Shallow and Deep Learning Principles

Zekâi Şen

Shallow and Deep Learning Principles

Scientific, Philosophical, and Logical Perspectives

Springer

Zekâi Şen
Istanbul Medipol University
Istanbul, Türkiye

ISBN 978-3-031-29554-6 ISBN 978-3-031-29555-3 (eBook)
https://doi.org/10.1007/978-3-031-29555-3

This Springer imprint is published by the registered company Springer Nature Switzerland AG
The registered company address is: Gewerbestrasse 11, 6330 Cham, Switzerland

RAHMAN VE RAHİM OLAN ALLAH'IN ADI İLE

IN THE NAME OF ALLAH, THE MOST MERCIFUL, THE MOST COMPASSIONATE

The following three harmoniously mixed components provide better scientific activities with peace and prosperity of humanity in freedom and mutual support

***Body fitness**: Physical training*
***Mind fitness**: Logic, mathematics, artificial intelligence trainings*
***Soul fitness**: Faith in Allah – God, spiritual training*

Preface

Although many numerical solution algorithms have been known since 1920 due to their extreme computational time requirement, they were at the shelves waiting for convenient future developments. With the emergence of the first computers after 1950, researchers began to use computers with simple programs with the Fortran language to solve boring problems quite quickly. As researchers have been interested in intelligent systems for the past three decades, the front runners were stochastic simulations with single input single output (SISO) architectural models, which later took the form of multiple input multiple output (MIMO) systems with input and output layers between the set of hidden layer nodes to mimic data to obtain the optimum prediction of known data within practicably acceptable limits. These were the pioneers of artificial intelligence (AI) systems that can make autonomic decisions and performances. Hence, the researchers were and can transfer decision process to AI system for more consistent and faster decision-making goals. Today these systems are used extensively almost in all convenient disciplines with care in the contents of artificial neural networks (ANNs), machine learning, and deep learning on probability, statistics, stochastic simulation, optimization techniques, genetic algorithms, convolution and recurrent networks, and similar methods. These methodologies are applied with supervised and unsupervised training methodologies. After evolution of AI methodologies through various stages, it became to be known as deep learning since the last decade.

This book presents the bases of shallow learning principles in terms of classical and well-known uncertainty principles based on the science philosophical and logical (two-value and fuzzy options) propositions and their conversion to mathematical equations for computer core writability. The shallow learning principles help researchers to understand the fundamental etymological, epistemological, and terminological contents of well-known and common numerical methodologies that pave way to better understanding of deep learning procedures. In fact, learning is a holistic procedure, but its classification as shallow and deep learning should be understood in terms of fuzzy logic, which provides overlapping between two learning methodologies. Unfortunately, many students, young scientists, and modeling beginners without sound background fundamental information, i.e., shallow learning information, may use ready software and get the solution results but cannot make proper interpretation. For this reason, it is recommended in this book that one should firmly dive into shallow learning principles with science philosophical and logical principles in order to swim safely in the deep learning ocean.

In addition to certain (deterministic) methods that provide the solution of various problems with strict rules (without any uncertainty), there are stochastic optimization methods that contain uncertainty to a certain extent. The common point of these two sets of methods is that they try to reach the solution by following a systematic way. This systematic path consists of successive steps along a trace. As a result of such a systematic approach, it cannot be decided that such techniques yield really the absolute optimization solutions. For this, it is necessary to apply some criteria with a series of trial-error sequential testing. This can cause a waste of time and still reach solutions that are not completely optimized.

With the method called genetic algorithms (GA), it is possible to reach a solution by elimination of the drawbacks mentioned above for classical methods. In this solution method, the area is scanned, not the line, and the solution is reached. One can summarize this with the phrase "defense the surface, not the line." The operation of GAs is completely random, with progress toward the goal. Although there are random elements in the method itself, GA can reach the absolute optimization solution in the shortest time.

In engineering and technology applications, the model of the investigated event is made, and the behavior patterns are determined. Parallel modeling methods have been developed with inspiration from the nervous system of the human brain and are called Artificial Neural Networks (ANN). One of the most important features of ANN, which includes many features beyond the models made by regression and stochastic methods that have been widely used until recent years and are still used in many parts of the world, is that it does not require some assumptions about the event or data at the beginning. For example, in all statistical-based methods, in addition to the necessity that the data or post-model errors fall into the normal (Gaussian) distribution, the parameters must be constant over time, parallel operations cannot be performed, the dependence is linear, etc. However, there are no such assumptions in ANNs, and for their implementation, at least three layers must be established together with cells, one of which is input, the other is output, and the third is hidden (intermediate) layer. For this, it is necessary to develop an architecture that will not model mathematical rules but only the action and response variables that control the event and the reactions that may occur within it.

It is a fact that technological developments are very rapid in the age we live in. Especially the developments in the computer world have become dizzying. These developments are speed in calculations, examination of the smallest (micro) and largest (macro) realms, acceleration of information production, etc. It has helped humanity in scientific and technological ways. The basis of this lays in the fact that computers process the scripts prepared for them very quickly.

Artificial intelligence (AI) is increasing its importance day by day in order to perform functions such as logic, rational approximation, perception, learning planning, mechanization and automation in a sustainable way. As in mathematics, it is possible for even better AI applications to emerge with the understanding of the principles of thought, philosophy of science, and ultimately rationality based on logic. AI continues its way if all kinds of living human intelligence are transferred to robots and machines within the framework of the rules of logic. One of the most important questions here, will AI devices replace human natural intelligence? The idea of the author of this book as many others is that such a situation could never arise.

Furthermore, machine learning ushered new and rapid development opportunities in scientific researches and technological developments. Such learning is based on computer technology through software with inter-disciplinary theoretical aspects such as probability, statistics, and algorithm developments in order to strengthen and enhance artificial intelligence attributes.

Under the umbrella of artificial intelligence is machine learning (ML), deep learning (DL), and artificial neural networks (ANN). It is possible to state that deep learning is a sub-branch from machine learning and machine learning is a sub-branch of deep learning. It is not possible to make crisp distinction between the machine learning and deep learning; machine learning has more algorithms that the other.

Deep learning (DL) is a mixture of artificial intelligence (AI) and machine learning (ML) concepts, which helps to pave the way to automatic scientific and technological developments, based on shallow learning principle improvements. In general, shallow learning (SL) principles are quite linear and based on descriptive conceptions leading to applicable formulations and algorithms; deep learning procedures are comparatively abstractive and complex and based on data mining statistical procedures leading to predictive models. Especially, large amounts of data collection, analyses, and interpretation are dealt with DL procedures in easier and faster manners. Traditional ML algorithms are linear compared with DL methodologies, which have hierarchically increasing developments, improvements, and abstractive complexities.

In this book, first the principles of philosophy, logic, mathematical, probabilistic, and statistical methodologies and similarities to other classical methods are explained and the reader's warming up about shallow leaning subjects and the principles are brought into consideration by giving the necessary clues. Furthermore, similarities between the two types of modeling are explained first by explaining the similar aspects of the classical shallow learning methods, which the reader knows beforehand or even if they do not know. In fact, it may be possible for a reader modeling according to classical systems to get used to ANNs and make applications in a short time. After a philosophical discussion of each model, explanations are made for rational reasons, taking into account possible similarities to previous methods based on philosophical and logical rules. With the given applications, efforts have been made to better understand them. One of the main purposes is to touch upon the AI issues that have sprouted throughout history and continued to increase exponentially, especially after 1950. In the meantime, it is to include the works presented on this subject by Muslim thinkers, philosophers, and scientists who are not rightly cited in the international literature.

Istanbul, Türkiye
25 August 2022

Zekâi Şen

Contents

Chapter 1
Introduction

1.1 General

The concept of deep learning has entered almost every higher education system in the world without much criticism and has become fashionable in places where basic conceptual understanding is insufficient, as ready-made software programs encourage obscure applications in the background. How can one learn and teach deeply without being aware of the shallow basic knowledge and information pair that is the root of all scientific disciplines and mathematical sciences? Rational scientific productions cannot be realized without basic definitions, concepts, assumptions, hypothesis, and related simplifications, which are largely in the field of shallow learning. For example, without traditional shallow learning principles, the effects of machine learning on deep learning procedures cannot be grasped objectively. It provides an introduction to the field of deep learning with shallower and narrower approaches, wonders, and sustainable end products, sometimes with conceptualizations. Deep learning provides strong performance in classification and optimization methodologies, provided that the shallow learning foundations are used coherently as cornerstones. Unfortunately, shallow learning is declining in popularity; this does not mean that deep learning methodologies are increasingly understood by everyone who uses deep learning approaches to problem solving.

A remarkable explanation was given about deep learning procedures, which are theoretically not easy to visualize and analyze. Accurate classification was lacking in large datasets through shallow learning methodologies for prediction, classification, pattern recognition, and optimization studies. Shallow learning methodologies are considered core solutions, and thus barriers when dealing with large datasets. Such data limitations in shallow learning lead to exploration of modern deep learning methodologies.

Z. Şen, *Shallow and Deep Learning Principles*,
https://doi.org/10.1007/978-3-031-29555-3_1

1.2 Historical Developments

Nearly a thousand years after the ancient Greek books and treasures of knowledge, a brand new era began to emerge in the name of Islam, this time with all the philosophical, logical, and rational prosperous resurrection and experimental activities. The original seeds of modern science have begun to take place in many disciplines with philosophical, logical, rational, and experimental interactive actions, mostly based on geometry, shape, form, algorithm, and map visualizations. As has been said before, there has never been a time in the history of civilization where geometry did not play a role. Even today, all imaginary screens are based on visualization, the first step in the history of science (Fig. 1.1). The emergence of algebra and algebraic equations enabled verbal relations to be expressed in the form of mathematical equations for the first time, and thus mathematics took the form of today's modern-time principles. Many mathematical expressions are solved by Al-Khwarizmi (780–850) algorithm based on geometry, but his name and Islamic civilization are not mentioned sufficiently even in the history of science. Sarton's (1950) book, which has contributed to the history of science, covering all civilizations, can be recommended for those who are deeply interested in the subject.

Mankind has been in constant interaction with his environment from his first day of existence and has obtained useful information through certain perceptions. Since the early ages, technological inventions have made more progress than science by developing some tools, equipment and devices that provide the comfort of human beings. However, with the emergence of writing 3000 years ago, it was possible to transfer the information obtained from generation to generation verbally.

1.3 Information and Knowledge Evolution Stages

There are various definitions of knowledge and information throughout the history of human civilizations. The main purpose of this section is to make scientific definitions as the final product after preliminary explanations. The simple definition of information is the communication between object and subject to obtain any rational knowledge about the subject's properties. This can be quite biased due to personal interest and perception. Information content has several stages such as perception, processing, transmission and storage in memory. Knowledge is more about the mental understanding of information because of its integration with other meaning content. At this stage, it is important to understand two terminological concepts as etymology and epistemology. The former deals linguistically with the root (origin) of words and their meaning have changed throughout history. Second, it concerns philosophy and the content of this book, which explores theories and fundamental foundations of meaningful knowledge, such as the scope and validity of the proposed meaningful content, and its relationship, if any, to natural

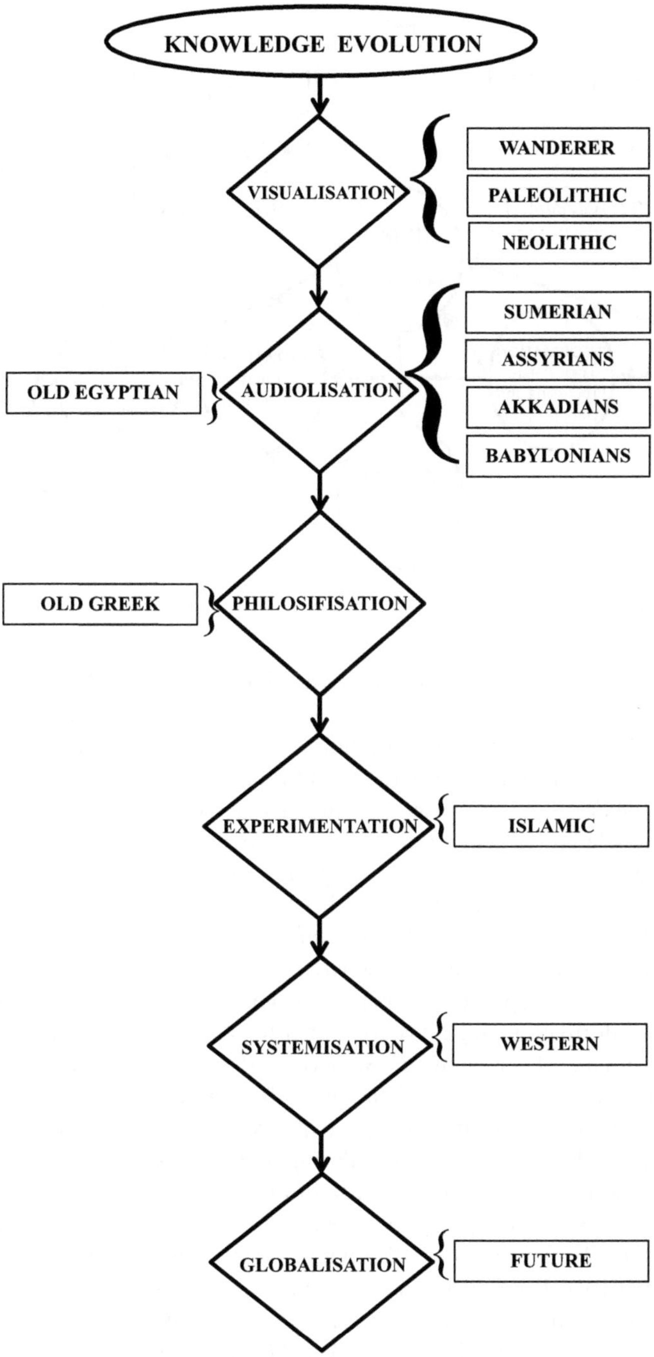

Fig. 1.1 Geometrical conceptualization time flow

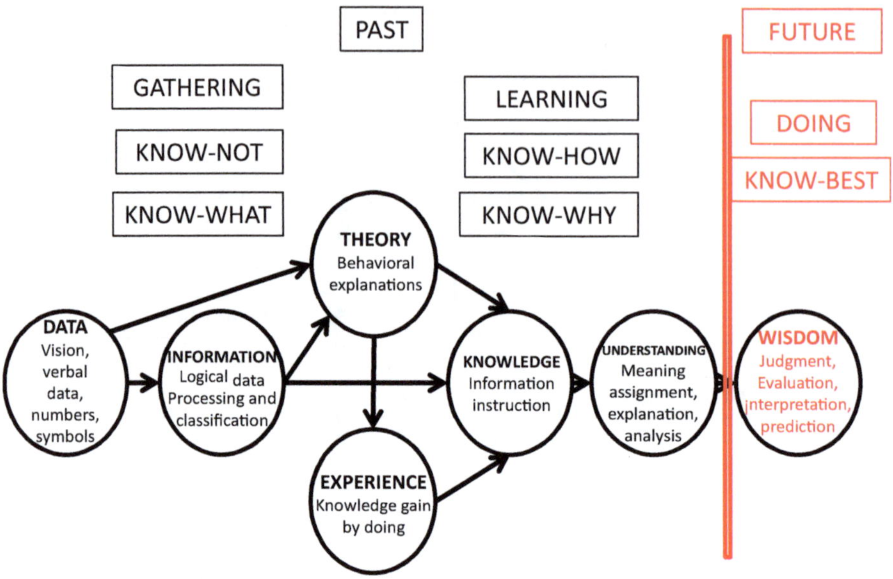

Fig. 1.2 Knowledge evolution stages

phenomena. Epistemology is perseverance and knowing things. It tries to distinguish between belief and rationality in terms of scope, methods, and validity.

In Fig. 1.2 knowledge and its evolution throughout human history are presented in two parts as past historical evolutions and future expectations.

The first stages of information and knowledge evolution were concerned with gathering knowledge with know-not then know-what without detailed knowledgeable meaning, but more simply knowing only. The first systematic gathering of information was through visualizations, because human beings tried to survive by hunting for food, mostly by sheltering in caves and clothing made from animal skins. The second stage of information gathering was through listening, through the transfer of visual concepts from experts to listeners with no prior experience. The audio-visual information stage has continued for thousands of years, starting with Ancient Egyptian, Mesopotamian (Sumerian, Babylon, Assyrian, Etruscan, and Hittite), Chinese, and Indian civilizations. After these two stages, the period of philosophical thinking and questioning expanded with rational thought based on philosophical and logical principles during Old Greek civilization. Behavioral explanations in the form of theories began to relate the mechanism of occurrences of an event to a set of principles, albeit speculatively, linguistically. The concepts of knowledge have begun to systematize with categorizations as well as thousands of years of experience from all previous civilizations. These developments have led people to think, learn and know better, why? and how?, thus to provide the principles of early shallow learning. Sarton (1884–1956) stated with a new dimension of experience that the Islamic civilization had many innovative developments based on the information and knowledge resources of previous civilizations. The last chain

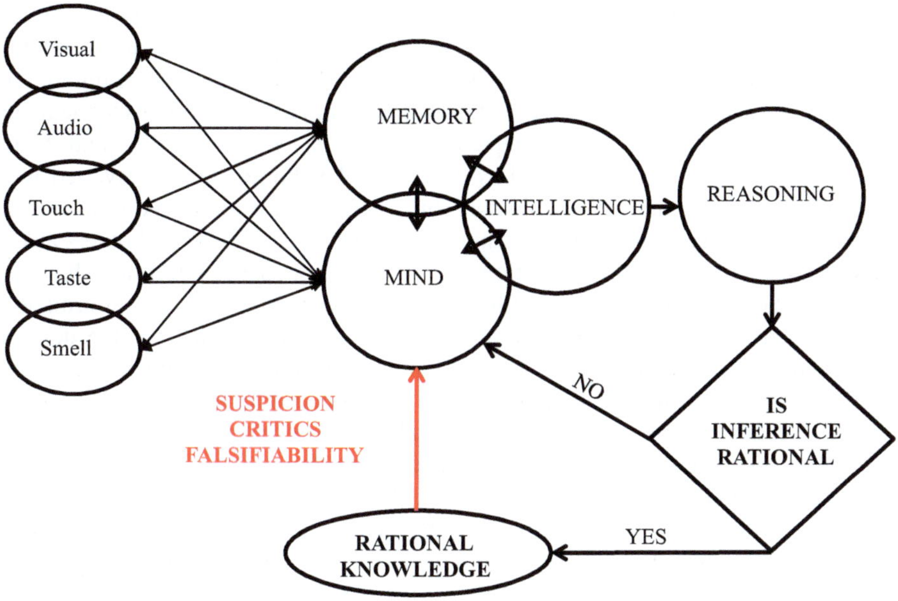

Fig. 1.3 Mind activity cycle

in the evolution of past information and knowledge assessment is the Western civilization that dominates these days. After all, a new era has begun to emerge as horizontal future aspirations, expectations, and aspirations for more innovative revolutionary information and deep learning knowledge phase. Deep learning is about scientific and technological reasoning, evaluation, assessment, interpretation, and prediction. All these latter activities were made possible by computer–human communication, which led to deeper artificial learning.

Figure 1.3 shows the natural human intelligent rational thinking stages in the form of mind cycle sustaining human wisdom activity and continuous thinking procedure.

Figure 1.3 shows the interactions between many facets of the human sense organs and their inner thought activities in the form of knowledge and information and the cycle of information acquisitions and outputs. The inputs of the sense organs are transferred to the mind, where they are first processed or stored in the memory. Intellect and memory interactions trigger human intelligence with reasoning, and therefore reasonable and rational inferences are reached with philosophical and logical tools. At this stage, verification of rationality or irrationality comes to the fore when doubt, critical review, and possible falsifiability assessments are effective. The success from these assessments provides scientifically acceptable information until it is deleted or improved through new and innovative cycles similar to stages in Fig. 1.3. Those who can provide a continuous critical evaluation of a problem may have the title of scientist.

In general, people's recognition, learning what they know and reasoning by criticizing what they have learned are called mind functions. With these features, the mind is a phenomenon that distinguishes humans from other living things. As will be explained below, concepts such as understanding, comprehension, intelligence, memory, and ideas all come with the mind. The process of determining the comparative and detailed differences of the signals from the sense organs to the brain and bringing them into a formal shape among themselves is called the mind. With the mind, the signals are processed as desired in the shortest time, converted into useful information, and by eliminating the human body as a feedback signal, isolating the useful ones from many signals, it benefits the person and others alike. A single person can perform an action that 10 or 100 people can accomplish. Only the mind helps to reveal useful information by organizing existing information among them. The mind reveals the necessary information to achieve the desired goal with practical and tireless information. In other words, there is no such thing as getting tired of the mind. It always does the task assigned to it. Other sense organs such as feet bear the brunt of the mindless head. Intelligence is an innate ability given to man by the creator and cannot be changed or transferred to another person.

People's minds are never equal, and just like fingerprints; everyone has slightly different mental abilities. In this respect, the Turkish proverb "mind is superior to mind" indicates that it would be better not only to love our own mind but also to benefit from the minds of others. It is possible to benefit from smart people by consulting and discussing common problems. If one is unable to overcome the problems, it is possible for the individual to seek advice from another person whom he respects, assuming that he is smarter than himself. As a result of realizing something, we can say that I lost my mind. Thus, it is possible for people to resemble each other according to their mental functions. As a result, because the way of reason is one, people agree with each other. When a person's body gets sick, they get better with some material medicine. However, it is understood that the mental illness of the individual begins with psychological disorder and its treatment is based on some spiritual suggestions. In this regard, religious beliefs play the most important role. Putting one's thought in a formal order is called the reasoning function. Re-enacting an event from the past is called coming to mind. Criticisms are conceptualized by accepting results and procedures that are reasonable in the end. It turns out that people who do whatever comes to their mind are thoughtless and unable to organize their mental functions. When the mind is preoccupied with something, the problem cannot be overcome, and the mind becomes obsessed with it.

Mind functionality cannot be measured concretely, but personality cannot be managed by some standard tests. It is dangerous to try to standardize minds, because the intelligence of those who speak of a mind is of paramount importance. The mind is never subjected to tax arithmetic operations. Accordingly, even if we do not know much about a class whether it has high or low significance, we can try to make better improvements through philosophical and logical thinking. Anybody who is willing to make decisions according to a systematic plan will be happy with smart people and to express decisively the knowledge learned and developed either in writing or orally.

A concept that we call "common sense" among the people is very diverse that can be presented and designed with a thought and consideration according to one's personal experience. It is necessary to investigate whether such common sense is always and everywhere correct. If it precedes reason, one prefers a technical aspect that is fundamental to morality, ethics, and virtue. As shown in Fig. 1.3, there is storage of previously obtained visual and auditory information in the memory filled with information that can be used for rational inferences. The information stored here is actively of passively perceived and memorized by the person. Almost all the information obtained throughout the history of humanity has been stored in human memory in the early ages. This information may be static from time to time and gains dynamism by importing more advanced ones; memory is always open to innovative knowledge development. For example, we can say that history is a science based entirely on memory. Information images of interesting events from childhood to old age are always kept in memory. Sometimes we say that some are mild or have lost their memory, and those in this situation are unable to make rational inferences using their minds. In a way, memory can also be thought of as a unit in which past knowledge, experiences, and emotions are recorded. Humans have genetically different memory capacities, but it is also possible to develop memory over time. In fact, the information stored in the memory plays a fundamental role in activating the mind, which is the function of the brain. By the way, there is a talent called intelligence based on superior or inferior views that may emerge from time to time in memory. No matter how superior a person's intelligence level is, he cannot improve his memory unless he works with more up-to-date and advanced information. Intelligence is a necessary but not enough condition for the development of memory.

In a way, memory can also be defined as the power to store information in the mind. Often, in a subconscious state, information is hidden and organized without thought, spontaneously stored in memory, and this can occur in relation to a thought event. Memories can be short-, medium-, and long term, as well as sensory and functional. In the form of a mixture of these, the information in the memory begins to play a role willingly and unwillingly.

One of the most important functions of memory is the ability to memorize a text and to repeat it automatically when the time comes. For example, memory can be used when a poem, a novel, a speech, or a religious act is done without thinking. In our culture, for example, those who memorize the Qur'an are called hafiz (memorizer). Even if the person does not know the meaning of something, he can memorize and repeat it by explaining it from the memory storage. For example, if a parrot repeats a phrase it has heard, it indicates that the parrot has a memory. Memory provides the ability to recall shallow and deep learning information principles. It should always be possible to disclose this information immediately, when necessary, without taking any action. A person with a strong memory should not forget the information and signals he gives in a short time with his sense organs. It can be easy to store information, especially for young people, as memory is new and rather empty. This accumulation process is called memorization for short. People who store information in their memory in this way are expected to transfer the information as it is, instead of processing that information and transforming it into other forms.

The process of detecting differences between the brain's existing information is called opinion or idea, which can change dynamically, and this change may be the result of new information processing by the mind and intelligence or the result of experience, logical friction with different communities and perhaps coercion. Reasonable discussion of an event can change one's mind by filtering out new information as a result of an impulse from one's inner world as persuasion that leads to a change of mind. It is possible to compare this to a factory with machines that process raw materials as inputs. These machines are like the mind of the human brain. In these machines, new materials can be produced with different properties and like the raw material mixture given for processing. When people do not have knowledge, they can say they have no idea, but how many people can say they do not have a mind? Obtaining an opinion among the public is equivalent to getting information from someone else on a subject that one does not know. To give an opinion is to express one's idea for the purpose of teaching. Speculation is expressing one's opinion about something or a situation from various perspectives. Knowledgeable and thoughtful people are intellectuals, and others are not well educated.

Reasoning and understanding are called cognition. Comprehending and keeping the results and information resulting from the fulfillment of all the mental functions listed above without causing any additional thoughts or doubts is called cognition. We can call it unconsciousness. The information that is perceived now is engraved in the minds like embroidery on a hard-to-wear stone, and all kinds of information persuasion methods can be used to convey that information to others and make them understand as well.

With the combination of all the above-mentioned brain functions in a harmonious shape and ratio any person can perceive the information and give them various crisp or fuzzy meanings in his mind. Thus, it becomes rather easy to extract new information from information mixtures. For the brain functions to work, it is necessary to arouse curiosity and desire. For this reason, it is necessary to determine the goal that the individual or society wants to achieve. These include the desire and even passion to dominate others, to be economically stronger, to be socially preferable, to occupy certain positions and control some centers, to serve people, and to enjoy solving their problems for the Creator, the homeland, and humanity. It is possible to set an innumerable variety of goals. One or more of these may become a target for some individuals or communities at different times and places. In order to achieve this, there may be some targets such as politics, bribery, and nepotism, which can be claimed to be made with reason and logic, but which gnaw at the society like ticks.

It is useful to make an analogy between the brain and the computer. No matter how much technological progress has been made so far, there are certainly inspirations for human intelligence from different beings, animal and inanimate. It is now possible to draw such analogies between the brain and computer functions. An individual's brain is a place where different brain functions emerge, as well as the environment in which the available information (data) in the computer processing unit is processed as software. It is not possible to operate the computer in the absence of a unit called disk operating system (DOS). Just as in the absence of the brain in the human body, the functions related to the mind cannot be performed. Here, the DOS unit in the computer corresponds to the mind in the brain. For the DOS unit to work, there are mathematical processors, that is, logical execution and memory units. In the event that a virus enters the computer, no operation can be performed or the operations performed do not give correct results, and similarly, if useless or bad or distorted information enters the human memory, the information to be obtained as a result of logic, intelligence and mental activities will be in error.

A person is exposed to different information about different views, ideas, ideologies, languages, religions, knowledge, science, politics, social events from time to time in his own world, and he thoughtlessly corrects some of them or tries to perceive them in a way that provides dynamisms by passing them through a mental filter. There can be no complete stability or full dynamism in any of the events. In other words, by collecting non-critical information and data on issues that can only be stored in the memory (Fig. 1.3) and discussed from time to time, relevant information can be communicated to others primarily based on language. These people do not use the productive function of their minds, they are happy to store information that can reach academic levels, but neither innovation nor invention can be made in the fields of science or technology.

In order to ensure the dynamism of the acquired information, the perceived knowledge must be kept in mind by prediction dynamism before it is filtered from the mind, and the only way to do this is to use the principles of science philosophy and logic and distinguish them from each other. One of God's greatest gifts to man is to have hearts that can be illuminated with reflections after the enlightenment of the mind. The premise of the most important tools in the mind–heart relationship is the "mind." As a matter of fact, one of the most important words of the Prophet Mohammad is as follows.

He who does not have a mind cannot have a religion.

Well, what are the conditions for the efficient use of the mind throughout human life? The answer involves language and thinking based on philosophy and logic.

1.4 Determinism Versus Uncertainty

There has been much debate and curiosity about natural phenomena that have occurred in the last century. These discussions included comparisons between the earth and atmospheric sciences and in physics, which inevitably led to the problem of determinism and indeterminism in nature (Leopold and Langbein 1963; Krauskopf 1968; Mann 1970).

The basis of scientific theories is the concept of "cause" and "effect" relationship with almost absolute certainty in scientific studies. Popper (1955), one of the modern philosophers of science said:

to give a causal explanation of a particular event means to deduce a statement that explains that event from two kinds of premises: from some universal laws and from some singular or particular statements there must be a very specific kind of connection between its conclusions and it must be deductive.

In this way, the conclusion necessarily follows from the premises. Before any mathematical formulation, the premises and the conclusion consist of verbal (linguistic) statements that can then be translated into forms such as flowcharts, methodological architectures, algorithms, graphs and diagrams even in sketches. At each step of the deductive argument, a logical conclusion can be drawn about the relationship between the statements. The concept of "law," on the other hand, is central to deductive explanation and thus to the precision of our knowledge about particular events.

With diversity in nature, human inability to understand all the causes and effects in physical and engineering systems, and practically every problem with insufficient data, every problem remains in an area of uncertainty to some extent. Even with large datasets, it is not possible to predict engineering events or natural phenomena inside of an error band. As a result of uncertainties, future events can never be fully predicted by researchers. Consequently, the researcher must consider uncertainty methods to assess the occurrence and quantity of certain events, and then try to determine the probabilities of their occurrence.

Poincare (1854–1912) discussed the problem of unpredictability in a non-technical way. For him, chance and determinism are associated with long-term unpredictability. He expressed this by saying that,

> a very small cause that escapes our notice determines an important effect that we cannot ignore, and then we say that this effect is due to coincidence.

Recently, the scientific evolution of methodologies has shown that as researchers try to clarify the boundaries of their interests, they become fuzzier with other areas of research. For example, when the hydrogeologist is trying to model groundwater pollution as one of humanity's modern challenges when it comes to water resources, they need information about the geological environment of aquifers (groundwater reservoirs), meteorological and atmospheric conditions for groundwater recharge, and social and traditional information for water consumption practices. Therefore, many common philosophies, logical foundations, methodologies, and approaches become common in different disciplines and in data processing, including shallow and deep learning methods. Scientists may use the same, similar, or different appropriate models, algorithms, or procedures for problem solving, depending on their area of expertise. Some common questions that may be asked by various research groups are discussed below. Many of these questions have been elucidated by Johnston (1989).

1. Is the event or phenomenon steady and deterministic over time with no random element?
2. If there are complexities, is it possible to reduce them to manageable form with a set of logical assumptions?
3. Are there any other variables that are closely related to the variable of interest?
4. Is it possible to set logical rules before resorting to mathematical equations for solution?
5. Are there relevant mathematical expressions with reliable results for the problem solution?

6. Are there nearly similar problems and their even approximate solutions in another discipline?
7. Do the homogeneity and isotropy assumptions help to simplify and solve the problem?
8. Does the problem need analytical, laboratory, or field work to achieve the desired conclusions?
9. Does the problem require inductive or deductive solution principles with integral variability characteristics?
10. What is the logical solution methodology? Analytical, probabilistic, chaotic, stochastic, artificial intelligence methodological, etc.?

1.4.1 Randomness

Random and randomness are probabilistic and statistical terms to describe any natural or engineering phenomenon that cannot be precisely predicted for a particular event. An illuminating definition of randomness is given by Parzen (1960) as follows:

> A random (or chance) phenomenon is an empirical phenomenon characterized by the property that its observation under a given set of conditions will not always lead to the same observed result (so there is no deterministic regularity), but different results in such a case, so there is a statistical regularity.

Probabilistic and statistical regularity refers to group and subgroup behavior of large datasets so that estimates for each group can be made more accurately than individual predictions. For example, given a long time series record of temperature in a place, it is confidently more likely to say that tomorrow will be warmer or colder than predicting the exact degree Celsius. As the entire text in this book suggests, statistical regularities are the result of some astronomical, natural, environmental, engineering, and social influences. For example, recent global climate change discussions are based on fossil fuel pollution of the lower atmosphere layer (troposphere) due to anthropogenic activities leading to greenhouse gas (GHG) emission concentrations. The impact of climate change has been expressed by different researchers and people, but the intensity of such a change cannot be predicted precisely even in the coming time periods. Statistical regularity means more complete unpredictability for single or individual events.

Deterministic phenomena are those in which, under a given set of conditions, the outcome of individual events can be predicted with complete precision if the necessary initial conditions are known. In the physical and astronomical sciences and engineering, the deterministic nature of phenomena is traditionally accepted. Therefore, the validity of sets of assumptions and initial conditions is necessary in the use of such approaches. In a way, deterministic scientific research with its idealization concepts, assumptions, and simplifications, yields results in the form of algorithms, procedures, or mathematical formulations that must be used with care taking into account restrictive conditions. The essence of determinism is idealization and assumptions so that indeterminate phenomena can be grasped and applied through mathematical procedures to work with analytical, probabilistic and statistical concepts. In a way, idealization and sets of assumptions remove the uncertainty components, transforming the uncertain phenomenon into a conceptually specific deterministic state. An important question to ask at this point is,

> is there no use in deterministic approaches in engineering and natural studies where events are uncertain?

The answer to this question is affirmative, because despite simplifying assumptions and idealizations, the skeleton of the indeterminate phenomenon is captured by deterministic methods.

1.5 Logic

The only way to keep the things explained in the above sections alive as a result of the activity in the mind or to make them more vivid is to turn to the environment of innovation and invention by transferring this knowledge and views through logic tool within the functional operation of thought. If we explain what the predicate of the word logic is, we can first understand that words and sentences are basic elements. Since words and sentences are also shaped by a language, it can be concluded that knowledge, information, innovation, and inventions are the property of not only educated people, but also anyone who tries to make their mind work with their thoughts can get their share. From this, we can conclude that it is not absolutely necessary to go through a formal education in order to be able to innovate and invent. If there is no philosophy of science and especially principles of logic that trigger thoughts, a formal education structure can only ensure that diploma and return practices take place in society in an imitative way, as shown in Fig. 1.1.

1.5.1 Bivalent (Crisp) Logic

The author of this book has been able to understand, throughout his life, that a purely formal and non-systematic education can be even more productive in the transfer of intellectual knowledge to others. The set of rules that enable human intelligence, reason, and functions to reach rational inferences is called logic rule base. Unfortunately, the fact that professions such as engineering that rely heavily on numbers, datasets, and formulations can only function with mathematical equations in formal education is a sign that innovation and invention may not take place in that society. One of the reasons and even the most important of the low number of patents on inventions in a society is the result of too much reliance on boring and formal entities such as formulas, algorithms, and ready software.

In order to reach the level of innovation and invention by making inferences with the rules of logic, it is recommended to follow the steps below, although not in order.

1. Taking or bringing to mind a phenomenon or event linguistically, philosophically, and logically.
2. Also, the mind must first know what the basic words (variables) of the phenomenon under study are. In the works written for this, there is a string of words called keywords.
3. In the analysis, it should be decided what the causes, namely the inputs, consequences and the outputs are. This means thinking about the relationship between inputs and outputs.
4. After determining the semantic loads (etymology) of cause–effect words, it is necessary to try to decide how the verbal relationship between each word of cause and effect is the two principles of thought that Allah (God) bestowed on every human being. These principles are,

    (a) The principle of direct or inverse proportionality,
    (b) Linearity or curvature (non-linearity) principle,

5. With the application of these two principles between each cause and effect words, non-numerical but verbal relations emerge. For example, to the question of what kind is the relationship between population and time, we arrive at the sentence "If time increases, population increases" with the first principle of thought. With the second principle, we can conclude that it will be curvilinear rather than linear. In fact, with these two principles, the situation of the mathematical relationship between population and time is revealed verbally,
6. Initial and boundary conditions should not be forgotten when making inferences. For example, if time is zero, the population can never be zero, and even this knowledge is useful in guiding the forms of mathematical relationships to infer functional form (line, parabolic, polynomial, exponential, logarithmic, etc.).

Relations have always been considered between all the cause (input) and effect (output) variables (words) in search for a mathematical equation base.

1.5.2 Fuzzy Logic

In the fuzzy logic problem, relations can become more thinkable by assigning adjectives to cause and effect words instead of holistic binary inferences (true-false, white-black, beautiful-ugly, 0–1, yes-no, etc.). For example, if we are asked what kind of relationship there is between force and acceleration, we can give a holistic answer as "If acceleration increases, then force increases." The words force and acceleration describe the situation holistically, but if the subordinates of these holistic words are taken into account, a fuzzy set of logic emerges. For example, if we think of force as "little" and "much," and acceleration as "small" and "large," we get a fuzzy situation because the adjective words of "little," "much," "small," and "large" provide uncertainty components instead of precision. The ambiguity implied by the words here should be understood as the first step towards fuzzy rules and inferences. Now, is there a relationship between force and acceleration? Not just one but four logic rules emerge as alternative answers given below.

1. IF force is "little" THEN acceleration is "small,"
2. IF force is "little" THEN acceleration is "large,"
3. IF force is "much" THEN acceleration is "small,"
4. IF force is "much" THEN acceleration is "large."

Here are four options. The reader must ignore the ones that do not make sense. Thus, it automatically turns out how many rule bases there can be from the sub-definitions of each cause and effect word, but not all of these rules make sense. With these rules, verbal (philosophical) and rational (logical) solutions of the situation studied without mathematical equations can be reached linguistically. If someone asks, they can say that mathematics means logic, so why not mention logical principles in the context of mathematics as foundation, but if mentioned it will lead to far more extensive innovations and inventions in mathematics and other related subjects. In fact, it may be a situation where a person who knows mathematics may not know the logic of the equations he is learning, but a person who knows the logic of the event can interpret and modify mathematical expressions for better improvements.

1.5.3 Bivalent-Fuzzy Distinction

Both Aristotelian crisp logic (plus-minus, black-and-white, something and its opposite) and Zadeh's (1973) fuzzy logic are expressed numerically by the numbers 0 and 1. In the case of strict logic the two alternatives are 0 and 1, exclusively, but in the case of fuzzy logic, any value between 0 and 1, inclusively. Likewise, true statements were coded as 1 and false statements as 0. As if there was no mixture of these two situations, that is, partly true or partly false, and precisely on the basis of human thought, and especially in philosophical thought, the middle case is excluded. For this reason, crisp logic is also known in the literature as the exclusion of middle case.

Fuzzy logic will even assign degrees to a scientific belief (degree of confirmation or falsification), accepting values between 0 and 1. It has been argued that scientific knowledge or theories can only be verified by logical positivists. The contradiction between the verifiability and falsifiability of scientific theories involves philosophical underpinnings that are blurred, but many philosophers have come to the conclusion with Aristotelian logic, which is quite contrary to the nature of scientific development. Although many scientists tried to solve a problem based on strict logic, considering the limitations of analytical and sometimes probabilistic scientific knowledge, the "fuzzy philosophy of science" did not enter the literature easily. Scientists cannot be completely objective in justifying scientific limitation or progress, but the components of the fuzzy are the impetus for the generation of new theories.

All scientific rule bases should be tested by the fuzzy inference engine that leads to fuzzy scientific results. In fact, scientific phenomena are inherently fuzzy, and especially the foundations of scientific philosophy implicitly contain fuzzy components. In science, the dogmatic nature of scientific knowledge or belief is the fruit of formal classical crisp logic, as if insensitive, whereas fuzzy logic is scientific for future improvements and knowledge production.

Since creation of man, logic became formalized and gained a binary (hot-cold, white-black, 0–1, etc.) style, and this situation continued dominantly almost until 70–80 years ago. Human thought never follows binary logic in everyday life, because there are always uncertainties. Although the rules of logic containing these uncertainties were stated by Farabi (870–950), one of the thinkers of Islamic philosophy, their rules were put forward by Lotfi Asker Zadeh only after 1960s. Today, when artificial intelligence is mentioned, in addition to binary logic, a more scientific and especially technology-sided thought emerged based on fuzzy logic and on its inferences.

1.6 Humans, Society, and Technology

These three terms, which are in harmonious solidarity with each other, show varying degrees of effectiveness in all activities, starting from family life, including natural and artificial intelligences (Fig. 1.4).

The origin (etymology) and semantic load (epistemology) of each word must be known. For example, creature is a phrase that is frequently used in every society that means God's creation. Creatures include also all inanimate beings, not just living things (plant, animal, and human). Without such a creation, neither man, nor society, nor science, nor technology studies could have emerged.

From the epistemological point of view, "forgetting" has a meaning, because people can forget a lot of information over time and therefore can use this information at appropriate times and places by reviewing and remembering. A person who trains himself to forget especially bad information can reach peace and happiness and instill these situations around him. Man is a three-dimensional creature because he has a body, mind, and soul. The body carries the mind and soul, but its handsomeness or beauty does not contribute much to the production of scientific knowledge and technology. A feature that other living things do not have, is the human conscience, which is actually a mixture of mind and soul in different proportions. Reason is necessary for the conscience, but not enough, because the soul is related to the non-scientific aspects of man and is a quality that has more weight in terms of conscience. The mainstay of science and technology studies is the mind, but the comfort of the mind is reflected in the human soul, which leads to a fluid path to more advanced and innovative scientific theories and technological developments.

God-given intelligence is one of the most important abilities that enable people to make rational inferences. There is no foolish person, but it is necessary to know the ways and methods of use according to his own power and ability. Throughout the

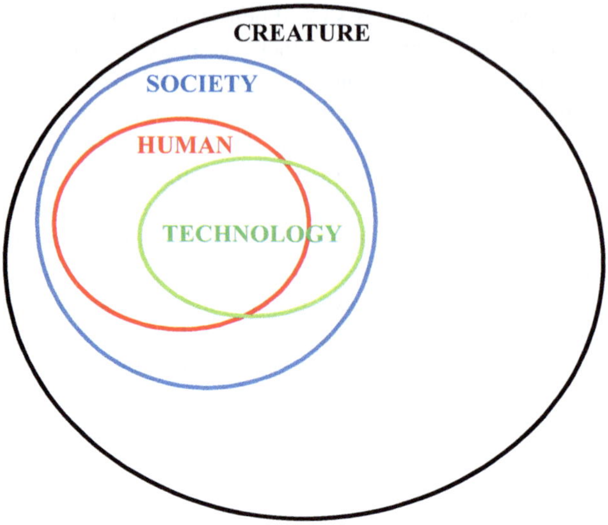

Fig. 1.4 Interrelationship between four words

history, mankind has continuously knowledgeably evolved and has led to the discovery of different tools and equipment with reasoning inferences. In a way, every era has its own intelligence. Thanks to these intelligences, different civilizations have been able to maintain their existence for a certain period of time.

Their intelligences can be used in social, economic, environment, energy, war and defense, health, hunting, nutrition, shelter, etc. There have been improvements to the levels that do not decrease over time, but slightly increase by using the available functionalities continuously. Human also gave rise to a sustainable lifestyle by attributing some of the intelligence uses of the tools and equipment he invented with natural intelligence.

Today, the word society is perceived as a group of people who come together with the same culture, tradition, customs, and common sense to understand each other on respectable levels. The society consists of individuals who come together with the information from the family and then the characteristics determined from the education system of primary and secondary schools. If 5% of higher education graduates can reach the current level of science and technology, that society can reach a modern level of development. The technological innovations can be realized by critically giving basic information that will feed the human mind and spirit, such as philosophy of science, logic inferences, history of science, and geometrical forms in education.

The term technology refers to the invention of devices, tools, software, and robotic machines that can make some aspects of human life easier, cheaper, practical, and automated. Those who discover such technologies from practical life and who do not have much higher education encounter a profession in the sense of inventor. Another aspect of technology is the name given to people who are highly educated and perform similar functions to scientists.

Considering formal education systems, social, economic, environmental, science, technology, etc., although the subjects are taught, the rules of science, philosophy, and logic are not given much importance. Knowing mathematical equations with symbols allows applications of the four arithmetic operations (addition, subtraction, multiplication, division), but it may not allow meaning and interpretation of these equations for further innovative developments. Although the so-called crisp (bivalent) logic and mathematics teaching take up more space than necessary in educational organization, the philosophy and logic of science that forms the basis of them is almost excluded. In numerical trainings, verbalism is used only for equations, formulas, algorithms and software explanations. With the philosophical thought and logic rules that must be given before the mathematical equations, it is almost impossible to achieve verbally similar results to the desired equations. Today, it is known that among the developers of most advanced technologies, there are many innovative and inventive people who have dropped out of university education or have never been to a university. The most obvious among the main reasons for this is that the principles of philosophy and logic in thought do not necessarily mean that they can be acquired through education. People who try to shape their thoughts according to the rules of philosophy and logic can achieve more productive science and technology results than those who have received classical education without the rules of philosophy and logic. In addition to the conclusions outlined above, the following points are advisory.

1. Oral expression and especially explanations, comments, and criticisms should be allowed in the mother tongue in education.
2. It is useful to explain the philosophy and logic of science that can lead to precise and plausible arithmetic in education departments that are numeric.
3. Philosophy, and especially philosophy of science, should be given great importance in education so that certain interpretations and forward looking views can emerge with confidence.
4. Valuing the verbal expressions obtained for the scientificity and rationality of the knowledge sets determined by philosophy according to the rules of logic.
5. In thinking, by using multiple options instead of only crisp ones, fuzzy logic rules should be reached in addition to the dull precise logic rules.

1.7 Education and Uncertainty

Education is a terminology used to enlighten others through a series of systematic lectures that are expected to provide students with a living space for reflection by certain key concepts. Of course, rational thinking, like education, is the essence of productive free thinking. In addition, there are three main aspects of education that must interactively contribute to productive and even emotionally stable end goals. Figure 1.5 illustrates these three components in their interactive lessons.

In an effective education system, education does not just mean the flow of information from teacher to student. There are situations in which information flows suddenly and quite unexpectedly from student to teacher in an effective

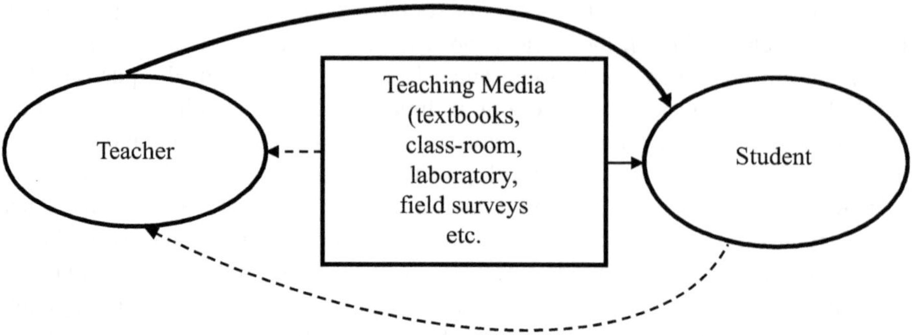

Fig. 1.5 Education system parts

education system, or new information emerges through mutual discussion. Unfortunately, very classical education systems operate in many parts of the world and especially in developing countries, they offer locally valid certificates in mass production, which provide unquestionable and vague information. The flaws in any traditional education system can be listed as follows:

1. Teachers, who are guided by state or customary rules that do not give freedom of productive thought. In such a system, logic means that the answer to any question is either black or white. This is the classic logical attitude towards problem solving.
2. Teaching environments, which can be called educational tools, can become indispensable organs over the years and are abused in a lively and quite dull way. Indeed, in non-native English-speaking communities, such devices can easily become showpieces for diverting students' attention to technological excursions rather than basic educational concepts.
3. There is an expectation of ready-made answers to textbook-style questions consisting of information shared without ambiguity by different learners and teachers.
4. Clear presentation of scientific concepts as if there is only one way to think with scientific certainty to solve problems.
5. Assumptions, hypotheses, and idealizations are common tools for the mind to grasp relevant phenomena, and therefore any scientific conclusion or equation is valid under certain conditions.

In a modern and innovative education system, uncertainty flexibility should be provided to almost all concepts, especially in higher education. The long history of science shows that not only freedom of thought is necessary for better progress, but also suspicion of scientific results. The word doubt itself gives to expectation and even uncertainty of scientific knowledge. For this reason, the basic points in a modern and innovative education system and the following points that are contrary to classical or traditional education should be considered.

1. In an innovative education system, traditional and classical elements should be minimized or even removed.
2. The teacher should not be completely dependent on educational tools and students should try to push the teacher to learn more about the limits of the material presented through discussions and questions.
3. It should be borne in mind that every scientific result is subject to uncertainty and doubt, and therefore, subject to further refinement leading to innovative ideas and changes.
4. In particular, keeping the principles of logic and philosophical foundations on the educational agenda by teachers so that every student can grasp the problem and approach it with their abilities.
5. At the higher education level, scientific thinking should focus on the falsifiability of results or theories rather than on certainty.

When all these points are considered together, it is possible to conclude that modern and innovative education should include philosophical thinking first and then logical arrangements that distinguishes illogicality in scientific education. This means that the conclusions reached are rationally acceptable with a certain degree of belief that is not entirely certain. Graduates must ensure that there is still room for them to make productive improvements, inventions, and scientific discoveries in the future. Otherwise, a classical and traditional education system with the principles of certainty in its results leaves no room for future developments, and therefore, those who graduate from these systems can only have certificates and dogmatic knowledge. However, with the advancing age, they may become frustrated, because the knowledge they have acquired during their education is not precise and changes for the better over time.

The traditional or classical education system can comfort teachers, but unfortunately, it kills the functioning of young minds in the hands of respected teachers in the society, especially in developing countries. In such education systems, teachers can be "mind killers" but are respected by higher authorities, where the wishes or representation of students is not considered significantly. The following points are suggested to improve the contents of the current education system.

1. Verbal knowledge bases should preferably be grasped through the mother tongue so that the subject can be discussed even with common people and a trigger idea can suddenly appear to achieve the final goal. Systematic education can provide a foundation in mathematics, algorithms, and physics. However, if the verbal background is unknown, new ideas and smart inferences may not be developed.
2. The unsystematic and sometimes random mix of knowledge and information should not be underestimated. There may be many hidden and useful sets of information that can be extracted from useful informative and illuminating pieces.
3. There are always possibilities for further improvement of existing or new ones, provided that questionable thinking is guided in scientific and technological approaches in any artificial intelligence (AI) production.

4. Entering the field of AI more broadly, explaining the problem at hand on the basis of logic, it is worth noting that no software was out of reach prior to clear and even recent principles of fuzzy logic (FL).
5. Any information learned from systematic education, teachers, books, the internet, or other means of communication should not be kept in mind without philosophy and criticism. Unfortunately, in many societies and countries, systematic education means filling the minds with mechanical and static information rather than effective and productive idea generation orientations.
6. After short or long discussions among team members, compatible partners should be selected for brainstorming conflicts and agreements,
7. Every technological product, computer hardware, and especially software technologies should be examined with the help of cognitive processes that enrich the human mind. Otherwise, general solutions may not be reached by being stuck in one of the three domains (soul, mind, body) rather than the field of AI.
8. The mind itself is the main natural machine without borders. But its action, production, and trigger fuel depend on collaboration with a reassessment of dominant intelligence concepts.
9. AI, which is considered valid instead of individual natural thinking, cannot lead to completely independent, productive innovative ideas. For this reason, it is recommended to have individuals with different ideas in a team. Otherwise, being of the same mind and thought is equivalent to a single individual,
10. Dynamically addressing expert systems with critical and questionable discussions, comments, and debates to generate better ideas instead of knowledge-based systems that provide statically transferable and memorable education.
11. After grasping the verbal foundations, numerical data information generation emerges, which needs various science functions such as mathematics, uncertainty techniques (probability, statistics, stochastic, chaotic), and expert system methods. However, the author suggests making simple graphical figures first to better interpret the characteristics of each dataset. Thus, their relationship with each other, their independence, dependency, or partial relationship can be revealed. These graphs, or any simple shape, are in the form of geometrical features that are more important to human intellect than mathematical equations. Because such simple and even explanatory figures provide very effective and rational information at the beginning, and then it may be possible to reach even the most complex mathematical equations.

1.8 Future Scientific Methodological Developments

The definition of science that can be understood by everyone today is the search for causes that investigate the reasons beyond the natural environment around man and artificial devices work. The historical investigation of these causalities with patience and difficulties has brought the scientific level and technological production that human beings have today. At the heart of all these efforts are inferences that are compatible with human thought and logic. The development of scientific ideas and methods by different cultures and people in different times and places continues uninterruptedly, and this will continue even more rapidly in the future (Sect. 1.3).

In general, human thought is concentrated in three groups: the idea of improving the relations of the individuals with the institutions and individuals of the society in which he lives, the idea of examining the events with physical objects outside himself, and finally the idea of making an evaluation. This inner group, which seems to be separate from each other here, is actually the thought process of the same person at the same time or at different times and reaching some inferences. However, although scientific and technological developments are achieved by considering the factual ones, it is never possible to completely isolate others from personal thought. The most important indicator of this is that in the development process of science, researchers and thinkers sometimes make generalizations by removing the inventions of researchers and thinkers from reality and then other people and researchers become the defenders of their own inventions. However, such thoughts and generalizations in the historical development of science have prevented scientific thought, which has a revolutionary structure, from freezing and decreasing its efficiency, but at least from diversifying and developing. In the past, during the development of science, two other groups opposed factual scientific knowledge and thought. In fact, even today, they think differently, and researchers do not deny that these three groups are always intertwined. After all, this is a result of instincts inherent in human nature and creation. However, people and societies that can keep these three thought groups in harmony have been able to reach today's advanced science and technology levels. Today, these three ideas clash from time to time in many countries of the world, causing these communities to advance their scientific and technological progress at a turtle speed. To put it briefly, ensuring that all kinds of free thoughts exist in the inner world of human beings is a necessary condition not only for scientific and technical development but also for the formation of a mentality that will enable them to sprout and develop by incorporating them. The history of science and humanity is full of many lived examples that can be taken as an example in this regard.

If these three ideas are not found in a harmonious ratio, even extremes may soon bring science to a breakthrough, but it cannot accelerate progress. Some dogmatic obsessions delayed the development of science for a while because of this over-factual thinking, in cases where only truth was given importance. The best example of this can be shown as the general validity of Newton's laws, which was accepted as the greatest revolution throughout the history of science until the nineteenth century and has now become a belief. According to this view, since it will be possible to explain all phenomena in the world and the universe with force, inertia and acceleration, as Simon de Laplace (1749–1827) emphasizes,

> *if a gifted person were to be given the positions and velocities of all objects in a single moment, this superior creature would be able to control these objects in time and from now on,* he can definitely foresee the topics that he can take.

Again, by the same person, we can determine the starting conditions of factual events in a healthy way and predict the future of any series of events and infer what its past was. Thus, the idea that the time and location of the cases can be calculated as independent parameters has prevailed. Those who agreed later suggested that their views could be applied to almost all other fields, and assumed that events, and especially the universe and those in it, were quite simple, completely controllable, distinct, purely mechanical, and uniform as if a mechanical clock were ticking. They believed their hypotheses and convinced others as well. Thus, some mechanicalness has been added to the pursuits of human thought. As a result, those who can think outside of this framework have not been looked after. For this example, what man does in science is the same thing he does.

The first objection to Newton's mechanistic view began in the early nineteenth century with the emergence of the laws of thermodynamics regarding heat problems that could not be explained by these mechanical laws. These heat laws led to a new scientific revolution by showing that Newton's view could not be valid in all areas. Naturally, it has been seen that the mechanistic and especially the positivistic thought founded by August Compte (1798–1857), were shaken. The most important view of the thermodynamic system of thought is that if the universe is likened to a machine, it will wear out over time, like every machine, and will no longer function as it used to a very simple but effective appearance. According to this, everything in the universe ages and this view was put forward by some thinkers in ancient times by using only reason. Among them, Platon (BC 478–423) claimed that the images of beings are perceived by people, in fact, there is an absolute truth called ideals for every being, and that these ideals do not change over time, but the images we perceive always change over time. Therefore, we can say and as a result, all scientific laws and principles valid today must change in the future. It is the divine behind these constant changes, and as a natural consequence of this, human beings will have to continue their efforts by constantly following the development of the laws of nature. At this point, it is meaningless for our world to change and form new continents we know as a result of today's scientific studies. For example, in the case of adjacent continents 400 million years ago, it is clear that the earth's climate and the circulation in the general troposphere were different from now. As a result, scientific rules and laws for meteorological and climatic features that were valid at that time are not valid today. It is not necessary to be a prophet to realize that today's scientific theses will not be valid in the future, given the expansion of the universe in remaining future times, as the lifespan of the earth, which is estimated to be approximately 15 billion years. For this reason, scientific findings should be perceived as a set of rules that are constantly evolving and changing both over time and due to incomplete information.

On the other hand, in the development of science and technology, the human thought system, namely philosophy, brought some views and led to the formation and development of today's science. Human nature always desires the good, the easy, the simple, the beautiful, the uniformity, and tries to understand many events. Meanwhile, he describes what he cannot simplify as ugly, obscure, or even useless, and becomes uninterested in it. As a matter of fact, one of the most dominant views in the development process of science is the principles and thoughts of certainty, which constitute the basis of the above mechanistic and positivistic views. In some societies and times, this view and its way of thinking have gone so far that those people and societies have become blind and even deaf from the thought of seeking certainty. Because they have come to believe in certain and under their control in a way that does not even allow scientific criticism. However, as in social events throughout history, free criticism, which is necessary for the scientific thought system to develop and produce knowledge, has been prevented under the name of being scientific. Sometimes this hindrance was unknowingly, and humanity took the medicine prepared as if it was on a prescription and swallowed the pill in a way that hinders scientific thought. The best examples of this are Euclidean geometry, which everyone knows very well. This geometry, contrary to human nature, deals only with ideal points, lines, planes and volumes. Euclid's assumptions were not criticized for nearly two thousand years or even if they were criticized by some, they were defended by others, preventing the discovery of the geometry necessary for the scientific progress of humanity. Although these obstacles are seen from time to time when other thoughts are thrown on them, it is also seen that scientific development does not progress due to some bigotry and dogmatization situations within itself. With Euclidean geometry, it is not possible to draw shapes in nature, such as a tree. This was only possible with the emergence of fractal geometry, which is developed towards the 1970s, and today, thanks to this geometry, it is almost indistinguishable from what humans do. Especially, starting from a very simple idea, aren't there any figures with decimal fractions between the quantities of 0, 1, 2, and 3, respectively, points, lines, planes, and volumes of Euclidean geometry (330–325 BC) as integers? The fractional (fractal) geometry originating in his principle was introduced by Mandelbrot (1924–2010), paving the way for the construction of artificial natures that would be almost indistinguishable from a natural landscape itself.

1.8.1 Shallow Learning

This learning system may imply the simple memorization of basic knowledge and information without the ability to use them rationally and actively to solve problems in context. However, in this book almost all the traditional basic methodologies in science and technology will be explained under the light of science philosophy and logical rule sets. Such a shallow enlightenment is necessary to establish deep learning foundations firmly. Shallow learning is the simplest form of machine learning and it provides basic methodologies to infer from a given set of numerical data, image, graph, figure, algorithm, and shapes. Different conventional methodologies are among the topics of shallow learning including probabilistic, statistical regression, clustering, classification, and reinforcement approach all of which depend on numerical data. Additionally, the Bernoulli Naïve Bayes, k-nearest neighbor, random forest and support vector machine procedures furnish the foundation of deep learning methodologies (Manikandan and Sivakumar 2018; Moe et al. 2019).

Recall and rote memorization are among the features of shallow learning activities, in general, without any modification of the existing methodologies or innovative scientific developments. If the shallow learning principles are not based on science philosophy and especially logical inferences, then the knowledge is bound to fade away from the memory or the knowledge passively remains in the memory.

1.8.2 Deep Learning

Deep learning methods have recently been found to show remarkable performance enhancement without extracting features or feature optimization (Krizhevsky et al. 2012). A modern way of knowledge gain after firm shallow learning principles is through deep learning, whose bases are artificial intelligence (AI), machine learning (ML), software learning (SL) and their execution in high speed computers. The application of these stages is dependent on the data availability, which is analyzed by means of logical rules, probabilistic, statistical methodologies for extraction of fine detailed structural features from available data for better estimation and prediction modeling. The very basis of deep learning is machine learning or proposal of a machine and its working principles through deep learning. Deep learning procedures learn from large data amounts by means of AI methodologies such as artificial neural networks (ANN). Human brain mimic is possible by deep learning through extensive data inputs, neural networks, and weights between layers and bias. Deep learning procedures are consistent of multiple layers in each interconnected nodes, which predict, classify, or optimize the internal structural features of given dataset by refinement of successive forward propagation functions. Deep learning algorithms are quite complex and there are different methodological applications such as the convolution neural networks (CNNs) and recurrent neural networks (RNNs), which are explained in detail in Chap. 9.

1.8.3 Shallow-Deep Learning Relations

In general, shallow learning procedures are regarded as machine learning methods that have reached almost their saturation levels to give birth for deep learning methodologies. Although there are computer modeling activities in shallow learning, the deep learning procedures are more frequently, intensively, and computationally oriented. In the approximation of any function, both shallow and deep learning methodologies are effective. For better accuracy, level deep learning procedures lead to more efficient results after extensive computational results with more representative and finer parameter sets and rather complex model structure. Preliminary expert views play significant role in the shallow learning procedure applications, whereas deep learning algorithms are capable of learning abstract numerical representations through different architectural structures of the models for better classification and optimization studies.

Shallow learning methods perform better than the deep learning approaches in cases of small data set availability leading to preliminary results with simpler and less computational requirements. Deep learning procedures help to process large datasets at finer scales, taking into account the cutting-edge shallow learning principles that lead to much better results. Deep learning subject is also machine learning algorithms but at higher level of complexity, effectiveness, and accuracy than shallow learning. Shallow learning is also concerned with machine learning but at a coarser scale.

In both the learning systems there are sequences of layers to learn, but in deep learning procedures the number of the layers is comparatively more and feature descriptions are more involved. Occasionally they are also referred to as feature extraction methods. The first layer in both learning procedures includes entrance of the available data and related information, which are processed internally through the internal layers leading to the outlet layer, which yields desired estimations, predictions, or results with more sensitive accuracy. The objective is to analyze available raw data to reach at valuable and logically appreciable valid deep learning results. As for the distinction of shallow learning from deep alternative there is not crisp boundary, but they intersect vaguely in a fuzzy manner. The first idea that comes into one's mind is that the deep learning is concerned with multiple hidden layers after the data (verbal or numerical) entrance layer. The theorem of universal approximation may imply that a single middle layer between the entrance and result layer is enough to solve problems at shallow learning domain, but this is not enough for deep learning solutions.

Goodfellow et al. (2013) stated in summary about the layers in deep learning the following passage.

> A feed forward network with a single layer is sufficient to represent any function, but the layer may be infeasibly large and may fail to learn and generalize correctly. In many circumstances, using deeper models can reduce the number of units required to represent the desired function and can reduce the amount of generalization error.

In the past, probabilistic and statistical models played a dominant role in the design of models based on numerical data for prediction purposes, and they are counted as shallow learning methods. Recently shallow and deep learning models started to show promise for extensive data analysis through intensive regression methodology.

Shallow learning desire comes into view mostly without or non-active rational thinking about the internal functional mechanism of the scientific methodologies. In general, shallow learning is understood in terms of memorization and ignorance of facts, but instead arguments. On the other hand, contrarily, deep learning involves data exploration in greater and finer details, understanding basic principles and to decide about the personal but scientific arguments concerning the methodologies. However, as will be explained in the next chapters of this book there is always interaction between the shallow and deep learning procedures. The aim of the book is to explain the fundamentals of scientific prediction, pattern recognition and prediction methodologies as shallow learning, which is then expanded to deep learning algorithms again scientific philosophically, logically, and rationally.

1.9 Innovation

Inventions imply advances in smart software, devices, scientific knowledge, and other issues in an obvious and attractive way. As for innovation, the adoption of some ideas, systems, scientific knowledge, fashion, art, and world views, by quoting them as they do not exist, is nothing but antiquity. To do something innovative can be achieved by taking what is said from somewhere, but after perception as rational thinking rather than transformation, and expressions in verbal knowledge about the its philosophy and then the logical rules and making their criticisms, new directions can be conquered leading to useful new information domains. Science and technology developments can be achieved with tangles.

If the information specified by a person or team is taken as it is, perhaps an innovative environment will be entered as a formative, but real and dynamic innovative horizons can be reached by revealing the better ones from the common thought information. Having received formal education about an event or phenomenon, even having a university degree or academic titles, cannot be said to be a novelty in what that person does. It is found that there are different elements in a biased roadmap as in Fig. 1.6.

When one looks at Fig. 1.6, one can talk about innovation if the inferences are at a more advanced level compared to previously known information and technology, and it is understood that this is trying to reveal new situations by entering the environment of events and facts with feedback or by entering the environment of reasoning, which is the engine of the generative thought dynamo with what they perceive. However, it turns out that the imitative results are starts of dead-end. It should not be forgotten that the most important device among the thought elements is the mother language. Who can perceive and, more importantly, digest shallow information on logical bases their thoughts are bound to generate new and innovative ideas in science and technology. From this point of view, the basis of the philosophy of science and the principles of logic are essential ingredients of education in general and university bachelor science education in particular. Unfortunately these two basic knowledge generative aspects are missing almost in every university world wise.

For the last three decades, the word "innovation" became frequently spelled not only in scientific research and technology domains but also in many branches of engineering and natural science circles. In general sense, innovation is rediscovery, modification, improvement, or development of an already known knowledge and information with addition of new visions or extensions that may lead to further methodological, technological, or research directions with additional economic benefits. Although, mathematics are thought at many education levels, the basis of expressions, equations, algorithms, or procedures are not founded on sound

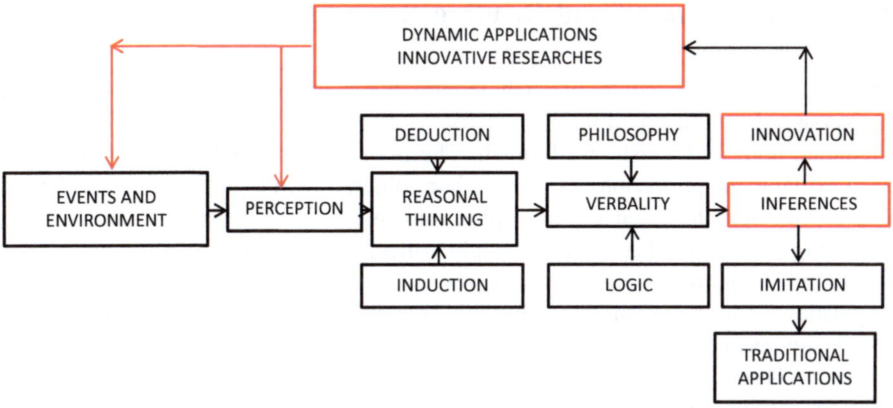

Fig. 1.6 Innovation flow

fundamentals. The basic fertile field for innovation should include philosophy and logical rules, which are missing in most education systems. Innovative ideas do not come through systematic education only, but even each individual who is suspicious about the scientific information can bring out better ideas provided that he has rational, logical, and approximate reasoning capabilities. Unfortunately, in the current education systems mathematical courses are abundant without geometrical backgrounds and principles.

1.9.1 Inventions

An important difference between inventions and improvement of something is that improvement does not mean innovation, because it is improving something more useful by making additions according to need. Constant reflection may not be necessary to invent. A mind that thinks and relies on examining many facts or events that comes before it, even at small scale, and its interpretation can suddenly ignite a situation by reflection and thus make an invention. This is also called discovery, and it is an invention to lift the veil in front of something that already exists and allow it to be seen by others. For example, criticizing Einstein's special and general theories of relativity may be an innovation, but it is not an invention. Introducing sub-particle (quantum) physics is a completely different invention rather than an innovation. From this, one can understand that the absence of innovation dimensions of inventions is also acceptable. In order to be able to invent, there is a need for thought, philosophy and logic, but these needs are productive thinking elements that must be activated when a certain invention idea comes to mind, not because of making an invention.

1.10 Book Content and Reading Recommendations

This book is the outcome of such lecture notes thought at Istanbul Technical University, Istanbul Medipol University, National and International Schools of Medicine both in Turkish and English. The book is composed of 10 chapters and each explains the philosophy of science intermingled with logical rule principles. In many countries, the science philosophy, logical inference principles, approximate reasoning, and rationality are not thought collectively in a plausible sequence. The author had experience and literature review and his own conception about the linguistic (verbal) explanations through the philosophical and logic principles, which are recently becoming more influential in the scientific and technological trainings. Throughout the book there are topics which encourage engineering experts to share common interest among different disciplines.

This book emphasizes the importance of innovative generative idea inventory that should be based on scientific philosophical, logical principles with accompaniment of engineering aspects that are available as inborn tools in any human being, who would like to invent new ideas, gadgets, instruments, robots, or at least their modifications to a certain extent. Among such human abilities are education dimension, which can inspire the person for additive new ideas to the existing ones, experiences and expert views, which can be gained by time. On these days innovative findings gain ratings and economic benefits almost in every aspects of life including even service sectors.

In all the disciplines there are different thoughts and thinking principles, where knowledge types and rationally wrought reasoning inferences are important. In this book, the following significant points are advised for medical philosophical and logical principles' applications in artificial intelligence (AI), machine learning (ML), and deep learning (DL) domains.

(a) Relationship between shallow and deep learning procedures,
(b) Basic and brief definitions related to AI, ML, and DL modeling aspects,
(c) AI history, types, and methods,
(d) Definition of philosophy and its effects on science philosophy,
(e) Logic principles and rules,
(f) Linguistic and numerical uncertainty types, inferences, and measures,
(g) Linguistic predicates and interferences in AI modeling,
(h) Crisp and fuzzy logic modeling principles based on sets,
(i) Probability and statistical modeling fundamentals,
(j) Regression techniques and restrictive assumptions,
(k) Human engineering,
(l) Mathematical modeling and functional interpretations,
(m) Optimization principles by genetic algorithm principles,
(n) Artificial neural networks as shallow learning stages,
(o) Prognosis, verbal, meaningful, educational searches,
(p) Fuzzy logic inferences system and modeling principles,
(q) Machine learning principles as the first step to deep learning,

(r) Deep learning requirements and methodological principles,
(s) Convolution and recurrent neural network modeling principles.

Chapter 1 provides brief information about the coverage content of all the chapters with some basic considerations in the basic information content. One of the main purposes of this chapter is to reflect the content of this book and to reconsider and reestablish educational basic knowledge and information on the basis of scientific, philosophical, and logical foundations leading to artificial intelligence (AI). Shallow and deep learning concepts and principles and their mutually inclusive relationships are explained.

Chapter 2 is concerned with preliminary and fundamental aspects of artificial intelligence (AI) with its historical evolution, types of intelligence, and AI basic modeling principles. Different versions of thought and thinking are explained so as to reach at plausible diagnoses with possible convenient treatment decisions all of which are explained to gain approximate reasoning abilities.

In Chap. 3, it is strongly emphasized that the very fundamental subjects for scientific progress are philosophy and logical principles, which should be priority courses for mathematical, physical, engineering, social, medical, financial, and all other disciplinary teaching and learning pair education system foundations. Intelligent functionality principles can be derived from these principles by considering various thinking principles as described in this chapter.

In Chap. 4, uncertainty types encountered in scientific studies are explained objectively concerning probability and statistical methodological uses in practical modeling and data description studies. It is tried to give the reader the skills and abilities to interpret uncertainty modeling procedures, problem solution, and validation on the bases of approximate reasoning and logical rules. Real-life uncertainties are avoided by a set of homogeneity, isotropy, stationarity, homoscedasticity, uniformity, linearity, and many other assumptions relative to the problem at hand. Basic probability and statistical parameters and methodologies are explained practically. Finally, trend identification methodologies are explained in detail.

Chapter 5 proposes a full explanation of the most commonly used mathematical functions linguistically followed by practical tips for their visual description and use in practical applications. For this purpose, previous chapters are used as a basis for scientific rational propositions in terms of "If. . . .Then. . . ." statements. In the chapter the conceptual, analytical, empirical, and numerical models are explained verbally leading to mathematical expressions. Also the importance of geometric fundamentals is explained for mathematical formulation derivations including differential equations.

Chapter 6 presents the principles of genetic algorithm (GA) for optimization (maximizing and minimizing) problems in numerical and random sequences with short time solutions than other mathematical optimization models. At the beginning of the chapter, the first historical appearance of decimal number system is explained in detail. It is shown that the principles of GA method are easy to understand it can reach the result with only arithmetic calculations without heavy mathematics. In any case, it is necessary to explain the verbal aspects of the problem first within the framework of philosophy, logic and rationality rules. Throughout the text, the reader ambition for the subject and self-development principles is taken into consideration by giving the necessary clues. Different application examples are given with numerical applications.

Chapter 7 deals with a comprehensive explanation of artificial neural network (ANN) architectural, mathematical, and procedural functionality by consideration of input, hidden, and output layer combinations. ANNs and especially linear perceptron in classification method through logistic regression presents the first examples of classification procedures as the first entrance into AI shallow learning field. Scientific and logical thinking fundamentals in ANN architectural construction for the problem at hand are explained on philosophical and logical aspects, which were touched upon in the first three chapters of this book. Various ANN modeling application disciplines and some practical applications are given.

Machine learning (ML) principles are discussed in Chap. 8, with support for deep learning software applications and shallow learning foundations. It is stated that ML depends on the mathematical procedures, but AI imitates and mimics human brain activities. ML depends on human thinking as shallow and deep learning skills and abilities. In the text, a roadmap is provided for data reliability and subsequently model output prediction validation by means of different loss function types for the best (least error) AI model results. Both crisp logic k-means and fuzzy logic c-means classification methodologies are explained with their philosophical, logical, and mathematical backgrounds.

Chapter 9 is concentrated on the deep learning (DL) methodological procedures as a mixture of artificial intelligence (AI) and machine learning (ML) rules that encompass all what have been explained in the previous chapters. Among the DL procedures two of them have weight as the convolution neural network (CNN) and recurrent neural network (RNN) that are explained in terms of model architecture. Although CNN has feed forward architectural structure in the secession of input, hidden, and output layers similar to the RNN method, the former has convolution, pooling, and fully connected series of hidden layer structure and the latter has in each hidden layer recurrence possibility for better modeling results. In the text, training, testing, and prediction stages of CNN and RNN are explained in a comparatively (Fig. 1.7).

1.11 Conclusions

The main content of this chapter is to provide fundamental gradients in the principles of shallow and deep learning through the assessment of knowledge and information based on different civilizations and centuries starting from the prehistoric era to the present. Various brain activities that are external (five sense organs) and internal (mind, memory, reasoning, intelligence and scientific rational approximate reasoning inference thinking paths) are explained. Foundations of scientific thinking for model building possibilities are explained as for the determinism versus uncertainty,

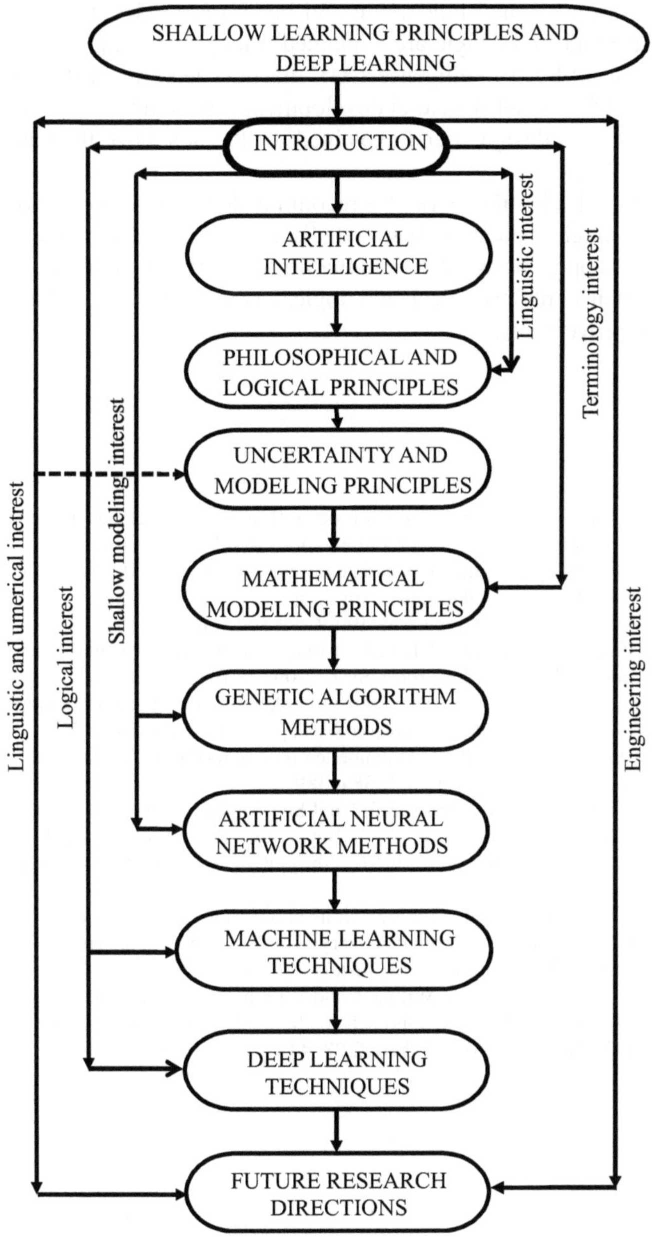

Fig. 1.7 Book content flowchart

randomness, fuzziness, and innovation and invention modes. Preliminary rules of crisp (bivalent) and fuzzy logic are explained briefly and comparatively for preparation to digest artificial intelligence (AI), machine learning (ML) and at the end deep learning (DL) model structural developments and solution algorithms. Finally, a roadmap has been designed and explained for the benefit of the reader about the content of this book.

For efficient AI scientific work, it is recommended that the content of this chapter be grasped through critical review and discussion in order to advance existing procedures. AI applications are accessible through off-the-shelf software, but even their inner science philosophical and logical components must be explored for innovative improvements.

References

Goodfellow I et al (2013) Joint training of deep Boltzmann machines for classification. In: International conference on learning representations: workshops track

Johnston RJ (1989) Philosophy, ideology and geography. In: Gregory D, Walford R (eds) Horizons in human geography. MacMillan Education, London

Krauskopf KB (1968) A tale of ten plutons. Bull Geol Soc Am 79:1–18

Krizhevsky A, Sutskever I, Hinton GE (2012) Imagenet classification with deep convolutional neural networks. Adv Neural Inf Proces Syst 5:1097–1105

Leopold LB, Langbein WB (1963) Association and indeterminacy in geomorphology. In: Albritton CC Jr (ed) The fabric of geology. Addison Wesley, Reading, pp 184–192

Manikandan R, Sivakumar DR (2018) Machine learning algorithms for text-documents classification: a review. Int J Acad Res Dev 3(2):384–389

Mann CJ (1970) Randomness in nature. Bull Geol Soc Am 81:95–104

Moe ZH et al (2019) Comparison of Naive Bayes and support vector machine classifiers on document classification. In: 2018 IEEE 7th global conference on consumer electronics (GCCE), pp 466–467

Parzen E (1960) Modern probability theory and its applications. Wiley, New York

Popper K (1955) The logic of scientific discovery. Routledge, New York, p 479

Sarton G (1884–1956) Introduction to the history of science (3 v. in 5). Carnegie Institution of Washington Publication no. 376. Williams and Wilkins, Co., Baltimore

Zadeh LA (1973) Outline of a new approach to the analysis of complex systems and decision processes. IEEE Trans Syst Man Cybern 2:28–44

Chapter 2
Artificial Intelligence

2.1 General

All human intelligence activities based on observation and imagination are transformed into viable and useful end products through conceptualizations based on science philosophy and logical principles. The history of science, and especially technology, is full of such intellectual inventions of gadgets, devices, and robots that have reached to this day. In a way, it is the uninterrupted accumulation of human intelligence from various civilizations with innovative, prestigious, and acceptable results. With tools from the prehistoric Paleolithic Stone Age human mind started to function with its first inventions as stone sharpeners, cutters, arrows, tools, etc. Such activities emerged when the human mind produced useful works with existing materials as stone, soil, wood copper, bronze, and iron. The main system is based on discovering and inventing concepts that are essentially historical (Newell 1982). So far, only a few works, such as "Thinking Machines," provide anything beyond this, and there is still no deliberate historiographical claim (McCorduck 1979).

Artificial Intelligence (AI) is a developing science field with the capacity to perform functions such as logic, which is increasing its importance in the direction of robotization (automation, the process of transforming a human into a robot) and mechanization. As with mathematics, even better AI applications are possible with an understanding of the principles of thought, the philosophy of science, and ultimately, rationality based on logic. AI continues on its way if all kinds of human intelligence are transferred to robots and machines within the framework of the logical rules. One of the most important questions here is, will AI devices replace human natural intelligence? The thought of the author of this book, like many others, is that such a situation would never arise. With the advanced capabilities and complexity of AI, systems are likely to be used more frequently in sectors such as economics, finance, medicine, energy, manufacturing, production, education, engineering, transportation, military, communication, and utilities. An important stage of AI is to inextricably linked human and machine.

Z. Şen, *Shallow and Deep Learning Principles*,
https://doi.org/10.1007/978-3-031-29555-3_2

Rational language rules have been used in the construction of the Egyptian pyramids. Even back then, natural intelligence was used through slavery instead of machines and robots. In fact, it can be said that the designs in these pyramids are made with reflection of natural intelligence (NI) as AI.

The indications for the first studies on artificial neural networks (ANNs), which are among the subjects of AI, were initiated by McCullogh and Pitts (1943). Initially, nerve cells use the fundamental mathematical modeling based on fixed-threshold logic elements. Later, Hebb (1949) developed a learning rule for ANN and caused this method to become widespread. A decade or so later, Rosenblatt (1958) took the so-called single linear perceptron model to an even higher level with the learning rule. In fact, although AI has existed in every period of history, its official introduction to the literature under this name was mentioned by McCarty (1960) in an international symposium. Ten years before that, Turing (1950) published an article on whether machines could influence humans. As he stated,

> While achievements have been made in modeling biological nervous systems, they still provide solutions to the complex problem of modeling intuition, consciousness, and emotion – which are integral parts of human intelligence.

He built his main reward system on discovering and inventing concepts that are essentially historical (Newell 1982). Until now, only a few works, such as "Thinking Machines," provide anything beyond this and still do not make deliberate historiographical claims (McCorduck 1979).

Especially after the 1990s, expert systems in the field of science started to be discussed and it was emphasized that much better results could be achieved by bringing together the views of different experts on a subject. Since more than one expert may have different experiences and knowledge in the same field of application, it is necessary to obtain and integrate information from more than one expert in order to generate an effective expert system. A Delphi-based approach is proposed to obtain information from multiple experts. Delphi was defined by Delbecq et al. (1975) as:

> A method of systematic solicitation and collection for decisions on a particular issue through a carefully designed series of sequential questionnaires interspersed with summary information and feedback from previously derived opinions and responses.

One of the most important issues developed in relation to AI is the work by Zadeh (1965, 1973, 1979), who developed fuzzy logic (FL) principles, which represent very closeness with natural logic instead of crisp (two-value) logic. These logical principles and rules, which were excluded for years after their first appearance in the West, have a very important place in AI applications today.

Considering the transition of human's natural soft abilities to some tools, software, robots, and machines, AI is a gift from the Creator to human. Mankind has been curious about AI since ancient times, albeit unconsciously. Early ideas and their mechanical means were either in the form of unrealized ideas, or were simple weapons, tools, and implements, or unimplemented drawings. The foundations of AI emerged in real and contemporary philosophical thought during the Ancient Greek period (Archimedes, BC ~ 159), in oral descriptive writings (Heron, AC ~ 50) in the early Roman period, and in medieval robotic mechanical drawings and explanations in the period of Islamic civilization (Abo-l Iz Al-Jazari, ~1200). In general, the foundations of AI thinking are the reduction and transformation of highly complex problems to human understandable levels with simple linguistic, symbolic, mathematical, algorithmic, and logistical solutions. AI studies may not only require formal training but may also depend on the individuality of group experiences, which can be expressed in linguistic terms that can be translated into mechanical activities through various machines. Today, AI explores human brain and mind functionality, machine learning, and healing machines and robots, whose main purpose is to serve humans.

Theoretical and practical studies and applications of AI have increased since the advent of digital computers, and these studies deal with uncertain, possible, probable, incomplete, and missing data situations through approximate reasoning models. Computers help visualize and implement AI configurations, provided that human intelligence can be translated into computer software through a series of assumptions and simplifications. Grammar and knowledge can be represented by modeling, reasoning, and decision-making procedures in addition to numerical data operations. The main source of AI is human brain functions that give rise to smart tools, gadgets, and machines aimed at improving social and economic life. Human intelligence is the main source of AI not only in smart vehicles and machines but also in any fundamental or applicable science and technology study. Scientific and technological advances cannot be achieved without the transformation of human intelligence into AI, which can appear in the form of formulations, equations, algorithms, software, models, and machines.

This chapter provides a brief explanation by referring to these forms in terms of AI. The main purpose of this chapter is to touch on AI issues that have sprouted throughout history and continued to increase exponentially especially after 1950. The foundations of the philosophy of science and the rules of logic are not fully covered in the international literature in an integrated way.

2.2 Artificial Intelligence History

The development of AI can be considered in two parts, pre-Renaissance and the recent period about 1940. The main purpose here is to reflect the historical development of AI missions from an unbiased perspective. Unfortunately, AI is considered the property of Western Civilization, but there are ancient civilizations where human natural intelligence (NI) could give rise to AI. This point has been documented and it has been understood that AI is a common feature of humanity as a whole.

2.2.1 Before Renaissance

Especially in early civilizations, technology developed quite independently of philosophical thoughts and scientific principles. In addition to foraging, people tried to take safe precautions against dangerous natural events by providing shelter, protection from wild animals, earthquakes, floods, droughts, fires, and extreme weather conditions. Different civilizations such as China, India, Mesopotamia, ancient Egypt, Greece, Islam and the West have contributed to technological and scientific developments with the principles of AI logically throughout history. Primitive technological developments in the historical origin of humanity have been identified in extensive archeological reports. Large-scale technological advances are detailed in the written documents by successive civilizations. Beginning with ancient Greece and after about the twelfth century, there were many researchers trying to use water power to operate some simple but effective devices for the service of humans. During the Hellenistic period, Heron (10–70) and Vitruvius (BC 80–15) of the Roman Empire were among the first thinkers to use water power for different human activities such as water transportation, water mill, and water clock. Most of the documents are lightly described in works of Heron from Alexandria and Vitruvius before the Islamic civilization. Unfortunately, there are no designs or drawings left from them, except for some scenarios related to the use of intelligence. On the other hand, Abou-l Iz Al-Jazari (1136–1206) of the twelfth century is the father of hydraulically powered robotics, who in many ways contributed significantly to the transition of scientific writing to the Western renaissance concerning basic root and fruit state of robotic, cybernetic, and artificial intelligence shoots. Abou-l Iz Al-Jazari has drawn many mechanically appropriate designs in his original handwritten Arabic book "Kitāb fi ma-'rifat al-Hiyal al-handasiyya (Arabic: ي صناعة الحيلالجامع بين العلم والعمل النافع ف), translated into English by Hill (1974, 1998) as "The Book of Knowledge of Ingenious Mechanical Devices."

Muslim thinker Abou-l Iz Al-Jazari wrote and explained, reviewing all the work of previous researchers from different cultures, along with his own drawings, designs, and devices for the use of water, animal, or wind power in the admirable automation species. However, although he is the pioneer of robotics and cybernetics not only in the western world, but also in the Islamic world of which he was a member, his name is not heard at all in education systems. His works were uncovered by German historians and engineers in the first quarter of the nineteenth century. Later, a British engineer named Hill (1974) translated his book from Arabic to English, summarizing modern automatic and robotic designs from the twelfth century. Nasr (1964) reviewed the technological, philosophical, and naturalistic views of Muslims thinkers. The first and most important work on Al-Jazari's biography was done by Wiedemann and Hauser (1915). The construction of Al-Jazari's devices is possible even today, because he has written clearly and in detail the common workings and every aspect of the elements.

Before Al-Jazari there were several Muslim thinkers, including the Banu Musa brothers, Harizmi and Radvan, who gave birth to technological ideas and tools through intellectual thoughts. The devices produced by these three brothers were later modified and used by Al-Jazari. These brothers evaluated everything presented to them by earlier researchers, especially Vitruvius and Heron.

Automata are the immediate ancestors of Europa's elaborate water clocks and were mechanically powered by water. Automaton making in Islam and later in

Europe was one of the factors that promoted them to intellectually develop rationalist and mechanistic explanations of natural phenomena, a highly fruitful attitude in the development of modern science. However, it was Al Jazari's monumental clock that showed off his most impressive automata sequences.

There were many full-scale machines that Al-Jazari described for raising water, and all of them contained features of great importance in the history of machine technology, and even today some of these parts are in common use in any machine. In the first robotic design, water power is used to raise the left and right hands of a robotic man on an elephant, as shown in Fig. 2.1. The water injection from the right pipe hits the left of the tool behind the elephant, raising the robotic man's left hand. This is the simplest initial design for robotics in the history of technology.

Another interesting machine from Al-Jazari that is important for today's AI technology is shown in Fig. 2.2. This machine has an important place in the development of steam engine and pump machines.

This system uses wind power, which is converted into mechanical energy by means of panels placed radially around a horizontal axis. A gear wheel is mounted on the same axle, which transfers the rotational motion to another wheel. Towards the outer edge of this wheel is a pin fixed perpendicular to the surface. This dowel went into a slot rod, the bottom end of which was rotated under the box. By turning this wheel the pin moves up and down, and as a result the rod moves from right to

Fig. 2.1 Robotic mean working by water injection

Fig. 2.2 Piston, cylinder, and valve

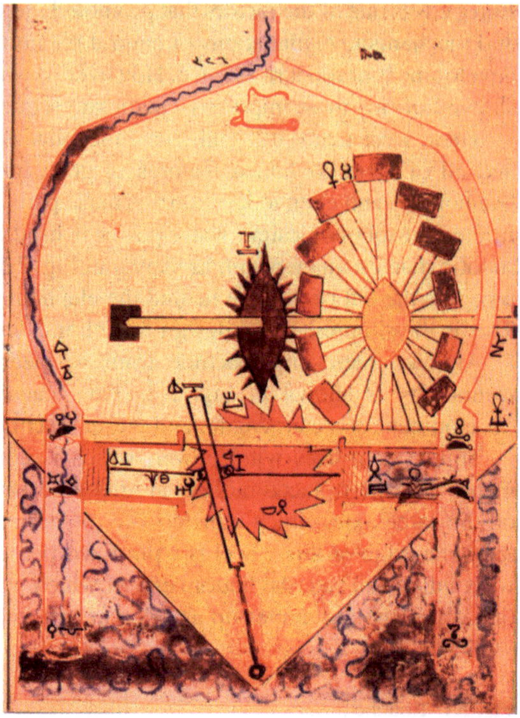

left and left to right. The other two rods are attached to the sides of the slit rod by ring and punch fittings, and a piston is attached to the end of each. Pistons move inside horizontal cylinders. Each of the pistons is equipped with a suction pipe descending into the water. The distribution pipes are reduced in diameter shortly after they are separated from the pistons and are interconnected to form a single outlet. As the split rod oscillates, one piston pumps and the other sucks. The pump working with the double-acting principle is a very early example of the back and forth piston movements.

His book contains comprehensive methods of automatic devices and the most up-to-date information on mechanics. It systematically includes the technological development of various devices and mechanisms with examples and in-depth intellectual explanations on automata and automation knowledge and scientific information. Hill (1974) highlights the importance of Al-Jazari's work in the history of engineering, cybernetics, and automation. Not only did he absorb non-Arab techniques, no work until the nineteenth century could replace his robotic designs. The influence of their inventions can be clearly seen in the subsequent design of steam engines and internal combustion engines, which paved the way for automatic control and other modern machinery. The influence of Al-Jazari's inventions is still felt in modern contemporary mechanical engineering (Hill 1998: II, pp. 231–2).

Even looking at the hand-drawn mechanical devices (Figs. 1.1 and 1.2) in his book and comparing them with today's robots makes one at first glance that he designed the water-powered machines as artificial intelligent productions. It is stated that his thoughts and drawings gave intuitive feelings towards mechanization in the nineteenth century (Sarton 1950).

2.2.2 *Recent History*

In the literature, the innovative AI principles of the Western Civilization, especially from 1940, are presented as if there were no AI devices proposed and designed by previous civilizations. There are at least two civilizations, like Ancient Greece, that did not contribute much, but there are actual designs based on AI thoughts of Aboul-l Iz Al-Jazari, which have cybernetic roots in Islamic Civilization.

Today, the term AI should also be understood in terms of the conceptualization of an intelligent robot or machine that automatically performs different functions, with its increasingly widespread development in engineering, social, economic, environmental, defense, agriculture, industry, transportation, and different sectors.

In fact, AI meets all the work done to study the logical foundations of natural human intelligence functions and to deduce their rules and transfer them to robots and machines. The designs at the end of these studies are to reveal a reasonable level of rationality. On the basis of AI studies, there is an approximate reasoning by harmoniously combining the set of logical relationships that arise in the natural way of thinking of humans. This is how AI is progressing and can therefore increase its interoperability over time by further elaboration and strengthening in rationality. AI development will evolve over time, focusing on the abilities of inventors and scientists in the public, but this development will never reach the level of human natural intelligence (NI). Emotional intelligence is happiness, joy, sadness, grief, fear, etc., that AI cannot detect thought modeling, which is an expression of human perceptions. In this respect, it is not a good idea to expect complete perfection in AI. But it can continue to evolve, especially through computer software programming.

Modern studies of AI began shortly after the Second World War, with the advent of computer technology and engineering in the 1950s. Digital computers provided a space for the simulation of natural phenomena in different disciplines (engineering, economic, social, etc.), and therefore, these early studies triggered human thinking in the direction of AI. A brief history of these recent developments is provided by Russell and Norvig (1995).

The first words about AI were uttered by John McCarthy in 1960. It is stated that a machine can do many movements and functions that a human can do. Today, many rational thoughts and behaviors can be scientifically simulated and modeled by programming computer software. With the improvement of these artificial behaviors made by machines over time, it has reached today's AI levels and will continue to develop from now on. Human intelligence is needed to solve all kinds of problems with scientific, rational, and logical rules, and machines and robots are needed to apply them. Since the first word of AI refers to machines and the second word refers to the intelligence of human mind, the first is an artificial and the second is a natural function. In the last 70–80 years, the following situations have played a role in the development of AI and continue to play a role. The following six items provide information from 1950 to recent years

1. The Birth of AI (1952–1956): Before the term AI was coined, there were developments in the cybernetic and neural field. The networks, which started to attract the attention of both, began to be felt in the scientific community and the public. In particular, the Dartmouth Conference held in 1956 served to identify the first cases in the field of AI. Thus, a roadmap emerged that led to the development of AI in the following years.
2. AI (1956–1974): In addition to solving algebra and geometric problems, computers of the time continued to develop AI a little further, for the first time speech qualities could be determined to some degree.
3. AI (1974–1980): In this period, the advantages and disadvantages of AI began to be mentioned. In this somewhat turbulent time period, AI methods were still able to continue to evolve.
4. AI (1980–1987): The gradual importation of expert systems to solve specific problems coincides with this period. Thus, verbal and survey-like expert opinions called Delphi, which were put forward by experts, were collected and started to be used in AI. It has always been the case that the rules of logic based on expert opinions, become functional as a whole. In this era, methods such as artificial neural networks (ANN), genetic algorithm (GA), and fuzzy logic (BM) have begun to emerge.
5. AI (1987–1993): With the use of desktop and laptop computers in the following years, an explosion occurred in the application of AI principles to different subjects, instead of computer centers that were not easily accessible by everyone, and this situation has continued to the present day. Later, between 1993 and 2000, new applications were developed in certain areas and the concept of "machine learning" emerged.
6. AI (2000–Today): In this period, especially the internet and later the web has successfully entered human life. Thus, a number of new concepts have emerged by both processing and transforming very large data sets such as "deep learning" and "data mining" into meaningful inferences.

2.2.3 AI Education

Although many research centers and universities are striving for AI, most of their work is initially based on some well-developed methods, machines, robots, etc., through approximate reasoning. However, many institutions, especially universities, do not have the philosophy of science and logic principles in their curricula, but have some mathematical variations (calculus, analytic, linear, differential and integral equations, uncertainty issues including probability, statistics, stochastic processes, chaotic methods), all have intelligent trigger components of thinking and critical appraisal software capabilities. This is why in many societies technology transfer is welcomed without being altered, enhanced, or replaced by better options.

Advances in AI have led to automation that seems far from human intelligence, but ultimately they tend to approach human innate ability similar to human intelligent thinking, reasoning, inference, and imagination. It is noted here that despite the overwhelming emphasis on mathematical principles in AI, geometry or shape visualizations are the main initiators of AI.

Here, some parts of human thought, concept and imagination, subjective and objective dimensions are partially examined. Some AI methods are mentioned, and especially the fuzzy logic (FL) inference system is explained in detail (Chap. 3), where not only mathematical symbols but also primary verbal logic rule foundations play the most important role in any AI activity.

The modern birth of the AI business originated as science fiction during World War II and was featured in the first software in the 1950s, "Can machines think?" it was Turing (1950) who asked. Then in 1955 the first modern AI program appeared, and the notable trend for it began after 1960. Today AI is seen as a discipline that has emerged in the past 4–5 decades with a mixture of various topics, in between mathematics and crisp (bivalent) logic, probability, statistics, stochastic, etc. Collectively, they seek to imitate and mimic human cognitive abilities. AI advances are closely linked to computational facilities that provide capabilities to tackle complex tasks previously unachievable with human intelligence alone.

2.3 Humans and Intelligence

All activities in this world can be thought through human thought, work, practice, and process. These activities are mental, social, educational, research and development, economic, environmental, military, health, etc. Each individual must engage him or her with some and perhaps more of these activities in order to achieve a personal purpose as a service for men in society or for the improvement and development of living standards. Although man has abilities as five sense organs that help to bring knowledge and information from outside (Chap. 1, Sect. 1.2), it is better to process them in a subjective or objective way. Figure 2.3 is a simple model for a person in the physical and to some extent spiritual environment.

The ultimate goal is to produce useful information, scientific knowledge, and technological tools that will help people live more comfortably. The input component of this figure is most influenced by the physical, social, and spiritual aspects of human life in the physical realities of life. For example, washing machines, dishwashers and similar service tools began to adhere to fuzzy logic principles for better social activities. Today, intelligence is considered given to express the physical developments of life, while other aspects of life such as social, cultural, and spiritual dimensions are neglected. Anyone who will be most satisfied with AI will also be satisfied with other supportive aspects of life. If there is a balance between the three input components, a peaceful and productive enlightenment and AI can emerge.

Other most important software is based on philosophy and logic principles. Enormous amount of knowledge, information, and practical skills can accumulate and remain unproductive agents in man's daily life as without any innovative developments. This is especially true if philosophical and logical aspirations are lacking in human thought. Especially, without the philosophy and logic of science, knowledge and information are like concrete without reinforcement. Every society, today and in future, seeks and encourages individuals with innovative ideas and encourages generations, at least as an improvement and development of existing facilities. AI science cannot exist without philosophical and logical thinking components. Also, for improvements, the AI must be "suspicious" of the gadget's behavior in our environment.

2.4 Intelligence Types

Intelligence provides the ability to rationally acquire knowledge and skill for productive application. It is the way to learn, understand, and provide new knowledge about the relevant event. In almost every successful operation of man's worldly activities, first of all, it is necessary to use the existence of the mind and to develop it and to set sail for generations of knowledge that can be innovative. We can say that the reason is to distinguish the good from the bad, starting from the basic knowledge that one has intelligence that provides a context between this information, because the origin (etymology) of the word mind means "to connect." It means expressing an opinion that comes to an approximate conclusion by passing through the human thought function of connections. Intelligence is the ability to see reality well and to produce solutions accordingly. The source of happiness for a smart person is success. In particular, reflecting the happiness of knowledge production to others and activating their minds adds happiness to happiness. Thus, it is possible to try to reach an advanced level of knowledge by criticizing existing scientific results.

God-given intelligence is one of the most important abilities that enable people to make wise inferences. There is no foolish person, but it is necessary to know the ways and methods, and to use each individual by triggering them in line with their own capacities and abilities. Throughout the history of humanity, humans have continuously provided the discovery of different tools and equipment with the inferences made by their own mind. In a way, each era has its own intelligence.

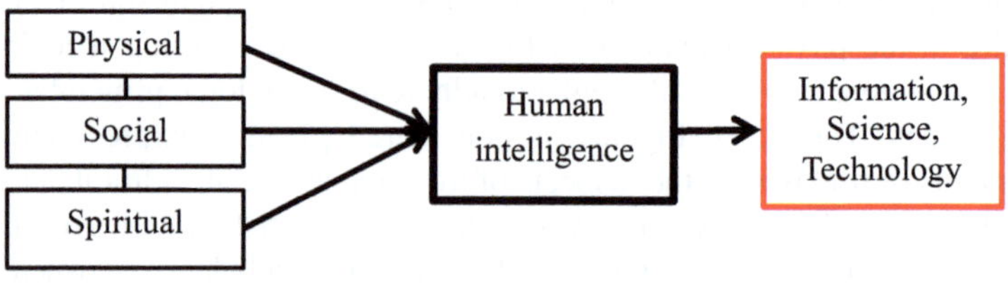

Fig. 2.3 Human activity models

Thanks to these intelligences, different civilizations have been able to maintain their existence for a certain period of time. By using them in continuous functions, they have improved the levels that do not decrease over time, but at least slightly increase. Man has also led to the emergence of a sustainable lifestyle by connecting some of his intelligence to the tools and equipment he invented. Unlike mind, intelligence is the ability to first understand an event of interest, grasp the existing connections for this, and find a rational solution to the problem by reasoning and making rational explanations. Intelligent people are also called scholars, and their understanding and comprehension skills are quite high. Even if a person does not solve an event himself, if they understand the details of the solution, they will be happy, and this feeling of happiness can lead them to success.

After what has been explained above, is an intelligent person in the form of a conceptual confusion? A question may arise. As mentioned earlier, each person has intelligence, but their degrees are different from each other. It is tried to raise the degrees of minds through education and training, and thus, according to the function chart in Fig. 1.1, the level of knowledge of the person increases, but the increase of the intelligence level of each person is quite different from each other. Although education contributes to intelligence, one's own curiosity, desire to learn, and interesting situations also play a role subjectively. There is no standard mind in the world, just as there is no standard intelligence. But intelligence, even if its measurements are limited, can break records over time, exceeding current limits.

Today, AI studies have begun to increase their impact on different sectors and continue to gain more importance. Similar definitions such as "gifted intelligence," "emotional intelligence," "cartesian intelligence," "active intelligence," and "practical intelligence" are importat. For example, those who have practical intelligence and even those who are not highly educated make many inventions called "inventor" with their practical intelligence. It is necessary to value them in a society, because intelligence is not only among those who are educated but also among those who make inventions according to themselves, but it is not appreciated because it does not have a title.

A smart person becomes happy by gaining the appreciation of those around him with his smart decisions and can open the minds of others or be watched and talked to with admiration. Intelligence begins to develop in the infancy and childhood and continues to increase up to a certain age, but declines with age. There is a decrease in the level of intelligence and, in parallel, a decrease in mental function.

The phenomenon of intelligence also has a technical content, so it is technically possible to measure it, whereas there is no technical phenomenon of the mind. In general, there are about 12 types of intelligences that are interrelated or chaotically related.

(1) *Visual Spatiotemporal Intelligence*: As described in Section 1.2, visual inspections provide an information gathering. The information that a person perceives and stores in his mind regarding the sense of sight is characterized as visual. The phrase of "I wouldn't believe it if I didn't see it with my own eyes" shows us how important visual information is for intelligence. If only verbal information is given while making a presentation, it is called chatting. A person can

perceive verbal information at a certain level of knowledge when he first hears it, but it cannot be said that he visualizes concrete information in his mind with every word he hears. Verbal information may remain abstract in minds, but its concretization is provided by visual information perceptions. Visual information can originate from reality, that is, from nature, or it can be perceived from virtual environments. The ancient Greek philosopher Aristotle (BC 384–322) claimed that he could see with a light coming out of the eye. According to this discourse of Aristotle, the eye must see in a dark environment. However, over 1300 years later, the Muslim physicist and the first master of optics Ibn al-Haytham (965–1040) saw that Aristotle's view was wrong, thanks to the rays reflecting off an object as a light beam emanating from a source helps to see.

Visually perceived information is also contained in memory called visual memory. The information in this memory is more meaningful than the verbal and is used consciously when appropriate. We can say that this is in the mind-open memory. Of course, the verbal mental infrastructure has common points with the visual mind. We cannot say that they are independent of each other. The fact that these two work together also means that memory is strengthened. Thus, both tangible and intangible information are kept together and intangibles are tried to be made more concrete over time. With the visualization of some imaginary visual shapes that are not visual or auditory, the human memory is given even more richness and efficiency. For example, feeling happy is neither a verbal nor a visual phenomenon, it is also called psychological comfort of the human soul.

As the basis of visual knowledge, it is recommended to use morphology instead of this term, which is called geometry but remains barren, because when geometry is mentioned, Euclidean (BC 330–325) plane geometry comes to mind according to the classical information given by education system. Man develops his mind, which is a wealth of verbal knowledge, together with a picture of morphology from everything in nature. In the early days, human beings increased the wealth of knowledge by making use of the objects in nature, then developed unknown geometric shapes to solve some scientific problems with the developing thought and built their scientific innovations on them. Muslim thinker Nasreddin Tusi (1202–1274) developed spherical geometry for the first time out of Euclidean geometry. Since necessity was the mother of inventions, when Muslims realized that they could not determine the qibla (direction from their place to Mecca) with the plane geometry, they discovered spherical geometry, which plays the most fundamental role in determining the lines of latitude and longitude today. The routes of the airplanes are determined accordingly.

Lobachevski (1792–1856) laid down the principles of hyperbolic geometry, which is named after him. Rieman (1879–195) realized the imaginary and concrete presentation of a geometry that would be useful in a physics calculation such as relativity in the West in the following years.

A very important conclusion to be drawn from what has been said above is how important it is to memorize verbal, visual, and figural information as messenger information set rather than numerical data in human thought. Numerical studies and mathematical equations and models have always been built on these foundations. It can be said that in the future, verbalism and morphology will inevitably open the way to enlightenment like a bulldozer. Although numerical data and mathematical principles are considered important in some branches today, there are seeds of verbalism, visuality, formality, and philosophical thought that will be mentioned later, and the principles and rules of logic that help to make rational inferences from it.

(2) *Verbal-Linguistic Intelligence*: Auditory hearing is perceived by ear, and three parts of the ear, namely outer, middle and inner ear, enable the person to hear. The scoop, which is the outer ear, allows the sound to be collected and transmitted to the middle ear, and snail-shaped structure in the inner ear to the nerves transfers the sound vibrations to the inner ear, and transmits the heard sound to the brain. The perception of the first information begins with the ability to hear, and thus, after a series of sensations, we automatically place in our minds the meaning of the words we heard by associating them with an object or figure, The main source of hearing is through the information in the conversations that everyone makes every day, and over time, the person can even make a meaning for himself from the tones in the speech, what is the meaning of the pronunciation of the word he hears. Is it a courtesy in the word used? Is it anger? Is it music? Lesson? Is it a recommendation? These recommendations are highly subjective and linguistically deducible. It is possible to understand what the event means from natural or artificial sounds. For example, we try to adjust our behavior according to audible information by instantly perceiving and making sense of events such as thunder, airplane noise, shouting, and the approach of a vehicle behind us. Each person's hearing ability is different from the other, and it is a fact that people, especially in advanced ages, suffer from hearing loss, albeit partially. Since incidental situations in the hearing ability may not be considered as important in terms of health, since they do not cause pain in the body. Voice announcements are always used to inform the public in cases such as disasters, wars, earthquakes. The memory in which verbal information is stored can be called verbal memory, but because some of what is in this memory cannot be embodied in the mind, forgetfulness often occurs. This memory may be filled with dull information called memorization, which is obtained by hearing and whose meaning is not even known, For example, memorizing poetry, memorizing the Qur'an and memorizing the meanings of the words and sentences in the mother tongue is an oral knowledge, but it cannot be productive, it can be communicated through advice and conversations. It should be understood that a person has to consult books called dictionaries to learn a word, term, concept, or definition that he does not know. Each language has its own syntax and rules that lie in grammar.

(3) *Intra-personal Intelligence*: These people are dependent on their own minds without being confused with interpersonal thoughts. They learn independently through reflections. They can explore their inner workings speculatively or rationally through emotional feelings. In this way, they wisely look inward for the shaping their feelings and motivations to reach the final solution to the problem. They can plan, run, activate, and manage intelligently with their own abilities. They think about what they want to achieve and how they can achieve the goal. What they can rely on and produce is not only verbal knowledge but also numerical information and data interpretations with verbal logical explanations and interpretations.

(4) *Inter-personal Intelligence*: In general, "people get along with language, animals with smell." This implies that people can understand each other linguistically through words and sentences. To overcome differences of opinion, linguistic communication helps to persuade each other. There is a distinction between verbal and numerical information, and when we look at the development processes of historical knowledge, verbal knowledge always takes place in the field of thought before numerical data. Verbal knowledge is possible not only with local or foreign languages but also with rhythmic movements of hand, arms, and body. Even today, texts filled with verbal expressions rather than numerical information play the most important role in books, scientific articles, projects, and research studies. In this respect, every profession has verbal knowledge and a wealth of information. Engineering terminology words and phrases help professionals understand each other without much discussion. Among the different fields of science, oral information is the most effective in any discipline because professionals communicate with clients as a preliminary access to solution. No matter how strong their expertise is, the expert tries to better understand the client's desire by asking questions. Whoever makes a request first of all tells the desires in their mind. Even in such requests, there is not enough explanation and information to fully understand the problem solution and the experts who listens to the same desires decides to make solution plans accordingly. Thus, the expert, who collects verbal information for solution, determines the appropriate means for the client and provides application possibility alternatives. Over the years, with the experience of hearing and speaking, words in every language gain meaning in our minds and become the common perception tools of the society. The person uses the language every day and therefore immediately grasps the meanings of common words and phrases. As a result of understanding, the mind acquires a habit of making sense of words effortlessly by hearing and reading them and allowing them to be passed on to others, if necessary. Each word is a symbol of different internal and external properties of real or imaginary objects. The community of subjects means nothing without the consciousness of the human mind. Just as a newly born child is named, a subject is named so that it is known. As there are different languages, the topics should have different names in the languages. People recognize that object with the words in their mother tongue and take an attitude accordingly.

(5) *Naturalistic Intelligence*: The most important influence in the germination of scientific ideas begun with the perception of images of objects encountered by people. Babies and toddlers who do not even know how to talk yet can boost their imaginations by making out of visually appearing objects, like a camera. By assigning the likeliness of different shapes in a single unit, they categorize with more general nomenclature and enrich the vocabulary by classifying them so that they reach their groups in the mind of the visible environment, giving them a name, and then giving another name to similar elements in that group separately. Enriched vocabulary tends to determine the relationships between them by positively influencing the mentality. Meanwhile, they try to process some words in their memory, not only visually but also by hearing and then try to make them concrete. Human civilizations follow each other since the creation of humanity to the modern age (see Fig. 1.3). As intelligent creatures, they have always maintained a rational relationship with their environment at the same time. With five sense organs, daily sheltering, feeding and dressing activities, and the search for raw materials are provided from the appropriate sources available in nature for the continuation of the human race (Sect. 1.2). Some animal and tree creatures are distinguished by their existence in different periods of the geological time scale, but the human race has been in continuity since creation seems to continue in the future, thanks to a gift of intelligence from Allah (God). Whatever natural conditions in the past, the human race could adapt to present conditions with intelligence, rationality, plausibility, and perseverance. Human natural intelligence (NI) persevered and found the least risky protection against even the most dangerous natural events such as floods, droughts, earthquakes, tornados, and alike. The light contemplations in these examples show that information is received through sense organs and subsequent pieces of information are transformed into meaningful and useful information through mental functional processes. NI has offered numerous alternatives and improvements throughout its many years of experience adaptation methods to comfort and embrace the natural environment, and hence the improvement in quality of survival has led to the continuous increase of NI. The backbone of NI is thinking about the visible and invisible aspects of life and generation of various thoughts. If the question is which sense organ helps human to grasp knowledge, it is undoubtedly the eyes that receive information from forms, shapes, geometries, sketches, and landscapes, especially in the early stages of human life. This type of knowledge acquisition is the most basic in the context of this book, called the visualization one of sources through images and figures. The reader should keep in mind that the geometry in this book eludes all forms and shapes that appear by sight or visualization with the mind's eye, even imaginary forms of that can never be translated into real knowledge without design prototypes in mind. For instance, all of the pictures and drawings on prehistoric cave walls relate to NI conceptual visualizations that were somehow converted into more ideal forms after eye examinations. Humans are born with five sense organs in the first stage of entry into this world, without a prior source of knowledge (Sect. 1.2, Fig. 1.2).

The passage of time is a dynamic and effective tool that provides various inspirational perceptions by these organs. The most effective and thought provoking examinations begin to emerge through the eye with visualization and the ear with hearing (auditory). While these two earlier inspirations leave their mark on the mind, the most outstanding are the dreams of visualization, which are the true sources of the sprouts of knowledge. This point supports the idea that human beings grasp prior knowledge with shapes and geometrical figures. Babies begin to smile by looking at the face of another person (mother, father, brother, alike). Although their names are not known exactly at such an original stage, it is possible to say that basic information is kept in mind with geometrical shapes that inspire learning what it is after a few repetitions. It is a common practice to stimulate the baby with different colored and sound toys to attract the attention. These primitive toys provide audio-visual physical sight and sound so that baby can concentrate on acquiring knowledge without any object and subject name. It is certain that the preferable stimulus is sight first and then sound. Therefore, it can be concluded that geometry and shapes in the forms of sketches, drawings, paints, plans, figures, graphs, diagrams, pictures, images, etc., are all informative linguistical information reflections of knowledge, discussion information, commentary, criticism all of which accumulate as a perception storage in mind for further physical, metaphysical, mathematical, and in general ontologically activities. The Aristotelian (BC 384–323) philosophy focuses on forms and their interior materials, as mentioned before, where forms include shape alternatives. In everyday seeing the surrounding shapes in different geometrical forms and thinking about their existences leads to better conceptualization of formations and in a better mood spiritually can be inspired, for example, by the natural landscape, some of which has multitude of shapes. In every residential area (cities, towns, villages) more ideal shapes of various colors are in front of us. The reader can compare these two positions and make a personal comment on the inspiration of the shapes as reason over spiritual, logical, social, and philosophical inspirations. This point indicates that the human mind is more inspired by shapes than other ideal geometry. Is there a day you do not come across geometric shapes? Therefore, the human mind is exposed to shapes and geometry much more than other phenomena. In this way, the brain makes more interpretations and calculations, statically or dynamically. This provides spatial exploration and the ability for approximate geometric reasoning. After all what have been suggested above, the most important conclusion is that because geometry is so important in NI, better understanding of its deep meaning, digestion of reasonable information, for a better understanding of many disciplines, opens the way of the "philosophy of geometry." This lies as a cornerstone for those who want to elaborate ideas in a very lively, rational, and sensible way. At this point, it should be reminded that the philosophy of geometry may not be necessary for many, but it provides an automatic mastery for those who want to dive into deep learning and establish roots of much disputable knowledge in any discipline. It is the key to learning and to teaching for better enlightment in any education system.

(6) *Musical-Rhythmic Intelligence*: Anyone equipped with musical intelligence can transmit the rhythms, sounds and patterns that give such characteristics. One can be engaged in fine arts by through productive thinking based on inspirations leading to rhythmical knowledge. Fine arts, including architectural designs, are also included this intelligence category.
(7) *Bodily Kinesthetic Intelligence*: This is about mind–body pair cooperation, joint action and coordination. This type of intelligence plays the most important role in athletic activations. The word kinaesthetic itself refers to the sense of mediation of muscular organs in harmonious cooperation with bodily movements and tensions. A person with bodily kinaesthetic intelligence can use the physical abilities of body physicality to deal with objects around him. It is possible to recognize body abilities and physical limits, and one can communicate with body language and actions to convey messages.
(8) *Philosophical-Logical Intelligence*: In general and simple sense, philosophy is defined in English as the love of knowledge and wisdom within the framework of materialism, while in Arabic it is called Hikmah, which also includes religious knowledge. Philosophy of science leaves out all spiritual, religious, and metaphysical concepts and focuses only on materialistic entities to reach rational and logical scientific inferences. Such a philosophy is known as the positivist philosophy of science. Positivist logicians place philosophy under science, which is not fully accepted by the author of this book. Since philosophical thought is not just the property of highly educated people but also a combination of knowledge and wisdom, ordinary people may also share such ways of understanding, discussing, and thinking to seek the ultimate truth, but this still remain vague, blind, and uncertain. Such obscure components enter the domain of philosophy in the form of doubt for a better and fairly clear understanding. Before any explanation, from the very beginning, the etymological and epistemological features of "philosophy" and "science" help the reader to grasp the contents of the next chapters. Unfortunately these words are often used even in academic circles, but without clear, concise, and satisfying insights. Especially the philosophy of science plays a big role in every subject for idea generations. Philosophy itself has five core divisions that are to some extent mutually inclusive: ontology, ethics, metaphysics, epistemology, and aesthetics. The main subject in the search for existence is ontology. Figure 2.4 represents each topic where ontology embraces all other branches of philosophy.

    Figure 2.4 shows interventions between different philosophical disciplines for philosophy professionals not interested in the philosophy of science. Ontology is the most comprehensive field of thinking regarding materialistic and nonmaterialistic (spiritual) figures, entities, properties, concepts, and the like. The nonmaterialistic part is limited to metaphysics. Ethics is a common area for every individual in any action, but there are also relative aspects for every civilization, society, culture, and religion in which moral identities play the most important role. Aesthetics is a very specific form of philosophy that may not strongly influence other philosophical activities. As one of the branches of

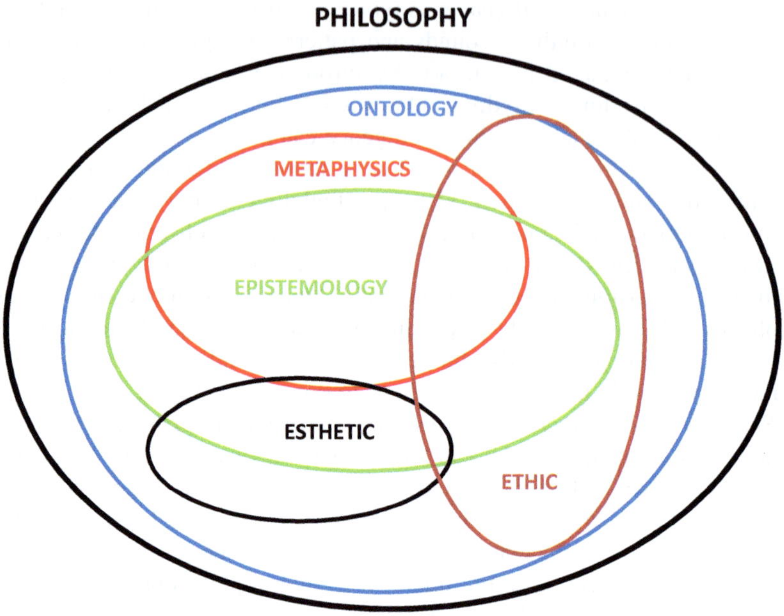

Fig. 2.4 General philosophical subjects

philosophy, epistemology encompasses the scope and meaning, understanding, and logical content of knowledge, and is therefore called the theory of knowledge in the form of scientific knowledge before it matures as (Chap. 3). It also deals with examining the nature, source, and validity of knowledge. To explore the epistemological world, two important questions can be asked: "How does one know and understand?" and "what is right or wrong?" As mentioned earlier, the effectiveness of thought content also falls under the umbrella of epistemology. The ultimate goal is to produce reliable, dependable, and valid information. Epistemology, "Can reality be known?" It seeks answers to a number of fundamental questions such as: Narrow skepticism is in which people claim that they do not have reliable information and that their search for truth is futile. This idea was well expressed by the Greek Sophist Gorgias (483–376 BC) who argued that nothing exists and that we cannot know it, if it exists [Jean-Paul Frederich Richter (1763–1825) in Foreign Literature, Science and Art Department. 12: 317. hdl: 2027/iau.31858055206621].

As the origin of the word, philosophy emerged from the combination of the Greek words "philo" and "sophia." In one sense, it can literally and clearly mean "love of knowledge," but epistemology has even deeper meanings. Philosophy is also concerned with abstract concepts and entities that include the freest, classical, and almost extraordinary thoughts, and therefore it first makes some inferences about them with rational approaches. In the past, it is accepted that all kinds of thought production was under philosophical issues until the eighteenth century. In this respect, philosophy is used as a concept that includes the sciences. Today, philosophy is perceived as separate from science. This is a potential error, because at the root of every thought is language, philosophy, and logic, which encompasses all rational reasoning and inferences. If one of them is not fully perceived, it is not possible for scientific knowledge to reach efficient productivity. It is thought that scientific subjects are no longer within the scope of philosophy as before, and with the emergence of many different branches in science, they are out of the scope of philosophy.

But philosophy is at the root of all sciences and arts as thought processes that are somewhat illogical at first, but gain rationality through logical rules and thus approaching even approximately acceptable results. It is unthinkable to satisfy people or satisfy the validity of science or art without philosophy. As a result of postgraduate studies, the highest academic title is "Ph.D.", namely "Philosophy of Doctorate". These words reveal the necessity for doctoral studies in every discipline to have philosophical foundations. It is essential for a person with a doctorate degree in a particular discipline to know the content of the subject rationally on a philosophical basis, not by memorization and stereotypes.

Intra-brain functions can be the most important, as it helps to classify and establish relationships between signals and information through logic, comparative, and critical operations. With the free thought of the mind, the unlimited horizons are chained by logic. In other words, logic means rational knowledge, mental activity within useful limits. A brain function whose logic is inconsistent or does not comply with the rules of logic cannot produce information and cannot make the necessary interpretations. Here, the organization used to produce knowledge or to bring different information together within certain rules is called logic and its rules are called logic rules. Thanks to these rules, logic has the task of optimizing memory usage by gathering information and extracting useful inferences from it and simply keeping it in memory. Even among the people, the phrase "There is reason and logic" shows that reason comes before logic. There can be no reason without logic, but there can be logic without reason. In this last situation, the mind stores unrelated information.

(9) *Scientific Intelligence*: The main purpose of scientific intelligence is to relate a reliable, easy body of knowledge and information contents. It also serves rational agreement between the scientists on the bases or rapid data assessment for easy rational interpretations and explorations. It takes three subjects from the philosophy segmental subjects, which are epistemology, aesthetics, and ethics without ontology and metaphysics.

(10) *Technological Intelligence*: When we look closely at the history of humanity, it is seen that the first intelligent activities were the technological studies. Because the first human being needed various tools and equipment to protect his lives against dangerous events. For this reason, they have produced technological tools in order to survive. First, they tried to activate their minds with geometric shapes and the shapes they saw or drew on the walls or the ground, then materials such as copper and bronze, and finally iron began to be used. In our modern studies, production of technological intelligence gains importance in successive stages as shown in Fig. 2.5.

(11) *Artificial Intelligence*: In the past, especially after the 1980s, "expert opinions" were not included in university curricula. Thus, even inventors without higher education were respected in society. The reason for this was the importance of verbal knowledge rather numerical information. The fundamental question is "whether verbal (linguistic) knowledge is the main source of digital intelligent inventions?" Human intelligence is infused with imagination and verbal inferences, which are then translated into formulations, algorithmic flowcharts, and computer software, all of which are products of human thoughts. Experts or expert system automations are far from human intelligence because there is an Arabic proverb that says "Even a donkey learns after repetitions." This does not mean that experts are like donkeys, but if the main functional aspects of the relevant topic are not known, this person cannot pass the knowledge on to others on the basis of philosophical, logical, and approximate reasoning. But even without such explanations, these experts are at the service of people with their practical and automatic abilities. These people cannot critically question the events they are examining. Their ultimate goals is not to move towards better AI methods, but are there such advances that they are trying to capitalize on? Real AI is strongly embodied in vision and science fiction, with the need to advance in fundamental research. This is tantamount to saying that AI is not just concerned with performance improvements in order to be able to model the world as a whole. The following points are important for philosophy to be a useful science.

1. Philosophy is the study of understanding.
2. Understanding is giving meaning to observations.
3. Meaning is obtained by applying a set of axioms to their observations.
4. Axioms are not instincts bestowed by nature in the generalization of an understanding, determining what man should or should do.
5. Reason is the process of applying axioms.
6. Reality is a meaningful worldview built with an understanding.
7. Wisdom is the habits adopted by an understanding dictated by the morality of knowledge.

AI is intelligence exhibited by machines or software. It is also the name of the academic field of study based on how to write computer software capable of intelligent behavior. Scientific method operations, analytical, stochastic, and various AI methods, which are the sole basis of such studies, can be addressed and followed, but this is not entirely true because there are emotional issues as well as such systematic and method approaches. It is also necessary to reveal fuzzy relationships between various variables that may be partially or somewhat related to the problem at hand. Below are a number of points for AI builds taken from the author's experience.

1. In order for the subject to be discussed even with many people, it must be grasped on the basis of oral knowledge and preferably through the mother tongue, a trigger idea may arise suddenly to achieve the final goal.

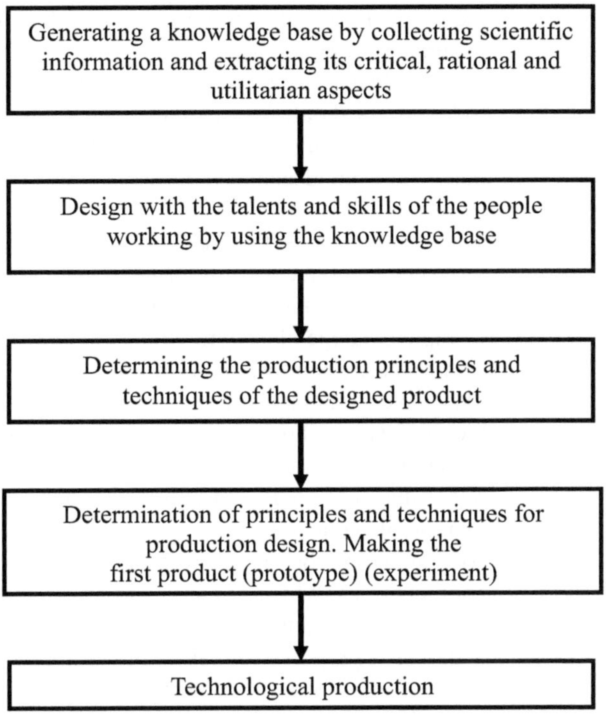

Fig. 2.5 Technological intelligence production stages

2. Systematic education can provide a basis in mathematics, physics, and algorithms, but if the verbal background is unknown, it may not develop new ideas and smart inferences.
3. The unsystematic and sometimes random mix of knowledge and information should not be underestimated, as useful information can have many hidden and undiscovered useful pieces of information from illuminating pieces.
4. In the generation of any AI, as can be seen from Fig. 1.2, there are possibilities for further improvement of existing or new ones, provided that questionable thinking is directed in any scientific and technological thinking.
5. It should be noted that no software could ever be achieved before the crisp and more recently fuzzy logic principles, which are increasingly entering the field of AI and explaining the problem at hand in a logical way.
6. Selection of suitable partners for disagreements and agreements after short or long discussions among team members.
7. Any information learned from systematic education training, teachers, books, the internet, or other means of communication should not be illogically and unreasonably kept in mind. Unfortunately, in many societies and countries, systematic education means filling the minds with

mechanical and static knowledge rather than dynamic and productive idea generation orientations.

8. Every technological product and computer hardware and especially software technologies should be examined with the help of cognitive processes that enrich the human mind. Otherwise, artificial intelligence with some integration of soul, mind and body, instead of math-based artificial intelligence, can be kept in human intelligence,
9. The mind itself is the main natural machine without borders, but its movement, production, and trigger fuel depend on cooperation with the reassessment of the dominant notions of intelligence ability,
10. Complete independence from other experts cannot lead to productive new smart ideas. Therefore, it is recommended that individuals with different ideas take part in a team.
11. Expert systems should be considered dynamically, with critical and questionable discussions, comments, and debates to generate better ideas, rather than knowledge-based systems that provide statically transferable, memorable education.
12. After grasping the verbal foundations, numerical data knowledge generation emerges, which requires various scientific functions such as mathematics, uncertainty techniques, and expert system methods, but the author first proposes to make simple graphic drawings to better interpret the properties of each dataset and hence it is possible to identify their interrelationships as their independence, dependency, or partial relationships. These graphs, or any simple shape, are in the form of visual geometric properties that are more important than mathematical equations because such simple and even descriptive shapes provide very effective rational information at the beginning and then even the most complex mathematical equations can be written down.

(12) *Natural Intelligence:* After all that has been described in the previous AI articles, it can be appreciated that AI is limited to human NI and therefore cannot produce human-like individuals, but many people can transfer work to AI tools or robots. AI is based on computer software, the more logical foundations there are in such programs, and the more effective robotic tools can be. There is a general trend as to whether AI will surpass human NI in the long run (Fig. 2.6). Such a feature is untenable, because even if NI has unlimited space, AI remains within the confines of NI. Perhaps the reason why many think AI can pass on NI is because AI productions can become uncontrollable as a result of some electronic contact or random behavior. For example, electricity is not dangerous to be at the service of man, but it can contact or malfunction, causing many inconveniences to man, such as fire, death, and other dangers. This does not mean that electricity has surpassed human producers.

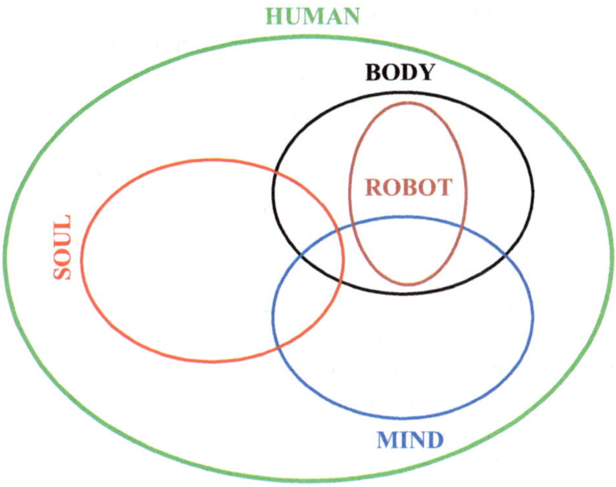

Fig. 2.6 Human and robot

(13) There is a trend lately all over the world that AI will surpass the NI and hence human life will be put in dangers that human intelligence cannot overcome; this is an unacceptable result. Although AI principles, methods and technological developments help to analyze environmental, engineering, social, economic, energy, service sectors and produce human beneficial results with optimum and maximum success, human NI cannot replace temporal and spatial characteristics. In particular, robotic, telecommnation, control system, image recognitions, data mining and many other activities can be oversimplified and atomized by AI, but AI productions cannot be expected to replace human intelligence in the long feature. So far, some AI methodologies have been touched upon and especially crisp and fuzzy logic inference systems will be explained in Chap. 3.

2.5 Artificial Intelligence Methods

As mentioned before, the term AI is directly related to the rules of logical inference, and a person who cannot use these rules will not be successful at this. First of all, the crisp (bivalent) logic as proposed by Aristotle (BC 384–323) is particularly successful in holistic inferences that lead to "symbolic logic," through mathematical equations and algebraic expressions first introduced by Muslim thinker Al-Khwarizm (780–850). By softening the sharp limitation of symbolic logic, various sub-methods have emerged in the direction of ANN concepts, genetic algorithms (GA), fuzzy logic (FL), and evolution calculations (EC) as more effective methods. Thus, "computational intelligence" was produced as a subclass of AI.

Computers, algorithms, and software simplify everyday tasks, and it is impossible to imagine how much of our lives could be managed without them. Is it possible to imagine how most methodologies, algorithms, and problem solving can be managed without manpower and intelligence? Mankind has lived with nature since its existence and learned many solutions by being inspired by nature. In particular, human beings have used natural resources to meet their needs and tried to struggle with natural disasters as much as possible. In addition, they tried to understand the events occurring in nature based on thoughts, mental activities, and emotions. In addition to the information and technology of the period, they also tried to examine the cause–effect relationships of events. During these studies, they developed many methods.

The methods studied have made unprecedented advances in the last few decades as a result of the rapid development of numerical computation based on advances in AI techniques after 1950. Some of the developed methods are inspired by living organisms. Genetic Algorithms (GA) and ANN can be given as examples of the methods that emerged by trying to express the functioning of human thinking organisms with mathematics (Şen 2004a, b; Öztemel 2016; Satman 2016).

The concepts of fuzzy set, logic, and system were introduced by Lotfi Asker Zadeh in 1965; his many years of work in the control field involved too many nonlinear equations to get the control he wanted, but resulted in complexity and difficulty solving. There has been some discontent in the literature with the introduction of fuzzy concepts. Among them, some researchers have embraced the idea of fuzziness and encouraged studies on this subject, but most of them were of the opposite opinion. Opponents argued that fuzzification does not comply with certain scientific principles and ii even contradicts science (Şen 2020).

2.5.1 Rational AI

It is a fact that technological developments are very rapid in the age we live in. Especially, the developments in the computer world have become dizzying. These developments are related to speed in calculations, examination of the smallest (micro) and largest (macro) realms, acceleration of knowledge production, etc. These helped humanity in many ways because computation facilities have tremendously boosted problem solving speeds.

The human brain is successful in seeing (visualization), speech (communication), error correction, shape image recognition in noisy environments with incomplete information. In fact, the human brain can produce verbally initial and efficient results than a computer. Nerves, known as electrochemical brain processing elements, respond in milliseconds (10^{-3}), while today's electronic technology products respond in nanoseconds (10^{-9}). Despite the fact that electronic technology products work 10^6 times faster, the reason why the brain works more efficiently than the computer in processing incomplete information and recognizing shapes has been a matter of curiosity for a long time. This curiosity led people to examine the working system of their own brains, and thus AI aspects began to emerge.

As a result of the examination of the brain and its tissue, it was determined that the difference was due to the information processing system. The striking result showed that the information processing of the nerves in the human brain is parallel. Based on this fact, scientists stepped up their work to ensure that information is processed in parallel, as in the human brain, so that it is as close as possible as to human behavior. Studies have focused on the design of methods, devices, and machines with the above-mentioned capabilities. Finally, methods such as GA and ANN were developed for the building intelligent systems, and thus AI for human intelligence modeling sprouted. Especially with the help of these methods, expert systems as a branch of AI have started to be used in many areas.

The real world is always complex, unless things are simplified by introducing clear scientific rules, and therefore the predictions are not absolutely correct even when firm thoughts and decisions are made. Information sources such as complexity and uncertainty, which appear in different forms in general, are called fuzzy sources (Chap. 4).

The more one learns about a system, the better he understands and the more complexities in the system are reduced, but not completely erased. If there is not enough data, the complexity of the systems includes too much uncertainty. In solutions, it is possible to make more meaningful and useful inferences from fuzzy input and output information by using fuzzy logic rules than crisp logic alternatives.

After 1970, fuzzy logic and system were given importance in the eastern world and especially in countries such as Japan, Korea, Malaysia, Singapore, and India. The use of fuzzy logic principles and rules in the construction and operation of technological devices is widely accepted all over the world today. The main reason why fuzzy logic is not widespread in western countries stems from the principle of certainty in binary logic, namely Aristotelian logic, which accepts two opposite alternatives symbolized by 1 or 0, as absolutely certain as true or false respectively. According to crisp (binary) logic, no other alternative is allowed between two values, as they are not absolute. Crisp logic is also called middle exclusion logic. In the West, the word fuzzy means insecurity. There is a belief in the East that beauty can be found even in the complexity or uncertainty. For example, even the establishment of necessary tolerance among people depends on such vague (ambiguous, highly personal) opinions.

2.5.2 Methodical AI

In engineering and technology applications, a model study of the problematic event is sought according to the event behavior patterns. Because this work is best done by the human brain, parallel processing modeling methods inspired by the brain working system gave birth to ANN as a part from AI procedures. One of the most important characteristics of ANN, which includes many features of regression and stochastic methods that are widely and still used in many parts of the world, requires some assumptions about the initial event behavior or data. For example, in all statistical-based methods, parameters must be constant over time, parallel operations cannot be performed, dependence is linear, homoscedasticity, etc. In practice, at least three successive layers are constructed: one for input, the other for output, and the third is an intermediate (hidden) layer. For this, it is necessary to develop an architecture that will model not only mathematical rules but also the action and response variables that control the event.

In this book, first of all, the shallow learning aspects of classical methods that the reader may or may not know beforehand are explained and the similarities of transition to deep learning methods are mentioned. In fact, it may be possible for the reader to model according to classical systems to get used to ANNs and make applications in a short time. First of all, after a short philosophical discussion, it is recommended to focus on the rational of the AI models, taking into account the similarity to the shallow methods learned earlier and the rules of logic.

2.6 Artificial Intelligence Methodologies

Models are the identification, imitation, and modification of solutions by translating human thoughts into a set of logical expressions and geometric shapes, then translating them into mathematics, and finally validation against real-life observations and, if possible, measurements. Each model is so understandable that real world problems (engineering, social, environmental, economic, health, energy, education, etc.) can be approached with human natural intelligence (NI). Various probabilistic, statistical, analytical, chaotic, and AI alternatives have emerged throughout the long history of modeling. All types of models are based on NI, which can be expressed in rational sentences that can be translated into mathematical symbols and thus mathematical analytical, empirical, and differential equations. Rational and logical expressions are very useful for communicating with computers through built-in models. The historical evolution of models also started in applications after the digital computer entered the field of scientific and technological researches. The first smart models were of the "black box" type, where input and output numerical data were available in the form of measurement data records. This caused the researchers to search "databases" to initiate model predictions. The first classical black box model has three components, as in Fig. 2.7, input, output, and transfer unit.

After obtaining the input and output databases, the researchers consider the transfer system to reproduce the output pattern from the input data. The transfer unit needs AI evaluations for the desired match between the two datasets. Unfortunately, at this stage many researchers are diving into existing shallow methodologies for transfer unit formation without their own intelligence contribution. In general, the transfer units are in the forms of probability, statistics, stochastic processes, analytical formulations, empirical equations, Fourier series, GA, ANN, analytical

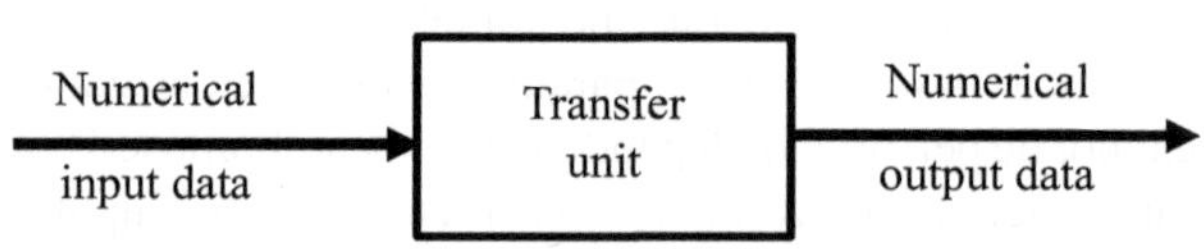

Fig. 2.7 Black box model components

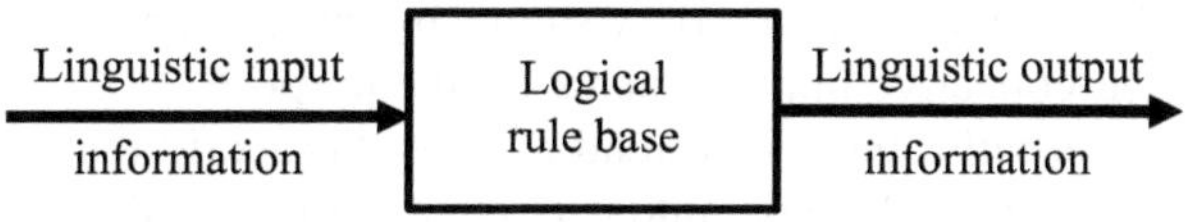

Fig. 2.8 Fuzzy model components

hierarchical methods, and many similar approaches. Such a shallow learning way does not activate human Intelligence; it gives ready-made answers within a certain error limit.

On the other hand, the active human mind, after learning the philosophical and logical foundations of each classical shallow learning methodology, may try to introduce another innovative modeling technique or at least suggest some modifications that imply the contribution of its intellectual ability. Such an application of human intelligence provides deep learning AI in model making.

Another AI methodology after the 1970s is fuzzy logic (FL) modeling (Zadeh 1965), which relies entirely on individual or group intelligence, because instead it requires a "database" and rational inferences that lead to a philosophical foundation and a set of constructive logical rule bases. Here, approximate reasoning plays the most active role for shallow learning AI inferences in light of human intelligence. The general parts of fuzzy modeling take the form shown in Fig. 2.8.

On the other hand, the active human mind, after learning the philosophical and logical foundations of each classical methodology, may try to introduce another innovative modeling technique, or at least suggest some modifications that imply the contribution of its intellectual ability. Such an application of human intelligence provides AI in model making. The first robotic tool based on human intelligence machine learning provides effective data mining procedures from which significantly applicable techniques can be extracted.

2.7 Natural and Artificial Intelligence Comparison

In general, intelligence includes the mental ability for absolute or approximate reasoning to solve problem and teaching tools such as robots for human service. The basic element of natural intelligence (NI) is the integration of cognitive abilities such as thought, reason, intellect, and wisdom in memory in a planned and organized way with observations from natural environment subjects based on a gift from creator to everyone. Every individual has an innate intelligence that can be provided through systematic and some chaotic training programs for innovative or advanced ideas, which are essential tools in the fields of scientific and technological prosperous development. The various NI degrees form the basis of tools, implementations, devices, machines, and practical robots.

With NI, a human tries to transfer the systematic knowledge and production mechanism of any action to other individuals or machines, in which case the subdivision of NI begins to take shape as AI. The fruitful area of AI is the principles of NI, which are strengthened by fuzzy degrees in different individuals, but do not allow the overlapping of wisdoms by integrating all the wisdom of individuals, but their collective improvements for the purpose.

Recently, there is a trend all over the world that AI will surpass NI and hence human life will be put into perils that human intelligence cannot overcome; this is an unacceptable result. Although AI principles, methods, and technological developments help to analyze engineering, environmental, social, economic, energy, and service sectors and produce human beneficial results with the optimum and maximum success, human NI cannot be replaced fully by AI procedures. In particular, robotic, telecommunications, control system, image recognition, machine learning, deep learning data mining, and many other activities can be oversimplified and atomized by AI, but AI productions cannot be expected to replace human NI in the long future.

One of the main aims of this book is to examine the subjective and objective dimensions of some human thought, concept, and imagination. Some of the AI methods are mentioned and especially the fuzzy logic (FL) inference system is explained in detail (Chap. 3), where not only mathematical symbols but also primary verbal logic rule foundations play the most important role in any AI activity.

After all that has been described in the previous chapters, it can be appreciated that AI is limited to human NI, and therefore, cannot produce human-like thinkers, but many humans can transfer the work to AI tools or robots. AI is based on computer software; the more logical foundations there are for such programs, the more effective can be the robotic tools. There is a general trend as to whether AI can surpass human AI in the long run. Such a feature is untenable as AI remains within the confines of NI. Perhaps the reason many people think AI can surpass NI is because AI productions can become uncontrollable as a result of some unintentional electronic error or random behavior. For example, electricity is not dangerous to humans, such as fire, earthquake, flood, drought and other hazards. This does not mean that electricity has surpassed human producers. Various aspects of human existence are given in Fig. 2.9.

Şen (2020) stated that there are three serial stages in any thought, namely imagination, visualization and idea generation as in Fig. 2.10. At each step, there is fuzzy uncertainty (incompleteness, fuzziness, uncertainty, etc.) that provides a space for slightly different decision-makings for the same problem. Each expert can reach the most appropriate solution depending on their ability, knowledge, cultural background, and experience (Şen 2014).

The imagining phase involves establishing appropriate assumptions for the problem at hand, and the visualization phase's purpose is to clarify shape (draft representation). Researchers often use different types of shapes (geometry, shapes, diagrams, graphs, functions, drawings, plans, etc.) to represent and defend assumptions. Justification of a set of assumptions is possible only by understanding the visual representation and should continue to refine and change the assumptions, if necessary. On the basis of assumptions, decision-makers act as philosophers, producing relevant ideas and their propagation to include new and even controversial ideas, so that other experts can transcend and further elaborate on basic assumptions. Arguments are expressed linguistically before any symbols and mathematical

Fig. 2.9 Humans and robot

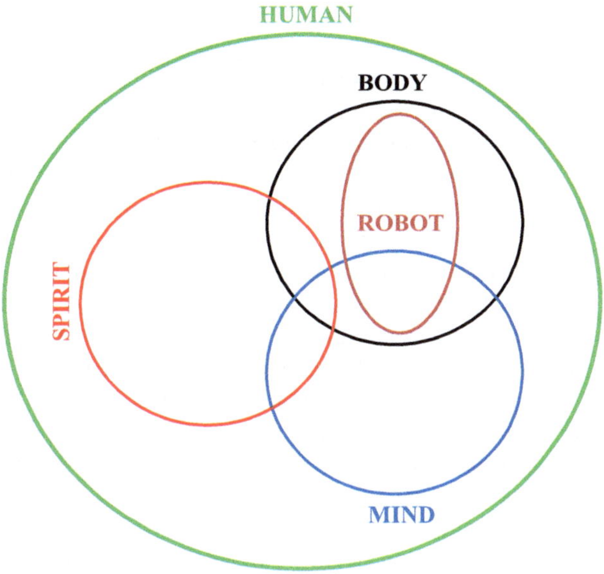

Fig. 2.10 Thought stages

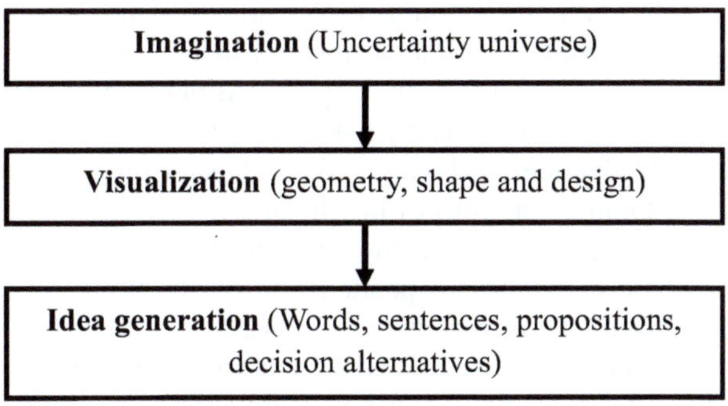

abstractions. In particular, the visualization phase is represented by algorithms, graphs, diagrams, charts, figures, etc., that contain an enormous amount of grammar.

2.8 AI Proposal

Below are a few points for AI productions drawn from the author's numerous experiences.

1. In order for the topic to be discussed even among ordinary people, it must be grasped on the basis of verbal knowledge and preferably through the mother tongue, so that an idea trigger may suddenly appear to achieve the final goal.

2. Systematic education can provide foundation in mathematics, algorithms, and physics, but cannot develop new ideas and intelligent inferences if the verbal background is grasped.
3. Irregular or random information should not be ignored, as useful information can be hidden and extracted from such sources.
4. As can be seen in Fig. 1.2 in Chap. 1, any AI generation has possibilities for further improvement of existing or new ones, provided that the questionable thinking is directed in any scientific and technological thinking.
5. Bearing in mind that before the crisp and even more recently fuzzy logic (FL) principles, which are increasingly entering the field of AI, no software can be achieved without logically explaining the problem at hand.
6. Selection of suitable partners for disagreements and agreements after short or long discussions among team members.
7. Information learned from systematic education, teachers, books, the internet, or other means of communication should not be illogical and unreasonably kept in mind. Unfortunately, in many societies and countries, systematic education means filling the minds with mechanical and static knowledge rather than dynamic and productive idea-generating orientations.
8. Every technological product and computer hardware, and especially, software technologies should be examined with the help of cognitive processes that enrich the human mind. Otherwise, instead of the AI, one can keep the mind with slightly overlapping AI in one of the three areas (spirit, mind, body).
9. The mind itself is the main natural machine without borders, but its movement, production, and trigger fuels depend on cooperation with a reassessment of dominant notions of intelligence ability.
10. One hand does not clap, two hands, and accordingly, individual natural thinking and complete independence from others cannot lead to productive new smart ideas. Therefore, it is recommended to have people with different ideas in a team, otherwise it is equivalent to one person if they all agree.
11. Expert systems should be handled dynamically, with critical and questionable discussions, comments and debates to generate better ideas, rather than knowledge-based systems that provide statically transferable, memorable education.
12. After grasping the verbal foundations, numerical data information generation that needs various scientific functions, including mathematics, uncertainty techniques, expert system methods, arises, but the author first proposes to make simple graphic drawings to better interpret the characteristics of each dataset, so that mutual relationships, independence, dependencies, or partial relationships can be revealed. These graphs, or any simple shapes, are in the form of geometric properties that are more important than mathematical equations, because such simple and even descriptive shapes initially provide very effective rational information and then even the most complex mathematical equations can be achieved.

2.9 AI in Science and Technology

It consists of harmonious and proportional connections of different human abilities, such as wisdom, reason, intellect, intuition, ambition, interest, culture, belief, scientific knowledge therapy, and similar gifts of God. Each individual has different intelligence scales and levels. No one is immune from these abilities. Intellectual intelligence can be at a static level without any research and development for the good of man in the society. Although the natural intelligence of man has existed since the creation of man and woman (Adam and Eve), it took centuries to reach today's intellectual level after a series of philosophical thoughts and logical propositions. Unfortunately, the reflection of human intelligence on science and technology has been overlooked in many disciplines and societies today, and as a result, these societies have lagged behind developed countries. This section provides a brief review and recommendations on the importance of AI from its historical perspective and methodological (models and machines) aspects. It is recommended that human intelligence sheds light on innovative inventions rather than classical, repetitive, and imitative approaches or at least partially replaces existing scientific and technologically available methodologies.

2.10 Misuses in Artificial Intelligence Studies

Many scientific papers are being published in an increasing rate, and many have the so called "scientific viruses" because they miss the basic rational and logical foundations. Scientific research needs appropriate methodologies such as analytical, numerical, probabilistic, statistical, and stochastic techniques that have been used in different aspects of scientific disciplines for nearly 200 years. Since the last 30 years, GA, ANN, and FL systems have started to take an effective role in modeling, simulation, prediction, and decision-making activities. AI methodologies are increasingly dominating the problem solving, simulation, and prediction processes. Many of these techniques require software for quick calculations, and at such a point misuse, misinterpretation, and misdirection begin as an integral part of many scientific papers in different journals. Unfortunately, such software causes scientific philosophy, rational thinking, and especially, logic to be overlooked. The main purpose of this section is to provide a basis for drawing attentions to such abuses of AI methodologies. In particular, the adaptive neuro-fuzzy inference system (ANFIS) approach is at the apex of misinterpretation.

The "publish or perish" savagery wrongly featured scientific publications with a purely mechanical flavor, even in most of the best journals. Especially in areas where the latest smart methodologies are used without logical care, but the mechanical demands through software have increased, the quality has decreased even though the publication rate has increased.

Early humans thought in a completely uncertain environment for their daily and vital activities. It is possible to say that early information and knowledge are concepts derived from frequent observations and experiences. For centuries, human thought received support from philosophical writings, drawings, calculations, logic, and finally mathematical calculations. Meanwhile, science especially after the Renaissance in the eighteenth century differs from philosophy in its axioms, hypotheses, laws, and final formulations. It is possible to say that with Newtonian classical physics, science has entered an almost completely deterministic world, where uncertainty is not even considered as scientific knowledge. However, today, there are uncertainty components in almost all branches of science, and many scientific deterministic foundations have begun to take the form of uncertainty in terms of probability, statistics, and stochastic, chaotic, and fuzzy alternatives. But others, such as geology and earth sciences, have never passed the determinism stage, but unfortunately determinism has influenced various education systems in many institutions all over the world. With the development of numerical uncertainty techniques such as probability, statistics, and stochastic principles, scientific advances in quantitative modeling have made rapid advances, but still have set aside sources of qualitative information and knowledge that can only be addressed with uncertainty principles.

Recently, famous philosophers and scientists have begun to articulate the uncertainty and fuzzy components that is the mainstay of scientific progress. For example, Russell and Norvig (1995) states,

> All conventional logic assumes that precise symbols are habitually used. Therefore, it is only applicable to an imaginary celestial existence, not to earthly life.

As for the fuzzy understanding Zadeh (1965) said:

> As the complexity of a system increases, our ability to make precise and yet meaningful statements about its behavior, beyond our precision and relevance (or relevance) become almost mutually exclusive properties.

As a tribute to Eastern thought, philosophical objects can be elevated with logical premises and inferences that lead to idea generation, as well as three basic mental activities, similar to Fig. 2.10, imagination, conceptualization, and then idealization. Man has interacted with nature, which provides the basic material of the chain of mental activity of man in the form of objects and events that change according to time and space. In the early stages of human history or in the childhood of any individual, these stages play a role at different rates and take their final shape with experience. Each of the chain elements in the thinking process contains uncertainty, because the stages of imagination, conceptualization, and idealization are highly subjective depending on the individuals. At any stage in the evolution of human thought, antecedents contain some degree of incompleteness, ambiguity, probability, possibility, and uncertainties. The derivation of mathematical structure, at every stage of physical or mechanical modeling, from the mental thinking process may seem certain, but even today, as a result of scientific developments, it is understood that there are at least partial uncertainties in every case, if not at the macro scale on a micro scale. It is clear today that mathematical conceptualization and idealization that leads to satisfactory mathematical construction of any physical reality is often an unrealistic requirement. As Einstein [2] stated,

> As long as the law of mathematics refers to reality, they are not exact. And they do not refer to the truth as much as they are sure.

At the most fundamental stages of mental thinking, activity objects are considered to be members or nonmembers of a certain or physically reasonable range of variability. This considers clusters containing possible consequences or basis of the phenomenon under investigation. In the formal sciences, such as physics and geology, these elements are almost invariably and automatically considered either entirely members of the set or entirely outside the same set. All mathematical equations in scientific disciplines are derived on the basis of crisp (bivalent, bivalent, Aristotelian) logic principles.

This section presented a brief description of the importance of AI in human life today and underlined that these techniques will gain more importance in the future. While everyone has NI, in recent years everyone has begun to load their personal intelligence activities on models that computers can deal with AI, more precisely on computer software-driven machines that can be broken down by computer programs into a simple or complex set logical rules. This chapter states that the first objective, physical, and mechanical artificial drawings and explanations emerged from the Islamic civilization, although some early versions were only in the form of general writings without reason and intellectual stimulation. AI emerges first in the form of models from NI and finally machines, including robots. NI continues to be explored more and more as there is no limit to the realm of thinking and idea generation, but in the realm of AI there are mostly mind and body functions that mostly lack spiritual content. AI can expand gradually and infrequently with a jump within the NI domain, the rate of deceleration or acceleration is not always predictable. Here, if possible, even inventors who do not have a systematic education, or as it has emerged throughout history, can use AI thinking opportunities.

2.11 Conclusions

Artificial intelligence (AI) shallow learning concepts are covered in this chapter, including Abou-l Iz Al-Jazari's original drawings of robotics and its historical development starting around 1200. With a contribution to the rapid execution of numerical modeling in computer technology from 1950 onwards, the shoots of AI began to mature in the last 70 years, especially accelerating from 1990 onwards. The relationship between human natural intelligence (NI) and AI is explained and it is recommended that AI remain within the confines of NI. AI types and procedures are discussed on rational and methodological grounds. About 12 intelligence alternatives are presented with etymological and epystemological philosophic contents. In the light of these issues, the education system is criticised and suggestions are also presented for future scientific thinking, method, and modeling purposes. Sequential thinking stages are stated as possibilities for imagination, visualization, and idea generation as well as critical interpretations and evaluations. Finally, some examples of misuses of AI conceptual and modeling stages are presented.

It is recommended that the reader not only benefits from AI but also other intelligence alternatives described in this chapter and develops their thinking skills and abilities accordingly.

References

Delbecq AL, Van de Ven AH, Gustafson DH (1975) Group techniques for program planning: a guide to nominal group and Delphi processes. Scott-Foresman

Hebb DO (1949) The organization of behavior. Wiley, New York

Hill D (1974) The book of knowledge of ingenious mechanical devices. D. Reidel Publishing Company, Dordrecht-Holland/Boston

Hill DR (1998) Studies in medieval Islamic technology. In: David A (ed) King. Ashgate, Variorum Collected Studies Series

McCarthy J (1960) Recursive functions of symbolic expressions and their computation by machine. I Commun ACM 3(4):184–195

McCorduck P (1979) Machines who think, Language, scenes, symbols and understanding. W. H. Freeman, San Francisco

Nasr SH (1964) Three Muslim sages. Mass, Cambridge

Newel A (1982) Intellectual issues in the history of artificial intelligence. Department of Computer Science Carnegie-Mellon University Pittsburgh, Pennsylvania, p 45

Öztemel E (2016) Yapay Sinir Ağları. (artificial neural networks). Papatya Bilim Yayını, Turkish, p 232

Rosenblatt F (1958) The perceptron: a probabilistic model for information storage and organization in the brain. Psychol Rev 65(6):386

Russel SJ, Norvig P (1995) Artificial intelligence a modern approach. Contributing writers: John F. Canny, Jitendra M. Malik, Douglas D. Edwards, Prentice Hall, Englewood Cliffs, ISBN 0-13-103805-2

Satman MH (2016) Genetik Algoritmalar (Genetic algorithms). Türkmen Kitapevi, Turkish, p 216

Şen Z (2004a) Yapay Sinir Ağları İlkeleri. (Artificial neural network principles). Su Vakfı Yayınları, Turkish, p 183

Şen Z (2004b) Genetik Algoritmalar ve En İyileme Yöntemleri. (Genetic algorithms and optimization methods). Su Vakfı Yayınları, Turkish, p 142

Şen Z (2014) Philosophical, logical and scientific perspectives in engineering. Springer, Heidelberg, p 260

Şen Z (2020) Bulanık Mantık İlkeleri ve Modelleme (Mühendislik ve Sosyal Bilimler) (Fuzzy logic principles and Modeling – engineering and social sciences). Su Vakfı Yayınları, Turkish, p 361

Turing AM (1950) Computing machinery and intelligence. Mind 49:433–460

Wiedemann E, Hauser F (1915) Über die Uhren in Bereich der Islamischen Kultur. In: Nova Acto Academiae Caesarae Leopoldino, (in German)

Zadeh LA (1965) Fuzzy sets. Inf Control 8:338–353

Zadeh LA (1973) Outline of a new approach to the analysis of complex systems and decision processes. IEEE Trans Syst Man Cybern 2:28–44

Zadeh LA (1979) A theory of approximate reasoning. In: Hayes JE, Michie D, Mikulich LI (eds) Machine Intelligence, vol 9. Elsevier, Amsterdam, pp 149–194

Chapter 3
Philosophical and Logical Principles in Science

3.1 General

Since the creation of men, imaginative and factual thinking abilities have played role steadily in the enlightenment along with accumulation of knowledge and information for betterment of a society without any materialistic cost at early stages. God's gifts of philosophy and logic together with the language provided dissemination of these enlightenment possessions in a society for further anthropogenic activities to deal with environmental, shelter, health, and food problems for sustainable life. For adaptable solutions, approximate reasoning principles are exploited based first on general philosophical imaginative and factual thoughts and then their filtering by means of logical principles for better rational idea generations. Technological improvements since Old Stone Age (Paleolithic) and the First Agricultural Age (Neolithic) centuries continued on the grounds of twin sisters philosophy and logic, and the human needs are satisfied with the physical inventions of new equipment, tools, and instruments starting from stone materials and went on through copper, bronze, iron, steel, and today's many new materials, but still philosophical and logical couple is in major role. Philosophy and logic continued their common existence in an intact manner up to almost eighteenth century imbedded with distinctive scientific ingredients.

Science also needs philosophical bases in terms of science philosophy, logic with systematic rational and methodological considerations. Thus, science excludes nonmaterialistic parts of ontology and completely denies metaphysical branch of philosophy in terms of positivistic views. However, even today science provides positivistic idea conquests from metaphysical philosophy domain. Recently, education systems have driven out philosophical and logical genuine principles, but they are thought more in the forms of rote and transporter manner, which may give rise to stagnant stability rather than active dynamism of philosophical, logical, and scientific sustainability.

Z. Şen, *Shallow and Deep Learning Principles*,
https://doi.org/10.1007/978-3-031-29555-3_3

This book emphasizes innovative education system, where the philosophy helps to solve confronted problems first linguistically in a rational manner prior to mathematics, systematic algorithms, and software. In this way, the reader may develop generative and analytical thinking skills and capabilities. The modern philosophy of science insists on falsifying existing scientific results, and consequently, there is always room for ambiguity, incompleteness, vagueness, and uncertainty in any scientific research activity. Innovative education systems should lean more towards the basic scientific philosophy of problem-solving with logical principles. However, many recent publications in the literature are software applications that do not provide the basic principles of linguistic and logical fundamentals.

Any innovative research should have improvement, modification, or innovative initial shallow learning soft logical bases for the advancement of science. On the other hand, in engineering, the application of scientific methodologies to various socioeconomic and civilizational activities must be justified with philosophical, logical, linguistic, aesthetical, and ethical components harmony. Otherwise, even applications can be mechanical, and the ultimate benefit may simply add one more line to the author's resume. It is noted here that some degree of inner exploration of methodological, procedural, and algorithmic techniques should be known at a grey level, rather than an outright mechanical work or use of off-the-shelf software. It is preferable to study the main components on philosophical, logical, linguistic, and scientific bases.

Not all knowledge and information are precise, but contain some element of uncertainty. In this book, a linguistic review of all knowledge and information is recommended through logical procedures prior to any mathematical modeling method. At this intersection, philosophical principles and logic propositions are the preliminary foundation stones for further developments.

For centuries, human thought has received support from scenarios, drawings, calculations, logic, and finally mathematical calculations. Meanwhile, science is separated from philosophy with its axioms, assumptions, laws, and formulations, especially after the eighteenth century. Today, it is understood that there are elements of uncertainty in almost all branches of science, and many scientific, commercial, and economic organizations try to make decisions under the concepts of uncertainty by taking into account some risk level. Such concepts include quantum physics, fractal geometry, chaos theory, and fuzzy model inference. Business, economics, management, and social sciences have never reached the stage of prominence. Advances in numerical uncertainty techniques such as probability, statistics, and stochastic principles (Chap. 4) can be found in almost all fields (engineering, sociology, economics, business management, etc.). It leads to quantitative scientific advances in these fields, but still leaves aside the sources of qualitative information and verbal knowledge that can be solved by fuzzy logic. Şen (2014) stated that prior to numerical modeling practices, there are three serial stages in any thought as imagination, visualization and idea generation as presented in Chap. 2, Fig. 2.9.

The main theme of this chapter is to furnish general and science philosophical aspects supported by logical rule bases to reach scientific derivations for better understanding of traditional shallow learning principles that are keys to enter deep learning arena including artificial intelligence (AI) and machine learning (ML). In this chapter, there is no mention about mathematical foundations, which will be dealt with through various methodologies in Chaps. 4 and 5.

3.2 Human Mind

Early knowledge and information were concepts derived from daily observations and experience. Human thinking had support from imaginations, scripts, drawings, logic, and primitive arithmetic calculations. The five sense organs provide information from the environment, and accordingly, decisions are based on logical and rational judgments (Chap. 1). The mind can deal with fuzzy impressions and concepts. After labeling each part with a "word" such as a noun, verb, or adjective, it breaks down the visible environmental reality into parts and categories, which are essential ingredients in categorization, analysis, and interpretations to reach reasonable consequences. These words have little to do with the unity of reality – a unity to which we all belong inextricably. Common words help to imagine the same or very similar objects in minds. The thinking world is made up of parts of senses, thoughts, and perceptions, which collectively serve to provide partial and thus distorted conceptual models of reality representing a perceived world in mind. It is not a world whose natural evolution brought us into being and where we are tied by the umbilical cord of vital and inseparable connections.

All conceptual models deal with parts of something as environments that are thought to use for what our ego-centered minds consider meaningful. Of course, there are overt and implicit interrelationships between these meaningful pieces for exploration of the human intellectual mind. Unsupervised or supervised (trained) minds in anything like scientists, engineers, economists, politicians, and philosophers are concerned with many distorted conceptual model adaptations to exert power and predict the evolving dynamics of reality that are beyond our ability to predict and control definitely. However, scientists and enthusiastic minds try to "do their best" to push reality into the meaningful beds of meaningless model reductions. In doing so, the mind is dependent on fragmentary information about reality.

The mind is faced with a dilemma, and therefore, chooses one thing and rejects the opposite according to crisp (two-value) logic, which trains the mind to think in true or false as a first approach to modeling reality. Such a distorted model of reality based on duality refers to the crisp or binary logic that was founded by Aristotle (BC 382–323). Although before Aristotle the human mind was based only on natural and innate logical principles, it is constrained by the choice of the principle of duality. The dualistic nature of the rational reasoning component of the mind is so strong that the mind alone cannot transcend it. So, the crisp logic does not consider vagueness, uncertainty, probability or possibility, because everything is either true or false. This type of thinking can trap the human mind in routines, stereotypes, prejudices, and habits that easily become a source of confusion, ultimately rendering one incapable of authentic experience. This is because human understanding is constantly filtered from already established mental patterns. Fanaticism is an extreme manifestation of such intense obscurity, because when man's ability tries to move beyond an established dogma, it is blocked completely.

On the other hand, even today all have vague, ambiguous, uncertain, possible and probable concepts and approaches in many works. The natural logic is broader and general than crisp logic, and therefore, is labeled as fuzzy logical thinking. In cases of fuzzy or probabilistic reasoning, it is possible to accept both opposites to some degree with intermediate alternatives. By following a fuzzy logic-based approach in thinking, one can agree to a certain degree with everything that others say, and this can easily lead one to conformity and indecision. When everyone is right, the uncritical acceptance of the fuzziness accompanies other people's thinking and makes it difficult to come up with one's own rational ideas. The polarity of opposites, contradictions, and conflicts of ideas provide the human mind with the necessary dynamics to transcend opposites. These dynamics manifest in the mind as an impulse to seek beyond the plane of conflict of opposites, and without such an impulse the mind can become stagnant, stuck in repetitions, or fascinated by illness.

3.3 Rational Thought Models and Reasoning

With the suggestions and inference rules explained in the previous sections, in order for a person to produce new ideas on the subject s/he is interested in, there must be a knowledge base or s/he must have gained it through life experience, education and difficulties. In rational thinking, there are basically two thinking modes as deductive and inductive. In both models, there is an intermediate layer thought system box connecting the inputs to outputs. In this layer, there are answers to the questions of how and why with intellectual activities.

When examining events, one can rise to a higher level that includes more information, either by making use of shallow knowledge or by categorizing them. On the contrary, one can reach the lower shallow levels of knowledge and learning, crumbling them with how and why questions. Thus, at first glance, the thought model systems can be divided into two; one going from the shallow to deep information and the other one from the deep to shallow information to understand firmly the basic conceptions, definitions and rationality. The first alternative is called induction and the other deduction. The inductive and deductive thought modeling explanations given below can also be viewed as a cause-effect relationship. If one tries to establish thought models that can help to explore these two different approaches, one understands that each have an input and an output layers. In order to be successful in modeling, thought is necessary, and to be successful in thinking, modeling is required. In a way, thought and modeling can be perceived as two sides of a coin. Great thinkers in history have been able to reveal the rules of thought with effective models. Man can be productive in the subject of interest, sometimes with one of these models, sometimes with the others, and sometimes with their mixtures. Unfortunately, models without science philosophy and logical rule base are perceived by rote or imitative way without knowing how and why they work. This trend seems to have increased even more these days, especially with the increase use of ready computer software.

3.3.1 Deductive

In the deductive system, high-level events are explained and everybody understands each part privileged by dividing into subcategories according to personal ability, and even be able to achieve induction by establishing relations between these categories (Fig. 3.1). The point to emphasize is that without thought models in any education system, if one tries to open the way for direct modeling such as in engineering, medicine and social sciences, he cannot be very successful. The shortest and shallow practical solutions are the most imitative, sterile and nonproductive.

Al-Farabi (870–950) is one of the Muslim thinkers, who categorized the science into two parts as absolute and probable. Among them, he described all those related to nature as possibilities. The most striking of absolute knowledge is "death." The knowledge that all living things will die is never denied, and from this point of view, as Al-Farabi says this knowledge is absolute and "completely" according to today's definition. Here, inferences made with logical rules based on such deductive information are called "deduction." An example of the rational inference to be reached as a result of reasoning by deduction can be given as follows:

1. *All living things are mortal.*
2. *The cherry tree is also alive.*
   IF that is THEN
3. *The cherry tree will also die.*

There are three stages here, the first of which is the definitive information, the second provides a sub-information, and the third provides a conclusion on the basis of previous information.

Comprehensive information cannot be known with such certainty at all times and places. Based on observations, experiments, and experiences in practical life, it can now be regarded as a complete knowledge according to the information load obtained up to that time. For example, the knowledge that "liquids are fluid" has a total character as a natural phenomenon.

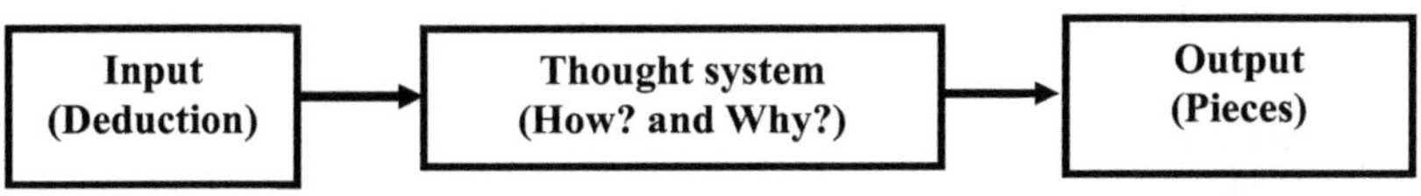

Fig. 3.1 Deduction system

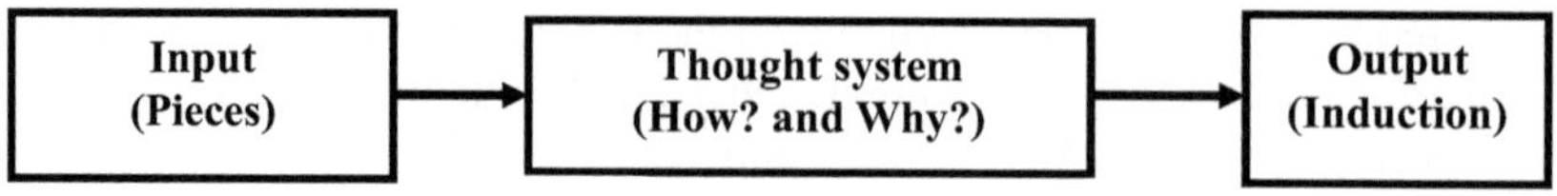

Fig. 3.2 Induction system

Some information counted as a whole takes on other forms over time. For example, when the first humans looked from the Earth, they regarded the planets revolving around the Earth as complete knowledge and put forward deductive inferences by reasoning. However, later, when the center of the solar system changed to the Sun, it became a complete knowledge that the planets revolve around the Sun today. Making rational inferences by using deductive logic has started in the depths of human history and continued until the last three to four centuries.

Since the information obtained as a result of inferences made by reasoning in this way causes clarity at shallow scales, it can also be called the "sub-information generation" way.

3.3.2 Inductive

Induction implies the thought model given in the education system of different countries, where thought system of many researchers, philosophers, and educators focus on. In this system, parts are explained to the recipients and with desire to reach initial shallow, but productive knowledge levels by combining them together. Figure 3.2 represents induction type of rational thinking model.

A student should be knowledgeable not only about the basic knowledge of his subject of interest, but also about other related subjects, so that s/he can get answers about how? and why? by combining closely related subjects? Today, education became stereotyped, separated by crisp boundaries, as experts have not many differences in opinions. As such, inductive education system is inefficient and unproductive. The ability of many young people to produce cognitive information is, so to speak, atrophied and injured. The inductive education model leads to a uniform thinking community.

In case the information is found piecemeal, it is called total arrival to reach rational inferences with the reasoning that emerges as if by trying to reach the whole by harmoniously gathering the ones that support each other meaningfully. Knowledge by repetition of the same experiment under the same environmental conditions with different objects may comply with a common law and hence it may be possible to reach a general conclusion. The most well-known example of this is the elongation of metals when heated, the following judgment is reached:

IF
Iron metal expands when heated.
Copper metal expands when heated.
Aluminum metal expands when heated.

Tin metal expands when heated.
Silver metal stretches when heated.
THEN
Metals extend when heating

Since the information obtained as a result of induction inferences made by reasoning in this way gives rise to enlightenment at higher scales, it can also be called the "meta-information generation" way.

3.3.3 Deductive and Inductive Conclusion

Some arguments are such that the truth of the premises is necessarily sufficient for the truth of the consequent. In the sense of the logical conclusion central to the existing tradition, such "imperative competence" distinguishes deductive validity from inductive validity. In inductively valid arguments, the (common) truth of the premises is very likely (but not necessary) to the truth of the consequent. An inductively valid argument is such that it makes the consequent of its premises more likely or more plausible, as is often argued (the conclusion may not be true given the common truth of the premises), the following argument deductively is not valid because the premises are not necessarily sufficient for the consequent chert can be heavy.

All the rocks observed so far have been heavy.
Chert is a rock.
Therefore, chert is heavy.

Distinctions can be made between different inductive arguments, which seem quite plausible and others are less plausible and hence a fuzzy implication enters the scene. There are many different ways to try and analyze the inductive result. One can consider the degree to which the premises make the consequent more probable (uncertain, probabilistic or fuzzy reading), or check whether the most normal conditions, where the premises are true also make the consequent true. The following three examples illustrate reasoning that does not lead from true premises to false consequent:

Ali analyzed whether Ahmed brought the samples and Ahmed brought the samples, so Ali analyzed it

Every geologist is talented and every hydrogeologist is a geologist, so every hydrogeologist is talented

The geoscientist is in the field, so one is in the field is geoscientist

In such cases, a thinker takes no epistemic risk by affirming the conditional claim that if the premises are true, the consequent is also true. The consequent immediately follows from the premises, with no further assumptions that might turn out to be false. In contrast, the following three examples illustrate reasoning that involves at least some risk of going wrong, from true premises to a false consequent:

Sand is soil and alluvial is soil, so alluvium can be cultivated

Not every living thing existed in the Precambrian, so not every living thing will exist in the long future.

Suppose the first statement, the alluvium can become uncultivable soil, which lags behind the demonstrative character. While the laws of nature may prevent immortality, the consequent of the last inference goes beyond its premise, even if in a sense it is foolish to resist the inference.

References to logical form arose in the context of attempts to say more about this intuitive distinction between impeccable inferences that invite metaphors of security and immediacy and inferences that risk slipping from truth to false. The motivations are both practical and theoretical. Experience teaches that an inference may seem safer than it originally was; and if one knew which inferences were risk-free, one could be more careful about where one risks making mistakes. This kind of vigilance can be valuable even if the risks are tolerable. As we shall see, claims about inference are also closely linked to claims about the nature of thought, its relation to language, and the possibility of ordinary language to "hide" the basic structure of thought. That is why we want to know that an inference (if any) is flawless. The most common recommendation is that certain inferences are safe thanks to their logical form, although the concepts of form have evolved and continue to evolve along with the concepts of logic and language.

3.3.4 Proportionality Rule

It is possible to appreciate a logical relationship, if any, between two variables holistically as a mind response to the following two subsequent questions:

1. Is there a possibility of probable relationship between the two variables.
2. In case of possibility, does the relationship appear according to directly or inversely proportionality principle. For example, is there a relationship between the temperature and heat? Of course, there is a relationship, which implies directly proportionality principle. As the temperature increases (decreases), the heat also increases (decreases).

3.3.5 Shape Rule

Another of the chains that must be critically thought is the principle of linearity, which has been adhered to in many problems. Almost all of the scientific laws contain the principle of linearity. The reason for this is man's most simple and linear thinking ability. Linearity in the natural events behaviors may be valid in short time durations and small spaces. However, there is nonlinearity for long periods and large spaces. Nonlinear numerical solutions in computers are affected by even one-millionth differences in initial conditions, which appear as chaotic behaviors

(Lorenz 1972). In this case, obvious equation solutions are found on completely chaotic trajectories called strange attractors. Thus, it is understood that science itself, which is thought to offer very clear results, can cause uncertainties even in terms of its methods. In order to further develop scientific studies and products in the future, it is useful to consider the following points briefly:

1. Scientific findings are not divine laws. It is necessary to forget never that all the events in nature and the findings that are regarded as definitive for now are in a state of development, and it is necessary to constantly strive for their better modeling representations.
2. The biggest mistake of a researcher is the thought that the present knowledge is firm and cannot be criticized. This may be true for geographical discoveries on earth, but false for scientific discoveries.
3. It should not be forgotten that throughout the history of science, it is necessary to try to generalize the results that have been stated as certain or to improve inferences according to changing conditions. It is necessary to try and find the falsifiability in scientific methodologies by criticisms.
4. In particular, "certainty in uncertainty," is known scientifically like chaotic behaviors, but obscurity will remain always in role. In the future, scientists will be challenged and perhaps excommunicated by those who cannot criticize science, believe in the inviolability of science, see it as a dogma and say everything is "scientifically proven."
5. It is necessary to reveal the limits of science in human nature and social life. Even if it is thought that there are no limits to the development of factual and artificial devices and robots in science, one should restlessly push towards exceeding these limits and conquering new knowledge.

3.4 Philosophy

In general, philosophy is a common thinking tool of humanity without limits and it provides full freedom to ponder about any event, phenomenon whether actual or imaginative in the extensive domain of ontology. Hence, ontology provides the first major subject of philosophy, which may not be known by this name, but each human being imagines some ideas and tries to visualize and understand its evolution and shape; then according to the self-capacity may reach far-sight views after meaningful inferences in the forms of estimations, predictions, or forecastings. There is no human mental activity that is without philosophical bases, because for thinking these features outside or inside environments of an individual urges one to come out with some thoughts, which can be transferable to other individuals logically first by means of language without any mathematical expressions.

Philosophy can be defined also as the accumulation of thoughts that gradually develop or fade as a result of thought interpretations to reach eventually a conclusion and explain it to other individuals and receive their criticism.

In the past, all kinds of thought production were accepted among the subjects of philosophy until the eighteenth century. In this respect, philosophy used to be a concept that also included the sciences. Today, philosophy is perceived as isolated from sciences. This is a big diversion, because at the root of every thought, philosophy should be taken as the subject that encompasses language, logic, proposition, and inferences. Today, it is thought that scientific subjects are not within the scope of philosophy as they used to be. However, philosophy as its basic definition is at the root of all sciences and arts as a systematic thought process.

The root of philosophy and science is the human desire to know. On the other hand, the root of technology depends on things that are necessary for the comfortability of human beings. Science has methodology, system, experiment, objectivity, logic, hypothesis, all of which are effective.

Philosophy, in short, means systematic thinking, reasoning with thought alternatives (deduction, induction, analogy) to make acceptable mind productions. Philosophical approaches to the questions were asked and sought answers even before the rules of logic were established prior to the Ancient Greek period about 2500 years ago. Philosophy has also scientific information content. It has five different branches as explained in the following subsections.

3.4.1 Ontology

Ontology is everything, whether physical or not, that we talk about is "Ontology." In fact, everything even the ones that do not exist are within the subject of ontology. For instance, "nil" is also an ontological existence, because we talk about it. If the ontological things are graspable and visible then they exist. Existing things may have dimensions that are too small to be seen with the naked eye or too large for humans to comprehend. Here, the fact of thoughts exist but cannot be considered materially and related to each other in a harmony. The existing thought is beyond understanding, but helps to put forward the views about the states of the existing beings. Today, even the most advanced scientific theories accept that human existence began after all inanimate beings. This understanding represents a common point not only in scientific circles, but also in holy religious books. On the other hand, in the evolution scenario put forward by materialistic thinkers, it was possible to go even further, in which human beings came into existence by chance from material beings, or that living beings became human through the development of simple and single-celled creatures over time. In short, all inanimate objects were primitive and existed even billions of years ago. Today, there are two different views in terms of existence that are simple enough for everyone to understand. The first is based on creation, and the other is the idea that the living things were subject to evolution originating from the nonliving things. Human ability is limited, but supported by science, technology and intelligence. As for the ontological branch of philosophy, the following points are important:

1. According to scientific understanding, scientist explains existing events and phenomena by laws and theories for their initial estimations, then these theories and laws must be verified by observations and experiments.
2. The scientist cannot touch on the hypothesis, observation, experiment, law, theory, and verification explicitly. These items get their existences through the philosophy of science.
3. Some of the philosophers let the metaphysical and speculative existences outside the science circle and in this manner they think that the scientific development will continue.
4. Philosophy and science use different approaches and methodologies from each other for the confirmation of existing things.
5. Philosophy of science tries to answer to "What is the science?" first on the ontological grounds.

3.4.2 Metaphysics

As another branch of philosophy, metaphysics concentrates on the original principles of being, reality, identity and change on real and imaginative time-space domain. It searches for causality, necessity and also possibility of real and even imaginative events including science, religion, social affairs, and alike. Its main concern is to relate mind functions to matters for better consciousness about the nature. Metaphysics is derived out from positivistic thinking scientific activities, but it still plays role in human affairs whether scientific or other. Its concern expands outside human understanding domain and tries to bring down some events beyond physics to physical world as science is concerned. Outside the scientific domain of influence, there are metaphysical functionalities in every human sense activity. For instance, although original physical composition and the after triggering of the Big Bang, there are many physical conceptualizations, but prior to triggering there are no scientifically valid explanations, but religious beliefs. The more the transfer from the metaphysics to physical domain by scientific works, the more are the scientific developments, advancements and innovative contributions.

3.4.3 Epistemology

As already mentioned in the previous chapters, although metaphysics tries to explain the origin and scope of nature, the meaningful explanation is possible by epistemology, which is equivalent to the theory of knowledge such that relationship of human mind with reality leading to plausible consequences. Truth, belief, and justification aspects are within the scope of epistemology. After perception and visualization of some existences, one provides informative knowledge about them by means of linguistic epistemological specifications. In general, epistemology is the theory of knowledge about its original source and collection of different types to reach at harmonious and detailed systematic knowledge status. Questioning of available knowledge is a concern of epistemology.

At the root of epistemology is etymology, which implies the root of each word and based on such knowledge then epistemology provides meaningful sentences, which open the way for further knowledge innovations and improvements. Empiricism and rationalism are two main branches of epistemology for knowledge generation functionalities. Epistemological explanations are open to critical thinking, debate, discussions and comments so as to improve the knowledge content about any phenomenon.

3.4.4 Aesthetics

Apart from the previous branches of philosophy, this is concerned with beauty and taste, which are rather subjective topics, but philosophy paves way to agreement among different individuals for common and better beautifulness. The interpretation and evaluation of art and artistic works fall within the domain of philosophical aesthetics, which urge for better levels of fine arts, architectural works, music and social affairs. It is possible to state that aesthetics is concerned with the individual's natural beauty enhancement. This is also concerned about something that looks good, and hence one feels comfortable. It means the pleasant, positive, or artful appearance of a subject. In a society, happiness comes with aesthetical feelings, appearances, hearings, and dialogs.

3.4.5 Ethics

Ethics is concerned with guidance and rules that are means of distinguishing between good and bad, moral and immoral, justice and injustice. Ethics is the collection of behavioral conducts that are acceptable commonly in a society or in a group of experts. It is equivalent to lawful conductance in a society. The philosophical aspect in it neither is concerned with nor without crisp boundaries but rather fuzzy and partially debatable concerns. Those with selflessness, justice, moral, self-respect, honesty, and loyalty are ethics and do not even need rules and regulations apart from the career respect and involvement. Decisions and actions can be conveyed to others with ethical conducts. Goodness, rightness, wrongness, and badness fall within the scope of ethics, and hence, a society may have peace and comfort without major problems.

3.5 Science

Scientific expressions are provided in a crisp logical manner as if there is only one way of thinking and solving the problems with certainty. Assumptions, hypotheses, and idealizations are the common means for mind to grasp the natural phenomena, and therefore, any scientific conclusion or equation is valid under certain idealization. It should be kept in mind that each scientific conclusion is subject to uncertainty and suspicion, and hence, to further refinements leading to innovative ideas and modifications. Prior to any concern, it is prime requisite to know what the fundamental principles of scientific criteria are.

At higher educational levels, scientific thinking must be geared towards the falsifiability of the conclusions or theories rather than exactness. General behavior of the phenomenon considered must be explained from different points of view on philosophical level, which indicates the significance of language in the planning and tackling of any problem.

During the presentation and definition stages of the problem, by all means the researcher must be open for various discussions, comments, and criticisms. The causative effects on the problem need to be identified with all possible details and verbal attachments of variables. Among the causative effects, a single variable of interest is depicted as the subject of the problem, and hence there are numbers of causative variables.

Logical propositions including premises among the subcategories of causative variables are constituted, and subsequently, each one of these premises is attached with sensible, rational, and logical consequent parts of the consequent variable.

3.5.1 *Phenomenological*

It is concerned with description, understanding, and interpretation of human experience to construct preliminary philosophical and logical principles towards scientific ends. Human experience is the main source of phenomenological thinking. This is a method for search of empirical relationships among various real events that may lead preliminary but theoretical thoughts to ripen scientifically valid theoretical conclusion. It paves the way towards the further scientific characteristics. Its grounds are physical philosophical pondering to reach rational consequences.

3.5.2 *Logical Foundation*

Any social, medical, cultural, economic, environmental, engineering discipline necessitates prior to numerical database, linguistic and verbal debates, which are interpretable by means of philosophical thinking and logical inference principles.

Authoritative researchers are based on ready laws and traditional rules, which may not allow for productive rational thinking. In such a research system, all types of answers are based on crisp (two-value) Aristotelian (BC 384–322) logic without any reference to Al-Farabi (870–950) probabilistic or Zadeh (1965) fuzzy logic, which are for uncertainty treatment methods. Research hardware, especially computer software, are given utmost priority as if they are indispensables, whereas human software thinking components such as science philosophy and logic are overlooked and even ignored at least partially. Hence, the road is paved towards memorization, transportation, copy and paste, and mechanical research system sustainability. Knowledge is taken from the books without doubt, critical questioning with certain conceptions without caring for any uncertainty ingredient. This implies acceptance of already established scientific paradigm by many researchers, who are referred to as normal researchers by Khun (1962) and make traditional scientific researches. The scientific content may be a small sample from a general population, and therefore, more coverage towards the population can be gained through common cooperation by distinctive researchers in a national of international team with similar aspirations. In this manner, some remaining blurriness in the explanation of the phenomenon can be reduced, and hence, the explanation and understanding of the phenomenon can cover even not experts in the subject area.

3.5.3 *Objectivity*

Epistemologically objectivity is one of the main characteristics of science, and hence, scientific findings and knowledge can be acceptable by many, under the bombardments of critical review and for betterment in information content expansion. Scientific knowledge is completely objective, which remains as it enters the domain of science, but possible to try and augment its objectivity degree further with new thoughts, improvements, and innovative research actions. Objectivity does not depend on beliefs or personal ideas. Favoritism does not enter the domain of science, because it is subjective. Scientific objective methodologies depend on unbiased experimentation without favorable interpretations that does not depend on evidence, and hence, the scientific knowledge by evolution takes refiner content.

3.5.4 *Testability*

All scientific knowledge must be testable according to scientific criteria and also repeatable under the same conditions at any time and location in the world. This requires experimentalism through technological tools that leads to measurements and they must accord with the scientific hypothesis and equations under a set of assumptions. The experiments help to testify the validity of any scientific hypothesis, and hence, many people can be convinced rationally in their area of scientific interest. Experimental details provide future possibility to modify existing information levels and get refined scientific knowledge. Any hypothesis in science must be verified by experimentation. If the concerned event is not testable, then it cannot enter the scientific scope domain.

3.5.5 Selectivity

In scientific research, there may be different alternatives that may give almost the same results. In this case, the best one can be selected according to scientific characteristics Sometimes, duality cannot be overcome in which case the selection may be according to convenience. For example, from scientific point of view, light is either in the form of wave or photon as a massless matter. In this case, depending on the purpose of the analysis one can select and depend on one of these two alternatives.

3.5.6 Falsification

The science and any related attributes to it are never completely verifiable but falsifiable, fuzzifiable, and hence, further developments in the form of prescience, traditional science, and occasional revolutionary science are valid at all times and spaces. Philosophy and science are twin brothers; each one needs the other for thought improvements, diagnosis, treatment, healing, and inferences from vague statements. Logical principles are like expert teachers for arriving at rational consequences even though they are approximate reasoning products. Popper (1955) proposed falsifiability principle for scientific evidence as already mentioned in Chap. 1. Scientific revolutionary structural basic concepts are presented by Kuhn (1962).

3.5.7 Restrictive Assumptions

A set of restrictive assumptions and isolation from the environment provides scientific information, forcing the problem into the world of certainty and ignoring all uncertainties. Experts, researchers, or decision makers multiply their final results by a factor of safety because of "overestimation" or "underestimation" possibility. Any factor of safety takes into account imperfections in the data or modeling calculations. None of the completely accurate formulations can be inferred from a set of simplifying assumptions, hypotheses, and laboratory (environmental) conditions. For example, the "factor of safety," as it is used in engineering and especially in civil engineering, includes the "factor of ignorance" due to the exclusion of all uncertain information about engineering design (Şen 2014). In mathematical modeling, the products are compared with the actual measurements, and as an initial judgment, if the relative error is less than ±5% or at the maximum ±10%, then the results are regarded as practically within the acceptable limits.

Reasoning is the most important human brain function that leads to generalizing ideas, methods, algorithms, and conclusions in addition to a continuous research and development process (R&D). The reasoning stage can be reached, provided that it is stimulating for the initial orientation of mental faculties due to some involvement such as science, engineering, trade, company, or business. The firing of reflection on a phenomenon comes with physical or mental effects that try to control the event of concern. These effects rely on the imagination of the event and the first drawings of the imagination with simple geometries or parts and connections between them (Şen 2014). In this way, ideas are crystallized and transmitted linguistically to other individuals, receiving their criticism, comments, suggestions, or support for the improvement of mental thought and final decision. Decision-making in any problem-solving is an important part of reasoning, because it is an application of inferring information about observable aspects of a case based on a set of information that may include different sources of uncertainty.

Especially, in the natural sciences, apart from field or laboratory measurements that lead to numerical data, experience is the most valuable and important piece of knowledge, which is easily accessible linguistically. Many experiences cannot be expressed by mathematical models, equations, or algorithms based on deterministic crisp (binary) logic. So the question to be answered: is it necessary to transform human thought into crisp logical principles, or to modify logical propositions and inferences to account for uncertainty? For example, in engineering and classical education systems, some assumptions are required before problem solution to avoid uncertainties (see Fig. 3.3).

The assumptions of homogeneity, isotropy, uniformity, linearity, that is, idealization of the reality, are necessary as the constraints of natural and engineering phenomena for their conceptual modeling. It is almost impossible to derive equations or to set up mathematical models without assumptions. Unfortunately, even today research institutions, especially in the engineering sciences, seek deterministic equations with basic assumptions, even though the related phenomena include uncertainties. Many disciplines seek to train their members through crisp logical principles (true or false) with the power of classical physical principles, mathematics, and statistical methodologies. It is time to change this view, with uncertain data treatment and artificial intelligence (AI) methodologies in the light of fuzzy logical principles. For this purpose, instead of teaching classical methodologies, an innovative research path should be based on philosophical thoughts and systematic form (shape) visualization with linguistically logical expressions.

All our actions in society, economics, administration, management, engineering, medicine, and the sciences take place in a complex world, where complexity often stems from vagueness in forms of uncertainty. Researchers and engineers subcon-

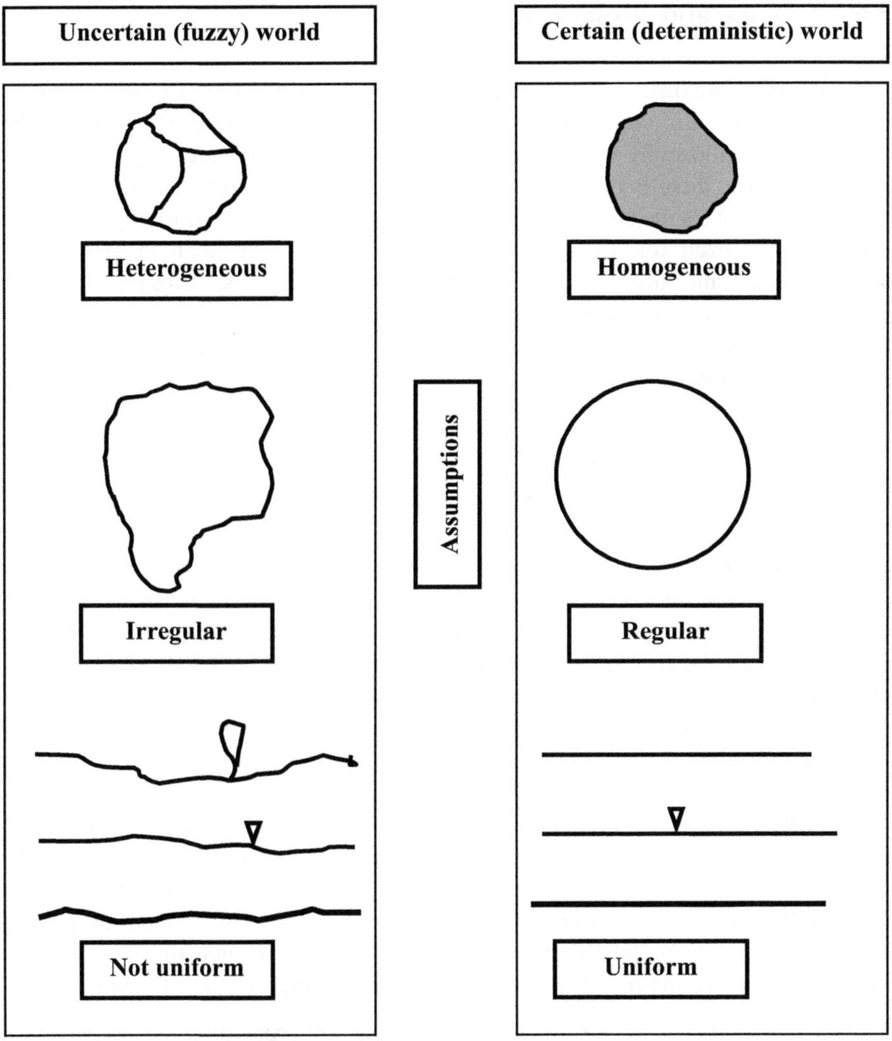

Fig. 3.3 Fuzziness versus determinism

sciously deal with problems of complexity and uncertainty, because these ubiquitous traits dominate the most natural, social, technical, and economic problems facing the human race. The only way computers can deal with complex and uncertain problems is through AI fuzzy logical principles, systematization, and control and decision-making procedures. Various uncertainty methodologies are presented in Chap. 4 including probability, statistics, and stochastic approaches.

3.6 Science and Philosophy

Although there are different definitions of philosophy, when it comes to philosophy of science, since the conclusions of science must be objective and testable, the inferences of philosophy of science cannot be valid always universally. According to Khun (1962), there is revolutionary scientific researching that changes the common paradigm. This means that the inferences as a result of the philosophy of science may remain frozen for a certain period of time; however, with the innovative opinions and contributions of a researcher or team members, it loses its validity and enters another stagnation period by replacing it with a better one, or by introducing a more evolved and matured state through additions or modifications to the previous one.

3.6.1 Philosophy of Science

Today, many graduates do not care closely about the principles of the philosophy of science, and therefore they can neither provide innovative studies nor advances in science and technology. It turns out that the minds that are not guided by the philosophy of science get stuck in the stagnant information level. However, the philosophy of science should allow information to produce new information in minds just like a power plant producing electricity. The first trigger for this is to try and ensure the maturation and development of philosophical thought within the framework of science.

The acronym Ph.D. (Philosophy of Doctorate) as the highest degree in scientific affairs means philosophy of science, which implies that the person who holds such a degree is capable to explain the specific knowledge that s/he has to anyone by taking into account their linguistically understanding level. The modern philosophy of science insists on the falsification of current scientific results, there are always rooms in any scientific research activity for the following uncertainty gradients:

1. *Ambiguity*: It has more than one meaning, interpretation or explanation concerning a word, sentence, phrase, conclusion, or solution, which is debatable. Any intention that is doubtful and suspectable in meaning with alternative intentional meaning is ambiguous.
2. *Vagueness*: This corresponds to unclear and imprecise language usage. It is just opposite of clarity and specificity. If any decision, description, or knowledge is not explained clearly, then it has vague information content and the meaning can be interpreted in a dubious manner. The words, sentences, and phrases without clear interpretation are blunt and cannot be expressible objectively.
3. *Imprecision*: Any measurement or thinking that has neither accuracy nor precision is imprecise. In expressions, there are confusions that lack correctness.

4. *Fuzziness*: The quality of anything that cannot be understood, heard, or spoken precisely is because there is some imprecision, incompleteness, vagueness, and ambiguity that exists in linguistic information and knowledge. In this chapter, very detailed explanations of fuzziness are presented especially through fuzzy logic principles.
5. *Uncertainty*: This term includes all the previous forms of unclearness in terms of skepticism, dubiety, doubt, lack of sureness, suspicion, complete or partial lack of conviction about knowledge.

Unfortunately, in many research institutions all over the world students are trained without basic science philosophy and logic. Thus, the end goal would be in a dreaded "publish or perish" ideology, without allowing existing methodologies to be improved or modified, but their rather blind use in many applications via software. For example, Artificial Neural Networks (ANNs) are tools for many researchers in matching a given input dataset to output data without logically knowing what happens between these two layers. Unfortunately, classical education systems are based on a multisystem, clear and organized framework based on crisp (two-value) logic in centuries with only two alternatives as something and its opposite like black and white, that is, true or false. However, real life and science have grey foregrounds and backgrounds in almost every corner of data and information sources. It is a great dilemma how to deal with grey, that is, uncertain information sources in order to arrive at scientific conclusions with acceptable and decisive logical principles. Such uncertainties can be dealt with probabilistic principles provided that there are numerical data (Chap. 4), but in the absence of numerical data then fuzzy logic (FL) principles with linguistically valid propositions and rather vague categories provide a solid scientific basis for the relevant phenomenon (Zadeh 1973). The first step is a genuine logical and ambiguous conceptualization of the phenomenon with cause and effect variables combined through crisp (holistically) or fuzzy (partially) logic propositions. Such an approach not only helps to logically visualize the relationships between different variables, but also provides a philosophical background concerning the mechanism of the phenomenon that can be presented linguistically without a mathematical treatment.

In order for the philosophy of science to develop and prosper within the framework of science for knowledge production, first of all, it is necessary to comprehend what the word "science" is (Sect. 3.5). Among the definitions of science made by different people, besides the philosophical ones, there are also definitions made due to some features, but a feature-based definition that is taken into account is always misleading. For this reason, until today, there is no single definition of science that is accepted by everyone. For example, a purely philosophical definition claims that science is a way of finding truth. If this is accepted as opaque information, then the first thing to be asked is whether this definition as a basis is what is the truth, and it is not possible to answer it fully. This definition brings with it many thoughts and questions other than the philosophy of science, and carries it to the field of general philosophy. One of the basic cornerstones in the definition of science is "matter"; thus a person who intends to do science studies has to accept this sentence in terms of philosophy of science, even if s/he cannot grasp fully the meaning of this sentence. When we say that matter does not come into existence and does not disappear from existence, the following sequence of information immediately emerges:

1. Since it is accepted that it does not exist out of nothing, nor does it disappear from existence, then the amount of matter is fixed.
2. If it does not exist out of nothing and does not disappear out of existence, then matter can change from one state to another. For example, water (liquid) can become solid (ice) or gas, but the amount remains the same according to the abovementioned principle.
3. Since the amount is constant, the total amount in the parts should be constant as well. For example, if one thinks of matter as a volume, this volume can turn into other volumes, but the material amount is always the same.

In the transformation of these points, which were taken out within the framework of the general philosophy above, into the philosophy of science, we must first consider that two words are absolutely necessary in the definition of science, which are

Science is materialistic
and
Matter undergoes change

The second proposition has the most weight, which implies temporal or spatial change of matter without additional generation or destruction, but conversion from one state to another. Hence, there are four dimensions as three spatial (x, y, and z) and temporal for material variations. Where there is no change, there is no scientific content. Philosophy of science can be entered after the acceptance of the philosophy principle as:

Matter does not come into existence from nothing and does not disappear from existence

In any scientific research, then, as a science inference, one can state the following general statement, which becomes imperative:

The difference between the initial state and the final state amounts of the substance is equal to the amount of matter between those two states

All verbal, linguistical, logical, and philosophical findings can be converted to mathematical expressions by representation of each variable symbolically, which are explained in different chapters in this book, especially in Chap. 5. For instance, if the amount of the matter is input, I, stage and output, O, stages then the change of matter amount is $\pm\Delta S$, hence, according to the continuity statement, one can write mathematically that

$$I - O = \pm \Delta S \tag{3.1}$$

This expression is referred to as the mass, momentum, energy conservation principle in physics, continuity equation in engineering, and budget in economics. The same philosophical concept behind this expression is valid, whatever the methodologies of shallow learning or deep learning domains are. In these learning procedures, most often the input information is available digitally as numerical data and also the output conditions and principles are set up, but in between the change of input information to output necessitates the structure of the variation component so that there is a balance between available input data and output data estimation or prediction. One who is capable to establish the transformation (change stage) of modeling without any rote and imitative manner can achieve successful research results. For such a capability, science, philosophical, and logical verbal reasoning principles pave the way for rational scientific successes. The philosophy of science should reject

engraving on the minds

and should be based on

implanting on the minds

because implementing means the knowledge sharing on the basis of philosophy, logic, and science.

3.6.2 Implications for the Philosophy of Science

Since philosophy is an unlimitedly invisible "mind" activity, it was functional in the past, functional presently, and will continue to function in the future. Particularly, in philosophy of science, it is necessary to make a categorization about the subjective variables such that researcher can consider it as interesting curious affair. As a simple example, if the event of rain is considered, at least a few of the factors that may cause rain must be considered verbally. Science and philosophical thoughts lead to rational conclusions after logical interpretations. As a philosophy of science, one can reach rules that are quite general, healthy, unbiased, criticizable, and acceptable by everyone. According to the philosophy of science, each thinker can verbally express what s/he thinks and then translates them into another language, geometry, and mathematics. While skepticism in general philosophy can oscillate widely, in science philosophy the field of skepticism is very narrow, because some shackles are put on the results, but still skepticism is there and not lacking. Science skepticism may develop innovation with normal thinking over time, but it rarely occurs in situations in sudden jumps as was in the history of science, which is referred to as revolutionary innovations. Even in any revolutionary innovation, skepticism persists. In order to think about the philosophy of science, it is necessary to have in mind the following four points:

1. The existence of a factual (material) existence (stone, earth, iron, mortar, water, air, fire, etc.).
2. Knowing the position of this entity in time and space.
3. The same asset has or is exposed to variability, that is, change.
4. To look for how and why the relationships between different factual entities are variable.

If there is effect between two or more variables, it is necessary to find out what kind of relationship exists between them. As a result of the objective knowledge accumulation, the philosophy of science ensures reaching the knowledge pools that increase over time. The information given by the philosophy of science is the product of the mind and the information obtained by thinking about a problem, and these can be categorized as rational science information. In order to specify the unknowns, a causality relationship should be sought between cause and effect variables by asking what the main reason is for determining the effect that causes the results. Another principle in the philosophy of science is the acceptance that the event will yield the same results under the same conditions, which is called the "principle of stationarity," and thus the uncertainty is completely excluded from rational inferences. For this dismissal, some assumptions have to be made so that the thought of science can move forward towards the philosophy of science. Science can never find the absolute truth and cannot reach it, which means that there is always an uncertainty in the scientific results, and therefore, it is necessary to make assumptions about the generation mechanism of the phenomena.

Another expectation is that every factual (material) event has a measure that can be expressed in numbers, which is called the "principle of measurability." In particular, with the acceptance that natural events do not give conflicting results with each other, an acceptance that we can call the "consistency principle." As it has been said before, there is a perception in some people and societies as if the knowledge of science cannot be doubted or falsified. This is the most dangerous thought against the development of science, because the results of science can be falsified at any time under all conditions, and for this reason, there is the "falsifiable principle," which is one of the most important acceptances in the philosophy of science (Popper 1955). People who do not remain impartially loyal to all of these listed principles cannot be called thinkers or scientists, but many people achieve the titles they want in academic life; after the last title, they have reached their level, which is "to hang their sieve on the wall," that is, without further research activities. The fact that the knowledge complies with the principles of science does not mean that it cannot be criticized or doubted, because in order for the knowledge in science to mature over time and yield even better results, it must always be kept under suspicion.

One of the most important features of science is that it does not comply with voting and especially does not have a "democratic" character. Throughout the history of science, thinkers who have been excluded from their societies due to their outlier knowledge and philosophy of science have affected the whole world even after death.

The aim of science philosophy is to reveal the relationships between cause and effect factors in general. If we put the mind in a box, and after processing the information that comes as input, we come up with the situation in Fig. 3.4 as a basic modeling concept.

Fig. 3.4 Mind function

In order for everyone to obtain information from the outside world under the same conditions, it must be assumed that the researcher exhibits a behavior completely independent of subjectivity. By looking at the people around him, s/he can take some information signals from them, make comments, be more curious about some of them than others, and make an effort to reach further details and reasons.

Observations are a way of obtaining information in the form of deduction, where details are not known immediately, and perhaps the conclusions will still be related to them without inner details. It opens the way for speculative knowledge extraction as in the ancient civilizations before Christ. However, if a problem is raised by asking "how" and "why" questions, this time the path of thought goes to the whole, where the principle of "causality" comes into play, which leads to results as in Fig. 3.4. Acquiring scientific knowledge with this kind of thinking and research is seen for the first time in the history of science in the studies first of ancient Greek and then Muslim thinkers and scientists during the Medieval Era. However, after knowing the detailed reasons, it is possible to make predictions about the event and produce scientific knowledge.

Another interpretation of Fig. 3.4 is a formal way of deriving the unknown (results) from the known (causes). In order for the unknown to be explained by the known, it is necessary to determine the correlation that may exist between them. For this reason, in the face of the question of "what is the cause of unknowns," the causative factors must first be determined verbally, which is referred to as the "causality" principle. Here, the information that the mind receives from the outside world, which is mostly verbal and numerical, is to process this information in the inner world through the function of reasoning, to produce verbal, figural, or numerical results. Instead of the mind in the center, other mechanisms generated by the mind can be substituted. These may include different mathematical operations, models, or simulations. Since the subject of this section is the philosophy of science and the first thinkers and scientists did not have the software, laboratories, and financial means like today, they always laid the foundation stones of today's science by making rational inferences from nature or an object with the mind as in Fig. 3.4. They were just doing philosophy and subsequent logical inferences similar to the philosophy of science as explained above.

In addition to the verbal and numerical information from the outside world as in Fig. 3.4, there is shape, taste, smell, hearing, etc., information also. For this, shapes can also be given in the Cartesian coordinate system, the first signs of which were stated verbally by the Muslim thinker Al-Biruni (973–1048), but geometrically shown by Rene Descartes (1596–1650), and this is very useful for mental activities and helps to visualize rational thoughts. This notation system is actually two-dimensional, that is, it serves to show how the relationship between one of the causes and one of the effects can be (Fig. 3.5).

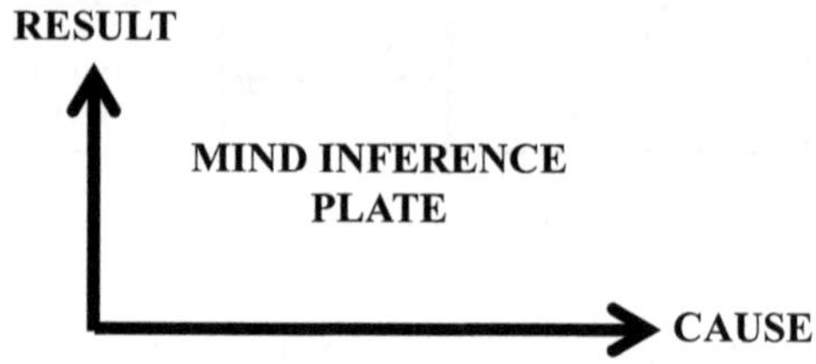

Fig. 3.5 Mind plane cause-and-result relationship

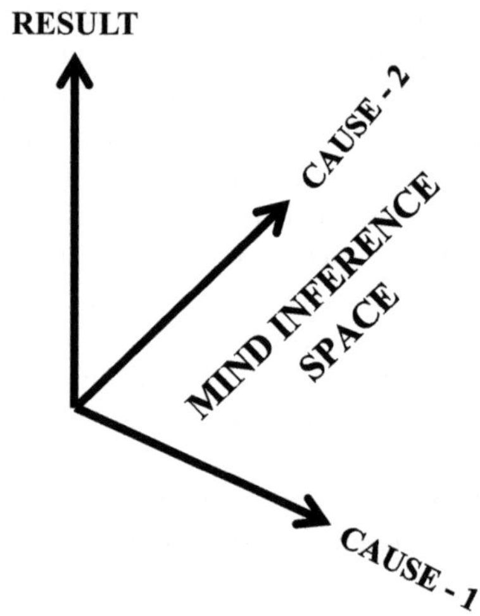

Fig. 3.6 Mind plate for two-cause-one-result relationship

When this figure is compared with Fig. 3.4, it is immediately clear that the two actually have the same arrangement. Figure 3.5 relates to the one-cause-one-effect case of Fig. 3.4. It can be said that Fig. 3.5 is more suitable for rational thought inferences, but Fig. 3.4 is more valid for today's complex model structures. Similar to Fig. 3.5, one can imagine a three-dimensional coordinate system also (Fig. 3.6).

In its simplest form, people can make rational inferences in the form of one-cause-one-effect, but some gifted people can even imagine what the reasoning pattern would look like in a two-cause-one-effect situation. After making verbal reasoning inferences as one-cause-one-effect, which is recommended here in terms of philosophy of science, one can arrive at two-cause-one-effect or generally many-cause-many-effect inferences, if necessary.

Although the spaces in Figs. 3.4 and 3.5 are filled by mathematical expressions, this book is based on enlightening the way of filling these spaces primarily verbally with the use of thought and reason at shallow learning level inferences, and then by logical interpretations, which are the tools for mathematical relationship search between the cause and effect variables.

3.7 Logic

Logic is a method of improving the ability of reasoning in order to understand the facts in a formal (systematic) way and to reveal the valid relationships between the causative and effective variables. Thus, one can say that logic is a guide to right thinking, and the information that passes through the scientific philosophical process provides results that will not be inconsistent. It serves to determine the provisions that do not constitute irrationality, because it is the function of proof, which includes the acceptance, idealization, and simplification that are consistent with the rationality. Thus, it is necessary to understand that the inferences made are approximations to the truth. Since the truth cannot be reached with such assumptions, it should be kept in mind that inferences in science are the result of approximate reasoning.

Throughout the history of science, the probabilistic logic, multiple logic, and various copies of it, which were said by Al-Farabi (870–950) after Aristotle (384–322), have emerged and although more healthy studies have been attempted according to these, the overwhelming superiority of the crisp (two-value) logic has not yet been overtopped. One interpretation of logical probability is that there are intermediate values of the events between two extreme situations (true and false). The most important saying of Prophet Mohammad on this subject is

The best of the works done is the one between the two extremes

On the other hand, according to Nasreddin Hodja (1208–1284) discourse, the logic of saying

Complainer and complained have both rights

represents in-between situation, the content of this saying is grey. Fuzzy logic is explained as having natural logic rules (Zadeh 1973) and the state of not being absolute true or false in every job can always be valid at all times.

Logic separates filters and identifies rational prescriptions from the ocean of philosophy. Its primary task is to set up systems and criteria for distinguishing rational from irrational, which express inferences, the processes whereby new assertions are produced from already established ones. It provides a mechanism for extension of knowledge.

Logic is a simple word, but its definition is unquestionably impossible and a precise definition cannot be reached. Many do not even try to provide a definition, and therefore, the definition of logic is rather vague in the literature. Different schools have given various definitions for logic. Chronologically, simple definitions of logic are arranged approximately as follows:

1. A means of distinguishing right from wrong (Averroes, 1126–1198).
2. The science of reasoning teaches the way of searching for unknown truth in connection with a thesis (Kilwardby 2015).
3. Art also guides the mind so that it does not err in the way of inference or knowing (Poinsot 1955).

4. The art of reasoning by knowing things (Antoine Arnauld, 1612–1694).
5. Correct use of reason in the pursuit of truth (Watts, 1736–1819).
6. The science as well as the art of reasoning (Whately, 1787–1863).
7. The science of the operations of the understanding subject to the estimation of evidence (Mill 1904).
8. The science of the laws of discursive thought (McCosh, 1811–1894).
9. The science of the most general laws of truth (Frege, 1848–1925).
10. The science that directs the operations of the mind to reach the truth (Joyce, 1908).
11. The branch of philosophy that deals with analyzing patterns of reasoning that draw a conclusion from a set of premises (Collins English Dictionary).
12. Official systematic study of valid inference and sound reasoning principles (Penguin Encyclopedia).

Instead of Aristotle's crisp (bivalent) logic, further and additional scientific and technological achievements in basic human thought took place by means of fuzzy logical principles (Zadeh 1999). Crisp logic allows to make inferences by deciding on one of the purely binary options such as true-false, yes-no, A-B, 0–1, white-black, etc. Unfortunately, this logic brings the person to the dual choice of "whose worldview is the same as mine" or "who does not," and when this is the case, it is not possible to talk about reconciliation, mutual tolerance, or dialog.

3.7.1 Logic Rules

Logic is the study of arguments, derived from the ancient Greek logos, which originally meant words or spoken arguments meaning thought or reason (logic). Its primary task is to establish systems and criteria for distinguishing rational arguments from irrational ones. Arguments express inferences, which process by which new claims are generated from already established ones. It is formal relations between newly produced claims and those already established, where "formal" means that the relations are independent of the claims. It is equally important to explore the validity of inference, including various possible validity definitions and the practical conditions for determination. Thus logic appears to play an important role in epistemology as it provides a mechanism for the expansion of knowledge. As a by-product, logic provides prescriptions for reasoning, that is, prescriptions for how humans as well as other intelligent beings, machines, and systems should reason. Since the mid-1800s, logic is widely studied in mathematics and even more recently in computer science. Logic as a science explores and classifies the structure of statements, arguments, designs, and schemes in which they are encoded. Therefore, the scope of logic can be very broad, including reasoning about causation, probability, and fuzziness.

Logic searches for meaningful propositions among others in a text or paragraph. Not all sentences have a logical structure, and only logical propositions lead to thinking, but the existence of interrelationships between various categories help to make a final decision. Therefore, it is necessary to have some guidelines for defining logical expressions in a particular text or structuring them in the process of thinking about some phenomena. The simplest way to search for a logical expression is to find one or more of the following logical words:

1. "AND" is the conjunction word that joins two categories or expressions such that both are included in the final decision. This is called "intercept" in classical set theory. We will use the phrase "ANDing" in this book. Since, this is taught as intersection in the textbooks, it is understood only as overlap area of two or more geometrically common Venn diagram sets, not the idea of logic.
2. "OR" is another conjunction word that also takes into account both categories and leads to a joint decision such that the parts of these two categories are the components of the deduction. In classical set theory, it is known as the "union" operation. It will be used as "ORing" in this book.
3. "NOT" is a negation of the original category. For example, if "tree" is the name of the category, "NOT a tree" includes anything that is not a tree. This is "complement" in classical set theory, but will be referred to as the "NOTing" in this book.
4. IF... (Premise, P)....THEN... (Consequent, C).... is a proposition that contains a very rigid logical expression that contains useful relationships between the input (premise) and output (consequent) categories. There are two parts here, one denoted by P between the words IF and THEN, implying the antecedent, input or prior knowledge; and the part denoted by C after THEN and containing the conclusion, output, result, final decision, or deduction based on the premise. As will be shown later, any expression of the form "IF.....THEN....." will be referred to as the logic rule for the relevant event.

A valid logical proposition is one whose consequent follows from its premise. It is also important to know in what sense the consequent (conclusion) follows from the premise. What is it for a conclusion to be a consequence of the premises? These questions are, in many ways, central to crisp and fuzzy logic systems.

There can be many different things that can be said about the same proposition, but most would agree that the proposition is valid if we are not ambiguous (if the terms mean the same thing in the premises and in the consequent), so the conclusion follows deductively from the structure of the propositions. This does not mean that the result is correct. Perhaps the premises are not true. However, if the premises are true, logically the conclusion is also true.

Crip logicians argue that fuzzy logic is unnecessary. It is stated that anything that uses fuzzy logic can be easily explained using classical logic. For example, true (white) and false (black) are discrete. Fuzzy logic contends that there can be grey versions between true and false. Classical logic says that the definition of terms is false as opposed to the true truth of the statement.

3.7.2 Elements of Logic

The general element of logic is language, and especially, mother tongue. The basic elements of logic are always words, terms, and propositions (sentences) that express a fact, situation, event, or phenomenon.

3.7.2.1 Bivalent (Crisp) Logic Words

The most basic parts of all languages in the world are syllables. With the combination of syllables, meaningful and meaningless words, terms, conjunctions, adjectives, slang expressions, curses, nice words, and words with rational and irrational meanings can emerge. In a way, if each syllable of every language is considered just like an integer (i.e., in the form of digits), one can then say that the basic structures of our mentality consist of syllables. While there may be single-syllable meaningful and meaningless words, rather multisyllable words appear. For example, in English, there are innumerable words with one syllable in general, for example, it is possible to say "throw, and, ask, hold, pass, come, stop, give, drink, eat" and many more words. We do not need to know the reasons for their first appearance or evolution in history.

Knowing that each word is loaded with two kinds of information helps to use the words stored in the mind dynamically. If these two information loads are not known, that word is condemned to remain mechanical and static in the mind by heart, which builds a wall against the ability to produce other information from the content of that word. The first of these information loads is to determine the "origin" (etymology) of the learned word, and the other is to determine what the "meaning load" (epistemology) is.

Although many words contain information, there are also words that do not contain much information and do not make sense on their own. They play an important role when they come between or before two other words and sentences. To give a meaning to these words, which cannot produce much perception in the mind in terms of logic; they are explained by the situations that occur as two circles in Fig. 3.7, each representing a word, partially overlap.

In this way, an object, fact, word, or sentence, which is described as two entities in the Fig. 3.7a, can be considered, and it is necessary to decide what kind of a logical operation to do between them at the end of this thought. As can be seen from the figure, there are three types of logic operations as mentioned earlier; they are ANDing (intersection), ORing (union), and NOTing (exclusion).

Similar to the logic basic operations that may arise as explained above in case of two subjects (A, B), the reader is recommended to analyze what kind of logical operations might occur if there are three (A, B, C) or more subjects. The reader who succeeds in this will be able to understand the basic operations of logic, so that s/he will be able to treat the analysis of more complex phenomena, which will be explained later, with logic as preparation for mathematics.

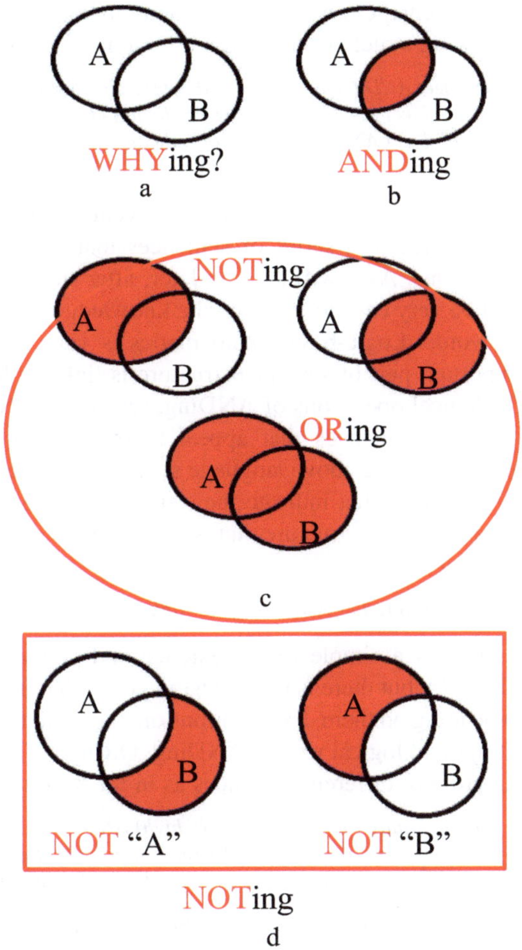

Fig. 3.7 (**a**) Two existences (**b**) ANDing (**c**) ORing (**d**) NOTing (exclusion)

3.7.3 Logic Sentences (Propositions)

Any language has its own rules called syntax (grammar) with names such as subject, object, and predicate (verb, adjective, conjunction, etc.). Among them, the subject actively and the object passively indicate the functions in a sentence structure. Proposition types indicate what kind of knowledge or relationship exists between the subject and the object. A simple sentence structure, say, in English is

Subject AND Object THEN Predicate

One can explain it with the ANDing logic operation. From this it is clear that only ANDing is present in a complete sentence. There may also be an ORing logic operation in a sentence, but this means expressing different option than the information content in the first sentence. Compound sentence structures emerge from simple sentences by using the logic operations ANDing, ORing and NOTing. The following structures are given as examples of compound sentences:

(Subject AND Object AND Predicate) AND (Subject AND Object AND Predicate)
(Subject AND Object AND Predicate) OR (Subject AND Object AND Predicate)
(Subject AND Object AND Predicate "(Subject AND Object AND Predicate) NOT
etc.

Thus, in order to explain an event or phenomenon, there is a necessity of coexistence in a series of sentences that can be connected with different logical operations. As it will be seen later, after such a structure is understood verbally, it will be very easy to translate the knowledge into mathematics with symbols instead of words. Logic-based mathematics is very interesting and cute. Thus, the most important part of sentence structure is linking the different elements together with the logical operations of ANDing, ORing, and NOTing.

Logic sentences that appeal to the human mind begin to emerge when the elements of the above language sentences (subject, object and premise) are replaced by situations that indicate the state of a phenomenon or variable. A simple logical statement has a structure that is connected to each other by at least one ANDing. For example,

Temperature low AND Pressure high

This is a simple logic sentence and connects some situations in nature as in Fig. 3.7b, but there is no provision yet so that the mind can make a logical inference depending on them. Explanations on the verdict and inference will be made later. For compound logical words, ANDing, ORing, or NOTing cases must be present in the structure of different sentences as in the following examples:

(Temperature low AND Pressure high) AND (Precipitation high AND high flow)
(Temperature low AND Pressure high) OR (High precipitation AND high flow)
(Temperature low AND Pressure high) (Precipitation high AND high flow) NOT

A person who understands the sentence structure in the mother tongue well, as a result of operating the rules of logic, can first filter the logical sentences by separating them from the others, and thus, a text of pages can be reduced to a very small number of paragraphs in terms of logic.

Since logic is used almost everywhere, and especially, in science and technology researches, when reading articles, papers and books, it should be questioned not as if reading a novel, but as to identify which sentences are logical. Underlining the logical sentences may lead to conclusions by considering them only in future researches and by means of inferences with more generative thoughts.

3.7.4 Propositions and Inferences

Two more words are involved in the structure of a propositional sentence that can help with logical inferences, namely, "IF" and "THEN." By identifying these two words in a sentence, it is understood that a logical proposition is in front of the reader. In the structure of a proposition, "causes" comes after the word IF, and "consequences," that are inferences, after the word THEN. The general form of a simple proposition is as follows:

IF "Causes" THEN "Consequences"

Another form of this in mathematical modeling procedures has the following form in terms of inputs and outputs:

IF "Inputs" THEN "Outputs"

To further clarify these simple propositional sentences, more concrete information as a condition (cause, input) can be placed after the word THEN. For example,

IF "Money is too much" THEN "spending may also be much"

Depending on the situation, different results (outputs) can be written as output in a simple proposition. Here, for someone who has no job, but loves to travel the logical proposition can be expressed as

IF "Money is too much" THEN "too much travel"

Simple logic propositions are popularly referred to as simply "cause-effect" relationships. The following proposition has more scientific content:

IF "Precipitation is heavy" THEN "There may be flooding"

Compound logic sentences for propositions and inferences also appear frequently. They are nothing, but a collection of simple sentences combined with the simple types of ANDing, ORing, and NOTing. The following is one of the examples:

IF "Precipitation is high AND Ground is permeable AND Vegetation is low" THEN "Excess water seeps into the ground"

In the field of science and technology, it is often possible to come across compound proposition and consequent inferences. Propositions may not explicitly contain IF-THEN words. For example,

Provided that I have money, I can spend more

It hides the IF-THEN within its body.

3.7.5 Logic Circuits

Especially in crisp logic options as true or false, which are completely exclusive and fully specific, say as "on" and "off" switch alternatives, there are numerous different options depending on the structure of a circuit. Circuits related to ANDing and ORing are shown in Fig. 3.8a–d, respectively.

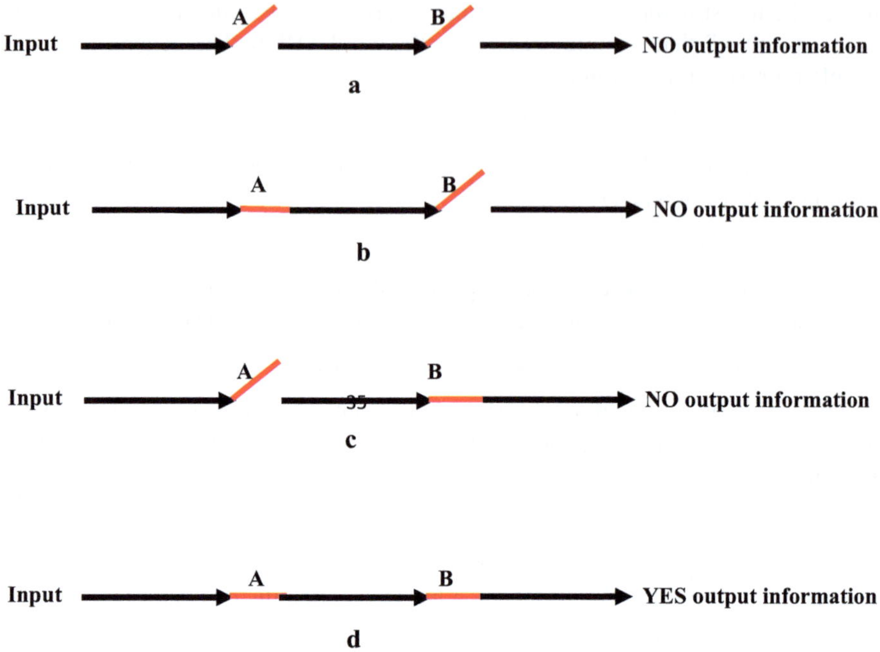

Fig. 3.8 ANDing circuits

As mentioned earlier according to crisp logic, there are two alternatives or sets in the use of ANDing or ORing words. Of these figures, two keys (A and B) are waiting for ANDing or ORing operations. The case in Fig. 3.8a is well appreciable, because both switches are open and therefore it is not possible to transfer knowledge from input to output. If only one switch is closed as in Fig. 3.8b, c, it is not possible to reach the output information. In the last option in Fig. 3.8d, if two switches are closed, that is, if they mean ANDing, it is understood that information is obtainable at the right-hand side as output.

In Fig. 3.9, some basic information can be reached either only by serial ANDing or parallel ORing logical words in propositions.

In case in Fig. 3.9a, if two parallel switches are open, it is not possible to obtain any information. However, this time, if one of the two switches is closed, as in Figs. 3.9b or c, information is obtainable. This information is partial, because it is only passed through one way. Full output information can be obtained when two switches are closed as in Fig. 3.9d.

The circuit shown in Fig. 3.10 has one input but two output points. It is recommended that the reader find different optional inferences using the ANDing, ORing, and NOTing logical rules to arrive output information.

The logic rules in the examination event may have a set of propositions that are different from each other.

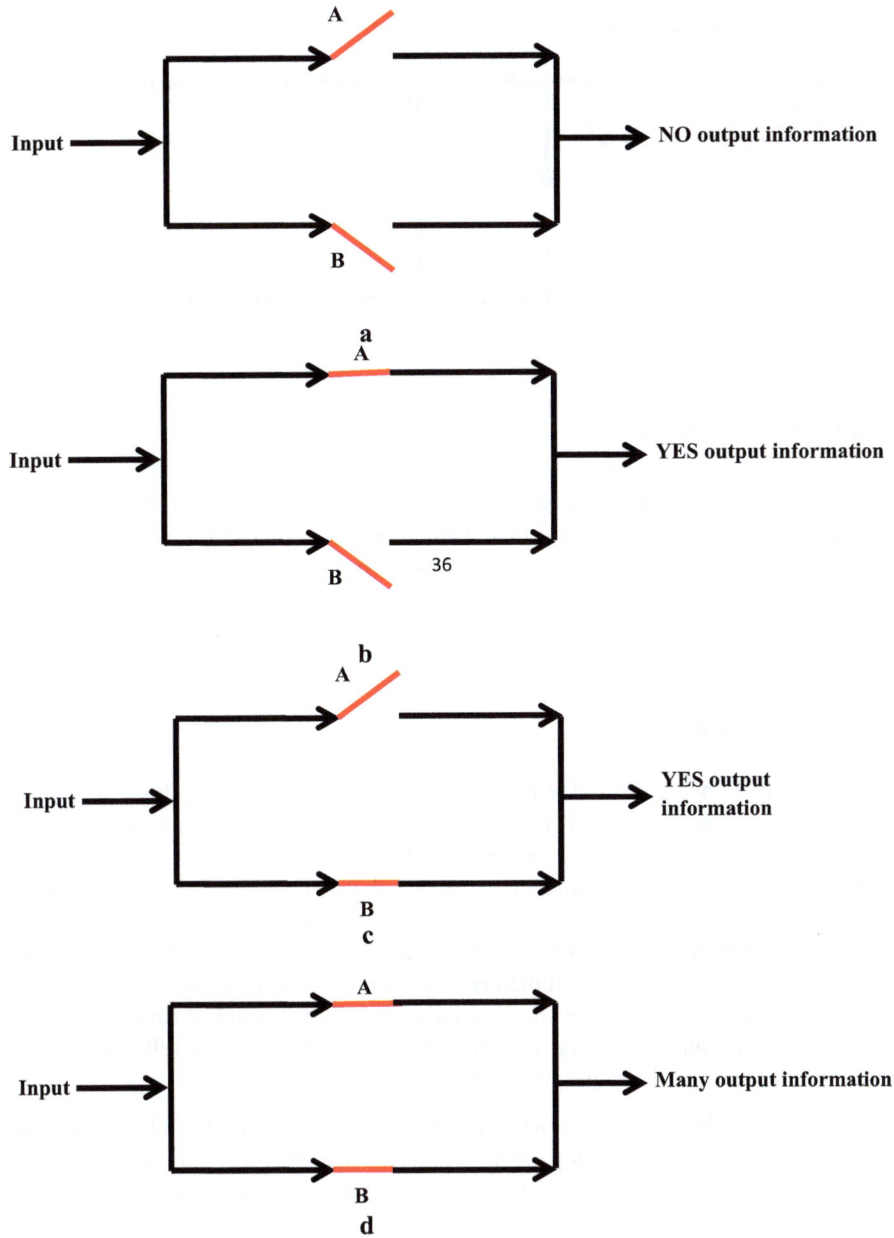

Fig. 3.9 ORing circuits

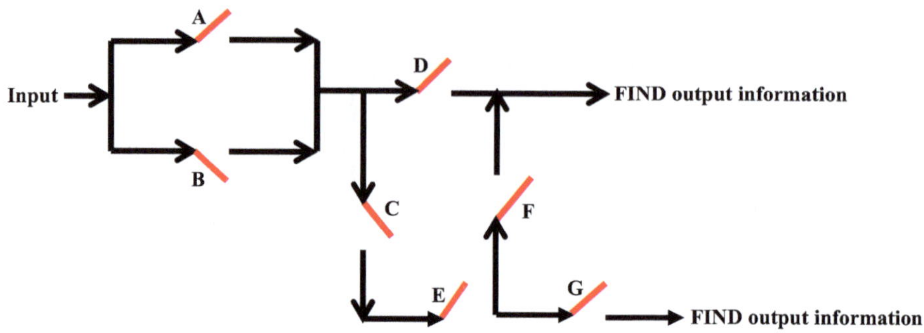

Fig. 3.10 Complex logic circuits

3.7.6 *Logic Types*

Logic is the main tool for rational inferences from ocean of knowledge and information sources based on philosophical thinking. During the human history, the thinking progress has evolved at various stages starting with the initial alternative by Aristotle (BC 384–322) as crisp logic. Later the logic took different alternatives depending on the necessity.

3.7.6.1 Crisp Logic

When the mind encounters a dilemma or duality, according to crisp logic it chooses one and rejects the opposite. It trains the mind to think in terms of true and false alternatives as the first approach to model reality. The dual nature of the rational reasoning component of the mind is so strong that reason alone cannot overcome it. The best that one can do is to reconcile opposites. Therefore, there is no vagueness, ambiguity or probability in the content of the two-value crisp logic, because everything is either completely true (white) or false (black). The classical true-false approach of thinking can easily trap the human mind in routines, stereotypes, prejudices, and habits that are unfit for real experiences. The following points are contrary to classical crisp training system:

1. The authorized knowledge must be employed with precautions and the solutions should not be completely dependent on knowledge gadgets, and through discussions and questions, the scientist should try to break the marginal frontiers of the crisp information.
2. Any scientific conclusion is subject to uncertainty and suspicion, and hence, further refinements are necessary for innovative ideas and modifications.
3. Logical principles, rules, and preliminary philosophical bases must be kept in the agenda by scientists so that each problem can be solved with expert contributions.
4. Scientific thinking must be geared towards the falsifiability of the conclusions or theories rather than readily acceptable conclusions.

3.7.6.2 Multiple Logic

This is many-value logic, where there are more than one alternative such as inclusion of the middle state, which is missing in two-value crisp logic. For fruitful research with innovative results, it is recommended to go beyond the restrictive rules of crisp logic and constraints, no matter how obscure (fuzzy) or obvious (crisp) the relevant phenomenon is. In this manner, there are many contradictory alternatives, which provide the decision on the most convenient one scientifically.

3.7.6.3 Probabilistic Logic

Scientific conclusions depend on propositions, which are logically valid rules of related phenomenon. These proportions are verbal or linguistic expressions, and thus involve ambiguity and vagueness in the initial philosophical thought. As more scientific evidence are obtained, rationally or empirically, the validity of these propositions increases or the uncertainty ratio decreases. In the philosophy of science, up to now, scientific statements are accepted as true with some degree of probability. However, objective probability attachment to scientific statements is a difficult task, and therefore, subjective (Bayesian) probabilistic statements are considered in many uncertain practical problem solutions (Chap. 4).

> *All traditional logic habitually assumes that precise symbols are being employed. It is, therefore, not applicable to this terrestrial life but only to an imagined celestial existence* (Bertrand Russes, 1948).

> *As the complexity of a system increases, our ability to make precise and yet significant statements about its behavior diminishes until a threshold is reached beyond which precision and significance (or relevance) become almost mutually exclusive characteristics.* (Zadeh 1973).

3.7.6.4 Symbolic Logic

This is another type of language, which translates philosophical and logical rational inferences into a system of symbolic language. All mathematical equations, expressions, and algorithms fall within the domain of symbolic logic, where logical ideas are expressible by symbols. Symbolic logic methodology manipulates rational ideas to mathematical expressions with arithmetical operations. In scientific modeling of any phenomenon, input and output variables are represented generally by alphabetic symbols, and hence, each mathematical equation renders independent input variables into an output. For instance, the phrase "two variables are directly and linearly related to each other" implies many well-known scientific laws as force (voltage, stress, heat) variable is related to mass (current, deformation, temperature) in the symbolic forms of $F = ma$, $V = RI$, $\sigma = E\varepsilon$, and $H = Dt$. Here, F, m, a, V, R, I, σ, E, ε,H, D, and t indicate force, mass, acceleration, voltage, resistance, current, stress, elasticity, strain, heat, diffusion coefficient, and temperature variation, respectively.

3.7.6.5 Logic, Engineering, and Computers

After having understood the fundamentals of logic, it is necessary to know the role of engineering in philosophical, logical, scientific, and technological aspects. Engineering is the application of scientific principles in order to transform natural resources into structures, machines, products, systems, and processes in the most efficient way (Page 341 of the main Encyclopedia Britannica). The whole of systematic studies that put the results of scientific studies and researches into technology and applications to meet the concrete needs of society is called engineering (8440-th page of the Great Larousse). An important question is "Is it the people who know today's engineers and remain at the technician level?" "Is it that there are people who are 'informed' and who can make new discoveries by constantly questioning what they know, leading themselves to critical thinking before anyone else, or who can make some changes for the betterment of those which have already been found?"

There is no human thought activity that does not contain elements of philosophical thought. Otherwise, it is not possible to talk about criticism, skepticism, innovation, and a constantly developing and spreading enlightenment. Since it is related to enlightenment in engineering, it should also have elements of philosophy. Today, unfortunately, it is sad to see that engineering education has become a professional career that does not include philosophy and logic issues and tries to serve only with the information perceived as it is in the forms of mathematical expressions. In Chap. 2, Fig. 2.5 shows significant stages of engineering work.

Logic and engineering should be interpreted freely, because engineering is about getting things done, usually building infrastructural things that accomplish a predetermined purpose. Logic is the domain of formal a priori truth that encompasses mathematics and everything that supports the construction and use of abstract or mathematical models that are crucial to engineering. Engineering is conceived as a discipline in which modeling techniques are increasingly mastered that allows the construction and evaluation of a design before its implementation is physically produced.

The increasingly dominant intellectual content of engineering problem-solving, modeling work, is basically pure logic. Software that supports these intellectual activities are more effective when built on solid logical foundations. The logical revolution, which is barely foreseen, derives from the same fundamental imperatives of how knowledge should be represented if we are able to manipulate it effectively? Digitization is a prerequisite for information process by computers, and it is used by engineers. What can be done about available information and numerical data depends on computers. A static image as a bitmap can be displayed, but manipulated less effectively than a representation of data with more structural information. A movie with a dynamic three-dimensional experience can be represented as a sequence of bitmaps, but more complex representations are required to allow interactive navigation. Ultimately, computers are needed to understand data manipulation and to reason about the behavior as well as the appearance for the system description. A sensible approach is to represent a system in an open-end way in terms of functionality that can offer and fit for the purposes of many different types of software integral workings together.

Logical procedure is a natural step in engineering knowledge representation that allows overt exploitation and manipulation. An exotic and unlikely development is in due course considerable an economic necessity, not only for engineering purposes, but also for education and entertainment, where the significance of models is undeniable.

3.8 Sets and Clusters

The concept of set must be understood very well in order to carry out all kinds of logical and especially fuzzy logical operations, because sets are structures that contain a group of word elements. Clusters are more concrete forms of concepts, terms, and definitions expressed in words in science studies (Dunn 1973). Concepts, terms, or definitions are based on an object in whole or in part. Holism includes all singulars related to that object. In partiality, a part of the wholeness is represented according to some characteristics. Here, the whole consisting of singular elements is called population set and the part of it is a subset. Set seems like an abstract concept that is thought to contain many elements, but it is possible to come across many concrete examples of this in daily life. For this, it is sufficient to look at the following examples.

3.8.1 Crisp Sets

When one says young people in Istanbul, a population or universal is meant in which all young people in the city are elements (members), and this is a holistic set or community cluster. In order to make the perception of such a community more concrete in the mind, one can ask the question, "Who is youth?" Since the answer to this is related to age, it is necessary to determine an age range that can be valid for the youth group. For example, that is why anyone between the ages of 18 and 23 is young. In order to better understand this inference, the words in Fig. 3.11 can be used.

In Fig. 3.11a, a youth society passing through the human mind in an abstract way is considered. One cannot know how many young people are there, and cannot doubt the existence of such a cluster. Since the word young immediately brings the state of age to mind, the appropriate numeric form of this set, which is abstract, is given in Fig. 3.11b, where the group of young people between the ages of 18 and 30 is expressed. Here, belonging degree to the set is represented by the number 1 and not belonging by 0. Thus, a bi-valued logical set emerges, and each person in this set is equally young without any distinction among them. Such sets are classical, fragile, and bivalent. It is called classical, because it represented by rectangle; it is called fragile because there is a sudden transition from 0 to 1 or vice versa, at its lower and upper limits; it is also bivalent, because there are two belonging degrees as 0 and 1.

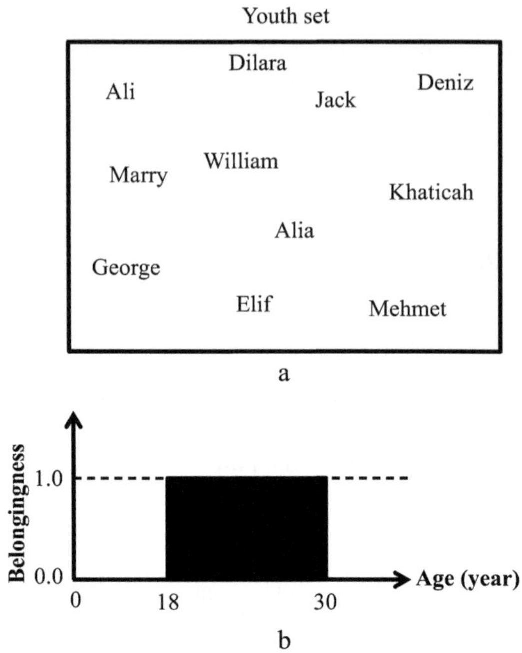

Fig. 3.11 Population set

3.8.1.1 Subsets

There are subsets of concept, term, and definitions, as well as subsets of a population set, which is larger in scale and impact than the subset. The youth population set given above has many subsets. For example, if we say "young people," it refers to a subset and its name is "young." If we say "almost young," another subset of the "young" emerges. In Fig. 3.12, where the population set is shown as a rectangle similar to Fig. 3.11, subsets are indicated as circles or ellipses. This geometric representation is called a Venn diagram. If we symbolically denote the population set as P and its subsets are A, B, C, D and many others that are not shown in this figure.

In this way, there are overlapping subsets (A, B, and C) as well as nonoverlapping discrete subsets (D).

In order for humans and computers to communicate, verbal logic expressions of humans need to be converted into numbers, which can be achieved by means of sets. The first studies in this direction were advanced by Boole (1815–1864). Thus, classical logic has become completely deterministic, where fuzziness and uncertainty from thoughts are completely excluded.

By denoting the belonging of an element to the set with 0 or 1, the quality differences between the elements disappear and the logic becomes restrictive, crisp, two-value, and absolutely certain (not fuzzy). For example, elements of a P plant set can be written as:

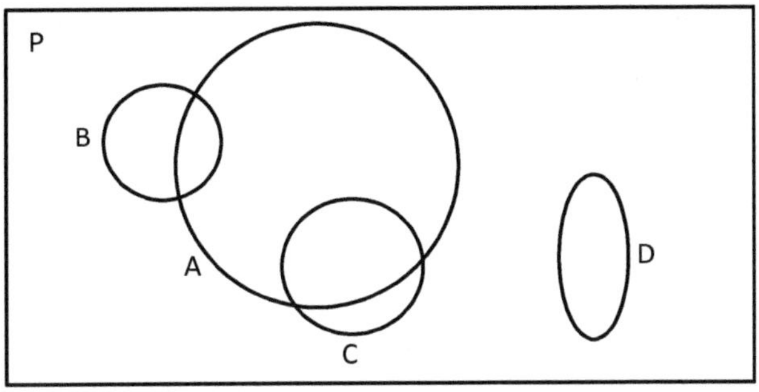

Fig. 3.12 Venn diagram representations of population and subsets

P = {Pear, Cherry, Watermelon, Fig, Grape, Garlic, Melon, Quince, Hemp, Chestnut, Rosehip, Lettuce}

It is known that each plant is very different from each other in terms of concept and quality. From here, we denote plants with the initials C as 1 and the others as 0.

$$P = \{0, 1, 0, 0, 0, 0, 0, 0, 0, 1, 0, 0\}$$

Hence, the community set expressed by numbers is obtained. In this set, the attributes among the plants have now completely disappeared, and the logic that we can call numeric does not give us any information other than the existence of two classes that start with the initial letter C (1 valued) and non-initiated (0 values).

As can be understood from the above explanations, crisp, that is, two-value logic is quite restrictive. The principles and rules of crisp logic have numerous benefits in the practical use of mathematical and computer operations and functions. For example, computers do not operate with decimal numbers, but on a binary (0 and 1) number system. The human enters decimal number data into the computer, and the computer first converts these numbers to binary logic (0 and 1 sequence) that it understands. The computer, which quickly performs operations in the binary number system, converts its outputs to the decimal system. The belongingness of the elements in the set, such as 0 and 1, is called the belongingness (characteristic) value, or membership degree (MD). The human and the computer jointly understand that the element with a MD of 1 belongs to the set considered. It brings up the display of sets as Venn diagrams in the Cartesian coordinate set by showing the states of belonging or not belonging to the set with a number each (see Fig. 3.11b).

3.8.2 Fuzzy Sets

Fuzzy logic assigns MDs to a scientific belief (degree of confirmation or falsification), accepting values between 0 and 1, inclusive. The contradiction between the verifiability and falsifiability of scientific theories involves philosophical underpinnings that are fuzzy, but many philosophers have concluded their work with crisp logic, which is completely contrary to the nature of scientific development. Although many philosophers of science try to solve this problem by considering the limitation of scientific knowledge and the possibility and sometimes the probability of scientific development, unfortunately "fuzzy philosophy of science" has not entered the literature for many years although it has been suggested many years ago by Zadeh (1973).

An explanation can be made regarding the limitation of scientific knowledge and fuzzy logic in scientific progress processes. Scientists cannot be completely objective in justifying scientific limitation or progress, but the components of the fuzziness are the impetus for the generation of new theories. All scientific and logical rule bases must be tested by fuzzy inference engine that leads to the fuzzy scientific field. Scientific phenomena are inherently fuzzy, and especially the foundations of scientific philosophy implicitly contain fuzzy components. In science, the dogmatic nature of scientific knowledge or belief, as if insensitive, is the fruit of formal crisp logic, whereas fuzzy logic is scientific for effective rational improvements and knowledge production in the future.

For fuzzy modeling, it is necessary to understand the operation of each membership function (MF) with data. Sets on the horizontal axis have the variability range related to the MD on the vertical axis. In a way, fuzzy functions are the conversion of data values to MDs or given MDs to data values. For example, in Fig. 3.13, the fuzzy "Low" category is exemplified by a trapezoidal fuzzy set near the origin. There are three data values that fall into the domain of the "Low" membership function (MF), and therefore, trigger (ignite) this function with three different MFs, as 0.3, 0.8, and 1, respectively.

This means that these three data values are all "low," but with different degrees of lowness. Someone with a membership rating of 1 is completely low, but a membership rating of 0.3 is at least "Low." This means that there are many MDs for many data values that fall within the "Low" category range. On the other hand, two data values on the right side do not fall within the range of this category, meaning they are outside the range with zero MDs as for the "Low" specification. Therefore, data for each category can be categorized as trigger and non-trigger. Non-triggers have zero MDs, but triggers have different degrees of membership ranging from 0 to 1. Likewise, Fig. 3.14 is for "High" fuzzy set with two triggering values and corresponding

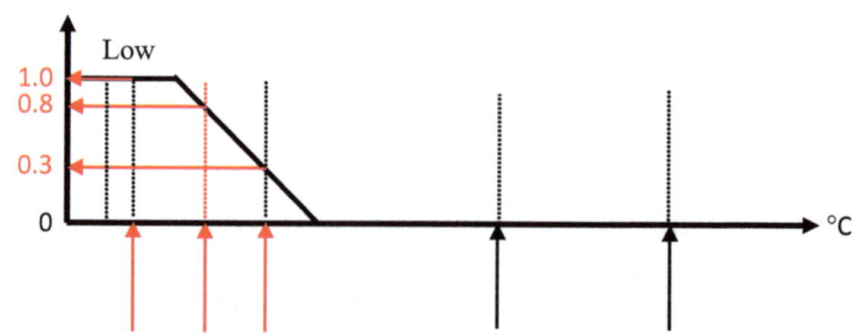

Fig. 3.13 "Low" fuzzy member and triggers

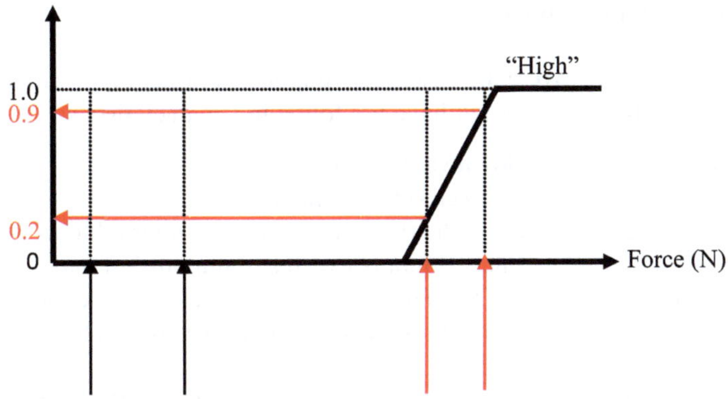

Fig. 3.14 The fuzzy "high" MF for force

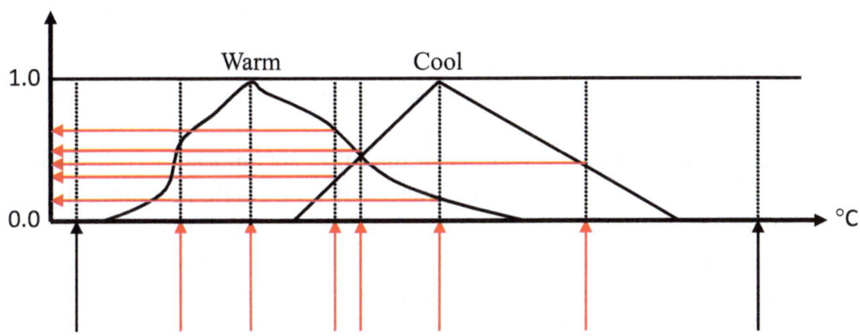

Fig. 3.15 Two overlapping fuzzy sets and firing

MDs on the vertical axis. Similar arguments apply for the force per unit area variable. There are two triggers and two non-triggers here.

MFs do not have to be of regular shapes such as triangles or trapezoids, but they can have any shape, provided that they decrease continuously after being equal to 1 on either sides or at least one side. Figure 3.15 illustrates such a situation with multiple data entries indicated by vertical arrows.

In this case, some input data trigger MFs, some do not. If the data value falls in the common area of two successive MFs, then each data value triggers not just one, but two MFs with different MD for each function. For the same data, one of these two degrees will be always greater than the other. These MDs can be considered to provide a measure of how significantly the data value triggers the two MFs. Of course, for the same data value, the MF with a higher MD is more important than the other.

3.8.2.1 Membership Functions

Membership functions (MFs) are essential components of the input variables in the forms of fuzzy sets, which should have the following characteristics:

1. They are in the forms of downward convex shapes.
2. Their MDs are confined between 0 and 1, inclusive.
3. The base of a fuzzy set lies on the 0 MD level and it is referred to as the support, which corresponds to an interval number with a minimum and maximum values.
4. The apex of fuzzy set has MD equal to 1 as a single point of in the form of another interval number, which falls within the support. These apex values are called as core of the set.
5. Input fuzzy sets cannot have any decreasing part between 0 and 1 MDs. Each fuzzy set has continuously increasing MDs starting from the extremes of support and reaching the core extremes on each side.
6. A fuzzy set with all these characteristics is labeled as "normal" fuzzy set. Otherwise it is non-normal and cannot be considered as input MF in fuzzy logic modeling studies.

Figure 3.16 presents all these characteristics collectively as normal and non-normal fuzzy sets.

The reader can appreciate the validity of the fuzzy sets in Fig. 3.16 frames under the light of the above six points the appearances of fuzzy logic modeling input MFs as normal or not.

In the following, the most used input fuzzy sets (MFs) are presented graphically for human and mathematically for communication with computers. In general there are two linear (triangular and trapezium) MFs, but there are several that are nonlinear.

(a) *Triangular MF*: Fig. 3.17 presents a normal triangular fuzzy set, which is most frequently used in many fuzzy logic modeling inference systems as input set. For mathematical expression derivation, the variable value on the horizontal axis is shown by x and its corresponding MD on the vertical axis as y. It has three parameters on the 0 MD axis as a, b, and c.

A general expression for y value can be simply obtained from the consideration of the two similar triangles on the left-hand side of the figure. Thus the proportionality of similar sides yields the following equation for the computers:

$$y = \begin{cases} 0 \ \text{for } x < a \\ \dfrac{x-a}{b-a} \ \text{for } a < x < b \\ \dfrac{x-b}{c-b} \ \text{for } b < x < c \\ 0 \ \text{for } x > c \end{cases} \tag{3.2}$$

A subclass hierarchy is defined to simplify the generation of fuzzy sets. The fuzzy sets that can be formed by this set of subclasses represent a fairly complete set of

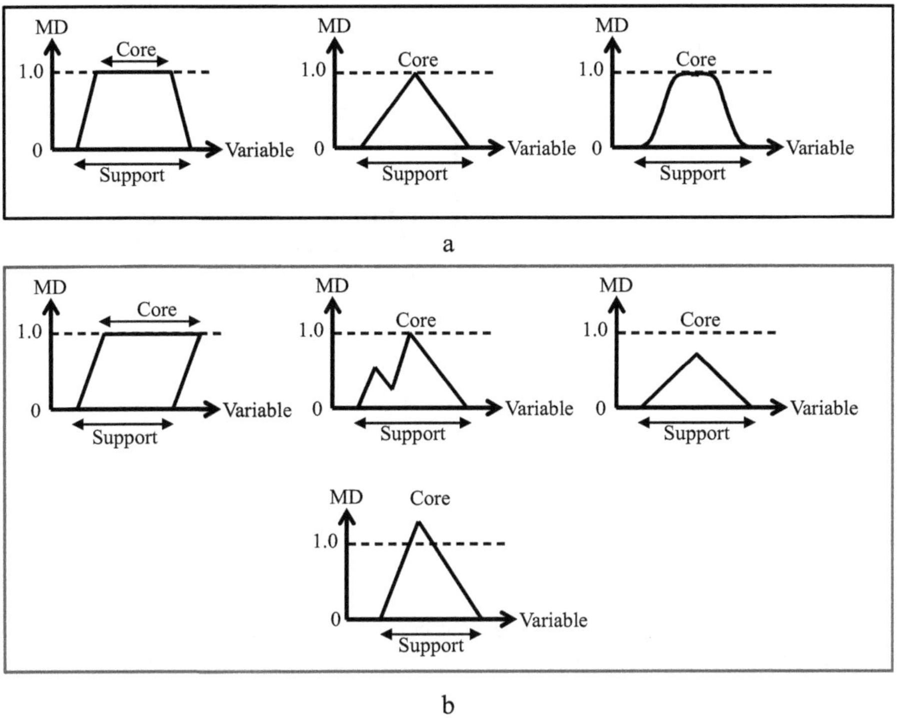

Fig. 3.16 Fuzzy sets (a) normal, (b) non-normal

Fig. 3.17 Triangular MF notations for mathematics

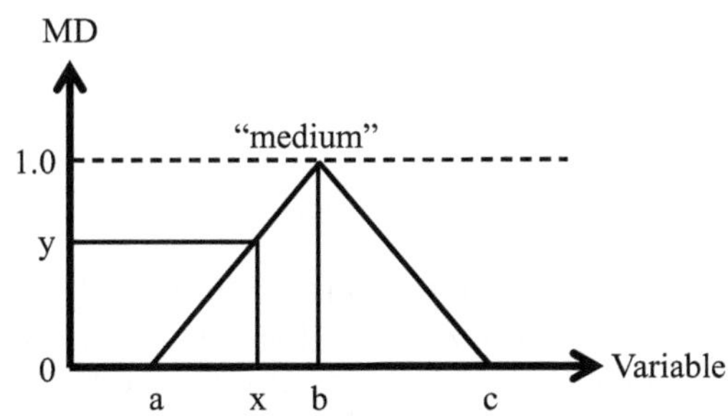

common shapes. The user can generate subclasses of other types to meet specific needs.

(b) *Trapezium MF*: This is most frequently used at the extreme ends of the variable considered and occasionally in the middle positions. Its notation is shown in Fig. 3.18 for mathematical expression derivation for computer communication. This MF has four parameters as a, b, c, and d.

Again consideration of the similar triangles on the left-hand side and then on the right-hand side of the figure yields the following equation:

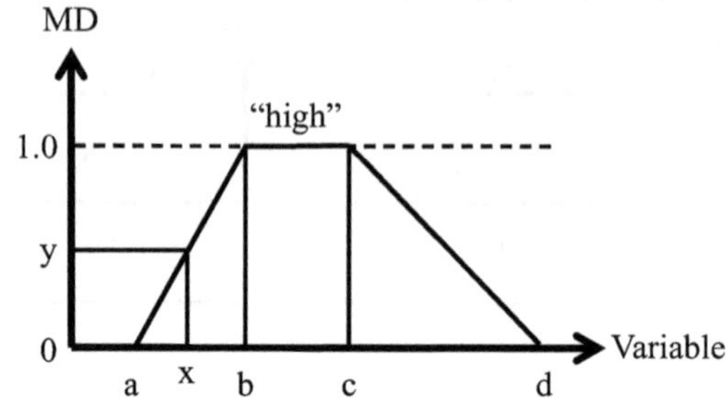

Fig. 3.18 Trapezium MF notations for mathematics

$$y = \begin{cases} 0 \ \text{for} \ x < a \\ \dfrac{x-a}{b-a} \ \text{for} \ a < x < b \\ 1 \ \text{for} \ b < c \\ \dfrac{x-c}{d-c} \ \text{for} \ b < x < c \\ 0 \ \text{for} \ x > d \end{cases} \tag{3.3}$$

(c) *Gaussian MF*: This has a symmetrical bell shape appearance with two parameters as a and b, which show the location and deviations from the location, respectively. Figure 3.19 is for the Gaussian MF.

The mathematical expression of this MF is very similar to the Gaussian probability distribution function and its expression is as follows:

$$y = e^{-\left(\frac{x-a}{b}\right)^2} \tag{3.4}$$

(d) *Sigmoid MF*: This is also dependent on two parameters *a* and *b* and is given graphically in Fig. 3.20.

The mathematical equation for this MF provides variation between 0 and 1 as required from fuzzy sets. There are numerous shapes for different values of a and b. In the figure, only two sets of a and b are presented. Depending on the sign of parameter *a,* the sigmoidal MF is inherently right or to left open and is therefore suitable for representing concepts such as "too large" or "too negative." More traditional looking MFs can be built by taking the product or difference of two different sigmoidal MFs.

(e) *Bell-shape MF*: The generalized bell function depends on three parameters *a, b,* and *c* as given as

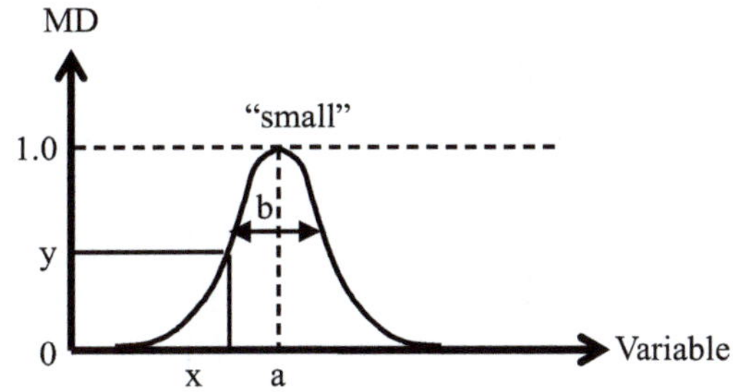

Fig. 3.19 Trapezium MF notations for mathematics

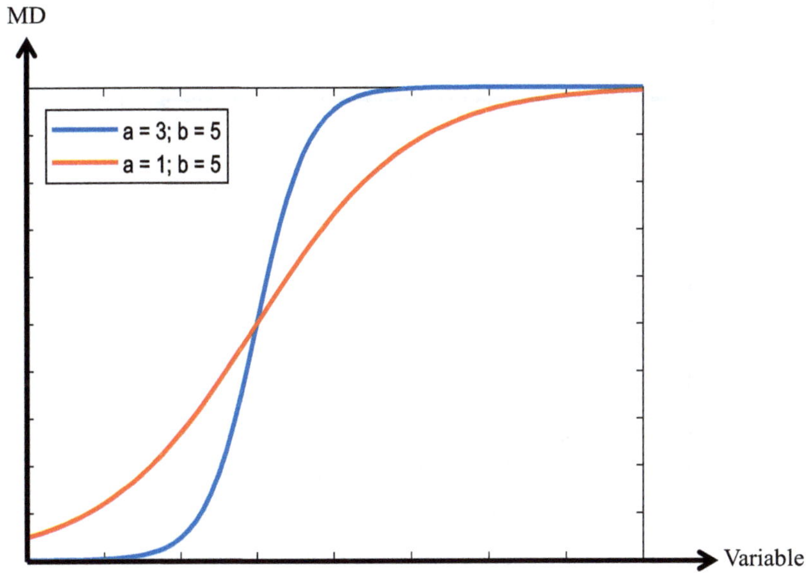

Fig. 3.20 Sigmoid MF notations for mathematics

$$y = \frac{1}{1 + \left|\frac{x - c}{a}\right|^{2b}} \tag{3.5}$$

where the parameter *b* is usually positive. The *c* parameter determines the center of the curve. Enter the parameter vector params, the second argument for gbellmf, as the vector whose entries are *a, b,* and *c*, respectively. The graph of this MF is given in Fig. 3.21.

Here c is for the location of the MF and b is the deviation measure around the location.

(f) *Two-piece Gaussian MF*: Its function is a combination of two of two parameters as was in the case of earlier explained Gaussian MF Eq. (3.4). Its various shapes are shown in Fig. 3.22.

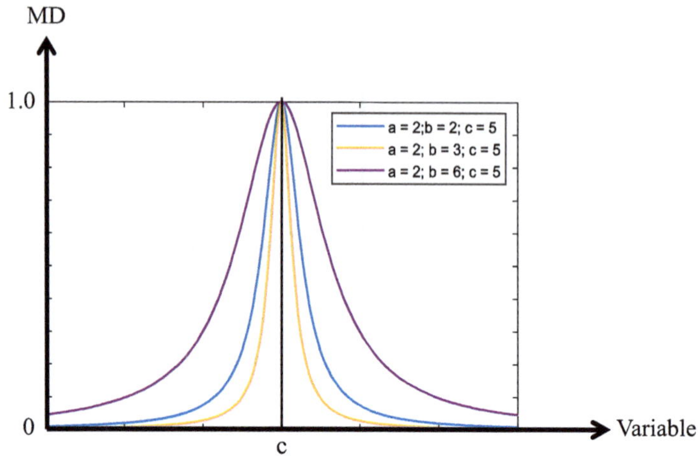

Fig. 3.21 Bell-shaped MF notations for mathematics

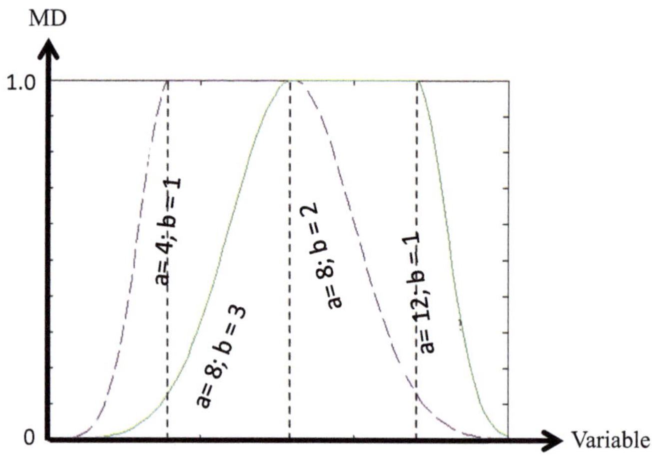

Fig. 3.22 Two-piece Gaussian MF notations for mathematics

The first (second) function, specified by a = 4; b = 1 (a = 12; b = 1), determines the shape of the leftmost (rightmost) curves. Whenever a < b, this MF reaches a maximum value of 1. Otherwise, the maximum value is less than 1 and such MFs are non-normal and not valid for input fuzzy set representations.

(g) *S-shaped MF*: This has also two parameters a and b and the parameters help locate the extremes of the sloped portion of the curve. The mathematical expression is given as follows:

$$y = \begin{cases} 0 \text{ for } x \leq a \\ 2\left(\dfrac{x-b}{b-a}\right)^2 \text{ for } a \leq x \; \dfrac{a+b}{2} \\ 1 - 2\left(\dfrac{x-b}{b-a}\right)^2 \text{ for } \dfrac{a+b}{2} \leq x \leq b \\ 1 \text{ for } x \geq b \end{cases} \tag{3.6}$$

According to this expression, the following shape appears for a = 1 and b = 8 as in Fig. 3.23:

(h) *Z-shaped MF*: Similar to the S-shape MF, it also has two parameters as a and b; its mathematical function is like the previous MF expression with interchange of the second and third lines as follows:

$$y = \begin{cases} 0 \text{ for } x \leq a \\ 1 - 2\left(\dfrac{x-b}{b-a}\right)^2 \text{ for } \dfrac{a+b}{2} \leq x \leq b \\ 2\left(\dfrac{x-b}{b-a}\right)^2 \text{ for } a \leq x \; \dfrac{a+b}{2} \\ 1 \text{ for } x \geq b \end{cases} \tag{3.7}$$

Figure 3.24 indicates the Z-shaped MF which has higher MDs on the left-hand side and decreases towards the right-hand side.

Determination methods are broadly divided into several categories, depending on the mode of application.

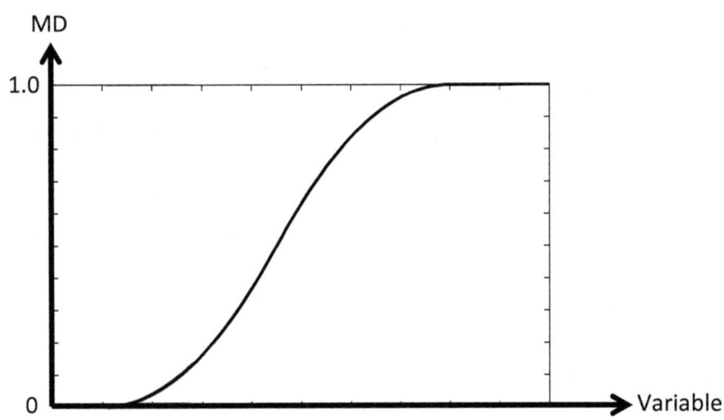

Fig. 3.23 Two-piece Gaussian MF notations for mathematics

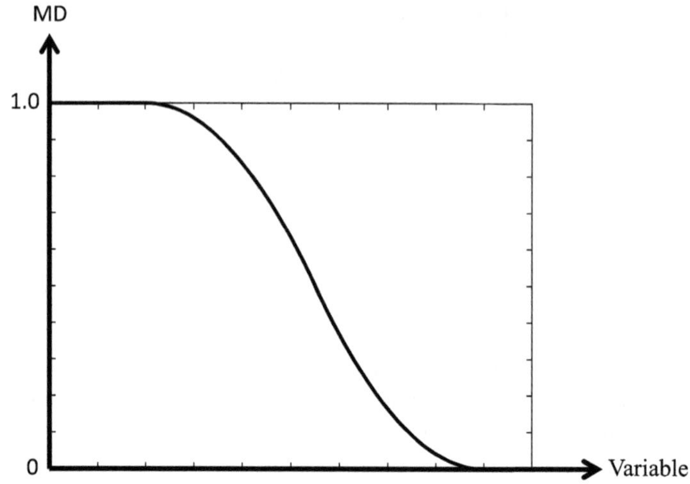

Fig. 3.24 Two-piece Gaussian MF notations for mathematics

1. Subjective evaluation and elicitation: Since fuzzy sets are generally intended to model people's cognitive states, they can be determined from simple or complex elicitation procedures. At the very least, subjects simply draw or otherwise specify different MF curves appropriate to a given problem. These subjects are typically experts in the problem domain. They are given a more constrained set of possible curves to choose from. Under more complex methods, users can be tested using psychological methods.
2. Ad hoc forms: Although there are many (quite endless) arrays of possible membership function (MF) forms, most true fuzzy control operations are drawn from a very small set of different curves, such as simple fuzzy forms. This simplifies the problem, for example, just choosing the central value and the slope on both sides.
3. Converted frequencies or probabilities: Sometimes information received in the form of frequency histograms or other probability curves is used as the basis to construct a MF (Chap. 4). There are several of possible conversion methods, each with its mathematical and methodological strengths and weaknesses. However, it should always be remembered that MFs are not necessarily probabilities.
4. Physical measurement: Many applications of fuzzy logic use physical measurement, but almost none directly measure the MD. Instead, a MF is provided by another method and then the individual MDs of the data are calculated from it.
5. Learning and adaptation: This question is what is the relationship between fuzzy truth values and probabilities? It should be answered in two ways.

   (a) How is fuzzy theory different from probability theory mathematically?
   (b) How does it differ in interpretation and application?

At the mathematical level, fuzzy values are often misunderstood as probabilities, or fuzzy logic is interpreted as a new way of handling probabilities, which is not the case. The minimum requirement for probabilities is that they be additive, must add to 1, or that the integral of the density curves must be 1 (Chap. 4).

This does not generally apply to MDs, although they can be determined by considering probability densities, there are other methods that have nothing to do with frequencies or probabilities.

Semantically, the distinction between fuzzy logic and probability theory has to do with the difference between probability concepts and degree of membership. Probability statements are about the likelihoods of outcomes. An event either happens or it does not and one can bet on it. With fuzziness, one cannot say with certainty whether an event has occurred and instead one is trying to model the extent to which an event has occurred.

The MF estimation problem can be considered as a pattern recognition problem (Bezdek 1981). Partitioning the samples of a trajectory into classes leads one to obtain the regions of attraction. In general, it is not possible to determine crisp boundaries between classes, and therefore fuzzy partitioning is a preferred methodology. At this point, there are two ways to proceed:

1. Using priori knowledge of the problem and suggesting reasonable class prototypes that model attraction regions by approximating the poles as shown in Fig. 4.3. MFs are calculated using an appropriate distance metric according to suitable analytical expression or to some objective function (which may or may not be optimized for the assumed prototypes).
2. An alternative approach is unsupervised pattern recognition, which results from a lack of precise knowledge about a physical process. The MFs are then estimated by an appropriate unsupervised fuzzy partitioning method.

3.8.2.2 Fuzzy Operations

Classical set theory defines three basic operations on sets, which are complement (NOTing), intersection (ANDing), and union (ORing) operations, and can be considered in relation to fuzzy sets. There are also many operations and manipulations that can be applied to fuzzy sets that are not applicable to crisp sets. Often there is a corresponding capability for fuzzy values with the constraint that the universe of discourse or range of x values must be observed for the fuzzy values, and for binary operations such as ANDing and ORing, the two fuzzy values must be the same fuzzy variable.

Fuzzy ANDing

A visual representation of the intersection of two sample fuzzy sets A and B is shown in Fig. 3.25. One set has trapezium and the other set has triangle set. As already explained in Sect. 3.7.2.1, ANDing operation should have simultaneous occurrence of two sets, which is mentioned as the intersection of the two sets. The ANDing operation of two fuzzy sets is defined such that the membership degree value at any x value is the minimum of the membership values of the two fuzzy sets, which are

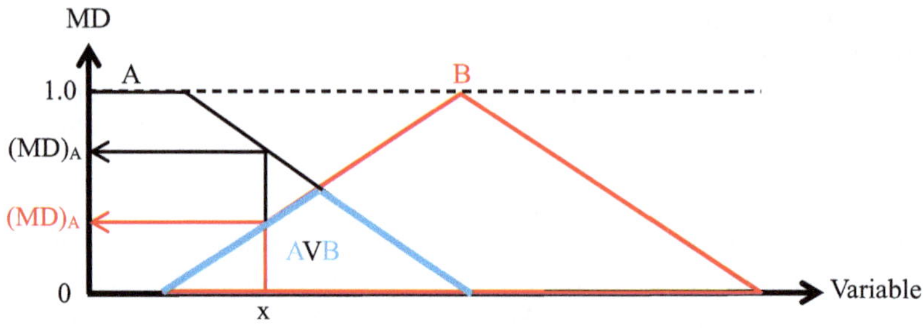

Fig. 3.25 Fuzzy ANDing operations

$(MD)_A$ and $(MD)_B$. As the x moves from left to right, these two MDs change their values and the ones with the minimum corresponds to the location of ANDing operation. This means that the ANDing MD should be obtained according to the following expression:

$$(MD)_{ANDing} = \min\left[(MD)_A, (MD)_B\right] \tag{3.8}$$

The resultant fuzzy operation is shown by blue boundaries in the same figure, which is a combination of MFs A and B. In fuzzy logic literature, ANDing operation sign is .

Under classical set theory, the ANDing of two sets is that which satisfies the conjunction of both the concepts represented by the two sets. However, under fuzzy set theory, an item may belong to both sets with differing MDs without having to be in the intersection. The intersection is therefore defined such that the memberships are the lower of the two sets.

Fuzzy ORing

The visual representation of the ORing of two fuzzy sets is depicted in Fig. 3.26. The ORing of two fuzzy sets is defined such that the MD value at any x value is the maximum of the membership values of the two fuzzy sets. Thus, similar to Eq. (3.8) one can write,

$$(MD)_{ORing} = \max\left[(MD)_A, (MD)_B\right] \tag{3.9}$$

According to the maximization procedure, the final Oring fuzzy set is shown by blue in the figure:

Following arguments similar to the ANDing operation, the ORing of two sets has memberships that are the larger of the two sets. The ORing operation is shown by V sign, thus the final fuzzy set is AVB.

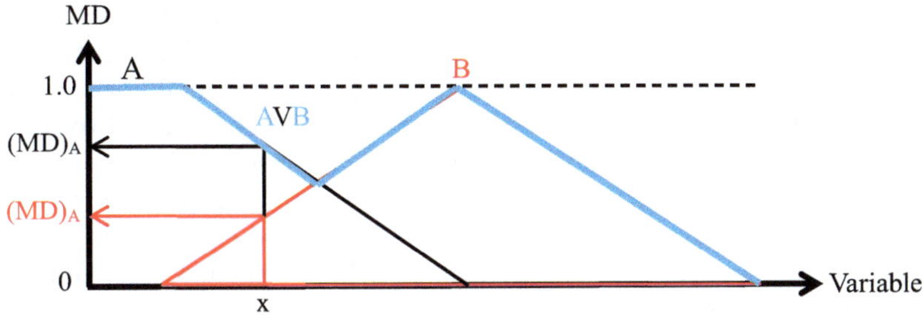

Fig. 3.26 Fuzzy ORing operations

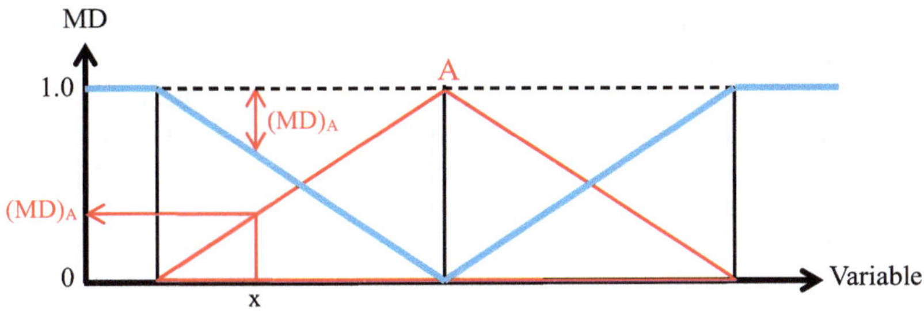

Fig. 3.27 Fuzzy NOTing operations

Fuzzy NOTing

It complements the MD, $(\mathrm{MD})_{\mathrm{A}}$ values of the set points in the fuzzy set according to the following mathematical expression:

$$(\mathrm{MD})_{\mathrm{NOTing}} = 1 - (\mathrm{MD})_{\mathrm{A}} \tag{3.10}$$

Given a fuzzy set A, the NOTing complement is shown symbolically as ¬A. The final NOTing fuzzy set is in blue lines in Fig. 3.27.

3.9 Fuzzy Logic Principles

On the contrary to crisp logic, everybody has vague, fuzzy, ambiguous, possible and probable concepts and approaches in our daily work. This natural logic is broader and more general than crisp (bivalent) logic and thus characterized as fuzzy thinking and reasoning in which it is possible to accept both oppositions with some degree of membership. By following the fuzzy logic-based approach in thinking, one can partially agree with everything, and this can easily push the expert into harmony and indecision (Dimitrov and Korotkich 2002).

Reasoning is based on graded concepts that everything is a matter of degree, that is, everything is fuzzy (flexibility). The theory of fuzzy logic took its general form in Zadeh's (1965) early publications. He wanted to generalize the traditional concept of a set through degrees of belongingness between 0 and 1, rather than a two-degree of belonging (equal to 1 or 0) and named them as membership degree (MD). Real situations are not clear and stable; therefore, they cannot be fully defined. A full description of a real system often requires more detailed data than a human can process and comprehend at the same time. This has been called the "principle of incompatibility" by Zadeh (1963), which means that the closer one looks at real-world problems, the more fuzzy appear they are.

All decisions must be expressed in language that can then be translated into universally used symbolic logic based on the principles of mathematics, statistics, or probability (Chaps. 4 and 5). This is why fuzzy logic is always followed by symbolic logic (including mathematics).

Subjectivity is very much in the perception phase of dependence on personal thoughts, and as one enters the field of visualization, subjectivities decrease and in the final stage objectivity ripens and becomes more logical, although still, some uncertainty (ambiguity, incompleteness, etc.) may remain. Accordingly, the final decision remains unclear. Uncertainties in physics and engineering can be minimized without too much harm to the final decision by a set of restrictive assumptions, but they always exist in engineering modeling, social, cultural, economic, commercial, and similar domains.

Classical scientific problem solution approaches work with crisp and organized numerical data on the basis of two-valued (white-black, on-off, yes-no, etc.) crisp logic. However, in every corner of daily life, scientific or nonscientific, approximate reasoning is valid with grey foregrounds and backgrounds in verbal information forms. It is a big dilemma how to deal with gray information for arriving at decisive deductions with crisp and deterministic principles. Fuzzy logic principles with linguistically valid propositions and vague categorizations provide a sound ground for the evaluation of such information. For a wider domain of reasoning and logical deduction possibilities, the preliminary step is fuzzy logic conceptualization of scientific axioms with uncertainties systematically in the antecedent and consequent parts of the propositions. Such an approach helps not only to visualize the relationships between different attributes, but furnishes a speculative detail about the systemic reasoning in scientific phenomena description, proposition, and axiomatic features by taking into consideration semantic and syntactic meanings within the etymological and epistemological contents of science philosophy. After all one can lead to linguistical deductions and by conversion to crisp logical mathematical formulation.

The concept of fuzzy logic was conceived by Zadeh (1965) and presented not as a control methodology, but as data processing by allowing partial membership (belongingness) sets rather than crisp sets as in the crisp (two-value) logic. This approach was not applied to control systems until the 1970s due to insufficient small-computer capability prior to that time. Zadeh (1921–2017) reasoned that people do not require precise numerical information input, and yet they are capable to control in highly adaptive manner. If feedback controllers could be programmed to accept noisy, imprecise and vague inputs, they would be much more effective and perhaps easier to implement.

Fuzzy logic provides a simple way to arrive at a definite conclusion based upon vague, ambiguous, imprecise, noisy, or incomplete input information. Fuzzy logic approach to control problems mimics how a person would make faster decisions.

3.9.1 Fuzziness in Daily Affairs

There are no fragile boundaries as mentioned. Uncertainty or indeterminacy principle, statement, articulated by Heisenberg (1927) provides that

> The position and the velocity of an object cannot both be measured exactly, at the same time.

This statement implies that human beings could never reach the exact truth in science and that uncertainty would remain at all times and places. So we live in an uncertain or better still a fuzzy and hazy world. Now let us ask, who exactly gives a definite answer when they say yes or no? Do you know exactly what the weather will be like tomorrow? Are the temperatures given on the television screens exact? Or are they approximations? After all one understands that there is always a fuzziness that is, uncertainty to some extent, between our lives and our pursuits. When it comes to uncertainty here, we should not think of numerical but especially verbal ones.

On the other hand, when fuzziness is mentioned, personal preferences also come into play. For example, liking colors, having a taste, giving a tip for a taste, giving help to a charity are all about personal preferences. In this respect, almost all the words and phrases used are fuzzy. For example, when "good" is said, there is a relativity that varies from person to person. For some, a "good" that is complete and ideal may be opposed to a "good" that is not so important to others. In this respect, the word goodness has a certain range of change in the human mind and thought. This means that there is no such thing as absolute goodness in this world. The word "good" denotes a fuzzy affair. If the word beautiful is perceived as giving the same sense of beauty to everyone, then a phenomenon with the word beautiful is always perceived exactly as the same. In this case, it is assumed that people's perceptions of the word beautiful are always at the same level, and they move away from the real world. However, it is an undeniable situation in real life that the word "beautiful" has a membership position between the words "ugly" and "very beautiful" that varies from person to person. If there are different MDs in a set, they are called fuzzy sets and their MDs can take any decimal number between 0 and 1. However, there is a case between 0 and 1 MF that most represents that word, or there are many adjacent cases that are smaller than this value over the valid set base. For example, the word "beautiful" has fragile and different fuzzy set options. By reflecting on each of these, the reader can, in a way, figure out the first differences between crisp and fuzzy sets with his (her) thoughts.

3.9.2 Fuzzy Logical Thinking Model

Unfortunately, classical education systems are based on very systematic, crisp, and organized framework based on more than twenty-three centuries old Aristotelian logic, which has only two alternatives like absolute true and absolute false. Many sciences have almost in every corner of information source grey foregrounds and backgrounds. It is a big dilemma how to deal with grey information sources in order to arrive at scientific conclusions with crisp and deterministic logical principles. Fuzzy logic principles with linguistically valid propositions and rather vague categorization provide a sound ground for the phenomenon concerned. The preliminary step is a genuine logical and uncertain conceptualization of the phenomenon with its causal (premise) and result (consequent) variables that are combined through the fuzzy logical propositions.

Such an approach not only helps to logically visualize the relationships between different variables, but also provides a philosophical background on the mechanism of the phenomenon that can be explained to anyone linguistically without mathematical treatment. In this book, it is emphasized that the basic science philosophy and fuzzy logic rationales in problem-solving in an innovative education system should be given linguistically before mathematics or systematic algorithms. In this way, the students may be able to develop their generative analytical thinking skills with the support of teachers, who have studied or at least worked in similar directions. Since the modern philosophy of science insists on falsifying existing scientific results, there is always room for ambiguity, inconsistence, vagueness and incompleteness in much scientific research activity. Innovative education systems should lean more towards the basic scientific philosophy of problem-solving with fuzzy logic principles.

Natural sciences are pretty self-explanatory, and there are always some partially overlapping results on the same phenomenon among different experts, depending on their background and experience. Differences of opinion open up potential questions in educational and research activities in the natural sciences. All these uncertainty concepts are the basis for further discussion as there is no systematic methodology for their assessment, control, and acceptability by different parties. The best that can be done is to use uncertainty techniques such as probability, statistics, and stochastic processes, but their application requires quite large numerical data (Chap. 4).

The perspective covered in this book is the use of innovative fuzzy logic principles in engineering and natural sciences' verbal modeling studies. In complex problem-solving, fuzzy systems are well suited to describe general and specific features. This book presents the basics of fuzzy thinking for problem-solving rather than scientific absolute accuracy. Fuzzy logic captures the quantitative and empirical knowledge of science; provides simultaneous simulation of multiple processes and nonlinear relationships by modeling principles.

The fuzzy concepts in understanding the problems that arise from the complexity of life cannot be resolved at the level of knowledge we have when these problems arise. But when our consciousness is raised to a higher level, the tension is reduced and problems are no longer a problem.

The philosophy of fuzzy thinking is based on granular concepts, which means that everything is a matter of degree, that is, soft (flexible). Zadeh (1963) wanted to generalize the traditional notion of a set and an expression to allow MDs and truth values, respectively. These efforts are attributed to complications that arise during modeling of the real world.

1. Real situations are not clear and stable; therefore they cannot be fully described.
2. A full description of a real system often requires far more detailed data than a human can recognize process and comprehend at the same time.

Zadeh's statement of dissonance principle is important; because it gives the message that the closer one looks at a real-world problem, the fuzzier becomes the solution.

3.9.3 The Need for Fuzzy Logic

The concept of Fuzzy Logic (FL) was devised by Zadeh (1921–2017), and presented not as a control methodology, but as a way of processing data by allowing partial set membership rather than exact set membership. This approach to nonmembership set theory was not applied to control systems until the 1970s due to insufficient small computer capacity. He thought that humans do not need precise, numerical input of information and yet have highly adaptive control ability. It would be much more efficient and perhaps easier to implement if feedback controllers could be programmed to accept noisy and imprecise inputs. In this context, FL is a problem-solving control system methodology that allows application in systems ranging from simple, small, embedded microcontrollers to large, networked, multichannel PC or workstation-based data acquisition and control systems. It can be implemented in hardware, software, or a combination of both. FL provides a simple way to arrive at a definitive conclusion based on uncertain, ambiguous, imprecise, noisy, or incomplete input information. FL approach to checking for problems mimics how a person would make decisions, but much faster.

> *The guiding principle of soft computing is to leverage precision, uncertainty, and partial truth tolerance to achieve traceability, robustness, and low resolution cost. What makes FL so important is that much of human reasoning and concept formation is linked to the use of fuzzy rules. By providing a systematic framework for computing with fuzzy rules, FL greatly increases the power of human reasoning Zadeh (1967).*

Among the beneficial aspects of fuzzy logic different than many methodologies can be stated along the following points:

1. FL systems use information efficiently; use all available evidence, extend to final clarification, solid to uncertain, incomplete or corrupt data.
2. FL encodes human expert knowledge/heuristic methods; common sense, easily interpreted, restrictions apply naturally.

3. FL systems are inexpensive, no training data required, no need for models or combined/conditional probability distributions.
4. Its design and implementation are relatively simple.
5. There is nothing fuzzy about FL.
6. FL is different from probability concepts.
7. Fuzzy sets are relatively easier to design than any other modeling.
8. Fuzzy systems are stable, easily adjustable, and conventionally verifiable.
9. FL no longer just handles control.
10. FL is a representation and reasoning process.

All the work we do in society, economics, management, administration, engineering, medicine, and science takes place in a complex world, where complexity often stems from uncertainty, because people can think and subconsciously deal with problems of complexity and uncertainty. Ubiquitous uncertainty dominates most scientific, social, technical, and economic problems. The only way computers deal with complex and uncertain problems is through fuzzy logical thinking, systematization, checking and decision-making procedures. FL affects many disciplines and provides technological advances in different fields such as washing machines, subway system operation, water resources management, automatic transmissions, factory operations, toasters, vacuum cleaners, and many more industrial fuzzy control processes. The number of fuzzy consumer products and fuzzy applications featuring new patents is growing rapidly.

3.9.4 Mind as the Source of Fuzziness

The biggest source of fuzziness is our mind. This powerful rational thinker never ceases to disassemble the whole of reality in order to analyze, classify, and label it, and then put it together or scrap it to shatter a world that has little to do with the unbreakable integrity of reality – a wholeness to which we inextricably belong.

The world we assemble from fragments of perceptions, sensations, and thoughts serves to provide partial, and therefore, distorted models of reality. These models represent a perceived world, a man-made world, not a world in which we are bound by the umbilical cord of vital and inseparable connections, whose natural evolution brought us into being. All of our models deal with parts of something we perceive as environments that are thought to be used for what our ego-centered minds consider meaningful.

An army of scientists, engineers, economists, politicians, and philosophers are involved in adapting many distorted models to predict and exert power over the evolving dynamics of reality. While we know that the complex dynamics of reality are beyond our ability to predict and control, we “do our best” to cripple reality so that meaningless models of reductionism can be pushed into impervious beds. Applications of such models have made both nature and social reality vulnerable; this is evident by the ecological catastrophes of today and the continued worsening of socioeconomic conditions for the largest and ever-increasing segment of society.

The mind can never go beyond duality – it either chooses something while rejecting its opposite (as in black-and-white thinking when using binary logic) or accepts both opposites to some degree (as in fuzzy thinking when using fuzzy or probabilistic reasoning). The dualistic nature of the rational reasoning component of the mind is so strong that the mind alone cannot transcend it; the best that can be done is to reconcile opposites.

By following a "true or false" approach to thinking (either A or not-A), we can easily become ensnared in routines, stereotypes, prejudices, and habits that eventually become a source of fuzzy that renders us incapable of authentic experience our "understanding" is constantly filtered from already established mental patterns. Fanaticism is an extreme manifestation of such intense obscurity, when man's ability to move beyond an established dogma is completely blocked.

By following a fuzzy logic-based approach to thinking ("both A and non-A" to some extent), we agree with everything others have said, and this can easily lead us to conformity and indecision. While everyone else is right, an uncritical acceptance of the fuzziness that accompanies other people's thinking makes it difficult for us to come up with our own generative ideas. It is the polarity of opposites – contradictions and conflicts of opinion – that provides the human mind with the necessary dynamics (forces and energies) to transcend opposites. These dynamics manifest in the mind as the urge to seek beyond the plane of conflict of opposites; without such an impulse, the mind can become stagnant, stuck in repetition, or be fascinated by delusional thoughts and dreams.

Fortunately, engineering applications of FL use a procedure called "defuzzification" to keep the degree of stability of a soft computation system at an effective operational level - the control mechanism of the fuzzy system is ready to generate non-fuzzy actions, that is, jumping to a simple crisp (binary) resolution at any time if necessary (Sect. 3.10). "Don't refuse anything! But don't stay with anything! Go beyond!" In our context, these words of wisdom say: "as you seek understanding, be prepared to go beyond logical rules and constraints, no matter how soft (fuzzy, probabilistic) or crisp (binary, deterministic)!"

3.9.5 Fuzzy Propositions

Every fuzzy proposition can consist of two parts, before and after the word THEN. The part before THEN is the premise (antecedent) segment, and the part after THEN is the consequent part. These binary fuzzy logical expressions are carefully divided into three, four, etc., parts by considering the sub-specifications of each variable. For example, let us say that the antecedent holistic variable for the problem has four specifications such as "Low," "Medium," "High," and "Extreme." If there are three holistic variables and each with four sub-specifications, there will be $4 \times 4 \times 4 = 64$ different combinations of these subcategories, and each of these combinations will be added, say, with the three subcategories of the holistic output variable.

Any relationship in fact leads to a conditional statement that if the state of some input variables are such and such then the output state may be in such a state. In this statement, two words provide the formal way of logical relationships, which are, as they were in the crisp logic case, IF and THEN, therefore, from now onwards any relationship in the FL terminology will be stated in the form of an IF-THEN proposition with fuzzy sets. Additionally, in any control or logical statement, human knowledge is also expressed in terms of fuzzy IF-THEN rules. These are conditional logical statements as

IF (antecedent fuzzy composition) THEN (consequent fuzzy specification)

In this manner, it is the connection or relationship between fuzzy input and output variables. Each logical statement implies a relationship between input and output variables and in the fuzzy terminology between the sub-domains of these variables.

The fuzzy statements can be considered as two types: *atomic fuzzy proposition* and *compound fuzzy proposition*. The former is a single statement and the latter is composed of few single statements, which are connected to each other in a serial manner by means of the three logical conjunctives as mentioned earlier in this chapter, which are "AND," "OR," and "NOT." These conjunctives correspond to complement, intersection, and union operations of classical sets, respectively. The atomic proposition has two parts separated by a verb in language. The first part before the verb is the fuzzy variable, v, and the second part is the atomic fuzzy word, w_f. Its general structure is

v *is* w_f

There are numerous atomic propositions in sciences such as

Rainfall is intensive
Force is strong
Energy is expensive
Bread is cheap
Grape is sweet
He is tall
etc., etc., etc.

A compound fuzzy proposition is a composition of atomic (simple) fuzzy propositions (s_1, s_2, s_3, ..), which are connected to each other by logical conjunctions. Its general form can be written as

s_1 *AND* s_2 *OR* s_3 *NOT* s_4

In the following are some of the compound fuzzy propositions. Their number can be expanded almost unlimitedly.

Rainfall is slow and recharge is little
Price is high and quality is good or taste is medium
She is tall and weight is heavy and hair is blonde

Note that the linguistic variables in a compound statement are not the same. They are also independent from each other. Furthermore, each atomic proposition in the compound proposition indicates partial variation domain of all the linguistic variables considered simultaneously with their respective fuzzy subsets. Therefore, any compound fuzzy proposition can be considered as fuzzy relations. The question is how to determine the MFs of these fuzzy relations? The following answers can be given under the light of what has been explained before.

1. The *logical connective* "AND" corresponds to intersection, and therefore, in the calculation of the MD of the compound proposition the intersection or ANDing operator starts the execution for a compound proposition as

   x is A AND y is B

   That is interpreted as the fuzzy relation $A \wedge B$ with MF as

$$\mu_{A \wedge B}(x, y) = t[\mu_A(x), \mu_B(y)] \tag{3.11}$$

   where t indicates t-conorms (Ross 1995),

2. If the logical connective is OR then the MDs can be computed according to union or "OR" connective. The general form of ORing compound proposition is

   X is A or y is B

   This should be interpreted as the fuzzy relation$A \vee B$with the corresponding MF as

$$\mu_{A \vee B}(x, y) = s[\mu_A(x), \mu_B(y)] \tag{3.12}$$

   where s implies on of the s-norms (Ross 1995).

3. Fuzzy complement operation is used for logical connective not in a compound proposition. In these propositions each "NOT" causes the replacement of each set by its complement.

Since nth fuzzy propositions are interpreted as fuzzy relations, then one would like to know how to interpret IF-THEN rules, that is, logical propositions. In general, a rule can be written as

IF p THEN q

where the prepositional variables can be either true (T) or false (F) in the classical crisp logic. Table 3.1 shows the final truth value deduced from the truth values of p and q.

In a fuzzy IF-THEN rule, p and q propositions are replaced with fuzzy propositions. Hence, the fuzzy IF-THEN rule can be interpreted by replacing the , $\vee$ *and* $\wedge$ operators corresponding to "NOT," "OR," and "AND" prepositions in the following equations with fuzzy complements, union, and intersection, respectively (Sect. 3.8.2.2):

Table 3.1 Truth values

p	q	IF p THEN q
T	T	T
T	F	F
F	T	T
F	F	T

$$\overline{p} \vee q \tag{3.13}$$

or

$$(p \wedge q) \vee \overline{p} \tag{3.14}$$

are equivalent. These two last equations both share the same truth table (Table 3.1).

There are wide varieties if fuzzy complements, fuzzy unions (s-norms), and fuzzy intersections (t-conorms) operators, a number of different interpretations of fuzzy IF-THEN rule are proposed in the literature (Ross, 1995). In order to summarize these operators collectively, p and q is replaced by fuzzy propositions (FP_1) and (FP_2) with the assumption that FP_1 is a fuzzy relation defined in $U = U_1 x U_2 x. \ldots. xU_n$, and FP_2 as $V = V_1 x V_2 x. \ldots .xV_n$. If the fuzzy relation is represented by Q, then the Zadeh (1973), Lukasiewicz (1878–1956) and Gödel (1906–1978) proposed that

$$\mu_{QZ}(x,y) = \max\left\{ \min\left[\mu_{FP_1}(x), \mu_{FP_2}(y)\right], 1 - \mu_{FP_1}(x)\right\} \tag{3.15}$$

$$\mu_{QL}(x,y) = \min\left[1, 1 - \mu_{FP_1}(x) + \mu_{FP_2}(y)\right] \tag{3.16}$$

$$\mu_{QG}(x,y) = \begin{cases} 1 & \text{if } \mu_{FP_1}(x) \leq \mu_{FP_2}(y) \\ \mu_{FP_2}(y) & \text{otherwise} \end{cases} \tag{3.17}$$

and

$$\mu_{QD}(x,y) = \max\left\{1 - \mu_{FP_1}(x), \mu_{FP_2}(y)\right\} \tag{3.18}$$

respectively. On the other hand, most frequently used Mamdani (1974) implications are the minimization of product operators as

$$\mu_{QM}(x,y) = \min\left[\mu_{FP_1}(x), \mu_{FP_2}(y)\right] \tag{3.19}$$

or

$$\mu_{QM}(x,y) = \mu_{FP_1}(x)\mu_{FP_2}(y) \tag{3.28}$$

respectively. These two implications are the most widely used ones in fuzzy systems and control studies. The human knowledge in terms of IF-THEN rules differs from each other, and therefore, different implication operations have emerged. The choice of the most convenient one depends on the type of problem and the expert view of the problem-solving.

3.9.6 Fuzzy Inferences System (FIS)

After verbal ambiguity (fuzziness) issues, fuzziness can be systematically represented by a set of logic rules with the basic principles of fuzzy logic of Zadeh (1967, 1973). Since the reader receives information with the two-valued crisp logic during the whole education, it is useful to learn the differences between it and fuzzy logic first. Many education systems educate students according to the crisp logic, but not according to even probabilistic logic of Al-Farabi (870–950), who is a Muslim philosopher and intellectual who brought probability to the agenda for the first time. Aristotle (bi-valued), Al-Farabi (probability), and Zadeh (fuzzy) logics all have situations that are grouped fundamentally by sets. According to crisp logic, all elements belonging to a set have the same degree of belonging to that set, and this is represented numerically by 1. This means that if it is an element belonging to that set, its MD is 1, otherwise it is 0. Thus, "right and left extreme values" called "excess and understatement" are considered to exist. Figure 3.28 shows the status of all three clusters.

As stated above in crisp logic, all elements in the set have a MD equal to 1; in Al-Farabi set, MDs vary between 0 and 1 intermittently; finally, in Zadeh sets there are steady increases between 0 and 1 and then decreases. Since the subject here is FL, the input sets must be normal as mentioned in Sect. 3.8.2.1.

Figure 3.29 shows classical and symmetrical and nonsymmetric fuzzy sets. It can be said that classical sets are always symmetrical because there is no difference in the MDs of the elements, because they all have MD equal to 1.

In this figure, we are talking about a set of all numbers between 2 and 4. While it is not possible to decide which of the numbers in this range is the most important or effective in classical logic, it is possible to determine the possibility and at least one member of the fuzzy sets has the most importance. While classical logic constitutes the basis of Western culture, fuzzy logic is a product of Eastern

The best of the things done is the one in the middle

which implies both probability and fuzzy logic sets. This means that, as can be seen in Fig. 3.29a, set elements do not have a clearly preferable situation. In this respect, crisp logic is also called the logic of the exclusion of the middle as mentioned earlier.

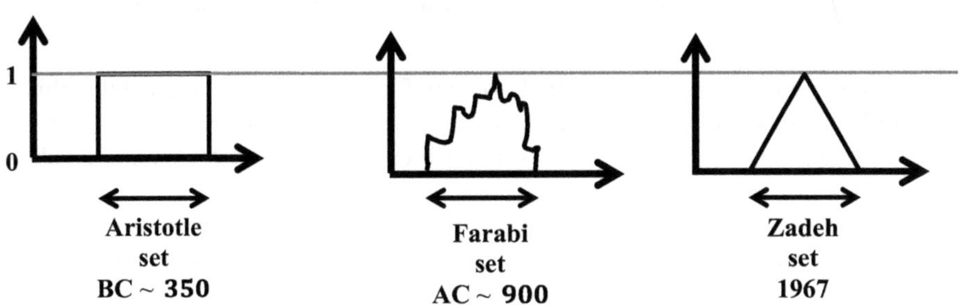

Fig. 3.28 Various logic sets

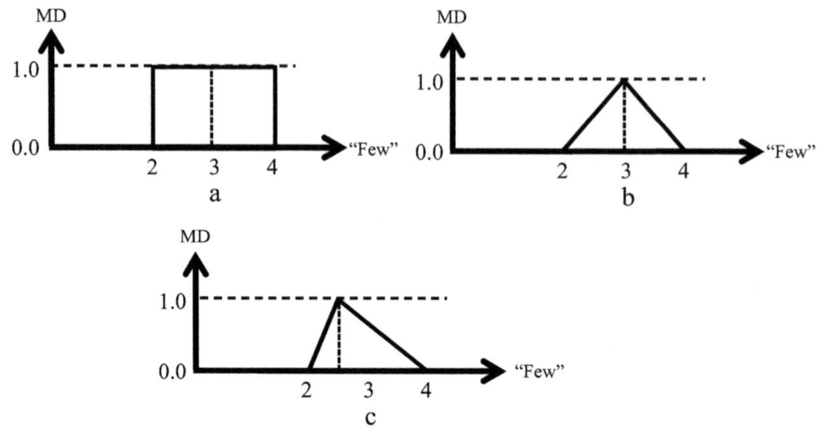

Fig. 3.29 Sets (**a**) classical, (**b**) symmetric fuzzy, (**c**) asymmetric fuzzy

Thus, it can be concluded that logic without middle preference has almost no meaning in daily life. There is a middle ground for almost everything in life.

The probability and degree of belonging of the elements in fuzzy sets have a number ranging from 0 to 1, so it is understood that there are inequalities, in the fuzzy logic sets. This shows that man is incomparably compatible with natural behavior patterns according to binary logic.

Fuzzy properties are also found in many scientific keywords (terminology), especially in common speech. For example, when the word "disease" is considered, there are many subsets among many different diseases that this word implies to crisp logic. Among the subclasses of the word disease there are "mild," "few," "little," "mild," "big," and "extreme" patient sets. Figure 3.30a shows their sets according to crisp binary logic, and even if there is a boundary separating them from each other, it is understood that these boundaries are sharp transition lines between adjacent classes. At the end of the questioning how a "little" disease can suddenly switch to a "mild" state during the healing process, it is concluded that it is more rational to develop it gradually rather than suddenly. In this case, FL clusters are shaped as in Fig. 3.30b.

It is possible to reach the following important conclusions by filtering these two logic types:

1. Considering the classical (bi-valued) logic in Fig. 3.30a, it is concluded that this is not plausible, since it is understood that the transition between subsets is sudden. In addition, it is strange that there is no difference in terms of MDs among subset elements.
2. If it is diagnosed that the disease is mostly in the "mild" group, it is seen that it does not differ from other subsets according to crisp logic. It is not fortunate that the disease degrees are equal to each other.

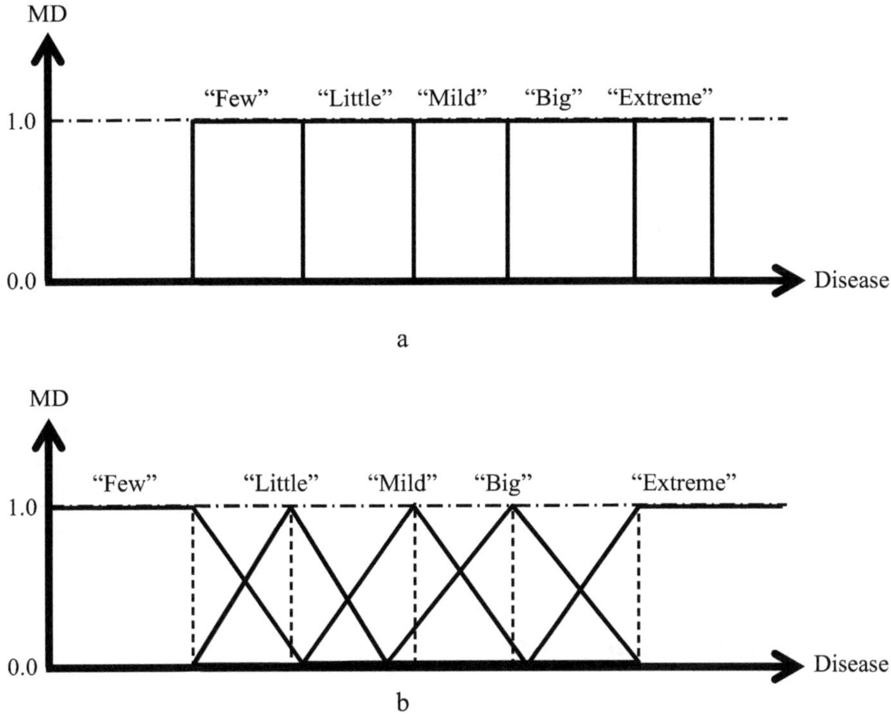

Fig. 3.30 Logical disease classification (**a**) classical, (**b**) fuzzy

3. If the disease level of those who fall into the same sub-disease cluster is the same, it is also foreseen to give the same dose to each of them, which is not a logical solution, because the immune system of each patient is different from each other.
4. According to the fuzzy set classification in Fig. 3.30b, a solution has been provided for each of the first three items.
5. According to the fuzzy set classification, it is understood that those in the same disease class have differences from each other.
6. It is a fact that a physician who makes his diagnosis according to classical logic cannot be as successful as one who performs his profession according to FL.

If a person is asked in daily life what are the differences between a disk, cylinder, and rod, s/he must look and visualize their shapes to answer the question. Otherwise, answering it by rote, mechanical, and transporter information may mislead anyone a lot. It is possible to see the differences between these three words in Fig. 3.31. It is possible to answer wisely by considering the geometric shape dimensions, say, two fuzzy words "length" and "diameter."

Classically, we encounter two fuzzy words for the disc, for example, "large" in diameter and "short in length." In addition to others, words containing fuzziness such as "small," "long," "medium" appear. The disc height is "short," the cylinder

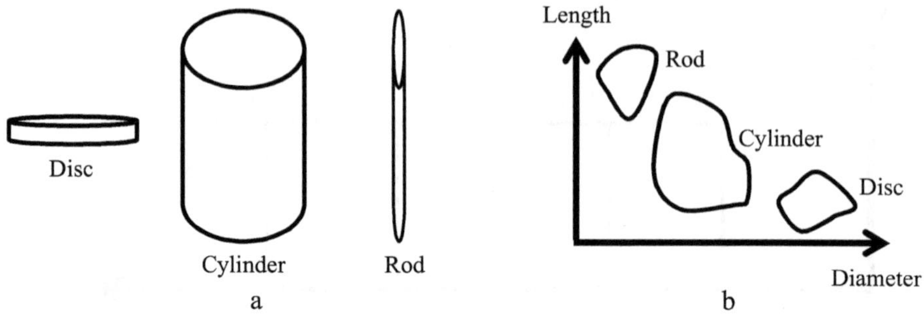

Fig. 3.31 Disc, cylinder, rod

height is "medium," but the rod has cylindrical shapes with a "long" height compared to others. The representative areas in Fig. 3.31b reveal that there are overlaps in diameter and length. However, overlaps are not allowed in crisp logic. Finding overlaps means the exclusion of crisp logic and the inclusion of FL principles.

3.9.7 Fuzzy Modeling Systems

In order to analyze a scientific problem, first of all, it is useful to make a design about its working mechanism. Otherwise, it will not be possible to introduce artificial intelligence (AI) functions by means of natural intelligence (NI) with the natural information in the rote, mechanic and transporter education system, and to load them into robots and machines. In order to construct a fuzzy AI system, the following points must be considered in order:

1. Before the thought of reaching the solution with the specific equations given by the education system in the full modeling of real events, an effort should be made to verbally visualize the preliminary information about the generation mechanism of the event, which is examined with a free thought without being captive to mathematical equations. Since the event to be examined is quite complex, it may not be possible to solve it with precise mathematical equations. In fact, every mathematical equation is valid in the light of certain hypothesis and a set of assumptions, but the operation of the event may not fully comply with these assumptions. For these reasons, it may naturally be appropriate for the researcher to prefer always to refer to methods that are approximate, but with solubility, even if they are not precise, among which FL principles are the most verbal and representative AI methods.
2. Mathematical equations express the studied event with approximation, if it is thought that they are really valid absolutely, this is biased and opposes that the mathematical equations are not quite correct.

3. Considering that the validity of the result obtained in natural and even engineering studies always carries a risk (earthquake, flood, strength calculations, etc.). It is therefore recommended to multiply the obtained results with a number greater than one in order to be on the safer side, and this is called the "safety factor." In fact, it is the coefficient of ignorance that covers the uncertainties.
4. Since "know-how" in the sense of "know why" is valid in today's information age, verbal information is used in all mathematics, physics, etc. modeling affairs. It should always be considered that it is more important than numerical information (Chap. 1).
5. For example, before an earthquake study risk calculation, the age of the buildings, the number of floors, material qualities, ground conditions, etc. situations need to be stated verbally. Thus, FL principles and especially systems are very useful in classifying buildings as "no damage," "low damage," "moderately damaged," "slightly damaged," and "demolishing."

We witness the FL in AI studies most commonly when a person learns to drive and obtains a driver's license. One learns to drive with FL principles without knowing any mathematical equations. Attempting to learn this with crisp (bivalent) logic increases the number of car accidents. A person can learn to drive only by perceiving verbal information. In this learning process, for example, the following two FL rules are very valid:

1. "IF speed is low THEN hit the throttle hard"

OR

2. "IF speed is high THEN accelerate less".

In general, classical model designs of classical and fuzzy logic rules are given separately as input (causes, variables) and output (results) in boxes in Fig. 3.32.

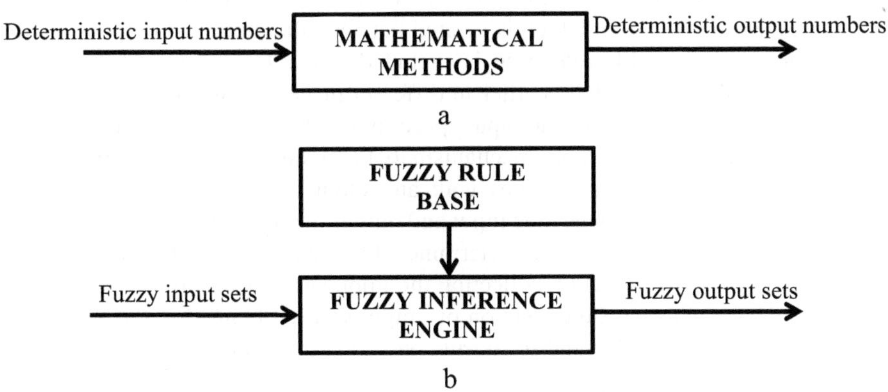

Fig. 3.32 Systems (**a**) classical, (**b**) fuzzy

Another important point is that the functionality of the classical system is made with hypotheses and assumptions, and the FL system is made with some logic rule bases. On the other hand, in classical systems, the necessity of inputs and outputs to be in the form of numerical data, in the FL system, these are expressed in numbers or verbally with words. In Fig. 3.32b, it is seen that there is a situation where mathematical models are replaced by rule bases and then by making these rule bases operative to obtain outputs. The following points can be drawn from the comparison of the two systems:

1. In the FL system, inputs and outputs are used to model situations involving verbal uncertainties only by means of logic rules, apart from mathematical principles; in the classical system, certainty and mathematical rules, namely equations, are essential.
2. Those who are accustomed to working in the classical system try to solve problems with mechanical approaches, with specific methods such as mathematical expressions (ordinary or partial differential, empirical, probability, statistics, stochastic, etc., equations) since they were trained in salience-based education. They try to make inferences by reaching solutions with mathematical equations and formulas, which are based on crisp logic principles, instead of logic rules.
3. Those working with the FL system, on the other hand, first of all determine the basic logic rules of the event they examine, and provide an approach with logic rules to a logical rule structure that already constitutes the basis of mathematical formulas. Thus, they can continue to walk in the field of AI with productive, healthy, peaceful, scientific, and innovative approaches that can be improved further.

The differences that arise in terms of the system can be detailed by considering the functions of the units below.

1. *Input*: This unit, which provides entrance to the systems, ensures that all known numbers and verbal information about the event are entered into the system.
2. *FL Rule Base*: This unit, which is located between the input and output units, is the part where the FL rule base is included that relates them. Here are the IF-THEN rules as described earlier in different places in this chapter. Each rule logically connects a part of the input space to a part of the output space.
3. *Fuzzy Inference Engine*: It is a mechanism that includes a collection of operations ensuring that the system behaves with an output by collecting all the partial relations established between the input and output fuzzy sets in the forms of fuzzy rule base. This engine is used to determine what kind of output the whole system will give under the inputs by collecting the implications of each rule.
4. *Output Unit*: Indicates the collection of output values obtained as a result of the interaction of knowledge and fuzzy rule bases via fuzzy inference engine.

In the abovementioned units, instead of the middle two, there is a middle unit in which mathematical equations are located in crisp logic systems.

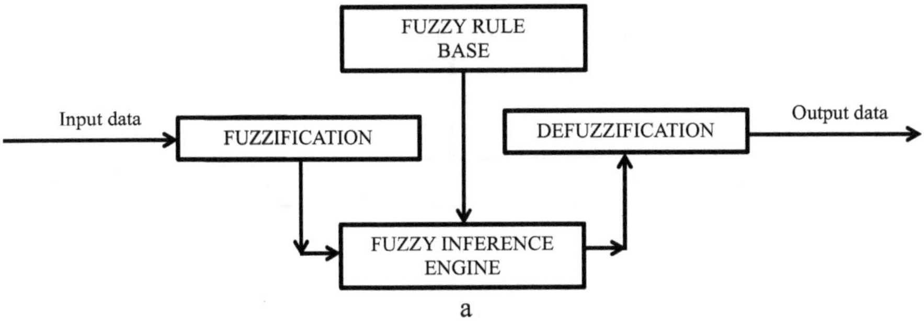

Fig. 3.33 Fuzzy logic general inference systems

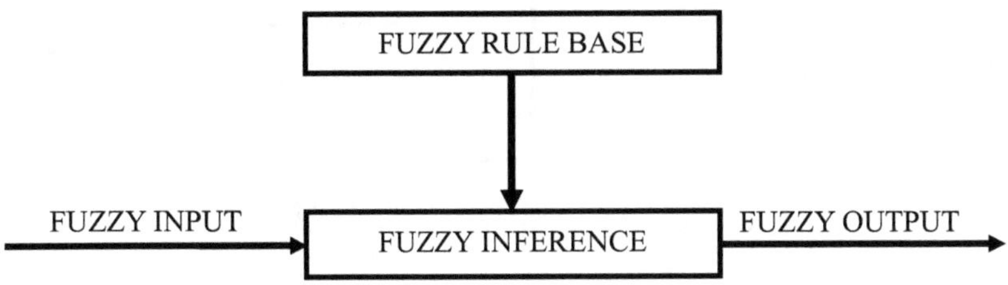

Fig. 3.34 Pure fuzzy systems

In the most general form, even when the inputs are significant numbers in the FL system, the conversion of them into verbal indefinite ones is called "fuzzification" and a system emerges as in Fig. 3.33. It is possible to see the functions of the fuzzy inputs and the FL rule base jointly and to have a fuzzy character in their outputs. The process of converting these fuzzy outputs back to numeric is called defuzzification.

3.9.7.1 Pure Fuzzy (Mamdani) System

In any fuzzy system, when compared to a classical counterpart, there should be two boxes instead of one, as in the classical system, since the system description cannot be provided only by mathematical equations or fuzzy logical propositions. The first of these boxes represents the fuzzy rule base collection, and the other is for triggering or inferring results, called an inference engine. Therefore, there are four boxes in the following fuzzy system (see Fig. 3.34):

3.9.7.2 Partial Fuzzy (Sugeno) System

The difference of this system from the previous one is that it has an arithmetic averaging procedure involved in the overall effect of the rule basis. The conceptual system model is shown in Fig. 3.35.

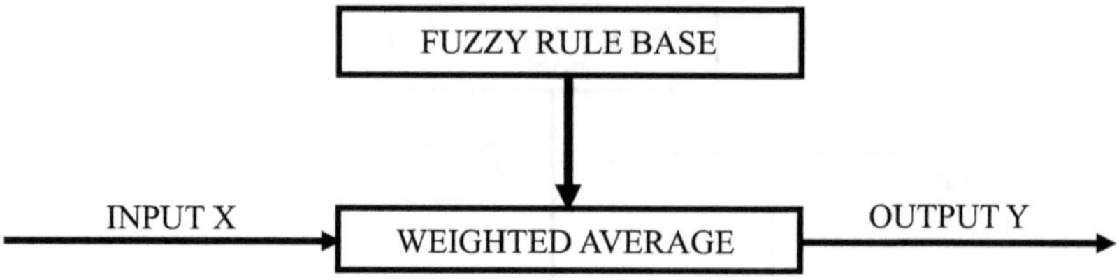

Fig. 3.35 Partial fuzzy systems

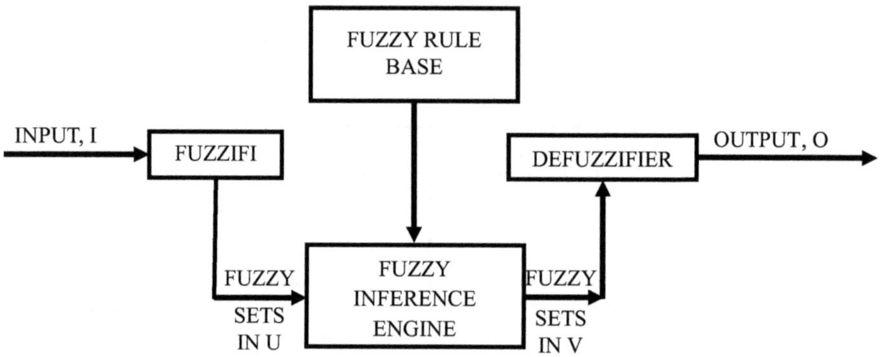

Fig. 3.36 General fuzzy system

3.9.7.3 General Fuzzy System

Its difference from the previous systems is that the inputs are fuzzified with a fuzzification unit and the outputs are turned into crisp forms by another unit, which is the defuzzification part. Hence, additional two more boxes are included in the conceptual model as in Fig. 3.36.

Two different types of fuzzy inference systems, namely, minimization and production are shown with triggering of sets in Fig. 3.37.

3.10 Defuzzification

Defuzzification is the reverse process of fuzzification. Converts output fuzzy sets of word descriptors to a real number. This may be necessary if a number needs to be output for use. A representative fuzzy result of the two rules is shown in Fig. 3.38.

Although there are different ways of defuzzification; quite simple methods are used in practical applications. It is intuitive that fuzzification and defuzzification are reversible. If defuzzification is to occur, this has effects on the shape of the MFs used to fuzzify the input variables into fuzzy sets used in the reasoning process, which ultimately results in defuzzification of an output fuzzy set.

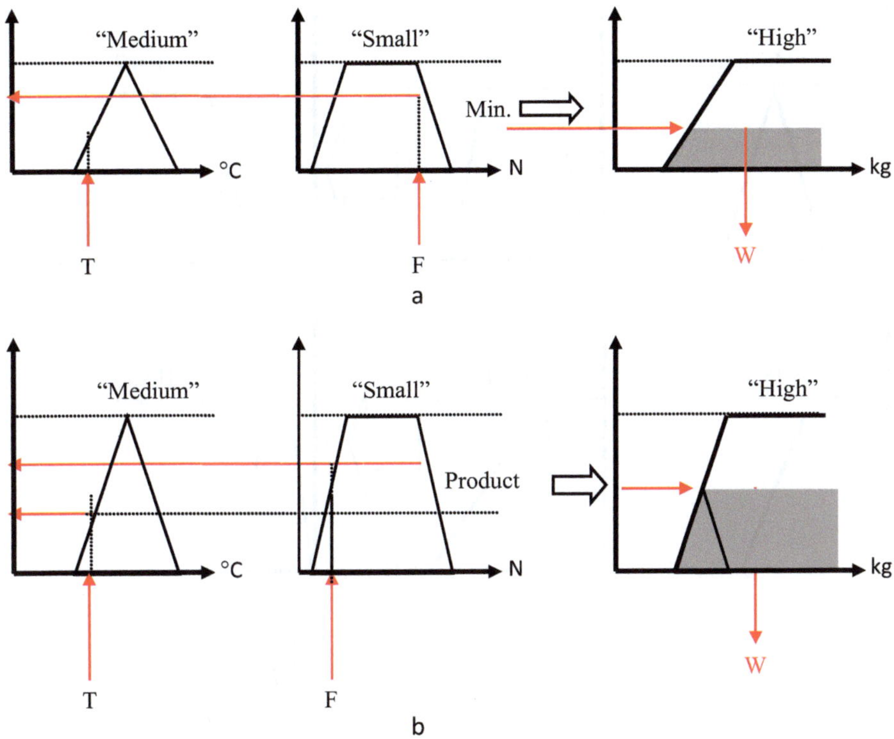

Fig. 3.37 Fuzzy rule triggering and inference (**a**) minimization, (**b**) multiplication

More generally, MFs are defined for the output fuzzy set and use the defuzzification command. Numerous defuzzification procedures are available for fuzzy control, but fuzzy reasoning applications can often be satisfied with few options (Ross 1995).

The best shape for an MF depends on its use. Flat-top MFs with adjacent functions and maximum overlap are usually best if the final output is non-numeric. On the other hand, if numeric output is desired, peak functions that intersect at half-full confidence are usually the easiest to manage.

The defuzzification process of a fuzzy set requires knowing the representative values corresponding to each fuzzy set member. The representative value for each fuzzy set member is multiplied by the degree in that member, the sum of the products, and the sum divided by the sum of the MDs. The options for representative value selection are the mean maximum (max) method and the center method. Alternatively, the rule can be written to implement almost any desired defuzzification scheme. As for the numbers that are fuzzy when defuzzification will occur, the MFs for the input fuzzy set should usually peak and intersect with half-full confidence.

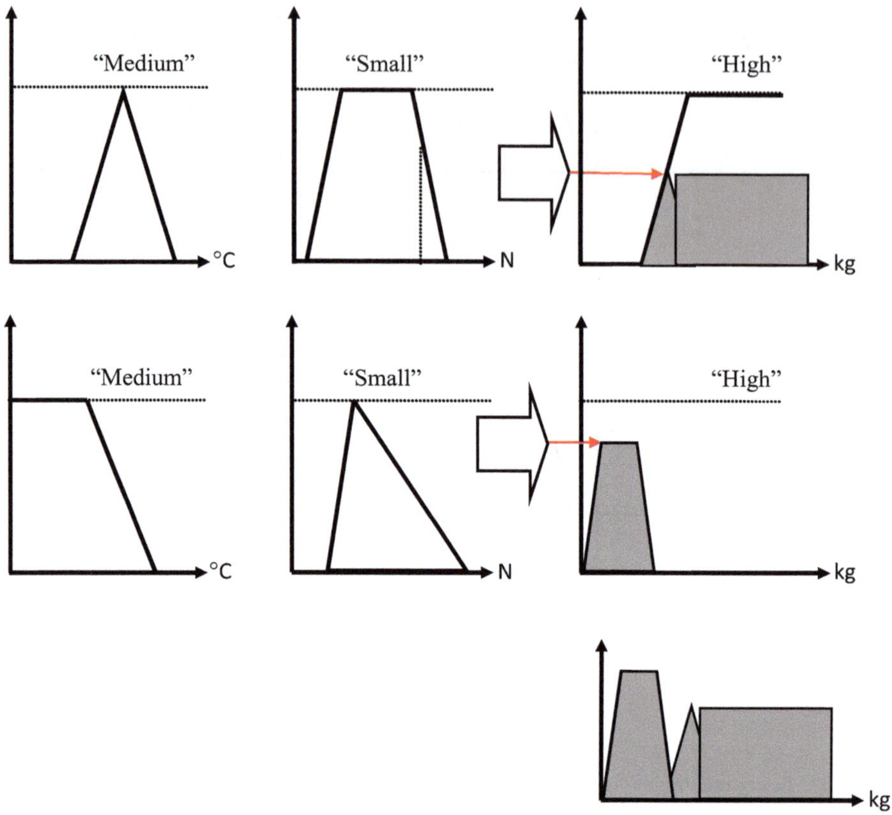

Fig. 3.38 Combination of fuzzy rules and fuzzy inference

There are different methods to get a single number as the output result. Since there is no procedural method for choosing which method is more appropriate, the most commonly used ones are the middle of the maximum, the center, and largest of the maximum.

After the final result as an irregular shape in Fig. 3.39, the final step in the approximate reasoning algorithm is defuzzification, that is, choosing an exact value for the output variable. In the Mamdani model, the fuzzy reasoning algorithm results in an output fuzzy set with certain MDs of possible numerical values of the output variable. Defuzzification compresses this information into an output value.

In most applications, it is recommended that the scientists look at the MDs of individual variable values and select one of them based on one of the criteria as "smallest maximal value," "largest maximal value," "center of gravity," "average of the range of maximum values," etc. One can even make his/her own choice according to the problem at hand. For example, in cases of economic crises studies, the minimum portion of the final fuzzy set may be acceptable as output of the

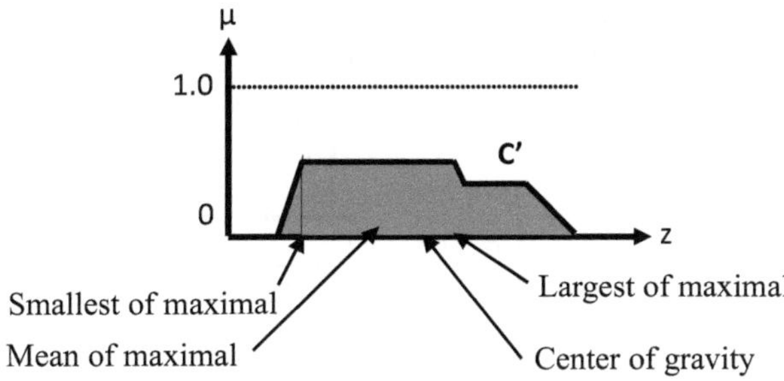

Fig. 3.39 Defuzzification of max-min composition

approximate reasoning in Fig. 3.39, where the defuzzification results in ascending order of magnitude will be "smallest of maximum," "mean of maximum," and "center of gravity" and "largest of greatest."

Defuzzification causes a large amount of information loss. Therefore, in any work, it is recommended that the scientist considers some other properties of the final compositional fuzzy set, in addition to the formal clarification methods. In classical expert systems, rules are derived only from human experts. However, in FL rule-based systems, rule generation is done by expert opinion in addition to numerical data. The combination of numerical data recorded by instruments and linguistic information revealed by experts is possible through fuzzy concepts and systems.

Fuzzy model development requires several iterations, where the first step defines the set of rules and corresponding input and output MFs. After execution of the model, the results are compared with the output data already available, and accordingly, either the MFs or the rules or both are revised for training with necessary corrections, if necessary. Then sequentially the model modifies rules and/or MF etc., for further training and test again for model validation. It is possible to divide the model building procedure into two successive parts, training and estimation. During the training phase, the previously mentioned MF or rule-based changes are completed. In the estimation phase, the output values are cross-validated with the existing data and again, if necessary, MF or rule base or both adjustments are applied. FL expert systems may not even require data for model building, including MF allocations and rule-based formulation. This stage is completely logical and is known as the expert opinion stage. Briefly, model development can be summarized as follows:

1. Expert knowledge is used to construct the model structure with relevant input and output variables in addition to the FL rule base. This step does not require any data and is therefore called the logical step in model construction. This model should be checked with numerical data from the relevant experiments.
2. Numerical data is handled in two parts as training and test stages. The existing dataset can be heuristically divided into two parts. In many studies, it is recommended to take about 70% of it as training data and 30% as test data and also in this book. MFs and rule bases are set as mentioned above during the training phase. Generally, some rules are deleted as they may not be triggered by existing data; some rules can be combined if they result in similar outputs; and still others can be changed.

Fig. 3.40 Weighted fuzzy defuzzification calculations

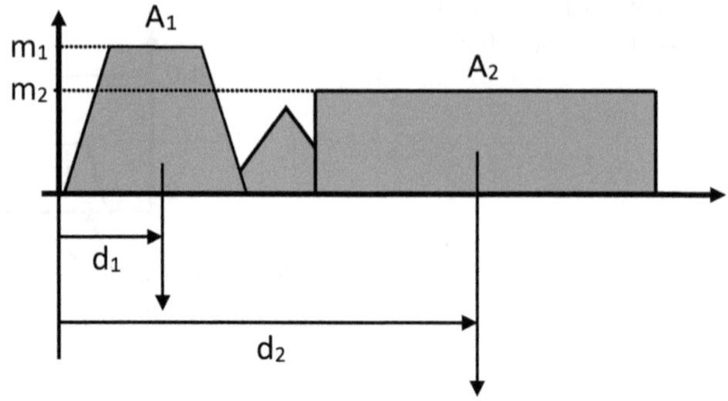

$$R_1 = (A_1 m_1 + A_2 m_2)/(A_1 + A_2)$$

$$R_1 = (A_1 d_1 + A_2 d_2)/(d_1 + d_2)$$

3. In the testing part, model cross-validation with prediction errors is observed. For example, it is possible to further tune MFs and rule bases on the basis of error minimization with the least squares procedure. If the overall forecast error is less than 10%, or preferably less than 5%, the model is adopted for forecasting.

Two different types of weighted average clarification are shown in Fig. 3.40. In the first, membership degrees and in the next, the areas of each fuzzy subset are used as weights.

3.11 Conclusions

The main concern of this chapter deals with the fundamental aspects of crisp and fuzzy logical words, phrases and sets, which are important parts in artificial intelligence, AI modeling studies. Philosophical and logical definitions, conceptions, and their rational meanings of science in innovative education systems provide the first stage of shallow and then deep learning linguistic modeling rules. It is stated that knowledge gains scientific garb if existing scientific expressions, methodologies, algorithms and models are considered as candidates for falsification to improve knowledge content. Deductive, inductive, analogy and their mixtures are explained in detail as rational thought models and approximate reasoning. The two mutually exclusive rational relationship-seeking principles have been given to man as proportionality (directly or inversely) and shape (linear or nonlinear), which is an innate grace by Allah (God). The difference between philosophy and scientific philosophy is explained on the effective branches of each concept. Crisp and fuzzy logic principle conjunctive words (AND; OR; NOT) and logical sentence structure such as IF....THEN.... propositions are discussed through graphs and numerical examples.

In the light of what is described in this chapter, it provides the reader with tools for development of productive and analytical thinking skills and capabilities.

References

Aydin AC, Tortum A, Yavuz M (2006) Prediction of concrete elastic modulus using adaptive neuro-fuzzy inference system. Civ Eng Environ Syst 23(4):295–309

Bezdek JC (1981) Pattern recognition with fuzzy objective function algorithms. Plenum Press, New York

Dimitrov V, Korotkich V (eds) (2002) Fuzzy logic: a framework for the new millennium. Physica Verlag, Heidelberg

Dunn JC (1973) A fuzzy relative of the ISODATA process and its use in detecting compact well-separated clusters. J Cybernetics 3:32–57

Fraile-Ardanuy J, Zufiria PJ (2007) Design and comparison of adaptive power system stabilizers based on neural fuzzy networks and genetic algorithms. Neurocomputing 70:2902–2912

Heisenberg W (1927) Über den anschaulichen Inhalt der quantentheoretischen Kinematik und Mechanik. Z Phys 43:172–198

Iphar M, Yavuz M, Ak H (2008) Prediction of ground vibrations resulting from the blasting operations in an open-pit mine by adaptive neuro-fuzzy inference system. Environ Geol. https://doi.org/10.1007/s00254-007-1143-6

Jang JSR (1993) ANFIS: adaptive-network-based fuzzy inference system. IEEE Trans Syst Man Cybern 23:665–685. https://doi.org/10.1109/21.256541

Jang JSR, Mizutani E (1996) Levenberg-Marquardt method for ANFIS learning. In: Biennial conference of the North American fuzzy information processing society—NAFIPS, pp 87–91

Kilwardby R (2015) Notulae libri Priorum, Ed. And trans. Paul Thom and John Scott Oxford

Kuhn TS (1962) The structure of scientific revolutions. University of Chicago Press, Chicago

Lorenz, E. N., (1972). Predictability: does the flap of a butterfly's wings in Brazil set off a tornado in Texas? 139th meeting of the American Association for the Advancement of Science., 29 December

Poinsot J (1955) The material logic of John of St. Thomas: basic treatises. Edited by John J. Glanville, G. Donald Hollenhorst Yves R. Simon. Translated by John J. Glanville, G. Donald Hollenhorst Yves R. Simon. University of Chicago Press, Chicago

Popper K (1955) The logic of scientific discovery. Routledge, New York, p 479

Reddy MJ, Mohanta DK (2007) A wavelet-neuro-fuzzy combined approach for digital relaying of transmission line faults. Int J Electr Power Energy Syst 29:669–678

Ross TJ (1995) Fuzzy logic with engineering applications. Addison Wesley

Singh TN, Kanchan R, Verma AK, Saigal K (2005) A comparative study of ANN and Neuro-fuzzy for the prediction of dynamic constant of rock mass. J Earth Syst Sci 114(1):75–86

Şen Z (2004) Fuzzy logic and system models in water sciences. Turkish Water Foundation, Istanbul

Şen Z (2006) Discussion "applying fuzzy theory and genetic algorithm to interpolate precipitation". J Hydrol 331:360–363

Şen Z (2007) Discussion "Takagi–Sugeno fuzzy inference system for modeling stage- discharge relationship". J Hydrol 337:242–243

Şen Z (2014) Philosophical, logical and scientific perspectives in engineering. Springer, Heidelberg, p 260

Zadeh LA (1965) Fuzzy sets. Inf Control 12:94–102

Zadeh LA (1973) Outline of a new approach to the analysis of complex systems and decision processes. IEEE Trans Syst Man Cybern 2:28–44

Zadeh LA (1999) Fuzzy sets as a basis for a theory of possibility. Fuzzy Sets Syst 100:9–34. https://doi.org/10.1016/S0165-0114(99)80004-9

Zio E, Gola G (2008) A neuro-fuzzy technique for fault diagnosis and its application to rotating machinery. Reliab Eng Syst Safety. https://doi.org/10.1016/j.ress.2007.03.040

Chapter 4
Uncertainty and Modeling Principles

4.1 General

There has been much discussion and curiosity especially about the natural phenomena. These discussions included comparisons between uncertainty in earth, atmospheric, and physical sciences which inevitably led to the problem of determinism and indeterminism in nature and engineering modeling aspects (Leopold and Langbein 1963; Krauskopf 1968; Mann 1970a, b). At the core of scientific theories and studies lies the concept of "cause" and "effect" relationship with absolute certainty (Chap. 3). According to Popper (1955), one of the modern philosophers of science,

> to give a causal explanation of a certain event means to deduce a statement explaining that event from two kinds of premises: from some universal laws, and from some singular or specific statements. We can call this certain initial conditions.

There must be a very special connection between the premises and the conclusions of a causal explanation (logical proposition), and it must be deductive. In this way, the conclusion necessarily follows from the premises. Before any mathematical formulation, the premises and the conclusion are made up of verbal (linguistic) statements. Each step of deductive argument must be justified by quoting a logical rule about the relationship between input and output variables. On the other hand, the concept of "law" is central to deductive explanation and thus to the precision of our knowledge about events.

Recently, the scientific evolution of methodologies has shown that as researchers try to clarify the boundaries of their interests, they become more blurred with other areas of research. For example, when the climatologist is trying to model climate change impact as one of humanity's modern challenges when it comes to water resources, they need information about the meteorological and atmospheric conditions for groundwater recharge, geological environment of aquifers, and social conditions. Therefore, many common philosophies, logical foundations, methodologies and approaches become common in different disciplines, and the data

Z. Şen, *Shallow and Deep Learning Principles*,
https://doi.org/10.1007/978-3-031-29555-3_4

processing is one of the most important topics that can be applied to the diversity of disciplines. The form of questions from earth, environmental, atmospheric, and engineering disciplines varies greatly, but solution algorithms may involve the same or at least similar procedures. Some of the common questions that may be asked by various research groups are summarized below. Many of these questions have been elucidated by Johnston (1989).

Any natural phenomenon or similar event occurs widely over a region and thus its recordings or observations in different locations raise some questions, for example, are there relationships between phenomena in various locations? In such a question, time seems to be frozen and the behavioral formation of the phenomenon in question over the field and between the locations is investigated. The answer to this question can be given descriptively in linguistic, subjective, and ambiguous terms that can be understood even by nonexperts. But their quantification requires objective methodologies, which is one of the purposes of the context in this book. Another question that can be stated right at the beginning of research in the earth, environment, atmospheric sciences, and engineering is whether places are different in terms of the phenomena occurrences? Such questions are the source of many people's interest in the subject.

The basis of any scientific thought includes ambiguities of various kinds, which can be broadly classified into linguistic and numerical forms. Everybody is exposed to linguistic ambiguity in the face of many aspects of social, financial, cultural, religious, scientific, administrative, and daily transactions in life. Such uncertainties make life alive, active, and prosperous in a way. Accordingly, each person has as many worlds of uncertainty as individual's current life. Such uncertainties that lead to knowledge production by both reducing the amount of uncertainty in previous information sources and leading to a new future uncertainty area or the problem being discussed may take another line of uncertainty along with other sources of uncertainty, including probability. Possibility is the term used to describe verbal uncertainty, imprecise, doubtful, incomplete information. A continuous series of uncertainties is encountered in daily life. For example, the very common question "How are you?" puts one in front of linguistic ambiguity, and ready-made answer is rarely definitive, and even the response "I am fine" may get the question as "Really? or "Are you sure?" and alike questions. Another example is "What will the weather be like tomorrow?" and this question may be vaguely waiting for an answer such as, "It will be fine," "It may be rainy," "It will be partly cloudy," and many other alternatives are possible. One can understand what is meant by language ambiguity if he thinks about sentences that he uses or encounters on daily basis. Based on common life, there are reflections of many words in any language. For example, in English words such as "slow," "almost," "about," "more or less," "quite a bit," "maybe," and similar ones express verbal ambiguity and questions, answers, phrases, allusions, sentences, accordingly includes a degree of uncertainty that can be intuitively appreciated depending on the context of the issue. While linguistic ambiguity is ubiquitous, there are always unknown and in some cases known limits to the relevant situation. Overcast has two extremes, cloudy and clear, and all other types classification or descriptions fall between these two extremes. Thus, as will be

Table 4.1 Certainty-uncertainty

Reality (uncertainty)	Scientific abstraction (certainty)
Complexity	Homogeneity
	Isotropy
	Heterogeneity
	Uniformity
Chaos	Homogeneity
Nonlinearity	Linearity
Turbulence	Isotropy

explained later in this chapter, if some logical numerical values are added to these two extremes, the others must take values between these two additions. The reader can suggest numerous examples of linguistic uncertainty from his/her daily life. Even today all ambiguities are in the form of numerical ambiguity character, but there are still linguistic ambiguities. At this stage, it can be argued that linguistic ambiguity cannot be completely avoided, which is a big issue. It is possible to say that science does not tell the absolute truth and therefore its results are discussed due to uncertainty. In classical scientific studies, linguistic ambiguities are avoided with a set of assumptions as in Table 4.1.

Besides linguistic ambiguity, there are also numerical ambiguities, which are used mostly by scientists, engineers, and many other careers for the quantification and notion of scale and size in accurate design. Some of the linguistic uncertainties are open-ended in their upper and lower constraints, but others imply approximate numbers that cannot be exact ends.

The main purpose of this chapter is to present methodologies for numerical uncertainties to model uncertain phenomena of any kind (social, economic, engineering, etc.). To use uncertainty techniques for any modeling or design purpose, the four basic arithmetical operations (addition, subtraction, multiplication and division) must be built on logical foundations. The basis of these arithmetical operations is logical operations such as ANDing, ORing, and NOTing, which are called logical operators (Chap. 3). Thus, it is confirmed once again that every operation has a linguistic basis, and then numerical characters become important in practical applications. If one does not understand linguistically what is going on in the problem that he is facing, then it is valid to point out that the only obvious way to deal with such problems is numerical solutions through mathematical formulations, which is very restrictive use of reason and generative thinking, but often almost blindly derived formulations as prescriptions that may not need yield viable solutions. Engineering education cares about clear expressions and, especially, formulations without further consideration of rather ambiguous sources of linguistic information.

There are many instances in scientific research and application affairs that question of unpredictability under the circumstances of uncertainty prevail. On this point, Poincare cited in Ruelle (1991) that

> A very small cause, which escapes us, determines a considerable effect, which we cannot ignore, and we then say that this effect is due to chance.

As mentioned in Chap. 3, in holistic systems based on crisp logic, there are simplicities, quite complexities, determinism and chance, which are quantifiable by individual measurements, but their measurement sequences perform uncertainties in most of the cases. There are many cases in any branch of science that rather than analytical and deterministic approaches the researcher is left with data measurement series from which scientifically plausible estimations, predictions, and forecasting must be deduced by means of some empirical uncertainty methodologies. It is possible to try and solve confronted problem after a set of restrictive assumptions, idealizations, and linearization concepts, but the result may not be valid due to high percentage of error. Among the model uncertainties are bias, discrepancy, approximation, lack of knowledge about the underlying problem, discrepancy between the model, and reality in addition to uncertainties in parameter estimations due to vague values and experimental uncontrollability. There is software about almost any complicated problem, but the model parameters are not certainly known, and therefore, most of the time applicators fiddle the model outputs to record measurements by parameter optimization or adjustment by any means. Complicated mathematical model software writing may have algorithmic uncertainty sources. In field or laboratory experimental works, there are always experimental uncertainties due to errors in observations, measurements, and indeterminism.

The measurement or model numerical output ambiguities can be assessed by theoretical probability distribution functions (PDFs) within intervals. The result comparison with measurement can be achieved by means of error bands which may be expressed in terms of root mean square, standard error of estimation, and percentage relative error limits ($\pm 5\%$ or $\pm 10\%$) (Benjamin and Cornell 1970). Widely used boxplots are based on normal (Gaussian) PDF on quartile range including median and outliers (Spear 1952; Tufte 1983).

Probability, statistics, and stochastic modeling procedures provide objectivities from uncertainty data sets within certain error limits. It is preliminary necessary that prior to the application any of these methodology, the researcher should have science philosophical, logical, and rational linguistical interpretations about description of concerned problem and appreciation of the arguments for predictions and interpretations (Chap. 3). It is also possible that the researcher may not have capability to predict specific events due to the intrinsic randomness (Mann 1970a, b). For instance, finite length time series should be regarded as random samples from the population of concerned event, and therefore, they have ensemble of replicates instead of a unique absolutely known behavioral feature. Probabilistic and statistical tools provide parametric information extraction from measurement set by means of percentages and meaningful parameters such as arithmetic average, standard deviation, skewness, and kurtosis. Each parameter is related to some physical behavior of concerned phenomenon. Hence, collection of several parameters provides in-depth information about the temporal or spatial occurrences of data.

The main purpose of this chapter is to provide fundamentals of probabilistic and statistical modeling and their usages in different scientific context under uncertainty umbrella. Each methodology is presented with fundamental assumptions along illustrative examples. The reader is encouraged to find additional information from some basic references (Kendall and Stuart 1979; Maity 2018).

In daily life, even if everybody wants to carry out his work according to a schedule, s/he realizes that daily work, which s/he expects to be evident by having to change his/her schedule in the face of unexpected situations, gains some uncertainty. The emergence of some of these uncertainties is not in the hands of man. Sometimes, one senses that if s/he shows a little disruption, it will have a random effect elsewhere in the program. Many of us experience that our programs are thought to be formal (systematic) but disrupted due to natural and social events and that we even suffer financial damage from this. In a way, life is full of uncertainties. Who can guarantee that we would not find ourselves in an accident when we drive car? No matter how careful we are in traffic with full consideration and awareness, if the other persons are careless and do not follow the traffic rules, we have a 50% chance of not having an accident. In half of the possible accidents, there is the danger of staying safe and even losing our lives on the other half. Every driver in traffic may feel that s/he may be involved in an accident, just like a player lost in games of chance. In countries with a good traffic system and where drivers try to obey the rules by using reason and logic, this ratio shifts towards the safe side and the danger rate decreases. For example, it can be converted to a 95% safety state. No matter how prescriptive the driver and traffic are, the effects of precipitation, temperature, fog, meteorological events, the slipperiness of the road, road repair, the effects of a pedestrian or living creature crossing the road, and unexpected uncertainty situations may suddenly occur despite these rules.

The fact that the account at home does not match the market in daily life is a proof that people live in an environment of uncertainty. The delay in the use of uncertainty methods has been due to the reluctance of people to learn and apply the methods with the indefinite adjective at the beginning, and that the result is not fully reliable if the result is reached at all. In general, in many societies, the principles of uncertainty, and the wording of probability, have been distrusted. Later, it was understood that the results of the uncertainty methods were more general, and their averages gave the results found by specific methods. Thus, it has been expressed in an objective mathematical language that it is possible to calculate with how much error the result can be trusted after application of specific methods.

In the last century, almost all the scientific revolutions have included the principles of uncertainty and the methods developed as a result of them. For example, at the end of the twentieth century, when physicists wanted to make a generalization by saying that there was nothing left to do in physics from now on, everything could be explained with the principles and laws of Newton and Einstein, which are deterministically certain methods, physicists dealing with molecules and subatomic structures concluded that deterministic methods could not describe these events. Their arrival caused unrest among physicists. In the following years, quantum physics emerged, which revolutionized the conventional classical, namely, Newtonian and Einsteinian physics. As mentioned in Chap. 3, Heisenberg (1927) stated that if humanity exists, the position and velocity of subatomic particles cannot be measured simultaneously without error. Hence, physics has become uncertain, and then, quantum physics is based on uncertainty methods such as probability and statistics.

In different disciplines and studies, there are different questions that probability methods can answer. It is not possible here for us to know and write down all the questions that can be answered with probabilistic principles. Some of these are given below. However, the reader can make use of the methods in this book by deciding on probable, which is, not deterministic events related to his/her subject.

(a) Evaluation of all kinds of luck-related events and predictions are made with probability calculations. For example, what is the probability of winning in each game? And how long should one continue the game so that s/he does not get hurt?
(b) The most important question that an insurance company is interested in is to know how much the danger (risk) of the event it is dealing with is, so that it can demand payment of dues from customers accordingly. The word hazard (risk) is literally a probability statement.
(c) How much and when are fluctuations in prices expected over time? What are the probabilities of certain expectations?
(d) What is the probability of earthquakes that may occur in a region and in what intensity? How many years apart do these probable events occur?
(e) What is the probability of an accident occurring between certain kilometers of a highway with high traffic?
(f) What are the limits of error in making weather forecasts? What is the probability of it being snowy tomorrow?
(g) While conducting oil exploration, if there are oil drillings that have been drilled before, deciding where it would be better to drill the next well still involves uncertainty. There are uncertainties in determining the probability of the next well to be shot dry or wet.
(h) Digitization of disturbing oscillations called interference or noise in electronic signals.
(i) Determination of whether the weather in agriculture will be rainy or dry for tomorrow and, if necessary, how many days there will be for rainy period.
(j) How many times the oil platforms built in the open seas can be exposed to the wind and waves that may arise in a given period and what is the danger (risk) of collapse?
(k) To be able to decide how the events with two or more options, which one will lead possible to a desirable result over time.
(l) Making use of random wind measurements in determining the direction and length of runways in airport designs.
(m) Concentrations of pollutants introduced into a river vary according to the amount of flow. In this respect, although the amount of pollutants given by the industry remains constant, the concentration varies randomly depending on the flow velocity and the amount of precipitation. At first glance, it comes to mind that during large flows a lot of pollutants can be introduced into the river. In this regard, uncertain precipitation and runoff measurements are used to adjust pollutant concentrations.

(n) Deterministic mathematical models are of no use in determining how many fish species are found in a lake. Probability calculations are used to calculate this approximately.

It is necessary to develop some criteria in order to measure not only the arithmetic mean but also how much oscillations around it, in random and uncertain investigations like the events mentioned above.

The purpose of this chapter is not to give the theory of probability in detail, but to provide the reader with insights to gain skills on the rules of reason and logic and simply how to construct probability models of natural and social events. Detailed information was provided by Şen (2002).

4.2 Percentages and Probability Principles

The concept of percentage is well known by everybody and many comparisons, predictions, bets, games of chance revolve around this understanding and, accordingly, the final product or result is expressed as a percentage. It reflects the proportional amount of some quantity with respect to a certain or known quantity; this is the ratio of the amount of some qualitative specification to the total amount of the same specification. The question is always "What is the percentage of a given quality in a given or known total or population? For example, "What is the percentage of Turkish students at University of London? "What is the chance of your son to succeed in the Istanbul university entrance examination? "What is the probability of rain tomorrow? Each one of these questions can be answered linguistically or numerically, depending on the purpose. If the main goal is to do more calculations or to use percentage information in numerical modeling techniques, the answer should be in percentages. Otherwise, the same questions may have verbal answers, but the recipient may add a percentage value subjectively or based on expert opinion. The percentage of Turkish students at the University of London may be "very low" or numerically 0.05 percent (0.05%). Of course, the numerical response is more objective and does not leave much room for further interpretation, but as "too low" the verbal response contains in itself ambiguity. In numerical response, the answer reflects the level of the percentage of Turkish students at the University of London, but the verbal response is two uncertainty sources: linguistic uncertainty and its subjective conversion to numerical uncertainty. A percentage can be defined as the ratio of a part of something to its total. In the light of this last sentence, a percentage can be defined as any shape and its part, which is shown in Fig. 4.1.

Each white shape labeled with T indicates the total area of variation or collection of objects in the form of a cluster, and black labeled patches are sharp subfields within the total area, and thus the black patches are smaller than the total white areas. If the question is what the percentage of black patches in each white area is, the answer is the ratio of the black area to total white area, and such a ratio has the following properties:

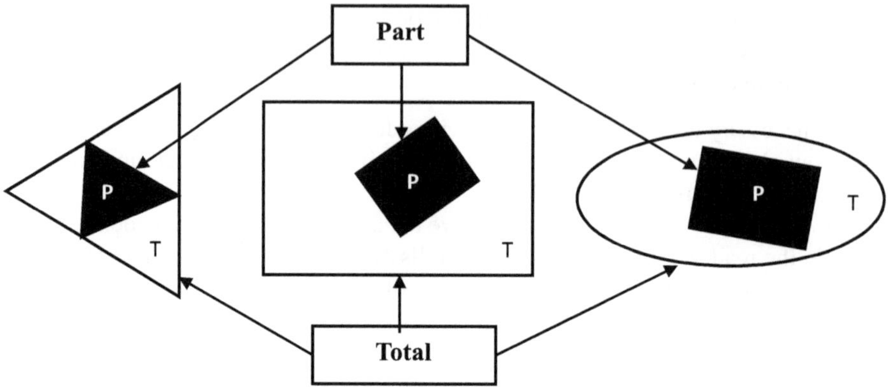

Fig. 4.1 Total and partial domains

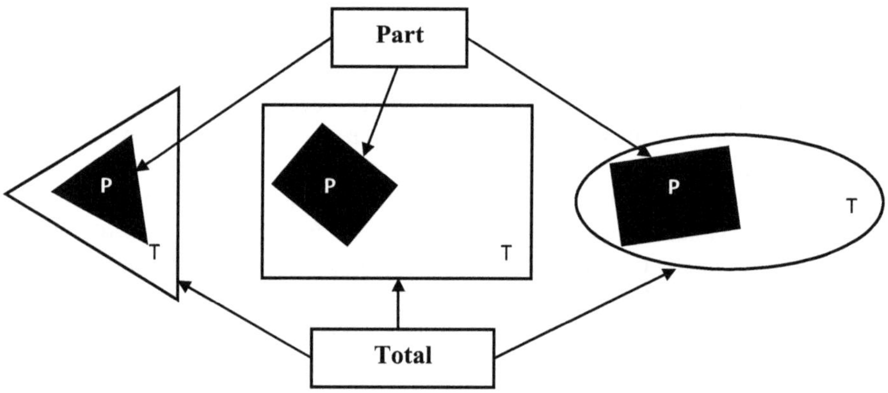

Fig. 4.2 Total and partial domains

1. The percentage defined in this way takes a value between 0 and 1 inclusive.
2. Regardless of the unit of patch and domain, it is unitless.
3. The percentage stays the same if it stays the same, but the black patches have the same extension at different locations as containers within the total domain. For example, the percentage definitions in Fig. 4.2 are the same as percentages in Fig. 4.1.

Comparing these two figures shows that the position of the patched area within the total area does not make any difference when it comes to percentage calculations. A property such as the equivalence of patched position can be called isotropy, and so there is one and only one percentage, which is independent of location but the percentage increases (decreases) with increasing (decreasing) the patched area within the total domain.

Example 1

It is also important to notice that the percentage definition is relative to total domain. In Figs. 4.1 and 4.2, there are two subareas within the total area, the black patched subarea and the remaining white subdomain; the sum of the two equals the total area. If either of these two subareas is considered as the main subdomain, the other will be supplementary domain. Likewise, the main percentage and the supplementary percentage can also be defined, and these two subfields have the following properties:

1. The two subdomains are crisp in the sense that there is clear-cut and precise boundary between the two.
2. Any point in any one of the subdomains does not belong to complementary domain.
3. Their sum is equal to the total area and hence the sum of percentages is equal to 1.
4. Such divergent and complementary subdomains are called mutually exclusive subdomains.

After all these points, it is possible to deduce the following statements regarding the two subfields that complement each other:

1. Percentage of any subdomain is equal to the percent of complementary subdomains subtracted from 1.
2. The sum of the two complementary subfields is equal to the probability of the total area, which is equal to 1.

Example 2

Now it is time to generalize the two complementary subdomains to n complementary subdomains, which are mutually exclusive but can be simultaneously contained within the total domain. Such general complementary subdomains are presented in Fig. 4.3.

Herein there are seven subdomains and they cover the whole domain and since any point in any part does not belong to any other part, again the subdomains are crisp and mutually exclusive, and all the seven parts are complementary. Each art has percentage coverage out of the total domain and the summation of these percentages is equal to 1.

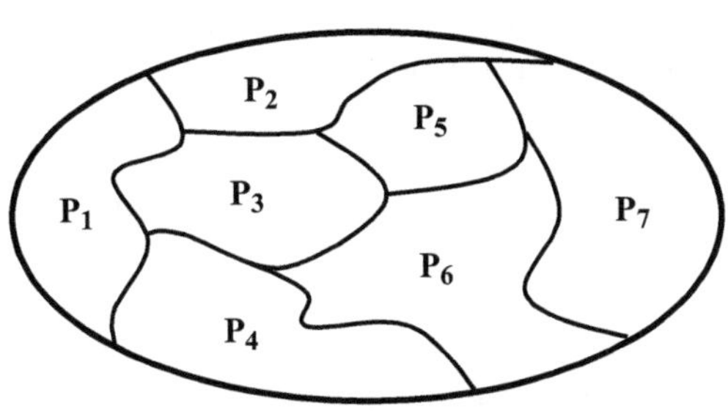

Fig. 4.3 Complementary and mutually exclusive subdomains

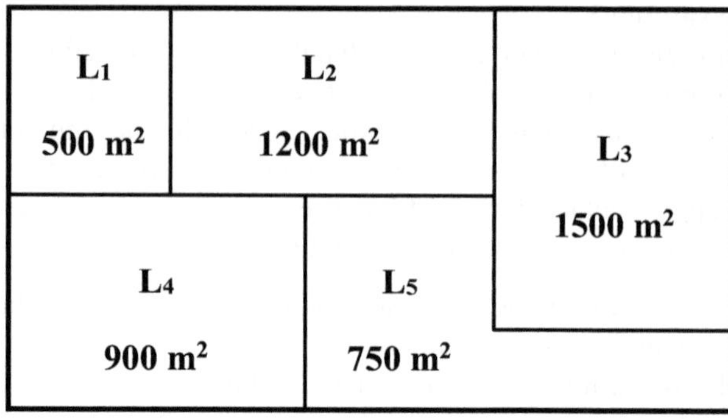

Fig. 4.4 Land use partition

Example 3

As in Fig. 4.4, in a region, there are five owners whose plots are labeled as L_1, L_2, L_3, L_4 and L_5 and those areas in m^2. So, calculate the percentage of each land owner property.

The total area is 4850 m^2 and therefore the percentages are $p_1 = 500/4850 = 0.1031$, $p_2 = 1200/4850 = 0.2474$, $p_3 = 1500/4850 = 0.3093$, $p_4 = 900/4850 = 0.1857$, $p_5 = 750/4850 = 0.545$, respectively. The sum of these percentages is 1, and while they are all percentages, they are fixed with the boundaries of land plots and do not change until there are some changes in the land boundaries. However, land masses are mutually exclusive, and this feature makes the percentage total equal to 1.

Example 4

The table below shows the 30 speed measurements at the exit of the Turkish Strait, Istanbul. After the necessary comments, find the percentages of exceeding and nonexceeding the speed of 120 mm/s.

Current velocity measurements are imprecise and vary with each taken sample and are never the same values and therefore there is an uncertainty due to random fluctuations. The variation of the speeds by the number of samples is shown in Fig. 4.5, which gives the random behavior at first glance.

As in the figure, 120 mm/s is a certain level, and corresponding to this level, all the measurements are now uncertain as there is no fixed trend of change. If the current velocity is to be considered as a range variable, the lower and upper limits are 3.6 mm/s and 156.3 mm/s, respectively as seen in Table 4.2. The question now is to decide the type of the range variable as explained in the previous section. As an assumption, based on the measurements given in Table 4.2, it is possible to say that the current variable can be considered a closed range, meaning that no lower than 3.6 mm/s is expected in any modeling or design work or greater than 153.2 mm/s. However, knowing the uncertainty in the measurements and keeping in mind those current velocity measurements at the same location in nature can vary randomly with time, it makes more sense to work on an open-range variable basis, where both hands can set records towards lower and higher values. Such an approach requires more methodology on how to deal with uncertain phenomena, such as current velocity measurements.

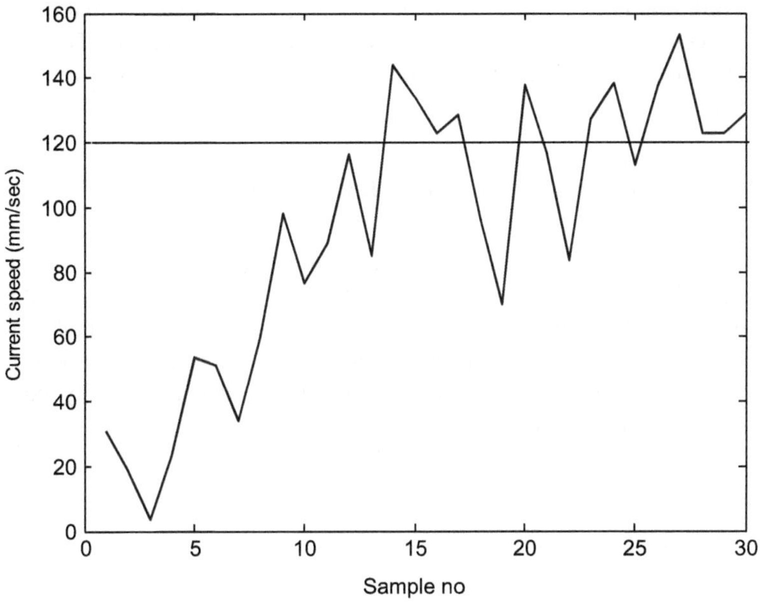

Fig. 4.5 Current speed variations

Table 4.2 Current speed measurements

Sample no.	Current speed (mm/s)	Sample no.	Current speed (mm/s)	Sample no.	Current speed (mm/s)
1	30,8	11	88,7	21	116,8
2	18,8	12	116,1	22	83,7
3	3,6	13	85,1	23	126,9
4	23,3	14	144,0	24	138,4
5	53,5	15	133,7	25	113,0
6	51,1	16	122,6	26	137,2
7	34,1	17	128,2	27	153,2
8	59,5	18	95,5	28	122,6
9	98,0	19	70,1	29	122,6
10	76,5	20	137,9	30	128,9

On the other hand, at the level of 120 mm/s, all data can be classified as smaller and larger at this level, and even precise data values in each class can be defined. For percentage calculation purposes, it is enough to know the number of data (not data values) that are smaller and greater than 120 mm/s. The total number of data is 30 and the number of data greater than 120 mm/s is 11 from Table 4.2 or Fig. 4.5 and hence the number of data less than 120 mm/s is 19. Accordingly, if the percentage of data is greater than 120 mm/s, it is 11/30 = 0.37, and therefore the percentage of smaller values is the complement of this value, which is 19/30 = 0.63, the summation of these two percentages is equal to 1.

It is worth noting that when calculating percentages, not their actual data values, but their occurrences compared to a given level are important. Therefore, the percentage can be defined as the ratio of the specified number of occurrences.

4.3 Probability Measures and Definitions

Although probability is equivalent to the everyday use of percentages, in practice the problem is how to define that percentage. For the definition of probability, it is necessary to have different categories for the same data. The fundamental question in any probability study is: What is the percentage of data for a class? This question can be answered in the following three ways, depending on data availability, conceptual model, enumeration, and personal belief about the phenomenon.

4.3.1 Frequency Definition

This is the most commonly used definition that requires the availability of measured data; otherwise, it cannot be used. In addition to data availability for classification, specific data values or design quantities should be considered. For example, given a data set X_i ($i = 1, 2, \ldots, n$) and a design level X_0, classical probability calculations can be performed easily. The significance of the design value is that it divides the given data into two parts, that is, the data values greater than the design value ($X_i > X_0$) and the others are smaller. If the number of greater data values is n_G, then the percentage or probability, p_G, can be calculated as

$$p_G = \frac{n_G}{n} \tag{4.1}$$

Similarly, the probability that the second class of data is smaller than the design value can be written as

$$p_S = \frac{n_S}{n} \tag{4.2}$$

This definition is also known as the relative frequency of classes. It is known that the summation is equal to one ($p_G + p_S = 1$). Instead of two classes there may be many, and the probability for each class can be calculated as the ratio between the numbers of data that falls within a class divided by the total number of data.

4.3.2 *Classical Definition*

This definition is based on the conceptual evolution of the phenomenon, and the percentage or probability can be defined even without data availability. Here again, different classes of outcomes of the phenomenon are considered. For example, the question of what the probability of the parent rock type is (igneous, sedimentary, and metamorphic) can be answered by considering these three categories, and igneous p_I, sedimentary, p_S, and metamorphic, p_m, rocks with equal probability up to 1/3. Again, the sum of the probabilities of these three categories is added to 1 as follows:

$$p_I + p_S + p_m = 1.0 \tag{4.3}$$

In general, the frequency definition of probability approaches the classical definition for very large data numbers.

4.3.3 *Subjective Definition*

In subjective interpretation, probability is considered as the degree of belief or quantitative judgment of an expert about the occurrence of a phenomenon. It is meaningful percentages to precipitation occurrences over the next few days. Thus, two individuals may have different subjective probabilities for an event without both necessarily being mistaken. However, the preferences of these two people are close to each other. Their recommendations depend on their long experience with the relevant phenomenon. Subjective probability does not mean that an individual is free to choose any number and calls it probability. Subjective probabilities should be consistent with the probability axioms in this chapter, and therefore, with the probability properties implied by these axioms.

4.4 Types of Probability

Any composite probability expression can be derived from these basic probability definitions. A conceptual understanding of these probability expressions provides derivatives of logical relationships between them. Such relationships are possible in the case of at least two different events or phenomena.

4.4.1 Common Probability

Basic definitions of probability relate to a phenomenon, but in practice there are often two or more events. The reciprocal treatment of these phenomena leads to some common definitions of probability. If two events are represented as X_i ($i = 1, 2, \ldots, n_X$) and Y_j ($j = 1, 2, \ldots, n_Y$), their joint probabilities can be defined considering the p_{XY} fundamental probability. However, for probability definition, classification level for each event is required as X_0 and Y_0, respectively. Each event can be divided into two categories as ($X_i > X_0$) with its complementary ($X_i > X_0$) and similarly ($Y_i > Y_0$) and ($Y_i > Y_0$). Based on these basic four categories, four common categories can be created as [($X_i > X_0$), ($Y_i > Y_0$)], [($X_i > X_0$), ($Y_i < Y_0$)], [($X_i < X_0$)., ($Y_i > Y_0$)] and [($X_i < X_0$), ($Y_i < Y_0$)]. Each pair represents a common category from two different events, and thus there are four different types of common categories and probability questions. For example, what is the common probability of the event [($X_i > X_0$), ($Y_i < Y_0$)]? The answer can be given according to the classical definition of probability, provided that the number of this common event is represented by n_{XY}.

$$p_{[(X_i \rangle X_0),(Y_i \langle Y_0)]} = \frac{n_{XY}}{n_X + n_Y} \tag{4.4}$$

4.4.2 Conditional Probability

Its definition also requires two different sequences, as in the case of common probability. However, the question is different as what the probability of a classification in is one of the events given a classification in the other event. Given the notations in the previous subsection, what is the probability of [($X_i > X_0$) and ($Y_i < Y_0$)] given ($Y_i < Y_0$)? This is the ratio of the [($X_i > X_0$) and ($Y_i < Y_0$)] event number to another event ($Y_i < Y_0$) number. In the expression [($X_i > X_0$) and ($Y_i < Y_0$)], the number is the simultaneous occurrence of events that can be represented as ($X_i > X_0$) and ($Y_i < Y_0$). If ($Y_i < Y_0$) is the number of events, the conditional probability is defined as

$$p_{[(X_i \rangle X)/[(Y_i \langle Y_0)]]} = \frac{n_{[(X_i \rangle X_0) \text{ and } (Y_i \langle Y_0)]}}{n_{(Y_i \langle Y_0)}} \tag{4.5}$$

4.4.3 *Marginal Probability*

This is also based on at least two events, and the question is, what percentage of one event class occurs when the other event is fully considered? This is equivalent to saying that the marginal probability of the class ($X_i > X_0$) occurs during the entire Y event regardless of its classes. So, notationally, this marginal probability can be written as

$$p_{(X_i)X_o)/Y} = \frac{n_{(X_i)_0)}}{n_Y} \tag{4.6}$$

4.5 Axioms of Probability

As shown in Fig. 4.6, the sample space and its constituent events do not have any numerical labels, but relationships are possible with numbers, axioms of probability, and their meanings.

The rules for the relative connection of events in the sample space are derived from logical considerations and there are basically three logical axioms of probability. These axioms can be loosely stated as

1. The probability of an event cannot be negative.
2. The probability of a sample space, that is, the composite event S, is equal to 1.
3. The probability of one or the other of two mutually exclusive events occurring is the sum of their two individual probabilities.

These axioms are the logical bases of the mathematical construction of probability. From these basic axioms, it is possible to deduce all the mathematical properties of probability. Many of these mathematical properties will be presented later in this chapter. Probability interpretations are found acceptable and useful in engineering, earth, environmental, and atmospheric sciences, in the same way as the particle-wave duality of the nature of electromagnetic radiation in quantum physics.

Fig. 4.6 Venn diagrams

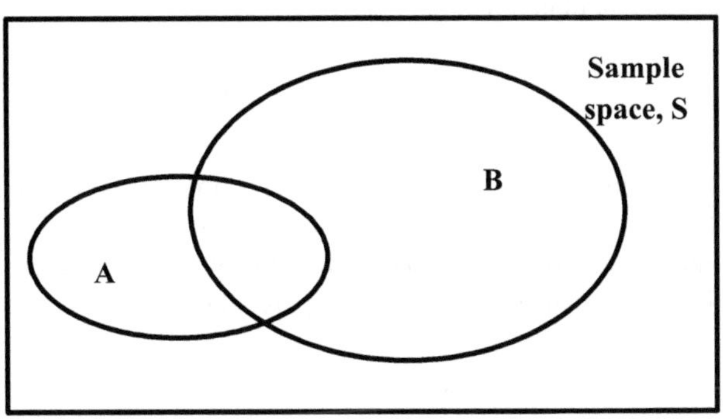

Venn diagrams are very useful tools for geometrically showing the probability of events. The area of the sample space in Fig. 4.1 is considered equal to a dimensionless unity, that is, 1, according to the second axiom. On the other hand, the first axiom says that no field can be negative. Finally, the third axiom says that the total area of the nonoverlapping parts is the sum of the areas of those parts. A set of mathematical probabilistic properties logically emerges from these axioms. All these features can be visualized with a Venn diagram.

4.5.1 Probability Dependence and Independence

Past or future values of any engineering, earth science, atmospheric or environmental variable often exhibit statistical dependence with their current value. Dependency can be defined as the presence of sequential effect between successive event occurrences of the same variable. In the case of positive (negative) dependence, large values of the variable tend to be followed by relatively large (small) values, and small values of the same variable tend to be followed by relatively small (large) values. In most geophysical events, this dependence along the time axis is positive. For example, if today's temperature is above average, tomorrow's temperature is expected to be above average.

To evaluate the serial dependence for any event, it is necessary to estimate the conditional probabilities of type P (event at present/event at previous time), where / stands for "given." This type of dependency coefficient has already been defined by Şen (1977). Rearranging the conditional probability expression in the equation leads to the following series of equations for two events E_1 and E_2:

$$P\{E_1 \cap E_2\} = P\{E_1/E_2\}p\{E_2\} = P\{E_2/E_1\}P\{E_1\} \tag{4.7}$$

Two events are said to be independent if the occurrence of one event does not affect the probability of the other.

In order to quantify the probabilities, for example, if one observes that an event is repeated n times under the same conditions, and the observed event occurs n_i times, the probability of this i-th item, called its relative frequency, is obtained by dividing the frequency (frequency) by the number of items, the probability of occurrence, P_{io}, is calculated as

$$P_{io} = \frac{n_n}{n} \tag{4.8}$$

The ratio defined in this way ranges from zero to one and it can never be negative. These values are called percentage, relative frequency, or probability of the event. Such ratios form the basis of probability theory. The probability that the same event does not occur is calculated as (percentage),

$$P_{in} = 1 - \frac{n_i}{n} \tag{4.9}$$

Here, P_{in} is called the complement of probability, P_{io}. Indeed, by summation of Eqs. (4.8) and (4.9), the following simple expression is obtained.

$$P_{io} + P_{in} = 1 \tag{4.10}$$

In this section, we try to find the consequential probabilities of the event, that is, the numbers that determine the priority of the class elements among themselves, by percentage weight. The numbers that give these weights for each item are called the probability criteria of that item.

Probability enables scientific interpretations of events that occur due to chance. Although the physics foundations of probability principles are quite meaningful and logical, they require care in making their calculations. To illustrate the possibility better, let us consider the following simple problem. Should you take an umbrella when going out tomorrow or not? For this, it is necessary to assume that there is a certain amount of precipitation, for example, 90% probability. Thus, an umbrella is taken in case of a 90% chance of precipitation. In this case, the probability of no precipitation (complementary) is 10%. Such fixed percentages in our daily life always help us to make some decisions. Generally, probabilities are expressed as a number between 0 and 100. In mathematics, this number is required to be between 0 and 1. Of these, the probability of 0 means that the event is impossible, and the probability of 1 means that it will happen. However, if the probability of an event falls between these two definite limit values, there is uncertainty. For example, it is understood that an event with a probability of 0.000001 may occur once in a million.

There are three different definitions that are independent of each other in finding the probability numbers for each element at the beginning and can be applied according to the situation. The first of these is the classical concept of probability. Here the consequences of the event are known as the finite number of elements. For example, the presence of seven different types of fish in a lake indicates that there are seven likelihoods (options) for each fish. The rate of occurrence of each of these seven elements (states) among themselves is 1/7. Thus, it is understood that the probability of each of the seven items can be taken as 1/7. Considering this, the classical definition of probability is made as the ratio of the number of items considered to the total number of items. The three basic features of probability are presented in Sect. 4.4 by three items stated there.

In the example of the fish species in the given lake, the probability value of each item (1/7) is equal. This means that the appearance of one species (item) in each fish catch has no advantage over the emergence of the other element; they are equally likely. However, in classical probability, the items may not always differ from each other. For example, the fish species in the lake are very similar, 2 and 3 species are denoted by 2. If 4, 5 and 6 species are denoted by 3 and 7 by 4, then at the beginning there are four sets of fish from seven species that are considered completely different from each other will be as.

$$T = \{1, 2, 2, 3, 3, 3, 4\}$$

In other words, there are four different elements in this set. Let us examine the classical probability values for such a lake. Since there are seven fish species in the lake, the total number is equal to 7. However, the elements represented by these faces are 4. After fishing at different times, the rate of 1 is 1/7, the rate of 2 is 2/7, 3 is 3/7 and, finally, 4 is 1/7. Compared to the previous example, although there are seven fish types, the probability values have changed. The three properties listed above have not changed and the sum of the probabilities is still equal to 1. The conclusion to be drawn from this is that in order to calculate classical probabilities, the number of possibilities and the number of elements must be known. In general, classical probability is defined as the ratio of the number of items to the number of likelihood (see Eqs. 4.8 and 4.9). In order to make a classical definition, the probabilities must be known exactly, which is impossible to determine for natural events. The number of possibilities can be infinite. The classical probability definition is valid for finite sets.

The possibilities of natural events are infinite. For example, expressing the amount of precipitation in a place as decimal number instead of discrete number as in fish species does not make the classical probability definition possible based on point probabilities. Here, a new definition must be made for probability calculation.

The set of all possible outcomes of an experiment is called the sampling set for that event. Among these results, one of them is called the element of that set and this element is called the sample. Thus, the set of possibilities is called the sample set of the event. In practice, records of precipitation or any event in previous years constitute an example from the population of that event. Precipitation is an infinite set containing all numbers between the smallest, P_s, and the biggest precipitation, P_b. For such sets to be useful in their applications, they must be divided into a finite number of subranges. The ratio defined by Eq. (4.8) is again a definition of probability, which is used frequently in practice and it is called the relative frequency definition of probability. This definition will serve as a basis for determining the statistical properties of the random variable, which will be seen in later sections of this chapter.

In order to be able to define possibilities given above, it is necessary to have numerical data. However, as a result of the experience about the event, one can predict the rate of occurrence of the same event for the future, that is, the probability. Experts who have a lot of experience on a subject can express their opinion about the probability values of the events related to that subject. In this way, the percentages determined by the experiences and opinions of the people are subjective probability, which is also widely used in practice.

4.5.2 Probability Assumptions

If a principal sampling space, S, and the probability of an event A in this space are denoted by P(A), it must conform to the following assumptions:

1. For each event, $0 < P(A) < 1$.
2. The probability of the universal set is equal to 1, so $P(S) = 1$.

3. If events A and B are incompatible, that is, they are two mutually exclusive events that do not have a common element then,

$$P(A \text{ and } B) = P(A) + P(B) \tag{4.11}$$

Thus, the probabilities of mutually exclusive events are summed. In order to prove this, let us observe that an experiment is performed n times, considering the relative frequency definition of probability, and as a result, events A and B occur n_1 and n_2 times. According to the relative frequency definition in n times repeated experiment, the incompatibly combined event A and B occurred $n_1 + n_2$ times:

$$P(A \text{ and } B) = (n_1 + n_2)/n = n_1/n + n_2/n = P(A) + P(B) \tag{4.12}$$

The property that both events cannot coexist in the same place and time is valid. This expression is not valid in case of congruent events. Like Eq. (4.12), it is possible to write for a set of incompatible events such as A_i (A_1, A_2, ..., A_n) the following expression:

$$P(A_1 \text{ and } A_2 \text{ and } \ldots \text{ and } A_n) = n_1/n + n_2/n + \ldots + n_n/n \tag{4.13}$$

Here, n_1, n_2, ..., n_n are the number of occurrences of each event.

In probability calculations, there is a product rule other than the rule of sum of the probabilities of incompatible events. In case the events are independent from each other, the product rule is valid. For example, if there is another B event that is simultaneous with the occurrence of an event A, the probability that these two events will occur together, if they do not physically affect each other, that is, if they are completely independent the probability expression is as follows:

$$P(A \text{ and } B) = P(A)P(B) = (n_1/n)(n_2/n) \tag{4.14}$$

If there is a sequence of event and they are completely independent from each other, then their common probability can be expressed as follows:

$$P(A_1 \text{ and } A_2 \text{ and } \ldots \text{ and } A_n) = (n_1/n)(n_2/n) \ \ldots \ (n_n/n) \tag{4.15}$$

4.6 Numerical Uncertainties

Uncertainties include probability theory, fuzzy logic, random set theory, rough sets, gray systems, etc. There is a lot of overlap and even confusion about the boundaries of each uncertainty principle, and many consider the boundaries to be precise, but not so in this book. Of course, the author's ideas do not mean that they are verifiable, but they can be falsifiable, if a better direction of improvement is developed

(Chap. 3). Every engineer and modeler should consider that exact solutions are not possible, and therefore, the results need to be questioned from different aspects. For example, the concept of "factor of safety" eliminates the possibility of philosophical issues entering engineering studies and is more of a factor of "ignorance"; ignorance in the sense of putting all the blame on this factor without further thinking and finding solutions in the field of uncertainty.

4.6.1 Uncertainty Definitions

As a general term, "uncertainty" means many different connotations, and experts understand it quite differently. According to some, it is a completely unknown and unpredictable pile of information, for others it is always there due to people's incomplete information and knowledge, and there is no way to get rid of it completely. Even in everyday life, individual's own explanations of the same phenomenon may overlap to a large extent with others, but there are differences, and therefore, these differences may not be understood by others, and therefore, remain unclear. The difference in expert opinions stems from the fact that initial information and knowledge are subjective concepts based on frequent observation and experience. In scientific activities, there are always philosophical thoughts with their own axioms, hypotheses, laws, and final formulations. With Newtonian classical physics, it is possible to say that science has entered an almost completely deterministic world, where uncertainty is not considered even among scientific knowledge. However, today, there are uncertainty components in almost all branches of science, and many scientific deterministic foundations have become uncertain in terms of random, probability, statistics, chaos, fractal, stochastic, quantum, and fuzzy inferences. On the other hand, environment, atmosphere, earth, and social sciences and the like have never passed the stage of complete determinism. Unfortunately, determinism dominates education systems in many institutions around the world. With the development of numerical uncertainty techniques such as probability, statistics, and stochastic processes, scientific advances in quantitative modeling have made rapid advancements, but still have set aside qualitative sources of information and knowledge that can only addressed be with fuzzy principles (Chap. 3).

Famous philosophers and scientists explain the uncertainty and fuzzy components that are the fundamental basis of scientific progress. For example, Russell (1924) stated:

> All conventional logic customarily assumes the use of precise symbols. Therefore, this is applicable only to an imaginary celestial existence and not to earthly life.

On the other hand, when it comes to verbal and linguistic fuzzy concepts, Zadeh (1965) said:

> As the complexity of a system increases, our ability to make precise and yet important statements about its behavior decreases, until a threshold is reached where precision and significance (or relevance) become almost mutually exclusive properties.

During the evolution of human thought, premises include elements of uncertainty such as ambiguity, incompleteness, probabilities, possibilities, and fuzziness. The inference of mathematical structure from the mental thinking process may seem certain, but even today, as a result of scientific developments, it is understood that there are at least parts of uncertainty, if not on the macroscale, at every stage of physical or mechanical modeling on a microscale. It is clear today that the mathematical conceptualization and idealization that lead to the satisfactory mathematical construction of any physical reality is often an unrealistic requirement. As Albert Einstein (1879–1955) noted,

> As long as the law of mathematics refers to truth, they are not exact. And they do not refer to the truth as much as they are sure.

This book has presented in Chap. 3 some uncertainty principles and the role of fuzzy logic principles comparatively. In addition to these stages after model adaptation and determination of its parameters, there is the "validation" stage, where the suitability of the selected model to the observation sequence is sought. Of course, the definition phase includes theoretically searching for a suitable model, estimating parameters for the model, and checking the suitability of the model. However, the most important stages in any study are identification followed by forecasting. In fact, these two stages follow each other.

4.7 Forecast: Estimation

The basic estimation work was done by Gauss in the early 1800s, who tried to fit the most appropriate curve over the scattering of points by using the least squares technique (LST) as a criterion that forms the basis of any uncertainty event evaluation in statistics without exception and stochastic process modeling. The successful implementation of the LST for almost two centuries is due to the following factors:

(i) Minimizing the sum of squared errors leads to a system of linear equations that is easy to solve and does not require extensive theory.
(ii) The sum of squares corresponds to various interpretations in many different contexts, as in engineering, physics, energy is expressed as the sum of squares; represents the moment of inertia in mechanics; in statistics, it provides the variance about the mean of fitted curve; and consequently, can be used as a measure of goodness of fit test.
(iii) The assumption of precisely clear analytical form to represent the observed data constitutes the basic application of the classical LST.
(iv) It is possible to apply LST to filtering problems without suggesting an explicit analytical expression. For example, a known differential equation may represent the related phenomenon. Likewise, the equilibrium and continuity equations are explicit expressions for a phenomenon in many disciplines.

(v) Wiener (1949) established a different implementation version of the LST, assuming certain statistical properties for useful signal and noise components of the observation sequences. The important difference of Wiener's approach lies in the fact that the useful and noisy parts are not characterized by analytical forms, but by statistical properties such as mean values being zero or returned to zero, and both serial and cross autocorrelations,

(vi) Carlton and Follin (1956) suggested the use of adaptive LST to reduce the computational burden after 1950.

To build a dynamic model for the simulation of any event, it is necessary to have a finite record of past observations. Given a historical record, the estimation process consists of calculating an estimate of the variable of interest at any time, whose position relative to the observation period gives rise to three types of estimation problems (Şen 1980):

(i) At any time, during the observation period, conditional estimation is referred to as "correction" in statistics or "interpolation" in mathematics.

(ii) Estimation of the state at the time of the last observation is called "filtering."

(iii) Specification of the variable estimation at a moment after the last observation, which is called "prediction" in the field of uncertainty or "extrapolation" in the field of mathematics.

4.8 Types of Uncertainty

Some researchers tentatively divide uncertainty into two broad groups: current uncertainty and future uncertainty; they pursue future forecasting models. Past records, if any, are considered together with the uncertainty in their structure, and the type of uncertainty is determined by probabilistic, statistical, and stochastic processes, and then future simulation studies are obtained through the addition of random components to the models such as Markov and Auto Regressive Integrated Moving Average (ARIMA), and similar processes (Box and Jenkins 1974). Past observations, with their embedded uncertainties, are valuable and indispensable information in these modeling estimates. Mankind is in the current uncertainty, whatever its problems and daily walks of life. Even daily agendas are full of current uncertainties, and hence, there is always a series of meetings and discussions to decide on a common solution; this may not even be exact, but it can reduce the uncertainty. The uncertain solution is already expressed by Heisenberg (1927) as indeterminate and by Popper (1955) as falsifiability principles (Chap. 3). Health problems are also among the daily uncertainties, and many diseases such as cancer seem to remain uncertain despite scientific and technological developments. All types of research involve uncertainty, and hence, researchers try to reduce the amount of uncertainty so that scientific advances can be achieved even on a microscale. The basis of thinking on philosophical issues and logical inferences all relate to aspects of uncertainty. We live every day with uncertain, incomplete, and dubious thoughts and feelings that are not quantitative but qualitative in the content of uncertainty and are fuzzy (Zadeh 1965). Thus, fuzziness is among the current uncertainty area that evolves over time.

The most questioned aspects of future uncertainty are randomness, probability occurrences, possibility, and risk additives as concepts and quantitative models of future uncertainty that are expected to lead to objective predictions within certain error limits.

Essentially, fuzzy, "grayness," "randomness," "roughness," and "uncertainty" are not related to past ambiguity. Mentally, if certain events are represented by exact numbers, let us say 1 for absolute precision and 0 for absolute uncertainty as in the crisp logic case (Chap. 3), then all fuzzy features will fall between these two limits. If the event is close to precision, its numbering will be close to one, otherwise close to zero.

Are "fuzzy," "grayness," "randomness," "roughness," and "uncertainty" different names for the same thing, or are they just fancy names for different things? In other words, what exactly do these terms refer to? At first, the answer is that these terms are related, but not synonymous, because each refers to a variety of phenomena. According to the philosopher Ludwig Josef Johann *Wittgenstein* (1889–1951), these can be defined not by necessary and enough conditions, but by family resemblances, in which some members of the family can be described with one or more other words.

Another question is, "Are fuzzy, grayness, randomness, roughness, and ambiguity mental or physical phenomena?" They are mental at first glance, but each one of them, like "almost red stone," "quite tall man," "greedy dog," etc., indicates some adjective suffixes in physical subjects.

There are many complex phenomena that occur around us and affect each other to an unknown extent. Therefore, "What are the relationships between or among these phenomena?" Here the answer is not clear, but fuzzy, because through rational thinking, fuzzy relations between the two phenomena can be reached. For example, the question is what is the relationship between the number of accidents and alcohol level? There are two variables here: "accident number" and "alcohol level." Rational thinking leads to a relationship of direct proportionality between the two and considers that zero-alcohol level does not mean zero accident counts, but the minimum number of accidents. All that has been said about the relationship between these two events can be illustrated as a model in Fig. 4.2a.

The second-level question is, "Is it possible for the relationship to be linear as in Fig. 4.7a?" If one thinks about two phenomena and the relationship between them, they will conclude that the relationship should be nonlinear as in Fig. 4.7b. So far everything seems complete and precise, but it should be noted that while the figural nonlinear relationship in Fig. 4.7b is acceptable, it is fuzzy because its exact position is not well established. There are many nonlinear curves like the one in Fig. 4.7b and not even one can be absolutely and completely defined, and therefore, there is still fuzziness in the final decision.

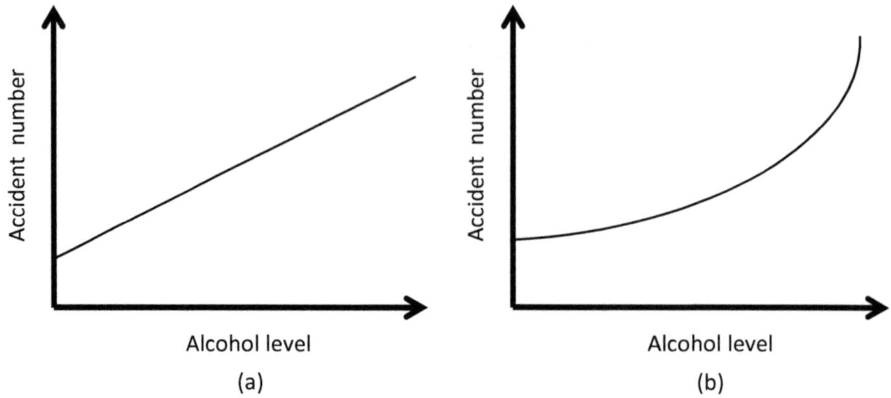

Fig. 4.7 Uncertain relationships

Another question is, "Under what conditions is a theory, procedure, algorithm, or model developed around any phenomenon of interest visibly more useful?" The answer will be in the domain of uncertainty, because if there is more than one method, the best can be determined based on the practically acceptable amount of error according to the same error criterion. The smaller the error, the better is the methodology. Mean absolute error (MAE), root mean square error (RMSE), relative error (RE), etc. are among the several commonly used error assessment criteria.

Another question is, "Can the relevant phenomenon be measured for the data collector helping to discover the internal formation structure of the event?" In the example given in Fig. 4.7, no measurement was made, but the mental experiment based on rational and logical reasoning nevertheless emerged as fuzzy with the relevant relationship. Similar rational and logical inferences can be used to investigate the mechanism of occurrence of the phenomenon. What is measured depends on what kind of problem the measurement can help solution. It is possible to draw some practical and empirical conclusions from the available data. For example, if there is time series data on an uncertain behavior of the relevant phenomenon, then one may be interested in whether the sequential data influence each other. This can be answered simply by treating the data graphs as "current" and "next," as in Fig. 4.8.

In this way, there are three alternatives, each with a distribution of data points. At first glance, it becomes clear that there is no deterministic and clear relationship between successive points, meaning that even in the data, there are uncertainty gradients that can have measurement errors to a certain degree, imperfect information and knowledge about the phenomenon; and finally, the behavior of the phenomenon may be erratic, etc. While Fig. 4.8a indicates the simplest form of linear relationship, in Fig. 4.8b, the relationship is nonlinear. Both relationships can be expressed by mathematical equations that turn the indefinite relationship into an absolute certainty. In other words, the application of any mathematical expression takes uncertainty out of the realm of certainty. However, in Fig. 4.3c, there is no type of mathematical relationship to get rid of ambiguity; instead there are three clusters with indefinite scatter in each.

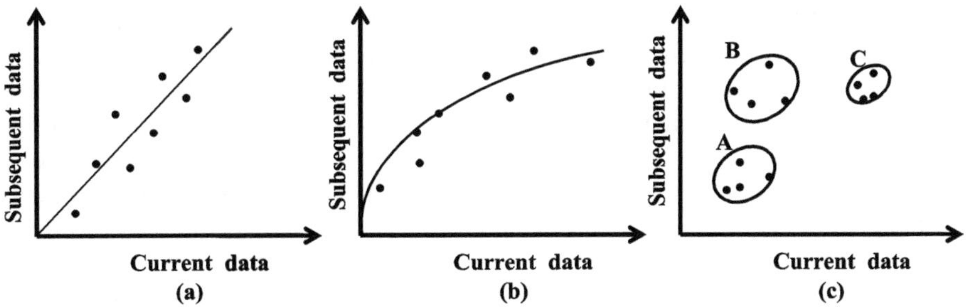

Fig. 4.8 Data-based empirical relationships

No study can accurately measure the relevant phenomenon, but researchers have the following question in mind: "Can the phenomenon in question (mental or physical) ever be accurately measured (quantified)? Without measurements, the relationships between physical phenomena can be described mentally, where the degree of precision depends on the nature of the problem and the relationship between the measurement values and how that value can help solve the problem one needs to solve.

4.8.1 Chaotic Uncertainty

The most important issue of any scientific study is the future prediction of the relevant phenomenon after the establishment of a reliable model. While forecasting relates to future unknown situations, the establishment phase is entirely dependent on past observations as numerical records of system variables. The essential step in any study is to identify the appropriate model and modify it to represent as closely as possible past observational sequences and model behavior. Not all developed models are successful in practical applications. In many areas of scientific forecasting, researchers are still behind in identifying a suitable model. With physical principles in addition to various simplifying assumptions, it is possible to explain many systems of empirical phenomena with ordinary and partial differential equations, but their application, for example, for forecasting purposes requires the measurement and definition of initial and boundary conditions spatiotemporally. Unfortunately, most of the cases investigated were not provided with reasonably enough data. In cases of unrealistic model availability or data scarcity, it is necessary to resort to some other simple but effective approaches. In most cases, deriving partial differential equations for the representation of the system concerned may seem simple, but its results are hampered by incompatibility of data with the model or lack of enough data.

In many applications, a simple but useful model is defined directly from the data, rather than partial differential equations and basic physical principles. Yule, for example, has proposed such simple models, considering the ordered structural behavior of the available data. Its purpose has been to treat data as a time series based on the principles of stochastic processes. Such a sequence was later recognized as one of the possible realizations among many possibilities unknown to the researcher.

Recently, chaotic behavior of dynamical systems has also exhibited random-like behavior quite different from classical randomness in stochastic processes. So, a question arises as to how to distinguish between chaotic and stochastic behaviors? Although the chaotic behavior shows a fundamental long-term pattern in the form of strange attractors in mean time, it suggests short-term forecasting possibilities (Lorenz 1972). Any data in time series format may appear to be a random series but may contain latent short-term coherences with several degrees of freedom. For example, while fluid flows represent chaotic behavior along the time axis, it shows a lower-dimensional attractor both theoretically and experimentally when constrained to a time-independent state-space representation. In the case of the existence of such a strange attractor for any phenomenon, time evolution results in a time series that hides chaotic remains.

A dynamical system is described by a phase-space diagram whose orbits show its evolution starting from an initial state. To enter this orbit, it is necessary to know the initial state quite precisely. Even different initial states, minute differences from each other, enter the strange attractor and after successive numerous steps, cover the whole strange attractor with chaotic transitions. Therefore, another deviation from classical time series is that successive time steps are equal in time series, in chaotic behavior the steps are random, but consecutive points stay on the same trajectory. In chaotic behavior, it is not enough to precisely determine the strange attractor, but it is necessary to model successive jumps that will remain on the attractor for predictions. Whatever the initial conditions, the orbits converge in a meaningful geometric pattern to a subspace within the entire space, then the attractor is captured. Otherwise, the behaviors are not chaotic, but stochastic unattractive. In short, strange attractors consist of points that occupy a small percentage of the phase space, but stochastic or completely random behavior randomly covers the entire phase space. Sub-coverage means that the dimension of the chaotic time series is less than the embedding dimension of the phase space.

4.9 Indeterminism

According to the concept of uncertainty, events (certain events or certain types of events) are not deterministically (causally) caused by previous events. It is the opposite of determinism and is about luck. This recalls the question, "Is the Heisenberg uncertainty principle relevant to the position taken by the vague, gray, randomness (probability), roughness and/or uncertainty experts?" The Heisenberg uncertainty principle is critical for events at the atomic level, while fuzzy events are attributed to larger scale events than atomic particles. Thus, Heisenberg's principle is not directly related to such phenomena, but the principle can be applied in a metaphorical sense.

As explained by Albert (1992, 1994), physicists have not found any special property of the electron that distorts its spin value by measuring the other. Therefore, the laws regarding the variation of returns by measurement cannot be deterministic. The theorist can introduce such uncertainty in probabilistic terms by simply referring to a pool of electrons. But s/he cannot formulate a deterministic law of the values of the two spins of a given electron. The uncertainty principle states that it is impossible to simultaneously determine the position and momentum of a particle. As soon as the experimenter knows the particle's location, the experimenter's instrument unpredictably affects the particle's momentum and vice versa. The principle is not about the impossibility of determining a reality due to the shortcomings of human means, but about the fact that reality does not consist of physical entities, each of which occupies a certain place and is characterized by certain properties. Bohm (1978, 1980) stated that matter constantly moves from the abstract space to the concrete one as the fuzzy of its state dissipates (Khalil 1997).

Physicists can only express the uncertainty of potential states in the probability distribution called Schrödinger's wave function when they consider many particles. Uncertainty is translated into certainty, that is, chance, which is expressed in the risk distribution, but only when physicists abandon the idea of predicting the behavior of a unique particle and instead focus on the representative particle, which is a fictional entity extracted from the average of the behavior of the multitudes (Khalil 1997). However, the resulting statistical description of the representative particle differs radically from the statistical description of the states resulting from the flip of a coin. There is no need to resort to "representative" coins to arrive at the probability distribution of their states. The probability distribution of the states of the coin does not result from being an individual, as in a quantum particle. Rather, the probability distribution is due to the observer's shortcomings. In principle, physicists can determine with certainty (100% confidence) whether a given shot will result in a head or tail. The result depends on the force and direction of the shot, air friction, the surface on which the coin falls, etc. Laplace (1951) stated that physicists cannot definitively determine the result. The only reason as de Laplace (1951) stated that physicists cannot determine the result with certainty.

The essence of the indeterminacy of chaos theory, interchangeably referred to here as "luck" or "risk," is essentially no different from the uncertainty of tossing a coin. Edward Lorenz (1963a, b, 1964) and Ruelle (1991), the pioneer of modern chaos theory, is a meteorologist. Weather was the optimal starting point, as its fluctuation challenged predictability over the centuries. Lorenz argued that the impossibility of non-probability prediction is due to "sensitivity to initial conditions" that are small enough to escape human perception.

A small change in an initial data can have a radically different outcome. Sensitivity to initial conditions occurs when the recursive feedback parameters are within a certain range. Lorenz half-jokingly stated that it is theoretically possible for a butterfly flapping its wings in Brazil to initiate a tornado in Texas via automatic feedback. The precise behavior of a storm structure can be predicted with improbable precision if the initial conditions are most precisely known as a Laplacian demon. But humans are not Laplacian demons who can keep track of all initial and alien variants. People must resort to a probabilistic form of estimation because of the astronomical cost of perfect knowledge.

As a result, chaos uncertainty (risk) is radically different from quantum uncertainty. Risk or chance results from incomplete information due to the observer's shortcomings. Conversely, uncertainty arises from incomplete information due to the nature of the object as a potential entity whose properties are not fully explained (Albert 1992). While the probability of chaos chance arises due to the size of the facts of a phenomenon, the probability of quantum uncertainty arises because the particle was not initially a definite, local phenomenon (Khalil 1997).

In fuzzy inference, many words can have ambiguous interpretations from many different perspectives. It is, therefore, necessary to bring together people from different disciplines interested in uncertainty principles to arrive at a more meaningful explanation for such words (Chap. 3). If researchers from the same discipline gather together to discuss "vocabulary content," they cannot enter a cycle of internal turmoil.

4.10 Uncertainty in Science

It is possible to say that early information and knowledge are concepts derived from frequent observations and experiences. For centuries, human thought received support from writings, drawings, calculations, logic, and finally mathematical calculations. Meanwhile, science differs from philosophy (Chap. 3) with its axioms, hypotheses, laws, and final formulations, especially after the Renaissance in the eighteenth century. With Newtonian classical physics, it is possible to say that science has entered an almost completely deterministic world, where uncertainty is not considered even among scientific knowledge. However, today, there are uncertainty components in almost all branches of science, and many scientific determinists have become obscure of fundamental fuzzy modifications. Such concepts include quantum physics, fractal geometry, chaos theory, and fuzzy logic principles. However, some, such as the geological sciences, have never passed the determinism stage, but unfortunately, deterministic education systems have also affected earth sciences education in many institutions all over the world. With the development of numerical uncertainty techniques such as probability, statistics, and stochastic principles, scientific advances in quantitative sciences have made rapid advances, but still set aside sources of qualitative information and knowledge that can only be dealt with fuzzy logic principles.

As a credit to Eastern thought, philosophical objects can be elevated with logical propositions and inferences that lead to idea generation, along with three basic mental activities, namely imagination, conceptualization, and then idealization. Since the existence of terrestrial life, man has interacted with nature, which provides the basic material for the chain of mental activity of man in the form of objects and events that change with time and space. In the early stages of human history or in the childhood of any individual, these stages play a role at different rates and take their final shape with experience. Each of the chain elements in the thinking process contains uncertainty, because the stages of imagination, conceptualization, and idealization are highly subjective depending on the individual's skills and capabilities. At any stage in the evolution of human thought, the antecedents contain some degree of ambiguity, vagueness, probabilities, possibilities, and uncertainties. The inference of mathematical structure from the mental thinking process may seem certain, but even today, as a result of scientific developments, it is understood that there are at least parts of uncertainty, if not at macroscale, but at every physical or mechanical stage of modeling on a microscale. It is clear today that the mathematical conceptualization and idealization that lead to the satisfactory mathematical construction of any physical reality is often an unrealistic requirement. As Einstein noted,

> As long as the law of mathematics refers to truth, they are not certain. And they do not refer to truth as much as they are sure.

In the most fundamental stages of mental thinking, activity objects are considered as members or nonmembers of a or physically reasonable range of variability. This considers sets containing possible consequences or basis of the phenomenon under investigation. In the formal sciences, such as physics, geology etc., these elements are almost invariably and automatically considered either entirely members of the set or outside the same set. Hence, the crisp logic of pairs of true or false; positive or negative; yes or no, black or white, one or zero, etc., is used for mathematical modeling of a phenomenon scientifically. However, Zadeh (1965) suggested continuity of membership grades between 0 and 1, inclusive, rather than a definitive assessment of membership. Therefore, fuzzy sets play an intuitively plausible philosophical basis at each stage of the chain of mental activity.

Poincaré (1904) has discussed the question of unpredictability in a nontechnical way. According to him, chance and determinism are reconciled by long-term unpredictability. He expressed this situation by saying that

> A very small cause, which escapes us, determines a considerable effect which we cannot ignore, and we then say that this effect is due to chance.

This point has been supported later by chaos theory, which is a branch of mathematics focusing on the behavior of dynamical systems that are highly sensitive to initial conditions, and any slight change in the initial or boundary conditions leads the solution to systematically uncertain domain, and one cannot estimate the values of the next step from previous numerical information.

Even though there are many deterministic modeling equations in use, especially for many events, the results may not capture absolute exactness due to the natural phenomena with internally inherited or externally excited random ingredients as already explained in the previous sections by numerous researchers. In practice, on the one hand, there are model uncertainties such as model bias, model discrepancy, model approximation, lack of knowledge about the underlying problem, discrepancy between the model and reality. On the other hand, there are also parameter uncertainties like model parameter influences, unknown values (exact, correct, best), and uncontrollability by experiment. Furthermore, numerical uncertainty, errors, approximations, and even mathematical model software writing for computer executions,

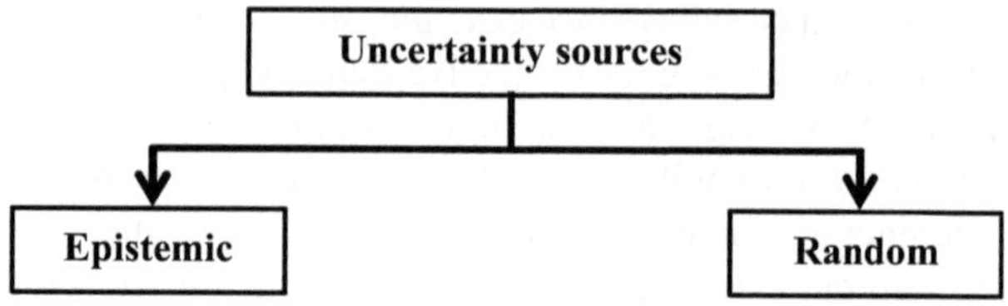

Fig. 4.9 Uncertainty sources

are among the algorithmic uncertainty sources. There are also experimental uncertainties including observational errors, experimental measurements variability, and indeterminism. Finally, there are interpolations uncertainties like the lack of available data, interpolation, or extrapolation results, and last but not the least, the methodological choice. It is possible to view uncertainty sources under two categories as in Fig. 4.9.

Epistemic (systematic uncertainty, known but impracticable principles, i.e., insufficient measurement or modeling, missing data, which can be alleviateable by better models and more accurate measurement) and random (statistical uncertainty, differences from each run, which cannot be eliminated by means of model improvement) uncertainty categories play role in the final production of any research study. Among the statistical uncertainties is visualization by common representations and probability distribution function (PDF) approximation. Whenever there is a sequence of measurements, scientists and researchers try to make the best estimate by calculating some quantity from the data on the assumption that some exact value exists considering that measurement is equal to the summation of the best estimate and an uncertainty term.

Measurement and synthetic generation data are not free of uncertainties, which are reflected in many disciplines by different modeling procedures. There may be different models from different research centers all over the world and their outputs constitute bundle of results, which are referred to collectively as ensemble. Ensembles are collection of multi-run simulations for parameters space exploration, mitigate model error, initial conditions cover ranges, and combination of multiple model outputs. They also help for collection of datasets, members, realizations, complete simulation run for each parameter set under input conditions. Ensembles might be multidimensional, spatial, and/or time domain simulate over many variables and support to assess many values for each variable or location. Uncertainty ambiguities are also available by means of numeric interval that is expected to contain data value with no assumptions about the probability distribution function (PDF) within the interval. Informational uncertainties are effective in many phenomena modeling that are known qualitative rather than quantitative information indication (Streit et al. 2008). Widely used boxplots are based on normal (Gaussian) PDF on quartile range, including median and outliers (Spear 1952; Tukey 1977).

Among the uncertainty methodologies the probability, statistics, and stochastic procedures have increased steadily during the last six decades in engineering, atmospheric, earth, and environmental sciences. Provided that uncertainty embedded numerical data are available in these domains, then the researchers seek to represent, treat, and interpret their data in an effectively quantitative manner. Any researcher with a firm mathematical background is able how s/he can obtain enough knowledge through probability, statistics, and stochastic processes, uncertainty methods after description of natural phenomena concerned and appreciation of the arguments for predictions and interpretations.

It is possible that the researcher may not have capability to predict specific events due to the intrinsic randomness (Mann 1970a, b). In the statistical sense, random and randomness are helpful to describe any phenomenon, which is unpredictable with any degree of uncertainty for a specific event and deterministic phenomena. The outcomes of individual events are predictable with complete certainty under any given set of circumstances, if the required initial and boundary conditions are known. The measurements of time series elements are only random samples from the natural event, and therefore, they have ensemble of replicates instead of a unique absolutely known behavioral feature. In fact, randomness is the ultimate and most profound physical concept of nature.

4.11 Importance of Statistics

Random and randomness are the two terms that are used in a statistical sense to describe any phenomenon, which is unpredictable with any degree of certainty for a specific event. An illuminating definition of randomness is provided by famous statistician Parzen (1960) as

> A random (or chance) phenomenon is an empirical phenomenon characterized by the property that its observation under a given set of circumstances does not always lead to the same observed outcome (so that there is no deterministic regularity) but rather to different outcomes in such a way that there is a statistical regularity.

The statistical regularity implies group and subgroup behaviors of many measurements so that the predictions can be made for each group more accurately than the individual predictions. For instance, provided that a long sequence of temperature observations is available at a location, it is then possible to say more confidently that the weather will be warm or cool or cold or hot tomorrow than specifying exactly by prediction the degree of centigrade. As will be explained in later sections, the statistical regularities are results of some astronomical, natural, environmental, and social effects. The global climate change discussions are based on the fossil fuel greenhouse gas (GHG) emission to the lower atmospheric layers due to anthropogenic activities. The climate change effect is expressed by different researchers and even common men, but the intensity of such a change cannot be determined with certainty over the coming time epochs. Statistical regularity implies further complete unpredictability for single or individual events.

The importance and implication of the statistical methodological applications in the engineering system planning, operation, estimation, projection, and simulation works are of utmost significance, because they digest uncertainty component in any measurement or ensemble spatial and time series objectively with reliable results. Statistics deals with uncertain data and identifies data features for practical applications. These methods help to design experimental data collection networks, registered data feature identifications, employment of correct analyses with effectively interpretable results. There is no any other way to assess the risks for future events except by means of the statistical techniques. They are helpful to deduct scientific discoveries, data-based proper decision-making and predictions. Statistics help to understand the phenomenon concerned more deeply based on the measurements.

Apart from the benefits from any statistical analysis, there are also pitfalls that should be kept in mind to get away from them. The overall quality of the statistical results depends on proper selection and measurements of data, sampling technique, sample size, and selection of the most convenient statistical methodology. Any slight deviation from these principles may lead to unreliable results even though one may think that s/he did the best. Among the pitfalls, the following points must be kept in mind during any statistical modeling.

1. *Bias effect*: Samples must be collected in such a way that any bias effect must be outside the procedure. For this purpose, the data must be checked against any bias such as systematic errors, mismeasurement, nonstandard measurement, which means that data measurements must be done by means of the same instruments and in any instrument check the new one (or the old one) must be calibrated with each other.
2. *Causality effect*: Prior to the decision on the methodology and its application, the researcher should have a sound knowledge about the causality, that is, the relationship between any pair of input and output variables. For this purpose, at least s/he should think about two principles deeply. One is the proportionality (directly or indirectly) relationship and then the next step is to get some impression about the linearity or nonlinearity about the relationships (Chap. 3). This is very essential for successful application of any statistical models such as single or multiple regressions that are very often used in many studies.
3. *Correctness effect*: Since most statistical analysis search for relationships, then one should be able to decide about one input or several input variables. In almost all statistical methodological models, practically the mean value is considered, but the best is to consider the mode value and at some instances the median value. The researcher may decide about this point by looking at the histogram of each variable. Another pitfall is linear fit to nonlinear relationships.
4. *Overgeneralization effect*: The results obtained from one set of data cannot be generalized to another set of data, because they may have different behaviors and hence different population models. One should be careful about the distinction of population behaviors in terms of their most suitable theoretical PDFs.
5. *Limitation effect*: Statistical inferences cannot be generalized easily, because they have limitations and one should appreciate the limitation features of any statistical models.
6. *Assumption effect*: In statistics, almost all methodologies have restrictive assumptions about the models, sample properties, and variables. In general, "central limit theorem" is taken into consideration, but its consideration has some specific conditions that should be cared for. Violation of one of the assumptions may result in misleading conclusions.

7. *Data effect*: If everything is applied correctly, the results may be incorrect, depending on a single measurement set. With different data sets the model must be run, and the results must be addressed to significance tests.
8. *Parameter effect*: There are different methodologies to estimate model parameters such as moments, maximum likelihood, and like procedures. However, in practical applications, the most trustable one is the least squares minimization, which provides the best result.

The points may be hidden in ready software, and therefore, the output from any software must be questioned under the light of these points prior to the reliable conclusion's acceptance.

4.12 Basic Questions Prior to Data Treatment

Statistical tools are means of extraction and condensation of information from assembled relevant data for different purposes. Data processing by the statistical tools depends on some restrictive assumptions or hypothesis. Although their use is rather straightforward and easy to apply, the user must have statistically practical information about each statistical methodology. Statistical tools are useful to describe the whole data set in a descriptive and better manner with plausible physical interpretation at the beginning. It is very important to keep in mind that any statistical parameter is attached with some physical behavior of the phenomenon, which gives rise to global values in terms of measurements. Further detailed analysis of the phenomenon can be achieved only through these preliminary examinations of the data set. Most often the interest is generally not in the individual values, but in the collective description of the phenomenon from which they are obtained. Examination of a data set in terms of its various statistical parameters may yield general ideas about joint characteristics of the variables. The statistical parameters represent the notable features of the data. The following questions are the most frequently encountered types in any data processing treatment whatever the type of available data.

1. What is the time average of a phenomenon at a desired location? What are the deviations from this average value? What are the possible levels of extreme values at this site and what are their risks of occurrence?
2. What is the general areal behavior of all the available stations in a region during a specific time period such as day, month, season, and year? Can these behaviors be shown in the form of maps?
3. What are the possible effective factors and how they are related to each other?
4. Is it possible, based on historic records, to make short-term future possible predictions from the available data?
5. Is the same type of variable data at different locations like each other behaviorally? Are they homogeneous or heterogeneous?

6. How one can calculate various possible periodical variations within a data set?
7. Are there trends (increasing or decreasing) within a given time series? If there is then how these can be detected and identified from the overall data sequence?

In order to answer to these questions rather than considering data values individually, relevant statistical parameters must be calculated for better deductions both verbally and numerically.

4.13 Simple Probability Principles

There is a long tradition in science and engineering to turn to probability theory when faced with a problem in which uncertainty plays an important role. This tradition was right when there were no tools for dealing with uncertainty other than those provided by probability theory, but today, this is no longer the case. What is not commonly realized is that there are many aspects of uncertainty that cannot be adequately addressed using standard probability theory. The prevailing view, let us call it adequate opinion, is that probability theory is all it takes to deal with any kind of uncertainty. In the words of Lindley (1923–2013), a well-known Bayesian, the belief of the competencies can be expressed as

> The only satisfactory definition of uncertainty is probability. By this I mean that every uncertainty expression must be in the form of a probability; that several uncertainties should be combined using probability rules; and that the calculus of probabilities is sufficient to handle all situations involving uncertainty... probability is the only logical definition of uncertainty and is sufficient for all problems involving uncertainty.

Anything that can be done with fuzzy logic, belief functions, upper and lower probabilities, or anything alternative to probability can be done better with probability.

There are those who feel that this gap reflects a fundamental inability of traditional mathematics—the mathematics of precisely defined points, functions, and sets, probability measures—to deal effectively with the analysis of biological systems. With such systems, which are often orders of magnitude more complex than man-made systems, we need a radically different kind of mathematics, the mathematics of fuzzy or cloudy quantities that cannot be defined in terms of probability distributions (Chap. 3). Indeed, the need for such mathematics is becoming more and more evident even in the field of inanimate systems, since in most practical situations a priori data is as imprecise as the criteria by which the performance of a man-made system is judged by specified or precisely known PDFs. (Zadeh 1965).

4.14 Statistical Principles

Our minds are preconditioned to Euclidean geometry, and as a result ordinary people must think of length, breadth, and depth in three-dimensional space in addition to time as the fourth dimension. All engineering, earth, environmental, and atmospheric sciences variables change along these four dimensions. If variations in time are not considered, it is said to be frozen in time and thus remains evolving along the time axis, but the variable in question has variations across space. A good example of such a change can be considered as geological events that change over million years. The iron content of a rock mass varies quite randomly from one point in the rock to the next, and spatial variation is therefore considered. Another example is the measurement of precipitation amounts in many meteorological stations at irregular locations spread over an area, that is, simultaneous measurement of precipitation amounts, again by freezing time to search for spatial variations, and hence, maps are prepared.

On the other hand, there are also temporal variations in natural phenomena. For such a change, it is enough to measure the event in a certain place, as in any weather station. Depending on whether the event is continuous over time, and hence, time series records can be obtained. A time series is the systematic measurement of any natural event at regular time intervals along the time axis. Depending on this time period, time series are called hourly, daily, weekly, monthly, or yearly time series.

Unlike time changes, it is not possible to think of space series in which records are kept at regular intervals, except in very special cases. For example, if a water sample is taken every 1 km along a river, the measurements provide a series of distances in terms of regularity. In fact, distance series are very limited, as if there was no such data series. On the other hand, depending on the relevance of the event, there are series that contain time data, but they are not time series due to the irregularity or randomness in the time intervals between successive occurrences of the same event. Flood, drought, and earthquake events correspond to such situations. In general, future behavior of these phenomena cannot be predicted exactly.

Any natural phenomenon develops in the four-dimensional field of human visualization, and as a result, its records must contain features of both temporal and spatial variability. Any record with this property is called spatiotemporal random variable.

4.15 Statistical Parameters

Parameter is a single value calculated from the whole of the available data value. It reflects a certain feature of the records in a single value. The position, dispersion, and many behaviors depend upon one or more constants called parameters. The calculation of parameters such as averages and measures of variability are related to statistical measures. A sample function (SF), that is, data series can be summarized by a set of parameters.

The statistical parameter is a function of all the data values in some way and any variation in any data value causes changes in the parameter estimation. The longer the SF, the smaller is the effect of data variation onto the parameter value. Each parameter gives answer to some question concerning the SF behavior. Some of the questions are as follows:

1. About what average level the SF indicates fluctuations?
2. What is the significance of deviations from the average value? And how they can be expressed for each SF?
3. How can one make the comparison of two or more SFs?
4. Are most data below the average or above or they are equal? If this is the case then the fluctuation of a SF can have symmetrical situation. If data numbers below and above average are not equal, then the SF is skewed.
5. Which of the data value(s) has the most frequently occurring case within the whole SF?
6. What is the data value that includes half of the data numbers above and the other half below?
7. Is there any effect of subsequent data values on each other? Or are they completely ineffective among themselves?
8. Is there any effect of time on data measurements? Is some part of data along the time axis a function of time? This implies existence of trends in SFs.
9. What are the extreme values and the range of data values change?
10. What portion or percentage of data falls within an interval of interest for some design purpose? What is the percentage of data that are greater (smaller) than a desired value?
11. Are there any geographic, astronomical, or environmental deterministic effects on the SF?
12. Is there any interference between two SFs?
13. Similar questions may be asked for a definite period (month, season) and subarea of interest.

In the following sequel, the parameters for answering such questions are presented with explanations.

4.15.1 Central Measure Parameters

Arrangement of different data values, X_i ($i = 1, 2, \ldots, n$), on a scaled axis shows rather randomly scatter with a dense central location as in Fig. 4.10. Hence, there is a central value that tells the general whereabouts of the data density location. In order to describe such dense portions of the data on the scaled axis, it is necessary to have measures that are referred to as the measures of location or central measurement parameters. Dense location is quantified as a single value of the measured variable, which in some way is a brief representative of the data.

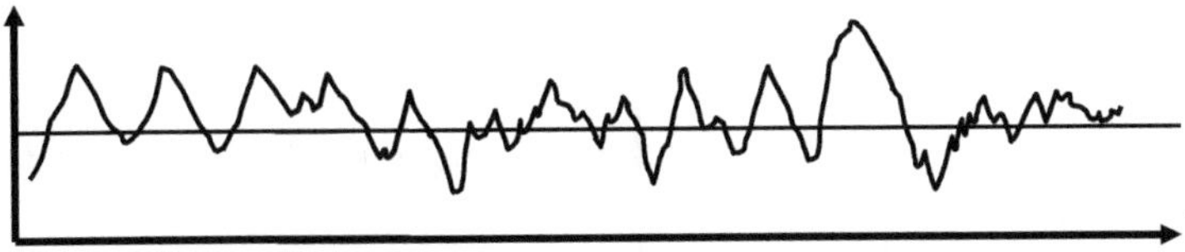

Fig. 4.10 Central level and deviations in sample function

This figure indicates the dense value and the data deviations from it with positive and negative values. In the following sequel, all the statistical parameters are defined by the dense value and the deviations from this value.

4.15.1.1 Arithmetic Mean

It shows the constant level around which the data values fluctuate. It is calculated as the summation of all data values divided by the number, n, of data as

$$\overline{X} = \frac{1}{n}\sum_{i=1}^{n} X_i \tag{4.16}$$

It implies that when the arithmetic average is multiplied with the number of data, $n\overline{X}$, the result is the summation of data. The arithmetic average uniforms all the data without giving any clue about the deviation scale. Eq. (4.16) gives the same weight to each data value without distinction. Hence, one or two extreme values may dominate the result, and therefore, the arithmetic average concept may not be the best representative central value for the whole data. For instance, the data samples in Fig. 4.11 cannot be represented by the arithmetic averages. In Fig. 4.11a, b the SF is not stable and there are few extreme values, which affect the arithmetic average calculation. The extreme values depart the arithmetic average from the dense region of the data variation. In Fig. 4.11c, there are two dense values of data variation and again the arithmetic average cannot represent whole data due to a sudden change. If there are seasonal fluctuations within the data structure as in Fig. 4.11d, again the arithmetic average cannot be representative. Finally, if there are increasing, decreasing linear or nonlinear trends in the data, arithmetic average must not be used as the representative central tendency value. All these examples indicate that prior to the use of the arithmetic average, it is necessary to inspect data visually in the form of SF and then decide about the feasibility of the arithmetic average usage. In many applications, the researchers do not take these points into consideration, and therefore, their conclusions are biased. One can deduct from all these discussions that in order to have a representative arithmetic average value the following points must be isolated from the data prior to its use:

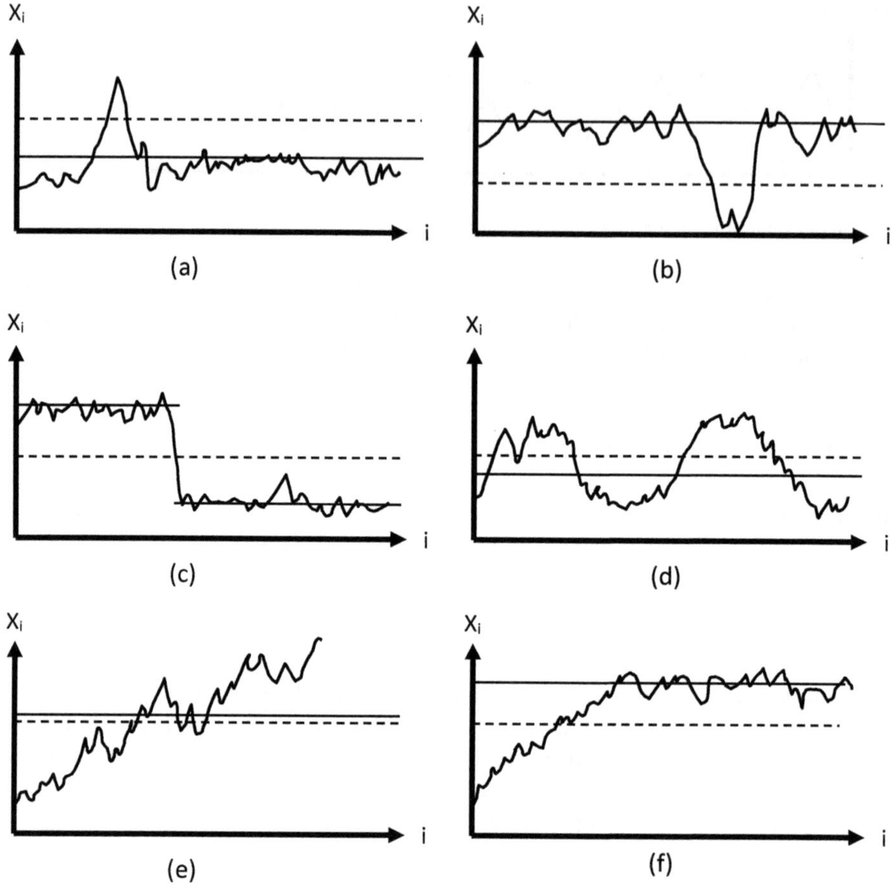

Fig. 4.11 Unreliable arithmetic averages

1. Extreme values.
2. Trends.
3. Sudden or gradual jumps.
4. Periodic fluctuations like seasonality.

The arithmetic average has the advantage of being simple and it is easy to interpret when the fluctuations around the dense value has a symmetric behavior. Otherwise, the arithmetic average must not be used in practical studies. Its virtue is more in the theoretical studies, where the fluctuations are considered as symmetric, which is reflected in a symmetric histogram or a normal (Gaussian) PDF as will be explained later in this chapter. This condition also implies first-order stationary data structure.

Other practically significant point about the arithmetic average is that the summation of deviations from this value is equal to zero whatever the data structure (including extreme values, trends, etc.). This can be written arithmetically from Eq. (4.16) as

$$\sum_{i=1}^{n}\left(X_i - \overline{X}\right) = 0 \tag{4.17}$$

Hence, the first measure of variability in the form of difference is encountered in this expression. Each term in this expression indicates the similarity or dissimilarity of data value to the arithmetic average value. The bigger (smaller) is the difference between any data value and the data arithmetic average, the bigger (smaller) the dissimilarity (similarity), that is, the bigger (smaller) is the variability.

Another practical view of the arithmetic average may be presented as a shifting operator as a result of subtraction from each data value the arithmetic average as in Eq. (4.17), because the whole data can be thought of varying without the geometrical change in the SF; it may be considered as fluctuation around the zero horizontal axes as shown in Fig. 4.12.

Figure 4.12a indicates the fluctuation of the data around its arithmetic mean whereas in Fig. 4.12b, the same sample function without any structural change fluctuates around the zero horizontal axes. The shifting procedure helps to work with small numbers, if the data have big values.

The consideration of the above-mentioned explanations leads to the following points about the calculation of the arithmetic average.

1. It is a linear transformation of data values to a single value (see Eq. 4.16).
2. If a constant value, C, is added or subtracted from each one of the data values and a new data series is obtained as $Y_i = X_i \pm C$, then the arithmetic average of this new series is $\overline{Y} = \overline{X} \pm C$. According to this rule, the arithmetic average after the shifting as $Y_i = X_i \pm \overline{X}$ becomes $\overline{Y} = \overline{X} \pm \overline{X} = 0$.

The graphical representation of such transformations is shown specifically in Fig. 4.13.

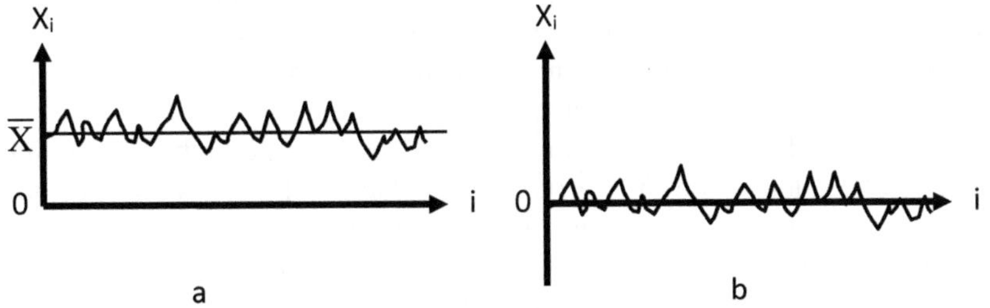

Fig. 4.12 Normal and shifted sample functions

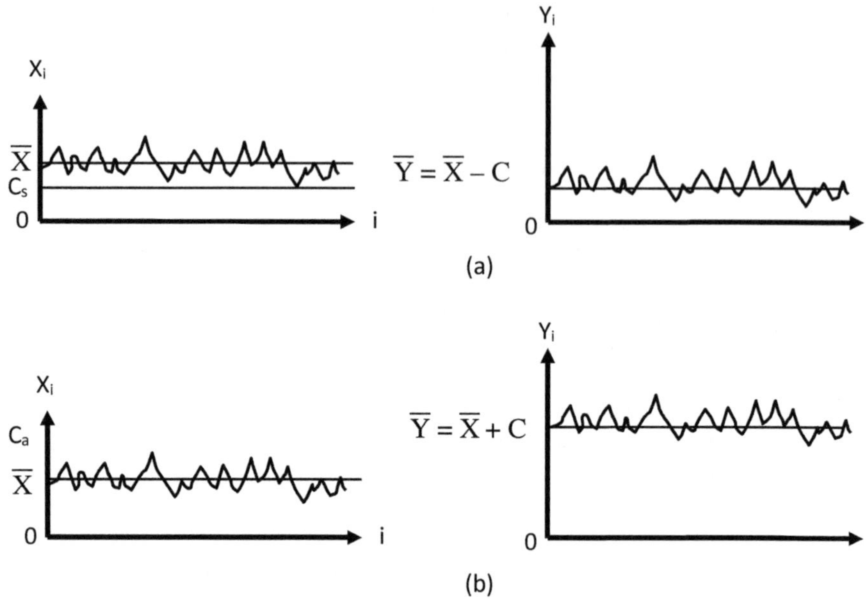

Fig. 4.13 General shifting operations (**a**) subtraction (**b**) addition

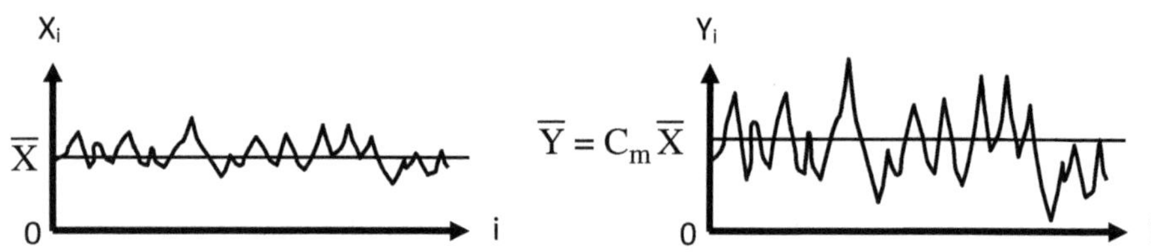

Fig. 4.14 Multiplication operation

If the data values are multiplied by a constant, C, leading to a new series as $Y_i = CX_i$, then the arithmetic average of the new series due to the linearity property of the arithmetic average calculation becomes $\overline{Y} = C\overline{X}$. As it is obvious from Fig. 4.14 after the multiplication with a constant value, the data fluctuations around its new arithmetic average level increases.

In the case of data division by a constant, C, as $Y_i = X_i/C$ the new arithmetic average can be calculated as $\overline{Y} = \overline{X}/C$. Similar shape to Fig. 4.14 will be obtained but with smaller deviations around the new arithmetic average value. Another benefit from the division operation is that the new data values become dimensionless so that different data types can be compared with each other.

Finally, provided that the transformation remains linear, it is possible to shift and scale any given data according to constant subtraction, C_s, and constant division, C_d, then the following formulation becomes valid:

$$Y_i = \frac{X_i - C_s}{C_d} \tag{4.18}$$

Its arithmetic average is

$$\overline{Y} = \frac{\overline{X} - C_s}{C_d} \tag{4.19}$$

This last transformation helps to obtain a new and dimensionless data series, which is referred to as statistical standardization and this expression will be used several times in this chapter.

4.15.1.2 Median

The dense data level may be sought based on data numbers rather than data values such that the number of data that falls above and below such constant value are equal, which is equal to n/2. If the number of data is odd then the median, X_{med}, value is equal to the value in the middle of the list of data that is arranged in increasing order. On the other hand, if the data number is even, then the arithmetic average of the two middle values is adopted as the median. The very definition of median value implies that the percentage of data values greater or smaller than median is equal to 50%. Especially, in the case of extreme value existence within the data, the median presents more reliable central measure because it is not affected by extreme values (see Fig. 4.15).

In this figure, since lower extremes are existent the arithmetic average value is drawn downwards whereas the median value is the sole representative for the dense level of data. The median has the same unit as the original data, and its comparison with the arithmetic average value provides the following practically useful information:

Fig. 4.15 Median value

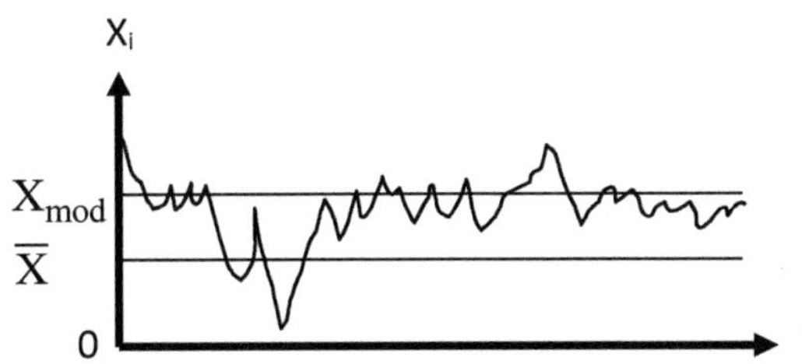

1. If the median is greater (smaller) than the arithmetic mean value, $X_{med} > \overline{X}$ ($X_{med} < \overline{X}$), then there are more (less) data above (below) the mean value than below (above) it. This implies that the dense data is above the arithmetic value. Both cases imply asymmetric nature of the data distribution within the variation range and the unreliability of the arithmetic average value.
2. If the median is equal to the average value theoretically, $X_{med} = \overline{X}$ then there is no difference in the use of the median or the arithmetic average, but due to its simpler calculation procedure, the arithmetic mean calculation is preferable. This condition also implies that the distribution (fluctuation) of data around arithmetic mean value is symmetrical and possibly also stationary if there are no trends within the data structure.

4.15.1.3 Mode

It is the most frequently occurring value in a sample or population either individually or within a certain subclass in the data variability domain. This should be the most preferable central measure of data in any application. It is not preferably directly due to difficulty in its calculation compared to previous central measure parameters.

If $X_{med} = \overline{X}$, then the mode is ought to be equal to these values. Otherwise, if $X_{med} > \overline{X}$ then $X_{mode} > X_{med} > \overline{X}$ or when $X_{med} < \overline{X}$ then $X_{mode} < X_{med} < \overline{X}$.

4.15.1.4 Percentiles

It has already been stated that the median value divides the data into two equal halves, that is, 50%. Likewise, it is possible to generalize this concept by asking the question: What is the percentage of data in more refiner percentages? If the refined percentage is 25%, it is called as quartiles, when 10% then the name is deciles; however, for 1% they are referred to as percentiles. This is a very fine data search and interpretation method that is not used very frequently in the practical studies. Instead, the frequency, relative frequency, and histogram concepts are employed, and these will be explained later in the chapter.

4.15.1.5 Statistical Weighted Average

This average attaches a certain weight to each data point and to a subgroup of samples. For example, in Fig. 4.11c, there are two different levels with each one having its representative arithmetic average, which is repeated in Fig. 4.16.

Let the arithmetic averages be $\overline{X}_1$ and $\overline{X}_2$ with their effective subgroup numbers of data as n_1 and n_2, respectively. In this case, the weighted average, $\overline{X}_w$, can be expressed as

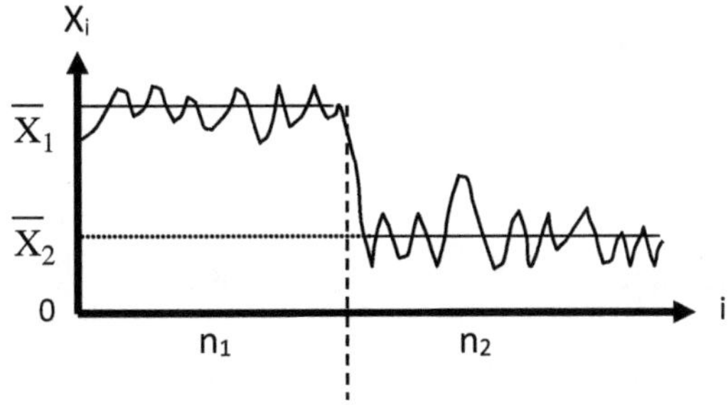

Fig. 4.16 Median value

$$\overline{X}_w = \frac{n_1\overline{X}_1 + n_2\overline{X}_2}{n_1 + n_2} \tag{4.20}$$

or

$$\overline{X}_w = \frac{n_1}{n_1 + n_2}\overline{X}_1 + \frac{n_2}{n_1 + n_2}\overline{X}_2$$

By defining the percentage of data as $p_1 = n_1/(n_1 + n_2)$ and $p_2 = n_2/(n_1 + n_2)$ the final form becomes

$$\overline{X}_w = p_1\overline{X}_1 + p_2\overline{X}_2 \tag{4.21}$$

which means that the weighted average may have the preference attachments according to the percentage (probability) of data occurrence out of the whole data set. Hence, if there are m sub-data groups, the weighted average can be generalized as

$$\overline{X}_w = p_1\overline{X}_1 + p_2\overline{X}_2 + \ldots + p_m\overline{X}_m \tag{4.22}$$

It is obvious that

$$p_1 + p_2 + \ldots + p_m = 1 \tag{4.23}$$

In practical studies, each one of the percentages can be given a different context depending on the problem at hand. For instance, in the calculation of areal average of spatial data, each measurement location should have an influence area as $A_1, A_2, \ldots, A_m$ as shown in Fig. 3.9.

4.15.1.6 Physical Weighted Average

Figure 4.17 shows a series of layers, where through each fluid heat quantity flows in the horizontal direction without any exchange between adjacent layers. The horizontal flow with total quantity, Q_H, movement is considered as output. Since no

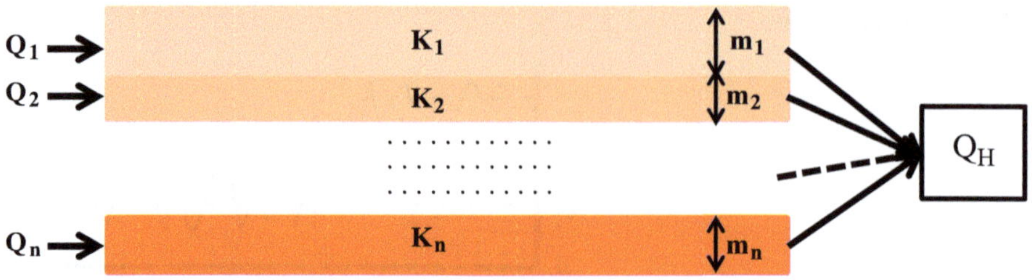

Fig. 4.17 Perpendicular fluid flows through the layers

quantity is gained or lost in passing through the various layers, the principle of mass conservation leads to the summation of quantities Q_1, Q_2, Q_3, … , Q_n in the individual layers as

$$Q_H = \sum_{i=1}^{n} Q_i \tag{4.24}$$

On the other hand, for unit width (w = 1) of the aquifer cross section the individual discharge is expressed as Darcy's law (Şen 1995),

$$Q_i = K_i m_i h \qquad (i = 1,\ 2,\ \ldots,\ n) \tag{4.25}$$

where K_i is the hydraulic conductivity of layer i and h is the hydraulic gradient along each layer.

Substitution of Eqs. (4.25) into Eq. (4.24) leads to

$$Q_H = h \sum_{i=1}^{n} K_i m_i \tag{4.26}$$

The specific discharge in the horizontal direction is $q_H = Q_H/A = Q_H/\sum_{i=1}^{n} m_i$, and therefore the substitution of Eq. (4.26) into this expression yields

$$q = \frac{\sum_{i=1}^{n} K_i m_i}{\sum_{i=1}^{n} m_i} \tag{4.27}$$

This last expression implies that the horizontal hydraulic conductivity is the weighted average of the individual hydraulic conductivities with layer thickness being the weights. It is interesting to notice that if the layers have the same thickness, then Eq. (4.27) takes the form as

$$q_H = \frac{1}{n} \sum_{i=1}^{N} K_i \tag{4.28}$$

in which n is the number of layers. This last expression implies that only in the case of equal layer thicknesses, the harmonic average is equivalent to the arithmetic average; otherwise, this statement is never valid even approximately for unequal layer thicknesses. If the thicknesses are not equal, the arithmetic averaging leads to under estimation.

4.15.1.7 Geometric Average

It is the nth root of the products of the n data values. Conversely, it is the exponential of the arithmetic average of the logarithms of data values, which can be written as

$$\overline{X}_G = \sqrt[n]{X_1 X_2 \ldots X_n} \tag{4.29}$$

By taking the logarithm of two sides, one can see that

$$\overline{X} = \ln\left(\overline{X}_G\right) \tag{4.30}$$

or

$$\overline{X}_G = \exp\left(\overline{X}\right) \tag{4.31}$$

Here, the arithmetic mean is expresses as

$$\overline{X} = \frac{1}{n}\left(\ln X_1 + \ln X_2 + \cdots + \ln X_n\right) \tag{4.32}$$

which implies that it is the arithmetic average of the logarithms of the data values. The following significant points are valid prior to the geometric mean calculation.

1. All the data values must be greater than zero. If among the data there is even one value close to zero, the geometric mean is not be preferable.
2. The geometric mean is always smaller than the statistical arithmetic mean, that is, $\overline{X}_G < \overline{X}$.

Another common area for the use of the geometric mean is when the logarithmically distributed data is transformed into a new variable with the hope that its distribution is normal (Gaussian) PDF. If the original data are X_i with logarithmic distribution, then $Y_i = \ln X_i$ will have a normal distribution function. Variables whose logarithms follow normal distribution are said to be log-normally distributed whose mean value is expressed by the geometric average.

4.15.1.8 Harmonic Average

This is used in the case of averaging process when the phenomenon concerned has unit with respect to some reference system like time or space and the requirement of the problem may necessitate its use. In order to familiarize the reader, let us consider the cases of vertical quantity, Q_V, flow through n sedimentary layer sequence as in Fig. 4.18. Let the thickness and hydraulic conductivities be $m_1, m_2, m_3, \ldots, m_n$ and $K_1, K_2, K_3, \ldots, K_n$, respectively, where n is the number of different layers in a multiple aquifer such as in Fig. 4.18. First the horizontal flow with total discharge Q_V is considered.

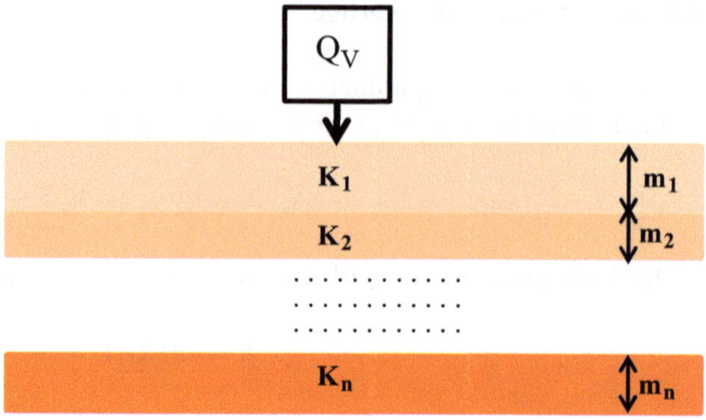

Fig. 4.18 Perpendicular fluid flows through the layers

Since no water is gained or lost in passing through the various layers, the principle of mass conservation can be used. In that, the total discharge is equal to the summation of discharge quantities Q_1' Q_2, Q_3, … ,Q_n in the individual layers. The overall hydraulic gradient is equal to the summation of the individual hydraulic heads divided by the total thickness (see Fig. 4.18).

$$i_V = \frac{H_V}{m} = \frac{\sum_{i=1}^{n} H_i}{\sum_{i=1}^{n} m_i} \tag{4.33}$$

Each layer allows passage of the same total vertical discharge Q_v, and therefore the specific discharge in each layer, q_v, is the same. The application of Darcy's law (Eq. 4.25) in each layer gives the individual head losses as

$$\begin{aligned} H_1 &= \frac{m_1}{K_1} q_v \\ H_2 &= \frac{m_2}{K_2} q_v \\ &\dots \\ H_n &= \frac{m_n}{K_{n1}} q_v \end{aligned} \tag{4.34}$$

Substitution of Eq. (4.34) into Eq. (4.32) yields

$$q_V = \frac{\sum_{i=1}^{n} m_i}{\sum_{i=1}^{n} \frac{m_i}{K_i}}$$

which is tantamount to saying that the vertical hydraulic conductivity, K_v, equation appears as follows:

$$K_V = \frac{\sum_{i=1}^{n} m_i}{\sum_{i=1}^{n} \frac{m_i}{K_i}} \tag{4.35}$$

Most frequently, multiple layers occur in the sedimentary rocks. In fact, they are often anisotropic with respect to hydraulic conductivity because they contain grains, which are not spherical but elongated in different shapes. During the deposition, these grains settle with their longest axes horizontally, and this usually causes the horizontal hydraulic conductivity to be greater that vertical conductivity in a single layer. However, when many layers are considered, then the bulk hydraulic conductivity of sediments is usually much greater than vertical counterpart. In order to prove this point from the previous equations by considering two layers only, the ratio K_H/K_v can be written as

$$\frac{K_H}{K_V} = \frac{m_1^2 + \left(\frac{K_1}{K_2} + \frac{K_2}{K_1}\right)m_1 m_2 + m_2^2}{m_1^2 + 2m_1 m_2 + m_1^2} \tag{4.36}$$

The hydraulic conductivity ratio terms in the denominator are always greater than 2, since if x is any number then x + (1/x) > 2. Consequently, always $K_H > K_v$. It is also true for alluvial deposits, which are usually composed of alternating layers or lenses of sand and gravel on occasional clays.

4.15.2 Deviation Parameters

These are also called as the dispersion parameters, which indicate the average deviation different powers from the arithmetic average. In all the deviation parameters, the statistically based arithmetic average is used as the reference for the deviation and hence the measure of variability.

4.15.2.1 Data Range

The range, R, is defined as the difference between the maximum, X_{max}, and minimum, X_{min}, data values:

$$R = X_{max} - X_{min} \tag{4.37}$$

The smaller the range, the better is the prediction of the variable concerning its future behaviors. All the data values are within the range 100%. The median, m, value divides the range into two equal parts and 50% of the data are above this value. As mentioned earlier, this is the probability of data values being greater or smaller from the median. This implies that the number of data within the intervals, $X_{max} - m$ and $m - X_{min}$ is the same and equal to n/2, where n is the number of data. The median divides the range into two equal halves as in Fig. 4.19.

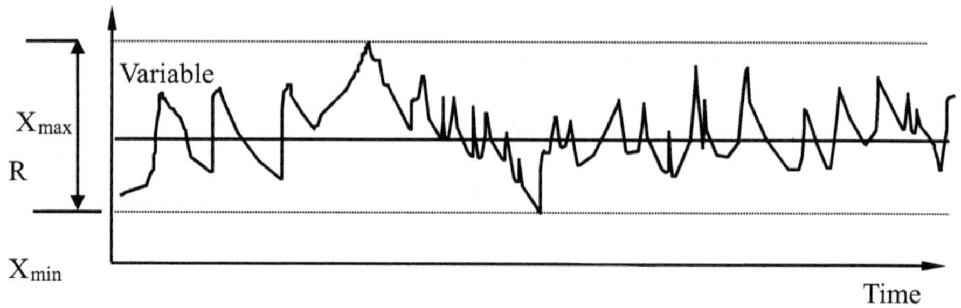

Fig. 4.19 Range

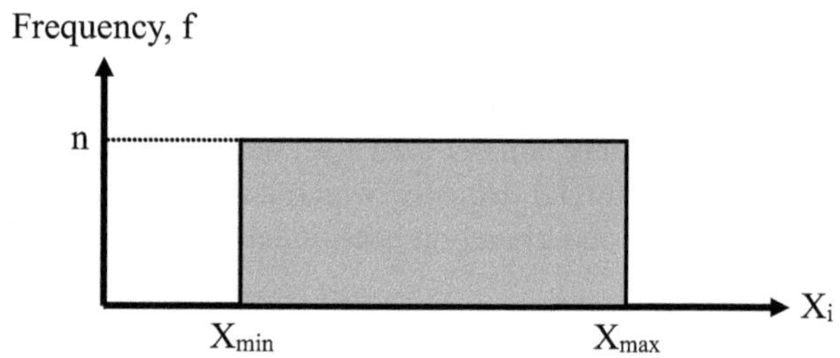

Fig. 4.20 Range-frequency diagrams

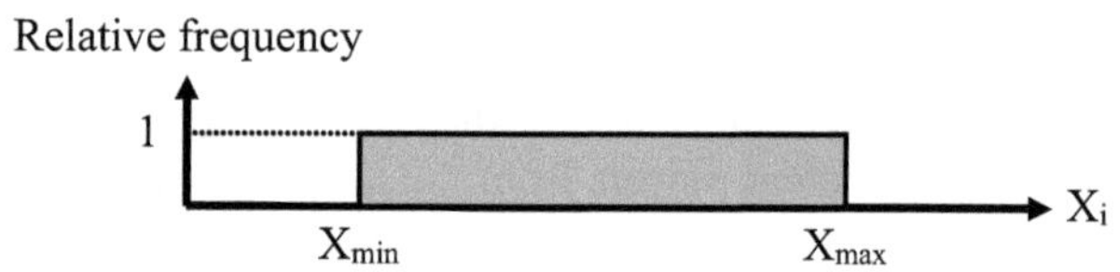

Fig. 4.21 Range-relative frequency diagrams

Since all the data are within this range, the number of data frequency is equal to n. This can be presented as a rectangle with base length, R, and height n according to crisp logic as in Fig. 4.20. Unfortunately, this frequency diagram does not tell anything about the internal data variability.

In order to have data number independent graph, the relative frequency is defined as the ratio of data number that falls within a range to the number of data. Since, here, the range is the data variability range, the relative frequency by this definition becomes equal to 1 and the corresponding relative frequency diagram is presented in Fig. 4.21.

4.15.2.2 Deviations

In statistical parameter definitions, the deviations are always defined from the arithmetic average value. It is already stated that the summation of the deviations is equal to zero (see Eq. 4.17). Hence, such a summation cannot be the measure of the deviations in the data sequence, since positive deviation summation is always equal to the negative deviations summation and then cancel each other. In order to have the simplest definition for the variation measure, it is possible to consider the summation of positive deviations and the absolute value of the negative deviations, which indicates that the smaller is this absolute summation the smaller is the data variability. On the other hand, it is also possible to measure the overall data variability by considering the maximum of the absolute deviations from the mean value. Again, the smaller is this maximum, the smaller is the data variability. Both definitions require the absolute value sign, which is rather difficult to consider in the mathematical derivation and integration operations.

4.15.2.3 Sum of Square Deviations

In order to alleviate the mathematical manipulations, it is convenient to consider the sum of the square deviations from the arithmetic average for the data variability. In practice, the purpose is to have as possible as small sum of square deviations, which is referred to as the least square analysis. The sum of squares, SS, is defined as

$$SS = \sum_{i=1}^{n} \left(X_i - \overline{X}\right)^2 \qquad (4.38)$$

The unit of SS is square of the unit of the original data.

4.15.2.4 Variance

The SS increases as the number of data increases, but it would be preferable to have the average measure of variation. Hence, the arithmetic average of the square deviations is referred to as the variance, V(X),

$$V(X) = \frac{1}{n} \sum_{i=1}^{n} \left(X_i - \overline{X}\right)^2 \qquad (4.39)$$

This is a nonlinear (power, parabolic) transformation of the data deviations from the arithmetic average. It is equal to zero provided that each data value is equal, which means that there is no data variability at all. The extreme values affect the variance in the same manner as it was for the arithmetic average value.

4.15.2.5 Skewness Coefficient

Like variance definition, it is possible to increase the power by one, and hence obtain various features of the data. Hence, the coefficient of skewness, $\gamma(X)$, is defined with third power as

$$\gamma(X) = \frac{1}{n} \sum_{i=1}^{n} \left(X_i - \overline{X}\right)^3 \qquad (4.40)$$

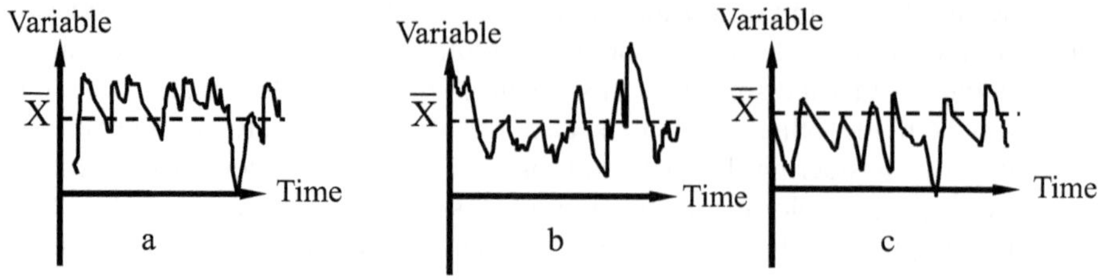

Fig. 4.22 Skewness coefficients (**a**) positive, (**b**) symmetric, (**c**) negative

This is also a nonlinear deviation transformation, which magnifies the effects of high deviations. The question of what percentage of data amount remains above and below the arithmetic average value can be answered by the skewness coefficient, γ, definition. This is simply defined as

$$\gamma = 1 - 2\frac{n_b}{n} \tag{4.41}$$

where n_b is the data number below the arithmetic average. This parameter varies between −1 and +1, and if it is equal to zero then there is no skewness in the SF. In any data sequence, if the number of data points above the arithmetic average is more than the below, this implies that extreme events will take place in the future. The above definition of skewness does not take into consideration whole the data values and the magnitudes of data but only the numbers. However, a better definition of skewness coefficient is given as follows:

$$\gamma = \frac{1}{n}\sum_{i=1}^{n}\left(X_i - \overline{X}\right)^3 \tag{4.42}$$

As explained before, like arithmetic average, the cubes of the deviations may have + or – values. Consequently, the skewness coefficient may be +, 0 or −. If equal to zero, this means that the fluctuations around the arithmetic average value are almost symmetric. In the case of positive (negative) skewness coefficient, the magnitude of deviations from the average is more dominant above (below) the arithmetic mean. The three positions of the skewness coefficient are presented in Fig. 4.22.

4.15.2.6 Kurtosis Coefficient

This statistical coefficient, δ(X), is defined as the fourth power of deviations from the arithmetic average:

$$\delta(X) = \frac{1}{n}\sum_{i=1}^{n}\left(X_i - \overline{X}\right)^4 \tag{4.43}$$

This indicates the spikiness of the data variability. The greater the kurtosis coefficient, the more concentrated is the data values around the arithmetic value.

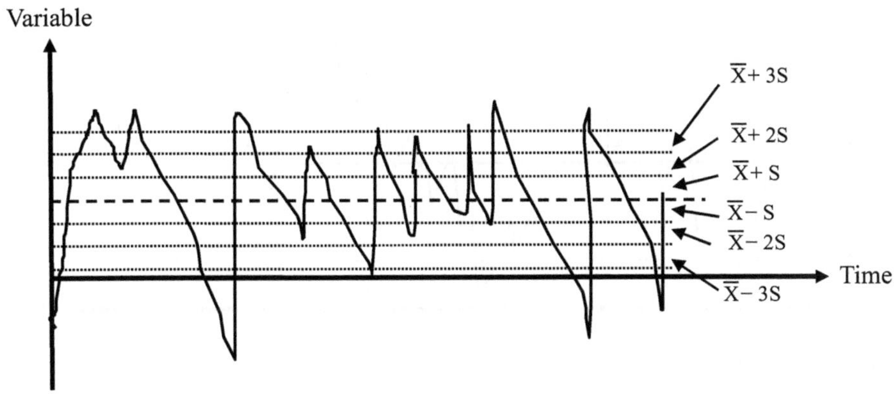

Fig. 4.23 Standard deviation

4.15.2.7 Standard Deviation

The variance has the unit of the original data unit square and therefore it is not possible to compare the data deviations from the arithmetic average with this statistical coefficient. In order to achieve this comparison, the square root of the variance can be taken, and it is the definition of standard deviation S(X) as

$$S(X) = \sqrt{\frac{1}{n}\sum_{i=1}^{n}\left(X_i - \overline{X}\right)^2} \tag{4.44}$$

The standard deviation gives an idea about the total deviation around the arithmetic average by considering 1, 2, and 3 standard deviation limits around the arithmetic average as in Fig. 4.23.

It is obvious that most of the data fall within the first standard deviation range, namely, between $\overline{X} - S_X$ and $\overline{X} + S_X$. Standard deviation has a very significant role in predictions. The smaller the standard deviation of a SF, the better will be the predictions.

In the statistical studies, the deviations from the arithmetic average for a given time series data $(X_i) - \overline{X}$ is very significant. Such deviations constitute the basic definitions of variance, covariance, correlation coefficients, and the regression method. It is possible to categorize the overall time series based on standard deviation distances above and below the arithmetic average as in Fig. 4.24.

In this figure, σ indicates the standard deviation of the population time series. With 1, 2, and 3 standard deviation limits, a given time series may be viewed in seven categories as in the following table with different specifications (Table 4.3).

In practice, most of the time series values fall within the normal limits with extreme values outside, above, and below normal extreme limits. In the case of normal (Gaussian) PDF, consideration of 1, 2, 3, and 4 standard deviation values around the arithmetic mean leads to the following numerical percentages.

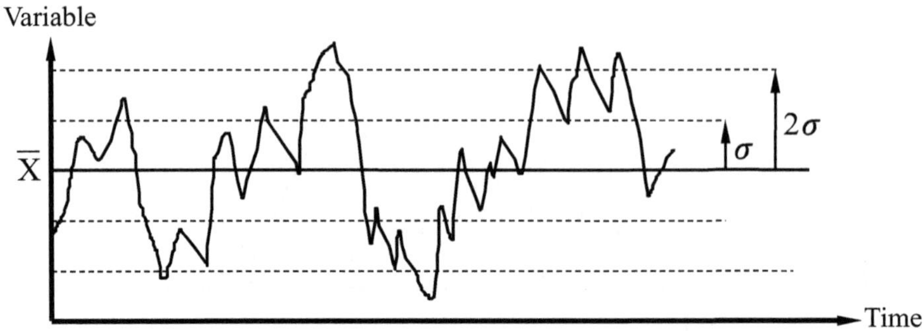

Fig. 4.24 Standard deviation truncations

Table 4.3 Truncation levels and specifications

Truncation	Specification
$\overline{\mathbf{X}} + 3\sigma < X_i$	Above normal extreme
$\overline{\mathbf{X}} + 3\sigma < X_i < \overline{\mathbf{X}} + 2\sigma$	Rather supernormal extreme
$\overline{\mathbf{X}} + 2\sigma < X_i < \overline{\mathbf{X}} + \sigma$	Above normal
$\overline{\mathbf{X}} + \sigma < X_i < \overline{\mathbf{X}} - \sigma$	Normal
$\overline{\mathbf{X}} - \sigma < X_i < \overline{\mathbf{X}} - 2\sigma$	Below normal
$\overline{\mathbf{X}} - 2\sigma < X_i < \overline{\mathbf{X}} - 3\sigma$	Rather subnormal extreme
$X_i < \overline{\mathbf{X}} - 3\sigma$	Below normal extreme

In interval,

$-1 < x_i < +1$, 68.269 percentage, that is, with probability 0.68269
$-2 < x_i < +2$, 95.450 percentage, that is, with probability 0.95450
$-3 < x_i < +3$, 99.730 percentage, that is, with probability 0.99730
$-4 < x_i < +4$, 99.994 percentage, that is, with probability 0.99994

In Fig. 4.25, a standard normal (Gaussian) PDF with arithmetic mean zero and unit variance is shown with the categorical division according to the standard deviation with these intervals.

4.15.2.8 Variation Coefficient

In order to be able to make comparison between two different unit data series, it is necessary to have dimensionless coefficient values. Based on the two simplest coefficients, namely, the arithmetic mean and the standard deviation, the coefficient of variation, $C_V(X)$ is defined as the standard deviation value per arithmetic average as

$$C_V(X) = \frac{S(X)}{\overline{X}} \tag{4.45}$$

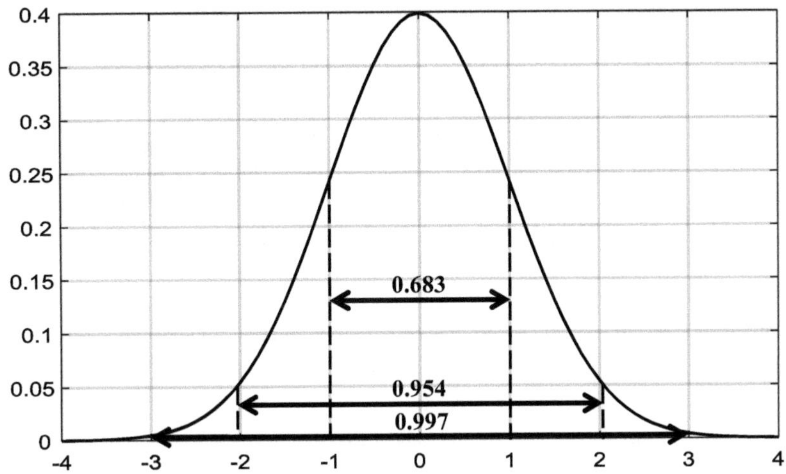

Fig. 4.25 Standard normal distributions

4.15.2.9 Standardized Variables

The coefficient of variation provides the comparison between two data series globally, but term-by-term comparison cannot be achieved, which is the most preferred procedure in the practical studies. In order to be able to have such an opportunity, the data values must be transformed into a dimensionless series like Eq. (4.18) as follows. The standardization operation is, in fact, the subtraction of the arithmetic average value from each data value and then their division by the standard deviation as.

$$\mathbf{x_i} = \frac{\mathbf{X_i} - \overline{\mathbf{X}}}{\mathbf{S_X}} \tag{4.46}$$

This standardized series has the following properties that should be cared for always:

1. The standardized variable has zero mean.
2. The standardized variable has unit variance and standard deviation.
3. It has positive and negative values with summation equal also to zero.
4. It is dimensionless.

One of the most frequently used concepts in data processing is the standard SF or series. Dimensionless property helps to compare two different data series with different original dimensions. For instance, in climatology, the comparison of temperature and rainfall series is not possible, but after the standardization operation, they can be compared directly.

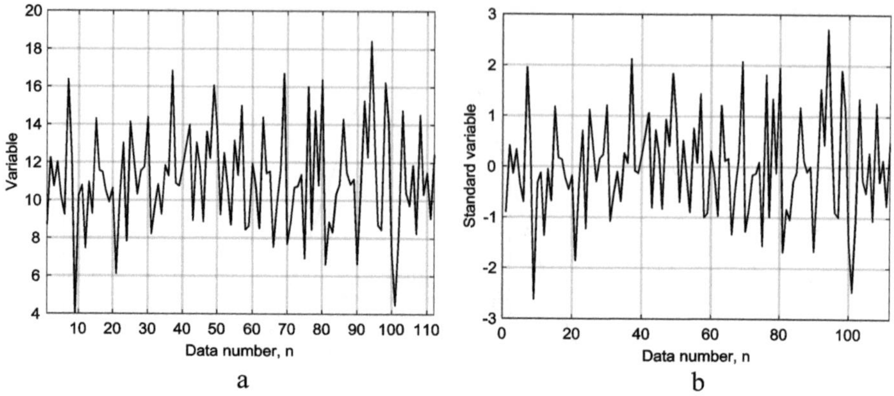

Fig. 4.26 Sample function (**a**) original, (**b**) standardized

4.15.2.10 Dimensionless Standard Series

The application of Eq. (4.46) to each SFs leads to standard SFs all with zeros arithmetic average and unit standard deviation, and they become comparable with each other. Figure 4.26 indicates the normal and standardized SFs of the same data series. Subtraction of arithmetic average from each data value shifts the horizontal axis to zero and division by standard deviation changes the scale. Finally, the normal and standardized SFs have the same shapes in different scales.

4.15.2.11 Skewness: Kurtosis Diagram

After the standardization of data, the two statistical parameters that remain nonstandardized are the skewness and kurtosis, which collectively provide a chart for the identification of possible probability distribution of the underlying data sequence. Gherghe has presented such a chart as in Fig. 4.27.

4.15.2.12 Engineering Truncations

Different from the statistical truncation, there are others that are useful for various human activities. In such truncations the comfort and benefit of humans are taken into consideration. These are referred to herein as the engineering truncations. For instance, for the comfort of humans the temperature must not be under 15 °C, and for the plant life below 7 °C. Daily water demand of Istanbul city is 3×10^6 m^3, and this is the truncation level for water supply projects to Istanbul city.

Fig. 4.27 Skewness-kurtosis relationships

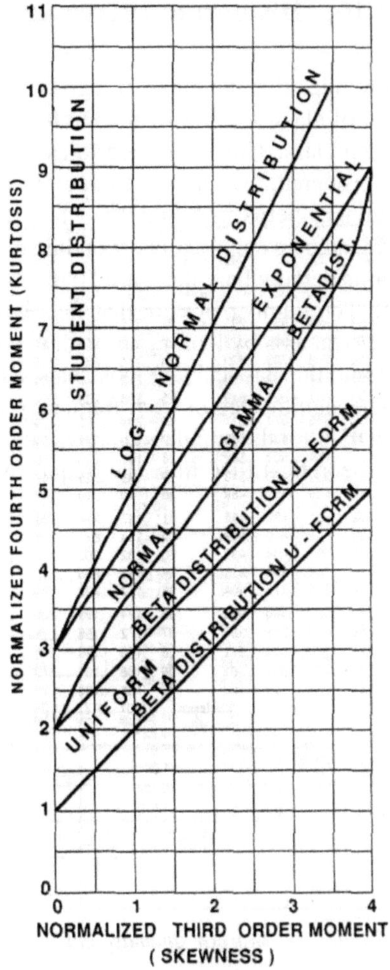

Table 4.4 SF truncation and specifications

	Temp.	Rainfall	Runoff	Humidity	Cloud	General
If $\mathbf{X_i} < \mathbf{X_0}$	Cold	Rainy	Dry	Humid	Open	Deficit
$\mathbf{X_i} > \mathbf{X_0}$	Hot	Non-rainy	Wet	N-humid	Close	Surplus

In the previous explanations, the values below and above of any truncation level are given in terms of numbers, percentages, or probabilities. However, as shown in Table 4.4, it is also possible to specify different phenomena with different words.

These bivariate specifications play important role in many diverse human engineering and social activities.

4.16 Histogram (Percentage Frequency Diagram)

Provided that the range of a given data set is known, it is possible to divide it into a set of classes to know the number and then percentage of the data that falls within each class. This is referred to as the frequency diagram, and when percentages are considered, then it is histogram. In general, the class number is chosen, any digital number between 5 and 15, which has almost been agreed commonly among the researchers empirically depending on experiences. In Figure 4.28a, seven classes are considered where there are 50 data values (Almazroui and Şen 2020).

The number of data that falls within each class is shown in Fig. 4.28b on the vertical axis by looking through the data from the left horizontally. In order to bring it into the classical textbook shapes, Fig. 4.28b is rotated 90° resulting in Fig. 4.28c. The significance of such frequency diagrams is for the identification of theoretical PDF fit for the population representation of the sample data set. Figure 4.29 indicates a theoretical PDF fit to the frequency diagram.

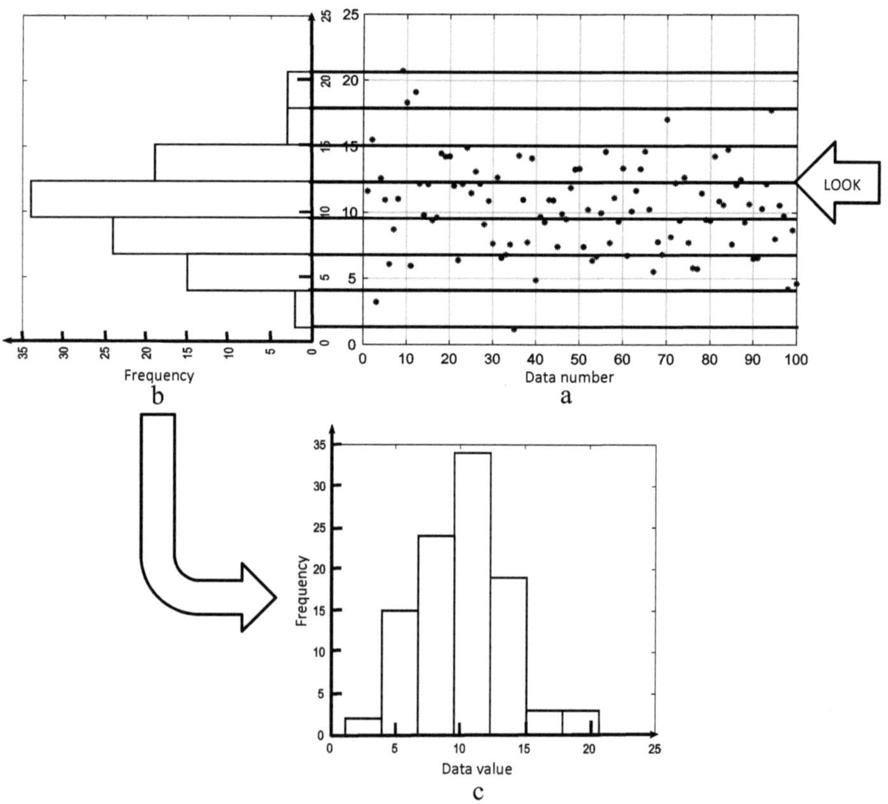

Fig. 4.28 Frequency diagram

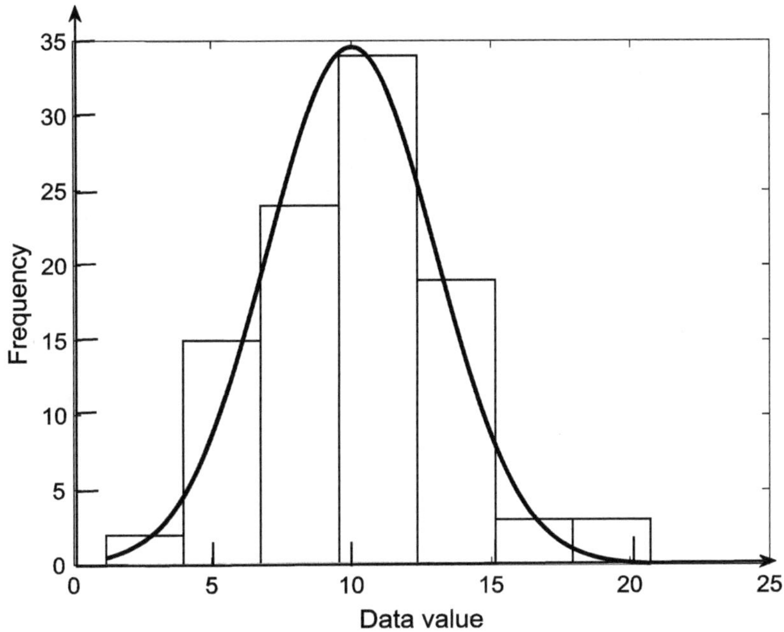

Fig. 4.29 Frequency diagram with theoretical PDF

4.16.1 Data Frequency

In the statistical data processing one of the most significant questions is the number of data that falls within a predetermined subinterval within the range. For instance, in practical studies, the frequencies of data within 1, 2, and 3 standard deviation, σ_X, distance above and below the average value, $\overline{X}$ are important in the overall data evaluation. These intervals are between $\overline{X} - \sigma_X$ and $\overline{X} + \sigma_X$; $\overline{X} - 2\sigma_X$ and $\overline{X} + 2\sigma_X$; $\overline{X} - 3\sigma_X$ and $\overline{X} + 3\sigma_X$. In any data processing, the number of data that falls within a subinterval in the range is very important, since it leads to the percentage, probability, or relative frequency of the data. In practice, the data range is divided into several adjacent equal length classes, and the number of data in each class is evaluated from the given data sequence. The following steps are necessary for such evaluations:

1. The total number of data in a given sample function (SF) or time series or data series within the range is n.
2. If the range is divided into two equal classes, the numbers of data that fall into these two categories which are referred to herein as the frequency, are not equal. Only in the case of symmetric SFs, the two classes will have equal frequencies. In skewed SFs, these two numbers are different from each other. If the frequencies in each class are n_1 and n_2 ($n_1 \neq n_2$), then it is necessary that $n_1 + n_2 = n$.

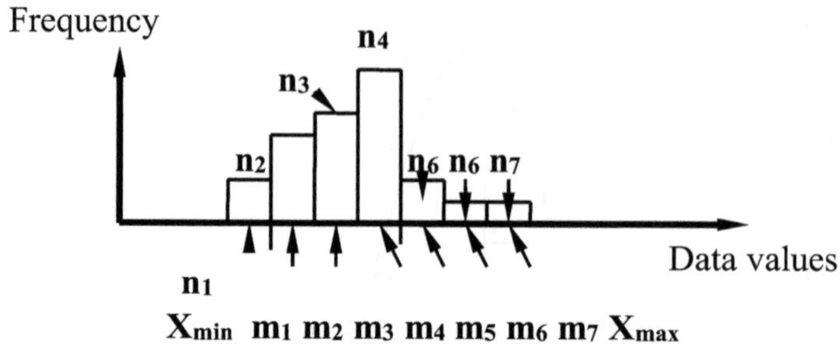

Fig. 4.30 Frequency diagram

3. If the data range, R, is divided into m classes, then the length, a, of each class can be calculated as

$$a = \frac{R}{m} \tag{4.47}$$

The sum of frequencies in each one of these classes is equal to the total number of data:

$$n_1 + n_2 + \ldots + n_m = n \tag{4.48}$$

If each class is represented by its mid-class value, then there will be m mid-class values as X_j (j = 1, 2, ... , m), with corresponding frequency values, n_i, (j = 1, 2, . . . , m). If these are plotted versus each other, a graph appears as in Fig. 4.30; it is named as the frequency diagram in statistics. In this evaluation, the key question is how to choose the number of classes, m. In practice, it is necessary that $5 < m < 15$. If the data number is between 10 and 20, the class number is adopted as 5, and for each 10 data number increase, m is increased by one. For instance, if n = 55, then according to this rule, m = 9.

Depending on the data behavior, there are different frequency diagrams in the practical applications. In Fig. 4.31, six of them are shown.

Frequency diagram is defined as the change of frequencies within the subclass midranges. It provides visually the change of frequency within the data values. The midpoints of each class are shown in Fig. 4.31. In Fig. 4.31e, almost all the frequencies are equal. Such a frequency diagram is referred to as the uniform frequency diagram. It implies completely random variation of the phenomenon concerned. On the other hand, in Fig. 4.31f, there are two maximum frequency values. This implies that in the data generation, there are two distinctive causes. In Fig. 4.31b, the frequency diagram is almost symmetrical. In this case, the mean, median, and mode values are expected as equal. The frequency diagrams in Fig. 4.31c, d are skewed to right and left, respectively. In the former case, mode ≤ median ≤ mean, but in the latter, mode > median > mean.

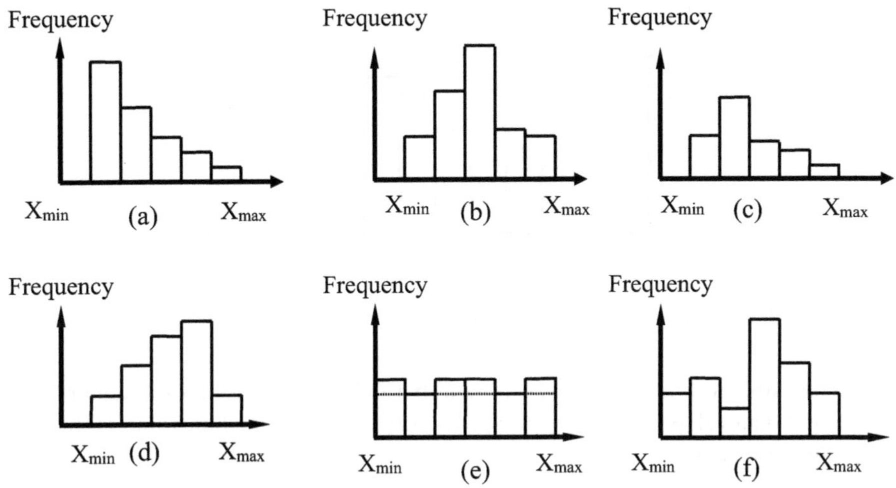

Fig. 4.31 Different frequency diagrams

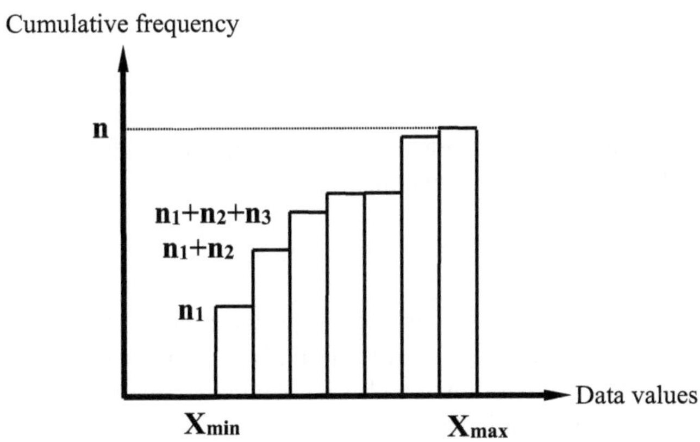

Fig. 4.32 Cumulative frequency diagrams

If the relative frequencies in Fig. 4.31 are successively added, then the cumulative frequency diagram is obtained as shown in Fig. 4.32.

The cumulative frequency diagram never decreases, and it always increases. If the same cumulative procedure is applied to frequency diagrams in Fig. 4.31, then similar cumulative frequency will emerge for each case. The benefits from the cumulative frequency diagram (CFD) are as follows:

1. The final value at the end of each CFD is equal to the number of data, n.
2. Any value on the vertical axis that corresponds to data value on the horizontal axis in a CFD is the number of data that are smaller than the adopted data on the horizontal axis.

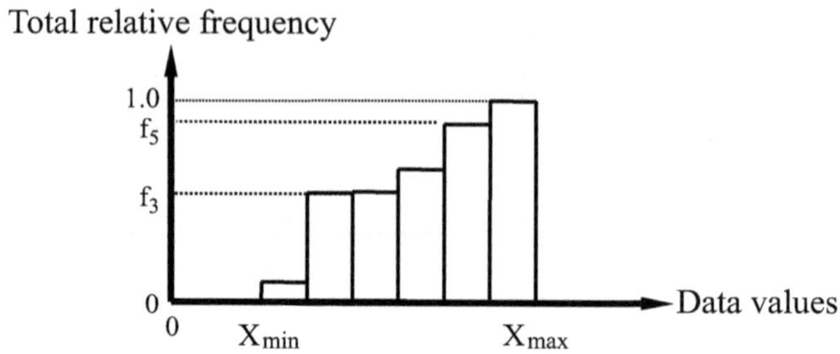

Fig. 4.33 Cumulative relative frequency diagrams

3. It is possible to find the number of data that falls within any desired data range.
4. The n/2 data value on the horizontal axis corresponds to median value of the data on the vertical axis.

It is possible to bring the vertical axis in any frequency or cumulative frequency diagram into a data number dependent form, by dividing the frequency numbers as appears in Eq. (4.48) by the total number of data, which yields

$$\frac{n_1}{n} + \frac{n_2}{n} + \ldots + \frac{n_m}{n} = 1 \tag{4.49}$$

If each term on the left-hand side is defined as the relative frequency, f_j ($j = 1, 2, \ldots, m$), then this last expression becomes

$$f_1 + f_2 + \ldots + f_m = 1 \tag{4.50}$$

The relative frequencies are also percentages or the probability values. Therefore, their values are confined between 0 and 1. Similarly, the CFD can be converted into a data number independent case also, and it is then called as the cumulative relative frequency diagram as shown in Fig. 4.33.

4.16.2 Subintervals and Parameters

If there are many data values, say more than 1000, the frequency of cumulative frequency diagram concept groups them at the maximum into 15 classes, and hence, there is a tremendous reduction in the data values. The question is then how to find the statistical parameters of data from such a grouped data? After the construction of frequency diagrams, there are two sequences that are significant for further calculations. These are:

$$x_1,\ x_2,\ \ldots,\ x_m\text{–class mid-points}$$

$$n_1,\ n_2,\ \ldots,\ n_m - \text{class frequencies}$$

or relative frequencies

$$f_1, \ f_2, \ \ldots, \ f_m\text{–class relative frequencies}$$

where $f_j = n_j/n$. In the case of two sequence existence, one can use the concept of weighted average which leads to

$$\overline{A}_a = \frac{\sum_{i=1}^{m} x_i n_i}{n} = \sum_{i=1}^{m} x_i f_i \tag{4.51}$$

This is equal to the arithmetic average of the given SF. In general, variance is the arithmetic average of the square of deviations from the mean value, and under the light of this definition, one can write that

$$V_a = \frac{\sum_{i=1}^{m} (x_i - \overline{x})^2 n_i}{n} = \sum_{i=1}^{m} (x_i - \overline{x})^2 f_i \tag{4.52}$$

Other statistical parameters can be calculated in the same way by considering weighted average concept.

4.17 Normal (Gaussian) Test

In the measurements of many uncertain events, the data show normal (Gaussian) PDF. Others can be transformed into a normal form by the application of various transformations (square root, logarithmic, etc.). Many significance tests in data processing are achieved through the normal PDF application. Therefore, one should know the properties of a normal distribution:

1. If the relative frequency distribution shows a symmetrical form, then the data are said to come from a possible normal PDF.
2. In a normal PDF, there is only a single peak (mode) value, which is approximately equal to the median and the arithmetic average values.
3. A normal PDF can be represented by two parameters only, namely, the arithmetic average and the standard deviation.
4. Majority of the data are close to the maximum frequency band on the symmetry axis. Towards the right and left, the relative frequency of the data decreases.
5. Theoretically, the area under a normal PDF, as in all PDFs, is equal to 1.
6. There are two extreme regions as tails on the right and left. These tails extend from $-\infty$ to $+\infty$, but in practical works, the smallest and the biggest values are finite. Mathematically, a normal distribution function is given mathematically as

$$f(X) = \frac{1}{\sqrt{2\pi}} \exp\left[-\frac{1}{2}\left(\frac{X-\mu}{\sigma}\right)^2\right] \tag{4.53}$$

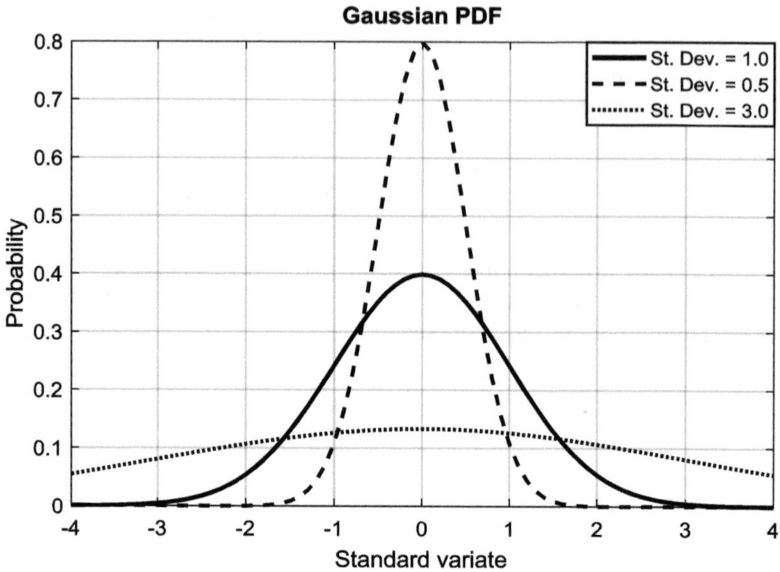

Fig. 4.34 Different normal distributions

where $\boldsymbol{\mu}$ and $\boldsymbol{\sigma}^2$ are the arithmetic average and the variance of the data. The geometrical exposition of this expression is given in Fig. 4.34.

Since the area calculation from this expression is mathematically impossible, the areas are calculated by numerical techniques and given in table form in many textbooks or available in the Internet. The areas in the table are from - ∞ up to the variable value, x. The subtraction of these areas from 1 leads to significance level values. By means of the table, if the significance level is given, then the corresponding area can be found. In order to use the table, it is necessary that after the data confirmation with the normal (Gaussian) PDF, the data values must be standardized such that the arithmetic average is equal to 0 and variance to 1. For a normal distribution test, the following steps must be executed:

1. The data sequence must be standardized using its arithmetic average and standard deviation value. This standard variant is referred to as the test quantity. Standardization gives rise to dimensionless data values such that they have zero mean and unit variance. The standard normal PDF has the following form with no explicit parameter values:

$$f(x) = \frac{1}{\sqrt{2\pi}} e^{-\frac{1}{2}x^2} \tag{4.54}$$

2. It is necessary to decide about the confidence interval. For this purpose, the significance level may be taken as 5% or in some approximate works as 10%.

The uses of uncertainty techniques such as the probability, statistics, and stochastic methods in sciences have increased rapidly since 1960s, and most of the researchers seek more training in these disciplines for dealing with uncertainty in a better quantitative way. Many professional journals, books, and technical reports in the science studies include significant parts on the uncertainty techniques in dealing with uncertain natural phenomena. Yet relatively few scientists and engineers in these disciplines have a strong background in school mathematics, and the question is then how they can obtain enough knowledge of uncertainty methods including probability, statistics, and stochastic processes in describing natural, social, and engineering phenomena and in appreciating the arguments, which they must read and then digest for successful applications in making predictions and interpretations.

Mann (1970a, b) stated that the inability to predict specific events may stem from the fact that nature is intrinsically random. According to him, random and randomness are used in a statistical sense to describe any phenomenon which is unpredictable with any degree of uncertainty for a specific event and deterministic phenomena, on the contrary, those in which outcomes of the individual events are predictable with complete certainty under any given set of circumstances, if the required initial conditions are known. In this approach, the nature is considered as random. Consequently, randomness has been suggested as the ultimate and most profound physical concept of nature. In an intrinsically random nature, predictions would be impossible, but the continued failure to predict does not entitle the issue of determinism.

Moreover, it is almost truest to claim that classical mechanics is not a deterministic theory, if the claim simply means that actual measurements confirm the predictions of the theory only approximately or only within certain statistically expressed limits. Any theory formulated, as in the classical mechanics, in terms of magnitudes capable of mathematically continuous variation must, in the nature of the case, be statistical and not quite deterministic in this sense. For the numerical values of physical magnitudes (such a velocity) obtained by experimental measurement never form a mathematically continuous series; any set of values so obtained will show some dispersion around values calculated from the theory. Nevertheless, a theory is labeled completely deterministic provided that its internal mechanism analysis shows that the theoretical state of a system at one instant determines logically a unique state of that system for any other instant. In this sense, and with respect to the theoretically defined mechanical states of systems, mechanics is unquestionably a deterministic theory. Consequently, as stated by Laplace (1747–1825) when a predictor is introduced into a philosophical discussion of determinism, it is not a human being but a "superhuman intelligence." Human predictors cannot settle the issue of determinism, because they are unable to predict physical events no matter what the world is really like.

4.18 Statistical Model Efficiency Formulations

In the literature, all model efficiency standard indicators are dependent on three basic statistical parameters among which are the arithmetic averages of the measurements, $\overline{M}$ and model predictions, $\overline{P}$; standard deviations, S_M and S_P and the cross-correlation, $C_{M,\,P}$ between measurement and model prediction sequences. Additionally, the regression line between the measurements and predictions has two parameters as intercept, I, or regression line central point ($\overline{M}$ and $\overline{P}$) coordinates and the slope, S.

The simple and necessary, but not enough mathematical efficiency measure is the bias, BI, which measures the distance between the measurements and model predictions as

$$BI = \frac{1}{n}\sum_{i=1}^{n}\left(P_i - M_i\right) = \overline{P} - \overline{M} \tag{4.55}$$

The ideal value for model efficiency is BI = 0; although this condition is necessary, but not enough. The second measure is the mean square error (MSE),

$$MSE = \frac{1}{n}\sum_{i=1}^{n}\left(P_i - M_i\right)^2 \tag{4.56}$$

The ideal value is MSE = 0, but this condition is not valid in any scientific model efficiency, because there are always random errors. This is the main reason why the best model MSE should have the minimum level among all other alternatives. Eq. (4.56) includes implicitly the standard deviations and the cross-correlation between measurements and predictions.

The Nash-Sutcliffe efficiency (NSE) measure includes the MSE with the standard deviation of the measurement data ratio as follows:

$$NSE = \frac{\sum_{i=1}^{n}\left(M_i - \overline{M}\right)^2 - \sum_{i=1}^{n}\left(P_i - M_i\right)^2}{\sum_{i=1}^{n}\left(M_i - \overline{M}\right)^2} = 1 - \frac{\sum_{i=1}^{n}\left(P_i - M_i\right)^2}{\sum_{i=1}^{n}\left(M_i - \overline{M}\right)^2} \tag{4.57}$$

The second term on the right-hand side is greater than 1, hence NSE has negative values. The ideal value of NSE is 1, but this is never verified in practical applications, and therefore, the closest is the value to 1, the better is the model efficiency.

As for the cross-correlation, CC, between the measurement and prediction can be calculated as

$$CC = \frac{\sum_{i=1}^{n}\left(M_i - \overline{M}\right)\left(P_i - \overline{P}\right)}{S_M S_P} \tag{4.58}$$

On the other hand, the straight-line regression intercept, I, and slope, S, values can be calculated according to the following expressions:

$$\overline{P} = S\overline{M} + I \tag{4.59}$$

and

$$S = \frac{n\sum_{i=1}^{n} P_i M_i - \sum_{i=1}^{n} P_i \sum_{i=1}^{n} M_i}{n\sum_{i=1}^{n} M_i^2 - \left(\sum_{i=1}^{n} M_i\right)^2} \tag{4.60}$$

respectively.

Apart from the above model efficiency measurements, there are others, which have been suggested for their rectification. One of the first versions is due to Willmott, who gave agreement index d as

$$d = 1 - \frac{\sum_{i=1}^{n} (P_i - M_i)^2}{\sum_{i=1}^{n} \left(|P_i'| + |M_i'|\right)^2} \tag{4.61}$$

where $P_i' = P_i - \overline{P}$ and $M_i' = M_i - \overline{M}$ are the deviations from the respective arithmetic averages. The expression in the dominator is referred to as the potential error (PE). The significance of d is that it measures the degree to which model predictions are error free, and its values varies between 0 and 1; where 1 represents the perfect agreement between the measurements and predictions, which is never possible in practical applications, and therefore, the researchers take the closest value to $d = 1$ as the model efficiency acceptance, but there is no criterion, which indicates objectively the limit value between acceptance and rejection, and hence, there appears subjectivity as in other efficiency measures. Equation (4.61) can be rewritten in terms of the MSE as follows:

$$d = 1 - \frac{nMSE}{PE} \tag{4.62}$$

Equations (4.55)–(4.62) include all the necessary numerical quantities that are useful in the construction of the visual inspection and numerical analysis method (VINAM) template (Şen et al. 2021) as will be explained and applied in the following sections.

4.19 Correlation Coefficient

In mathematics, when two variables are related with each other, their variation or graph on a Cartesian coordinate system does not appear as a horizontal or vertical line, but rather a line with many slopes at each curve point. The simplest form of dependence is the linear one, which is always used in the statistical or stochastically modeling works. Hence, in order to find the quantitative representation of the dependence, the SF must be plotted on the Cartesian coordinate axis in some manner. If there are two different SFs, then they can be plotted versus each other, and hence the graphical relationship between these two variables can be visualized. Consequently, by assessing with necked eye the scatter of the points, one can appreciate whether the dependence is high, small, or none. In the case of scatter points lie around a regular trend, then the dependence is strong, otherwise it is weak.

4.19.1 Pearson Correlation

It is valid for linear correlation measurement between two variables. The basis of the methodology is either without any shift or some lag shifts as lag-1, lag-2, lag-3, etc. Pearson correlation can be used serially within a single time series or crosswise between two time series. In Fig. 4.35, Lag-2 serial correlation coefficient calculation mechanism is shown.

Whatever is the lag value, the corresponding pairs are multiplied, and hence, a multiplication sequence is obtained as $X_1X_3, X_2X_4, \ldots, X_{n-2}X_n$. In the calculation of correlation coefficient for sampling error minimization, the maximum number of lag-k is taken as one third of the sample length, n (k = 1, 2, . . . , n/3). In practice, the standardized time series are taken into consideration. Based on the standardized time series with zero mean and unit variance, the arithmetic average of this sequence is the Pearson serial correlation coefficient value. For lag-k the Pearson correlation coefficient, ρ_k, can be written as follows:

$$\rho_k = \frac{1}{n-k}\sum_{i=k}^{n-k} x_i x_{i+k} \tag{4.63}$$

where x_i is the standardized time series. This can be interpreted as the arithmetic average of dot product of two time series. This indicates the linear dependence because of the straight-line fit through the scatter of points in Fig. 4.36.

In this figure, the slope of the straight line is equivalent with the Pearson correlation coefficient. The arithmetic average and the standard deviation of the lag-one correlation coefficient theoretical distribution are given by Anderson (1942) as

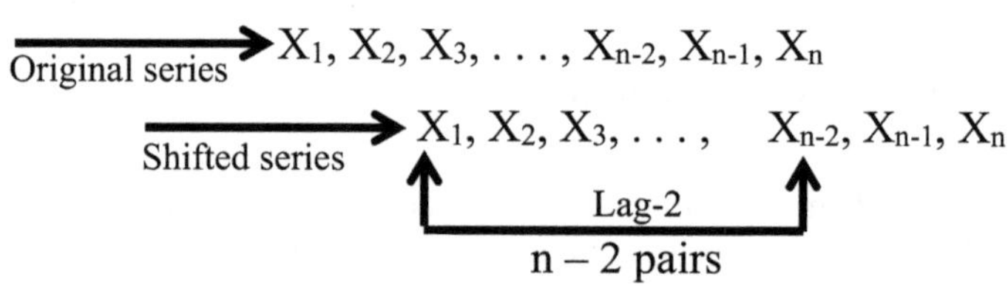

Fig. 4.35 Lag-2 shifts

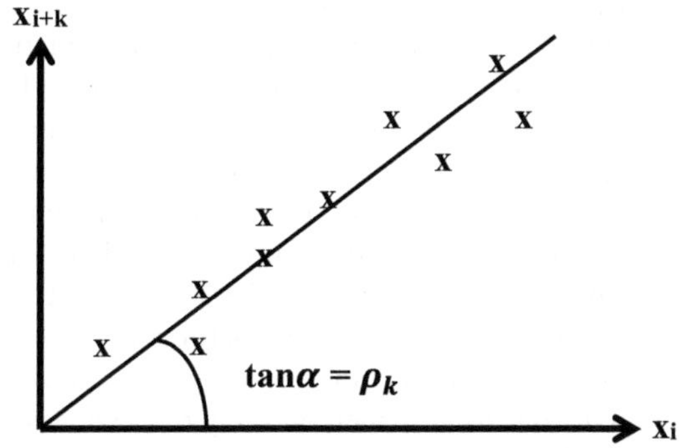

Fig. 4.36 Pearson correlation coefficients

$$\overline{\rho} = -\frac{1}{(n-1)} \tag{4.64}$$

and

$$V_{\rho} = \frac{1}{(n-1)} \tag{4.65}$$

respectively. One implicit assumption that comes from the theoretical formulation is that the sample first-order correlation coefficient has the normal (Gaussian) PDF with mean $\overline{\boldsymbol{\rho}}$ and variance, $\mathbf{V}_{\boldsymbol{\rho}}$.

The cross-correlation coefficient, ρ_c, between two different time series (X_i and Y_i, i = 1, 2, . . . , n) can be defined like the serial correlation function as follows:

$$\boldsymbol{\rho}_c = \frac{\mathbf{1}}{\mathbf{n}}\sum_{i=1}^{n}\left(\frac{\mathbf{X}_i - \overline{\mathbf{X}}}{\boldsymbol{\sigma}_X}\right)\left(\frac{\mathbf{Y}_i - \overline{\mathbf{Y}}}{\boldsymbol{\sigma}_Y}\right) \tag{4.66}$$

where $\overline{X}$ and $\overline{Y}$ are the arithmetic averages of each series and likewise σ_X and σ_Y are the respective standard deviations. The correlation coefficient is referred to as the Pearson correlation and it assumes values from −1 to +1. In the mid-value is zero correlation coefficient corresponding to complete independence. Otherwise, values close to +1(−1) implies positive, that is, directly proportional (negatively, i.e., inversely) strong correlations. After all what have been explained above, one can come out with the following important points:

1. In case of nonlinear relationship between two variables, the Pearson correlation coefficient definition cannot be applicable, because it is valid only for linear dependence cases.
2. Eqs. (4.64) and (4.6) provide bias values if there are one or more extreme values in the sample function, and therefore, the Pearson correlation coefficient cannot be representative of the dependence.

3. If the PDF of the data is away from the normal (Gaussian) PDF, then the Pearson correlation coefficient cannot be representative for the records. It is necessary to transform non-normal PDFs to normal case for the Pearson correlation coefficient calculation, but in this case the correlation coefficient cannot be genuinely representative of the original time series. Even the reverse transformation does not guarantee that the correlation coefficient is equal to the original value. In the literature, there is the autorun correlation coefficient definition, which is not dependent on the type of PDF (Şen 1977).
4. The Pearson correlation coefficient necessitates variance constancy, that is, homoscedasticity; otherwise it is not representative.
5. In the uncertainty methodology of fuzziness (Ross 1995), the Pearson correlation coefficient cannot be used, because it is dependent on a single number, but not on verbal and linguistic data.

Prior to the application of the Pearson correlation coefficient to any time series, these points need to be checked. Anderson (1942) provided a significance test for the acceptance of significant Pearson correlation coefficient. As stated by Feller (1967), it is by no means a general measure of dependence, because it involves many assumptions, but they are mostly not considered in many applications.

The correlation coefficient takes values between -1 and $+ 1$. The closer the coefficient to zero, the more random, that is, independent are the two SFs, otherwise, close values to +1 (-1) implies positively (negatively) strong correlations. Positive correlation means proportionality and negative value shows disproportionality. In the case of positive dependence, high (low) values follow high (low) values whereas in the case of negative dependence high (low) values follow low (high) values. The dependence that is calculated through Eq. (4.63) is the serial correlation or autocorrelation coefficient. It assumes values between -1 and $+ 1$ and the intermediate values are given specific explanation as stated in Table 4.5. One should not memorize this table, because they are the subjective opinion of this author. Other authors may deviate from the specifications given in this table owing to their experiences, but the deviations are not significant in practical works. This correlation coefficient is the measure of linear dependence between two variables. If the correlation is not linear, then these definitions are invalid. In the complete positive dependence ($\rho_p = +1$), the two SFs are directly proportional to each other, otherwise for complete negative dependence case ($\rho_p = -1$) they are inversely related. In both cases, there is a complete mathematical relationship between the two variables. Prior to any calculation according to Eq. (4.66), it is very useful to visualize the scatter diagram between the two variables on a Cartesian system. If the relationship from the scatter diagram gives the impression of linearity, only then Eq. (4.66) must be used for numerical calculations. Otherwise, direct numerical calculations may lead to unnecessary conclusions without any benefit. Verbal specifications of various correlation coefficient values are given in Table 4.5.

The following points are the deficiencies of the Pearson correlation coefficient concept in practical applications:

1. Even if the correlation is not linear, the correlation value calculation according to Eq. (4.63) will appear between -1 and $+1$. This may not have logical and physical meaning, because it is valid only for linear relationships.

Table 4.5 Correlation coefficient classes

Numerical value intervals	Linguistic interpretations
$\rho_p = -1.0$	Completely negative dependence
$-1.0 < \rho_p < -0.9$	Strong negative dependence
$-0.9 < \rho_p < -0.7$	Quite negative dependence
$-0.7 < \rho_p < -0.5$	Weak negative dependence
$-0.5 < \rho_p < -0.3$	Very weak negative dependence
$-0.3 < \rho_p < -0.1$	Insignificant negative dependence
$\rho_p = 0.0$	Complete independence
$0.1 < \rho_p < 0.3$	Insignificant positive dependence
$0.3 < \rho_p < 0.5$	Very weak positive dependence
$0.5 < \rho_p < 0.7$	Weak positive dependence
$0.7 < \rho_p < 0.9$	Strong positive dependence
$\rho_p = 1.0$	Complete positive dependence

2. If there are one or more extreme values in the SF, then these values effect Eq. (4.63) in such a manner, the correlation coefficient appears biased and/or unrepresentative.
3. The data must abide with a normal PDF, otherwise the correlation coefficient is not meaningful.
4. For meaningful and reliable correlation coefficient, the standard deviation of the data must be constant and finite, that is, homoscedasticity.
5. Correlation coefficient definitions in Eqs. (4.63) and (4.66) cannot be used for verbal and linguistic data.
6. If data is transformed by any means, the correlation coefficient of the transformed data will not be the same as the original data. Even the reverse transformation does not guarantee that the correlation coefficient is equal to the observed correlation value.

After all what has been said above, it is obvious that the domain of the Pearson correlation coefficient is rather restrictive, and prior to its use, all the necessary assumptions must be cared for its validity. The significance for this correlation coefficient can be achieved by Anderson test.

Most empirical research belongs clearly to one of those two general categories. In correlation research one does not (or at least try not to) influence any variables, but only measure them and look for relations (correlations) between some set of variables. In experimental research, we manipulate some variables and then measure the effects of this manipulation on other variables; for example, a researcher might artificially increase blood pressure and then record cholesterol level. Data analysis in experimental research also comes down to calculating "correlations"between variables, specifically, those manipulated and those affected by the manipulation. However, experimental data may potentially provide qualitatively better information. Only experimental data can conclusively demonstrate causal relations between variables.

4.19.2 Nonparametric Correlation Coefficient

In parametric statistics, the correlation coefficient is named as the Pearson correlation and it is defined as the product moment as in Eq. (4.63). In the nonparametric statistics domain, the analogous to the Pearson correlation is the Spearman's rank correlation coefficient. Pearson correlation coefficient requires that both variables should comply by the normal PDs, which are not the case in many scientific studies.

4.19.2.1 Spearman's Rank Correlation

This is a nonparametric correlation calculation procedure, where given time series are ranked separately in ascending order with ranks $R(X_i)$ and $R(Y_i)$, respectively. Spearman correlation coefficient does not require normal (Gaussian) PDF of the data and it is robust to extreme values. There is a perfect correlation between the two variables, if ranks in both series are equivalence at each instant. Otherwise, if there are differences between corresponding ranks then Spearman correlation is defined mathematically as the sum of the differences. The Spearman's rank coefficients are then scaled down to between -1 (perfect negative correlation) and +1 (perfect positive correlation). In the midway, there is a value equal to zero indicating no correlation. The following steps are for this rank correlation calculation.

1. The null hypothesis, H_0, adapts the complete independence as zero correlation coefficient, $\rho_h = 0$.
2. Other than zero rank correlation, values are for the alternative hypothesis, H_A.
3. The test statistic, ρ_s, is given in terms of ranks and the data number, n, in each set:

$$\rho_s = 1 - 6\frac{\sum_{i=1}^{n}[R(X_i) - R(Y_i)]}{n(n^2 - 1)} \tag{4.67}$$

Contrary to Pearson correlation coefficient, ρ_s does not mean linear relationship indicator.

4.19.2.2 Autorun (Şen) Test

It is generally accepted that if a series is dependent, it means that the high (low) values tend to follow high (low) values. This statement shall be interpreted differently for a dependent series; specifically, periods of surplus (positive run) and deficit (negative run) spells tend to be greater than that in the case of an independent series (Feller 1967). Such a property is termed "persistence." Scientists have made various attempts to measure "persistence" mainly by three procedures, namely, autocorrelation function, spectral analysis, and rescaled-range analysis (Hurst 1951). Surplus and deficit spells are directly related to run properties as explained by various researchers (Mood 1940; Swed and Eisenhart 1943; Şen 1976).

Autocorrelation analysis is a means of measuring linear dependence between any two observations. As stated by Feller (1967), it is by no means a general measure of dependence, because it involves all the assumptions stated in the previous section. On the other hand, spectral analyses of sequential patterns in series by the classical periodogram method have been in use for many years. Finally, the rescaled-range analysis, which was introduced by Hurst (1951, 1956), has the advantage of being comparatively more robust than any other technique, and it is not very sensitive to the marginal PDF.

Şen (1979) presented a special case of the autorun coefficient based on the median value, which corresponds to exceedance probability of $p = 0.50$. The generalized autorun coefficient is considered for any probability level $0 < p < 1$. In the following, the definition of autorun methodology is briefly reviewed in addition to suggestion of new formulations.

Joint and conditional probabilities are also measuring of dependence (Sects. 4.4.2 and 4.5.1). In general, a joint probability is equal to the multiplication of a conditional probability by a marginal probability (Feller 1967). If x_i and x_{i-k} are two dependent events with joint PDF, P (x_i, x_{i-k}), their conditional PDF is denoted by P (x_i/x_{i-k}). Herein, k is referred to as the lag and indicates the time difference between the two events if P (x_{i-k}) is the marginal PDF of event x_{i-k}, then the following probability statement is valid between these two probabilities as.

$$P(x_i, x_{i-k}) = P\left(\frac{x_i}{x_{i-k}}\right) P(x_{i-k}) \tag{4.68}$$

Furthermore, the conditional probability is defined by Şen (1976) as the autorun coefficient, $r = P(x_i/x_{i-k})$, and hence, the Eq. (4.68) yields

$$r_k = \frac{P(x_{i-k}, x_i)}{P(x_i)} \tag{4.69}$$

In more detail, if a time series is truncated at an arbitrary constant level, x_0, and then this expression can be defined in a more illuminating way in terms of exceedance probabilities as

$$r_k = \frac{P(x_{i-k} > x_0,\ \ x_i > x_0)}{P(x > x_0)} \tag{4.70}$$

A special case has been defined by Şen (1976) for the median, m, truncation level for which $P(x_i > m) = 0.5$. Herein, $p = P(x_i > m)$ is the exceedance probability. The non-exceedance probability, is $q = 1 - p$. Hence, generally, the autorun coefficient definition becomes

$$r_k = \frac{P(x_{i-k} > m, x_i > m)}{p} \tag{4.71}$$

The necessary information for evaluating the population autorun coefficient is the joint PDF of observations lag-k apart. Unfortunately, in practice, the population PDF of a variable is not available, but instead, a sequence of measurements may be available. Hence, sample estimates of the autorun coefficients must be estimated from the measurement data.

An estimate, $\widehat{r}_k$, of r_k can simply be proposed by considering Eq. (4.71) together with the classical definition of probability in textbooks. According to this definition, the probability, $P(x_i)$, is found by counting the total number, n_x, of occurrences, exceedance in the autorun case, and consequently, one can define simply that.

$$P(x_i > x_0) = n_x/n \tag{4.72}$$

In other words, $P(x_i)$ is the ratio of the favorable alternatives to the total number, n, of alternatives. On the other hand, in the case of a joint event ($x_i > m$, $x_{i-k} > m$), in a sequence of n observations, n – k possible alternatives exist for two observations lag–k apart being simultaneously greater than m. As a result, if the number of joint events in a given sequence of length n is n_k, then from Eq. (4.72) one can obtain

$$P\,(x_i > m, x_{i-k} > m) = \frac{n_k}{n-k} \tag{4.73}$$

The substitution of which into Eq. (4.71) leads to the small sample estimate of r_k as

$$\widehat{r}_k = \frac{n_k}{p(n-k)} \tag{4.74}$$

The numerator of this expression is an integer-valued random variable, whereas the denominator is a fixed value for given n, k, and p. Hence, the random characteristics of the estimate $\widehat{r}_k$ can be obtained from the characteristics of random variable, n_k. For instance, if both the expected value $E(n_k)$ and the variance $V(n_k)$, of n_k are known, then the expectation and variance of $\widehat{r}_k$ could be evaluated, respectively, through the following expressions:

$$E\,(\widehat{r}_k) = \frac{E(n_k)}{p(n-k)} \tag{4.75}$$

and

$$V(\widehat{r}_k) = \frac{V(n_k)}{pq(n-k)^2} \tag{4.76}$$

Based on the frequency interpretation of the probability, the estimate $\widehat{r}_k$ can be calculated by successive execution of the following steps:

1. The exceedance probability, p, and its corresponding truncation level x_0 is calculated from a given sequence $x_1, x_2, \ldots, x_n$.
2. The series is truncated at the x_0 level, hence, sequences of surpluses ($x_i > x_0$) and deficits ($x_i \leq x_0$) are identified.
3. The number, n_k, of overlapping successive surplus pairs (observations lag-k apart) are counted.
4. The estimate of r_k is then calculated from Eq. (4.74).

It is interesting to point out at this stage that the calculations in the four steps are all distribution-free. When the length of historic data theoretically approaches infinity, then $\widehat{r}_k$ becomes

$$r_k = \lim_{n \to \infty} \frac{2n_k}{n-k} = 2P(x_i > m, \quad x_{i-k} > m) \tag{4.77}$$

Contrary to the autocorrelation, the autorun analysis does not distort the dependence structure of the sequence considered. Additionally, the autorun coefficients are easier to calculate than the autocorrelation coefficients. Furthermore, autorun analysis is directly related to run properties, which play an effective role in various engineering problems. Extensive simulation studies using a first-order Markov process were carried out and the autorun function of the synthetic sequences was calculated according to Eq. (4.75).

4.20 Classical Regression Techniques

The essence of many statistical forecasting depends on the principle that the summation of the prediction error squares should be the least. Herein, the prediction error is defined as the difference between the observed and the predicted value. Simple revision of the statistical least squares technique is presented in the following sequel. Detailed account for the weather predictors and earth scientists can be found in textbooks by Koch and Link (1971) and Davis (2002). There are six restrictive assumptions in the regression equation parameter estimations that should be taken into consideration prior to any application.

1. *Linearity*: Regression technique fits a straight-line trend through a scatter of data points, and correlation analysis test for the "goodness-of-fit" of this line. Clearly, if the trend cannot be represented by a straight line, regression analysis will not portray it accurately. In the case of a definite trend the use cross-correlation is necessary, and it brings the linearity restriction by definition.

2. *Normality*: It is widely assumed that use of the linear regression model requires that the variables have normal (Gaussian) distributions. Although, the requirement is not that the raw data be normally distributed, but it is that the conditional distribution of the residuals should be normally distributed. If the conditional distribution is normal, then it is almost certain that the distributions of dependent and independent variables are also normally distributed. Thus, it is necessary to test if the data are normally distributed in order to inquire as to whether a prerequisite for normal conditional distribution exists.
3. *Means of conditional distributions*: For every value of independent variable, the mean of the differences between the measured and predicted global dependent variable values obtained by Eq. (4.17) must be zero. If it is not, the coefficients of the regression equation (**a** and **b**) are biased estimates. Furthermore, the implication of major departure from this assumption is that the trend in the scatter diagram is not linear.
4. *Homoscedasticity*: This means equal variances in the conditional distributions and it is an important assumption. If it is not satisfied, then the regression equation coefficients (**a** and **b**) may be severely biased. In order to test for homoscedasticity, the data must be subdivided into three or more groups and the variance of each group must be calculated. If there is significant difference between any of these variances, then the data has homoscedasticity.
5. *Autocorrelation*: The crux of this assumption is that the value of each observation on the independent variable is independent of all the values of all others, so that one cannot predict the value of variable at time, say i, if one knows variable value at time, $i - 1$. There are two interpretations as to the importance of this assumption; one is substantively logical and the other is statistically logical. The statistical interpretation of autocorrelation relates to the linearity assumption.
6. *Lack of measurement error*: This assumption requires that both dependent and independent variable measurements are without error. If this is not the case, and the magnitude of the error is not known, then the coefficients of the regression equation may be biased to an extent that cannot be estimated.

All what have been explained above leave suspicions in the regression coefficient estimations, if the necessary tests are not performed and the data are not prepared for the requirements? In practical studies, all over the world, researchers most often do not care or even think in these restrictive assumptions and consequently the coefficient estimations might remain biased. Even the amount of the global bias is not known and therefore bias correction procedures cannot be defined and applied. Hence, the parameter estimates remain under suspicion. In order to avoid all these restrictive assumptions, rather than the application of procedural regression analysis to data with a set of restrictive assumptions, it may be preferable to try and preserve only the arithmetic averages and variances of the dependent and independent variable data in many practical studies. After all, the arithmetic averages and variances are the most significant statistical parameters in any design work.

4.20.1 Scatter Diagrams

If the question is the search for the type of relationship between two variables, the practical answer can only be given by plotting the corresponding values of two sample functions, SFs, on a Cartesian coordinate system. Consequently, the two SFs' data values give rise to scatter points as presented in Fig. 4.37, which provides visual inspection, preliminary feeling, and consideration of the relationship type between two variables. Such coordinate systems with data points are referred to as the scatter diagram in the statistics terminology.

Comparison of the scatter diagram with the functional relationships visually provides the first opinion about the type of deterministic relation form (mathematical functions), which shows the general trend between the two SF values.

The simplest of the possible relationships is the straight-line form, and it is used very frequently in different disciplines.

$$Y = a_{yx} + b_{yx}X \tag{4.78}$$

This model is also referred to as the simple regression technique since herein X is regressed on Y. In any actual prediction model, there is more than one predictor, but the ideas for simple linear regression can be generalized easily to multiple linear regression. The representation of Eq. (4.78) on a Cartesian coordinate system yields to a straight line in the mathematical sense, but scatter of points in the statistical sense, where each one of these points is associated with the data pairs (X_i, Y_i) for n data pairs (i = 1,2, ... , n). The relative position of the straight line must be

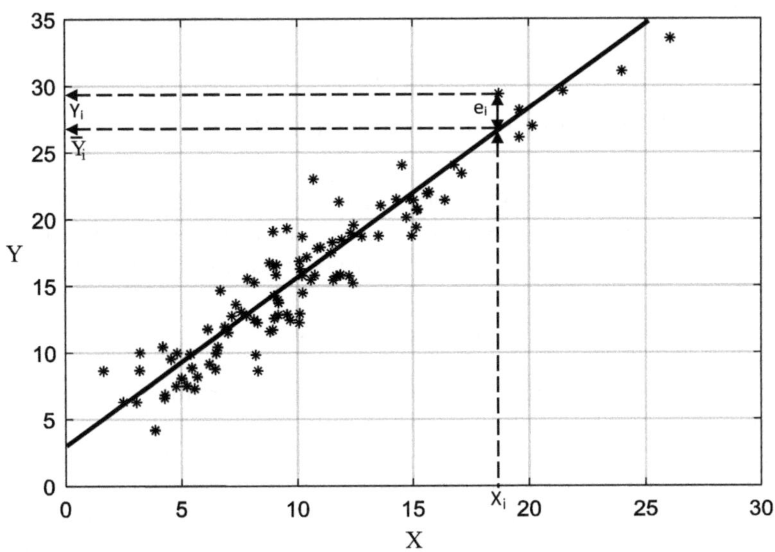

Fig. 4.37 Scatter diagram

determined in such a way that the summation of squared deviation of each point from this straight line must be the smallest possible. Herein, the deviation is synonymously used as the error. As shown in Fig. 4.37, these deviations from the straight line may be decided as the horizontal, vertical, or perpendicular distances. However, in practical studies most often sum of vertical deviation squares are minimized to fix the regression line through the scatter diagram. The choice of the sum of the squared-error criteria is convenient not only that it is necessarily for the best model, but also it is mathematically tractable for differentiations. Thus, the scatter diagrams are very significant for the identification of functional relationship between the variables. In any functional relationship such as in Eq. (4.78), there are parameters as a_{yx} and b_{yx}. The subscript of the parameters as yx indicates that the Y variable (predictand) is predicted from X predictor variable. After the decision of visual best relationship form, it is important to determine or estimate the model parameters from the available SF data values. The principle in the estimation is that the error (between any predicted, $\hat{Y}_i$ and measured, Y_i) sum squares value is the minimum. The parameters are dependent on the SFs' data, which can be expressed implicitly as.

$$a_{yx} = f_1(X, Y) \tag{4.79}$$

and

$$b_{yx} = f_2(X, Y) \tag{4.80}$$

Finally, the whole question is now how to obtain the explicit formulation of the model parameters. For this purpose, the regression procedure is used which will be explained in the following sequel.

4.20.2 Mathematical Linear Regression Model

In this section, only straight-line model is considered. The regression procedure is the search procedure for the explicit expressions of Eqs. (4.79) and (4.80). In order to grasp the question more closely, let Eq. (4.78) be written for the i-th data pairs as.

$$Y_i = a_{YX} + b_{YX} X_i \tag{4.81}$$

In Fig. 4.38, different straight-line forms are shown for different sets of model parameters.

In Fig. 4.38a, there is quite steep slope. This implies that b_{yx} parameter in Eq. (4.81) has a big value. In this manner, increase in X causes increase in Y. However, in Fig. 4.38b, the situation is just the opposite and increase in X value gives rise to decrease in Y. The former corresponds to a positive and the latter to a negative dependence. Hence, b_{yx} parameter expresses the slope of the straight

Fig. 4.38 Straight lines for different parameter sets

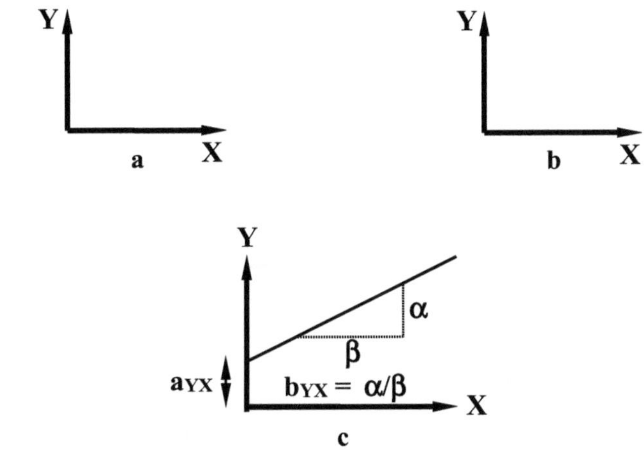

line, and it is represented geometrically in Fig. 4.38c. For some β distance along the X-axis, the corresponding vertical distance on the Y-axis is **α**, and hence the slope can be expressed as.

$$b_{YX} = \alpha/\beta \tag{4.82}$$

If $\beta = 1$, then the α value on the Y-axis gives directly the slope of the straight line which is equal to b_{yx}. This parameter is called as the regression coefficient. It is the change in Y dependent variable corresponding to each unit increment in X independent variable. In Figure 4.37c, a_{yx} corresponds to the ordinate of intercept point on the Y-axis. Up to now, the regression parameters (a_{yx} and b_{yx}) are explained in a mathematical manner completely independent of SF data values. It is already stated that these parameters should be determined such that the sum of deviations (error) squares is minimum.

4.20.3 Statistical Linear Regression Model

The straight-line parameters must be estimated from the best fitting through the scatter diagram shown in Fig. 4.32a. This means that the fitted straight line must be close as much as possible to overall scatter points. The statement "as much as possible" in this sentence implies that the variance of the points from the straight line must have its minimum value. In general, in any classical straight-line model fitting, the deviations are adopted as the vertical errors that are parallel to Y-axis as shown in Fig. 4.37a. Hence, the minimization of the total sum of error squares based on n data points can be expressed mathematically as

$$\mathrm{Min}\sum_{i=1}^{n}\left(\hat{Y}_i - Y\right)^2 \tag{4.82}$$

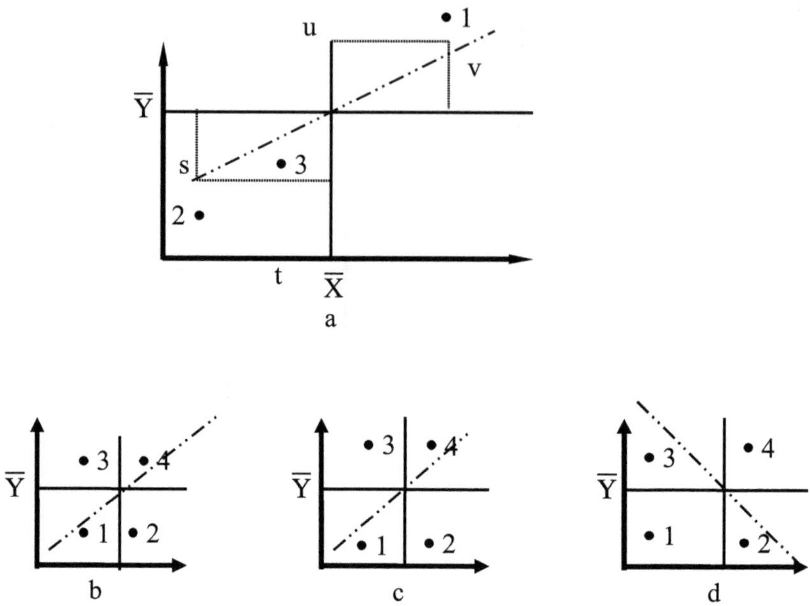

Fig. 4.39 Various deviations

where $\hat{Y}_i$ shows the predicted Y value on the straight line corresponding to i-th independent data, X_i. The expression in Eq. (4.82) is known as the least square procedure in the regression methodology. The main subject in any regression procedure is the relationship between the variances of dependent and independent variables (Y_i and X_i). With this information, let us concentrate on various Y_i and X_i points in Fig. 4.38. For better understanding, after the arithmetic averages, $\overline{X}$ and $\overline{Y}$ of the two variables, the contributions, u, v, s, and t deviations of points 1 and 2 to the overall variances, namely, $S_X{}^2$ and $S_Y{}^2$ are shown in Fig. 4.39a.

In Fig. 4.39a, the contribution from point 1 to $\mathbf{S_X{}^2}$ and $\mathbf{S_Y{}^2}$ are $\mathbf{u^2}$ and $\mathbf{v^2}$, respectively. In this manner, points 1 and 2 contributes significantly to variances $\mathbf{S_X{}^2}$ and $\mathbf{S_Y{}^2}$ because of their comparatively faraway locations from the arithmetic averages $\overline{\mathbf{X}}$ and $\overline{\mathbf{Y}}$ whereas point 3 has very little contribution. In order to commonly account for these contributions collectively, it is necessary to develop the concept of covariance. In general, covariance is defined as the average value of products of deviations from the averages. Hence,

$$\mathrm{Cov}(X, Y) = \frac{1}{n}\sum_{i=1}^{n}\left(X_i - \overline{X}\right)\left(Y_i - \overline{Y}\right) \tag{4.83}$$

For instance, in Fig. 4.33a, the covariance contributions of points 1 and 2 are

$$\left(X_1 - \overline{X}\right)\left(Y_1 - \overline{Y}\right) = uv$$
$$\left(X_2 - \overline{X}\right)\left(Y_{21} - \overline{Y}\right) = ts$$

respectively. For these two points from Eq. (4.83), the covariance can be expressed as

$$\mathrm{Cov}(X, Y) = 0.5(uv + ts)$$

In Fig. 4.39b, c, d, the regression line slope appears as the ratio between this covariance and the variance, $\mathbf{S_X}^2$, of independent variable as

$$b_{YX} = \frac{\mathrm{Cov}(X, Y)}{S_X^2} \tag{4.84}$$

It is obvious from Fig. 4.39 that the overall regression line crosses through the weight point of all $\mathbf{X_i}$ and $\mathbf{Y_i}$ data corresponding to point ($\overline{X}$, $\overline{Y}$). With this information at hand, the intercept point ordinate of the straight line on the Y-axis, the parameter $\mathbf{a_{YX}}$ can be easily calculated leading to

$$\overline{Y} = a_{YX} + b_{YX}\overline{X} \tag{4.85}$$

and

$$a_{YX} = \overline{Y} - b_{YX}\overline{X} \tag{4.86}$$

The regression parameters must be found based on "the least sum of error squares." For this purpose, the basic straight line can be rewritten for the i-th data point by taking into consideration the error term h_i as

$$Y_i = a_{YX} + b_{YX} + e_i \tag{4.87}$$

Hence, the error for i-th data value becomes as

$$e_i = Y_i - (a_{YX} - b_{YX}) \tag{4.88}$$

Furthermore, the error square is

$$e_i^2 = [Y_i - (a_{YX} - b_{YX})]^2 \tag{4.89}$$

Finally, the sum of error squares over all the available data becomes

$$H_T = \sum_{i=1}^{n} e_i^2 = \sum_{i=1}^{n} [Y_i - (a_{YX} - b_{YX}X_i)]^2 \tag{4.90}$$

In order to minimize this expression, mathematically, it is necessary to take the partial derivatives with respect to unknowns (herein the unknowns are a_{yx} and b_{yx}) and then equated to zero as follows:

$$\frac{\partial H_T}{\partial a_{YX}} = \sum_{i=1}^{n} 2[Y_i - (a_{YX} + b_{YX}X_i)](-1) = 0 \tag{4.91}$$

and

$$\frac{\partial H_T}{\partial b_{YX}} = \sum_{i=1}^{n} 2[Y_i - (a_{YX} + b_{YX}X_i)](-X_i) = 0 \tag{4.92}$$

and after the simplification,

$$\sum_{i=1}^{n} Y_i = na_{YX} + b_{YX}\sum_{i=1}^{n} X_i \tag{4.93}$$

and

$$\sum_{i=1}^{n} Y_iX_i = a_{YX}\sum_{i=1}^{n} X_i + b_{YX}\sum_{i=1}^{n} X_i^2 \tag{4.94}$$

Division of these equations by the number of data, n, leads to expressions that can be written in terms of arithmetic averages as

$$\overline{Y} = a_{YX} + b_{YX}\overline{X} \tag{4.95}$$

and

$$\overline{YX} = a_{YX}\overline{X} + b_{YX}\overline{X^2} \tag{4.96}$$

It is obvious that Eq. (4.95) is equivalent to previously obtained Eq. (4.83) and on the other hand, substitution of a_{yx} from Eq. (4.95) into Eq. (4.96) leads after the necessary algebraic manipulations to

$$b_{YX} = \frac{\overline{YX} - \overline{X}\overline{X}}{\overline{X^2} - \overline{X}^2} \tag{4.97}$$

This should have the similar interpretation with Eq. (4.97).

4.20.4 Least Squares

The essence of many statistical weather forecasting depends on this principle that the summation of the prediction error squares should be the least. Herein, the prediction error is defined as the difference between the observed and the predicted value. Simple revision of the statistical least squares technique is presented in the following sequel. Detailed account for the weather predictors and earth sciences can be found in textbooks by Wilks (1995), Neter et al. (1996), and Fuller (1996).

4.20.5 Simple Linear Regression Procedure

In this title, linearity means the simplest linear relationship between any two variables. One of these variables is called "independent" or "predictor" variable and the other is the "dependent" or "predictand" variable. In general, the predictand, say Y, is expressed in terms of the predictor, X, as.

$$Y = a + bX \tag{4.98}$$

where a and b are the simple model coefficients. This model is also referred to as the simple regression technique since herein X is regressed on Y. In any actual prediction model, there are more predictors but the ideas for simple linear regression generalize easily to this complex case of multiple linear regression. The representation of Eq. (4.98) on a Cartesian coordinate system yields to a straight line in the mathematical sense but scatter of points in the statistical sense where each one of these points are associated with the data pairs (X_i, Y_i) for n data pairs (i = 1,2, . . . , n). The relative position of the straight line must be determined in such a way that the summation of squared-deviation of each point from this straight line must be the smallest possible. Herein, the deviation is synonymously used as the error. As shown in Fig. 4.33, these deviations from the straight line may be decided as the vertical, horizontal, or perpendicular distances. Although, theoretically, the best straight-line situation is obtained by considering the perpendicular distances as errors, however, their calculations require time and effort therefore in practical studies most often vertical deviations are minimized to fix the regression line through the regression line. Vertical deviations correspond to the prediction of the predictand given the value of the predictor. It is the choice of the squared-error criterion that is the basis of the least squares regression. The choice of the sum of the squared-error criteria is convenient not only that it is necessarily the best but also it is mathematically tractable for the differentiations. As shown in Fig. 4.31, assumption that the model parameters, namely, a and b are known, the prediction of Y given the predictor value X_i yields a point on the straight line as, Y' which is expressed simple from Eq. (4.98) as.

$$Y' = a + bX_i \tag{4.99}$$

The vertical distance between the data point Y_i and the point on the line, Y' corresponding to the same X_i value of the predictor is the error amount e_i which is also referred to as the residual and it is defined objectively as.

$$e_i = Y_i - Y' \tag{4.100}$$

Hence, corresponding to each data pair (X_i, Y_i), there is a separate error amount, e_i. Depending on the data point position as to whether above or below the straight line in Fig. 4.36, the error term has positive or negative sign, respectively. This is the usual convention in statistics but is opposite to what is often seen in the atmospheric sciences, where forecasts lower than the observations (the line being below the point) are regarded as having negative errors, and vice versa (Wilks 1995).Since definition of the best line fitting is based on the squared residuals, the sign convention is not important. The regression equation with error and the data value can be written as.

$$Y_i = a + bX_i + e_i \tag{4.101}$$

This expression says that the true value of the predictand is equal to the summation of the predicted value and the residual given the predictor value. Again, the same expression gives the error and its square and the summation of whole the squared-error attached with the given data becomes

$$T = \sum_{i=1}^{n} (e_i)^2 = \sum_{i=1}^{n} (Y_i - Y©)^2 = \sum_{i=1}^{n} [Y_i - (a + bX_i)]^2 \tag{4.102}$$

where T shows the total squares residuals. This equation is the basis for determining the model parameter expressions in terms of the given data values. For this purpose, it is necessary to differentiate the total square residuals term with respect to a and b and equate these two expressions to zero. If the necessary algebraic simplifications are made, then finally, it is possible to obtain

$$a = \overline{Y} - b\overline{X} \tag{4.103}$$

and

$$b = \frac{n\sum_{i=1}^{n} X_i Y_i - \sum_{i=1}^{n} X_i \sum_{i=1}^{n} Y_i}{n\sum_{i=1}^{n} (X_i)^2 - \left(\sum_{i=1}^{n} X_i\right)^2} \tag{4.104}$$

It must be noticed that the slope parameter, b, is equivalent to the Pearson correlation coefficient and intercept parameter, a, in Eq. (4.102) says that the regression straight line passes through the centroid of the data scatter. This is tantamount to saying that the regression line passes through the arithmetic averages of predictand and predictor. Hence, the calculation of the intercept parameter is very simple and requires only the calculation of the arithmetic averages. The calculation of the slope parameter can be simplified if the standard variates of the predictand and predictor sequence are considered. Let us consider that these two standard variates are x_i and y_i, then the slope parameter simplifies to

$$b = \frac{1}{n} \sum_{i=1}^{n} x_i y_i = \overline{xy} \tag{4.105}$$

which verbally says that the slope parameter is equal to the arithmetic average of the cross multiplication of the standard predictand and predictor variates.

4.20.6 Residual Properties

Fitting the best straight line through the scatter of data points necessitates the mathematical procedures as explained above. For a successful application of the regression method, the residuals should have some properties. Among the most significant of these are that the summation of the residuals is equal to zero by definition and it is expected that the variance of the residuals is constant, that is, does not change with the value of the predictor. Furthermore, in any regression technique, the residuals should also be independently distributed. Often, the additional assumption is made that these residuals follow a Gaussian distribution.

Since the summation of the residuals is equal to zero, equivalently, the arithmetic average of the residuals is also equal to zero. This statement does not give any idea about the deviations around the regression line and therefore average deviations can be assessed by the standard deviation of the residuals. Residuals are expected to scatter randomly about some mean value. Therefore, a given residual, positive or negative, large or small is by assumption equally likely to occur at any part of the regression line. The distribution of residuals is less spread out which means that they have smaller variance than the distribution of Y, reflecting less uncertainty about Y if a corresponding X value is known. In order to make statistical inferences in the regression setting, it is necessary to estimate the constant variance estimation of the residuals from the sample of residuals as defined from Eq. (4.100). Since the sample arithmetic average of the residuals is equal to zero in the calculation of residual variance, two degrees of freedom (because a and b have been estimated) must be taken into consideration and hence

$$V(e) = \frac{1}{n-2}\sum_{i=1}^{n} e_i^2 \tag{4.106}$$

In any regression model the total variance is equal to the summation of two different variances, namely, variance of the regression predictions and the variance of the errors. Since, each one of these is variances, then the factor 1/n cancels each other and finally the following statement becomes valid. That is that the sum of squares of the total data, SST, is equal to the summation of the sum of squares of the regression prediction, SSR and the sum of squares of the errors. This can be expressed as.

$$SST = SSR + SSE \tag{4.107}$$

This expression also means that the variation of the predictand, Y, and a partitioning of that variation between the portion represented by the regression and the unrepresented portion ascribed to the variation of the residuals. The explicit forms of the terms in Eq. (4.107) can be written as follows:

$$SST = \sum_{i=1}^{n} Y_i^2 - n\overline{Y}^2 \tag{4.108}$$

$$\mathrm{SSR}=\sum\nolimits_{i=1}^{n}\left[\mathrm{Y©}-\overline{\mathrm{Y}}\right]^{2} \tag{4.109}$$

and

$$\mathrm{SSE}=\sum\nolimits_{i=1}^{n}e_i^2 \tag{4.110}$$

By rearranging Eq. (4.107), the error variance can also be defined as

$$\begin{aligned} V(e) &= \frac{1}{n-2}(\mathrm{SST}-\mathrm{SSR}) \\ &= \frac{1}{n-2}\left\{\sum\nolimits_{i=1}^{n}Y_i^2 - n\overline{Y}^2 - b^2\left[\sum\nolimits_{i=1}^{n}X_i^2 - n\overline{X}^2\right]\right\} \end{aligned} \tag{4.111}$$

In the theoretical background of the regression methodology, there are six basic assumptions as mentioned earlier in Sect. 4.19.

4.21 Cluster Regression Analysis

This analysis methodology was proposed by Şen et al. as an alternative to the classical approach, but with clustering structure along the linear model. In the continuation below, much of the text is reflections from this article.

As questions of gradual (trend) or sudden (shifts) climatic change have received attention in recent years (rising concerns about the effects of greenhouse gasses, GHG, on the climate); most of the research is based on temperature and precipitation data but in this section lake level change records are taken for practical application of the cluster regression analysis, which can be applied to any other time series records even in industry. This point has been evaluated in detail by Slivitzky and Mathier (1993). Along this line of research, Hubert et al. (1989), Vannitsem and Demaree (1991), and Sneyers (1992) used statistical methods to show that temperature, pressure, and runoff series in Africa and Europe have changed several times over this and previous centuries. On the other hand, as noted by Slivitzky and Mathier (1993), most modeling of levels and runoff series in the Great Lakes assumed stationarity of time series using the autoregressive-moving average (ARIMA) processes introduced by Box and Jenkins (1974). It is assumed that lake level fluctuations do not have stationarity property, and therefore, classical models such as ARIMA processes cannot reliably stimulate lake levels.

Multivariate models using monthly lake variability failed to adequately reproduce statistical characteristics and persistence of watershed resources (Loucks 1989; Iruine and Eberthardt 1992). Spectral analysis of water levels pointed to the possibility of significant trends in lake level hydrological variables (Privalsky 1990; Kite 1990). Almost all these scientific studies have relied heavily on the existence of an autocorrelation coefficient as an indicator of long-term persistence in lake level time series. However, many researchers have shown that shifts in mean lake level can introduce unrealistic and spurious autocorrelations. This is the main reason why classical statistical models often fail to reproduce the statistical properties. However, Mathier et al. (1992) were able to adequately reproduce the statistical properties of a shifting-mean model. In this section, a cluster, linear regression model is developed and then used to simulate monthly lake level fluctuations in a way that retains statistical properties and correlation coefficient. The method is applied to water level fluctuations in Lake Van in eastern Turkey.

4.21.1 *Study Area*

Lake Van, the world's largest soda lake, is located on the Anatolian high plateau in eastern Turkey (38.5°N and 43°E) (Fig. 4.40). The Lake Van region experiences very harsh winters with temperatures often below 0 °C. Most precipitation falls in the form of snow in winter, and heavy rains occur in late spring.

During spring snowmelt, high runoff rates occur, during which more than 80% of the annual discharge reaches the lake. The summer period (July to September) is hot and dry with an average temperature of 20 °C. The daily temperature changes are about 20 °C. Lake Van has a large drainage basin of 12,500 km^2. The lake surface currently averages 3600 km^2 (Kempe et al. 1978). The surface is at an altitude of about 1650 m above sea level. The lake is surrounded by hills and mountains reaching 4000 m.

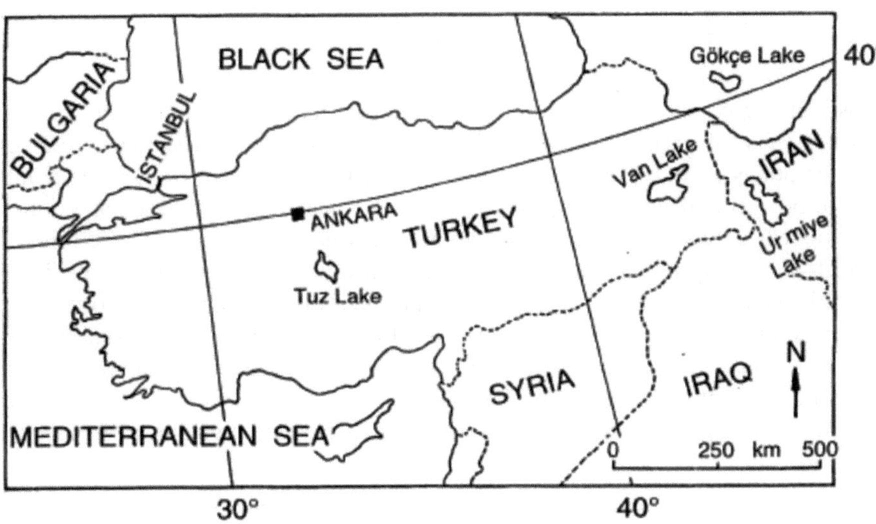

Fig. 4.40 Location map

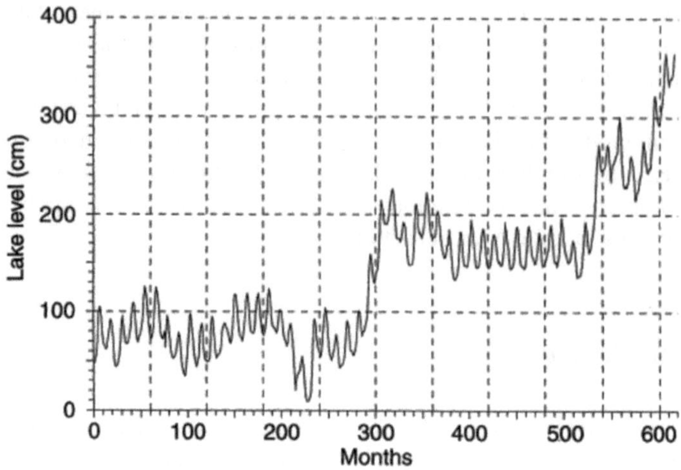

Fig. 4.41 Lake Van level fluctuations (1944–1994)

Lake Van has been exposed to a net water level rise of about 2 m in the last decade and as a result, the low, flooded areas along the coast are now causing problems for local administrators, governmental officials, irrigation activities, and people's property. Figure 4.41 shows the 50-year lake level fluctuations between 1944 and 1994 from January to June each year, the water level rises and falls in the second half of the year.

4.21.2 Cluster Regression Model

Classical regression analysis has several assumptions about the normality and independence of residuals. Also, an implied assumption that is skipped from the considerations in most regression line implementations is that the scatter points should be evenly distributed around a line. Unfortunately, this assumption is often overlooked, especially if the scatter diagram has not been drawn. If the original records are homogeneous and stable, a uniform distribution of points along the line is possible without any shifts, trends, or seasonality. If there are level shifts over time, the scatterplot will contain clusters of points along the regression line. Confirmation of such clusters is evident in Fig. 4.42, which shows a lake level lag-one scatter diagram for monthly records from Lake Van.

The following conclusions can be drawn from the interpretation of the scatter diagram:

(a) The lag-one scatter diagram shows a general straight-line relationship between successive lake level occurrences. The existence of such a line corresponds to the first-order autocorrelation coefficient in monthly lake level time series. Therefore, persistence at lake level is maintained with this straight line.

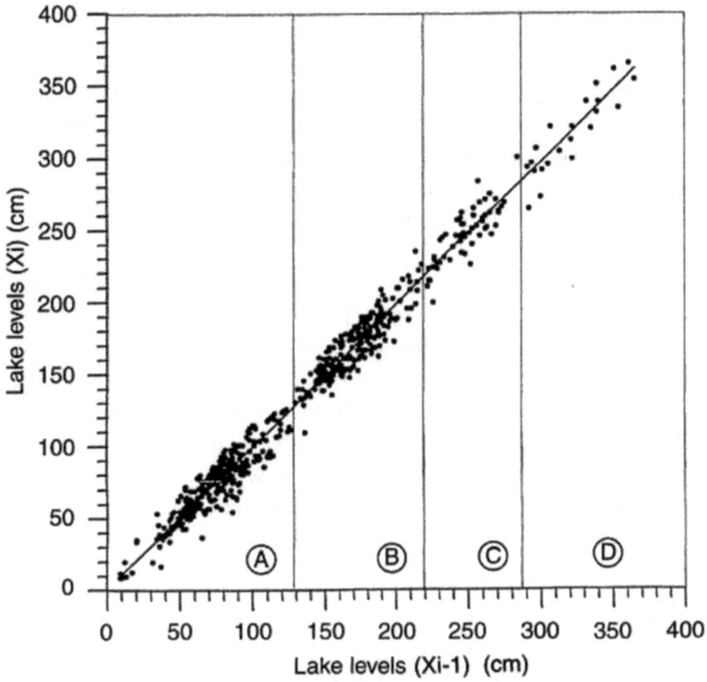

Fig. 4.42 Lag-one water level fluctuations and cluster boundaries

(b) The scatter of points around the straight line is constrained within a narrow band, meaning that the forecast of near-future levels cannot differ much from the current level provided there is no shifts in the data.
(c) There are different cluster regions along the straight line. Such clusters are not expected in the classical regression approach, but the presence of these clusters transforms the classical regression analysis into a cluster regression analysis, the basis of which will be presented later in this section. Separate clusters correspond to periods of shifted lake level.
(d) Classical regression analysis provides a basis for estimating water levels by runoff, but in the cluster regression line approach, reliable estimates are possible only provided the probability of cluster formation is considered. Here, questions arise about which cluster should be taken in the predictions for the? Should the future prediction stay in the same cluster? Any transition from one cluster to another means a shift in the water level. Therefore, one needs to know the transition probabilities between the various clusters. Cluster regression shows not only the autocorrelation coefficient but also the influence domain of each cluster along the horizontal axis as A, B, C, and D as shown in Fig. 4.42. Influences domains help calculate transition probabilities between the clusters from the original water level records.
(e) For any current cluster, it is possible to predict future normal lake levels using the regression line equation.

The following steps are required for reliable estimates through cluster regression analysis:

1. To decide which domain of influence (A, B, C or D) should be considered first, a uniform distribution function taking random values between 0 and 400 cm.
2. Generate a uniformly distributed random number and accordingly decide on the next cluster, considering influence domains. For example, if the uniformly distributed random number is 272 then the current cluster from Fig. 4.42 influence domain C will be.
3. Generate another uniformly distributed random number and if the level remains within the same cluster, use the regression equation to estimate. Otherwise, take the average water level value in the new cluster. The new level will be considered the midpoint of the cluster areas in Fig. 4.42. A better estimation would be based on regenerating the random variable from a uniform distribution constrained within the variation space of each cluster. Also, the value found in this way will be added to a random residual value. This will then provide the basis of the future water level estimates within the same cluster area.

4.21.3 Application

Cluster regression approach was applied here to the recorded water level fluctuations of Lake Van. For this purpose, various lag scatter points of successive levels are first plotted in the Figs. 4.43 and 4.44. The overview of these figures implies the applicability of the cluster regression equation steps as mentioned in the previous section. All figures have straight lines and the transition boundaries between clusters A, B, C, and D are given in Table 4.6 in addition to the boundaries of each cluster at different lags up to 9 years.

Here, A is considered as a low lake-level cluster, where only transitions from low level to low level are allowed. B and C refer to lower medium and upper medium level clusters and, finally, D is the cluster that includes highest levels only.

It is obvious from Table 4.6 that the transition limits between A and B, and B and C are practically constant on average for all lags and equal to 129 and 219, respectively. However, the upper limit transition between C and D increases with the increase in the lag value. The difference between the first and ninth lags has a relative error percentage of $100 \times (308 - 285)/308 = 7.4$ which may be regarded as small for practical purposes.

Scatter diagrams in Figs. 4.43 and 4.44 gives the following specific comments for fluctuations in Lake Van level.

(a) The scatter diagrams have four clusters with the denser point concentration in cluster A representing low water level followed by low water levels. Extreme values of water level fluctuations have the least frequency of occurrences in cluster D.

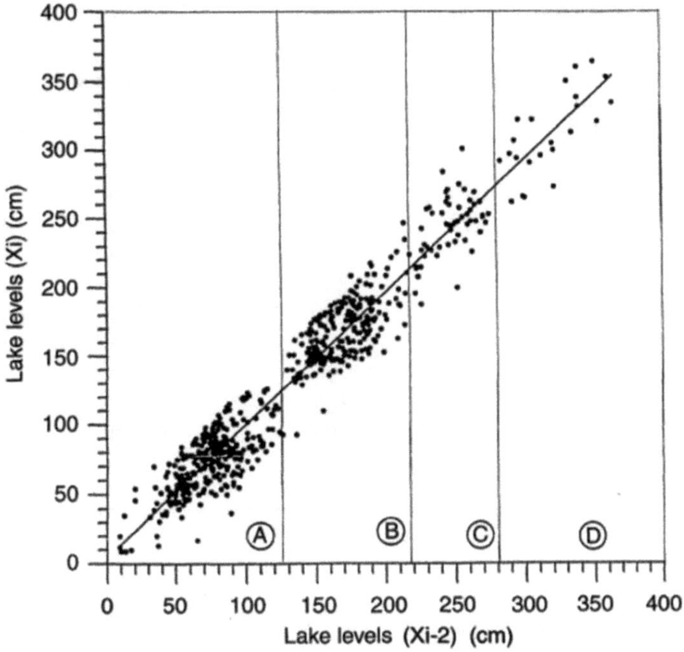

Fig. 4.43 Lag-two water level fluctuations and cluster boundaries

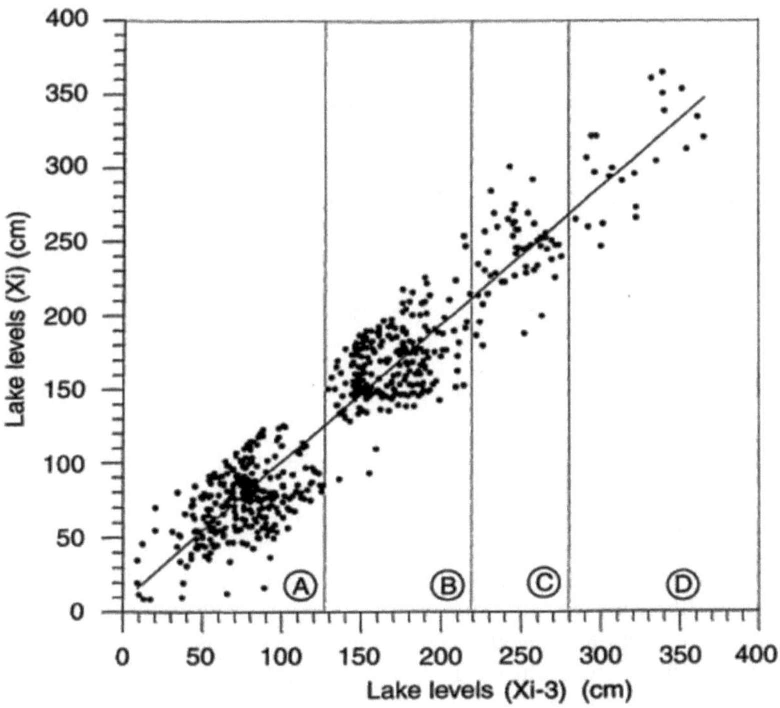

Fig. 4.44 Lag-three water level fluctuations and cluster boundaries

Table 4.6 Cluster regression boundaries and coefficients

Lag	Transboundary values			Regression coefficients	
	A-B	B-C	C-D	a	b
1	130	220	> 285	0.985	1.459
2	125	218	> 280	0.960	4.564
3	129	215	> 280	0.930	8.239
4	130	219	> 281	0.901	11.725
5	128	222	> 280	0.878	14.486
6	128	212	> 287	0.862	16.293
7	130	221	> 296	0.853	17.114
8	132	225	> 302	0.852	16.835
9	131	222	> 308	0.858	15.532
Average	129	219			

(b) Regardless of the lag value, the points in the scatter diagram deviate from the regression line within a narrow band. This indicates that once the water level is within a cluster, it will be relatively very likely to remain within this cluster, as will be discussed later in this section. Also, transitions between clusters are expected to occur quite rarely, and indeed only between the adjacent clusters.

(c) In any of the scatter diagrams, the transition of water level from one cluster to another nonadjacent cluster is not possible. This can be verified from the computed transition matrix elements, since there are no elements along the main and the two diagonals. Here, only lag-one regression line will be considered to model lake levels while considering transition probabilities between adjacent clusters. Monthly level time series data for Lake Van from 1944 to 1994 yields a lag-one transition probability matrix, [M] as follows:

$$[M] = \begin{array}{c} \\ A \\ B \\ C \\ D \end{array} \begin{array}{c} \begin{array}{cccc} A & B & C & D \end{array} \\ \begin{bmatrix} 291 & 2 & 0 & 0 \\ 1 & 233 & 4 & 0 \\ 0 & 3 & 54 & 2 \\ 0 & 0 & 1 & 20 \end{bmatrix} \end{array} \quad (4.112)$$

The diagonal values in this matrix are the number of passes within each cluster. For example, there are 291 transitions from low to low levels within cluster A. In the same matrix, intercluster transitions occur quite rarely along diagonals, for example as four transitions from cluster B to C. In classical stochastic processes, the calculation of transition matrix elements is based on the basic assumption that the process is time reversible. This is equivalent to saying that transitions as A → B is the same as B → A. As a result, the resulting matrix should be symmetrical. However, in the proposed cluster regression method, only one-way transitions into the future along the time axis are allowed. This means that the transition along the time axis is irreversible. As a result, the transition matrix is not symmetrical. Accordingly, the matrix in Eq. (4.1) is not symmetric; the transition from C to B is 3, not 4. Zero values next to the off diagonals indicate that water levels can only move to adjacent clusters. Therefore, possible transitions are A, B, C, and D only. For example, the transition to cluster C is possible four times from B, 54 times from previous C, and only once from D with no transition from A (hence a total of 59 transitions). Column values indicate the transition from other clusters to the cluster under consideration and the transition probabilities can be calculated by dividing each value in the column by the column total. Therefore, the transition probability matrix [P] is from Eq. (4.112) as.

$$[P] = \begin{array}{c} \\ A \\ B \\ C \\ D \end{array} \begin{array}{c} \begin{array}{cccc} A & B & C & D \end{array} \\ \begin{bmatrix} 0.9932 & 0.0068 & 0 & 0 \\ 0.0042 & 0.9790 & 0.0168 & 0 \\ 0 & 0.0508 & 0.9152 & 0.0339 \\ 0 & 0 & 0.0476 & 0.9524 \end{bmatrix}_{(2)} \end{array} \quad (4.113)$$

The linear regression line associating two successive water levels, namely, Wi and W (i-1), can be obtained from the cluster scatter diagram in Fig. 4.44 as

$$W_i = 0.9858W_{i-1} + 1.45918 + \epsilon_i \quad (4.114)$$

where ϵ_i denotes vertical random deviations from the regression line. Theoretically, these random deviations should have a normal (Gaussian) distribution function for the regression line to be valid, and Fig. 4.45 shows that they are normally distributed. To adopt Eq. (4.114) estimates with cluster scatters, it is essential to consider the following steps:

(a) Since the most frequently occurring water levels are confined in cluster A, the initial state W_0 is randomly chosen from the actual water levels in this cluster.
(b) The decision of whether there is a transition to the next cluster is provided by the transition probabilities given in matrix [P] in Eq. (4.113). Transitions take place according to the following rules:

   1. The transition to cluster A is possible only from cluster B or the level remains in the same cluster. From the transition matrix, their probabilities are 0.9932 and 0.0042 and their sum equals 1.0. In order to decide which of these two clusters will be effective in the next time step, it is necessary to generate a uniform random number, n_i, ranging from zero and one.

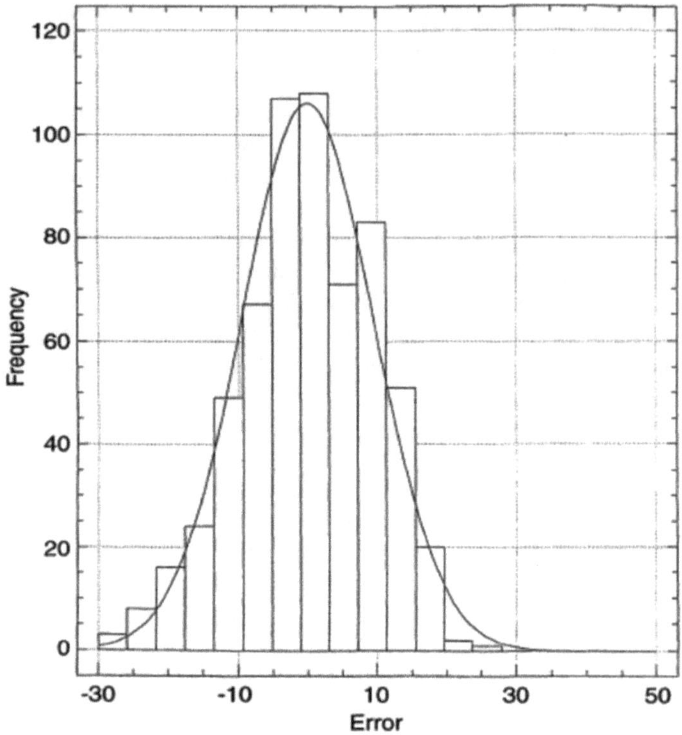

Fig. 4.45 Regression error distribution function

If ni < 0.9932, the water level will remain within cluster A, otherwise for 0.9932 < ni < 1.0 a transition from cluster A to B. In the first case, after generating a normally distributed random number, the new water level value, ϵ_i, is generated using the cluster regression model in Eq. (4.114). However, in the latter case, the water level will be chosen randomly from the range of water levels for cluster B.

2. At transition from two adjacent clusters (A or C)to cluster B can occur at any time. The transitional probabilities from A and C are 0.0068 and 0.0508, respectively, and the complementary probability of 0.9790 remains in cluster B. Now the transition decision to B will have three independent regions of uniform distribution, that is, if 0 < ni <0.0068 a transition from A to B occurs or 0.0068 < ni <0.9858 when the water level stays within cluster B, and finally, 0.9858 < ni, <1.0 a transition from C to B occurs. If the water level stays within cluster B, it is produced as a normal variable ϵ_i and the regression expression in Eq. (4.114) is used to predict the next water level. In transition states, the water level is randomly displayed from the current levels.

(c) Transitions to cluster C show a similar mechanism to cluster B, with different transition probabilities but the same generating mechanism.

(d) Finally, besides staying in the same cluster, only transition from cluster C to cluster D is possible. The application of all these procedures and steps to monthly water level changes in Lake Van results in the development of the synthetic transition matrix $[M_s]$.

$$[M_s] = \begin{array}{c} \\ A \\ B \\ C \\ D \end{array} \begin{array}{c} \begin{array}{cccc} A & B & C & D \end{array} \\ \begin{bmatrix} 287 & 2 & 0 & 0 \\ 1 & 230 & 3 & 0 \\ 0 & 4 & 56 & 2 \\ 0 & 0 & 0 & 19 \end{bmatrix} \end{array} \quad (4.115)$$

Comparison of corresponding elements is shown between two matrices in Eqs. (4.112) and (4.113) differ with a relative error of less than 5%. This indicates that the conservation of transition numbers as probabilities at predicted lake levels is indistinguishable from actual water level data. Synthetic cluster scatter diagram derived from the use of Eqs. (4.113) and (4.114) is shown in Fig. 4.46 where the regression line has the following form:

$$W_i = 0.978 W_{i-1} + 1.47 + \varepsilon_I \quad (4.116)$$

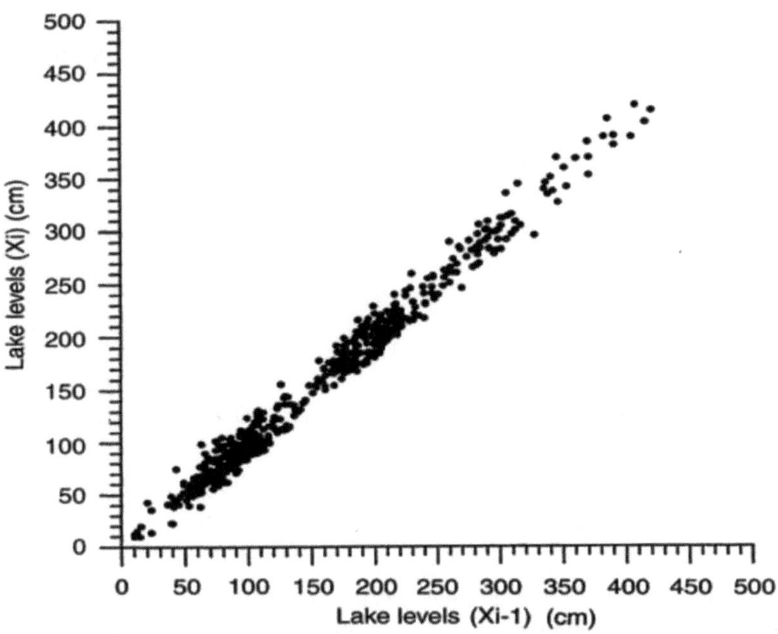

Fig. 4.46 Synthetic lag-one water level fluctuations

Again, comparing this expression with Eq. (4.114) shows that the corresponding coefficients vary with a relative error of less than 5%. In other words, the autocorrelation coefficient in the estimation of water levels is maintained despite shifts from the original data.

4.22 Trend Identification Methodologies

One of the most important techniques in different disciplines is the identification of possible systematic variations in the form of a trend tendency within a given time series. As mentioned earlier, any time series record has, in general, three distinctive parts as the periodic (seasonality), systematic (trend), and stochastic (uncertainty) components. Trend methods cover most of the common concern of researchers and practitioners in different discipline studies related to the non-stationarity of natural events. In all the methods, a time series is considered with n number of measurements notationally as X_i ($i = 1, 2, \ldots, n$).

4.22.1 Mann-Kendal (MK) Test

This is a nonparametric trend test for identification of a holistic and monotonic straight line through the whole record length. It is first suggested by Mann and later with some modification by Kendall (1975) and additional improvements by Hirsch and Slack (1984) with seasonality component. Kendall's tau is used for its calculation, which measures association between two ranked samples provided that the data have independent structure. The MK test statistic, S, is defined as

$$S = \sum_{i=1}^{n-1} \sum_{j=i+1}^{n} \text{sign}(X_j - X_i) \tag{4.117}$$

where

$$\text{sign} = \begin{cases} +1 \text{ if } (X_j - X_i) > 0 \\ 0 \text{ if } (X_j - X_i) > 0 \\ -1 \text{ if } (X_j - X_i) < 0 \end{cases}$$

The PDF of S should have a normal (Gaussian) PDF with the long-term arithmetic average equal to zero and variance as

$$V(S) = \frac{1}{18}\left[n(n-1)(2n+5) - \sum_{k=1}^{m} t_k(t_k - 1)(2t_k + 5)\right] \tag{4.118}$$

where k is the number of tied sets and t_k is the size of k-th group. The test statistics can be obtained by standardization of S variable with parameters, which leads to

$$Z = \begin{cases} \frac{S-1}{\sqrt{V(S)}}, & \text{if } S > 0 \\ 0 & \text{if } S = 0 \\ \frac{S+1}{\sqrt{V(S)}}, & \text{if } S < 0 \end{cases} \quad (4.119)$$

Based on these statistics, MK test hypothesis decision is given in many textbooks (Naghettini 2016). The slope of the trend can be estimated by the regression line, but better by the Sen procedure as will be explained in the subsequent section. Prior to its application, the following assumptions must be satisfied, otherwise its results are questionable.

1. The measurements should have independent structure (at least the first-order Pearson correlation must be practically zero) with identical PDF.
2. The number of data should not be less than 30 to avoid biased estimates.
3. Normally, (Gaussian) PDF yields better results according to the theory, but this is not very restrictive assumption.

This trend methodology is applicable even though there might be missing values in the records, but the performance of the test is adversely affectable by such events.

4.22.2 Sen Slope (SS)

MK trend test has no slope calculation method but tells only whether any trend exists in a time series. Sen proposed a simple but effective way of slope calculation by considering all the possible slopes between any pair of data, and then the median of these slopes yields a single slope, S, value as.

$$S = \text{median}\left(\frac{X_j - X_i}{j - i}\right)(j > i) \quad (4.120)$$

The MK monotonic trend mathematical formulation is in the form of a linear function with slope, S, and a crossing point as the half time of the record and the arithmetic average of the data values.

4.22.3 Regression Method (RM)

This is a parametric method for trend identification. It is commonly and frequently used methodology in the statistical evaluations as already explained in Sect. 4.8. Given a time series, a linear regression line is fitted to the data, which provides the systematic change of average data values by time. This method is used quite frequently for trend identification, but again without consideration of the underlying restrictive assumptions as already explained in Sect. 4.19.

4.22.4 Spearman's Rho Test (SR)

This is another nonparametric method, which compares two time series as to identify their similarity. For this purpose, both series are ranked in ascending orders and then Spearman's Rho is calculated depending on the comparison of the ordered data values according to the following expression (Spearman 1904). Since the application of this methodology requires two separate time series, for its consideration for trend identification a given time series is divided into two parts as "early" and "late" time durations. This method is identical with the Spearman's rank correlation as explained in Sect. 4.18.

4.22.5 Pettitt Change Point Test

Pettitt (1979) has considered that a given time series has a change point at time instant, τ, if the PDF prior and posterior parts to this point have different types. This method does not require that the prior identification of the time instant at which the change is supposed to have occurred. Its null hypothesis states no change point in time series. Like MK test, the sign matrix elements are defined as follows:

$$D_{i,j} = \text{sign}\left(X_i - X_j\right) \tag{4.121}$$

The sum of sub-matrices, $U_{t,n}$, of the matrix $D_{i,j}$ results in the Pettitt statistic as

$$U_{t,n} = \sum_{i=1}^{t} \sum_{j=t+1}^{n} D_{i,j} \tag{4.122}$$

This statistics is computed for t values from 1 to n, which can be achieved by the following expression (Rybski and Neumann 2011) as

$$U_{t,n} = U_{t-1,n} + \sum_{j=1}^{n} \text{sign}\left(X_t - X_j\right) \tag{4.36}$$

The test is two-tailed and the Pettitt statistic for final decision is given by the following expression:

$$K_n = \max_{1 \le t \le n} \left|U_{t,n}\right| \tag{4.123}$$

More detailed information about the change point search in a given time series has been presented by Şen (2017) through the successive average method (SAM).

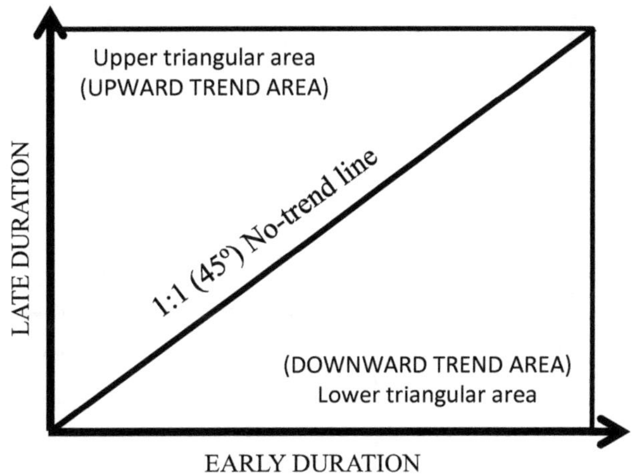

Fig. 4.47 ITA template

4.22.6 Innovative Trend Analysis (ITA)

This is the most recent technique that gives way first to subjective visual inspection followed by objective numerical analysis for trend analysis. Like the Spearman's Rho test, ITA also considers the early and late parts of a given time series. Each one of the half series is ranked in ascending order and then the early part on the horizontal axis is plotted against the late duration part, which provides the scatter of points. The scatter diagram is given equal axis form with the same scales on each axis. Such a scatter diagram is referred to as the ITA template, which has 1:1 (45°) straight line as the no-trend diagonal. The upper (lower) triangular area is for the increasing (decreasing) trend occurrences as in Fig. 4.47.

As shown also by Folland et al. (2001), the classical indication of climate change is given with relative locations of two probability distribution functions (PDFs) with a shift toward hot or cold temperature direction to indicate monotonic increasing and decreasing trend lines as in Fig. 4.48.

The corresponding representative ITA templates are below each PDF climate change graphs. Donat and Alexander by using respective PDFs of their 60-year data using two 30-year periods show that the PDF of the variable has significantly shifted towards hotter values in the latter period compared to the first period. This is tantamount to saying that the time series of 60-year duration (1950–2010) is divided into two equal parts as 1950–1980 and 1981–2010, which is the same concept as already mentioned by Şen (2012, 2014). A close inspection of this figure leads to the following major points that are important for the methodological developments in future applications:

1. Two different PDFs imply two different time series (X_1, X_2, ... , $X_{n/2}$) and ($X_{n/2+1}$, $X_{n/2+2}$, ... , X_n) in terms of two halves of a given time series (X_1, X_2, ... , X_n) with n number of data. Each half with arithmetic average values, $\overline{X}_1$ m_1 and m_2, respectively.

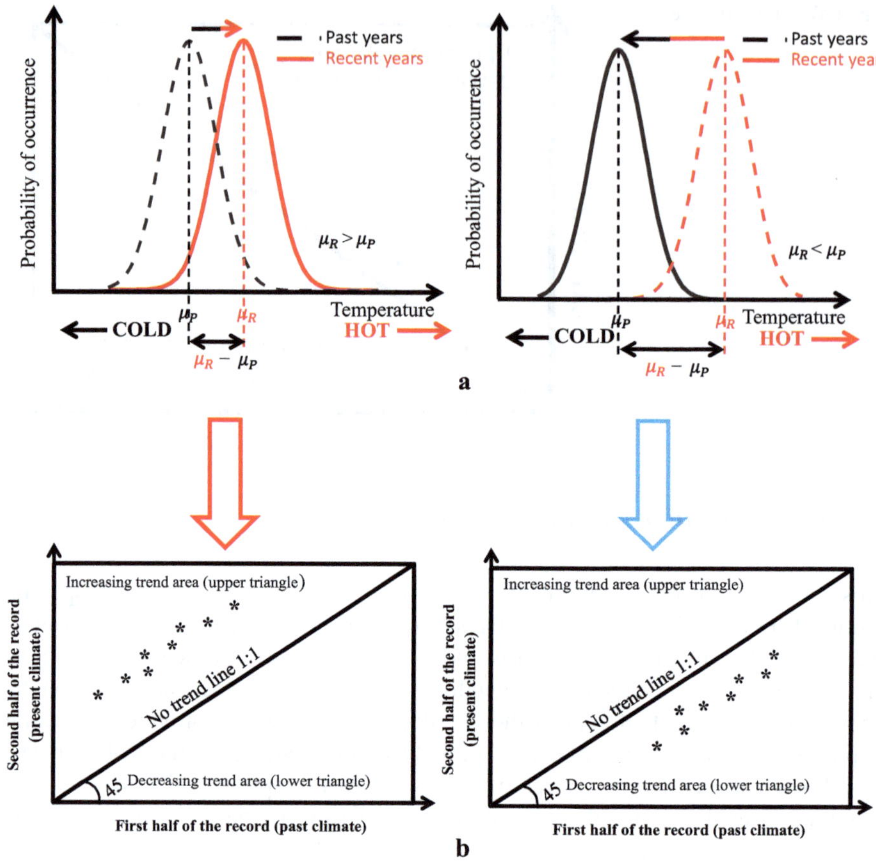

Fig. 4.48 PDF and ITA trends (**a**) increasing, (**b**) decreasing

2. If the shapes of the PDFs of the two halve time series are the same without any shift in time, then time series has the stationarity property, that is, statistical parameters do not change with time. This also implies that there is no trend in given time series.
3. For trend detection rather than the consideration of the PDFs, it is equivalently valid to compare the arithmetic averages. In case of significant difference between them, one can conclude that there is a trend in the given time series. The trend slope is related to the difference in the arithmetic averages, $\pm\Delta\mu$, as shown in Chap. 4, Sect. 4.22.6, Fig. 4.48. The measure of significant difference can be expressed through the relative error percentage, α, which is defined as follows:

$$\alpha = \pm 100\frac{\Delta\mu}{\mu} \tag{4.124}$$

If $\alpha < \pm 5\%$, then there is no significant trend.

4. It is also important to know the time shift, Δt, between the two PDFs, which practically corresponds to the time series shift time. The change of arithmetic mean per time is equal to the trend slope, S_μ, as

$$S_\mu^{\pm} = \frac{\mu - (\mu \pm \Delta\mu)}{\Delta t} = \pm \frac{\Delta\mu}{\Delta t} \quad (4.125)$$

Considering the innovative trend methodology (Şen 2012, 2014), the practical correspondence of the arithmetic mean change with time can be logically defined from the time series on the right-hand side in Fig. 4.40 as follows:

$$S_m = \frac{m_2 - m_1}{\left(\frac{n}{2}\right)} \quad (4.126)$$

where m_1 and m_2 are the arithmetic means of the halves and n is the number of data in a given time series. Depending on the sign in Eq. (4.127), it can be decided whether the trend has increasing (positive) or decreasing (negative) behavior. The trend slope in Eq. (4.12) is an alternative to the classical Sen slope, S, calculation procedure in Mann-Kendal trend analysis.

$$S = \text{median}\left(\frac{X_i - X_j}{i - j}\right) \quad \text{for} \quad (i > j) \quad (4.127)$$

where X_i (X_j) is the *i*-th (*j*-th) time-series value. Philosophically and logically, there is a conceptual difference between this, and the suggested alternative slope formulation proposed in Eq. (4.127).

Standard deviation relative positions to A (B and C PDFs) are indicators of variability in a given time series. Significant difference between the standard deviations based on relative error percentage, again like Eq. (4.124), implies variability in the time series structure. The change in standard deviation per time period is the definition of the slope of variability and thus logically like Eq. (4.122) it is possible to express the slope of variability as

$$S_\sigma^{\pm} = \frac{\sigma - (\sigma \pm \Delta\sigma)}{\Delta t} = \mp \frac{\Delta\sigma}{\Delta t} \quad (4.128)$$

Practically speaking, the slope of variability, S_v, can be written logically like Eq. (4.3), considering the two standard deviations, s_1 and s_2, as shown in the right-hand time series for the first and the second halves in Fig. 4.2, as

$$Sv = \frac{s_2 - s_1}{\left(\frac{n}{2}\right)} \quad (4.129)$$

Equations (4.126) and (4.129) can be mathematically misinterpreted as $n \to \infty$ considering the slopes to be zero. This is not a valid statement because each of these formulations defines the slope per time interval.

4.23 Future Directions and Recommendations

Although there are statistical techniques that are helpful in data processing to identify the holistic features of a given time series, many of them have fundamental restrictive assumptions that may be relieved by means of future methodological developments or with close care about these assumptions. In the following are brief set of suggestions and recommendations for refined statistical calculations.

1. Prior to the application of any statistical method, the applicant should know the assumptions and try to act accordingly for reliable calculations and conclusions.
2. It is recommended strongly that prior to any statistical methodological indulgent to look for visual inspection of data through the scatter diagrams, which provide linguistic information, which directs for convenient statistical methodology choice.
3. Statistical tests are concerned with significance tests, which depend on certain PDFs, and therefore, prior to the statistical methodological applications, it is necessary to search the most suitable PDF and compare it with the basic assumptions. For instance, in some methods the normal (Gaussian) PDF is a must.
4. In using ready software for the statistical data processing purposes, the researcher should have a sound background about the statistical method used, otherwise s/he will not be able to make plausible interpretations, but only classical, mechanical, and commonly known knowledge can be reached.
5. Unfortunately, in some circles, statistics is regarded as one of liars, which is an indication of statistical ignorance, because statistics are sound and reliable way to extract meaningful interpretations leading to practically applicable conclusions.
6. It must be kept in mind that each statistical parameter, test, and methodology has logical fundamentals and once they are not grasped then one cannot fall into statistical pitfall with invalid interpretations.

The use of statistical techniques is very essential in data processing works, especially for meteorology, climate change, and hydrology literature with plausible efficiency, which is missing in many publications due to careless concern about the basic restrictive assumptions and limitations. This chapter provides a common interest for statistical technique employments in data processing with logical, rational, physical, and plausible principles. In the text, reasons for statistical methodological usage are presented with explanations that it is the most convenient numerical technique to treat especially uncertainties in a given data set. In meteorology, climatology, hydrology, and water resources, research exclusion of the statistical methods is impossible, and therefore, the practitioner and researchers alike must take into consideration the basic science philosophical, logical, and practical aspects of the methods. In this chapter, the basic statistical parameters are explained in detail with mathematical and physical applications. The most extensively used statistical methodologies, such as in the regression technique and especially trend identification methods, are explained in the simplest manner. Finally, a list of recommendations for the present and future researches are presented for statistical methodologies and observance of these principles is expected to pave the research way in the best possible manner.

4.24 Conclusions

Everyday life and even physical phenomena reside in the world of uncertainty in precise components that can be approximated with various tools such as probability, statistics, logical rules, mathematical equations, and artificial intelligence (AI) or many other modeling styles. It is stated that the best models are those with minimum uncertainty components. Uncertainties take place in the form of randomness, vagueness, ambiguity, complexity, blur, fuzzy, chaos, nonlinearity, and turbulence. In analytical modeling, such uncertainty components are removed from visualization and conceptualization through many assumptions such as homogeneity, isotropy, uniformity, stationarity, linearity, homoscedasticity, and idealization. Probability and statistical methodologies are explained simply as shallow learning principles, which are the foundations for furthermore complex modeling alternatives. In this chapter, classical regression and trend determination methodologies are explained with examples.

The reader is encouraged to pay attention to key assumptions and simplification considerations such as data reliability, model building, and minimization of error between model output and measurement value, if supervised learning procedure is employed.

References

Albert DZ (1992) Quantum mechanics and experience. Harvard University Press, Cambridge, MA

Albert DZ (1994) Bohm's alternative to quantum mechanics. Sci Am 70(5):58–67

Almazroui M, Şen Z (2020) Trend analyses methodologies in hydro-meteorological records. Earth Syst Environ 4(1):713. https://doi.org/10.1007/s41748-020-00190-6

Anderson RL (1942) Distribution of the serial correlation coefficient. Ann Math Stat 13:1–13

Benjamin JR, Cornell CA (1970) Probability, statistics, and decision for civil engineers. McGraw-Hill, New York

Bohm D (1978) The implicate order: a new order for physics. Process Stud 8(2):73–102

Bohm D (1980) Wholeness and the implicate order. Routledge and Kegan Paul, London

Box GEP, Jenkins GM (1974) Time series analysis forecasting and control. Holden-Day, San Francisco, Calif., pp 489

Carlton AG, Follin JW (1956) Recent developments in fixed and adaptive filtering, Agardograph, vol 21

Davis JC (2002) Statistics and data analysis, 3rd edn. Wiley, New York, p 257

Feller W (1967) An introduction to probability theory and its applications, vol I, p 527

Folland CK, Rayner NA, Brown SJ, Smith TM, Shen SSP, Parker DE, Macadam I (2001) Global temperature change and its uncertainties since 1861. Geohys Res Lett 28:2621–2624

Fuller WA (1996) Introduction to statistical time series, 2nd edn. Wiley/UVC, New York, p 689

Heisenberg W (1927) Über den anschaulichen Inhalt der quantentheoretischen Kinematik und Mechanik. Z Phys 43:172–198

Hirsch RM, Slack JR (1984) Non-parametric trend test for seasonal data with serial dependence. Water Resour Res 20:727–732. https://doi.org/10.1029/WR020i006p00727

Hubert P, Carbomnel JD, Chaouche A (1989) Segmentation des series hydrometeoraloques: ap plication a des series de precip-itation et de debits de L'Afrique de L'ouest. J Hydrol 110:349–367

Hurst HE (1951) Long-term storage capacity of reservoirs. Trans Am Soc Civil Eng 116:770–808

Hurst HE (1956) Methods of using long-term storage in reservoirs. Proc Inst Civil Eng 5:519

Iruine KN, Eberthardt AK (1992) Multiplicative seasonal ARIMA models for Lake Erie and Lake Ontario water levels. Water Reso Bull 28(3):385–396

Johnston RJ (1989) Environmental problems: nature, economy and state. Masquarie University, Sydney

Kempe S, Khoo F, Guerleyik Y (1978) Hydrography of lake Van and its drainage area. In: Degen ET, Kurtman F (eds) The geology of Lake Van. The Mineral Research and Exploration Institute of Turkey, pp 30–45. Rep. 169

Kendall MG (1975) Rank correlation methods, 4th edn. Charles Griffin, London

Kendall MG, Stuart A (1979) The advanced theory of statistics. Butler and Turner Ltd., London

Khalil EL (1997) Chaos theory versus Heisenberg's uncertainty: risk, uncertainty and economic theory. Am Econ 41(2):27–40

Kite V (1990) Time series analysis of Lake Erie levels. In: Hartmann HC, Donalhue MJ (eds) Proc Great Lakes water level forecasting and statistics symposium. Great Lake Commission, Ann Arbor, pp 265–277

Koch GS, Link RF (1971) The coefficient of variation; a guide to the sampling of ore deposits. Econ Geol 66(2):293–301

Krauskopf KB (1968) A tale of ten plutons. Bull Geol Soc Am 79:1–18

Laplace PS (1951) A philosophical essay on probability. Dover, New York

Leopold LB, Langbein WB (1963) Association and indeterminacy in geomorphology. In: Albritton CC Jr (ed) The fabric of geology. Addison Wesley, Reading, pp 184–192

Lorenz EN (1963a) Deterministic non-periodic flow. J Atmos Sci 20:130–141

Lorenz EN (1963b) The mechanics of vacillation. J Atmos Sci 20:448–464

Lorenz EN (1964) The problem of deducing the climate from the governing equations. Tellus 16: 1–11

Lorenz EN (1972) Predictability: does the flap of a butterfly's wings in Brazil set off a tornado in Texas? Paper presented at: American Association for the Advancement of Science

Loucks ED (1989) Modeling the Great Lakes hydrologic-hydraulic system. PhD Thesis, University of Wisconsin, Madison

Maity R (2018) Statistical methods in hydrology and hydro climatology. Springer, Singapore

Mann CJ (1970a) Randomness in nature. Bull Geol Soc Am 81:95–104

Mann HB (1970b) Non-parametric tests against trend. Econ Soc 3:245–259

Mathier L, Fagherazzi L, Rasson JC, Bobee B (1992) Great Lakes net basin supply simulation by a stochastic approach. INRS-Eau Rapp Scienti®que 362, INRS-Eau, Sainte-Foy, 95 pp.

Mood AM (1940) The distribution theory of runs. Ann Math Statist 11:427–432

Naghettini M (2016) Fundamentals of statistical hydrology. Springer. ISBN: 978-3-319-43561-9

Neter J, Wasserman W, Kunter MH (1996) Applied linear regression model, 4th edn. IRWIN Book Team, USA

Parzen E (1960) Modern probability theory and its applications. Wiley, New York, p 464

Pettitt AN (1979) A non-parametric approach to the change-point problem. Appl Statist 28:126–135

Poincaré H (1904) L'état actuel et l'avenir de la physique mathématique. Bull Sci Math 28:302–324. http://henripoincarepapers.univ-lorraine.fr/chp/hp-df/hp1904bs.pdf

Popper K (1955) The logic of scientific discovery. Routledge, New York, 479 pp

Privalsky V (1990) Statistical analysis and predictability of Lake Erie water level variations. In: Hartmann HC, Donalhue MJ (eds) Proc Great Lakes water level forecasting and statistics symposium. Great Lake Commission, Ann Arbor, pp 255–264

Ross TJ (1995) Fuzzy logic with engineering applications. Addison Wesley

Ruelle D (1991) Chance and chaos. Princeton University Press, Princeton

Russell B (1924) Icarus, or the future of science. E. P. Dutton

Rybski D, Neumann J (2011) A review on the Pettitt test. In: Kropp J, Schellnhuber HJ (eds) In extremis. Springer, Berlin, pp 202–213

Şen Z (1976) Wet and dry periods of annual flow series. ASCE J Hydraul Div 102(HY10): 1503–1514
Şen Z (1977) Run-sums of annual flow series. J Hydrol 35:311–324
Şen Z (1979) Application of the autorun test to hydrologic data. J Hydrol 42:1–7
Şen Z (1980) The numerical calculation of extreme wet and dry periods in hydrological time series. Bull Hydrol Sci 25:135–142
Şen Z (1995) Applied hydrogeology for scientists and engineers. CRC Lewis Publiahers, Boca Raton, 444
Şen Z (2002) İhtimaller Hesabı Prensipleri (Probabilty calculation principles). Bilge Kültür Sanat, Istanbul (in Turkish)
Şen Z (2012) Innovative trend analysis methodology. ASCE J Hydrol Eng 17(9):1042–1046
Şen Z (2014) Trend identification simulation and application. ASCE J Hydrol Eng 19(3):635–642
Şen Z (2017) Innovative trend significance test and applications. Theor Appl Climatol 127(3): 939–947
Şen Z, Şişman E, Kızılöz B (2021) A new innovative method for model efficiency performance. Water Sci Technol Water Supply 22(3):589–601. https://doi.org/10.2166/ws.2021.245
Slivitzky M, Mathier L (1993) Climatic changes during the 20th century on the Laurentian Great Lakes and their impacts on hydrologic regime. NATO Advanced Study Institutes, Deauaille
Sneyers R (1992) On the use of statistical analysis for the objective determination of climate change. Meteorol Z 1:247–256
Spear ME (1952) Charting statistics. McGraw-Hill
Spearman C (1904) The proof and measurement of association between two things. Am J Psychol 15(1):72–101. https://doi.org/10.2307/1412159
Streit T, Pham B, Brown R (2008) A spreadsheet approach to facilitate visualization of uncertainty in information. IEEE Trans Vis Comput Graph 14(1):61–72
Swed FS, Eisenhart C (1943) Tables for testing randomness of grouping in a sequence of alternatives. Ann Math Stat XIV(1):66–87
Tufte E (1983) The visual display of quantitative information. Graphics Press
Tukey JW (1977) Exploratory data analysis. Addison-Wesley
Vannitsem S, Demaree G (1991) Detection et modelisation des secheresses an Sahel-proposition d'une nouvelle methodologie. Hydrol Cont 6(2):155–171
Wiener N (1949) The extrapolation, interpolation and smoothing of stationary time series. Wiley, New York
Wilks DS (1995) Statistical methods in the atmospheric sciences. Academic Press, pp 160–176
Zadeh LA (1965) Fuzzy sets. Information and control, vol 8, pp 338–352. Part I, pp 519–542

Chapter 5
Mathematical Modeling Principles

5.1 General

As explained in the previous chapters, science philosophical views and logical principles are always the common arena of all thoughts and ideas shared by researchers. No scientific objectivity, generalization, selectivity, or testability is taken into consideration during philosophical thoughts. Subsequently, logic helps to generate more rational productive sub-arenas by filtration of thoughts. Logic is not the art of eloquence, but the art of making intelligent scientific inferences from conversations, debates, and discussions. This art is used to search for the interaction between the factors affecting it and the effect that occurs during the examination of an event. The causes and consequences of the verbal conclusions lead to rational mathematical expressions. Logic is like a necessary tool that is not only in science, but it is open to anyone who tries to make rational inferences. Logical principles constitute foundations of science, because scientific inferences are rational and can be tested only with facts in the forms of observations, experiments, or measurements. Under the light of the falsification principle, the testability of scientific inferences can be validated, and hence it is always possible to reach results that are closer to reality. Scientific testability is possible only with the strict application of logical rules. From the explanations in the previous chapters, it is understood that logic can survive without scientific principles, but that science can never gain functionality and innovation outside the rules of logic.

Anyone can explain thoughts to others by philosophical and daily conversation, writing even without appreciation of the logic. From this point, there is no difference between a shepherd and a highly educated person. Logic is everyone's property, but mathematics, which is dependent on logic, is a privilege and a choice by the rational thinkers and scientists. When the history of science is considered, logical principle applications come before mathematics, and although there was logic even in the primitive civilizations, it was written in a formal way and explained for the first time in Ancient Greek civilization to mean "tool" by Aristotle (382–323 BC). Today, in

Z. Şen, *Shallow and Deep Learning Principles*,
https://doi.org/10.1007/978-3-031-29555-3_5

almost every country, even if it is voluntary or involuntary, logic is always tied to crisp logic (Chap. 3) leading to mathematical expressions and places, and the idea of fuzzy or uncertainty logic is blocked in this way to some extent. However, logic operations and functions always existed in every stage and subject of daily life.

Mathematics is a field of thought dealing with symbolic logic, numbers, and sets establishing a relationship in the form of equations at the end of scientific studies; it is used to translate logic inferences from a language to a collection of symbols in a formal way, starting from philosophical thought through equations, algorithms, and methods in the forms of theoretical and practical symbolic expressions. First simple arithmetical operations sprouted in the time of Mesopotamia (between the Tigris and Euphrates rivers), Sumer and Babylon, and Ancient Egyptian civilizations along the Nile River in the 4000–5000 years before Christ, and it has reached its present state with increasing knowledge by time. Today, there is almost no study, and especially scientific development, that mathematics does not enter or support. Mathematics is a subject that contains the philosophical and especially the logical foundations of the event or phenomenon studied within the scientific circles. According to the definition given by Yıldırım (1988), accuracy in mathematics is logical, not factual. Mathematics gives the translation into symbolic language of the conclusions reached by understanding the internal and external generation mechanism of the phenomena, which are the subject of scientific investigation, according to the principles of logic.

Mathematics prior to germination started with the use of numbers for measuring, weighing, trading, construction, etc., in a very simple way to face daily practical life problems and in different works. Such situations have played a role in revealing the useful ones through some verbal relations. Thus, it is understood that mathematical development always has a relationship with logic and that logical inferences must be verbally understood before entering the field of mathematics. Instead of perceiving abstract mathematical expressions, a person who wants to do mathematical studies should first have a good verbal understanding of the logical principles, rules, and inferences that constitute the foundations of mathematics. However, in solely practice-oriented professions such as engineering, mathematical formulas are used in practical life perhaps without entering the logical principles, because they are readily imbedded into mathematical expressions, whether they are analytical or empirical.

There is also logic that has been tried to be made independent of human language, which is called symbolic logic that also constitutes the foundation of mathematical expressions. When a proposition is made in any language, it is translated into other languages, the way the words are spoken changes, but the logic remains the same. In order to make this invariance language-independent, the antecedents, successors, and conjunctions such as "and," "or," "not" (Chap. 3) that show the relationship between them are all represented by letters and some arithmetic operation symbols as addition, +, subtraction, −, division, / and multiplication, ×. Thus, a logic language like the language of mathematics emerges, which is called symbolic logic. This logic has no difference in that it is crisp logic (bivalent, true-false) and assumes variables holistically. In terms of being based on a binary basis, all the extracted models give approximate results due to the assumptions (Chap. 3).

In this chapter, in order to reach mathematical expressions, first of all, it is recommended to analyze the principles and rules of logic thoroughly and to reveal the meaningful words and sentences in the world of thought, especially those that contain IF-THEN propositions, that is, cause-effect relationships, and to reveal the logical consistency of the facts with observations, field and laboratory works or mind experiments as already explained in Chap. 3 in detail. After the establishment of logic rules verbally, one can encounter mathematical expressions by translating them into symbolic equations. All these preliminary features are presented in the previous chapter.

5.2 Conceptual Models

After our minds label each descriptive part of a phenomenon with a "word" such as a noun, pronoun, or adjective, it divides the visible environmental reality into parts and categories for precise inferences, producing vague impressions and concepts. These words have little to do with the unity of reality; a unity to which we all belong inextricably. Common word etymologies help to imagine the same or very similar objects in minds. Words are the foundation of feelings, thoughts, and perceptions. They collectively serve to present partial, and therefore, distorted conceptual models of reality that represent the perceived world produced by the human mind. The natural environment is a world in which we are connected through situations made up of vital and inseparable connections that make us exist (Dimitrov and Korotkich 2002).

All conceptual models deal with parts of something perceived by the human mind as the environment that is thought to be used for what our ego-centered minds consider meaningful. There are, of course, overt and implicit interrelationships between these meaningful pieces that exist for the exploration of the human mind workings. In any subject such as science, engineering, economics, politics, and philosophy, unsupervised or supervised (trained) minds study conceptual models that contain many uncertainties to predict and apply power over the evolving dynamics of reality.

5.2.1 Knowledge and Information

Education or personal experiences help one to become aware or understand the acquired subject in the form of knowledge. It provides objective and relevant information for conclusions. Information is the basic concept related to the elements of any event that cannot be systematically brought into knowledge. For example, it is possible to observe and then know that the probability of precipitation occurs when the sky is cloudy; this is basic knowledge. However, why it occurs is another source of knowledge that requires a more systematic connection between basic information in order to become knowledge. There are different sources of information. Information is grasped through sense organs and processed mentally in order to reach a better result. Early humans thought contemplated in a completely uncertain environment for their daily and vital activities. It is possible to say that early information and knowledge are concepts derived from observation and experience.

5.2.2 Observation

Every individual observes many small and large-scale events in daily life. One then tries to make some comments without having any numerical, arithmetic, or mathematical basis. All depend on mind functionality, logic, and personal interests. Initial results are linguistic, and therefore, all subjects to some degree include fuzziness. Observations enable one to put their impressions about the observed phenomenon into linguistic expressions. For example, a villager may observe that in his area the river flow has a pattern, where high (low) flows follow high (low) flows. In this way, it implies a positive correlation between subsequent flow quantities.

5.2.3 Experience

Experience means that some behavior of any event occurs frequently, and therefore, the observer learns by repeating the basic information. For example, learning our mother tongue goes through constant repetition. After repetitions, one gets used to the taste of local dishes that give experience to our taste buds. Likewise, working in the same area of interest, one is experienced over time, and this means that if one observes the same phenomenon repeatedly, one can automatically come up with a set of linguistic propositions.

5.2.4 Audiovisual

Another way of acquiring knowledge and information is through listening and hearing, which are the most common sources of information and knowledge in our age (Chap. 1). For example, over 95% of the information and knowledge throughout the entire education system is obtained by listening to teacher. In this way, unfortunately, one can depend on already organized knowledge, and if there is no critical discussion collectively or individually, then learning process will be a routine procedure with the production of certificates, but there will be no scientific idea improvements and developments. This type of learning is more dangerous than the previously mentioned observation and experience gains.

5.3 Mathematics

Mathematics, which is the art of solving problems, tries to reach the result by using the features of number or morphology by acting on the ways of reasoning. The way of expressing the objective information obtained from verbal reasoning that results with symbols is called mathematics, and this method constitutes the basis of science and technology studies.

In the body of mathematics, firstly, the information obtained by intuition is filtered through logic, and after correctness and conceptual ground are gained, there is analysis, generalization, and finally a formal mathematical structure. The importance of logic in mathematics is in terms of proving mathematical propositions. The foundations of mathematics are the search for basic concepts and logical structure. Thus, it is possible to reach the singularity of human knowledge. The most important mathematical concepts are numbers, definitions, shapes, sets, functions, algorithms, equations, and mathematical proofs. Mathematics principles are used in all scientific studies, but what mathematics itself means is not known to many people who use it. We can say that mathematics is the science of quantities. It provides scientific structure, order, and relation that are dependent on arithmetic, measuring, and shape description of objects.

There is reasoning in each thought, but this thought should be transformed into a form that can be perceived by everyone, and the reasoning should be given objectivity not subjectivity. In general, there are much more reasoning thoughts and inferences in everyday life that may not have that much logic in them.

The development process of mathematics has reached its current level by passing through consistent, formal, logical, and rational ways. There is no branch of science that separates from the body of mathematics and gains prestige. Mathematics, unlike other branches of science, has become a huge tree, its structure always increasing in the process. The basis of mathematics having a very robust and unshakable structure is that the rules of logic work strictly. These rules are primarily based on rational foundations that emerge as a result of sound thinking and reasoning, even if it is approximate. The logic, which is the basis of mathematics, can always maintain its accuracy by putting the suggestions made on a solid basis after various comparisons. Mathematics is a constantly fertile science in terms of generating new rules that will shed light on different subjects from the set of definite rules it deduces based on logic. The principles of right thought enter the field of logic, and with the application of these principles, mathematics continues to play an increasing role in the service of humanity through science and technology.

The German logician Gottlob Frege (1848–1925), the founder of contemporary logic and philosophy, said,

Mathematics is the application area of logic.

Based on his view, he thought that mathematics could be built on the axiomatic system of logic.

In the origins of logic and later mathematics, thoughts in human mind activities are the most important first movement and production elements. Developments in mathematics are achieved after the ideas are matured by transferring them to others through language, the origin (etymology), and semantic loads (epistemology) of the words in the language are visualized in detail, and the rational meanings of the proposed sentences are perceived. Before thoughts reach the field of mathematics, they must pass through the filter of logic and gain accuracy and organization. There are different ways to enter logical thinking, which has accuracy and organization, into these patterns. From here, one can say that logic is a cornerstone of science that triggers the human thought activities. Unlike sentences in daily conversations, logical propositions contain rational sentences that connect causes to effects. In order the public to perceive it easily in each proposition "IF .... THEN ......" parts must be explicitly or implicitly present in its structure (Chap. 3). From this, it is understood that it is necessary to consider the sentences in the form of propositions first for logic and then for mathematics. The effort of making rational inferences in this set of propositions, which are logical sentences, is called reasoning. In the reasoning process, connecting the results to the causes in a concrete way must first be obtained verbally.

5.3.1 Arithmetic Operations

It has already been stated that all the rules of logic form the basis of the mathematical principles. In general, it is possible to generalize them further, primarily by arithmetic operations (addition, subtraction, multiplication, and division). What are the logic rules of arithmetic operations? ANDing? ORing? or NOTing? Does the reader have any prior thoughts on these issues? Or did s/he perceive them only as arithmetic operations by heart?

Now let us assume that 2 + 2 equals 4. What is the logical equivalent of the "plus" sign here? Is this sign ANDing? ORing? For this, let us choose one of the following according to the logic rules explained in Chap. 3.

1. *2 OR 2 is four!*
2. *NOT 2 but 2 is four!*
3. *2 AND 2 are four!*

When the first option is considered, three independent results emerge; the first of which is 2, the second is again 2, and the third is 4 (Fig. 3.7c), which of these is acceptable? Of course, the last one, but one may not always be able to decide which one is valid since with ORing one will always have three options. The NOTing operation in the second option does not make any sense; it is completely excluded as it is not logical. In the third case, ANDing always yields only one result as 4. So, the logical equivalent of the "plus" sign is ANDing (Fig. 3.7d).

As a result, it is a fact that all signs in mathematical equations based on crisp logic rules are ANDing. If there is more than one result in the solutions, then ORing may be involved depending on the circumstances.

5.3.2 *Logical Relationships*

In order to reach generalizations by making inferences in mathematics, it is enough to make two different inquiries that complement each other. If these two successive relationship determination rules are applied on a logical basis, it becomes possible to make inferences of all the basic equations that have been seen throughout the educational systems. One does not even need to be educated to know these two relationship building rules. These are the basic thoughts that everyone has for the human mind to work. Plainly, both rules are put into practice simply and easily with questioning between two different words, facts, or events, as well as between rules of logic. By questioning according to these two rules, one can reach a conclusion about the type and form of the relationship between two objects. These two rules have been touched upon in Chap. 3 in detail.

1. *The proportionality rule*: This rule is the principle of questioning how there can be a relationship between the two variables under consideration. There are two different answers that can be given. One can name these as "direct" or "inverse" proportionality. In simple terms, direct (inverse) proportionality indicates that if one variable increases, the other will also increase (decrease).
2. *Linearity rule*: It is useful to make an inference by questioning whether the relationship type determined by the previous rule will be in the form of a straight line (linearity) or a curve (nonlinearity).

In fact, these two rules serve to reveal logically the shape (function) of how there can be a relationship between two different variables. Since there are two different options in each rule, when these are matched, four different alternatives emerge, and there is no other option for rational inferences. The use of only the names of both variables indicates that each of them is considered in its entirety without subsets, and thus the operation with the combination of these rules is nothing but a deduction inference. Figure 5.1 shows four different pairings (a, b, c, and d) related to proportionality and relationship geometry.

If the variable on the horizontal axis, input, I, and on the vertical axis output, O, variables are shown on the Cartesian axis set of the four alternatives, the situation corresponding to each of them is shown in Fig. 5.2 with the letters in Fig. 5.1.

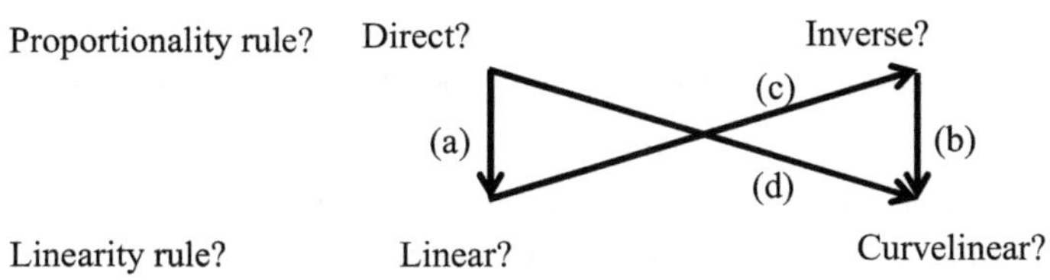

Fig. 5.1 Joint proportionality and linearity rules alternatives

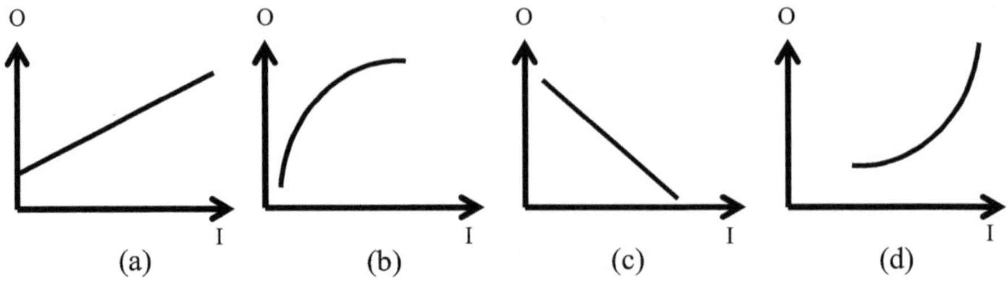

Fig. 5.2 Joint proportionality-linearity shapes

5.3.3 Equations

In the backyard of each equation, there is a rational thought, and then the rules of logic can originate from that thought. In the study of an event, just like trying to fish in the richness of thought, as much as there are intelligent logical inferences in the ocean of thoughts, keeping them and obtaining the equations is verbal information. The equation is a mathematical expression and it has two sides. In general, there are output (result) variables on the left and input (reason) variables connected with logic propositions on the right. If the variable that causes an event is x and its output is y, the simplest mathematical equation is given in the form below.

$$y = f(x) \tag{5.1}$$

This implicit expression, which makes sense to mathematicians and logicians, may imply similar information for those working in practice, but the explicit expression of this equation must be revealed especially in engineering applications. Many inputs to this equation are x, z, u, v, … , and hence, the most general expression is obtained in the case of multiple variables as

$$y = f(x, z, u, v, \ldots.) \tag{5.2}$$

In the case of these implicit equations, attention is paid neither to the relations between the input and output variables, nor to the relations between the input variables themselves. The important thing here is that only the input variables are thought to influence the output variable, but neither the proportion (direct, inverse) nor the shape (linear, curvilinear) of this effect is given importance, it is written only in closed form. It should be known that mathematical equations appear in science and engineering studies, and the following points should be considered in their derivation:

1. Based on an event or phenomenon that may be the basis of science.
2. To determine at least one of the variables that may affect this event or phenomenon.

3. The time variability of the variables, if any, should be considered first.
4. The spatial variability of the variables, if any, should also be considered.
5. Considering the variability of variables both in time and space.

Making inferences between the output and input variables is possible by considering the geometrical shapes of mathematical proportionality and relations with the rules of logic. After interrogating the input variables separately with the output variable in the previous steps, the final equation is obtained by combining them collectively.

It should not be forgotten that there are idealizations and linearization with some assumptions and hypothesis in the backyard of every equation.

5.4 Geometry and Algebra

Al-Khwarizmi (780–850) invented the algebraic equations, and hence the first time in the history of philosophy-science and mathematics equality expressions took place. He has also suggested the solution of the second-order degree polynomial root solution by geometric principles as shown in Fig. 5.3. It is left to the reader to follow the conceptual modeling steps in this figure to reach the final root solution as

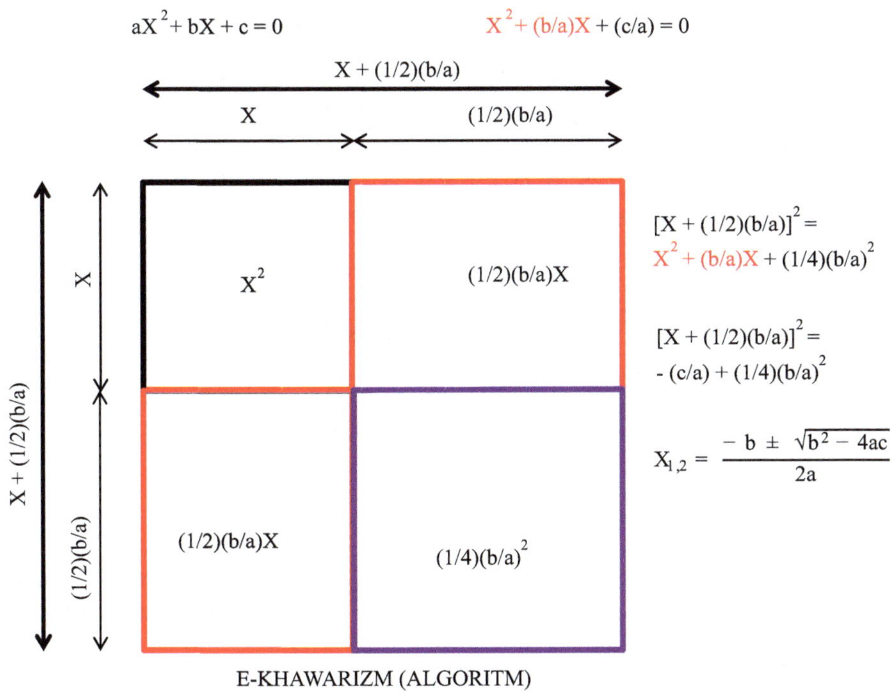

Fig. 5.3 Al-Khawarizm's second-order equation root finding geometric procedure

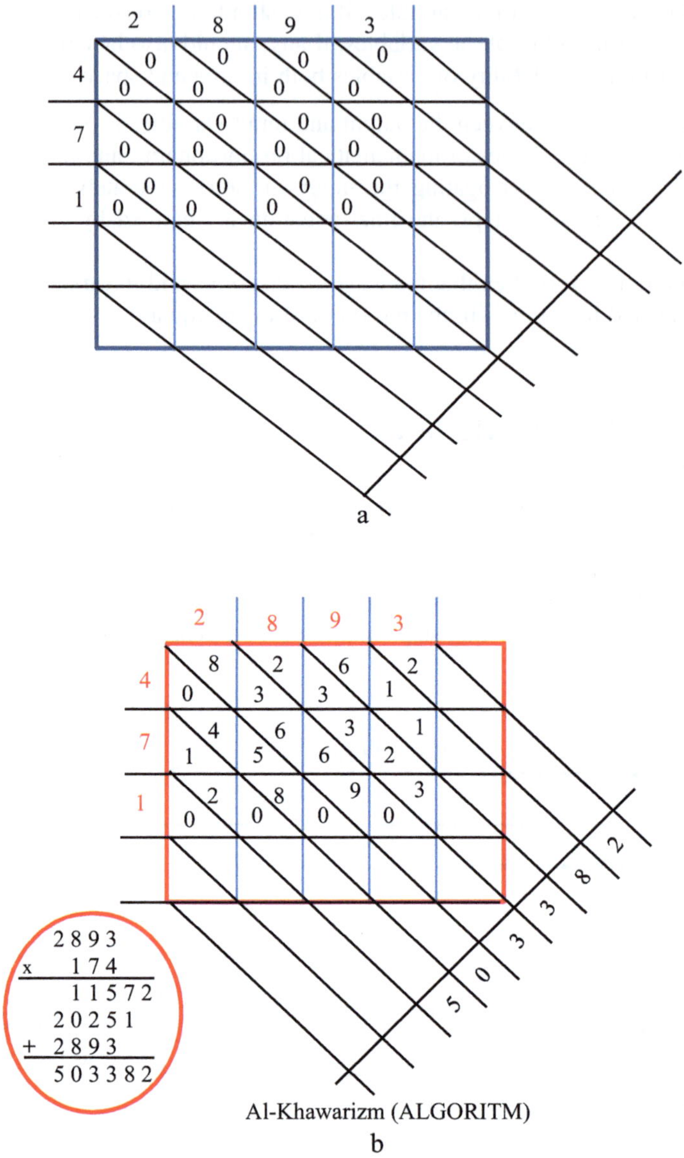

Fig. 5.4 Al-Khawarizm multiplication geometry

we are asked to memorize today in the education system without any background of logical reasoning. This figure is the first proof that mathematical equation solutions are based on geometrical sketch principles.

Along the same geometrical design principle, he also provided in detail the multiplication procedure as suggested in the following templates in Fig. 5.4a, b. In this template, multiplication of two digits one from upper horizontal and another one

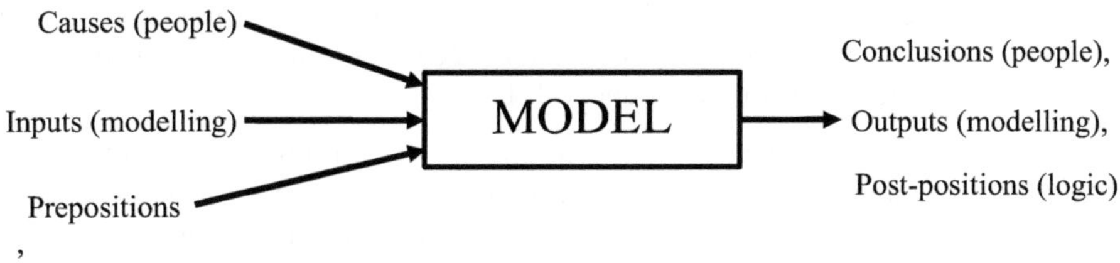

Fig. 5.5 Black-box models

from the left-hand side vertical ends up with two decimal numbers. Decimal cell is written on the lower half on the common square and the other in the upper triangle as, for example, multiplication of $4 \times 8 = 32$; its appearance in Fig. 5.4 is obvious. The reader can contemplate and fill the other squares corresponding to the other digital pair's multiplication. Finally, the summation of the numbers within the main diagonal directions leads to the result as in Fig. 5.4b.

5.5 Modeling Principles

Prior to any modeling, one should ask the following questions and then consider the black-box conceptual model elements as in Fig. 5.5.

1. Which event(s) to model?
2. What is the expected output?
3. What are the reasons (causatives) and variables?

In general, although it is independent of professions, there are models of thought that are shaped according to the profession and simplified by perceiving experts' training of the subjects initially at shallow but rational teaching approach with restrictive assumptions, which vary according to the rate of dealing with natural and social events. No matter how many uncontrollable events there are, some assumptions are made in order to analyze the situations of facilitator, simplification, and approximation to the ideal case. These assumptions are preserved in different designs. For this reason, whether it is a deductive or inductive model, much emphasis is placed on cause-effect relationships. The models are quite limited, and their validity may not be universal, but the situation may change by time. So, even in the simplest modeling procedure, one tries to draw conclusions from causes (see Figs. 3.2 and 3.3). In a modeling process, the following three situations occur differently from each other:

(a) If the inputs of the model are unknown, but the description of the actual behavior and the outputs are known, it is called filtering or smoothing,
(b) If the model's inputs and outputs are known and only the behavior pattern is determined, this is called model identification. In a way, one can say that the reasons are analyzed in detail in the light of the results.

(c) If the model is known to be defined by its inputs, this type of modeling is also called predictive modeling. Here, various scenarios can be examined easily about what the results might be under different reasons.

In general, the mathematical models in practice and in this book help to make estimations and predictions. If a model is without some restrictive assumptions, then it cannot be successful, because there are always incompleteness, vagueness, fuzziness, and uncertainty in natural and artificial events. In particular, the geometry of the event and the mathematical approximations are among the initial ideas leading to assumptions. The assumptions simplify not only plausibility of the overall event behavior, but also aid to reach early results that can be improved later by release of some of the initial assumptions, which are necessary to reduce uncertainty, and hence, to reduce possible complexities to the level of human grasp and then derive the mathematical equations.

Among the geometrical assumptions, the most ideal and simple shapes are always preferable (line or circle if line, square or circle if area, sphere, cylinder or rectangular prism if volume). At the beginning of the nineteenth century, Rutherford (1871–1937) was able to visualize the atom as electrons circulating in circular trajectories around a nucleus. This is not the case, but it helps to visualize event, and hence, calculations yield close or identical results to the measurements. The most interesting assumption made about geometry in modeling many matter-related phenomena is a rectangular prism with very small dimensions as in Fig. 5.6. This matter is either solid or liquid or gaseous.

Unlike the thinking models in Chap. 3, the matter is variable along the perpendicular x, y, and z axes system according to the Cartesian-Euclidean geometry. This rectangular prism is also called as "control volume" and is a simple model that is at the core of many analytical studies. This rectangular prism is always at the root of the

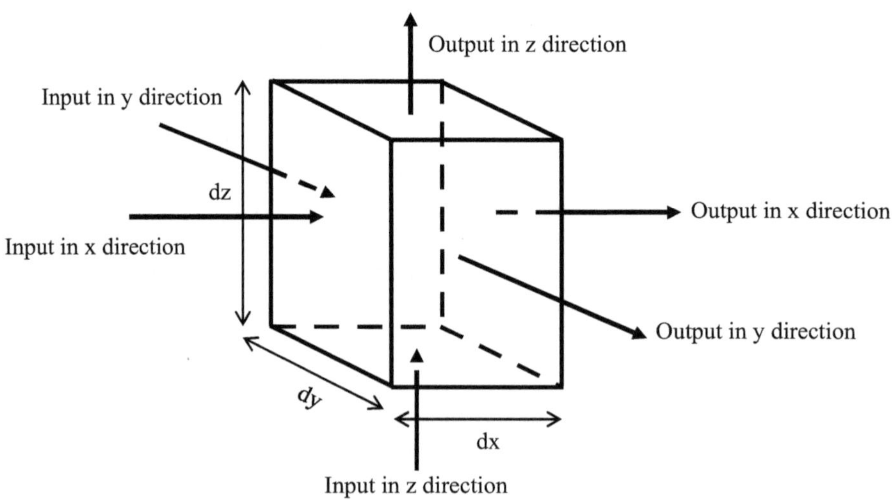

Fig. 5.6 Volume representation

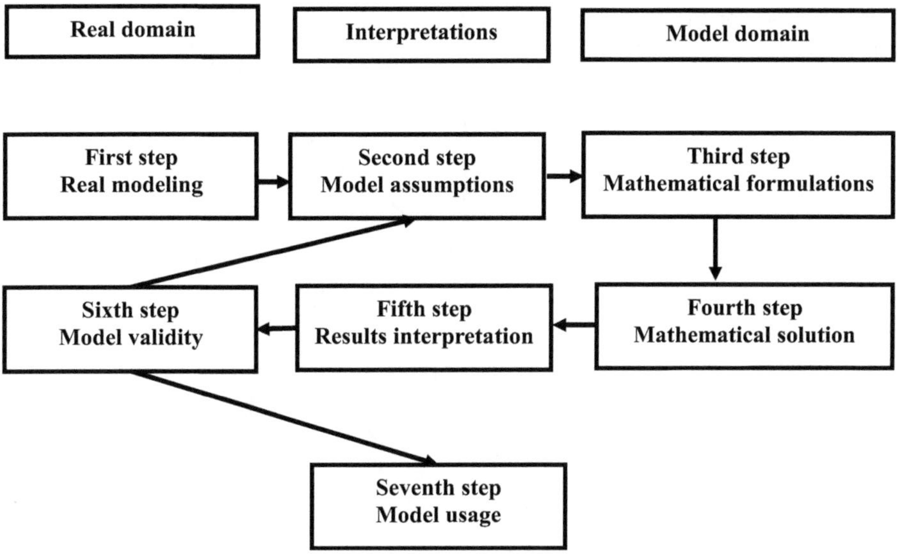

Fig. 5.7 Modeling stages

quadratic differential equations obtained by combining conservation (mass, energy, and momentum) and body (Hooke, Darcy, Ohm, Fourier, Hubble, etc.) equations.

The purpose of modeling is to explain the observation and measurement of internal generation mechanisms in a simple and mechanical way to obtain predictions for decision-making. In order to achieve the final product in a systematic way, it is appropriate to translate the real and concrete world behaviors into mathematical symbols in the abstract way. For modeling, first, the variables of the problem are defined and then the possible relationships among them are investigated. All simplifying assumptions and the relationship among variables form first the basis of mathematical modeling. Thus, thoughts about the phenomenon under investigation, the way they operate and behave are transformed into relevant mathematical equations, which play major role in the solution by reasoning. After the final solution is found, the reliability of the prediction is checked again with rational approaches, their interpretations, and then with measurements. Interpretations of the result mean checking the real-world inferences from the abstract mathematical world along the steps in Fig. 5.7. At this stage, the degree of accuracy of the theoretically obtained model results by comparing them with those available from observations and measurements in the real world.

For model validity control, it is recommended first to use the relative error concept. If the model prediction and measurement results are R_m and R_M, respectively, then the relative error,α, is expressed as a percentage:

$$\alpha = 100\frac{R_M - R_m}{R_M} \tag{5.3}$$

In practical applications, it is desirable that the relative error be less than $\pm 5\%$ or $\pm 10\%$. If there are many measurements and model predictions then the arithmetic

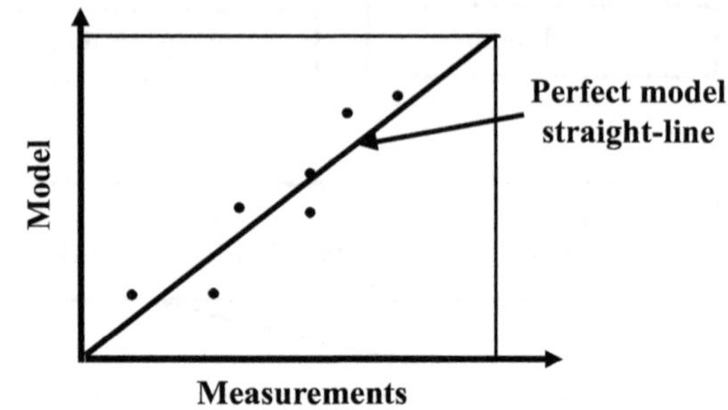

Fig. 5.8 Model prediction-measurement scatter diagram

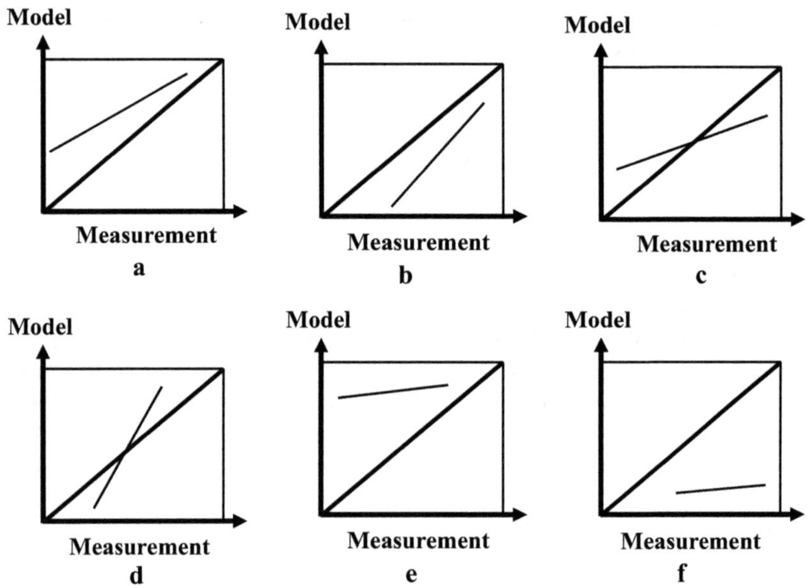

Fig. 5.9 Measurement and model inconvenience

average of the relative errors should be less than these limits. It is possible to adjust the model results with corresponding measurements on a Cartesian coordinate system such that the scatter of plots should fall around a 45° line as in Fig. 5.8.

There are different deviation possibilities from the perfect model straight line as in Fig. 5.9. For example, it is possible to encounter situations, where the model over (under) predicts as in Fig. 5.9a, b, respectively. Similar situations appear in Fig. 5.9e, f, the cases in Figure 5.9c, d corresponds to partial overestimation and underestimation.

If good predictions cannot be obtained from the model, the simplifying assumption made to move from the real to the modeling world and the relations among the variables should be reconsidered. Remodeling is continued by importing the necessary improvements, which are possible to revise the model structure step by step. This is continued until the model predicts the measurements within the limits of an appropriate percent relative error (±5 or ±10).

In scientific researches, one is interested in relating measurements to model prediction outputs, which may be in the form of single-input-single-output (SISO) or various versions as multi-input-multi-output (MIMO) models. In the literature, there is a set of standard coefficients that provides agreement between measurements and predictions through a set of single parameter values. In most of the cases, authors report that their model is suitably based on the comparison of statistical parameters composed of measurement and model output data series by considering one or few of the well-established agreement or association metrics among which the most commonly used, accepted, and recommended ones are bias (BI), Percent Bias (PBI), coefficient of determination (R^2), mean square error (MSE) or root mean square error (RMSE), correlation coefficient (CC), Nash-Sutcliffe efficiency (NSE), and index of agreement (d) (Pearson 1895; Nash & Sutcliffe 1970; Willmott 1981; Gupta et al. 2002; Santhi et al. 2001; Van Liew et al. 2007; Moriasi et al. 2007; Tian et al. 2015; Zhang et al. 2016; Dariane and Azimi 2018). In the literature, there are also other versions as the modified index of agreement (Legates and McCabe 1999), Prediction efficiency (P_e) as explicated by Santhi et al. (2001), persistence model efficiency (PME) (Gupta et al. 2002), RMSE-observation standard deviation ratio (RSR) as given by Moriasi et al. (2007), and Kling-Gupta efficiency (KGE) measured by Gupta et al. (2009).

As stated by Willmott (1981), almost all the models have elusive predictions, which cannot be covered by the model efficiency measures, and even, in general, by significance tests. Freedman et al. (1978) have mentioned that the statistical significance tests are concepts that must be viewed with skepticism. Along the same line, Willmott (1981) stated that it may be appropriate to test an agreement measure and report the value of a test statistics at a significance level, but the distinction between the significant and insignificant levels is completely unjustified. For instance, if the significance level is adapted as 0.05, then what are the differences, say, among 0.049, 0.048, 0.047, and 0.051, 0.052, and 0.053? Additionally, such significance levels depend on the number of data for depiction of the most suitable theoretical probability distribution function (PDF). Even though ASCE (1993) accentuate the need to explicitly define model evaluation criteria, no widely accepted guidance has been established, but a few performance ratings and specific statistics have been used (Saleh et al. 2000; Santhi et al. 2001; Bracmort et al. 2006; Van Liew et al. 2007).

For a more objective assessment of model efficiency, calibration and validation, measurement association, and comparison visual inspections must be preliminary conditions for better insights, interpretations, and model modification possibilities. The basic statistical parameters such as arithmetic averages, standard deviations, regression relationship between the measurement (independent variable) and model prediction (dependent variable) data through the scatter diagram are very important ingredients even for visual inspection to identify systematic and random components. Unfortunately, most often the model efficiency measure is obtained by available software, which does not provide any informative visual inspection and assessment. Even though there is extensive literature on model calibration and validation, it is difficult to compare the modeling results (Moriasi et al. 2012). Numerous models of calibration and validation have been the subject of discussion by scientists and experts (ASCE 1993; Van Der Keur 2001; Li et al. 2009; Moriasi et al. 2012; Harmel et al. 2013; Ritter and Muñoz-Carpena 2013; Pfannerstill et al. 2014; Larabi et al. 2018; Rujner et al. 2018; Swathi et al. 2019). Şen et al. (2021) presented visual inspection and numerical analysis methodology (VINAM) for effective model efficiency and ideal validation, and if necessary, modification or calibration of the model predictions to comply with the measurements.

5.5.1 Square Graph for Model Output Justification

For visual inspection of the model predictions associated with the measurements, one can regard the measurements as independent variables, model predictions as dependent variables, and plot them on a coordinate system. In the case of an ideal match between the two series, one expects that they fall on the 1:1 (45°) straight line, which appears as the diagonal straight line on the square template graph as in Fig. 5.10 like Figs. 5.7 and 5.8.

This straight line divide the square area into two half triangles with the upper (lower) one representing model overestimation (underestimation) domain provided that all scatter points fall completely in either one of these triangles. It is also possible that the scatter points may have positions partially in each triangle in which case the model has some points as overestimation or underestimation. The mathematical expression of the ideal model efficiency case as 1:1 line is

$$P_i = M_i \tag{5.4}$$

In Fig. 5.11 a and b coefficients correspond to the minimum and maximum values among the measurements and predictions. The most significant feature of a square template is its ability to reflect almost all the previously defined efficiency criteria properties in a single graph. For instance, in Fig. 5.10a, the scatter of points is shown in the upper triangular area (overestimation), but there is no linear trend between the measurements and prediction scatter points, which are randomly distributed. This provides the message that the model is not capable of representing the measurements at all. It is necessary to try and model the measurements with another suitable model, which must yield at least some consistency among the scatter points.

For instance, in Fig. 5.11b, the scatter points have a linear tendency, which is the first indication that the model for predictions is suitable, because the scatters are around a regression line. The following features are the most important information pieces in this figure.

1. The centroid point ($\overline{M}$ and $\overline{P}$) on the regression line is at a distance, D, from the ideal prediction line.

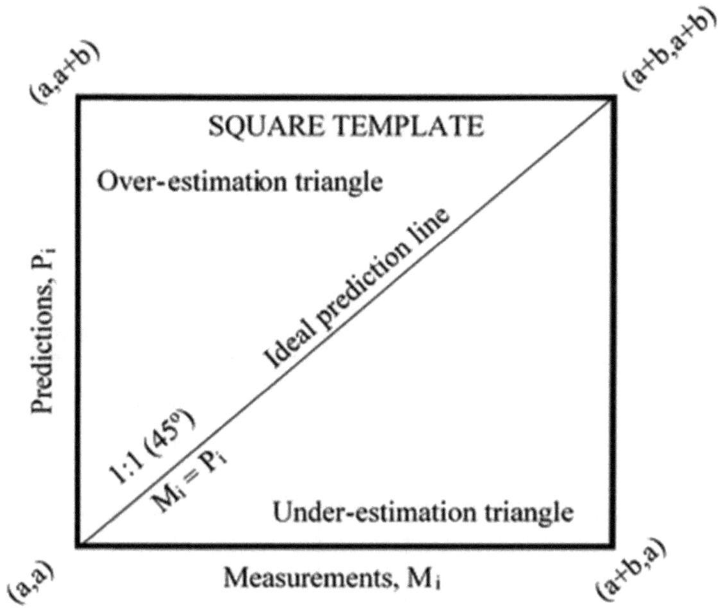

Fig. 5.10 Square templates for VINAM

2. The same straight line has a slope, S, with the horizontal axis, the value of which can be calculated.
3. The model regression straight line in the figure has an intercept, I, on the vertical axis. It also crosses from the centroid ($\overline{M}$ and $\overline{P}$) point.
4. The straight line passes through the centroid point.
5. After the regression straight-line expression determination, one can calculate the vertical deviations of each scatter point from ideal prediction line, which constitute the error sequence, ε_i.

Tian et al. (2015) suggested representing the straight line without any visual explanation as follows with notations in this article,

$$P_i = SM_i + I + \varepsilon_i \tag{5.5}$$

where ε_i represents deviations from the perfect line. This is exactly the reflection of the regression line in Fig. 5.11b. In this paper, S is referred to as the rotational error and I as shift error. It is obvious from Fig. 5.11b that each one of these is systematic deviations from the ideal prediction line. No need to say, Figure 5.10c is the underestimation alternative of Fig. 5.11b.

Figure 5.11d, e is for partial model overestimation (underestimation) case, where the central point of the regression line centroid coincides with the ideal prediction

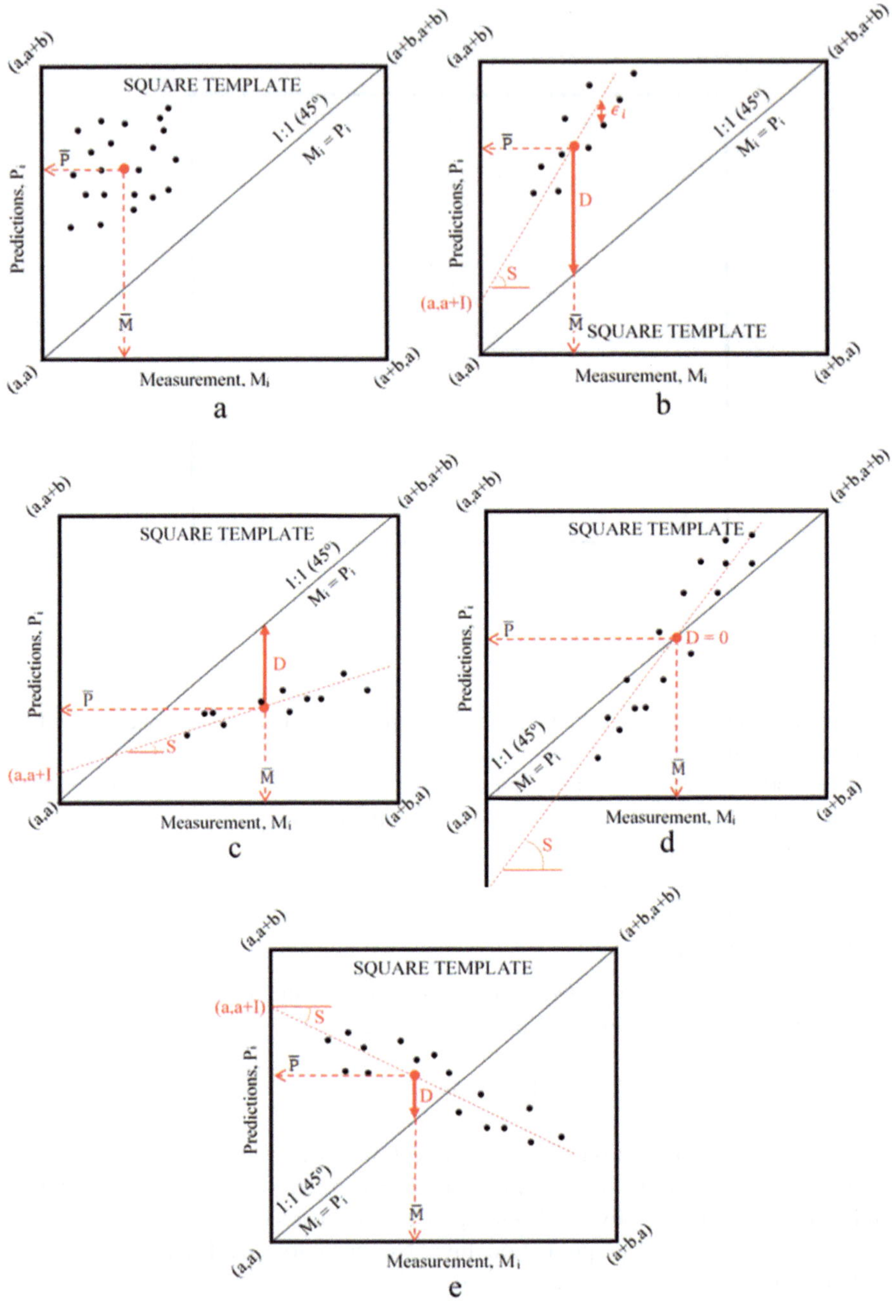

Fig. 5.11 Measurement and prediction scatter points in the various VINAM templates

line (D = 0), and away from the ideal line, respectively. In the former case, there is no shift error, and for the other case everything is self-explanatory under the light of the above explanations

According to the suggested template and algorithm, the measurement data is accepted as constant, and it is tried to approach systematically the predictions to these measurement data or to define the recalibration operation between the obtained model results and the measurement for the best and optimum efficiency model. When the measurement data accepted as an independent variable is shown on the horizontal axis, and the prediction data as a dependent variable on the vertical axis, the rotation and translation can be achieved by performing mathematical operations sequentially. In this way, a model prediction or calibration that is closer to actual measurement values is made by reducing systematic errors. In this case, a distortion occurs due to the change in the vertical distances with the ideal line. This result is unavoidable to make better predictions. Because of the new approach in this study, the optimization of total vertical changes has been achieved by taking all available data into account. The positive contribution of the suggested method is seen by controlling the obtained results through the six different performance indicators.

5.5.2 *Model Modification*

After all explanations in the previous section, an important question is "is it possible to improve the model performance, and how to increase its efficiency?" The best and optimum efficiency is possible after shift and rotation operations on the VINAM template regression lines. The following steps are necessary for arriving at the best solution:

1. Shifting operation of the central regression point vertically such that it sits on the ideal prediction line (1:1). Only vertical shifts are possible for keeping the measurements as they are.
2. After the shifting, the regression line is rotated according to the rotation angle as (1-S), so that the regression line coincides with the ideal prediction line (1:1).
3. These two operations are preferable if there is no other choice to get the VINAM regression line to coincide with the ideal prediction line.

In shifting operation, there is no problem, because the whole scatter points are moved by the amount of D downwards or upwards. The shifting operation mathematical expression is

$$P_i' = P_i \pm D \tag{5.6}$$

As for the rotation operation, the horizontal locations of each scatter point must remain the same, so as not to disturb measurement values. Such a rotation can be achieved by means of the following expression where P_i'' is for final data.

$$P_i'' = P_i - I - SM_i \tag{5.7}$$

The method suggested in this study aims to improve the model performances by reducing the differences between the systematic errors in the model prediction results and the predictive measurements. When the accuracy and reliability levels are analyzed, it is seen that there are systematic and random errors between the predictions and measurements. These errors vary depending on certain factors such as the experience of the modeler, the data quality, and the selected modeling methodology. The error evaluation can be made according to the ideal line given in the square template described in Fig. 5.8 which is frequently preferred in studies, in addition to various performance indicators. When the first results of model studies are evaluated, it is seen that different alternatives may arise (Fig. 5.11). The vertical differences of the prediction results compose of the consistent difference between the mean values of the measurement and prediction, the angle between the 1:1 ideal line expected to be between the model and the prediction, and the regression line obtained according to the least-squares between the measurements and model predictions, and finally random differences. These three components show the quality, accuracy, and reliability of the model. The model performance depending on the first two components can be improved significantly through the suggested method. The model design needs to be revised to improve the model performance by developing the third component prediction.

5.5.3 *Discussions*

In the Appendix-A, the necessary software is given for the application of all VINAM steps. The applications of the VINAM procedure are presented for two well-known models, which are artificial neural network (ANN) and adaptive network-based fuzzy inference system (ANFIS). These applications are based on the water losses measurement in potable water distribution systems, for which water loss predictions are among the most important issues of water stress control (Şişman and Kizilöz 2020). The most important component in the evaluation of a water distribution system with regard to water losses is the non-revenue water (Kizilöz and Şişman 2021). Jang and Choi built a model to calculate the non-revenue water (NRW) ratio of Incheon, Republic of Korea, by means of ANN methodology. When the best model was examined, R^2 was obtained as 0.397. It is seen that the models can be improved when the measurement values and model projections scatter plots appear along a regression line, as already explained in Fig. 5.9.

The NRW ratio estimates are modeled through ANN and ANFIS for Kocaeli district, Turkey, and the implementation of suggested VINAM method is carried out on similar model outputs for further improvements. A total of eight models (four ANN and four ANFIS) with nine input measurements are developed through the modeling procedures. Water demand quantity, domestic water storage tank, number of network failure, number of service connection failure and failure ratio, network length, water meter, number of junctions, and mean pipe diameter are the model input parameters. All models are validated by VINAM approach, and the model efficiency evaluations are carried out through statistical values according to BI, MSE, CC, R^2, d, and NSE.

For ANN model performance, 55% of the available data is arranged as training, 35% as validation and 10% as testing. These models are developed with one hidden layer including four neurons and feed forward back propagation training procedure with support of Levenberg-Marquardt back propagation algorithm (Rahman et al. 2019; Coulibaly et al. 2000; Kermani et al. 2005; Kızılöz et al. 2015; Şişman and Kizilöz 2020). As for the ANFIS model implementation, 66% of the obtained data are taken as training and the remaining 34% as validation (testing) purpose. For this model, various membership functions (MFs) are considered as triangular (Trimf), Gaussian bell-shaped (gbellmf), and trapezium (trammf) with "low," "medium," and "high" linguistic terms. The statistical properties of input components and model outputs are given in Table 5.1 for ANN and ANFIS models.

The resultant VINAM graphs are presented in Fig. 5.12 for ANN model versions with the model's efficiency classical and VINAM improvements in Table 5.2.

In this study, the NRW prediction rate of the selected model was calculated over nine different parameters through the ANN and ANFIS methodologies. The performance indicator results of NRW ratio predictions are made through three different combinations of input parameters given in Table 5.3, and it seems that the model results are not at the desired level. On the other hand, when the NRW ratio

Table 5.1 Input-output parameters

Model parameters			Range	Unit
Input	Water demand quantity	WDQ	315.445–2.844.526	m^3
	Domestic water storage tank	DWST	4000–85.901	m^3
	Number of network failure	NNF	34–628	Number
	Number of service connection failure	NSCF	23–541	Number
	Failure ratio	FR	0.01–3.43	–
	Network length	NL	306–1600	Km
	Water meter	WM	15.124–160.135	Number
	Number of junctions	NJ	9.616–52.565	Number
	Mean pipe diameter	MPD	108–159	Mm
Output	Non-revenue water ratio	NRW	0.13–0.54	–

predictions are analyzed through the VINAM graphs as in Fig. 5.13, it is seen that the combinations determined by certain systematic errors can make good predictions. So, a considerable improvement has been achieved by calibrating the models (according to the ideal line) over the classical approaches through the suggested methodology. It is possible to predict the NRW ratios with specific levels that can be accepted with only three parameters, and evaluate the network losses over three parameters such as WDQ, WM, FR.

The second application graph links to the measurements and ANFIS models VVA diagram with Fig. 5.13. The classical efficiency and VINAM improvements are available in Table 5.3.

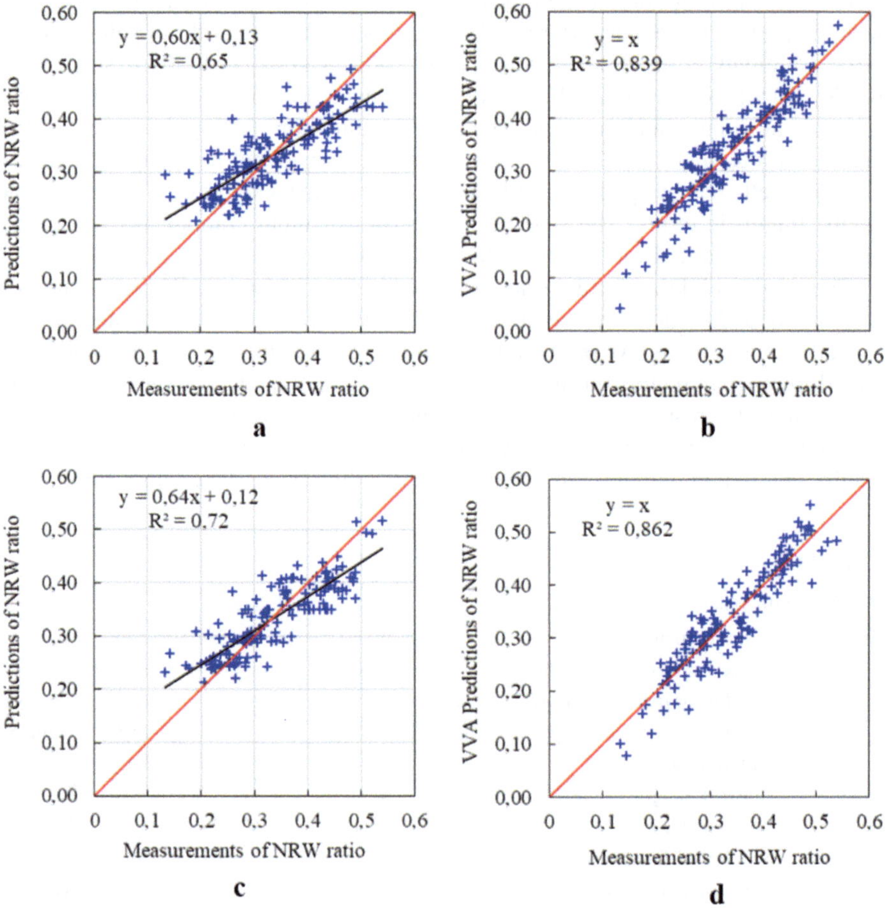

Fig. 5.12 ANN classical (**a, c, e, g**) and VINAM (**b, d, f, h**) approach

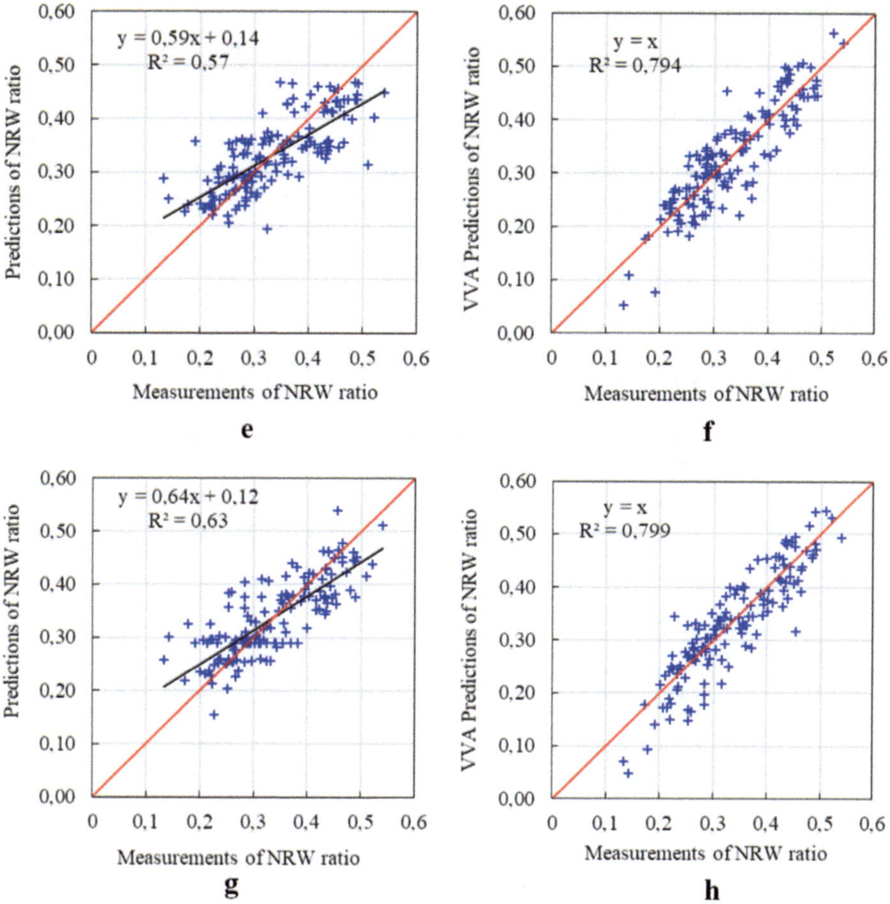

Fig. 5.12 (continued)

Table 5.2 ANN and VINAM ANN models result

Model No	Input combinations	R^2	MSE	NSE	BI	CC	d
ANN 1	WDQ – MPD – NNF	0.65	0.0028	0.648	−0.0014	0.808	0.874
ANN 2	WDQ – NL – NNF	0.72	0.0022	0.715	−0.0024	0.851	0.901
ANN 3	WDQ – WM – FR	0.57	0.0034	0.572	−0.0022	0.757	0.851
ANN 4	WDQ – DWST – FR	0.63	0.0290	0.625	0.0018	0.791	0.877
VINAMANN 1	WDQ – MPD – NNF	0.84	0.0015	0.809	−0.0004	0.916	0.955
VINAMANN 2	WDQ – NL – NNF	0.86	0.0013	0.839	−0.0004	0.928	0.962
VINAMANN 3	WDQ – WM – FR	0.79	0.0020	0.740	−0.0004	0.891	0.940
VINAMANN 4	WDQ – DWST – FR	0.80	0.0020	0.745	−0.0049	0.894	0.941

Table 5.3 ANFIS models with three inputs

Model No	Input combinations	R^2	MSE	NSE	BI	CC	d
ANFIS 1	WDQ – MPD – NNF	0.52	0.0051	0.523	−0.0014	0.724	0.825
ANFIS 2	WDQ – NL – NNF	0.007	0.0138	−0.292	0.0086	0.258	0.536
ANFIS 3	WDQ – WM – FR	0.24	0.0081	0.238	0.0031	0.491	0.621
ANFIS 4	WDQ – DWST – FR	0.15	0.0093	0.123	−0.0011	0.384	0.575
VINAM_ANFIS 1	WDQ – MPD – NNF	0.79	0.0028	0.738	−0.0012	0.89	0.939
VINAM_ANFIS 2	WDQ – NL – NNF	0.607	0.0075	0.296	−0.0224	0.779	0.856
VINAM_ANFIS 3	WDQ – WM – FR	0.823	0.0024	0.776	−0.0095	0.907	0.948
VINAM_ANFIS 4	WDQ – DWST – FR	0.804	0.0026	0.756	−0.002	0.897	0.944

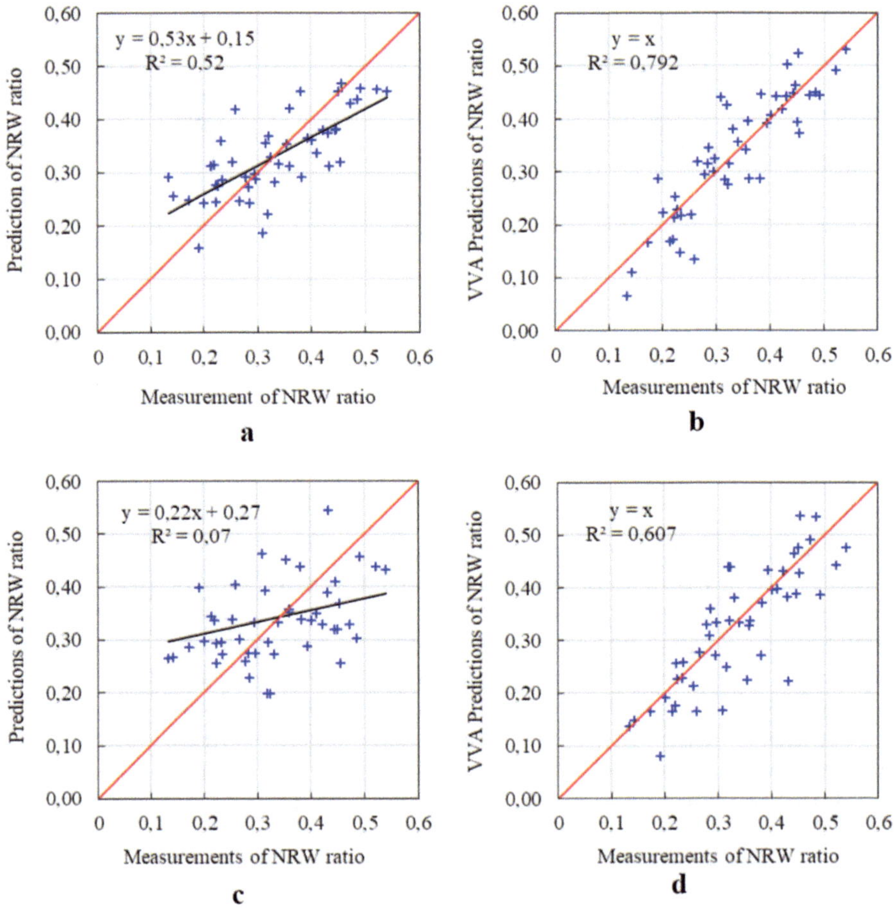

Fig. 5.13 ANFIS classical (**a, c, e, g**) and VINAM (**b, d, f, h**) approach

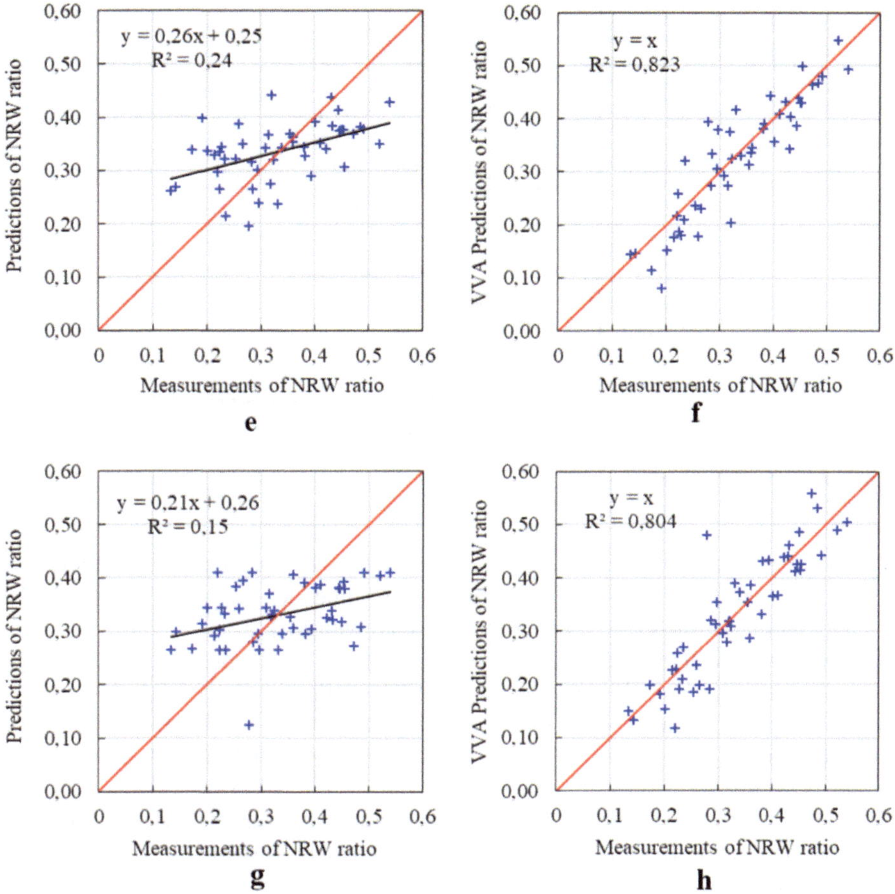

Fig. 5.13 (continued)

5.6 Equation with Logic

The leading way to reach mathematical equations in the simplest, fastest, economical and trouble-free way is to find clear expressions of equations that can at least approximate, considering the logic and mathematical rules mentioned before. For this, the reader must be preoccupied with a problem. Here, there is an example of logically deriving a basic equation for population growth that is attractive for all. Since, the goal is to obtain an equation in mathematics; the verbal logic information must be obtained beforehand (Chaps. 2, 3). If only the change in population growth over time is taken as a basis, making the following assumptions helps to obtain the mathematical equation after the simplest logic rules.

1. If a time is taken as the beginning for the population, then an initial population must be found at this time. Let us denote the start time with t_o and the current population amount with $P(t_o)$.

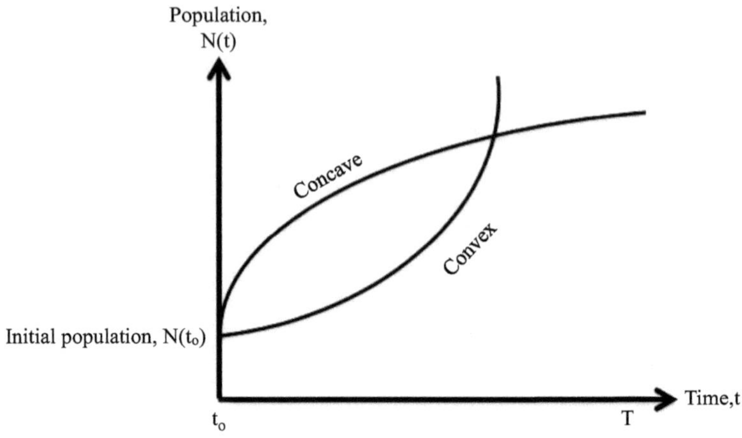

Fig. 5.14 Population model

2. How the population may change over time during the next T period? If there was no war, epidemic, or migration during this time, let the value of the initial population at time t ($t < T$) denote its change with P(t).
3. Since there will normally be population growth, the expression $P(t) > P(t_o)$ indicates that population growth continues over time.
4. How will this increase in the population over time continue? It should be answered whether it will be linear or curvilinear. Rational thinkers say this will never be linear.
5. Whether the curvilinear increase will be concave or convex is also decided by rational thought. Here, it should be decided that the increase in the population will increase or decrease gradually over time.
6. In Fig. 5.14, concave and convex shape curvatures are given together. It is left to the reader to decide on one of them.

If the mathematical equations of two different curves, concave and convex, as reached after all the above steps are questioned, it will be understood that this is an exponential function. Şen (2002) recommended the rule of thumb for the exponential function that the curve cuts an axis and the other end extends asymptotically. Even if this rule is not known, if the curves in Fig. 3.9 are shown to a mathematician, the answer to be taken is the following clear mathematical expression:

$$N(t) = ae^{bt} \tag{5.8}$$

There are two unknown constants in this equation, a and b. Since the population is $N(t_o)$ for the initial time, $a = N(t_0)e^{-bt0}$ is obtained as a result of a simple arithmetic operation by substituting the values of t_o and $N(t_o)$ in the previous equation. By substituting this into Eq. (5.8), one can obtain,

$$N(t) = N(t_o)e^{b(t-to)} \tag{5.9}$$

Thus, the only unknown constant is b, and its determination requires annual data showing the change of population over time.

The previously explained logic and mathematical rules were used to obtain the population equation given above. A reader who has correctly perceived the steps described here can now at least reach the geometry of the changes in time, space or time-space variation of the event or phenomenon, which s/he examines with logic and simple mathematical rules as in Fig. 5.14, that is, how the output from the input variable changes. It is not necessary to be a mathematician for this, but it is necessary to perceive logic and simple mathematical rules in a dynamic way.

5.6.1 Equation by Experiment

In the previous section, the relationship between the input and output variables was found from the constant initial condition that provided the equality. In some cases, the proportionality and relationship can be determined rationally, but the constant coefficient necessary for the establishment of full equality cannot be determined by the rules of reason and logic. In such cases, experimentation is necessary to determine the equality constant. To give an example of this situation, for example, when a certain tensile or compressive force is applied to an elastic rod, the elongation or shortening (which is called deformation ratio, or strain) is directly proportional to the force. Considering the cross-sectional area of the bar perpendicular to the applied force, if the force per unit area is defined as stress, it is concluded that stress and strain are directly proportional. Verbally, the following propositional sentence is reached.

As the stress increases, the strain also increases in direct proportion.

It should be noted that the words IF-THEN (Chaps. 3 and 4), which indicate the judgment, are hidden in this last statement. In order to write the mathematical expression for this, if stress and strain are denoted by the symbols σ and ε, respectively, then the proportionality can be expressed as follows:

$$\sigma \; \alpha \; \varepsilon \tag{5.10}$$

where α implies proportionality property. As such, this expression is not an equation; it is merely the symbolic form of the verbal proposition. To convert this expression into equality, a coefficient must be imported so that the equation takes the form of equality below:

$$\sigma = \mathrm{E}\varepsilon \tag{5.11}$$

Here, E is a constant that provides the balance between the two variables. In material engineering it is referred to as the elasticity modulus. This constant is determined by logical thinking, but in order to determine its numerical value, at least one stress, σ_1 and strain, ε_1, measurement must be known. In this case, the numerical value of E is calculated as $E_1 = \sigma_1/\varepsilon_1$. In the case of many stress and strain measurements, the arithmetic means simply taken after obtaining different E_i values for each tool can be regarded as the numerical value of $\overline{E}$.

5.6.2 *Extracting Equations from Data*

It is not possible to determine some events with reason and logic approaches and rules as explained above. What kind of a relationship exists between even the simplest two variables is hidden within the measurement or experiment data. In particular, the direct or inverse proportionality of the relationship can be found with the rules of logic. On the other hand, being able to determine the form of the relationship between those two variables from the data is called the method of "extracting equations from the data," which is known as empirical formulation. For this, it is necessary to know the numerical data of the input (I_1, I_2, . . ., I_n) and the corresponding output (O_1, O_2, . . ., O_n) from measuring tool records. If these two series are marked on two sets of axes perpendicular to each other in the I-O plane, a group of points called a scatter diagram emerges. By reviewing these points together, either there is no relationship between the two variables, or the shape of the relationship can be determined. Figure 5.15 shows some of the most general of the different options that may arise.

The units of the input and output variables do not need to be the same in the mathematical equation extraction proposed in this section. Dimensionality is not sought in such equations based on experience (i.e., data, empirical). The extraction of mathematical expressions from the scatter diagrams given in Fig. 5.15 is given in detail below.

In Fig. 5.15a, the dots are scattered in such a way that nothing can be seen that can be discerned between them. As a result, it is concluded that there is no relationship between the input and output variables (words) on the horizontal and vertical axes. There is no mathematical relationship here, because a verbal inference that complies with the principles and rules of logic cannot be reached. Such data scatters can be subjected to cluster analysis for depicting possible groups (Chap. 4).

In Fig. 5.14b, one understands that there is a direct proportionality between I and O, since one can come to logic rule that IF the variable I increases THEN the variable O also increases. One wonders if this straight ratio is along a straight line or has a curvature. The answer to the question can be determined by eye without doing any mathematical operations, and a straight line passing through these points can be assigned. The right one assigned varies from person to person, but they are very close to each other. Thus, if mathematical principles are not involved, a ball of truth based on subjectivity emerges between people. However, in mathematics, an equation form has been determined as

$$O = a + bI \tag{5.12}$$

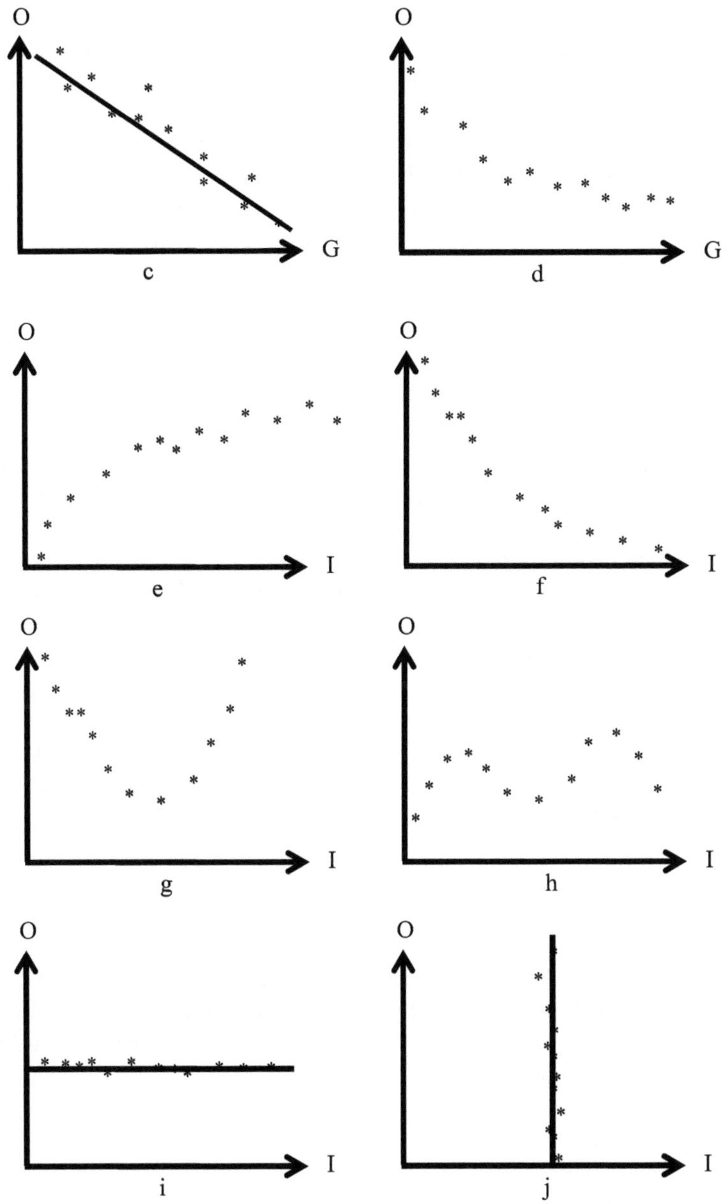

Fig. 5.15 Equation extractions from data

In order to obtain a and b coefficients from the available I and O data, a common line equation accepted by everyone is reached by calculating with the method of linear regression (finding the most appropriate objective relationship between two variables) (Chap. 4).

The scattering of the dots indicates that the O decreases with the increase of the I variable visually (see Fig. 5.15c). This means that there is an inversely proportional relationship between I and O. Is this inverse proportion relationship along the straight line? or is there a curvature? Everyone concludes that the answer is along the straight line, but subjectively it is a mathematical expression as

$$O = a - bI \tag{5.13}$$

The coefficient a here indicates the O value of the intercept point on the O axis, and b indicates the slope of the line.

In Fig. 5.14d, the scattering of the points between I and O again shows that there is an inverse proportion. Since it is also observed that the inverse ratio has curvature, it is verbally concluded that it can be represented by a suitable curve, not a straight line. When examined in more detail, it is understood that this curve intersects the O axis but is almost parallel to I axis on its left side. Close inspector can subjectively determine the most suitable curve to pass through here as a cluster (bundle) of curves, each of which is close to each other. For subjectivity to gain objectivity, it is necessary to introduce mathematical operations. A person who knows that a curve that intersects one axis and shows parallelism (asymptote) to the other will be a decreasing exponential (exponential) mathematical expression, will find that this is a mathematical equation as follows (Şen 2011) where a and b are coefficients to be calculated from numeric al data.

$$O = ae^{-bI} \tag{5.14}$$

In the Fig. 3.11e scatter diagram, it is observed that there is direct proportion and curvature. In addition, it is understood to be an exponential curve, as it intersects the horizontal axis and is parallel to it in the last parts. In this case, the form of the mathematical equation is obtained by replacing the minus in the previous equation with plus for the direct proportion.

$$O = ae^{bI} \tag{5.15}$$

The scattering points in Fig. 5.15f show parallelism or rather tangency (asymptote) to both axes, the proportion is reversed, and there is curvature. With this much information, it is learned that it will be a curve like a hyperbola (parallel to or tangent to two axes) (Şen 2011). This verbal information includes logic principles and rules. The hyperbola expression that corresponds to this verbalism in the mathematical world:

$$O = \frac{a}{I^b} \tag{5.16}$$

It is seen from Fig. 5.15g that the scattering points are arranged in a formal way along the two arms, and they appear in reverse on one arm and in correct proportions on the other arm. As such, there is also a valley (or it can be a peak) point in this way. It is observed among the observations that both arms are constantly increasing with a decreasing slope. There are also logic rules hidden here. It is recommended that the reader come up with these by considering the logical principles and rules explained earlier. A mathematical expression with these properties is called a parabola (quadratic polynomial), and its mathematical symbols are given below.

$$O = a + bI + cI^2 \tag{5.17}$$

Figure 5.15h shows the scattering of the wavy dots. It is concluded that the mathematical expression of this consists of sine and cosine curves.

The scattering of the points in Fig. 5.14i are parallel to the horizontal axis with no slope. This implies that there is a constant relationship between I and O variables, or the O outputs are connected to I inputs by a constant. In the last Fig. 5.15j, unlike the previous one, O is bound to I by a constant.

5.6.3 Extracting Equations from Dimensions

In the previous section, it has been explained that even in the absence of unit, it is possible to obtain equations that can be used in practical applications with experience (empirically, experimentally) from the measurement data at hand. In some cases, from complex units, which may arise from the basic definition of a term, the equations associated with it can also be derived. For example, someone who knows the definition of a term called discharge as the volume of fluid (water, gas) per time, t, can also derive some basic equations from it. If the discharge is Q, the volume is V, and the time is t, the verbal definition becomes a mathematical equation as

$$Q = \frac{V}{t} \tag{5.18}$$

The unit of this definition is m^3/s; it is possible to divide this unit into possible subunits, for example, $m^2 \times (m/s)$; Since the two units here are physically area A and velocity v, hence, the definition in the previous equation takes the following expressional shape:

$$Q = Av \tag{5.19}$$

Thus, a brand-new equation is obtained.

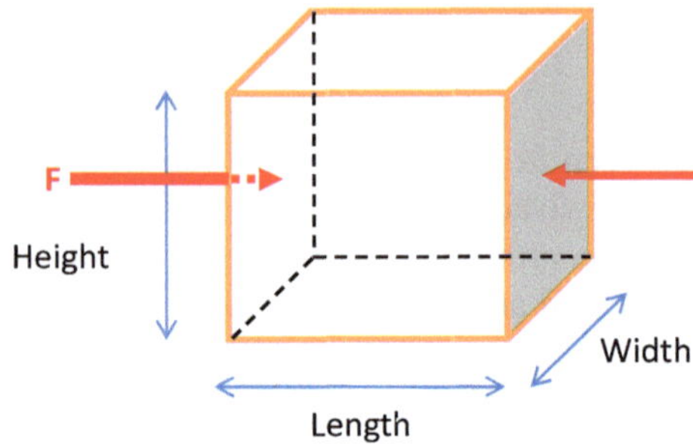

Fig. 5.16 Elastic but incompressible rectangular prism

5.7 Logical Mathematical Derivations

The principle, which is called "conservation of mass" in the scientific world, is referred to by different names such as "balance equation," "continuity equation," or "budget equation" (Chap. 3). At the root of all scientific equations are verbal expressions of simple thought and logic principles and rules. Here, an inference will be presented that will appeal to those who do not have much education, but who want to reach rational conclusions by thinking with reason, logic, and skeptical criticism.

There are numerous benefits in considering the geometry of the phenomenon or event, which is examined by observation, experiment, or imagination, especially in order to reach scientific results. For this, consider an elastic but incompressible rectangular prism with width, height, and length, as shown in Fig. 5.16. Here, the meaning of the "incompressible" principle among the public is that its volume will not change after subjection to a force as in the figure.

If the deformable prism is suddenly (to the exclusion of time) subjected to compression by a force, F, in the direction shown, then logically the shape will change, but the volume will not. It is thought that a shortening will occur in the direction in which the force is applied, and extensions will occur in other directions. The amounts of elongation and shortening are different from each other. Shortening in the length direction leads to a decrease in volume along that direction; it causes an increase in volume in the width and height directions. There are volume changes in the directions, but no change in the total volume. If one makes these verbal expressions a little tidier, s/he arrives at an addition operation with the following sentences.

Volume change in length direction + Volume change in width direction + Volume change in height direction = 0

Writing it in another way turns into a verbal mathematical expression in the frame below with aggregation.

Volume change along length
Volume change along width
Volume change along height
+ ____________________
ZERO

The above expression, which has been verbally explained and obtained only by using the principles and rules of reason and logic, is a kind of mathematical equation. In order to translate this result, which is reached with logic, into mathematics, some symbols must be used. If length is denoted as l, width w, height h, volume V, and change, that is, difference by d symbols, changes in height, width, and height are expressed as dV/dl, dV/dw, and dV/dh, respectively. So, the above verbal explanations appear in the form of the following symbolic equation:

$$\frac{dV}{dl}+\frac{dV}{dw}+\frac{dV}{dh}=0 \tag{5.20}$$

Again, one of the mathematical assumptions is that if an event depends on more than one dimension (where volume depends on height, width and height), the symbol ∂ should be used for partial change indications instead of d, and accordingly, the previous mathematical equation becomes

$$\frac{\partial V}{\partial l}+\frac{\partial V}{\partial w}+\frac{\partial V}{\partial h}=0 \tag{5.21}$$

In the above mind experiment, if the force acting on the prism is applied slowly over time, not suddenly, a change in volume over time also comes into play. Like the previous explanations, if one includes time, which is the fourth dimension, instead of three dimensions, into the volume change work, at the end t to show time, the final expression takes the following form:

$$\frac{\partial V}{\partial l}+\frac{\partial V}{\partial w}+\frac{\partial V}{\partial h}+\frac{\partial V}{\partial t}=0 \tag{5.22}$$

There is nothing logically missing in this last statement, and one can perceive it as just right. However, the length, width, and height are dimensions of space and are thought to be different from the time dimension, and hence, it turns out that we have added apples and pears in this last equation. Apples and pears cannot be added, but in science they can be added rationally. For this, it is necessary to either convert apples into pears or pears into apples. Since the first three terms of the last mathematical expression show the change of volume with a length, that is, volume changes in unit lengths, that is, in derivatives, and the fourth term shows the change in volume per unit time, the units of volume changes with respect to space and time are different from each other. A coefficient is needed to convert one of the different units to the other. If such a coefficient is denoted by α, the most accurate mathematical equation converted from time unit to space unit takes the following form.

$$\frac{\partial V}{\partial l}+\frac{\partial V}{\partial w}+\frac{\partial V}{\partial h}+\alpha\frac{\partial V}{\partial t}=0 \tag{5.23}$$

It is advised that before mathematics, logical principles, rules, and verbal inference could be made, and then mathematical expressions could be reached by symbolizing words and sentences.

5.7.1 *Logic Modeling of Electrical Circuits*

Considering the initial explanations of the principles of crisp logic (Chap. 2), it is possible for the inferences to be either true or false. In general, engineering and natural science formulations are dependent on crisp logic principles as explained in the previous sections, most often the events include vagueness, imprecision, and uncertainty, and therefore, fuzzy logic is more suitable to digest the uncertainties first verbally and then with fuzzy logic inferences numerically (Chap. 2). Although there are numerous procedures to deal with numerical data such as probability and statistical methodologies, but most often researchers do not care their basic assumptions and apply the available convenient formulations directly. We can represent the principle that logic inferences are verifiable or falsifiable with a circuit shown in Fig. 5.17. There is a switch that enables communication between two points, such as A and B. If this switch is closed, yes, if it is open, there will be no. We can think of this as the presence or absence of electric current between points A and B.

However, it is very common in practice that more complex circuits arise by increasing the number of switches given in series (Fig. 5.18), in parallel (Fig. 5.19), and both in series and parallel (Fig. 5.20). In this case, there are situations where the electrical circuits work reliably or not (danger, risk).

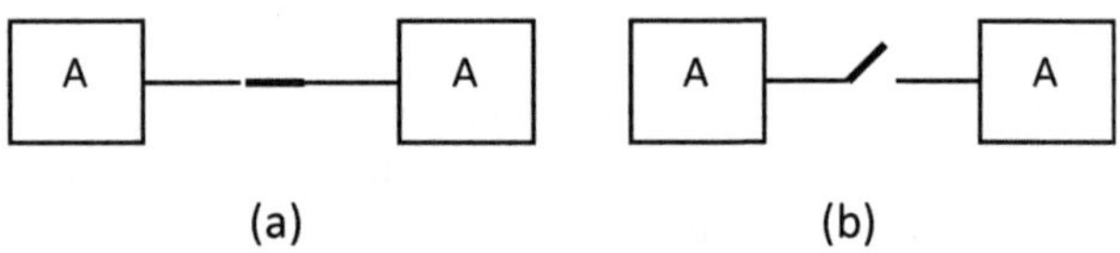

Fig. 5.17 Logical inference principles (**a**) verifiable, (**b**) falsifiable

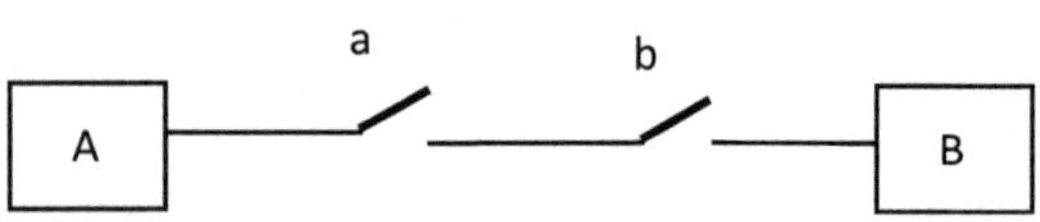

Fig. 5.18 Serial connected keys

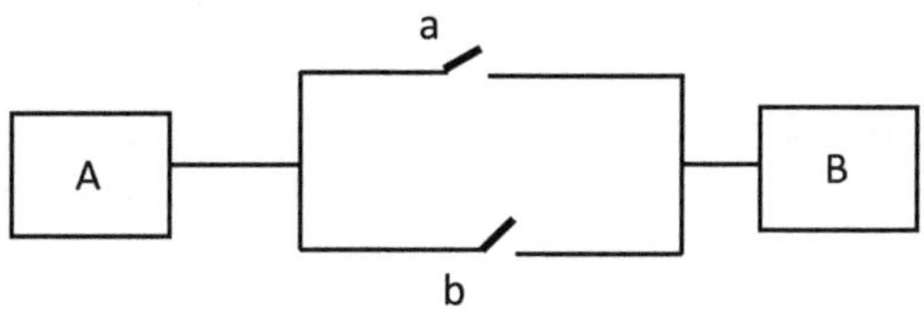

Fig. 5.19 Parallel connected keys

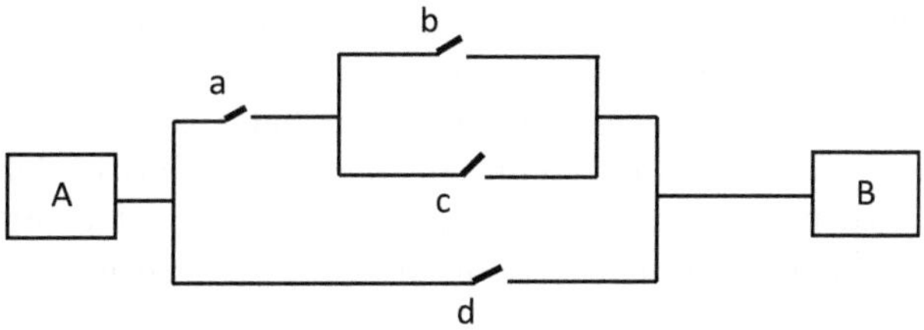

Fig. 5.20 Serial and parallel connected keys

5.8 Risk and Reliability

with lines or in anyIn practice verification is regarded as "reliability" and falsification as "risk." It should be considered that the switches in each of the above circuits are manufactured by the same company and that their reliability, R, in percentage, that is, exceedance probability or risk, r, is complementary events. When these risk and reliability amounts are expressed as percentages (likelihoods), their sum must be equal to 1. The following expression is valid simultaneously for each key.

$$R + r = 1 \tag{5.24}$$

Taking advantage of these basic reliability and risk possibilities, both switches must be closed (verified, that is, ANDing) at the same time for current to flow from A to B through the series-connected switches shown in Figs. 3.8 (Chap. 3). Since closing one of them will not have any effect on closing the other, which is, closing the switches does not affect each other, that is, they are independent, and hence, the multiplication rule of probabilities is applicable. Accordingly, the probability of current flowing through two series-connected switches given is

$$R_{1-2} = R_1 R_2 \tag{5.25}$$

If there are m switches with similar characteristics in series between points A and B, the probability of electric current flowing through this circuit can be calculated by generalizing the previous expression in the most general manner as

$$R_{1-2-\ldots.m} = R_1 R_2 \ldots\ldots R_m \tag{5.26}$$

According to the situation of the parallel connection of switches given in Fig. 3.9a–d (Chap. 3), there are three possible situations for electric current to pass from point A to point B. The first of these is when only switch a is closed, the second is only switch b is closed, and the third is when both switches are closed at the same time. If one, two, or all three of these conditions are present, an electric current flows from A to B. For this reason, since only the switch a must be closed, that is, verifiable, for the system in Fig. 3.9 (Chap. 3) to function, the probability of this situation is that if a is closed and b is open do not affect each other, the probability of multiplication R(1-R) can be written according to the independence principle of probability. Similarly, if b is closed but a is open, electric current will still flow through the system and the combined probability of this is R(1-R). In the last case, since both keys are closed, it is found from the principle of multiplication of probabilities, as a result of the principle of independence of probability calculations. In that case, the possibility of the system to pass electric current always and situations, with the help of the ORing of these three incompatible situations and the addition principle of probability calculations,

$$R_{OR} = R(1-R) + R(1-R) + RR = 2R(1-R) + R^2 \tag{5.27}$$

Finally, in Fig. 3.15 (Chap. 3), there are cases, where switches b and c are connected in parallel with switch a in series, and the case that occurs when switches b and c are connected in parallel with switch d again. A valid probability model can be written for such a complex system by considering the parallel and serial connections, respectively, according to the two-state probability calculations described previously. Here are a few options that may arise first, and the rest is left to the reader to complete:

(a) ANDing of a and b.
(b) ORing b and c with a.
(c) ANDing of a, b, c, and d.
(d) ANDing of a and d.
(e) Others.

Assume that all keys are manufactured by the same company to the same quality. As a first step, if the b and c switches are connected in parallel, according to the Eq. (5.26) if they have a combined probability, it is possible that they can transfer influx to the B point. This combined possibility

$$R_{bc} = 2R(1-R) + R^2 \tag{5.28}$$

Let us show it with notation. Now, the idea that switch a is connected in series with this parallel system (switches b and c), using probability products according to Eq. (5.24), the following expression is reached:

$$R_{abc} = R\left[2R(1-R) + R^2\right] \tag{5.29}$$

Since this last obtained probability value (keys a, b and c) will be connected in parallel with key d, considering Eq. (5.28), hence one can obtain a rather complex expression,

$$\begin{aligned} R_{abcd} = {} & R\left[2\,R\,(1-R) + R^2\right](1-R) + \left\{1 - R\left[2\,R\,(1-R) + R^2\right]\right\} \\ & R + R\left[2R(1-R) + R^2\right] R \end{aligned} \tag{5.30}$$

Although this last statement may seem complicated, if the steps in its derivation are followed wisely, the reader can easily deduce probability models for current flow even through more complex circuits in the form of such circuits.

5.9 The Logic of Mathematical Functions

What is explained in this section consists of the subjects explained in the book "Science and Scientific Research Principles" written in Turkish by Şen (2011).

As a result of the mathematics education received in secondary and higher education, students take the shapes of some functions and the symbolic expressions of the functions that belong to them mechanically, but "I wonder what their logical foundations are?" Can it be placed in the rote thought system at the end of a questioning? For example, they may know that derivative and integral calculations are series of mechanical operations, but can they also answer the question of what are their logical bases? You can find the answers to all these questions in the explanations below. In the absence of reasonableness (rationality), criticism (uncertainty), doubt, and perception of the principles and definitions of logic, which is the basis of the work, at the end of education, a person falls into the path of memorization, dullness, dogmatism, and conveying without being aware of it. Below logical-based explanations of some basic mathematical figures (geometry, functions) are explained in a way that supports dynamic rationality. The author of this book presents from his experience the necessity of perceiving information given to the student with its reasons. In the field of mathematics, the general names of the equations taken during secondary and higher education stages and most often used in practical life are given in Fig. 5.21.

The reader should consider the plane geometry environment for all the cases described below. This means that shapes exist in one-, two-, and three-dimensional spaces, that is, Euclidian geometrical domain. However, there are also decimal spaces with its name in the last 40 years as fractional (fractal) geometry and has been left out of this chapter (Mandelbrot 1982).

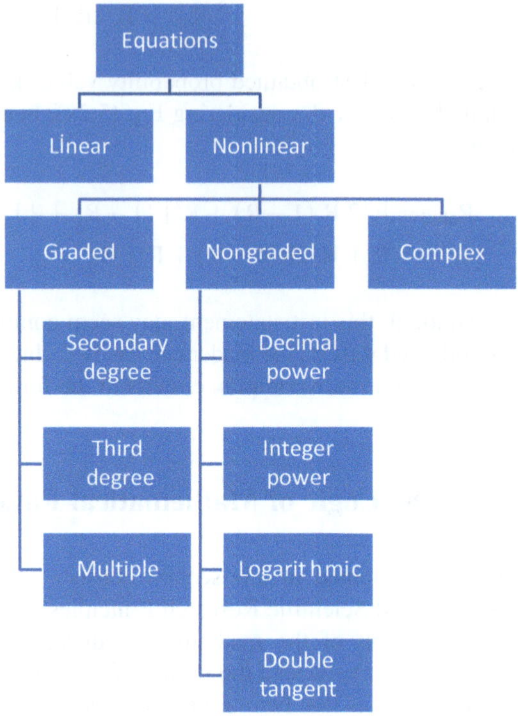

Fig. 5.21 Some mathematical names

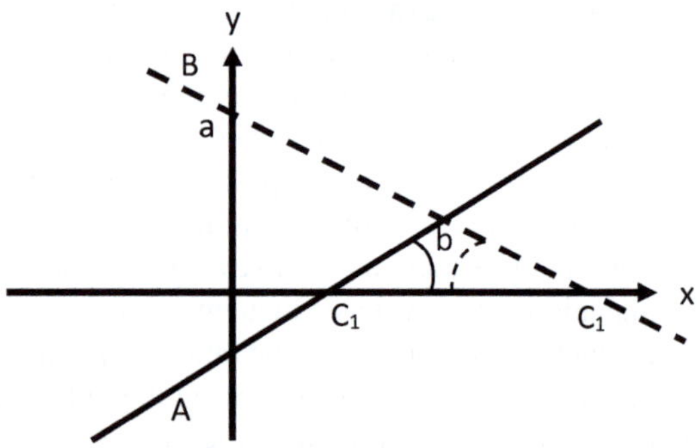

Fig. 5.22 Straight-line shape

5.9.1 *Straight Line*

Every person, educated or not, thinks that knowing two points will be enough to draw a line. This means that two points are enough to draw the line. A straight line passing through points A and B is given in Fig. 5.22. The logical relationship between at least two variables is directly proportional and linear, and in general the expression is

$$y = a + bx \tag{5.31}$$

Here, x is input variable in modeling (reason in philosophy of science, independent variable in mathematics, predicator variable in engineering), and y is called output variable in modeling (result in philosophy of science, dependent variable in mathematics, prediction variable in engineering).

There are two constant values in a straight-line equation. Of these, a, indicates the intercept value on the y-axis when $x = 0$, and b indicates the slope of the straight line. One can also look at this slope as an increase in y versus an increase in x. Since Eq. (5.32) has a root corresponding to $y = 0$) $C_1 = -a/b$, in terms of root,

$$y = b(x - C_1) \tag{5.32}$$

One can also write as if the constant is $a > 0$ ($a < 0$), the geometric (function) situation A (B) appears in Fig. 5.21.

All the physical law equations (Newton, Ohm, Hubble, Hooke, Darcy, etc.) put forward in science so far are expressed as a straight-line equation with $a = 0$. So, one can look at the equation "$y = bx$" as the equation of all these laws with different verbal expression for each variable. The most general form of the previous equation is if there are n input variables ($x_1, x_2, x_3, \ldots x_n$) in n-dimensional space it is a single point. Such input variables are the main input variables in different artificial neural network (ANN) architectures (see Chap. 7).

$$y = a + b_{x1}x_1 + b_{x2}x_2 + b_{x3}x_3 + \cdots\cdots + b_{xn}x_n = a + \sum_{i=1}^{n} b_{xi}x_i \tag{5.33}$$

Here, the values denoted by each b_{xi} indicate the amount of slope that ith variable shares among others. Geometrically, if this expression contains only the case up to the variable x_1 (like Eq. 5.32), it represents a straight line in two-dimensional space. If the part up to the x_2 term is concerned, it gives the plane in three-dimensional space. In general, it is a plane that one cannot visualize in mind in n-dimensional space up to the x_n term, but that one can think and express a direct proportional and linear relationship in every direction.

$$y = a + b_{xi}x_i \quad (i = 1, 2, \ldots, n) \tag{5.34}$$

with lines or in any three-dimensional space as

$$y = a + b_{xi}x_i + b_{xj}x_j \quad (i \neq j; i = 1, 2, \ldots, n \text{ and } j = 1, 2, \ldots, n) \tag{5.35}$$

Again, we can visualize it in our minds with planes that can be perceived as cross-sections. It is enough to look at Fig. 5.23 to visualize this last statement with symbols.

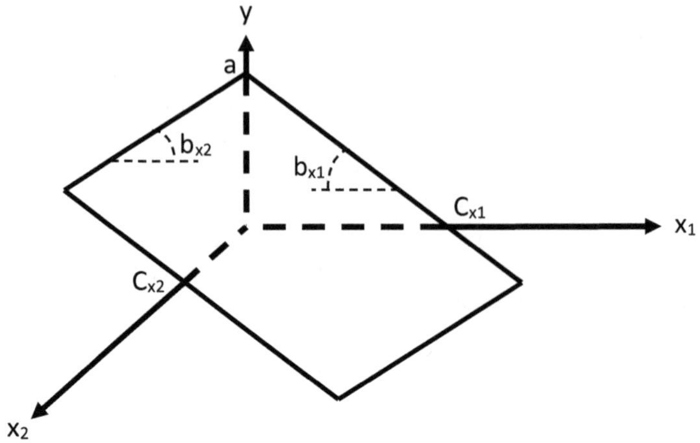

Fig. 5.23 Plane geometry in 3-dimansion

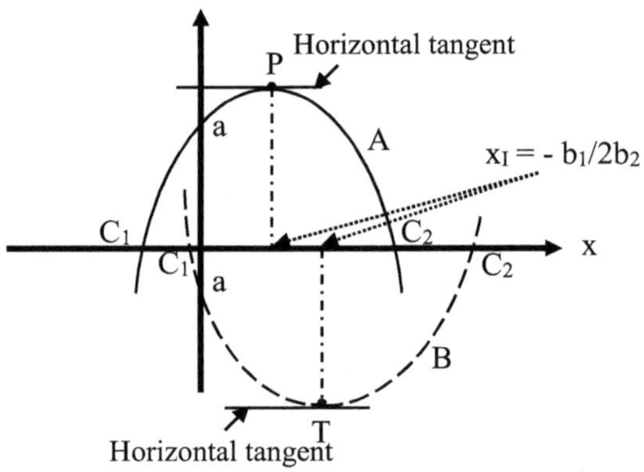

Fig. 5.24 Second-degree equations (parabola)

In addition, the intersection points on the x_1 and x_2 axes, which can be considered as a kind of roots from Eq. (5.36), appear as $C_{x1} = -a/b_{x1}$ and $C_{x2} = -a/b_{x2}$, respectively (Fig. 5.23). Also, Eq. (5.34) is written in terms of roots of n.

$$y = a\sum_{i=1}^{n}(x\text{-}C_i) \quad (5.36)$$

5.9.2 *Quadratic Curve (Parabola)*

It is a situation that arises by adding another term with an integer exponent to the straight-line equation given above (Fig. 5.24).

$$y = a + b_1 x + b_2 x^2 \quad (5.37)$$

All variables are the same as what was said before for the straight-line equation. Here, a new quadratic term is preceded by a slope constant, b_2.

This equation has two roots, C_1 and C_2, which intersect the x-horizontal axis. Equation in terms of roots can also be written as

$$y = a(x - C_1)(x - C_2) \quad (5.38)$$

In the figure, graph A (B) shows the constant $a > 0$ ($a < 0$) state, which corresponds to a peak, P, and a trough, T, respectively. The practical meaning of this is that there is one turning point (peak, P or trough, T) in quadratic expressions (Fig. 5.24). The characteristic of the turning point is that there is a horizontal tangent, that is, the slope is equal to zero at these points.

Since the derivative means slope as a basic definition, the derivative of dy/dx must be equal to zero here. By simply taking the derivative of Eq. (5.38)

$$\frac{dy}{dx} = b_1 + 2b_2 x \quad (5.39)$$

By setting this derivative equal to zero, we obtain the position of the projection of the apex or trough point on the horizontal axis. The value of the projection point is $x_1 = -b_1/2b_2$ (Fig. 5.24). Verbal information is obtained about the position of the peak or trough point to the right or left of the vertical axis, depending on whether this value is plus or minus. Eq. (5.40) can be converted into a straight equation that people can easily understand, and it can be given the form of Eq. (5.32). For this, by describing the derivative with another variable such as t obtained

$$t = b_1 + 2b_2 x \quad (5.40)$$

As a result of comparing this with Eq. (3.21), verbal perceptions of b_1 and b_2 slopes can be reached with all the interpretations given there.

5.9.3 Cubic Curve

By adding the next integer exponential variable term to the quadratic equation, its mathematical expression is obtained as follows:

$$y = a + b_1 x + b_2 x^2 + b_3 x^3 \quad (5.41)$$

Figure 5.25 shows the geometry (function) of this expression and the states of the required symbols. In general, it is known to have three roots. Accordingly, writing the previous equation in terms of roots takes the form,

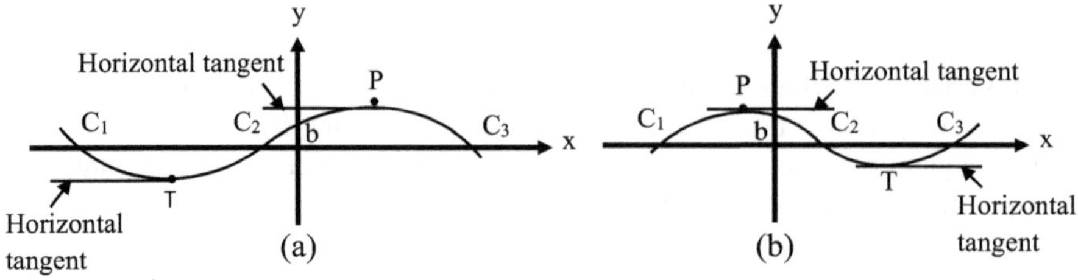

Fig. 5.25 Third-degree shapes

Table 5.4 Degree turning point number relationship

Degree of equation	Turning point number	Linearization derivative number
1	0	0
2	1	First derivative
3	2	Second derivative

$$y = a(x - C_1)(x - C_2)(x - C_3) \tag{5.42}$$

Figure 5.25a, b geometric shapes appear, respectively, according to the positive or negative value of a, which is called the shape parameter. As can be seen from this figure, there are successive peaks and troughs in the geometry of Eq. (5.42), that is, two turning points.

As a result of comparing this with the first- and second-order figures given before, the rule in Table 5.4 can be deduced and then generalized.

Additional information can be obtained from the landmark concept. Since the definition of turning point is given above as the derivative being equal to zero, by taking the derivative of Eq. (5.42) a quadratic equation is obtained:

$$t = b_1 + 2b_2x + 3b_3x^2 \tag{5.43}$$

Interpretation of this can be done by comparing it with the quadratic equation given above. Thus, verbal meanings can be given to the slopes of b_1, b_2 and b_3 in this equation. This situation is left to the reader. In another case left to the reader, by taking the derivative of Eq. (5.44) again

$$u = 2b_2 + 6b_3x \tag{5.44}$$

By obtaining the expression, a straight-line equation is reached as a result. Since u is the derivative of the derivative, it is called the quadratic derivative, d^2y/dx^2, in terms of the principal variables. As a result of comparing this with Eq. (5.32), necessary verbal meanings can be attributed to b_2 and b_3 in u-x space this time.

Another rule scan be taken from here; mathematics is to take consecutive derivatives to linearize equations with integer exponents. Then the basic equation can be linearized by taking consecutive derivatives up to the turning point.

5.9.4 *Multi-degree Curve (Polynomial)*

Equation with exponent up to nth integer terms, which is the most general form of the figures given in the previous steps, is written as follows:

$$y = a + b_{x1}x + b_{x2}x^2 + b_{x3}x^3 + \ldots + b_{xn}x^n \tag{5.45}$$

With the generalization of Table 5.4, the rule that it is necessary to have n-1 inflection points in the shape (geometry) of an nth degree multi-degree equation emerges from the relation "Number of inflection points = (Greatest exponent-1)." According to this, for example, the general and root-based mathematics of a fifth-degree equation are in order.

$$y = a + b_{x1}x + b_{x2}x^2 + b_{x3}x^3 + b_{x4}x^4 + b_{x5}x^5 \tag{5.46}$$

and

$$y = a(x–C_1)(x–C_2)(x–C_3)(x–C_4)(x–C_5) \tag{5.47}$$

The forms (according to the principal and derivatives) and the verbal expressions implied by this last equation are left to the reader. The reader is strongly encouraged to do so. For the reader who does this, difficulties in verbalizing such equations or verbalizing the given equations disappear and the reader reaches scientific happiness as rational information plays a role in the mind.

5.9.5 *Equation with Decimal Exponent (Power Function)*

One of the mathematical symbols frequently used in practical applications is the decimal form of the integer power of $y = ax^2$, which is the simplest form of the quadratic equation.

$$y = ax^b \tag{5.48}$$

In Fig. 5.26, the geometry (function) of the values ü takes with respect to 2 is shown. For $b > 2$ values of the basic quadratic (parabola) equation, there are closings like an accordion, whereas there are openings for $b < 2$. Here, a coefficient can be called a scalar, because the scale can be changed according to its large and small values. Although a can be shown in the figure, it is not possible to show b.

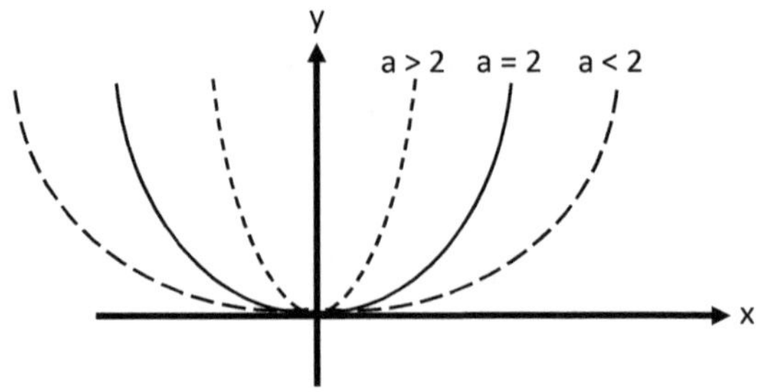

Fig. 5.26 Decimal power shapes

The linearization of this frequently used equation cannot be obtained by taking the derivative and its successive derivatives, as in the previous ones, because b is decimal. For the linearization of Eq. (5.49), it is possible to do it simply by defining a new variable such as u = bx and transforming it into y = au. Such transformations do not give verbal information about what b means.

On the other hand, taking the logarithm of both sides of Eq. (5.49) gives a clue to the new state linearity.

$$\log\ y = \log\ a + b\log\ x \tag{5.49}$$

Since the logarithm of the scale constant u and a will be constant for the new variable u, which is obtained by taking the logarithm of the variable y on the right side of this expression, if one defines it as a new variable to the log(x), which is the new equivalent of A and x, the simplest linear equation desired is the desired linear equation takes the form:

$$u = A + bv \tag{5.50}$$

As a result of comparing this with Eq. (5.32), the slope of the line obtained by marking y and x values with logarithms of b on the vertical axis and if the y variable with logarithms of A is equal to zero, b = −A/v or b on the v axis. It can be deduced that it must be equal to –A/log(x) and from this it will be A = −alog(x).

If one needs to develop a method for the practical solution of the above operations, it would be useful to go back to Eq. (5.50). This equation shows that a straight line with slope b and cutoff value b = log(a)/log(x) on the horizontal axis will emerge in [log(x), log(y)] logarithmic space, not in (x, y) normal space. Therefore, it is concluded that if the axes are taken directly as logarithmic straight-line geometry will be obtained without any difficulty. Figure 5.27 shows a set of orthogonal axes with both axes logarithmic.

On this paper, there is an order that is shown at intervals of 10 on both axes and is divided into six subintervals within it. According to Eq. (5.50), an example is given for the straight lines and the inversely proportional lines that have two options in the figure. Here, a value is shown as the slope in the figure. Since one cannot give y a zero value in order to understand what the value of a in the equation means,

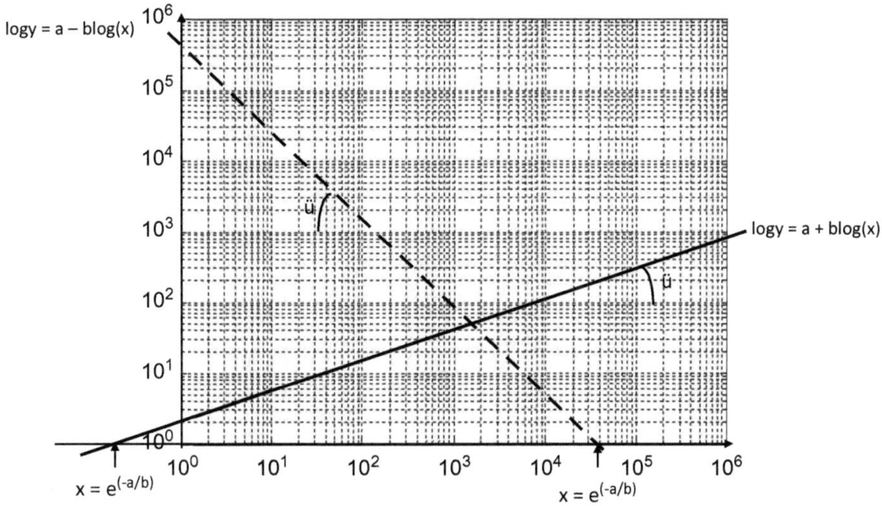

Fig. 5.27 Double logarithmic axes

considering log(y) = 1, since y = 100 on the vertical axis, the intersection points of the lines on the 100-horizontal line on the vertical axis are from a + blog(x) = 0 to x = e(−a/b) is obtained.

While marking on this paper, the values are given to the intervals in the horizontal and vertical axes by looking at the minimum and maximum values of the values at hand on both axes, writing them as the power of 10 and looking at the exponent of 10. For example, suppose the smallest (largest) value of x data is 0.012 (2457.8) and the smallest (largest) value of y data is 27.9 (379.8). To mark data on double-logarithmic paper, the data must be written as powers of 10. Thus, as a result of evaluating the largest and smallest values of the x and y axes according to base 10, 1.2 × 10–2 (2.4578 × 103) and 2.79 × 101 (3.798 × 102) are found.

Example Trace the x and y data given in Table 5.5 on the double logarithmic paper given in Fig. 5.28.

5.9.6 Exponential Curve

Another constantly increasing or decreasing curve, which is frequently encountered in practice, is the exponential curve that intersects one of the axes and is tangent to the other at very large values (if there is an asymptote at infinity, as if there is an infinity reality) as follows:

$$y = ae^{-bx} \tag{5.51}$$

Table 5.5 x and y data

x	y	x	y
1	3,2	1×10^0	$3{,}2 \times 10^0$
3	82,944	3×10^0	$82{,}944 \times 10^0$
5	11,06998	5×10^0	$11{,}06998 \times 10^1$
7	12,7713	7×10^0	$127{,}713 \times 10^1$
9	14,6277	9×10^0	$146{,}277 \times 10^1$
11	20,3283	$1{,}1 \times 10^1$	$203{,}283 \times 10^1$
13	24,9082	$1{,}3 \times 10^1$	$249{,}082 \times 10^1$
15	23,3904	$1{,}5 \times 10^1$	$233{,}904 \times 10^1$
17	26,7909	$1{,}7 \times 10^1$	$267{,}909 \times 10^1$
19	29,1216	$1{,}9 \times 10^1$	$291{,}216 \times 10^1$
21	30,3917	$2{,}1 \times 10^1$	$303{,}917 \times 10^1$
23	31,6082	$2{,}3 \times 10^1$	$316{,}082 \times 10^1$

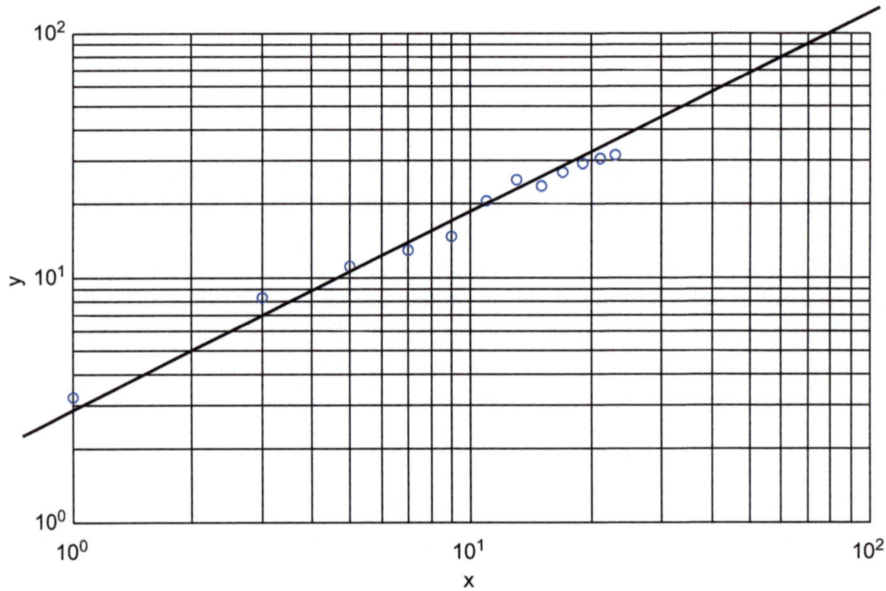

Fig. 5.28 Plotting and scatter diagram on a double logarithmic paper

Here, a is the scaling constant and b is the branching constant. For continuity in verbal expression, since there is information about being tangent to another axis at large values when cutting one axis, different sets of curves emerge with constants shown in Fig. 5.29.

Of the se, the geometry (function) set in Fig. 5.29a is the curve collection expressed by Eq. (5.52). Equations can be found for others in a similar way. For example, the mathematical expression for Fig. 5.29b is

$$x = ae^{-by} \tag{5.52}$$

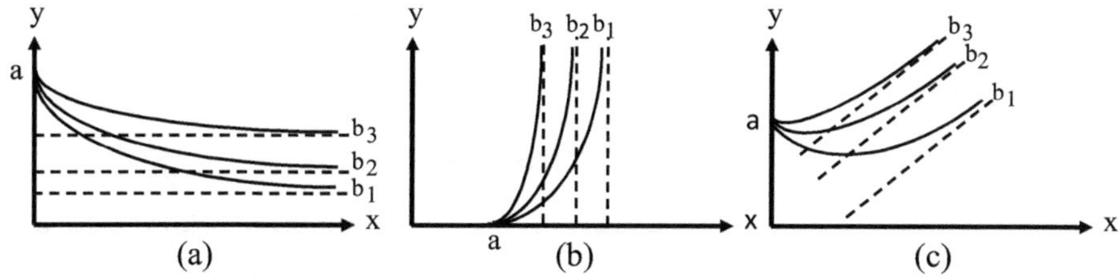

Fig. 5.29 Exponential shapes

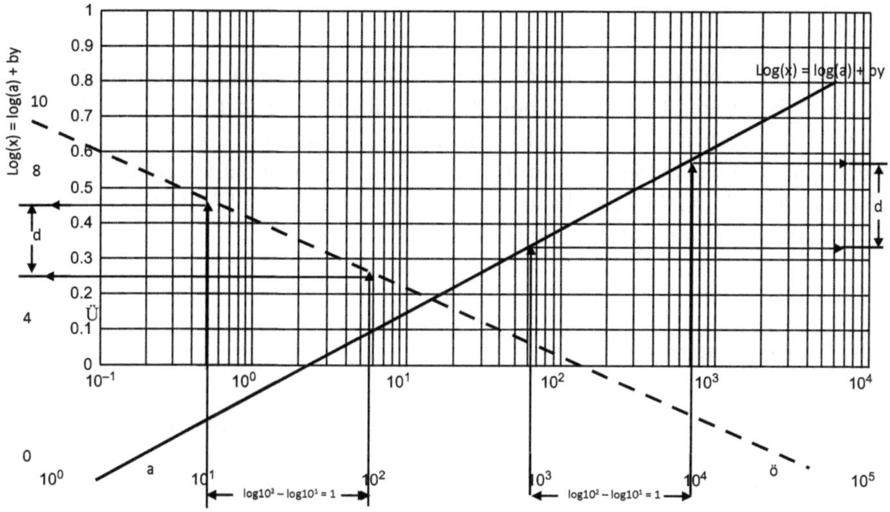

Fig. 5.30 Exponential shapes on semilogarithmic paper

The linearization of this expression can be done as logarithm, like the previous curve. In this situation,

$$\log(x) = \log(a) - by \tag{5.53}$$

If defined here as $u = \log(y)$ and $A = \log(a)$

$$u = A - by \tag{5.54}$$

This is the correct equation. Comparing this with Eq. (5.32) provides the necessary verbal expressions. Since only the logarithm of y is taken from the variables according to Eq. (5.54), the lines obtained according to the plus and minus sign of d by showing it on semilogarithmic paper are shown in Fig. 5.30.

Since b shows the slope of the line in semilogarithmic space, this slope is obtained by the ratio of the opposite side to the adjacent side from the general slope definition.

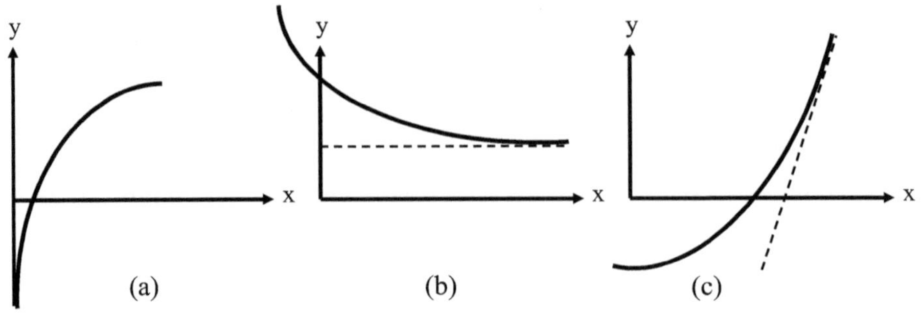

Fig. 5.31 Logarithmic shapes

For slope in semilogarithmic space, one complete revolution is taken on the logarithmic axis. Its length is, for example, $\log 10^2 - \log 10^1 = 1$ and so on (e.g., $\log 10^5 - \log 10^4 = 1$ etc.); it always equals 1. In Fig. 3.25, $\log 10^2 - \log 10^1$ and $\log 10^4 - \log 10^3$ are shown. Differences in the vertical axis corresponding to these circuits give the slope d. On the other hand, by taking $y = 0$ in Eq. (5.53), $\log(x) = \log(a)$ or $x = a$ corresponding to this value is obtained.

5.9.7 *Logarithmic Curve*

At large values of an axis it is tangent (asymptote at infinity), curves that take increasingly large values at large values of the other axis perpendicular to it and do not pass through the origin of the perpendicular axes are called logarithmic curves and the mathematics of this verbal definition appears symbolically in the following expression:

$$y = a + b \ \log(x) \tag{5.55}$$

The most important difference of this model from the exponential one is that it does not pass through the zero point and the variable x cannot take a negative value. The different geometry (function) of this curve according to the verbal definition is shown in Fig. 5.31.

5.9.8 *Double Asymptotic Curve (Hyperbola)*

Verbally, it is the name given to curves along two axes that are tangent to them at large values, and its simplest mathematics (tangent to two perpendicular axes at large values of the axes) can be given by the following symbols (Fig. 5.32):

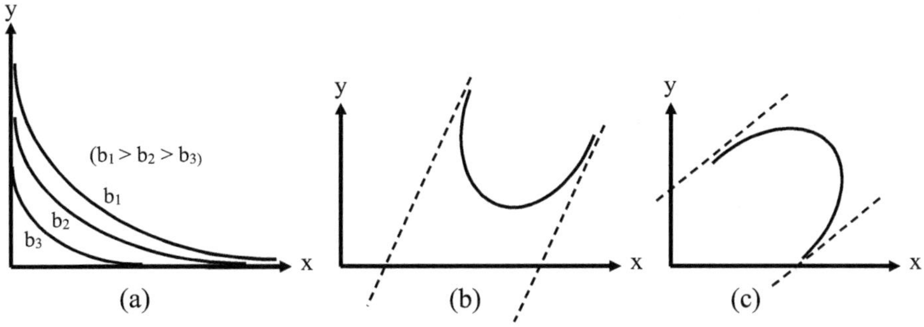

Fig. 5.32 Double-tangency shapes

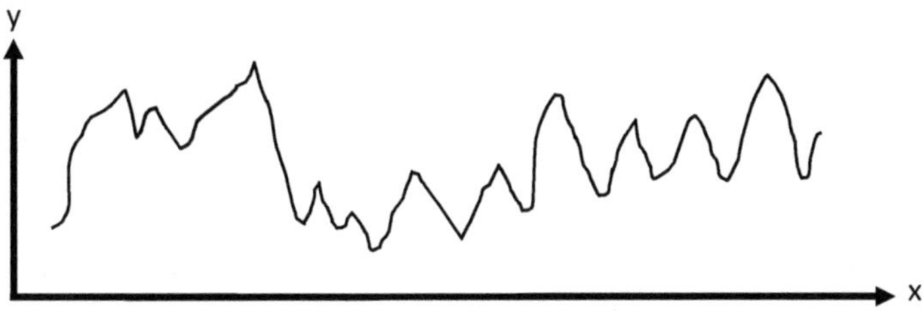

Fig. 5.33 Complex shape

$$y = \frac{d}{x} \tag{5.56}$$

In this expression, d provides a cluster curve as the distribution coefficient. With the decrease (growth) of this coefficient, the curve tangent to the two axes at large values gets smaller (grows).

5.9.9 Complex Curve

The multi-order curve equation described above can be used to model very complex shapes. By looking at this, milestones are counted so that many degrees of mathematical symbols can be applied; one less than that (Fig. 5.32). In this case, with the inclusion of very detailed milestones, the number of terms in the equation may increase, making it very difficult to manage. Instead, even the most complex shapes can be modeled by superimposing sine and cosine waves of different scales and amplitudes, which are more convenient, fast, and easy (Fig. 5.33).

5.10 Mathematics Logic and Language

As mentioned before, the basis of mathematics is logic and the expression of logical ideas is possible with language. The importance of binary and fuzzy logic principles is explained in Chap. 3.

5.10.1 From Language to Mathematical

In previous sections, the benefits of making inferences from the geometry after establishing its shape by visualizing the event to be examined in the mind with the ways of reason and intuition are mentioned in order to produce scientific knowledge. In such inferences, deduction or induction was used and it was emphasized that the main inferences are very important. The verbalization of some events causes the emergence of verbal rules in the examinations. In such studies, it should be tried to extract the shape information (geometry) of the investigated event. The first thing that comes to mind is to investigate the variability in the two-dimensional space, which is called a function, which is taught to us throughout the education as morphology. Even if it is the first time that the reader hears that functions are geometric shapes, s/he can make healthy and wise scientific inferences by paying attention to this point in his future studies. For a phenomenon to be scientific, the following points must coexist:

1. The substance of the phenomenon (air, water, stone, soil, concrete, iron, wood, etc.)
2. Change of phenomenon behavior with time and/or location (distance, area, or volume).

The meaning of the second point is desirable to have an output of interest under the influence of many factors in the emergence or behavior of the investigated event and to predict this from the inputs. Before such a prediction, it is necessary to understand the time and location variability of the event linguistically. For this reason, the first thing to be done is to try to visualize the time and/or location variability of each of the desired output or input variables, if possible. If we denote the output variable as O, time t, distance D, area A, and volume V, shapes (relationships) are searched in the two-variable fields in Fig. 5.34.

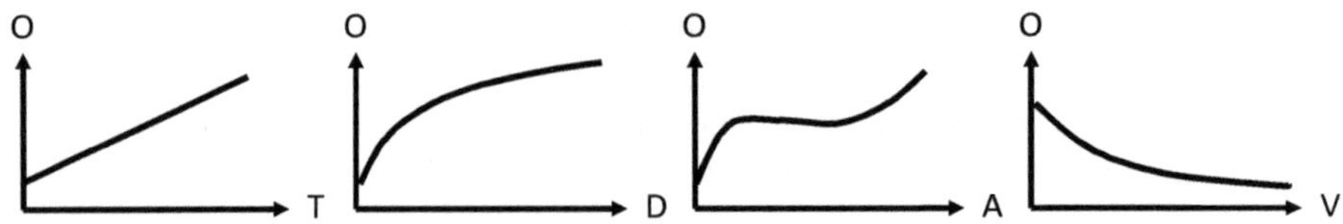

Fig. 5.34 Time and shape variations

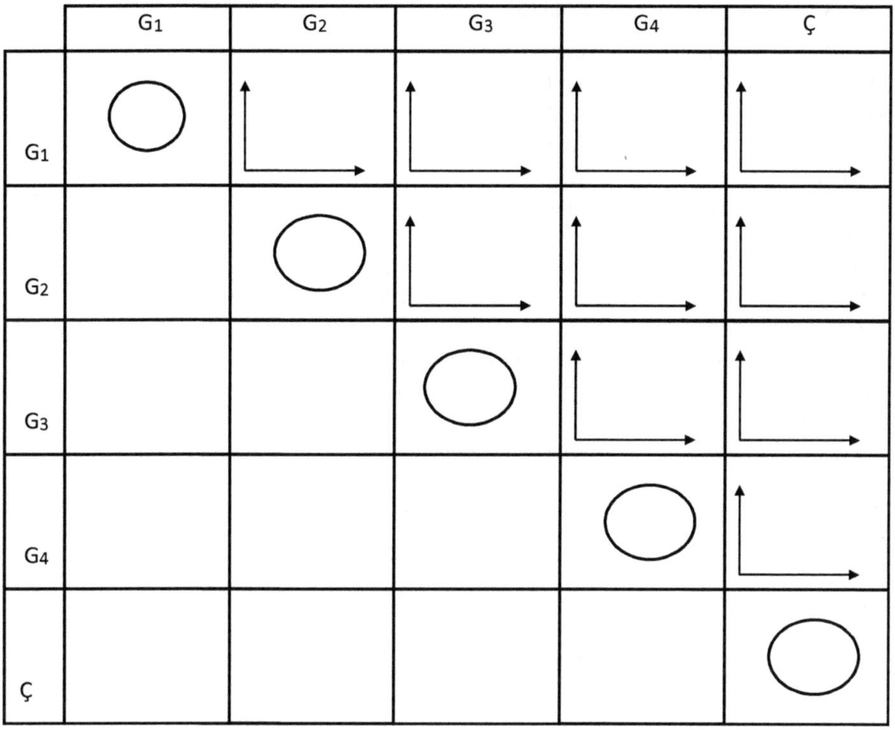

Fig. 5.35 Binary relationship matches

As a second verbal inquiry approach while doing research, it is useful to show it as a rational binary relationship match (matrix) between input and output variables. This matching matrix is similar to the one outlined in the preliminary modeling section, but it is important to see how the two relationships take shape (Fig. 5.35).

Examples in two-form environment, it is now time to translate them into mathematics. For this, first, it is necessary to know the linguistic of the mathematical symbols. Table 5.6 shows these translations.

5.10.2 From Mathematics to Language

Contrary to the mathematical expressions reached in the previous section, being able to translate from mathematics to a language provides access to additional detailed information in future studies. When a given mathematical formula is considered, it is as if deductive inference is used since it is a matter of going into details. Such scientific translations into a language have both verbal and rational visual benefits.

Table 5.6 English-mathematics dictionary

English	Mathematics	Explanations
Plus	$+$	ANDing
Minus	$-$	ANDing
Multiplication	$\times$	ANDing
Division	$/$	ANDing
Ordinary derivative	Dy/dx	Slope, y variation per unit x variation
Partial derivative	$\partial y/\partial x$	Conditional slope (for y variation all variations are accepted as constant except x variation)
Square root	$\sqrt{}$	
Power	x^a	Power or exponential
Ordinary logarithm	Log(x)	Ordinary logarithm of x
Natural logarithm	Ln(x)	Natural logarithm of x
Absolute value	$\lvert x \rvert$	Absolute value of x
Bigger than	$x > y$	X is bigger than y
Less than	$x < y$	X is smaller than y
Equality	$x = y$	X is equal to y
Approximately	$x \approx y$	X is equal to y approximately
Error	$\pm$ 5%	Plus-minus 5% error
	∇	
Infinite	∞	There is no limit
Difference	Δx	Difference in x
Pi number	π	3.14
Limit	$\text{Lim}(x \to y)$	X approaches y at large values

The person who will translate from mathematics must first know linguistically the meanings of the symbols in the formulas. In addition, it should be known what some signs (derivative, integral, etc.) mean (Table 5.6).

Example 5.1 Translate the following equation into English and graphically show the relationships between each of the input and the output variables. Let us assume that in this equation p represents the pressure, t the temperature, h the humidity, v the wind speed, and finally r the amount of rain. Knowing the names of these symbols is very important in order to show how the given expression will be translated into English. It is straight away possible to write these variables in an implicit mathematical form as follows:

$$r = f(p, t, h, v) \tag{5.57}$$

The first step in explicitly verbalizing this equation is to reveal the relations between the output variable r and the input variables on the right side by considering the other input variables as constant. For this, we need to apply the two criteria mentioned earlier (proportionality and geometry), keeping in mind. One of them is direct or inverse proportionality, and the second is linearity (linearity) or curvature (nonlinearity).

1. r-p relationship: There is direct proportionality and curvature, which is in the form of a power function since p has an exponent other than 1.
2. r-t relationship: There is direct proportionality and linearity. The linearity here is due to the power of t being equal to 1.
3. r-h relationship: There is inverse proportionality and curvature. Curvature arises from two sources. The first is a hyperbola-shaped curvature since h is in the denominator. The second is an additional curvature from the square root in addition to the previous curvature.
4. r-v relationship: Inverse proportionality and linearity are valid. It is inversely proportional to the minus sign in front of it. This means that as v increases, r decreases.

With these points, the linguistical logic principles of the equation are reached and verbal information is obtained about the types of different relations. Using these verbal relations, it is possible to write the general form of the given equation at least as follows:

$$y = a\frac{x^b u}{z^c} - v \tag{5.58}$$

Here, the coefficient a is called the equalization constant. On the other hand, b and c are expressions of curvature, their numerical values are inconceivable. Therefore, experimental studies, measurements, and numerical values (data) are needed to determine b and c values. From the equation given in addition to the above points, one can deduce what the verbal relations between the input's variables are based on the same idea. For this, it will be enough to consider that the other input and the output variables remain constant, except for the two input variables considered (Fig. 5.34). Thus, the verbal logic expressions of the binary relations between p, t, and v are given below:

1. r-p relationship: They are in a direct proportional and curvilinear relationship.
2. r-t relationship: They are in a directly proportional linear relationship.
3. r-v relationship: They are inversely proportional but in a linear relationship.

We can make the model relationship match of the given equation as shown in Fig. 5.36. From this, it is understood how the shapes of the relations, that is, the geometric (function) appearances, are. All verbal relationships given above are shown here visually.

Example 5.2 Heat conduction, groundwater flow, pollutant distribution, etc. The equation expressing the changes of events in very small control volumes was derived in the previous section by using the verbal reasoning principles. The general form of

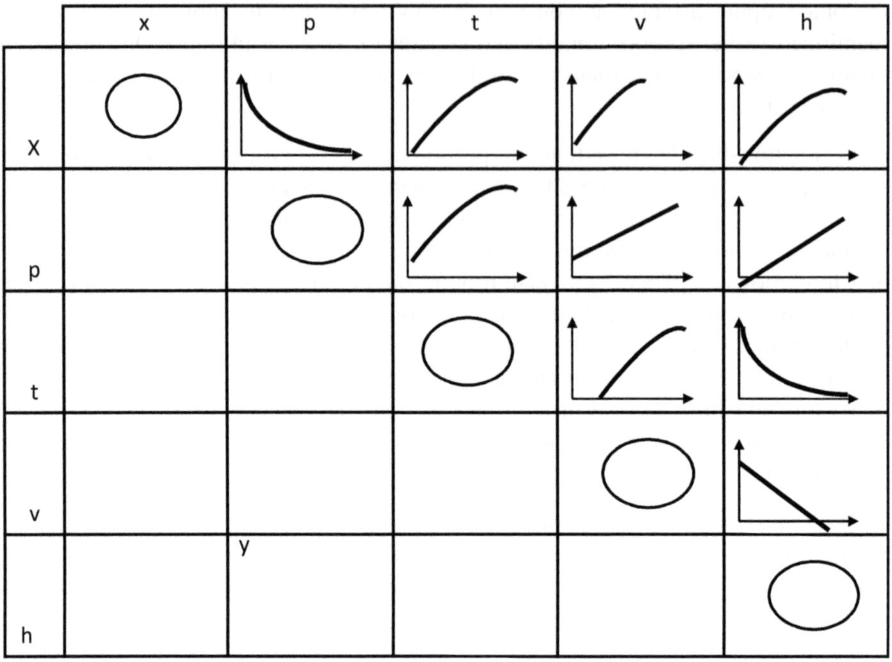

Fig. 5.36 Equation (5.58) relationship matches

the differential equation representing these events is as follows. If we assume that T is temperature and t is time, what kind of verbal information does this equation give us?

$$\frac{\partial^2 T}{\partial x^2}+\frac{\partial^2 T}{\partial y^2}+\frac{\partial^2 T}{\partial z^2}=\alpha\frac{\partial T}{\partial t} \tag{5.59}$$

By writing this expression in more detail, the following situation is obtained:

$$\frac{\partial}{\partial x}\left(\frac{\partial T}{\partial x}\right)+\frac{\partial}{\partial y}\left(\frac{\partial T}{\partial y}\right)+\frac{\partial}{\partial z}\left(\frac{\partial T}{\partial z}\right)=\alpha\frac{\partial T}{\partial t} \tag{5.60}$$

It is understood that the parts in parentheses are partial derivatives as explained in Table 5.6. We can explain the verbal expressions of the left-hand brackets as the slopes of the temperature according to the x, y, and z directions, or the changes in the unit direction lengths. If we show the slopes according to their direction with the symbols s_x, s_y, and s_z, respectively, and the slope according to the time on the right with the s_t symbol, the previous expression will be reduced to shape,

$$\frac{\partial s_x}{\partial x} + \frac{\partial s_y}{\partial y} + \frac{\partial s_z}{\partial z} = \alpha s_t \tag{5.61}$$

From this last case, it becomes clear that each term on the left-hand side means the slopes of the slopes.

If the event is multiple (heterogeneous), since the features will be different in each direction, generalization can be made by introducing different coefficients in three directions. In this case, the following equation can be easily written as

$$C_x \frac{\partial^2 T}{\partial x^2} + C_y \frac{\partial^2 T}{\partial y^2} + C_z \frac{\partial^2 T}{\partial z^2} = \alpha \frac{\partial T}{\partial t} \tag{5.62}$$

Here, C_x, C_y, and C_z represent the coefficients related to the additional variability of the investigated event in the x, y, and z directions, respectively. In addition to the variability of the event itself, these coefficients show the situations related to the variability of the environment in which that event occurs. For example, each of them is conductivity coefficient in heat-related problems, hydraulic permeability coefficient in groundwater investigations, diffusion coefficient in air pollution, and elasticity coefficient in strength of materials subjects.

5.11 Mathematical Models

Mathematics is based on the rules of science philosophy and logic. Mathematics consists of symbolic expressions found with idealizations in one's own world, even outside of nonphysical events in a strictly definite way. For the first time in history, the Muslim thinker Al-Khwarizmi-Algorithm (780–850) symbolized systematically thoughts and provided the solution of problems through algebraic equalities for the first time in the history of science. With the algebraic solutions, he put forward symbolic logic, which constitutes the bases of mathematical developments through different analysis methods today. Symbolic and especially algebraic expressions called functions are nothing more than mathematical models. The essence of modeling is to reveal the geometry of the event in some way and express it logically, visually, or algebraically. In fact, all functions taught abstractly in mathematics represent geometry. Every mathematical function serves to express a variable in terms of other variables. It is possible to plot their graphs as dependent versus independent variables. Every graph drawn gives the relationship between the variables, either as a curve or as a surface.

The different types of functions we learn in mathematics classes all have phenomena that they can represent physically. The models to be extracted in this book are based on classical functions known from mathematics. Therefore, in this section, the properties of mathematical functions will be reviewed that may be of practical importance in terms of modeling.

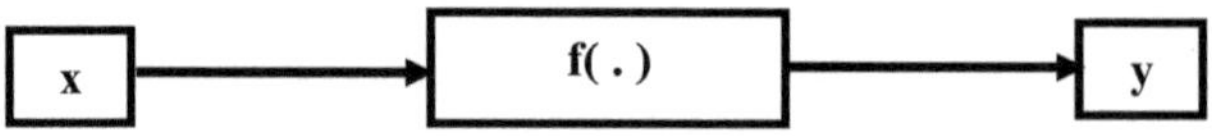

Fig. 5.37 Implicit mathematical model

5.11.1 Closed Mathematics Models

Expressing a variable y in terms of another variable x in its most closed and simple form in mathematics is displayed as

$$y = f(x) \tag{5.63}$$

Such an expression symbolically can represent multivariate geometric shapes in terms of modeling. First, if x is viewed as an input variable and y as an output variable, this symbolic spelling literally means transforming an input variable into an output variable. In Fig. 5.37 a model emerges that can, in general, be given for all such implicit mathematical equations. Here, the conversion is done with the f(.) mechanism.

It is thought that all implicit equations in mathematics can be represented by such a model. Such a closed model, while important to the mathematician, is hardly useful to practitioners. In the implicit form, it is not clear how the system converts input variables into output estimations.

It is always clear that engineers, social scientists, or researchers want to know that statistical regression models, in which determinations can be made by using measurement values, include empiricism and rationality. It is not always possible to reveal the explicit states of implicit mathematical expressions with rationality and reasoning about the investigated event. For this, the person must have assimilation of some basic mathematical equations that s/he has learned so far, not only symbolically, but by rational-logical interpretations. This point is overlooked in pure mathematics lessons. Whereas, mathematical expressions are useful and usable especially in engineering, physics, etc. to the extent that they represent an event.

5.11.2 Explicit Mathematics Models

All the different functions that we learned so far in mathematics are written clearly and their derivatives and integrals can be taken if necessary. For example, a person who does not know or not always consider that derivative means slope, and integral means area or volume cannot benefit from explicit mathematical models, nor can s/he derive productive information from them. It should be known what each equation written in mathematics means verbally. If the equation can be associated with physical events, the importance of mathematics in applications will increase. In addition, if one knows what the geometry of that equation is and how it can be

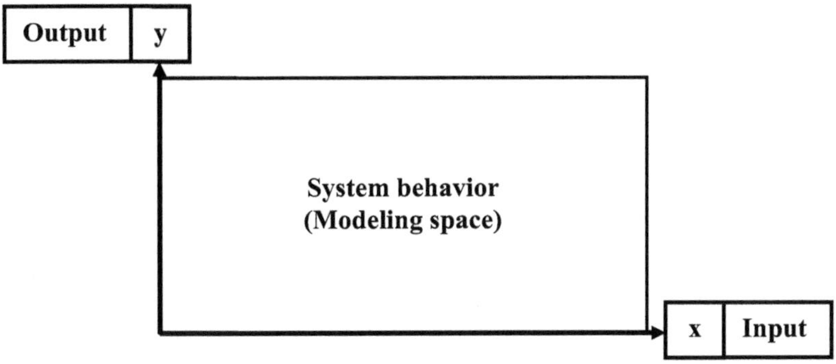

Fig. 5.38 Cartesian modeling

completed with verbal expressions, different chemistry, physics, engineering, sociology, etc., it may be possible to fit the events on the subject or rather to express these events with that mathematical equation.

In mathematics, one should not expect any modeling benefit from them unless explicitly symbolic writing of equations is supported by verbal expressions. At least in explicit expressions, it is necessary to know a form of system behavior that converts input variables, parameters and independent variables into dependent variables. Explicit mathematical models are visualized by considering two or three sets of axes perpendicular to each other. If there are two correlated variables, x independent and y dependent, it has become commonplace since about 1600 to represent them in a perpendicular two-axis system and they are known as Cartesian axes, because they were first propounded by Rene de Carte (1596–1650). Such a modeling situation is shown in Fig. 5.38.

Along this axis, the values of that variable are written and read. In mathematics, the axes are called abscissa and ordinate, respectively. Each of the basic mathematical equations taken in high school education necessarily gives the solution of an idealized event or behavior in engineering, natural or social sciences. In another sense, each of these is a mathematical model with three components as input, system behavior, and output.

5.11.2.1 Linear Model Through Origin

In its simplest form, it represents the verbal statement that as one of the two variables increases, the other will increase or decrease linearly. If the independent variable is x and the dependent is y, then equation shows the simplest version of the linear model.

$$y = ax \tag{5.64}$$

The representation of this in the Cartesian coordinate system is given in Fig. 5.38. From this figure, one can deduce the following significant points.

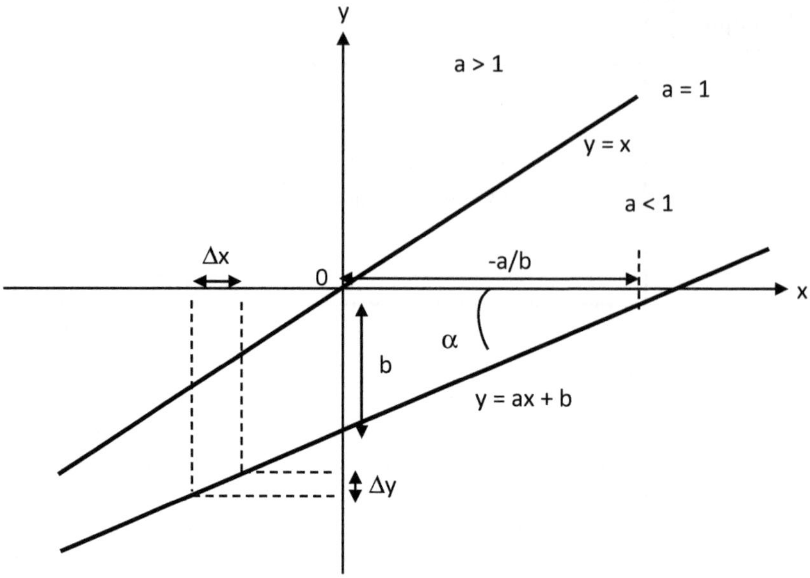

Fig. 5.39 Straight-line model

(a) The increasing (decreasing) values of the variable x correspond to the decreasing (increasing) values of y.
(b) Independent variable, x, takes the value zero, and the dependent variable y takes the value zero.
(c) If x overgrows (shrinks) y overgrows (shrinks).
(d) The constant difference between two consecutive values of x corresponds to a constant difference value in y. This difference is placed on the x-axis, the corresponding difference of y always remains the same. The difference on the vertical axis, corresponding to the difference Δx, on the horizontal axis, is Δy. The ratio of Δy to Δx gives the value of parameter a in Eq. (5.64). This is briefly called the moment of slope of the curve. So, the slope of this line is

$$a = \Delta y / \Delta x \tag{5.65}$$

In the simplest straight-line model given in Eq. (5.64), a can also be called the proportionality coefficient. Thus, one can say that the x and y variables change in directly proportional manner.
(e) The constant in Eq. (5.64) is called its sole parameter. As the parameter value changes, the y values corresponding to x change according to the parameter ratio. Straight-line models valid for different parameter values are given in Fig. 5.39.

In this equation, a = 1 means y = x. This line divides the field of the Cartesian coordinate system into two parts, one a > 1 and the other a < 1. The linear model in Eq. (5.67) does not necessarily have to pass through the origin. A more general version of Eq. (5.68) is by adding b constant to its right-hand side. Thus, the most general linear mathematical model becomes as

$$y = ax + b \tag{5.66}$$

Here, the second parameter, b, can take any value and if x = 0 (i.e., on the y-axis) it represents the abscissa of the point, where the line intersects the y-axis (see Fig. 5.38).

It also explains other features of the linear model, which has benefits for modeling physics phenomena. The parameter a in the model is also called the scaling parameter, apart from the slope. Indeed, parameter a has a magnifying effect if the argument x (a < 1) is diminutive (a > 1). The b parameter in the model serves to shift the value whose scale is set; in this respect, the b parameter can also be named as the shift parameter. On the other hand, the parameter a, defined as slope in Eq. (5.68), can be written a = tanα since, according to this equation, the opposite side is the ratio of the opposite side, αy, to the adjacent side in a right triangle, αx. Here α denotes the angle of inclination of the straight model in degrees.

Now, how can the linear models be used in scientific studies? Let us try to find the answer to the question. In the past, researchers have used the linear model in many branches of science, especially in its simplest form in Eq. (5.64). One can list a few of them as follows:

(a) In physics, force, F, is directly proportional to acceleration, a. The direct proportionality coefficient is called the constant mass, m, in physics. Thus, the law of motion first advanced by Newton,

$$F = ma$$

It is a linear mathematical model. The verbal expression of this is that if the mass is constant, the force changes linearly and proportionally with the acceleration. The expression 'linear changes' is necessary in the right-hand side of this expression. Just the word proportionality can represent many mathematical equations. For example, $f = m\sqrt{a}$, $F = ma^{0.2}$, $F = ma^3$, etc., are not valid for this law. In all expressions, force and acceleration are directly proportional, but in each of them, force does not change linearly with acceleration.

(b) In earth sciences, groundwater velocity, v, is directly proportional to hydraulic slope, i, and has linear variation. Here, the proportionality coefficient denotes the permeability coefficient of the geological rock, k, and the law is

$$v = ki$$

(c) In materials science, stress, σ, is proportional to strain, ε, and stress varies linearly with strain. The proportionality coefficient is here called the modulus of elasticity E; like this Hooke's law has been stated:

$$\sigma = E\varepsilon$$

Except for these mathematical linear models, all other models are nonlinear. This means that in the Cartesian coordinate set, the geometric figure between y and x in the form of two variables is not straight but shows curvature.

5.11.3 Polynomial Models

The most widely used and simple to understand mathematical models are x-independent and y-dependent variables is in the form as

$$y = a_0 + a_1 x + a_2 x^2 + a_3 x^3 + \ldots\ldots\ldots\ldots\ldots + a_n x^n \tag{5.67}$$

This is called an nth order polynomial. Here, a_i ($i = 0, 1, 2, \ldots, n$) denotes the polynomial coefficients. To be a polynomial, n must be an integer greater than 1. There are special cases of this general polynomial that are simple but used in practice. The most important of these is the straight-line equation, which is explained in the previous section and emerges by considering the first two terms on the right side of Eq. (5.69).

In addition, it includes various simple nonlinear models of this polynomial. Knowing their clear characteristics and graphic shapes very well by the reader is very useful in applying the approach determined as the experimental approach, which is obtainable by measurements made on various scientific subjects. Here, we will focus on the shapes and some meaningful properties of these mathematical models, not on derivatives and differentials, which are important in mathematics.

5.11.3.1 Parabola Model

It is the simplest nonlinear model obtained by considering the three terms on the right side of the general polynomial equation:

$$y = a_0 + a_1 x + a_2 x^2 \tag{5.68}$$

This model has three parameters, which can be calculated from the x and y data obtained by experiment or observation by means of the least square's method, which will be explained in the next chapter. Unlike the parabola equation, can also be

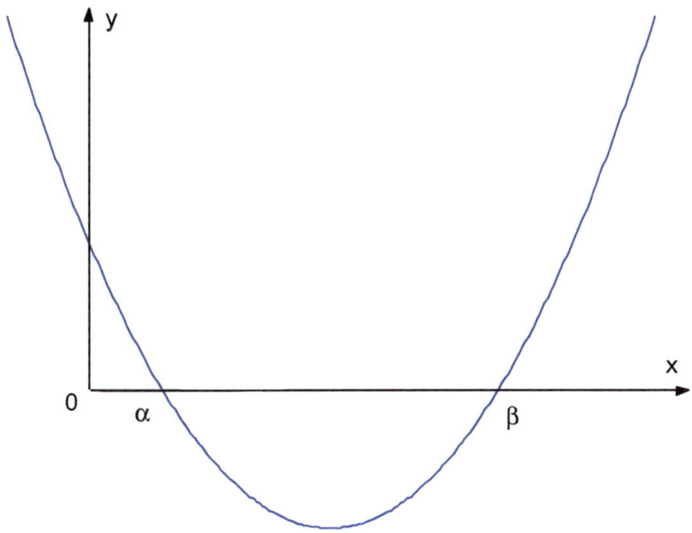

Fig. 5.40 Parabola with trough

written as in this software, the values of α and β represent the roots of the parabola for $y = 0$.

$$Y = a_2(x - \alpha)(x - \beta) \tag{5.69}$$

It will be later in this chapter how these roots were found by Al-Khwarizmi-Algorithm (780–850) with a geometrically rational model. One can summarize the information about the shape of the parabola according to the parameters in Eq. (5.69) in the following points.

(a) In general, if the degree of a polynomial is n, its graphical representation has a total of $(n - 1)$ inflection points. Since the parabola is of the second order, there is one inflection point indicating that it has reached either the largest or the smallest value.
(b) If $a_2 > 0$ and $\alpha < \beta$, the y's corresponding to the interval $\alpha < x < \beta$ are all negative. In this case, as y goes to ∞, x goes to ∞ or y goes to −∞ (Fig. 5.40),

(c) If $a_2 < 0$ and $\alpha < \beta$, then all y values in the range $\alpha < x < \beta$ are plus sign. Here, as y goes to −∞, x goes to ∞ or x goes to −∞ again. The parabola graph in this case is shown in Fig. 5.41.

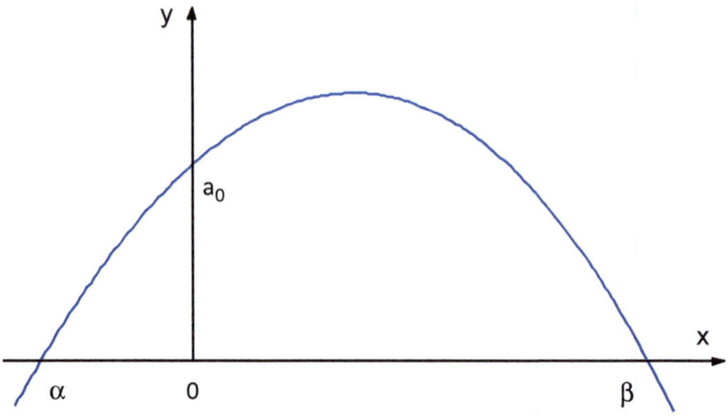

Fig. 5.41 Parabola with peak

5.12 Conclusions

Mathematical models are derived after the geometrical conceptualization of the relevant event, or vice versa, in the case of mathematical equations, it should be able to visualize its background linguistically in terms of philosophical and logical principles of science. First, it is important to build a conceptual model with interrelated input and output parts through mathematical (probabilistic, statistical or analytical) expressions. Geometrical conceptualization and design are of great importance in the solution of mathematical equations. For example, the second-order polynomial root solution was presented by El-Khawarizm (Algorithm) about 1200 years ago, and this is explained extensively in this chapter in addition to other algebraic equations. The arithmetic operation of multiplication is also illustrated by the basic geometric design and operation, the design of the square mesh and main diagonal elements. Even analytical formulations derivation of differential equations is presented linguistically and at the end the corresponding equation is derived. The validation and calibration procedures are shown in seven steps that can be improved by translation and rotation procedures in case of systematic errors. Commonly used mathematical basic equation linguistic interpretations are presented so that the reader can remember the type of equation at first glance.

In the Appendix A the necessary software is given for the application of all VINAM steps.

Appendix A: VINAM Matlab Software

```
function [PDG,MSER,MSEG,MSEP] = VINAM(ME,MO)
% This program is written by Zekâi Şen on 7 June 2019
% ME    : Measurement time series
% MO    : Model output time series
% MI    : Model improvement
% PDG   : Pseudo-rotational shift data
% MESR  : Regression mean-error-square
% MESG  : Global shift mean-error-square
% MESP  : Pseudo-rotational shift mean-error-square
n=length(ME);
%
% Regression equation between the measurements and model output
%
b= (mean(ME.*MO)-mean(ME)*mean(MO))/(mean(ME.*ME)-mean(ME)*mean(ME));
a=mean(MO)-b*mean(ME);
% Scatter measurement versus model output
figure
scatter(ME,MO,'k*')
hold on
xlabel('Measurement data')
ylabel('Model output')
title('SQUARE TEMPLATE GRAPG (STG)')
grid on
box on
hold on
% Adjust square plot framework
mME=min(ME);
MME=max(ME);
mMO=min(MO);
MMO=max(MO);
m=min(mME,mMO);
M=max(MME,MMO);
% Plot the regression line
X=mME:0.1:MME;
Y=a+b*X;
plot(X,Y,'Color','k','LineWidth',2)
mm=mean(ME); % Measurement mean
om=mean(MO); % Model mean
scatter(mm,om,'kd')
axis([m M m M]);
line([m M],[m M],'Color','r','LineWidth',2);
legend('Regression data','Regression line','Regresssion centroid point',...
    'Perfect efficiency line','Location','Northwest')
%
% Mean Square Error (MSE) calculation for regression
%
RP=a+b*ME;    % Regression points corresponding to measurements
MSER=(1/n)*sum(RP-ME)^2;     % MSE for regression
%
% MODEL GLOBAL SHIFTING OPERATION
%
figure
D=mm-om; % SHIFTINH AMOUNT. IF POSITVE (NEGATIVE) DOWNWARD (UPWARD)
if D > 0
    MOS=MO+D; % MOdel output upward shift
    mos=mean(MOS);
    scatter(ME,MOS,'b*')
    hold on
    YY=a+b*X-D;
    plot(X,YY,'Color','b','LineWidth',2)
    YYY=MOS-a-b*ME-D;
    PDG=ME-YYY;     % Point pseudo-rotational shift data
else
    MOS=MO-D; % MOdel output downward shift
    mos=mean(MOS);
    scatter(ME,MOS,'b*')
    hold on
```

..........continued..........

```
    YY=a+b*X+D;
    plot(X,YY,'Color','b','LineWidth',2)
    YYY=MOS-(a+b*ME+D);
    PDG=ME-YYY;    % Point pseudo-rotational shift data
end
hold on
xlabel('Measurements')
ylabel('Global shift model output')
title('SQUARE TEMPLATE GRAPG (STG)')
grid on
box on
scatter(mm,mos,'bd')
axis([m M]);
line([m M],[m M],'Color','r','LineWidth',2);
legend('Global shift data','Global shift line','Global shift centroid',...
    'Perfect efficiency line','Location','Northwest')
%
% Mean Square Error (MSE) calculation for global shift
%
GP=a+b*ME-D;   % Global shift points corresponding to measurements
MSEG=(1/n)*sum(GP-MOS)^2;    % MSE for global shift
% MODEL POINT PSEUDO-ROTATIONAL SHIFTINGS OPERATION
figure
scatter(ME,PDG,'r*')    % Final scatter points
%scatter(ME,MOS,'r*')    % Final scatter pointshold on
xlabel('Measurements')
ylabel('Pseudo-rotational data')
title('SQUARE TEMPLATE GRAPG (STG)')
grid on
box on
hold on
m=min(mME,mMO);
M=max(MME,MMO);
axis([m M M]);
line([m M],[m M],'Color','r','LineWidth',2);
mpdg=mean(PDG); % Pseudo-rotational data mean
scatter(mm,mpdg,'rd')
legend('Pseudo-rotational data','Perfect efficiency line','Location','Northwest')
%
% Mean Square Error (MSE) calculation for point shift
%
MSEP=(1/n)*sum(PDG-ME)^2;    % MSE for point pseudo-rotational shift
%
% ALL CASES ARE IN ONE FIGURE
%
figure
scatter(ME,MO,'k*')
hold on
xlabel('Measurements')
ylabel('Pseudo-rotational data')
title('SQUARE TEMPLATE GRAPG (STG)')
grid on
box on
plot(X,Y,'Color','k','LineWidth',2)
scatter(mm,om,'kd')
scatter(ME,MOS,'b*')
plot(X,YY,'Color','b','LineWidth',2)
scatter(mm,mos,'bd')
scatter(ME,PDG','r*')    % Final scatter points
%scatter(ME,MOS','r*')    % Final scatter pointsaxis([m M]);
line([m M],[m M],'Color','r','LineWidth',2);
legend('Data','Regression line','Regresssion centroid point',...
    'Global shift data','Global shift line','Global shift centroid',...
    'Points shift data','Perfect efficiency line','Location','Northwest')
end
```

References

ASCE (1993) Criteria for evaluation of watershed models (definition of criteria for evaluation of watershed models of the watershed management committee, irrigation and drainage division). J Irrig Drain Eng 119(3):429–442

Bracmort KS, Arabi M, Frankenberger JR, Engel BA, Arnold JG (2006) Modeling long-term water quality impact of structural BMPs. Trans ASABE 49(2):367–374

Coulibaly P, Anctil F, Bobée B (2000) Daily reservoir inflow forecasting using artificial neural networks with stopped training approach. J Hydrol 230(3–4):44–257

Dariane AB, Azimi S (2018) Streamflow forecasting by combining neural networks and fuzzy models using advanced methods of input variable selection. J Hydroinf 20(2):520–532

Dimitrov V, Korotkich V (2002) Fuzzy logic a framework for the new millennium. Springer, Heidelberg, p 249

Freedman D, Purves R, Pisani R (1978) Statistics. W.W. Norton and Co, New York

Gupta HV, Sorooshian S, Yapo PO (2002) Status of automatic calibration for hydrologic models: comparison with multilevel expert calibration. J Hydrol Eng 4(2):135–143

Gupta HV, Kling H, Yilmaz KK, Martinez GF (2009) Decomposition of the mean squared error and NSE performance criteria: implications for improving hydrological modelling. J Hydrol 377(1–2):80–91

Harmel RD, Smith PK, Migliaccio KW (2013) Modifying goodness-of-fit indicators to incorporate both measurement and model uncertainty in model calibration and validation. Trans ASABE 53(1):55–63

Kermani BG, Schiffman SS, Nagle HT (2005) Performance of the Levenberg-Marquardt neural network training method in electronic nose applications. Sensors Actuators B Chem 110(1): 13–22

Kizilöz B, ŞiŞman E (2021) Exceedance probabilities of non-revenue water and performance analysis. Int J Environ Sci Technol:1–12. https://doi.org/10.1007/s13762-020-03018-y

Larabi S, St-Hilaire A, Chebana F, Latraverse M (2018) Using functional data analysis to calibrate and evaluate hydrological model performance. J Hydrol Eng 23(7):1–12

Legates DR, McCabe GJ (1999) Evaluating the use of 'goodness-of-fit'measures in hydrologic and hydroclimatic model validation. Water Resour Res 35(1):233–241

Li Z, Liu W, Zhang X, Zheng F (2009) Impacts of land use change and climate variability on hydrology in an agricultural catchment on the loess plateau of China. J Hydrol 377(1–2):35–42

Mandelbrot BB (1982) The fractal geometry and nature. WH Freeman and Company, San Francisco

Moriasi DN, Wilson BN, Douglas-Mankin KR, Arnold JG, Gowda PH (2012) Hydrologic and water quality models: use, calibration, and validation. Trans ASABE 55(4):1241–1247

Moriasi DN, Arnold JG, Van Liew MW, Bingner RL, Harmel RD, Veith TL (2007) Model evaluation guidelines for systematic quantification of accuracy in watershed simulations. Trans ASABE 50(3):885–900

Nash JE, Sutcliffe JV (1970) River flow forecasting through conceptual models part I - a discussion of principles. J Hydrol 10(3):282–290

Pearson K (1895) Notes on regression and inheritance in the case of two parents. Proc R Soc Lond 58(347–352):240–242

Pfannerstill M, Guse B, Fohrer N (2014) Smart low flow signature metrics for an improved overall performance evaluation of hydrological models. J Hydrol 510:447–458

Rahman M, Ningsheng C, Islam MM, Dewan A, Iqbal J, Washakh RMA, Shufeng T (2019) Flood susceptibility assessment in Bangladesh using machine learning and multi-criteria decision analysis. Earth Syst Environ 3:123

Ritter A, Muñoz-Carpena R (2013) Performance evaluation of hydrological models: statistical significance for reducing subjectivity in goodness-of-fit assessments. J Hydrol 480:33–45

Rujner H, Leonhardt G, Marsalek J, Viklander M (2018) High-resolution modelling of the grass swale response to runoff inflows with Mike SHE. J Hydrol 562:411–422

Saleh A, Arnold JG, Gassman PW, Hauck LM, Rosenthal WD, Williams JR, McFarland AMS (2000) Application of SWAT for the upper north Bosque River watershed. Trans Am Soc Agric Eng 43(5):1077–1087

Santhi C, Arnold JG, Williams JR, Dugas WA, Srinivasan R, Hauck LM (2001) Validation of the SWAT model on a large river basin with point and nonpoint sources. J Am Water Resour Assoc 37(5):1169–1188

Swathi V, Srinivasa Raju K, Varma MR, Sai Veena S (2019) Automatic calibration of SWMM using NSGA-III and the effects of delineation scale on an urban catchment. J Hydroinf 21(5): 781–797

Şen Z (2002) Bilimsel Düşünce ve Matematik Modelleme İlkeleri (scientific thinking and mathematical modeling principles). Su Vafı Yayınları 184. (in Turkish)

Şen Z (2011) Bilim ve Bilimsel Araştırma İlkeleri (science and scientific research principles). Su Vakfı Yayınları, p 201. (in Turkish)

Şen Z, Şişman E, Kızılöz B (2021) A new innovative method for model efficiency performance. Water Sci Technol Water Supply 22(3). https://doi.org/10.2166/ws.2021.245

ŞiŞman E, Kizilöz B (2020) Artificial neural network system analysis and Kriging methodology for estimation of non-revenue water ratio. Water Sci Technol Water Supply 20(5):1871–1883

Tian Y, Nearing GS, Peters-Lidard CD, Harrison KW, Tang L (2015) Performance metrics, error modeling, and uncertainty quantification. Mon Weather Rev 144(2):607–613

Van Der Keur P, Hansen S, Schelde K, Thomsen A (2001) Modification of DAISY SVATmodel for potential use of remotely sensed data. Agric For Meteorol 106(3):215–231

Van Liew MW, Veith TL, Bosch DD, Arnold JG (2007) Suitability of SWAT for the conservation effects assessment project: comparison on USDA agricultural research service watersheds. J Hydrol Eng 12(2):173–189

Willmott CJ (1981) On the validation of models. Phys Geogr 2(2):184–194

Yıldırım C (1988) Matematiksel Düşünce (mathematical thinking). Remzi Kitabevi, İstanbul. (in Turkish)

Zhang R, Moreira M, Corte-Real J (2016) Multi-objective calibration of the physically based, spatially distributed SHETRAN hydrological model. J Hydroinf 18(3):428–445

Chapter 6
Genetic Algorithm

6.1 General

In the last four decades, interesting methods have been developed for modeling, simulation, optimization, and prediction of future behaviors in various subjects, inspired by the way natural events operate and behave. Among these, the genetic behavior patterns of living things played a very important role in the emergence of genetic algorithm (GA) methodology. In general, the use of GA, which is suitable for solving optimization (maximizing and minimizing) problems, can reach much better solutions in a short time in numerical and random orders compared to other methods. The terms "numerical pattern" and "random order" here refer to the most important features of recently developed modern methods including artificial intelligence (AI).

It is not possible to find or approach the solution of every problem with analytical or mathematical methods. For such solutions, it is necessary to determine the mathematical expression of the problem at hand, but not every problem can be expressible through mathematical functions. In such cases, the direction was taken towards the numerical algorithm's development. Among the most important pioneers of these are the differential and integral equation numerical solution computations by converting them into finite difference, elements, and boundary discretization formulations. With the increase in computer possibilities after 1950, numerical analysis has since become the current solution algorithms. The solution space geometry of these numerical methods must have a strict pattern (rectangular or square grid, finite elements in the form of triangles or polygons). Solutions can be reached after such an organization and systemization. However, strict pattern means introduction of many redundant grid points or subfields into the calculation procedures without significant contribution to final solution. In order to avoid this, the final solution is approached in a random order. Such algorithms serve to reach the sought-after solutions more dominantly, effectively, and quickly. In this chapter, the most modern of these, GAs, will be explained in detail. The very basic principles of GA have already been explained in a Turkish book written by Şen (2004).

Z. Şen, *Shallow and Deep Learning Principles*,
https://doi.org/10.1007/978-3-031-29555-3_6

6.2 Decimal Number System

In the nineteenth century, a binary number system based only on the numbers 0 and 1 was developed in a branch of mathematics called Boolean (1815–1864) algebra. Until then, in the historical development of mathematics, mankind understood that the closest number system to human nature is the decimal system. The existence of 10 fingers on human hands and feet has also led to the development of such a number system. It is very easy to generate numbers with unknown endings based on 10 digits from 1 to 9 including zero. Despite all the convenience of this number system, it is necessary to learn especially the multiplication table by systematic memorizing while performing arithmetic operations. If one looks at the multiplication of each number in the multiplication table with the other as a rule, approximately $9 \times 8/2 = 36$ rules should be memorized. One of these rules is, for example, $8 \times 7 = 56$. A person goes to memorize this multiplication table at a younger age and over time, by repeating similar operations he assimilates it and reaches the maturity of applying the rules without pondering. In a way, it becomes automatic in remembering these basic rules after applications repeatedly. After multiplication stage, with the division operation development, two- or more-digit numbers can be operated arithmetically on a plain paper with a pencil under the light of these rules. Hence, all kinds of numerical operations need four basic arithmetic rules (addition, subtraction, multiplication, and division).

It is not possible to pass without mentioning the founder and father of algebra, Al-Khwarizmi (780–840), one of the great Islamic scholars and his name in Latin pronunciation is Algorithm. This name is also given to the scientific procedure called algorithm, which is the basis of the main theoretical and applied activities in computers software programming today (Ifrah 1994). Thus, Al-Khwarizmi left a considerable mark and an inexhaustible effect on the whole world. One of the most important operations that this scientist brought to humanity is to transform systematically the product of the 36 basic two-number multiplications into an algorithmic framework considering the digits in today's sense. While doing this, he explained his thoughts using geometry as in Fig. 6.1. For example, consider the product of the numbers 9 and 75. Al-Khwarizmi explained the answer as follows. Here, the digits in each number are considered one by one and multiplied one by one. This means that two numbers are matched in a certain way. Such basic and simple mappings are at the core of GA operations today in another way. For general products, one can first show the multiplication table of the two basic numbers from 0 to 9 as a table in Fig. 6.1. Here, the first row contains the base decimal numbers from left to right and the first column from bottom to top.

The product of any two integers can be represented easily using the system in this table. For example, if one wants to see what 7×8 is, he can find the answer to the multiplication at the intersection of the row and column by considering 7 in the first row and 8 in the first column from the multiplication table. The answer to the question is shown in Fig. 6.2 (see also Chap. 5).

	0	1	2	3	4	5	6	7	8	9
9	0/0	0/9	1/8	2/7	3/6	4/5	5/4	6/3	7/2	8/1
8	0/0	0/8	1/6	2/4	3/2	4/0	4/8	5/6	6/4	7/2
7	0/0	0/7	1/4	2/1	2/8	3/5	4/2	4/9	5/6	6/3
6	0/0	0/6	1/2	1/8	2/4	3/0	3/6	4/2	4/8	5/4
5	0/0	0/5	1/0	1/5	2/0	2/5	3/0	3/5	4/0	4/5
4	0/0	0/4	0/8	1/2	1/6	2/0	2/4	2/8	3/2	3/6
3	0/0	0/3	0/6	0/9	1/2	1/5	1/8	2/1	2/4	2/7
2	0/0	0/2	0/4	0/6	0/8	1/0	1/2	1/4	1/6	1/8
1	0/0	0/1	0/2	0/3	0/4	0/5	0/6	0/7	0/8	0/9
0	0/0	0/0	0/0	0/0	0/0	0/0	0/0	0/0	0/0	0/0

Fig. 6.1 Multiplication table

The answer is obtained by considering the square corresponding to the numbers 7 and 8 in the multiplication table in Fig. 6.1. The two-decimal digit as answer 56 is within the cross small square diagonal and the two digits are below and above the main diagonal. The same system is valid for the multiplication of other integers. Accordingly, the reader can immediately find the answer to any two-digit multiplication from the multiplication table in Fig. 6.1.

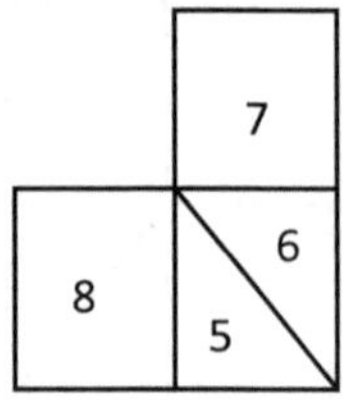

Fig. 6.2 Al-Khwarizmi pattern of multiplication of 8 × 7

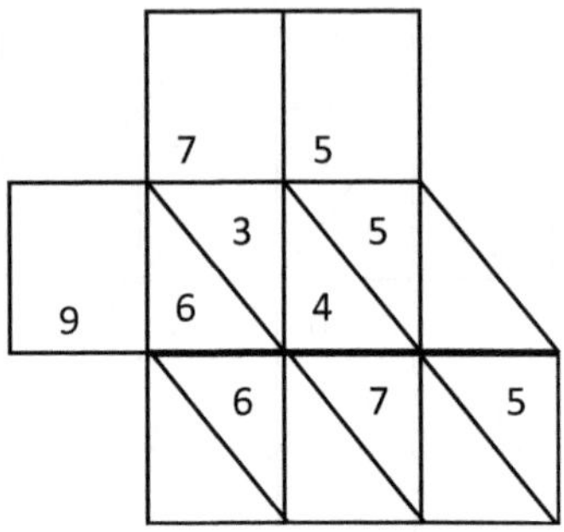

Fig. 6.3 Al-Khwarizmi pattern of multiplication 75 × 9

If one of the numbers to be multiplied has 2 digits and the other has 1 digit, for example, let's explain how the table in Fig. 6.3 can be used to multiply the number 75 and 9. For this, the diagonal multiplication results corresponding to the numbers 7 and 5 are written as rows (columns) side by side (one below the other). If we consider that the numbers 75 are written side by side in the first line, the multiplication process is given as follows, including the number to be multiplied (9) in the first column.

To find the result, if the numbers between the diagonals at the top of the thick line are added starting from the right side, the answer is found in the line below the line in Fig. 6.3. Now, let's try to find the product of 9 × 75 according to Al-Khwarizmi representation with the change of order. Here, if the number 9 forms the column of Al-Khwarizmi show, two lines are needed for the numbers 7 and 5. For this, the question arises whether the number 75 should be written from bottom to top or from top to bottom. Both cases are shown in Fig. 6.4. However, it is understood that the correct answer will be obtained in the order in Fig. 6.4b, that is, by writing the number 75 from bottom to top.

After memorizing the bottom-up rule, it is now possible to multiply numbers with all kinds of digits. For example, let us show how to multiply the number 97 × 78 according to Al-Khwarizmi notation. Again, by making use of the multiplication table in Fig. 6.1, the Al-Khwarizmi software in Fig. 6.5 emerges.

This multiplication system, first shown by Al-Khwarizmi, also led to the emergence of today's multiplication operation. To explain this, let's consider the Al-Khwarizmi representation in Fig. 6.5 in parts. For this, 8 × 97 and then 7 × 97 products are separately shown in Fig. 6.6.

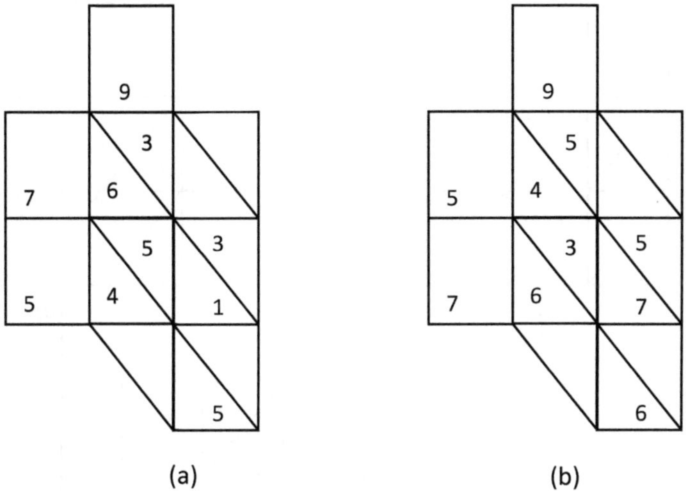

Fig. 6.4 Al-Khwarizmi pattern of multiplication 75 × 9

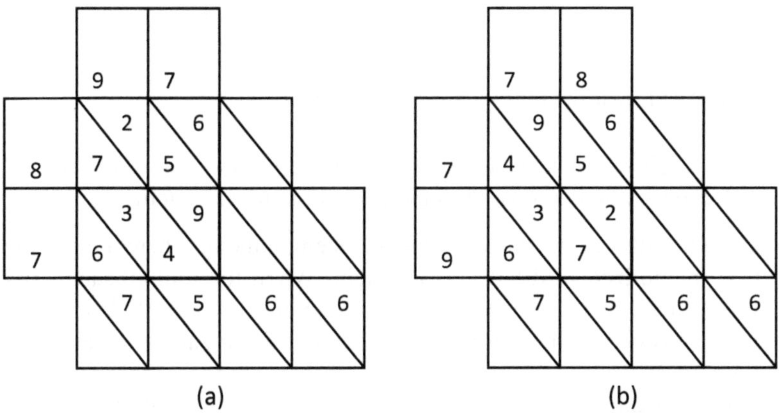

Fig. 6.5 Al-Khwarizmi pattern of multiplication 97 × 78

To obtain the valid result, the multiplication of 78 by 8 is written in the ones digit and 97 on a line, then the result of multiplying the same number by the digit in the tens digit (7) is added by shifting one digit to the left below it.

6.3 Binary Number System

People accustomed to the decimal number system for centuries saw the binary number system consisting of only 0 and 1 digits of Boolean arithmetic as a fantasy and unnecessary work. For this reason, the binary system remained buried in the

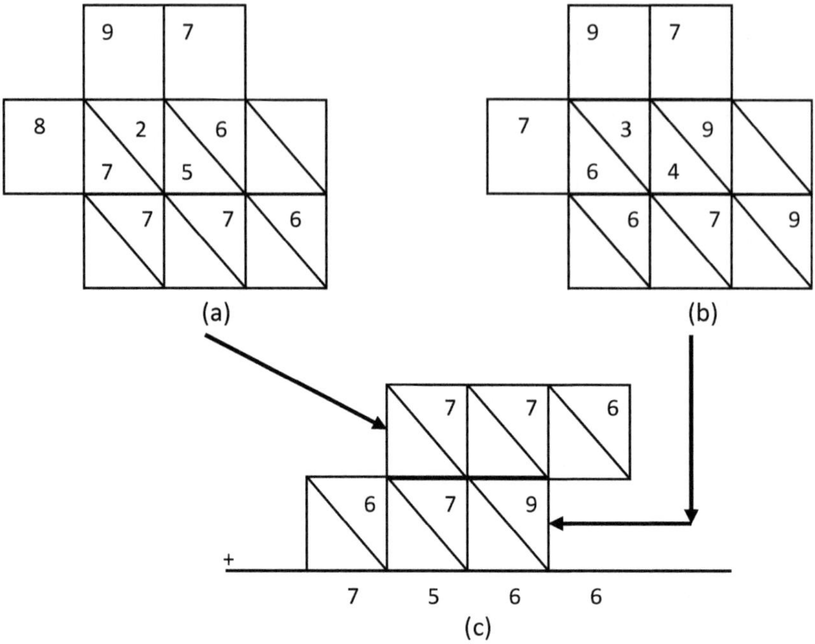

Fig. 6.6 Partial Al-Khwarizmi pattern of multiplication 97 × 78

dark for years. Then, it has been understood that the binary number system has the following benefits when compared to the decimal number system.

1. There are no rules in the binary number system, although there are 36 rules that must be memorized in the decimal system. When the multiplication rules are mentioned here, there are 2 rules only. These are $0 \times 1 = 0$ $(1 \pm 0 = 1)$ and $1 \times 1 = 1$. However, since these do not require memorization, it can be said that there are no rules.
2. There is no need for paper and pen to process, only to write the result of the process.
3. There are only addition and multiplication operations that can be executed very easily in binary number system as opposed to 4 comprehensive arithmetic operations in decimal number system.

The binary number system has some drawbacks compared to the decimal. These are not very undesirable situations from a practical point of view. For example, in the expression of a quantity in the binary number system, the number of digits consisting of 0s and 1s is high. The decimal number 91 consists of 1011011 digits (7 digits) in the binary number system. Another problem stems from human habit. Since people are accustomed to the decimal number system, they find it a hassle to change their habit to the binary system and cannot or will not want to make this transition.

However, there is a binary order in the rules that one may call crisp (bivalent) logic, which states fundamentally that everything is expressible with its opposite.

For example, plus–minus (not almost plus), beautiful–ugly (not almost beautiful), yes–no (not almost yes), too little (not too much), and long–short (not tall) are all within the bivalent logic domain. These correspond to the principle that there is no middle value, but extremes only. There is always a dilemma, such as, or simply “”’something-nothing.” If bivalent logic is adapted then any person simply must adjust all thoughts according to these two options. There are already two crisp logic operations. These are known as AND (ANDing) or (ORing) (Chap. 3).

To understand the relationship between decimal and binary number systems, it is necessary to grasp what each number digit means. For example, one can express the number 432 verbally as “4 hundreds,” “3 tens,” and “2 ones.” Since the words “hundred,” “ten,” and “oneness” here can all be written as $4 \times 10^2 + 3 \times 10^1 + 2 \times 10^0$ based on decimal system, where x symbol is identical to the multiplication in arithmetic. If one considers this system, the number 1098 will be written accordingly as follows.

$$1 \times 10^3 + 0 \times 10^2 + 9 \times 10^1 + 8 \times 10^0$$

On the other hand, a number given in detail, for example,

$$5 \times 10^4 + 9 \times 10^3 + 0 \times 10^2 + 5 \times 10^1 + 0 \times 10^0$$

this has its decimal software as 59050 for short. Another point that must not be forgotten here is that the powers of 10 are integers, and they must continuously decrease from the largest number to the smallest integer 0.

Now, taking advantage of this short writing of decimal numbers, the system of decreasing powers from a suitable largest integer to the smallest integer 0 should remain the same in number powers according to the binary system, with the number 2 being the base. In the binary number system, power numbers with base 2 will be preceded by the number 0 or 1. According to this

$$1 \times 2^5 + 1 \times 2^4 + 0 \times 2^3 + 1 \times 2^2 + 0 \times 2^1 + 0 \times 2^0$$

Its detailed expansion is briefly written as “110100” according to the binary system. In fact, this number is calculated by performing the necessary mathematical operations in open software.

$$1 \times 32 + 1 \times 16 + 0 \times 8 + 1 \times 4 + 0 \times 2 + 0 \times 1 = 52$$

this is equivalent to the decimal number. So, the binary equivalent of the decimal number 52 is “110100.”

Conversions between decimal and binary number systems form the basis of GA methods. It is recommended that the reader switch from one to the other with the numbers he chooses in head, so that he can practically grasp the transition from

Table 6.1 Comparison of decimal and binary number systems

Decimal	Binary
0	0000
1	0001
2	0010
3	0011
4	0100
5	0101
6	0110
7	0111
8	1000
9	1001

decimal to binary or from binary to decimal. He can even develop simple algorithm software by programming in a software language he knows. The equivalents of 10 digits in the decimal number system in the binary system are given in Table 6.1.

Numbers in the binary system can be viewed as finite-length sequences of 0s and 1s. Using this chart, performing arithmetic operations of different numbers in the binary system easily leads one to the same result in the decimal system. The rules of arithmetic operation in every number system are the same. For example, application of the arithmetic multiplication rules to 5, i.e. "101" and 6, and "110" in the binary number system yields the following operational structure.

```
   101
x  110
------
   000
  101
 101
+------
 11110
```

The number "11110" in the binary system corresponds to the number 30 in the decimal system. This is equivalent to the decimal number system $5 \times 6 = 30$ products. The result is the same in both systems, but the symbols are different. As can be seen from the cross multiplication above, there is only the multiplication of the numbers 1 and 0. Although, the binary system is very easy even for humans, but their final solutions lead to a long sequence of 1s and 0s. Without the need to memorize a detailed multiplication table, that is, the multiplication rules, as in the decimal system, it is possible to get to the result quickly by only performing operations in the form of the product of 0 and 1. If "1000" (8) and "1001" (9) are added as an example for addition.

```
  1000
+ 1011
------
 10011
```

Here again, starting from the right, the consecutive digits are collected. If the sum of the digits is in the form of 1 + 1, since the total is greater than 1, after 0 is written, the number 1 is written in front of it by saying "have 1 at hand." Thus, 1 + 1 = 10, which corresponds to the decimal number 2, as can be seen from Table 6.1. By using this rule of "have 1 at hand", the sum of the binary system numbers "1000" and "1001" are obtained as "10001." If attention is paid, the number obtained is one digit more than the collected numbers. The decimal alternative of the binary number "10001" is 17 (8 + 9 = 17). Thus, one concludes that the operation is correct from our knowledge in the decimal number system.

From the explanations above, operations in the binary system are carried out according to the rules of the decimal number system. Computers as fully obedient porters are command-taking robots that are designed and programmed to perform operations according to the binary number system. There must be a unit that will provide the exchange between humans who are accustomed to the decimal system, which can be considered the external world, and computers that work according to the binary system, which one can call the inner world. It is also called a "converter," meaning a unit that translates from decimal to binary or vice versa. So, in the first stage, computers convert decimal numbers given by humans into binary system. After that, it performs the necessary arithmetic and mathematical operations according to the binary system and outputs the results in a format that can be understood by the decimal world.

Such a converter system is also necessary for genetic algorithms (GAs), and therefore, computers are more compatible and efficient for GA analysis. Like decimal and binary number systems, it is also possible to generate different systems based on each number. For instance, Sumerians about 3000 years ago have developed 6-based numbering systems, but it is not used in daily mathematical or modeling procedures. However, today 6-base system and its many folds appear in different forms such as, it is known that 60 s is equal to 1 min and 60 min is equal to 1 h. On the other hand, it is widely used that 360° also corresponds to the circumference of a circle. Furthermore, the numerical summation of three angles of a triangle is assumed to be equal to 180, which is 30-fold of 6.

6.4 Random Numbers

The numbers falling into the solution space of a problem can be chosen either systematically or randomly. By now most of us are accustomed to always systematic organization. There is a belief that if formal numbers are not used in scientific studies, the results will not be reliable. Natural events are organized in medium

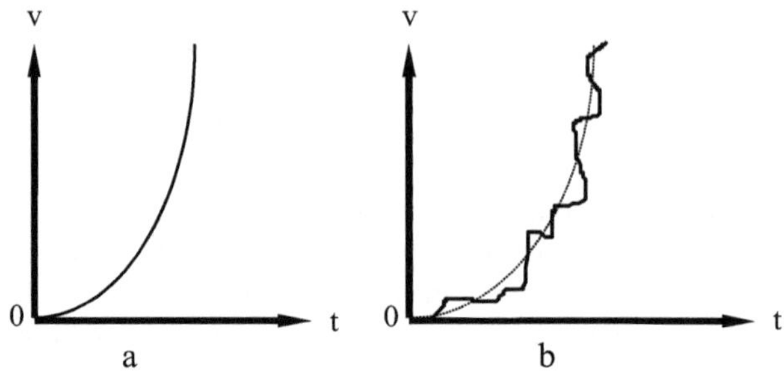

Fig. 6.7 Variable (**a**) non-random, (**b**) random. (Şen 2002)

scales, perhaps in terms of facilitating human perception, but in very small and especially subatomic scales, everything proceeds randomly according to human perception. Here we have to say that there are degrees of randomness as well. For example, when and where an earthquake will occur with what intensity; how many deaths it will cause; how many buildings will be destroyed, uninhabitable, etc. It is not possible to know in advance what will happen. The numerical values of events appear randomly even at large scales in nature, but this has no role in digitizing or revealing that event as a fact. Thus, one can say that natural, social, economic, psychological, medical, engineering, etc., events have randomness.

All variables in mathematics or classical physics involve certainty, and the relationship between variables can be calculated precisely by means of a suitable function. For example, when velocity, v, and time, t, are given by the function, $v = gt^2$ (where g is a gravitational constant), one finds exactly what v is for any desired value of t. The geometric view of such a relation is given in Fig. 6.7a.

If the geometric view between two variables includes random oscillations as in Fig. 6.7b, precise calculations cannot be made since there is no exact relationship, and even if the value at one time is known, the next one cannot be predicted with certainty. Variables that have this randomness characteristic with quite complex geometry are random. Graphs showing the change of these variables according to the time or space axis as in Fig. 6.7b are given different names as signal, time series, or sample function (Şen 2002).

The outcome of uncertain events has many possibilities, probabilities, and random numerical values. For example, among such events are wind load values to which an engineering structure will be exposed during its lifetime; the number of accidents between certain kilometers along an expressway in traffic; the number of deputies a political party will issue in the general elections; the number of children born in a day; the height of sea waves; the position and momentum of subatomic particles; the amount of water that will come to the dams monthly; the most severe risky event in the next 10-year; intensity of the earthquake on the Richter scale, etc. Many events have random behavior and, as a result, randomly measurable variables. The simple case of this is the probability that the value of probability is random number, which is called state randomness. Days with or without precipitation, with or without fog, with or without sun; whether the borehole is dry or wet; wins, losses,

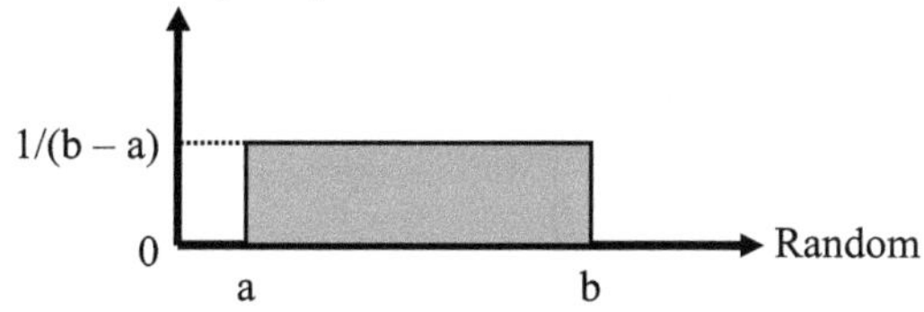

Fig. 6.8 Uniformly distributed random numbers between a and b

or score in football matches; whether the trade is profitable or unprofitable, etc. Each of these states is an example of randomness states.

Finally, it is the case that although the states (likelihoods) are definite, only their quantities are random. For example, if a coin is tossed, it is a tail or a head in the probability space, and if it is tossed 95 times, the sum of heads and tails is known, but they are random numbers since it is not known how many tosses will end up as heads. For uncertain events that vary over time, for example, the days considered as likelihood in a series of fog measurements are date specific; however, it cannot be known exactly how much amount will appear on which day. This randomness of the third kind is called intensity or quantity randomness. An example of quantity randomness is the uncertainty of how much rain will fall on a rainy day. There are always holistic variables such as temperature and humidity every day. Their daily intensity, that is, how much their amount will be, has a random variability. In this respect, variables such as temperature and humidity are also called random variables.

In the GA analysis method, the state and size of the randomness are used. It is necessary to generate random numbers to use during the operation of this algorithm. Since many software today freely provide programs for generating random numbers, it will not be discussed here how random numbers are generated. The relative frequency distributions of random numbers need to be considered in some cases. In GA analysis, two types are used, one with a uniform (uniform) and the other with a normal (Gaussian) probability distribution function (PDF). The commonly used type is the uniform one, which represents that all real numbers between two numbers a and b are chosen with equal probability, with no advantage (selection advantage) of one over the other. The relative frequency diagrams (histograms) of such random numbers consist of a rectangle containing the random numbers between the given lower and upper limits (a and b) but having a height equal to 1/(b – a), as shown in Fig. 6.8. The area under this rectangle is equal to 1 on definition.

Another random number distribution occasionally used in GA operations is numbers that theoretically take values between minus and plus infinity and oscillate around a mean, μ, with a certain standard deviation, σ. These are called numbers with a normal (Gaussian) PDF. If the mean value is equal to zero and the standard deviation is equal to one, the relative frequency (probability) distribution of these numbers is given in Fig. 6.9 as a standard bell shape curve. As it can be seen, the probabilities of the normal numbers are not equal. For those close to the mean, there are big probabilities, but as one goes from there towards the tails, there are small probabilities. Approaching the tails means that the random number takes extreme (extreme) values.

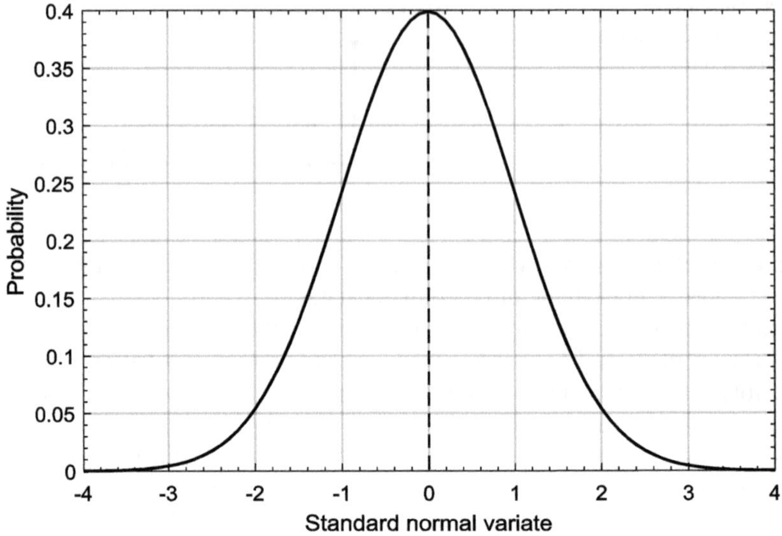

Fig. 6.9 Normally distributed random numbers (zero mean and unit standard deviation)

This is called the standard random number relative frequency. However, non-standard random numbers, S_r, whose mean, μ, is different from zero and standard deviation, σ, is also different from one, can be generated from standard normal variable, s_r, according to the following simple equation.

$$S_r = \mu + \sigma s_r \tag{6.1}$$

6.4.1 Random Selections

While performing operations with genetic numbers, which will be defined below, in GA analysis sections, random selections must be made from a random group of numbers or among the digits of numbers. Here, the random numbers described above are used in the random selections to be made among a group of numbers. Random selections are made in one of three ways as card, roulette, or stochastic methods.

6.4.1.1 Cards Drawing

This method has been a very common approach in times or places where computers were not available. One can compare this to playing cards. Each of the collection of numbers is written on cards of the same size, shape, and color. Then, if this collection of cards is randomly shuffled, what one can call a pool of random numbers emerges. Here, the number on a piece of cards to be drawn randomly from this pool represents the random number. If one does not want similar numbers to be found in the draws, the drawn number is not returned to the pool so that the drawn number will not be drawn again. In this way, a series of random numbers is obtained by continuing the second, third, and desired number of draws. There are no identical numbers in such a sequence. However, here, the number of draws cannot be more than the number of cards put into the pool.

A second way is to continue drawing from as many numbers as the initial number of cards each time, after the drawn card is put back into the pool. This put-back shot will have identical numbers. Drawing can be continued here if desired. In this way, one can list the drawbacks in the process of determining random numbers by drawing cards from the pool as follows.

1. It is not possible to obtain numbers other than the numbers written on the cards in the pool. For example, if the amounts of accidents that have occurred in the past are written on a card and put into the pool, and if an interpretation is made as if the future accident is predicted by random selections, it is assumed that the future accident will not differ from the past values.
2. If the cards are not put back into the pool, the probability (chance) of each card will be different from each other.
3. When the cards are put in place, the probability of drawing each card will be the same.
4. It is not possible to adjust the degree of randomness in both selection types. However, as has been said before, there is some degree of dependency even in randomness.
5. No matter how big or small the numbers written on the cards are, the probability of each being chosen is the same, so one does not have an advantage over the other. However, in solving real problems, numbers will be preferred according to their proximity to the solution point. In every optimization analysis, like GA, large numbers should be preferred for maximizing and small numbers for minimizing.

In order to eliminate many of the drawbacks listed here, the roulette wheel selection method is used as explained below.

6.4.1.2 Roulette Wheel

Here, the selection principle is based on spinning a wheel and waiting for it to stop randomly in a slice. Here, the slices of the roulette wheel have replaced the cards in the previous selection procedure. Each slice is adjusted according to the importance of a number in the ensemble of numbers. Wheel circumference length, L, and desired numbers N_1, N_2, . . . , N_n, should satisfy the following equation.

Fig. 6.10 Roulette wheel

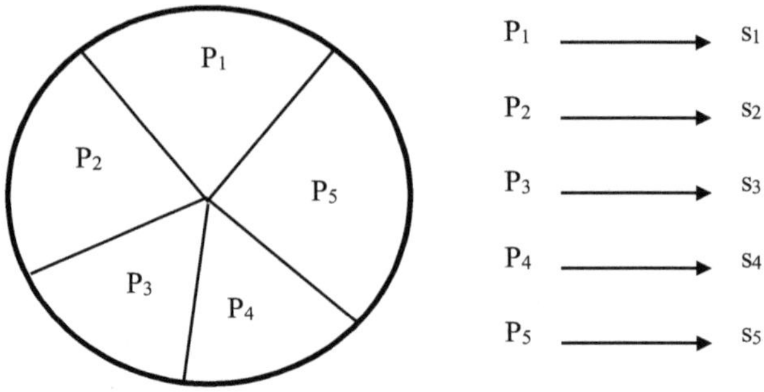

$$N_1 + N_2 + \cdots + N_n = L \tag{6.2}$$

After the division of both sides by L one can obtain the following equation.

$$P_1 + P_2 + \cdots + P_n = 1 \tag{6.3}$$

Hence, the probability ($P_1, P_2, \ldots, P_n$) of each number being selected appears. In this case, the standard length of the roulette wheel is equal to 1. Roulette wheel perimeter is divided into n slices according to probability values. For example, if $n = 5$, there are 5 different slices (s_1, s_2, s_3, s_4, s_5) as representatively shown in Fig. 6.10.

The selection style with the roulette wheel is also called the stochastic selection method. Here, each of the probability values can be thought of as getting a slice from the pie. After that, a random number between 0 and 1 is generated and the number in which this random number falls is selected.

For a better understanding of the subject, 10 numbers and the probability of being selected from them are shown in Table 6.2. These probabilities show the size of each number's share of the pie. The probability of being selected is obtained by dividing the vigor degrees (numbers) by the total vigor.

For example, if a selection is desired for a collection of numbers with 6 elements, first 6 independent random numbers are generated from a uniform distribution between 0 and 1. Suppose that these random numbers are $r_1 = 0.81$; $r_2 = 0.32$; $r_3 = 0.96$; $r_4 = 0.01$; $r_5 = 0.65$ and $r_6 = 0.42$. Using these random numbers and the last row of data in Table 6.2, which decision points will be selected is decided from the standard length of the expansion roulette wheel in Fig. 6.11.

Accordingly, the numbers 7, 2, 9, 1, 5, and 3 were chosen as the members of the population. The numbers corresponding to these sequences are again 0.6; 0.7; 3.0; 1.2; 2.1; from 1.2 and 1.9 (Table 6.2).

6.4.1.3 Stochastic Choice

This selection method has zero bias and minimal spread. Here, as in Fig. 6.11, each decision point is represented one after the other by a line segment (see Fig. 6.12). Points with equal intervals equal to the number to be selections are selected after the

Table 6.2 Numbers and probability of being selected

Sequence number	1	2	3	4	5	6	7	8	9	10
Number	3.0	2.1	1.9	1.6	1.2	1.5	0.6	0.3	0.7	0.4
Selection probability	0.225	0.158	0.143	0.120	0.090	0.113	0.045	0.023	0.053	0.030
Cumulative probability	0.225	0.383	0.526	0.646	0.736	0.849	0.894	0.917	0.9700	1.000

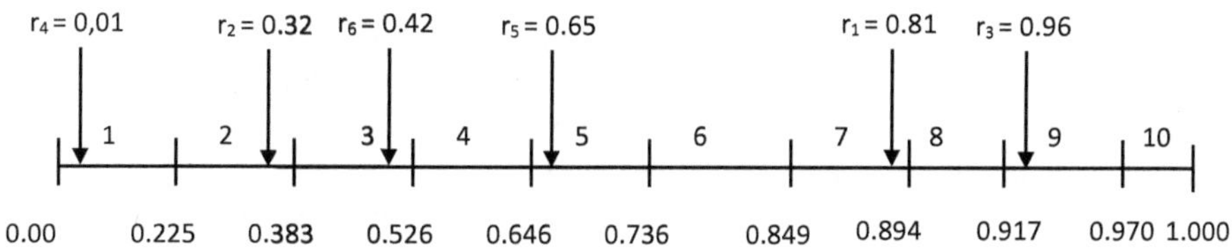

Fig. 6.11 Roulette wheel selection

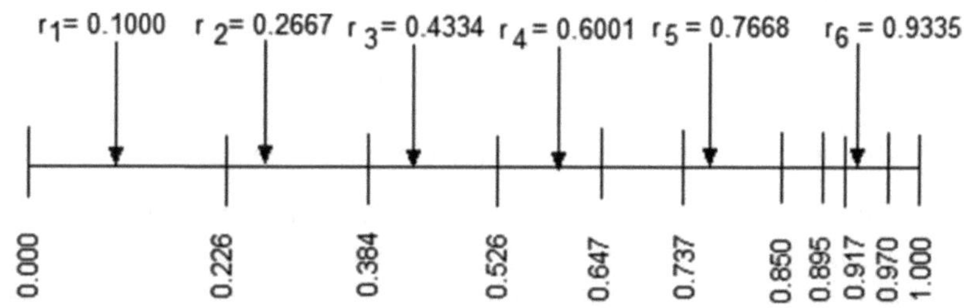

Fig. 6.12 Stochastic population sampling

first random point. The number corresponding to the interval in which each equally spaced point falls is taken as the member of the population. If n points are to be selected, the position of the first point is a random number between 0 and 1/n, and after being randomly selected, the intermediate distances of the equally spaced points are placed with 1/n. Similar to the above, the structure in Fig. 6.12 appears when the equal distance between the points is selected as 0.1 from an independent uniform PDF with the equal distance between the points in the range of $1/6 = 0.167$ and (0, 0.167) for the selection of 6 numbers, similar to the above.

The numbers corresponding to the other random number positions are found by adding 0.1667 each time after the random number was initially chosen as 0.1 are 1, 2, 3, 4, 6, and 9. The corresponding numbers from Table 6.2 are 3.0, 2.1, 1.9, 1.6, 1.5, and 0.7.

6.5 Genetic Numbers

After the GA is decided to solve the problem with one of the decimal or binary number systems described above prior to any application, the solution needs some self-processing by making use of random number generation. For this, a basic structure called the genetic number system must be thoroughly understood at the beginning of the work. For the genetic number system, it is useful first to know the best solution structure of the problem. Each optimization problem has the following components.

1. Variables represent the problem and are called decision variables,
2. Constraints specifying the limits of change of these variables specify the valid solution (decision) space limits of the problem,
3. Constants are the invariants of the problem,
4. A goal function consisting of a mixture of decision variables and constants,
5. An optimum solution algorithm that will optimize (minimize or maximize) this function.

In this chapter, since the optimization method will be GA, the previous steps should be prepared in a structure suitable for it. Although the decision variables are continuous characters in the mathematical sense (such as x, y, z), since GA methods will solve the problem numerically, they must be represented by numbers. There are four number types with special names in the GA number system.

1. In the GA number system, each digit of the decision variable numbers will be called a "digit" in this chapter. So, in GA numbers, as in other number systems, the smallest element is digit. Each digit has either 1 or 0 in the binary number system, and only one of the 10 digits between 0 and 9 in the decimal number system. Thus, each decision variable is encoded according to a number system. In general, the coding format for GA analysis is based on the binary number system; because this is the same as the working system of the computer.
2. The structure that emerges when the digits come together in a series is called "gene" in GA terminology. Accordingly, the numerical value of each decision variable constitutes a gene. There are as many genes as there are decision variables in a problem. The length of the sequence in the gene, that is, the number of digits, is called the length of that gene.
3. The sequence of genes that emerges by arranging them one after the other is called a "chromosome." From this, it is understood that each of the decision variables of the problem in a chromosome are collectively in a series.
4. All the numbers formed by the combination of two or more chromosomes are called GA population.

In a genetic number construction, the order of the digits in each digit of the gene is important. Each digit has its special significance. The GA digitization (mutation) operation, which will be explained in Sect. 6.14, works based on gene digits. The order of the genes in a chromosome structure is not important for GA analysis. It is only necessary to know which decision variable represents each gene in the chromosome. The order of the chromosomes in the randomly selected initial population is also irrelevant. It will also be explained in Sect. 6.3 that while searching for a solution to a problem with the GA operation, all genetic numbers will undergo an evolution during the processes, that is, their values will be renewed step by step and eventually optimization will come to a stop by approaching the solution. From a philosophical point of view, digits, genes, and chromosomes need to be processed for the better by GA principles.

For a good understanding of GA numbers, let us say one wants to buy by a certain amount of money, P, food, F, and clothing, C. Here, the unknown variables are F and C. Each of these is limited to 0 and P. Mathematically, spending all the money on these two species can be expressed as,

$$P = F + C$$

where the restrictions are $0 < F < P$ and $0 < Y < P$. If the unit prices of the food and cloth are P_F and P_C, respectively, then the total spending, S, is,

$$S = P_F F + P_C C$$

It is called the target function, and its optimization is desirable. This is a linear target function and there are classical linear optimization methods that can be used without going to the GA method for its analysis. For GA, curvilinear (non-linear) target functions would be more appropriate such as the following.

$$S = P_F F^{0.5} + P_C C^2$$

Let us assume that initial amount of money is 1050 Euro. Now, in order to solve the problem according to GA, the genetic number system given above with 4 steps should be considered. One wants to solve the problem according to the decimal number system that we are used to as human beings. Since the solution will be numerical, one must first determine the boundaries of the solution space of the decision variables F and C. In the example at hand, this space is limited to $0 < F < 1050$ and $0 < C < 1050$. The solution must be a pair of numbers (F and C) in this space. For example, if we say $F = 528$ Euro, and $C = 3$ Euro, with the head in this space, they are a case for a solution since they are in the valid space.

Here, the word "throws away" means choosing a random number to stay in this space after defining the boundaries of the solution space. Thus, at the very beginning, randomness is encountered for the first time in revealing the genetic numbers that will constitute a GA infrastructure. As will be explained later, this is an important point in GA analysis. In the light of what has been said about genetic numbers above by considering $F_1 = 528$ Euro, and $C_1 = 3$ Euro, one understands that the decision variable F is 3 digits and C is 1 digit. The subscripts of 1 in the variables indicate that they are the first set of solutions in the solution space. There are two genes here, 528 and 3.

After making the digit and gene explanations, let us look at the chromosome that will constitute the solution to the problem. Again, according to the definition above, the chromosome appears as 5283 the sequence of genes or as 3528, because the order is not important. From now on, chromosomes will be represented symbolically in big brackets { }. Then the chromosome of the problem at hand is chosen as either {5283} or {3528}. The first with symbols is $\{F_1C_1\}$ and the second is $\{C_1F_1\}$. The gene order within the accepted chromosome should never be changed during GA operations.

Now, to form the population that can be a solution, by throwing it out of the solution space again, for example, $F_2 = 132$, $C_2 = 913$. If three more candidate teams, such as $F_3 = 1$, and $C_3 = 71$, and $F_4 = 19$, $C_4 = 623$, are selected, one understands that they emerged from 3, 3, 1, 2, 2, 3 digits as genes, respectively. From these three pairs, one can get three more chromosomes. At this stage, the order of the decision variables in the chromosome structure must be the same as the one selected in the first chromosome. If the chromosome genetic number is formed in the first chromosome with the F decision variable and then the C decision variable, {132913}, {171} and {19623} chromosome genetic numbers are written from the genes of these new sets, respectively. Using these genetic numbers at hand, one can write the number of the genetic population. [ ] brackets will be used for population representation. Within these brackets, each of the chromosomes will be lined up one under the other, not side by side. It indicates that each line targeted from this constitutes an option to the solution of the problem at hand. Then, with the chromosome structures as designed above and the information at hand, the genetic population can be presented as follows.

$$\begin{bmatrix} 5283 \\ 132913 \\ 171 \\ 19623 \end{bmatrix}$$

Thus, the genetic numbers of the problem related to spending money (gene, chromosome, population) are determined. After all these explanations, some problems that come to mind are as follows.

1. How long should the gene lengths be?
2. Should each of the genes be of different lengths (different digit number)? Or is it better if they all have the same number of digits?
3. Is it helpful to make headshots in the solution space?
4. How many chromosomes should be in the population genetic number?

In order to answer the first of these, it is necessary to know the solution space limits of the problem. Note that integers are used in the example above. If decimal numbers are used, the question of how many digits gains even more importance. The number of digits is determined according to the sensitivity expected from the solution and the meaning of the digits. In the example given above, if it is to be worked with integers, it is understood that the genes will have a maximum of four digits. In the case of decimal, how many more digits to take can be decided by physical meaning. For example, if one of the decision variables is 5.3241567 m in meters, the numbers beyond the three digits after the point have no practical meaning. For this reason, it is desirable that the decision variable contains at most three decimal digits after the dot, because in practice, perhaps the mm digit may be needed. If the height of a hill is to be estimated, perhaps 1 or 2 digits should be taken after the decimal point. If the decision size is the distance of the stars, no digits should be taken after the point. From this, it is concluded that the reader should decide on the number of digits according to the scale of the problem he is dealing with. In the meantime, it should be noted that with the increase in the number of digits, the computing time will also increase.

In answer to the second question above, if one wishes to stay within a systematic, he should be careful that all genes are of the same length, that is, from the same number of digits. However, if some decision variables are detected more sensitively than others, then different gene lengths can be taken. As a recommendation here, it is useful to have the same length of genes if there is not very specific expert knowledge. Thus, the chromosomes are of the same length and there is no confusion.

Chromosomes are the members that will constitute the population required for GA analysis. If N_{cro} chromosomes and N_{gen} genes in each chromosome are considered in a population, the whole population can be represented as a $N_{cro} \times N_{gen}$ population matrix together with its chromosomes and genes. This much,

$$T = \begin{bmatrix} g_{1,1} & g_{1,2} & g_{1,3} & \cdots & g_{1,N_{gen}} \\ g_{2,1} & g_{2,2} & g_{2,3} & \cdots & g_{2,N_{gen}} \\ g_{3,1} & g_{3,2} & g_{3,3} & \cdots & g_{3,N_{gen}} \\ \vdots & \vdots & \vdots & \cdots & \vdots \\ g_{N_{kro},1} & g_{N_{kro},2} & g_{N_{kro},3} & \cdots & g_{N_{kro},N_{gen}} \end{bmatrix} \tag{6.4}$$

Thus, one can represent the chromosomes in the same shape as a matrix. Each element of this matrix corresponds to a gene. Also, each row denotes a chromosome (member).

Overthinking in the solution space is very useful so that random sampling is done in that space. Man's head is synonymous with computer generating random numbers. Therefore, random number operations are needed in addition to decimal or binary number systems in GA solutions. While GA operations are explained in Sect. 6.3, it will be understood that random number operations are also needed in other ways.

The necessity of finding how many chromosomes with genetic character in the population is a factor that will affect the GA resolution time and sensitivity. There is no strict rule that can be given here. However, the points where this will change depending on the size of the solution space and the smoothness of the target function will be discussed in the next sections. There is no specific general rule that everyone can use for every problem. Better decisions can be made as time and knowledge, skills, and expertise increase. There is no rule that the more the chromosomes, the better is the solution. However, it can be recommended that the number of chromosomes should not be lower than 15–20.

6.5.1 Genetic Algorithm Data Structure

When trying to solve a problem with GA, the following five consecutive steps should be completed preliminarily.

1. Should try to reach detailed preliminary information about what the problem is, and verbal and especially numerical data must be gathered as much as possible,
2. Deciding on the chromosome structures of the variables that may be related to the problem to be solved with GA and preparing data,
3. To prepare all the infrastructure that may be necessary to pass from decision variables to chromosomes,
4. To determine the analytical expression of the target function related to the problem depending on the decision variables,
5. To determine the transformation from the target function that will be used to calculate the degrees that will come against each decision set (chromosome).

During the preparatory studies, researches can be done in the series or parallel forms. After collecting verbal and numerical information, it should be tried to determine how they are logically related.

6.5.1.1 Digital Converter

If it is decided to work with the binary number system, it is necessary to consider at the outset a converter to ensure that the decision variables are human-understandable decimals, binary number system understood by GA, or vice versa. It is a digital translation processor that will enable the agreement between the human and the GA program.

6.5.1.2 Target Function

Such a function is the target value that corresponds to the values, which the decision variables take jointly in the decision space in the decimal system. While the decision variables of the target functions are decimal numbers in simple optimization problems, they can be vectors in multi-target optimization problems. It should be noted that the values of the target function are not the same as the degrees of vigor (fitness) even for the same decision variables (Sect. 6.3). In general, n target values are calculated from the target function. These n values can be represented as a column matrix in the following form.

$$T = \begin{bmatrix} t_{1,1} \\ t_{2,1} \\ t_{3,1} \\ \vdots \\ t_{n,1} \end{bmatrix} \tag{6.5}$$

In fact, each row here belongs to a chromosome. Equation (6.5), which includes chromosomes, is the column equivalent of the matrix shown as a gene in Eq. (6.4).

6.5.1.3 Fitness Function

The fitness values are obtained from the target function either by a scaling or ranking (from largest to smallest) operation. Since each member will have only one degree of fitness, they are represented by a column matrix shown below. While the target values represented by Eq. (6.5) are values with minus signs, the degrees of fitness are always numbers with a plus sign.

$$D = \begin{bmatrix} d_1 \\ d_2 \\ d_3 \\ \vdots \\ d_N \end{bmatrix} \tag{6.6}$$

If the target function only gives plus-signed values, no additional vigor function is required. How the fitness function is derived from the target function is explained in Sect. 6.3.

6.6 Methods of Optimization

Human beings have been living together with nature since the depths of history. In this unity, nature has been a source of inspiration for people in many ways. For example, the construction of the airplane and the invention of the radar are among the human technological findings inspirational from the natural events. Technological developments have been realized as a result of examining nature and living things in nature. During the study of nature, many methods have been developed to solve some problems. GA is one of these methods, but they are also the newest.

Holland (1975), working on machine learning, tried to perform genetic operations in living things in a virtual environment and explained the effectiveness of these processes. Interest in GAs, which was initially thought to be of no practical use, increased with the Young Researcher award given by the National Science Foundation for the doctoral thesis of Goldberg, a student of Holland, and the publication of his classic work 4 years later (Goldberg 1989). In this section, classical optimization (CO) methods and GA will be explained comparatively.

6.6.1 Definition and Characteristics of Genetic Algorithms

Since it is known that this algorithm was developed with inspiration from the theory of evolution, first, there is a need for "creatures" that can make such an evolution. These creatures must be able to give birth and then die. They leave some of the behaviors they have after birth as a residue in newborns. It is also necessary to measure the abilities of these creatures, which one can call their abilities, and which are stated in this book as fitness. However, thanks to the values of fitness, privilege is understood among newborns. Among them, the strong ones survive and enter the next population, and some of the weak ones can form stronger creatures by merging with another creature of their own kind. With the union of two creatures in a genetic population, not one but two new generations are formed. In the GA approach, a new generation emerges when one of the digits in the genes described in the previous section changes its number. From this it is understood that fertility in GA analysis is different from that in human life.

Every born creature has a mixed inheritance from its creators. In GAs, it is tried to provide the unions that can have the best inheritance. Some of these creatures, after living for a long time, merge with others to acquire new characteristics (become stronger) and may reach the next population (generations). Thus, with the renewal of each population, stronger creatures emerge, and the population becomes healthier. This means getting closer to the numerical optimization point. During this evolution, there are very few or no creatures that can survive without any residue from other creatures.

The GA method works on the principles of evolutionary theory and seeks to find the best solution or solutions for a given problem. It does this by starting from many initial points in the decision variable space and progressing collectively towards the best direction with a series of parallel operations. At these points in the decision space, there is no need for information other than the degrees of fitness. The development of the points in the society, working in parallel, towards optimization is provided by the principles of randomness.

The essence of GAs is based on the rules of natural selection and genetics. These rules should be perceived as the survival of the creatures that are most adapted to the environment and the elimination of those that cannot adapt. GA methodology is an optimization approach that aims to search for the best by using these two rules together.

GA requires simple calculations, but their efficiency is not diminished by their simplicity. It does not require the continuity and differentiability as many other optimization methods. Heavy mathematical knowledge is not used in its applications. Unlike random search methods, GAs use probability principles as a tool to perform genetic operations in decision variable space. Here are the differences between GAs and other traditional methods.

1. GA uses decision variables in its analysis by encoding (converting to binary number system) according to the genetic number system explained in the previous sections. In this number system, the genes of the decision variables collectively represent a point in the decision space.
2. In GA, instead of one point, it is a collection of points at the same time. As a result of the development of this population with the evolution of the GA, the optimization solution is reached. During this evolution the system is not stuck to the local optima.
3. During the GA evolution, the target function values at the points specified by the decision variables are used. Since, there is no need for derivative and integral operations, there is no need to make some classical assumptions with initial and boundary conditions.
4. GA evolution operations are based on uncertainty (randomness, probability) rules, not certainty. Selection operations are made in the light of probability principles (see Chap. 4).

Apart from the classical methods counted for optimization, simpler, less mathematical and objective GA approaches can be used. Random approaches in GA ensure that the final solution is not stuck with local best solutions. According to the study of Adeli and Hung (1995), there are five steps in GA, which are coding, initial conditions, fitness measure, evolution performance measurement, and working parameters. GAs is based on genetic mechanics principles in the development of natural events. Initially, the solution candidates can be genetically developed randomly under a certain set of rules, and the conclusion can be reached by moving towards the best solution (Goldberg 1989). GAs differ from classical methods in the following points (Buckles and Petry 1992).

1. GAs, during their development, try to reach the best solution by making use of the information available up to that time in the decision space,
2. GAs have embedded parallel computing method in their structure. In this way, they also approach step by step towards the best point in the solution space, starting from many points,
3. GAs are random algorithms. It uses probability principles to reach its conclusions. Accordingly, a random number generator is needed during the studies,
4. GAs operate on many solution candidates simultaneously. During this process sequence, it has a mechanism that moves towards even better solutions, thanks to the information it collects from neighboring and previous points and random operations.

6.6.1.1 Optimization

Whoever thinks about an event collects different information and generates ideas, and for optimization moves towards the solution. As the information increases, the event begins to be understood better. Similarly, as an engineer or scientist increases his knowledge about the subjects he is dealing with, he can find or reveal different solutions to the same problem. There are several algorithms as analysis methods developed to find the best of these solutions as soon as possible, and determination of the best among them depends on the situation and conditions. The best example of this is differential equations with uncertain boundary and initial conditions. For example, the analytical equation for solving a heat, h, problem is, in general, with α being a constant appears in the form,

$$\frac{\partial^2 h}{\partial x^2}+\frac{\partial^2 h}{\partial y^2}+\frac{\partial^2 h}{\partial z^2}=\alpha\frac{\partial h}{\partial t} \tag{6.7}$$

However, it is necessary to know the initial and boundary conditions in order to determine the most appropriate, that is, the best solution, according to the situation. Even if one may know these conditions, different questions such as “Is this optimum solution?” or “Is there any other solution to this problem, if there is, how can one reach the best choice between them?” may constantly occupy the agenda. It is natural to have such questions. If not, the conclusions may be natural and static.

Here, one needs to reveal what he understands or wants from optimization. In general, optimization appears in the form of minimization or maximization. However, beyond these, optimizations can be considered that can be made to reach a desired average or a specific target. In trade, the trader tries to maximize his profit and the buyer tries to minimize his expenses. Thus, one can say that there may be a competition between different optimization methods, and even that it always exists. Optimum one can be obtained with very specific information or it may have to be done with preliminary information that has not yet fully crystallized, that is, incomplete or not certain. In general, clear information is used in simple optimization methods. As a result, if the data containing some uncertainty are definite (deterministic), a solution can be reached with optimization methods. In real life, nothing is completely certain. For this reason, while optimization methods containing uncertainty were used with probability, statistics, and stochastic processes, today they are more modernly made with GA.

The word “best” in the optimization has relative meaning and implies that the solution to the problem being tackled includes not one but more than one options. The value of each solution (economic, social or technological) differs from each other depending on the situation. It varies according to the acceptances made, the accepted tolerance (significance) limits, and the method used. In a way, the best solution may even depend on the opinion involved in solving the problem.

6.6.1.2 Optimization Stages

Optimization is all the efforts to make an occupation even better than it can be. After researchers and engineers visualize an event in their minds, they try to develop it with optimization to achieve even better results depending on time and possibilities. The optimization of an event is its development in line with one of its maximization or minimization aspects. After determining the desired target, optimization is the search for solutions that approach the target in a direction that can be the most perfect, by considering the additional information that can be obtained by utilizing the existing information and data. Even in daily life, each person works different activities for optimization. For example, even a person's daily life according to a certain diet by paying attention to their food in order to adjust their weight can be considered a method of optimization. In the absence of a certain method, people's thoughts, actions, and their attempts to make optimization are relative. Instead of a single solution to a problem, there may be many solutions. In this case, in order to obtain the best solution, a single solution can be reached with optimization as a result of subjecting each of them to a set of rules on the same scale. Today, the optimization process is the most effective, most useful, efficient and appropriate methodology consisting of three parts (see Fig. 6.13). For the optimization process, it is necessary to have prior knowledge of the first and last links of the chain.

If a person makes an optimization by reviewing the situation when he gets new opportunities, he can safely come to better positions every day than the previous one. The saying of Prophet Mohammad also points to the optimization process.

He who has two days equal is in loss

Since a person who does not optimize will not have a rational target, he may go towards the worst by squandering the opportunities at hand, perhaps without even realizing them. Even on the worst road there must be a specific goal. Otherwise, the result can be extremely sad and unproductive to repeat.

After determining the possibilities in the first stage in Fig. 6.13 and the target in the last stage, it is necessary to make an adjustment between these two in the middle stage. For this, verbal information and necessary thoughts must be processed quickly, and especially in order to benefit from computers today, it is necessary to symbolize, that is, to write mathematical rules. All these rules are written and transferred to the computer and all the processes to be done for the solution of optimization are called the optimization process. When the word is approached from mathematics, it is understood that it is necessary to benefit from methods that are a set of rules. After the verbal expressions are converted into mathematical equations (Chap. 5), the derivatives can be taken and the optimization (minimization or

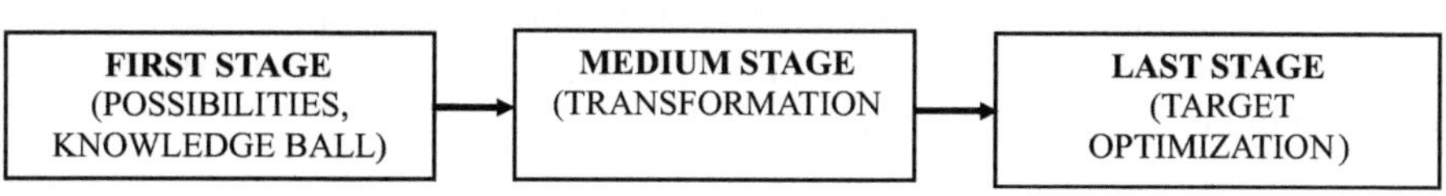

Fig. 6.13 Optimization stages

maximization) operation can be performed. Since it is not possible to use mathematical operations in case the event cannot be expressed with continuous mathematical functions, the development of numerically working methods has been encountered in the last 100 years. The most up-to-date of these is the modern optimization method called GA. This method has many advantages over others. Optimization can be defined as a method to transform input information into target outputs as shown in Fig. 6.13. It is useful to consider outputs as costs for a better understanding of optimization processes. Because every day, people show their most sensitive thoughts on money-related issues and constantly try to make optimization transactions on this issue. The following Turkish phrase is nothing more than an optimization warning

> Stretch your feet according to your quilt

However, today this saying is replaced by a modern but without any optimization based on the credit card as.

> Stretch your feet according to your credit card

Optimization means that a problem has more than one solution that are not equal, but that are very close to each other. Optimization analysis is closely related to the following points.

1. The best solution is relative to the problem at hand.
2. The best solution is also relative to the chosen method.
3. The best solution is also relative to the accepted approximation tolerance.
4. The best solution is also relative to the skill, knowledge, and approach of the researcher trying to solve the problem.

Identification of optimization solution includes morals such as education, opinion, backing, and siphoning and as immoral fatigue, insomnia, and tension body. It may also depend on spiritual functions such as religion and conscience. The solution for a basketball player is to get the ball into the crucible. This inlet has a very different shape because the diameter of the ball is smaller than the diameter of the crucible. Here, the tolerance depends on the greatest cross-sectional area of the ball being in different positions within the crucible area.

The best route for a person's daily commute is not always the physically shortest. If the goal is to reach the job as soon as possible, the solution to the optimization process is different. The weather may be snowy, misty, foggy, rainy, windy etc., and the best solution varies depending on weather pattern. In cases of accident or terrorism, there are different solutions. Therefore, optimization also shows what kind of analysis of a problem is valid under different conditions. The solution to the questions of which projects a country or company should prioritize and rank in the next 5-year planning can be found by the optimization process. In designs, optimization analysis is made by considering multiple targets such as the lightest, cheapest, and most durable and demanded material.

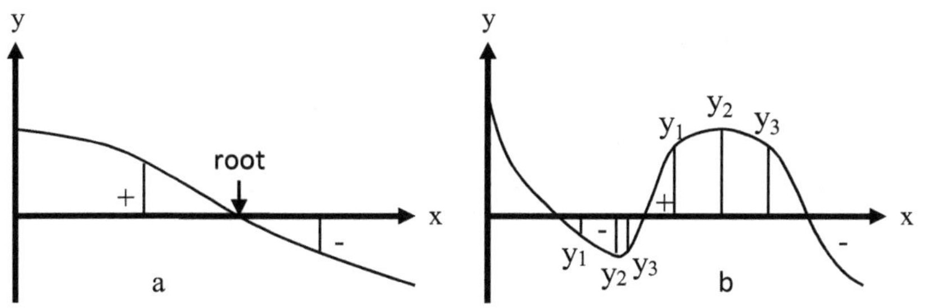

Fig. 6.14 Root finding operation

Mathematical approximate root-finding studies, which can be given as an example to the optimization operation in mathematics, are based on the principle that the root must be between a minus and a plus value close to each other. By approximating these minus and plus values on condition that their signs remain constant, when the minus and plus values that are practically close to the root are reached, it is said that the optimization analysis is in this range (see Fig. 6.14). If the largest (smallest) point is three consecutive rows of values $y_1 < y_2 > y_3$ ($y_1 > y_2 < y_3$), the root is sought from y_1 to y_3. It is necessary to know the relative positions of three consecutive values with minus-plus-minus or plus-minus-plus signs.

These are the verbal expressions necessary to find the root and the extreme (largest and smallest) points. In the light of these statements, numerical optimization methods are developed and analyzed. On the other hand, if the mathematical expression of the function whose root is to be found is known, it is possible to calculate the necessary roots precisely by setting it equal to zero. However, the marked approaches described above are general and apply to all situations. In order to work with the GA optimization process, the target function or initial information does not need to be expressed with mathematical functions. Accordingly, GA will not require a derivative operation as in mathematical optimization.

One of the most important problems encountered during the optimization process is whether the solution found is local or global. In practice, it is desirable to find the absolute optimization solution. As will be explained later, GA is free from such problems, because it achieves its analysis by randomly scanning an area, not by following a certain trace formally. Here, too, it is seen how valid is the "line defense, not the surface defense." Indeed, GA is not a linear trace; it can reach the absolute target in a very short time without falling into the local swamps by sampling from all possible points in an area.

Another dead-end of classical optimization methods is that the approximations compute linearly over finite small intervals. Some scientific tricks are used to reach the result in curvilinear (non-linear) optimization problems. However, the cost of such a trick is too high in terms of time and volume of transactions. As seen in Fig. 6.15, it is necessary to keep the stride length very small in order to closely follow a nonlinear trace. The smaller the step, the more sampling points are tracked

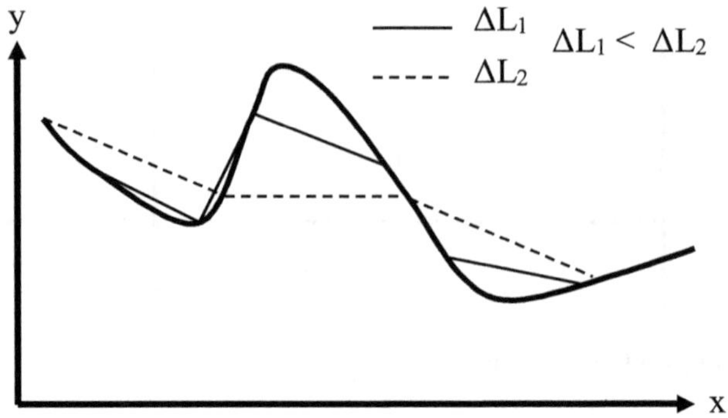

Fig. 6.15 Straight-line approximation to nonlinear trace

on the target function. Thus, even very convoluted places of the target function are represented with a more frequent sampling. This means an increase in the number of steps and, in parallel, the number of transactions. In other words, such studies are time-consuming. Although many people do not care, it can be concluded that if the computer time used is considered in monetary terms, very expensive solutions will be obtained.

Since there is no trace tracking in the GA method, it takes relatively little time to reach a conclusion despite the large step lengths. In classical methods, this step length is generally kept constant during searches for optimization analysis. In GA solutions, the step lengths change randomly.

6.6.1.3 What Is Optimization?

In one's daily life, more than one opportunity may come into the way. One may have to decide on one of them by evaluation in the form of different options. In fact, one always does optimization first albeit verbally. Life is driven by events that are nonlinear, chaotic, uncertain, and sometimes even accidental. Even a small deviation or fluctuation in the initial conditions can cause huge differences in future solutions (Lorenz (1963). For example, failure to fulfill a promise on time can greatly affect a person's later behavior or schedule. This influence can affect one's earnings and even health. For this reason, it is necessary to act in a planned and programmed manner in order to arrive at the best solution in all cases. Everyone solves or tries to solve the problem of how long sleep he needs at night so that one can be more vigorous the next day. How much work, a student who prepares for exams should study, is also an optimization problem. In all these optimization thoughts and problems, the person has input and expectation output information. The entry should contain some information so that the solution to the problem can be known approximately. In the method part, there is a function called target or fitness (vigor). Output has the best solution selected. Here, the target may be a function that needs to be minimized. To get the maximization of something, it is enough to change its sign.

After such a sign change, the minimization becomes a maximization operation. For example, maximizing x^2 in the range $-1 < x < +1$ is equivalent to minimizing the opposite sign $(-x^2)$ of this function in the same range.

Finding the roots of a quadratic equation given in mathematics or zeroing is a method of optimization (Chap. 5). Since they are mathematically based, they do not contain uncertainty and their solutions are certain. To find the logical zero point, as explained in the previous section, there must be a minus (plus) value to the right of that point and a plus (minus) value to the left. Therefore, two points are needed in the reset. In maximizing or minimizing, at least three points values should be compared with each other (see Fig. 6.14b). This chapter will always talk about minimizing, not maximizing. Anyway, the reader now understands that maximizing is minimizing by a sign difference.

It is also very important to determine that the result obtained in the optimization analysis is not the local but the absolute best point. A minimizing point found may be local. This also needs to be checked. Either all local smallest points are found, and the absolute smallest point is reached by minimizing again, or it is possible to automatically find the absolute minimum point in the structure of the developed optimization algorithm without leaving any room for such a second optimization. Here, the GA method is used to find the absolute minimizing point directly.

6.6.1.4 Optimization Classes

It is possible to perform an optimization operation with one of six different classes that are compatible with each other.

1. *Trial and Error Optimization*: In this type of optimizations, a solution candidate point is first obtained in the solution space of the problem by giving systematic or random values to the decision variables. Then, it is searched whether it is possible to improve this point. If all the values of the decision variables in the solution space can be considered, the smallest value to be chosen among them gives the minimization solution of the problem. For example, in search for the value of $\sqrt{2}$, first approximate convenient values out of head are given and it is checked whether the square of which is equal to root 2 or not. Since obtaining the exact equality requires a lot of trial and error, a solution is sought by accepting a certain percentage of error. When this error starts to stay within the limits, it means that the solution has been obtained. Today, this type of optimization operation is used in computers to take the square root of a number. This type of optimization method, which was put forward by the Islamic scholar Al-Kereci (953–1029) is used as a ready-made sub-program function in computers software even today. With the SQRT command, one automatically takes the square root of a desired number without knowing what the basis is. However, El-Kereci used the Al-Khwarizmi algebra mentioned earlier in this chapter to solve this problem.

First, he said x to the unknown square root, and since its square must be equal to the desired number, for example 3, he cleverly wrote the equation $x^2 = 3$ as $x = 3/x$, ensuring that the same decision variable is found on both sides of the equation. So, this last equation can only be solved by trial and error. To solve this equation, if one first considers $x = 1$, substituting this in the equation gives $1 = 3/1 = 3$, which is never an equality, but an inequality ($1 < 3$). So, what should take the value of x as the second step to approach equality? If one looks at the main equation ($x = 3/x$) again, it will be understood that the new value will need to be greater than 1. For example, if $x = 2$, the result is $2 = 3/2 = 1.5$. This is not equality, but an inequality of the form $2 > 1.5$. However, since the direction of the inequality changes according to the previous inequality, it is expected that the square root solution will be a number between 1 and 2. If $x = 1.7$ is chosen again, the result is $1.7 = 3/1.7 = 1.76$, so the desired equality is almost reached. If a few more choices are made in the light of what has been said, the result will be more precise. For the exact solution, if two decimals are considered after the point, $x = 1.73$ absolute solution is obtained. Since the optimization process is to choose the most suitable one among the various solutions, such a selection is obtained by trial and error rather than formality. As a matter of fact, by learning the necessary values, judgments, and knowledge through trial and error throughout human life, he can reach the best (appropriate) solution quite easily with experience,

2. *Multidimensional Optimization*: As explained in the previous step, if there is only one decision variable, the solution of the problem is obtained by trial and error without requiring much time. However, as the size increases, the analysis time also increases. In this respect, efforts should be made to reduce the dimensions of multidimensional optimization studies as much as possible and preferably to transform them into one-dimensional cases. The number of variables can be reduced by finding the relationships between the factors in the emergence of a problem. For this, regression, principle components, Fourier, spectral, etc., classical analysis methods can be used,
3. *Dynamic Optimization*: Here, the solution of the problem depends on time. As time advances, changes are expected in the solution. Otherwise, time-independent static optimizations are in question. For example, it is the determination of the best transportation route between a person's workplace and his working point. First, the shortest path between two points is found from the map, and this is called the static optimization solution. This solution is correct in time-independent situations. However, as many people think independently of time and come to this shortest route, the experience and knowledge of the increase or decrease in traffic at certain hours gives it dynamism. In this case, static optimization does not give the optimal solution. The solution requires the consideration of factors such as traffic, accidents, weather conditions, which are time-dependent, that is, affect the commute from home to work over time,
4. *Discrete Optimization*: If the decision variables of a problem take discrete, that is, integer values, the decision variables space includes finite points. For optimization, each of these can be considered individually. However, the fact that the variables are continuous requires an infinite number of cases to be considered.

Discrete-valued optimizations can also be viewed as combination calculations, since the number of best cases is tried to be determined from a finite number of cases. For example, one of the optimization methods can be used by transforming the variable into a discontinuous one to maximize a continuous function over a given finite range. If there are 5 different weather and 3 different ground conditions that will pose a danger in a place; if both are effective, one will have $5 \times 3 = 15$ options. Among these options, the most dangerous one can be found with the discrete optimization process,

5. *Restricted Optimization*: In an optimization problem, it is also the case that the ranges of change of the constant and decision variables are limited. In real life, nothing is eternal. Finite quantities impose constraints on optimization algorithms. In this respect, we can look at optimization methods from two perspectives either restricted or unconstrained. Mostly engineering problems are solved with constrained optimization algorithms,
6. *Optimization with Initial Parameter Value*: In many cases, the values of the initial decision variables must be determined somehow for the optimization methods to reach a solution. These methods easily get stuck in local minima, and it does not take long to get there. These are classical optimization methods and are generally based on differential and integral calculations.

6.7 Least Minimization Methods

The most important aspect of these methods is to search for the point where a given problem will have the smallest value in the solution space. The solution lies in a space that has as many dimensions as the number of variables. Considering that the latitude, longitude, and altitude of all points in a region are given for a better understanding of the method, it is understood that a three-dimensional solution space should be considered with these data. In order to minimize in such an area, a person would have to walk from a random point to work and looking around continuously to the less elevated points. The position is a point of minimization when the walker realizes that he has finally reached a point where the elevation can no longer decrease. Are there points with smaller elevations in the desired solution area? For this, is it possible for the person with bionic eyes to see the back sides through the mountains surrounding the smallest elevated point, and according at the local smallest point or at the smallest point of the general solution area should be able to decide. Since such a talent is not found in all creatures, scientific methods must be rationally developed in a way that they can see beyond the mountains and take them into account. In general, since all classical and early optimization methods are based on this type of walking, their solutions are often local smallest solution points. If the terrain is completely in the shape of a bowl, the walker understands that the point he arrives at is the solution point with the lowest elevation overall.

In practical studies, the finding of an optimization solution may not be achieved with certainty. In many cases, one may have to settle for solutions that are closest to the best solution. The reasons for this may be that there is a lot of time and transaction volume, and the decision variables of the problem are many and complex relationships. In all optimization analyzes, there are variables that describe the decision variable or solution space domain, and the target variable, which is a function of these variables. In addition to these, there may be physical and mathematical constraints for decision variables. Starting from a set of optimization decision variables, first corresponding target value is calculated, then re-determining the target value by sequentially renewing the decision variables to bring this target variable closer to the optimization point. In this way, one travels in the decision space along a path whose route is not predetermined. As every road or journey has an end, optimization travel must end somehow. The absolute end really comes down to getting 100% to the best point; but this is nothing more than a theoretical expectation. Therefore, a criterion must be established for the optimization travel to stop. This criterion is either to limit the number of steps, to make a fuzzy decision with expert opinion, or to objectively require that the difference between the target values in successive progressions be smaller than a predetermined amount or percentage. Considering all that has been said, in general, the logical flow of operations in an optimization study is done according to the scheme shown in Fig. 6.16.

If the stopping criterion is not met, the newly selected decision variables are always circulated in the valid (with constraints) decision space and target function

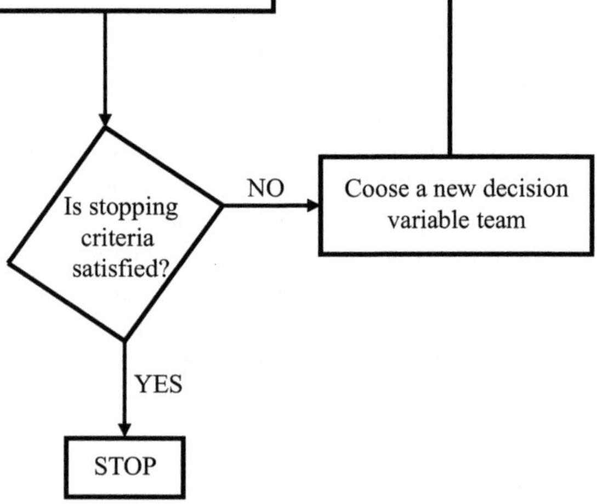

Fig. 6.16 Optimization flow charts

calculations are made. Provided that the decision space stays in its valid region, even if the set of decision variables is chosen in a formal or random way, the important thing is the new value of the target function and its status relative to the previous target function. If the new target value is smaller than the previous one (here assuming that optimization is minimization), reselection of the next set of decision variables continues. If the new target value is greater than the previous one, then either continue with the new decision team randomly or choose the decision team by traveling in different directions from the direction specified by the decision variables. In summary, the common point in all optimization methods is that the decision set is chosen independently of the target function, even if it is random, on the condition that it stays in the valid decision space, and then the target function value is calculated depending on this.

There can be many minimum points in the decision space, one of which is necessarily the best in terms of the value of the target function. When almost all the optimization methods are stuck in one of them, it is not possible to get out of there formally. The most useful method for this is to choose a random decision set. From this point of view, GAs can reach the absolute smallest point among the other optimization methods, most effectively and without getting stuck at the local smallest points. In particular, the maximum gradient descent (minimization) or similar hill climb (maximization) methods get stuck at local minimization points.

6.7.1 *Completely Systematic Research Method*

These methods, which are very difficult due to being systematic and following a certain set of rules, are based on the idea that minimization can be made by dividing the objective function into very small decision variable sampling parts and comparing the value among them. In this approach, since many intermediate examples that do not directly contribute to the solution are tested, it takes a lot of time, effort, and patience. For example, if the target surface is in the form of a trigonometric function, there will be too many valleys, hills, and ridges in the solution area. In order to find the absolute minimum point, the picture of the objective function in the entire solution space must be known. This situation can be observed visually in three dimensions. For example, the goal function

$$f(x, y) = x \sin(1.5x) + 2.3 \cos(y) \tag{6.8}$$

The complex solution surface that this will give if taken is shown in Fig. 6.17. Here, the f(x, y) value is calculated as the product of the number of intervals in the x and y directions. For example, since 100 sub-intervals are taken along the x and y axes as in Fig. 6.17, $100 \times 100 = 10000$ f(x, y) values should be calculated. Here, among all these points, the ones with the smallest f(x, y) values are an option to the solution.

The vast majority of those out of these options have been calculated for nothing. All points need to be explored in this approach. In order to better understand the

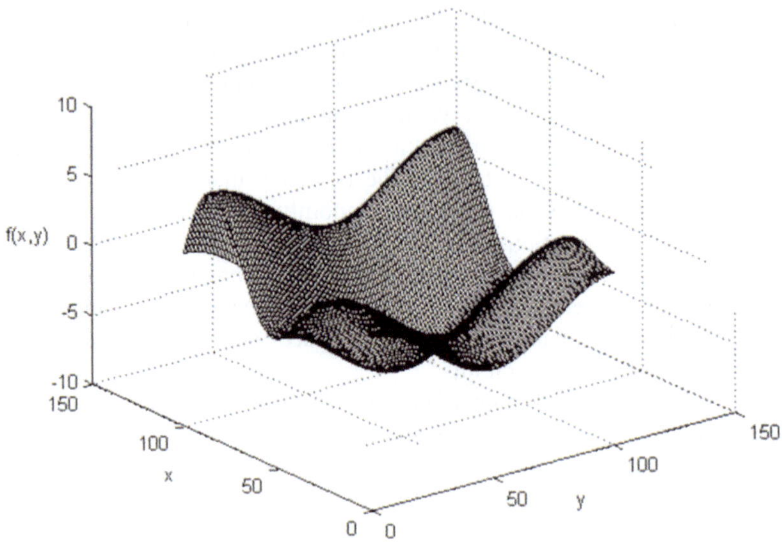

Fig. 6.17 Three-dimensional presentation of Eq. (6.8)

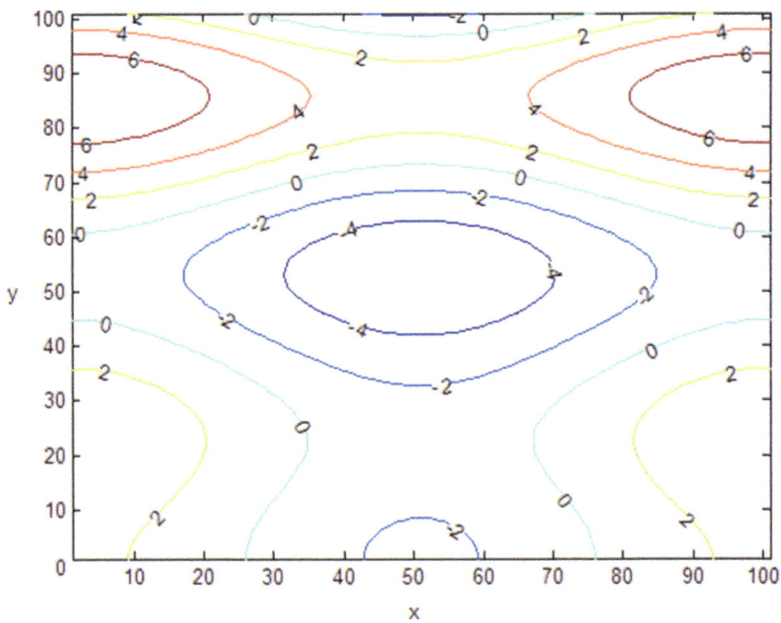

Fig. 6.18 Equal value lines for Eq. (6.8)

situation, the map shown in Fig. 6.18 is obtained by drawing the equivalent lines of f (x, y) values since it is three-dimensional. Now the situation is transformed into finding the smallest raised point on a topography map.

By examining this equivalent map, it is calculated that the smallest f (x, y) value is –5.5094 corresponding to the f(51, 51) point. It is not possible to use such a visual approach for 4 or more-dimensional minimization problems of the overall search method. The time spent to find the absolute smallest solution point of these methods can be so much that one loses patience. Graphs and maps are very useful for visual inspections, but up to three dimensions and variables. In case of more variables, numerical methods should be used.

It is not always possible to find a mathematical expression like the one above among the variables in order to perform the optimization process in a formal way in research. In practical studies, the relationship between the three variables is often discrete. For example, contour maps can be drawn as a result of knowing the elevations at different points in geography according to latitude and longitude. Similarly, three different but simultaneously measured variables can be used instead of topographic elevation. For example, from the Calcium, (Ca), Magnesium, (Mg), and Bicarbonate (HCO_3) measurements made in relation to water quality in one place, the three-dimensional surface and two-dimensional equivalent bicarbonate values of HCO_3 based on two other variables are shown in Figs. 6.19 and 6.20, respectively. These figures provide important visually information about how bicarbonate concentration changes according to Ca and Mg measurements. In addition, it is possible to visually and approximately determine which Ca and Mg values correspond to the largest absolute and local values of HCO_3. Here, the HCO_3 target variable Ca and Mg act as decision variables.

By plotting iso-bicarbonate curves as shown in the figure, many inferences can be made about the variation of bicarbonate by comparisons using the relative positions of the curves. Such maps give, at first glance, the first visual information about where the optimization points (largest and smallest) are. In Fig. 6.20, the absolute

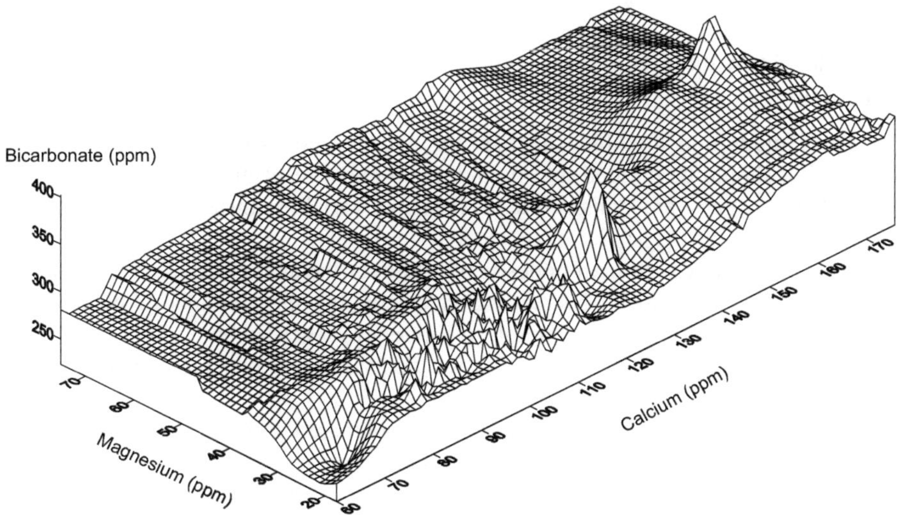

Fig. 6.19 Bicarbonate concentration surface in three dimensions

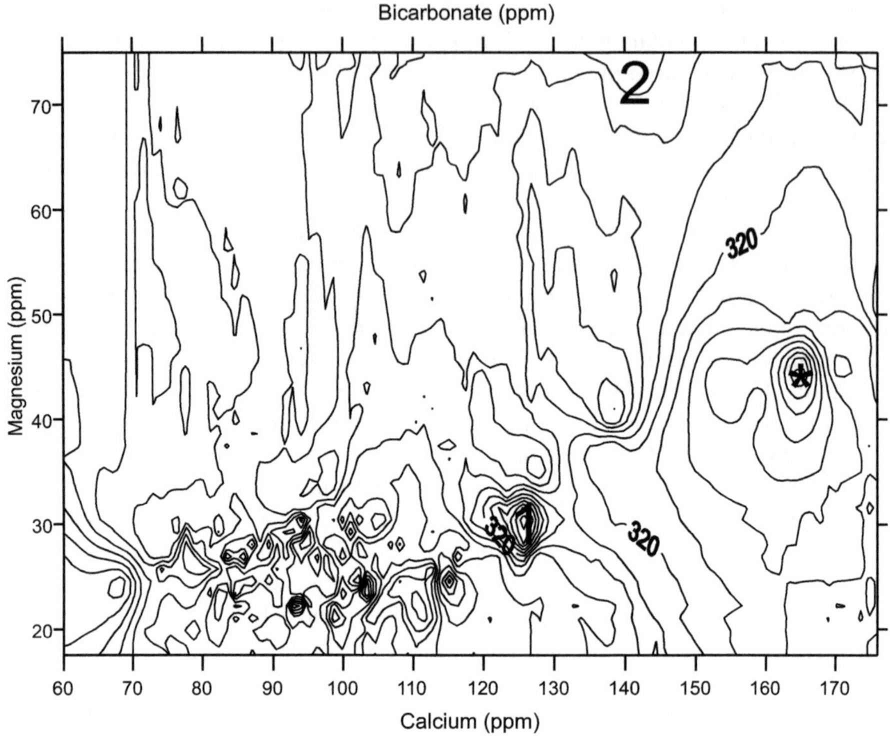

Fig. 6.20 Bicarbonate concentration surface in two dimensions

maximum bicarbonate value is shown with an asterisk. Local maximum bicarbonate values are marked with numbers 1 and 2. The following steps must be completed in order to obtain the optimization points by transferring such a map to the digital environment.

1. The latitude and longitude degrees (or two decision variables) of the area of interest should be divided into m and n sub-parts, respectively. In Fig. 6.20, Ca and Mg values are considered as decision variables instead of latitude and longitude,
2. Thus, a network with mxn nodes is passed over the area of decision variables,
3. This map was digitized by reading or calculating the amounts of bicarbonate concentration as the target variable separately for each node.
4. The data of these three variables (two decisions and one target variable) are entered into the computer,
5. The value at each decision (node) point is compared with the others in an order and the desired optimization value is found,
6. If there are k decision variables instead of two variables such as latitude and longitude, each of them has n_1, n_2, n_3, . . . ,n_l; $n_1 \times n_2 \times n_3 \ldots xn_k$, in the case of providing a multi-space network by dividing it into n_k sub-ranges nodes and

decision points are obtained. Even for these 3–5 variables, a large number is reached. In this case, since maps with more than three decision variables cannot be perceived visually by humans, it will be very difficult to find optimization solutions. For this, it is necessary to develop numerical methods. The most advanced method in this regard is GA.

It is another problem that the optimization analysis to be obtained as a result of the application of the completely systematic research method depends on the length of the lower interval. The smaller the interval lengths, the greater are the number of nodes, the greater the computational volume. This is practically undesirable. In addition, if the appropriate sub-interval lengths cannot be chosen from the beginning, the absolute optimization solution point cannot be found, and even some local solutions may not be fully obtained. Starting from large interval lengths, first the regions where absolute optimization analysis can be found, are determined. Then it seems plausible to rescan these locally with smaller sub-interval length networks. These processes are both tiring and require a lot of time and effort.

On the other hand, this method constitutes the basis of all optimization methods. It is too cumbersome and tiring in terms of considering the node points one by one. Many nodes do not directly contribute to the optimization analysis. Then, the answers to the question of how one should choose the points that can contribute to the optimization process can be listed as follows:

1. Instead of systematically scanning the area over all nodal points, should it scan in certain directions? If so, how should these directions be chosen? Line optimization methods have been developed in response to this. The most famous of these are the steepest descending or climbing methods,
2. Considering all points systematically, a starting point should be systematically chosen for comparisons. In practice, in general, the optimization analysis is achieved by making comparisons starting from the top left side of the grid. However, starting from any point of the grid, the same result will be reached. This raises the question of whether it should be started randomly. If there is a region preferred by the researcher based on his experience, it is important to start from there in terms of reducing the calculations,
3. Should it take different lengths each time instead of keeping the sub-range lengths constant? Should different lengths be taken for different variables? The answer to all these questions is yes. However, in this case, the question arises of how the interval lengths should be decided,
4. Is it enough for the sub-range lengths to be different? In addition, would it be more beneficial to change their direction? Yes, changing the direction allows the method to turn to target points in a shorter time,
5. Another question is, would it be more appropriate to determine the number of nodes and their locations systematically, instead of determining them randomly or irregularly? The answer is yes, and the best method for this is GA.

A general question arising from the above points is whether some randomness should be imported into the method in order to reach the optimization analysis in the shortest and most reliable way instead of going systematically. Indeed, during the GA, optimization analysis processes, there is a lot of randomness and the inclusion of expert opinion in the analysis process.

6.7.2 Analytical Optimization

If a problem can be expressed mathematically in terms of the decision variables of the target function, optimization analysis can be done with differential calculation methods. This is identical to finding the extreme (smallest or largest) value points of a given function. The mathematical expressions used here are, in general as, follows.

$$\frac{\partial f}{\partial x} = 0 \tag{6.9}$$

According to the topics in this book, f represents the decision variable in the target function x. The meaning of this is that the slope of the target function is horizontal, that is, its angle is zero. Indeed, in the case of the continuous variable, the slopes at the largest (peak) and smallest (valley) points are equal to zero. How many decision variables there are, so like Eq. (6.9), the partial derivative of the target function with respect to each decision variable is taken and set to zero. Thus, the set of non-linear equations resulting from taking the slope of the target function equal to zero must be solved simultaneously. The search is made around the points that are likely to be peaks or troughs. The shortcomings of this method are as follows.

1. Target functions must be differentiable: Otherwise, it is impossible to reach a solution by analytical method. However, as there are discontinuities from time to time in the data obtained from real life, there is also noise (interference, error) in the data,
2. Non-linear sets of equations are very difficult to solve. The solution can be reached within the initial and boundary conditions. It is also possible that the existing solution might be chaotic depending on the initial values (Chap. 4),
3. Since the search will be done around points that are likely to be peaks or troughs, this method may work well for geometries such as Fig. 6.21a, which has only one trough, but may not work well for Fig. 6.21b. The method can be stuck to local best instead of absolute best (smallest). The method proceeds from a local point of view.

Differential-integral calculus offers very attractive solutions in finding the smallest point of different target functions. As it is easily known from the education training even at the high school level, in case of a decision variable, the target function is set to zero by taking the first derivative of that function in order to find its smallest point (Eq. 6.9). Thus, the found point gives the value of the decision variable and the target function. If the second derivative is greater than zero, it is a minimization point.

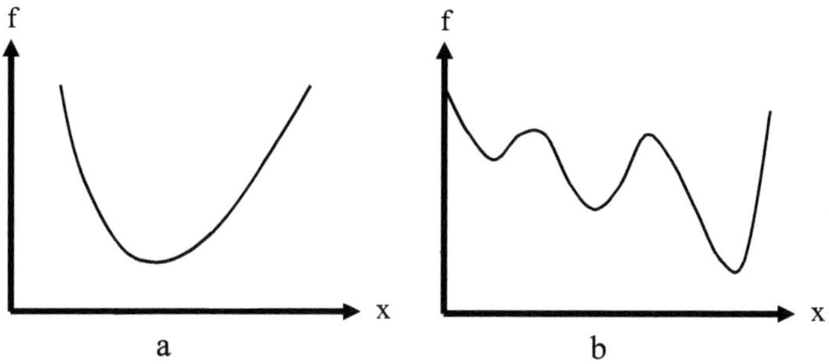

Fig. 6.21 Extreme values

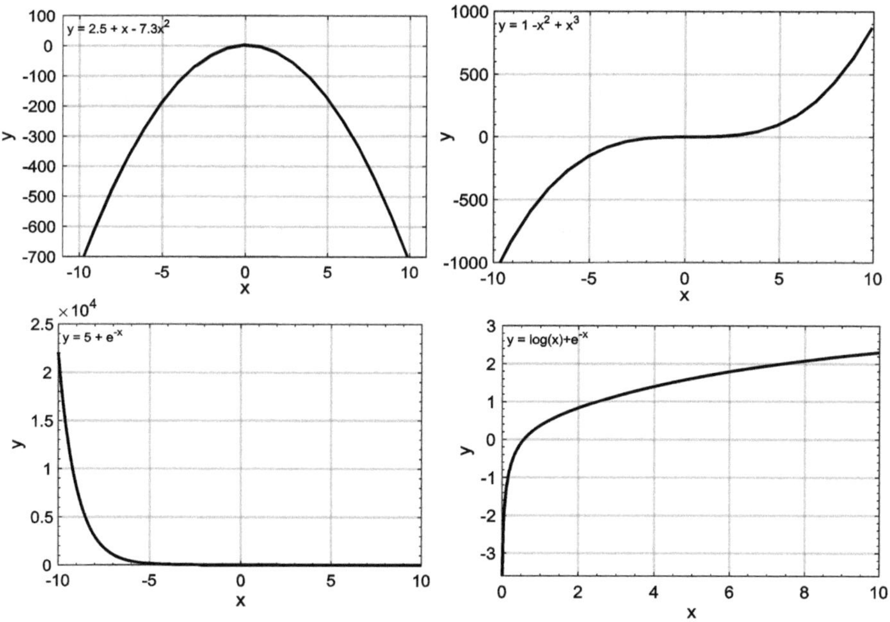

Fig. 6.22 Some continuous functions

$$\frac{\partial^2 f}{\partial x^2} \geq 0 \tag{6.10}$$

However, in order to use the analytical solution method, either the given equations must be continuous or, if they are discrete, the most appropriate continuous curves must be found by fitting them. Some of these are shown in Fig. 6.22.

In the case of many decision variables, the first order partial derivatives are taken with the idea that all other decision variables are fixed and equated to zero, and the location of the smallest target function point in the solution space is determined by substituting the values of these decision variables in the target function.

If the mathematical expression of the objective function, f, depends on the decision variables x and y are given as follows,

$$f = 1.2 + 3.5yx^3 – 2.4x^2y^2 \tag{6.11}$$

then the slope expressions in the x (y fixed) and y (x fixed) directions are,

$$\frac{\partial f}{\partial x} = 10.5yx^2 - 4.8xy^2 = 0 \tag{6.12}$$

and

$$\frac{\partial f}{\partial y} = 3.5x^3 - 4.8x^2y = 0 \tag{6.13}$$

respectively. With the simultaneous solution of these equations, y = 3.5/4.8 = 0.729 and x = (4.8/10.5) × 0.729 = 0.33 decision variables are found. At this point f has the optimization solution. However, in order to understand whether it is the largest or the smallest, it is necessary to decide by looking at the second derivatives. The second derivatives are available in the following forms in the same direction.

$$\frac{\partial^2 f}{\partial x^2} = 21yx - 4.8y^2 \tag{6.14}$$

and

$$\frac{\partial^2 f}{\partial y^2} = -4.8x^2 \tag{6.15}$$

Since the sum after substituting the calculated x and y values above has a minus sign, it is decided that the optimization solution is maximization.

6.7.3 Steepest Ascend (Descent) Method

This optimization search can be applied directly. In this form of search, first an estimate is made about the location of the minimization point. Starting from an arbitrarily (or randomly) selected initial decision variable point, the search continues. From this point one proceeds in the steepest possible slope direction to find the local best point. The shortcomings of this method are:

1. The method proceeds from a local point of view. So, it gives good results for Fig. 6.22a but can be plugged into local best for Fig. 6.22b,

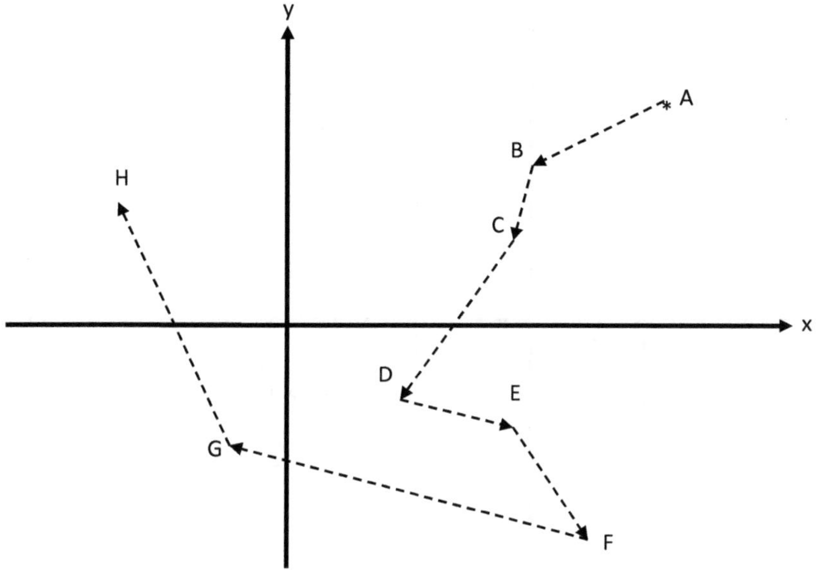

Fig. 6.23 Optimization line by direction changes

2. Target functions must be of differentiable type. Otherwise, this method becomes unable to search.

However, the GA method is an approach that provides a continuous descent towards the smallest point without requiring derivatives, as in the analytical method, in parallel with the computer support.

Most minimization methods are based on line minimization. The solution starts from a randomly chosen point in the decision space. There are infinite directions passing through this point. One can collect these in two main classes, one is variable and the other is parallel line tracing. After a direction is determined from these, it is continued decreasingly until the target function starts to increase on the solution surface line along that direction. At the point before the target function starts to increase, the direction is changed, and the new solution direction is tried to be determined. In Fig. 6.23, starting from point A, the direction has been changed in successive goings to points B, C, D, E, F, G, and H. If the direction that started at A is continued after B, the target function value starts to increase again on this track. Thus, the solution continues until it proceeds to the smallest point solution. The location of the smallest solution point in the decision space in Fig. 6.23 is indicated as H. For the algorithm to work quickly, the solution direction must be chosen appropriately. Here, the steepest descent (ascent) direction is selected according to the shape of the target function surface; the general decision is given from directions in different directions from each point.

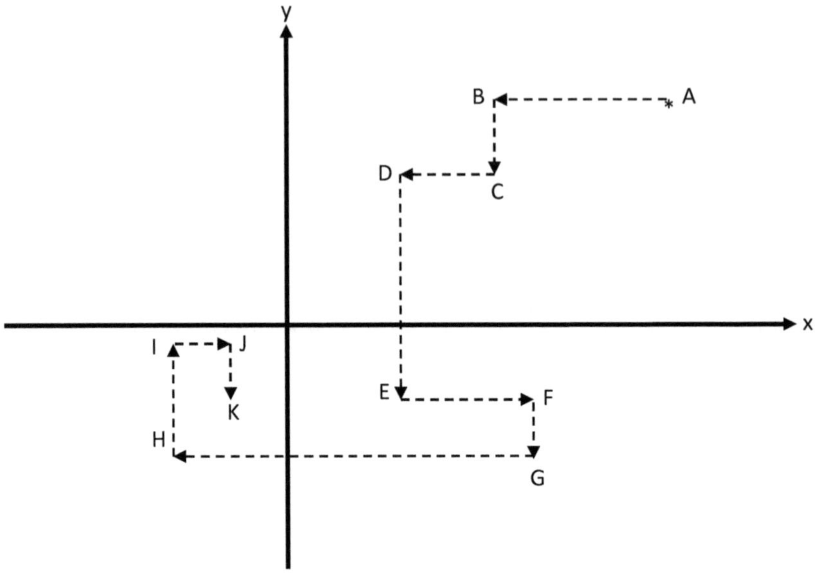

Fig. 6.24 Optimization line along two constant directions

In the most primitive version of this method, after choosing a random point, the objective function is minimized by changing only one of the variables. Changing only one of the decision variables and keeping the others constant requires the direction to be parallel to the axis of the changed decision variable. Then, by changing the other variables, this minimization process, which one can call parallel axis, is continued until the smallest point is found. After all the variables have taken their order, the variables are changed one by one periodically in order (see Fig. 6.23). In general, this method is slow and requires a lot of time to reach the solution. If there are two independent variables such as x and y that controls the target functions, changing them in parallel to the Cartesian coordinate axes, respectively, is shown in Fig. 6.24.

Here, in the decision variables space, after starting the optimization process from a point denoted by A, minimization points are searched consecutively first in the x and then in the parallel directions to the y axes. Here, the end of each arrow in Fig. 6.24 gives the smallest value of the target function when the line is followed in the direction of that arrow. When it is desired to go further from these points, the way to maximization is taken as the opposite of minimization. In this way, successive minimization points are A, B, . . . and after obtaining J, point K is reached, which is even smaller than all points. This is where the decision variables x and y have the smallest target function value. If the direction is changed again at this point, the desired minimization point has been reached since there will be a small displacement that is practically negligible.

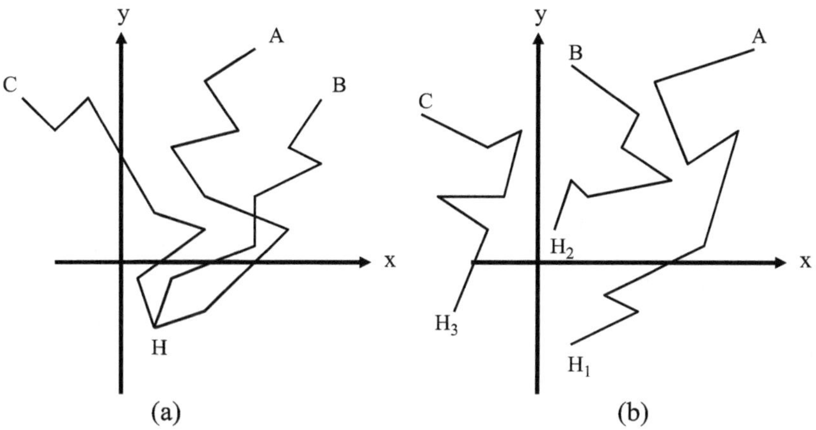

Fig. 6.25 Local and global optimization solutions

However, it should be suspected that this point is the absolute smallest point in the entire decision space. Because, by continuing the optimization process along the consecutive lines, local minimum values can also be reached. In order to be sure of this, a few random starting points in the decision space should be chosen and the optimization operations should be continued in a similar way. If all the considered line optimization operations result in the same point, it can be concluded that the absolute minimization point has been reached. The greater the number of line optimization operations with different starting points, the more confidently the absolute smallest point can be reached. Repeating (serial) repetition of such optimization operations with many starting points makes it is even slower to achieve the absolute best point in an already very slow and time-consuming method. In Fig. 6.25, the results of the line optimization method obtained with different starting points are shown.

Since the same resolution point, H, is always reached from different starting points (A, B, and C) in Fig. 6.25a, it can be inferred that this point may be the global minimization point. However, in Fig. 6.25b, since three different minimization points such as H_1, H_2, and H_3 are reached in variable-direction solutions, respectively; there is no absolute solution point here. However, it can be said that the smallest target value point is closer to the absolute smallest point in these three different line solutions. In order to obtain the true absolute smallest point, the line optimization process must be continued by selecting several starting points. However, repetitive application of such a method to the same problem is tiring for the researcher and very time-consuming for the computer. The important points that can be drawn from here can be listed as follows.

1. Optimization methods with the steepest slope must reach the conclusion by making progress on the line,

2. In order to find the absolute solution with these methods, it is necessary to apply the same analysis method repeatedly from different points,
3. These methods are both tedious and time-consuming,
4. In sensitive cases, an absolute solution with enough health cannot be obtained due to rounding errors.

A method that can overcome these drawbacks should be put into use. Such a method should make progress in the field, not on the line, should not be tedious and time-consuming, should ensure to catch the absolute solution with a series of parallel operations, and ultimately achieve the desired precision. The methodology to include all these features is the GA method, which is the main subject of the next sections in this chapter.

The trace of each initial decision set in the decision space is independent of each other. Because of this independence principle, if the probability of a trace not finding the absolute smallest point is P, the probability of n traces is P_n.

$$P_n = 1\text{–}P^n \tag{6.16}$$

The larger n is, the higher the probability of finding the absolute smallest point, since $P_n < 1$. If, in the valid decision space, the investigated optimization problem has N best points, the probability of finding the absolute point is $P = (N - 1)/N$ in the absence of other information. From here, to find the absolute best point with 0.90 probability it is necessary to satisfy the following inequality.

$$\left[1 - \left(\frac{N-1}{N}\right)^n\right] \geq 0.9 \tag{6.17}$$

Hence, one can obtain,

$$n \geq \frac{0.9}{\log N - \log(N-1)} \tag{6.18}$$

Accordingly, if the 10 smallest points are in a valid decision space, it is understood that $n = 21$ independent initial decision sets will need to be selected in order to find the absolute best point with a probability of 0.90. In the GA optimization method, it is necessary to choose independent decision sets (chromosomes) at the beginning.

A question that comes to mind is, if one takes these different initial decision sets at the same time, and starting from these points, can he reach the absolute smallest point with simultaneous, that is, parallel calculations? The answer is yes. However, this greatest slope cannot be done with the stroke method, even if it is tried to be done, it takes too much time due to serial operations. This is where the GA optimization method comes to the researcher's aid; because in this method, the absolute smallest point can be reached by parallel computation based on the randomly selected initial decision sets.

6.8 Simulated Annealing (SA) Method

In the largest descending or ascending (hill-climbing) methods, the next step is taken in the direction of the greatest slope, and the best solution is reached if there is only one valley (hill). However, when the number of valleys (hills) is high, one often cannot reach the absolute best solution, if in a wrong valley (hill). In this case, Kirkpatrick (1983) developed the optimization method with simulated annealing (SA) like the energy change during gradual cooling (annealing) after metal melting.

The SA optimization method is based on the idea that the iron cools slowly (minimizing) or heats up (maximizing) like the annealing and forging of the iron. This method is like minimizing the energy probability distribution function (PDF) for the formation of crystals if a metal is heated above its melting point and then left to cool slowly on its own. This crystal structure consists of millions of nodes in a regular grid. They complete the minimization process naturally by arranging in the most appropriate way in an orderly and harmonious way among themselves. If the natural cooling is accelerated, very amorphous situations will arise, and good order and harmony will not be seen, because the energy is frozen higher than the best. The key to crystal formation is to allow the variable temperature to decrease or control appropriately.

Similarly, in the optimization algorithm, after the initial prediction states are selected, we can call the changing of the parameters "heating" them. Here, the "target function" represents the energy level of the substance. The crucial point of the SA method is the addition of a control variable called the temperature variable, which will lead the solution to the optimization. The task of this variable is to adjust the approach speed to the smallest point of the target function. This control variable determines the stride length in such a way that at the outset, the algorithm forces large variations in variable values to occur. Changes over time move the algorithm away from the optimization solution. This allows the algorithm to search for new regions of the decision space. After a certain number of repetitions, the value of the control parameter is decreased. This allows smaller steps in the variables. By slowly decreasing the value of the control parameter, the algorithm reaches the valley with the absolute smallest point before reaching the smallest point.

In case of slow cooling (annealing) of a molten metal, the energy decreases evenly at different temperatures. Since the velocities and kinetic energies of metal molecules at high degrees are high, different patterns can emerge even with the smallest interventions. The best stable structures are obtained by step annealing. In sudden cooling, which neglects the annealing process, fragile structures are reached, which is of little use. The annealing process can be viewed as the processes of maximizing the force between the molecules and minimizing the fragility with the least energy. The SA optimization method can be viewed as a probabilistic approximation of the serial solutions to be overhauled by the largest slope descent or raise method. Here, the best suitable solutions are reached with a randomly selected decision set at a given temperature of an energy (target) function, which is the essence of the problem. The acceptance of the move from the existing decision set to a new decision variable depends on the current temperature and the resulting energy change. Deciding on this course and making the next similar steps depends on the reduction of the temperature parameter continuously. In the operation of the SA optimization method, on the one hand, the temperature parameter is reduced continuously; on the other hand, it is to decide which new decision set to go to. Thus, in the SA optimization method, there are two stages, one to make the move and the other to decide whether this move will be accepted or not.

The selection of a new decision set is always done in close vicinity of the previous decision variable. If the decision set has 6 digits consisting of the integers 0 and 1, for example (1 0 0 1 1 0), only one of these digits is replaced with the opposite value, for example (1 0 1 1 1 0), staying around the previous team. The digit change here is the same as the digit change (mutation) in GA method. If this single-digit substitution is done consecutively, it is possible to move from one team to nearby teams. This digit change operation depends neither on the temperature parameter nor on the energy (target) function. As said above, it is nothing but navigating the valid decision space independently of them. In fact, this is like the digit replacement (mutation) operation in the GA formal optimization system, which will be explained later (Sect. 6.9.11).

Transitions in the SA method is considered in the case of energy reduction, like the gradient reduction in the largest gradient descent method. However, sometimes there is a possibility of acceptance in transitions that increase energy. Depending on the acceptance probability, P, the temperature parameter, T, and the energy levels old, E_O and new, E_N, is calculated as,

$$P = e^{(E_O - E_N)/T} \tag{6.19}$$

The higher the temperature, the greater is the probability of acceptance. This point represents the melting of natural metals, easily changing their shape at high temperatures, and subsequent annealing. An increase in the temperature parameter also means an increase in the noise in the processes. This possibility also grows with the shrinking of the difference between the old and new energy levels. Although the probability of each decision set to be a solution is equal at high temperatures, low-energy decision sets have a higher probability value at low temperatures. As the processes continue, the saturation state is reached, as if the probability of the decision team depends only on its energy. Since checking whether this state is reached or not requires heavy calculations and long-time durations, in practical applications, processes are continued for longer periods with the decrease in temperature. If operations are performed for a long time at a temperature level, this algorithm finally gives teams that fit the Boltzmann distribution, and the mathematical expression for this is the probability that the energy of the current state j is Ej, is calculated as,

$$P_j = \frac{e^{-E_j/T}}{\sum_{i=1}^{n} e^{-E_j/T}} \tag{6.20}$$

When the SA optimization method reaches the absolute best solution, the temperature in the Boltzmann distribution becomes equal to 0 and probability to 1. Unfortunately, it takes a very long time and a very slow temperature reduction in order to reach this conclusion. If a local minimum solution point is reached, it cannot be guaranteed to reach the absolute optimization point with this method with a small number of steps. To improve this point, instead of the exponential temperature cooling explained in the SA optimization method, the previous temperature is reduced by a number less than 1. Instead of extremely slow cooling, the best solution point among them is selected by using the SA optimization method, fast cooling at the beginning, and the largest slope descent method at the end.

6.8.1 Application

It is an optimization problem to divide the 6-node network given in Fig. 6.26 into two parts with equal sides. Each node here can be expressed as numbers in the binary system with as many digits as the node. For example, since there are 5 nodes, the second, third and fourth nodes (100011) form a part with the symbol 0, and the others form the second part. Its split into two parts means that there are always 3 0s and 3 1s in such a number sequence. Since the condition of having equal number of sides is required here, according to the example given, the energy of the 0 part is $E_0 = 4$, and the 1 part is $E_1 = 7$. This means that in the case of the (100011) solution considered, the probability of moving to higher energy states according to Eq. (6.19) is,

Fig. 6.26 Given net

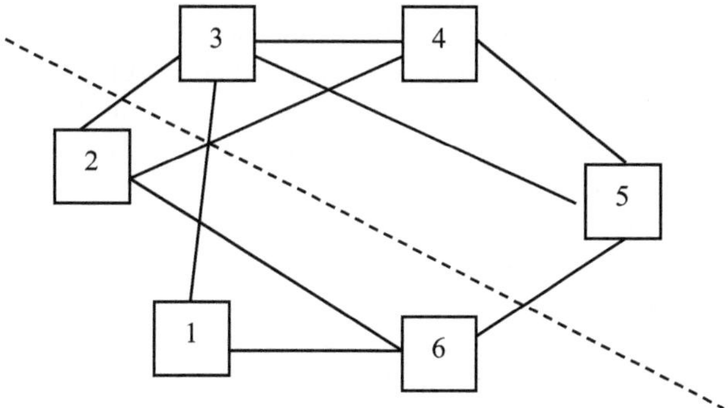

$$P = e^{(4-7)/T} \tag{6.21}$$

Since there must be two equal parts continuously (100011), this condition is kept in place by making one of the 0s 1 and one of the 1s 0 at each step from the initial value. Since the energy level will be different for each option, the probability of acceptance of motion will also change. Accordingly, the teams after the starting (100011) decision team can be (010011), (001011), (000111), (110001), (101001), (100101), (110010), (101010), and (100110).

Considering that the decision set at a time is (010011), its energy is $E_O = 6$. If the transition to (100011) is considered, its energy will be $E_N = 4 < E_O = 6$. If the team (100110) goes, $E_N = 8$ and the probability of its acceptance will depend on the temperature.

$$P = e^{(6-8)/T} = e^{-2/T} \tag{6.22}$$

If the temperature is chosen as 50, this equation gives the probability of acceptance $P = 0.96$. On the other hand, from the same equation for $T = 10$, $P = 0.96$ and for $T = 1$, $P = 0.13$. From here, it is seen that the solution set with higher energy level is passed with high probability at high temperatures.

6.8.2 Random Optimization Methods

The target value of the initial decision set is selected randomly in the valid decision space, without any formal research structure. In the next step, it is compared with the target value of another randomly selected decision set. If the next target value is smaller than the previous one, the previous one is forgotten, and the next best point search is based on the new decision set and its target value. If the target value of the newly selected decision set is greater than the previous one, forgetting this newly selected decision variable and target value, again randomly calculating another point of the valid decision space and its target value as a result of the comparison with the previous one, it is decided which one will be forgotten and the process will continue. Thus, in random serial operations, optimization is performed at multiple points of the valid decision space in succession, and the best absolute point can be reached at the end. The criticism of this method is that it is the increase in the transaction volume and time, as there is always a random point in the decision space and there is no connection between them. The lack of a clear method does not make it a desirable method. This method, which corresponds to the completely random sampling of the decision space in practice, is called the Monte Carlo method. Any method that has some degree of formality (systematism) in its structure is preferable than one that is completely random.

If the solution space of this random method is m-dimensional, and the decision variables space is based only on the integers of 0 and 1, there are 2m different and randomly selected cases. The probability of reaching the absolute best point after n steps, is given as

$$P = 1 - (1 - 1/2^m)^n \tag{6.23}$$

Accordingly, if $m = 10$, $n > 2356$ must be in order to get the absolute best point with a probability of $P = 0.90$. This even exceeds $2^{10} = 1024$ memory. For this reason, it is practically not possible to undertake optimization with the completely random decision team selection method.

In the GA optimization method, on the other hand, the initial decision teams are randomly selected at the same time, and since each team is taken to the next points of the decision space with a relationship between them, the calculation limits are extremely reduced and remain well below the practical limits.

In order to improve the random optimization method a little, instead of choosing the decision sets in successive steps with equal probability across all boundaries of the decision space, the computational volume will be reduced if it is chosen more likely to stay close to an obtained best point. Here, the focus is on its area around a local best point rather than the entire decision. Thus, the absolute best solution can be reached by randomly searching around the local smallest points obtained in the previous steps, rather than arriving at the absolute best point sequentially and forgetting the previous ones.

6.9 Binary Genetic Algorithms (GA)

In the numerical optimization methods described in the previous sections, sequential transitions to the next smaller target value points are provided in different ways, generally starting from one point. Their local smallest points are often captured. In the last 35–40 years, algorithms have emerged with effective and new aspects. The most recent of these is the GA approach, which is based on the theory of evolution and natural selection. Here, randomly distributed starting points are first interspersed in the solution space. Based on these, it is tried to reach the decision variable points closer to the target by making use of the evolutionary process theory procedures. GA performs the optimization process by using probability methods. In a way, this is like taking random samples from different points of the solution surface. However, the randomness process here is very different from the classical sense. A systematic numerical evolution process has been added to the classical randomness (Sect. 6.8) whose sequential and unique evolution processes have been described in the previous section. An evolution is achieved in the decision space such that points at any stage will get better (smaller target function values) to the next solution candidate points. The individuals of this evolution are the genetic number system as already explained in the first sections of this chapter.

6.9.1 *Benefits and Consequences of Genetic Algorithm (GA)*

There are different subsets representing biological development within the GAs used to model highly complex target functions. In GA, a population of different individuals (chromosomes) is allowed first. This ensemble changes the goal function with certain rules so that it becomes even more optimizing. While some chromosomes die and leave the process, others continue their lives even more healthy. Some of the benefits in solving optimization problems with the genetic algorithm are as follows.

1. Optimization can be done with discrete or continuous variables,
2. There is no need for derivatives,
3. In the solution space, research is started from many points in a wide area at the same time,
4. Optimization operations can be done with many variables,
5. It is very suitable for parallel calculations,
6. It can do optimization even in the case of target functions with too many extreme (largest and smallest, extreme) values,
7. Can jump over local minima,
8. It can give not only the absolute best solution, but even a list of the best solutions,
9. Makes optimization in the coding world by coding decision variables,
10. It works with numbers produced according to the genetic number system. These can be experimental data or analytical functions.

One should not conclude from here that GAs should be used in every optimization problem. Classical optimization methods, which are mathematical, should be preferred in order to reach fast and simple solutions in the presence of smooth target functions and few variables.

The GA or evolution algorithm is a method inspired by the theory of evolution in order to find the most appropriate solution to an optimization problem as soon as possible. The problem is tried to approach the most suitable solution by using the evolution processes between the decision variables encoded with the genetic number sequences and the solution options. The GA solution algorithm is very different from the classical optimization methods. GAs are based on random sampling and are therefore a non-deterministic method. Uncertainty methods are used in GAs and there is no clarity in the operation of the algorithm. Here, one can define uncertainty as probability, game of chance or stochastic. As a natural consequence of this, when the same GA model is used at different times for the same problem, it may give slightly different results. However, linear, dynamic, or curvilinear (non-linear) formal optimization methods are all obvious approaches. There is no randomness in their operation, and their conclusions are always the same under the same problems and conditions.

One of the drawbacks of GAs is that the best solution is relative among known solutions. It may not allow checking whether the solution reached is the best. For this reason, GAs is best used when it is not known exactly what the solution might be. Naturally, evolution (genetic) algorithms do not know when to stop for the best solution. A stopping criterion must be set. One understands that the GA decision space will be scanned completely randomly, although in classical optimization methods a direction or systematic trail must be followed. Here, one can list some of the advantages of this random optimization research method over others.

1. While GA evaluates at many points simultaneously and in parallel in the decision space, this situation appears as a single point and direction in classical methods,
2. GA operations never require calculations such as mathematical derivatives and integrals, and all calculations are based on arithmetic operations. During the scanning of the decision space, the only important effects are the target function and the resulting fitness,
3. Probability and stochastic transitions are used during GA operations, not deterministic,
4. In GA, variables can have their binary coded equivalents instead of their actual decimal values. However, it has been developed in GAs working with decimal values.

The GA method yields too many viable solutions to a given problem. It is up to the discretion of the researcher to choose the most suitable one among them. It is recommended to use the GA approach instead of the methods that can be stuck at local optimization points.

The simplest GA mechanism was presented by Goldberg (1989). The general structure of this is given briefly with the flow chart in Fig. 6.27. In this diagram, the symbol t is used to express the change of population over time. The initial population is P(0), and the population at time t is P(t).

In summary, there are the following steps in a GA.

1. Representation of possible solutions of the problem with genetic numbers,
2. The generation of a new population by changing the initial solution population,

Fig. 6.27 Simple GA steps

```
GA  steps
Start
        t = 0;
        Satart P(t);
        Evaluate P(t);
        In the mean time
        Start
                t = t+1;
                Select P(t) from P(t – 1);
                Renew couples from P(t);
                Evaluate P(t);
        End
End
```

3. A function that determines the target of the problem and an evaluation of fitness to be obtained from it,
4. A set of genetic operations (crossover, mutation, etc.) to improve the structure of the solutions,
5. Calculation of variable values used in GA,

Many large-scale optimization problems can be solved with GA in approximately a short time. GAs is a stochastic technique that performs natural genetic inheritance behaviors theoretically and artificially on computers. GAs are different from classical optimization methods in that they collect directed and stochastic search features. These algorithms start from a point and do not follow a trail until they find the best solution in the decision space. Instead, they enter the decision space randomly from many points, and the decision points develop by following the rules of the theory of evolution among themselves, generating new populations that have more fitness, that is, approaching the solution, and reach the best solution. Each population is a form of the previous one, according to the theory of evolution. The closer the decision point in any population is to a solution, the more vigorous it is and the more likely it is to live accordingly. Thus, that decision point (chromosome) can continue its life in later populations. According to the steps given in Fig. 6.27, at one iteration time, GA has a population containing candidates such as T(t) that can be a solution. The fitness value is calculated for each chromosome. The fitness values are a measure of whether the chromosomes can survive in the next evolution of society.

6.9.2 *Definition of GAs*

Inspired by the biological evolution theory, GA methods seek the best solution by bombarding the solution area stochastically, that is, randomly. In order to reach the solution, first a collection of points is taken randomly in the decision variable space, then, by making matches between these points in the light of the rules to be shown, some members of the population disappear, and new ones come in their place. By joining the newcomers to the population, it is ensured that that population is healthier than before, that is, closer to the goal. With the application of the necessary genetic processes among the members of this population, a new population is obtained, so to speak, that is more fit, that is, closer to the goal. Thus, the target is approached from many directions and short paths, not along one direction or path as explained in the previous sections. Each member of this population is represented by a coded genetic number sequence. In the first sections of this article (Sect. 6.3), these sequences of numbers are called chromosomes. Since GAs generally work with binary (0 and 1) number systems, their space is called binary and people's is called decimal space system. Although the coding of GAs is done according to the binary number system, it can also be done in the decimal number system, as will be explained in Sect. 6.4.1.

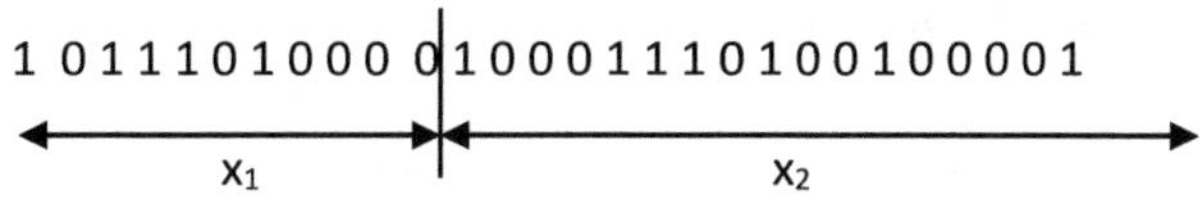

Fig. 6.28 A chromosome structure

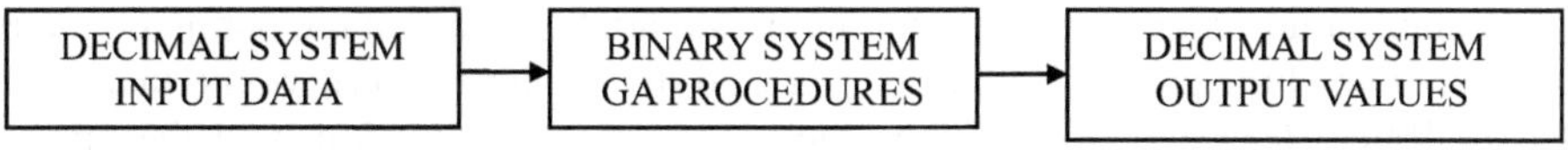

Fig. 6.29 Decimal-binary systems and GA

However, the binary system is mostly used due to its ease of calculations. If there are two variables such as x_1 and x_2 while solving a problem, their chromosome structure in the base-two system is given in Fig. 6.28.

In this chromosome, x_1 and x_2 are represented by 10 and 18 digits, respectively. These digits, which are the smallest building blocks of chromosomes, are also called "bits." The sequence of numbers 0 and 1 representing each of the two decision variables in the chromosome in Fig. 6.28 is also called a gene, like biology studies. Chromosomes are made up of genes. The chromosome given above makes no sense on its own. In Fig. 6.28, the chromosome is given in binary number system. For this number to be understood by humans, the values of the x_1 and x_2 variables in the chromosome must be converted to the decimal system, $x_1 = 744$ and $x_2 = 729993$. During GA studies, the inputs and outputs are in the decimal GA, while the inner workings of the GA are in the binary number system space (see Fig. 6.29).

Chromosomes containing encoded numbers are used during GA operations. After the decision variables are coded in this way by means of chromosomes, the fitness of each chromosome should be calculated separately in the population. There should be a target function to calculate the fitness of the members. The target function is chosen according to the type of problem at hand. Chromosomes that are members of the population according to the goal function criteria either continue their lives or they must leave that population. The most important feature of the target function is that it helps to select members that will cause more vigorous members to occur in the population by matching.

At each renewal stage, the chromosomes in the population receive degrees of fitness according to their target function before doing any work. These grades are then used for the reproduction of more vigorous chromosomes. Those with high fitness are more likely to remain in population for many generations, but those with low fitness are less likely to stay in that population. The fitness of each member is expressed as a relative ratio in the fitness of the whole population.

They decide that some members will stay in the future population, because of their strong genes by GA processes and making operations on the genes of the chromosomes. They are prepared for the next population by performing GA operations (evolution) between two or more genetic number pairs.

6.9.3 Binary GA

Figure 6.30 summarizes the similarities between biological and binary number evolutions. It starts with startup populations that both have random members. Let it be assumed that each of the binary numbers shown on the left corresponds to a population of horses. If the best neighing horses are to remain in the population and the rest to be eliminated, this appears as a kind of optimization criterion (Haupt and Haupt 1998). These horses continue their lives in the population with the binary numbers that are assumed to be related. Two of this population are randomly selected and exchanged among them, resulting in two new horses. It is possible that these horses can be expected to neigh even better.

Their parents were of the kind that can neigh well. Each of the new population inherits some of the characteristics of the mother and father. Newly born horses that neigh better exclude two from the population that neigh worse. The same number of new horses is assumed to occur so that the population remains the same size (same

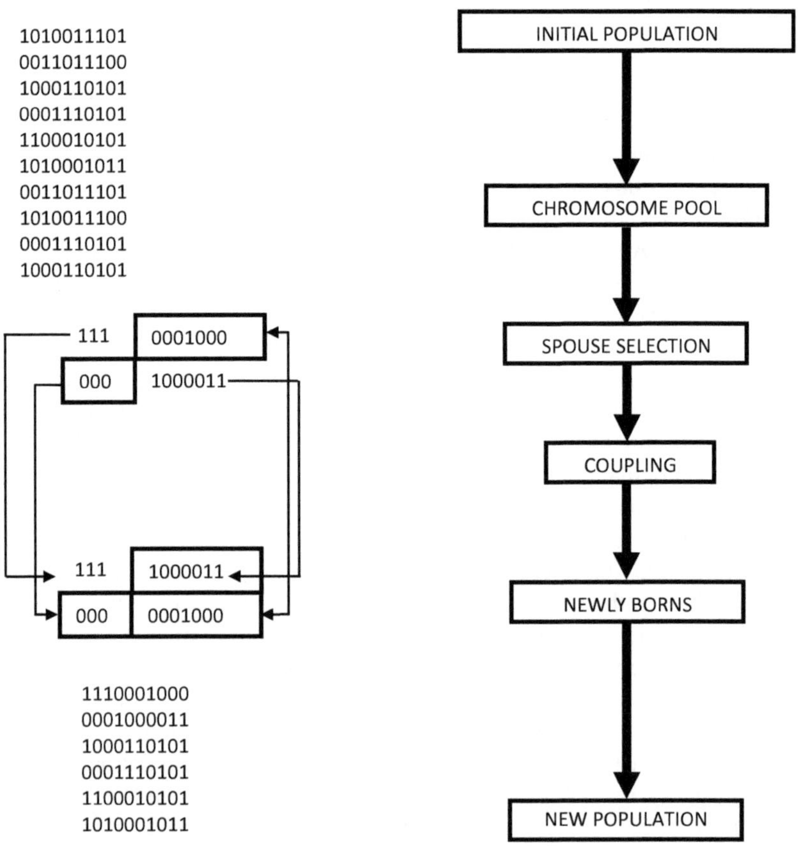

Fig. 6.30 Numerical and biological GA similarities

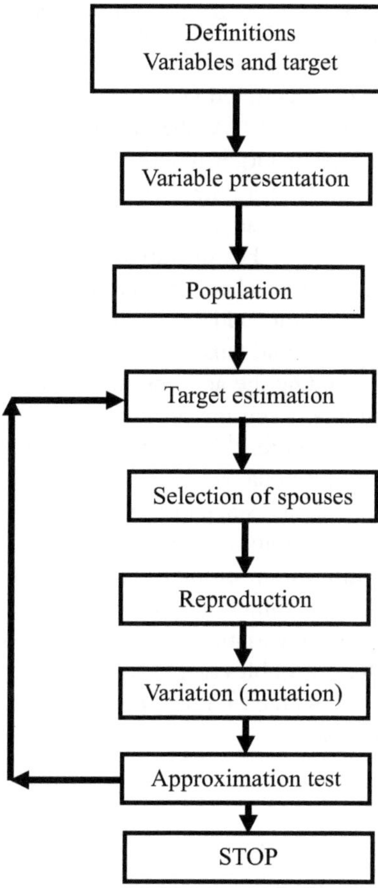

Fig. 6.31 Binary number system-based GA flow diagram

number of chromosomes). If this regeneration process is continued successively, a much better neighing horse population can be obtained.

As in other optimization methods, optimization analysis in GAs starts with the replacement of the selected decision variables in the target function. The processes are continued in this way, and finally, the best solution is terminated if its approximation remains within a fixed error limit. Apart from these, GAs is very different from classical optimization methods. In Fig. 6.31, the current steps to perform GA optimization on a computer are shown in flow. Topics within each box will be explained in detail as parts of this chapter.

The target function is a complex surface with peaks, valleys, ridges, and troughs in the space determined by the variables in general. In the three-parameter case, this is like a topographic map. The optimization process is solved by finding one of the most economical of the valleys. Optimization methods to find peaks deal with the maximization of the target function. For example, finding the summit of Mount Ararat, the highest peak in Turkey, is similar.

6.9.4 Selection of Variables and Target Function

The target function generates an output surface by incorporating the variables of the problem. In GA, the variants correspond to the genes on the chromosomes (Sect. 6.1). The target function can be a mathematical expression, as well as the results obtained from the experiments or surfaces consisting of systematic rules. The goal here is to change the variables in such a way that the optimization point can be found at the end. For example, when filling the tub with water before taking a bath, the target is reached with the optimization of the difference between the actual water temperature and the desired temperature. Here, the question is, how much should the hot and cold taps be opened variably? In this case, experimental results are obtained by finding the appropriate temperature for the feelings. There is a constant relationship between the opening amounts of the taps and the achievement of the desired temperature. When the desired temperature changes then the opening amount of the taps also changes.

The first stage of GAs starts with the generation of the chromosome representing the variable sequence for which the best solution values are desired. It is an optimization problem to try to obtain the values that make the target function of these variables, h ( ), the best. In Sect. 6.1, it is explained that the value of each of the decision variables of a triangle is called a gene. Provided that a problem has d_1, d_2, . . d_{Ngen} variables (gens), a chromosome is written as a sequence with as many elements as C, with each one consisting of 0 and 1, and hence chromosome can be written as follows.

$$C = \left[d_1, d_2, \ldots, p_{N_{gen}}\right] \tag{6.24}$$

For example, if the goal is to find the highest peak in an area, then there will be two variables, one for latitude, Lat, and the other for longitude, Lon, takes the form (Ngen = 2)

$$C = [\mathrm{Lat}, \mathrm{Log}] \tag{6.25}$$

There is a target value, T, which corresponds to each chromosome and is found from its target function. The definition of this value is given by the following equation.

$$T = t(C) = f\left(d_1, d_2, \ldots, d_{Ngen}\right) \tag{6.26}$$

If one wants to find the highest peak in a locality, the problem is reduced to a minimization by writing the elevations, y, as negative values. For this, briefly one can write as,

$$t(\text{Lon}, \text{Lat}) = -y \tag{6.27}$$

In general, the target function gets smaller and smaller values, just as when maximizing the distance, a car travels relative to the fuel it burns. In this case, the model maker must decide which of the expert variables has the highest priority. Too many variables prevent GA from working well. Sometimes considering the most necessary variables can be found with expert judgment or a few trial and error calculations. Almost all the optimization problems have lower and upper limits of the variables. The dimensions of the case under investigation should be well decided. If a horse-size dimension is expected, it should be known that it will not exceed 2–3 m. It is very beneficial for the person to have realistic predictions about almost all the variables and values in the problem to be optimized.

Some variants of optimization problems may be interdependent. For example, a person's height and weight are generally linked. Similarly, runoff amounts depend on precipitation. Here, classical minimization algorithms described in the previous section should be used rather than GA in optimization problems whose variables are interdependent. GAs is mostly used in solving optimization problems with moderate or weak dependent variables. If the dependency is too great, pure random optimization analyzes are appropriate.

The most difficult step in an optimization problem is the determination of the target function. The question here is the goal one? or more? will depend on the variable. For example, in the manufacture of a car, is the formula t = (distance traveled)/(fuel expended) the best target function? Otherwise, the overall total value of the target, if the investment cost of car manufacturing, i, the operating cost, c, depends on the car weight, w, and its volume, v (Haupt and Haupt 1998) as,

$$t(w, v) = i + c \tag{6.28}$$

If both investment and operating costs are significant, working with standard costs is appropriate in optimization problems. Considering the smallest (i_{min}, c_{min}) and largest (c_{max}, c_{max}) values of the variables in standard cost, t, is defined as,

$$t = (i - i_{min})/(i_{max} - i_{min}) + (c - c_{min})/(c_{max} - c_{min}) \tag{6.29}$$

This standard cost now equals a number between 0 and 2. In general, standard costs include investment weight a_p and operation ac should be considered. Here, $a_i + a_c = 1$.

$$h = a_p(i - i_{min})/(i_{max} - i_{min}) + a_c(c - c_{min})/(c_{max} - c_{min}) \tag{6.30}$$

Sometimes the target function can be extremely complex and time-consuming to calculate. For this reason, it is necessary to try to make solutions in such a way that there is little need for calculation. Since the target function calculations are based on chromosomes, identical chromosomes should be avoided. In order to avoid twin chromosomes, first, identical chromosomes should be eliminated during the establishment of the initial population. Identical chromosomes are common in base-number GAs. Such a situation is rare with decimal GAs, which will be explained in Sect. 6.4. To ensure that each chromosome is different, let's consider the following 8 chromosomes. If the first three digits are desired to be different from each other, since the number of combinations is $2^3 = 8$, it is possible to arrange the remaining 5 digits completely randomly, after ensuring that the first three digits are different from each other by combination calculation deterministically. Even if the same sequences occur by chance in this random ordering, the chromosomes are different from each other because the first three digits are different from each other.

000101010
001101010
010010101
011001100
100110011
101000111
110111000
111011011

By ensuring that the first three digits of these are different from each other, all the chromosomes in the starting population are different from each other. Here, it is ensured that only the first genes differ from each other. For it to differ from one another in its other genes, the point is that if a chromosome consists of 3 genes with three digits in each, this time in $2^3 = 8$ different sequences, the first digit of each gene can be adjusted to be different so that all chromosomes are completely different from each other. Here, since the first digit of each gene is desired to be strictly different from each other, the following chromosome structures are obtained. The first digit of each strictly set gene is shown in italics.

*1*01*1*10*1*10
*1*01*1*10*0*10
*1*01*0*10*1*10
*1*01*0*10*0*10
*0*01*1*10*1*10
*0*01*1*10*0*10
*0*01*0*10*1*10
*0*01*0*10*0*10

Thus, two or more calculations of the same chromosomes in the initial population were prevented in the target function calculation. This means that it is not guaranteed that the same ones will appear in later population chromosomes.

In order not to recalculate the target function with the same chromosomes, target calculations are made with the new chromosomes that appear in the next stages and the target values of the unchanged chromosomes remain the same. This means that as new chromosomes emerge, they choose the way of calculating their target values.

This approach is applied in situations, where the calculation of the target function is not difficult and time-consuming.

Another approach is to know that the newly emerged chromosome is different from the previous chromosomes. If target value calculations take more time than searching for identical chromosomes in the population, the chromosomes should first be checked to see if they are the same, otherwise, this path is not followed. Only the different chromosomes that appear at any stage of the GA calculations should be calculated on target.

Another way to simplify cost calculations is to simplify the target function itself. For example, target and GA calculations are made on a coarsely spaced network, and precision calculations are made by increasing the resolution of the network with the information obtained from this. In the first step, a coarsely meshed target function and chromosome calculations are performed. Calculations are continued by increasing the resolution as the next steps are passed. Considering a rough mesh first, the first indications of more precise mesh-spaced solutions are obtained with less computations in a short time. For example, while a 100×100 solution area is used in the first stages, a solution can be made by gradually considering 500×500 networks in the same solution area in later stages.

There are 4 different values as parameters in the structure of GAs. The first of these is the number of options that will generate the population. The number of genes in each option must then be determined. In addition to these, these four parameters are completed together with the crossover and mutation probability values. Generally, high (such as 0.7) probabilities are chosen for crossover, but low (0.001) probabilities for digit switching.

There is also an environment, where the GA event occurs with the vigor of each option. It should be known or stated that at least some options have superiority over others, even verbally. The necessary steps in the operation of the GA can be explained as follows.

1. During each iteration (circuit) choose n random options called chromosomes,
2. Calculate the vigor of each of them,
3. Randomly select two of the options with the roulette wheel,
4. By matching these two options among themselves with the crossover probability, two new options emerge,
5. Apply the digit change operation with the probability of digit change to each of these new options,
6. Replace the old population with a population emerging from new options.

6.9.5 Target Function and Vigor Measurement

A target function, t(x), is necessary in order to indicate how close each of the population chromosomes circulating at different points in the decision space is to the solution sought. It is desirable to reach the small or large values of the target function according to the direction of the optimization, that is, whether it is minimization or maximization. In order to calculate the relative vigor of each chromosome (member) in the GA population, their absolute (only plus value) vigor values, that is, the values they take in the target function are needed. It is necessary to somehow obtain a function, v(x), which will give the degrees of vigor from the target function. The closed relationship between these two functions; one can show it as,

$$v(x) = F[t(x)] \tag{6.31}$$

Here, the F transformation should be such that the values of the function giving the vigor degrees must be positive for all decision variables. The explicit form of this last statement, with $v(x_i)$ being the vigor value of the i-th decision variable, can be calculated as follows.

$$v(x_i) = \frac{t(x_i)}{\sum_{j=1}^{N_U} t(x_j)} \tag{6.32}$$

The fact that all the target function values have plus signs, the degree of vigor gives the percentage (probability) of that decision variable in the population. This can be called relative vigor. The biggest drawback of this definition is that it cannot consider the minus sign values of the target function. In order to eliminate the shortcomings that may arise in the target function, the explicit expression of another vigor assignment is linear, given in the following form.

$$v(x) = at(x) + b \tag{6.33}$$

Here, a and b are the scaling and translation coefficients, respectively. The sign of a is taken as minus in case of minimization of optimization and plus in case of maximization. However, the value of b is to ensure that the vigor coefficients that will emerge are not negative. The operation in Eq. (6.33) is also called linear scaling.

Another proposed approach is to give the target function values of the vigor degrees according to the order in which they are arranged from the smallest to the largest.

6.9.6 *Representation of Variables*

Binary GAs operates in a very large but finite solution space. With this feature, GA analysis is very useful when the variables take a finite number of values. If a variable is continuous, it must be discretized before GA applications. For this, the range of variation is divided into a finite number of sub-intervals. The value falling on any of these ranges is considered the midpoint of the range. The division here is like obtaining classes in histogram studies (Chap. 4). This type of study is the most appropriate in optimization methods. The amount of error that may occur is also the largest. This minimization of the biggest errors ensures that the smaller ones are even smaller. It has already been said that the sequence of digits representing each variable in a binary system is called a gene (Sect. 6.1). Mathematical formula for binary number encoding or back-coding of a given d_n variable is achieved according to the following formulation.

$$d_{norm} = (d_n - d_{min}) / (d_{max} - d_{min}) \tag{6.34}$$

Here, d_{min} and d_{max} are the smallest and largest values of the variable values (genes), respectively. This last statement is also called the standardization operation. The given variable value is thus standardized to fall between 0 and 1. To find the binary opposite of this value in the binary number system the following steps must be followed.

1. First, it is decided how many digits the binary number system will work with. In the case of sticking to a system, binary numbers are $2, 2^2, 2^3, \ldots, 2^n$ digits. Here, as the upper number increases, it is worked with a more detailed and healthy number system. GA applications usually work with $M = 2^3 = 8$ or $M = 2^4 = 16$-digit numbers. The number of digits does not necessarily have to be a coefficient of 2.
2. Since the interval of change of the standard variable obtained from Eq. (6.34) is 1, its division by M and the length of the interval corresponding to each digit of the binary number system is found as,

$$\Delta a = \frac{1}{M} \tag{6.35}$$

3. The lower and upper limit values of the standard variable sub-ranges are calculated and written in the first column of Table 6.3. The first range limits are 0 – 1/M, the second 1/M – 2/M, the third 2/M – 3/M, and the M-th (1 – 1/M) – M. Results for M = 8 are given in Table 6.3.

Table 6.3 Converter between decimal and binary numbers

Class boundaries	Interval sequence	Binary number
0 – 1/8	0	000
1/8 – 2/8	1	001
2/8 – 3/8	2	010
3/8 – 4/8	3	011
4/8 – 5/8	4	100
5/8 – 6/8	5	101
6/8 – 7/8	6	110
7/8 – 1	7	111

4. If one writes the others in order to give the number 0 to the first range, 7 range numbers, (M – 1), are found in the last range. These values are written in the second column of the table. Thus, there is mutual agreement between the decimal numbers in the first column and the decimal numbers in the second column.
5. The third column includes the binary system equivalents of the decimal numbers in the second column.
6. The first and third columns show each other what the standard data values correspond to in the binary number system. Thus, all standard values are represented by binary numbers.
7. There are many values falling within the corresponding range in the first column to the binary number in the third column. In other words, there is no one-to-one agreement. To resolve this, it should be assumed that either the upper limit, middle value, or lower limit of the range corresponds to the binary number in the first column. In practical studies, this agreement is the middle value.
8. By accepting the median value, it is understood that the GA results working with the binary system are only an approximation of the decimal number system.
9. As the number of digits M in the binary number system increases, the approximation specified in the previous step decreases, but the calculation volume increases greatly.

GA generally works with a binary number system, but the target function requires a decimal number system. For example, for a GA optimization analysis, considering that there are three different variables (Ngen = 3) and each of them has 10 digits in the binary number system (N digit = 10), and in the appropriate chromosome becomes

C = [**1111001001***0011011111*..........0000101010]

The string of bold digits is gen1, the string of italic digits is gen2, etc., and the last 10-digit string indicates N_{gen}. In such a chromosome, there are binary numbers 0 or 1, which are called "digits," up to $N_{bit}N_{gen} = 10N_{gen}$.

To find the location of the highest peak in a place, take 128 × 128 elevation points. The latitude, Lat, and longitude, Lon, variable values required to find the peak here appear as two genes. If one considers each of these as N digits = 7 digits, different values can be considered thanks to 2^7 combinations for each of the Lat and Lon values. Accordingly, in the chromosome to be taken randomly, there are 7 digits for each variable binary numbers and 14 digits in the chromosome. For example,

C = [**1110001***1000110*]

The first seven of them are Lat and the remainder is the representation of Lon variable value (gen) in binary number system.

6.9.7 Initial Population

As said earlier, GAs start with at least around 15–20, and preferably even more, chromosomes, and this represents the initial population. If N_{cho} of chromosomes are found in this population, a matrix of size $N_{chr}xN_{gen}xN_{bit}$ is written, consisting of elements 0 and 1, (see Sect. 6.3). If this is called the startup population, SPOP, matrix is obtained.

$$\mathrm{SPOP} = \mathrm{Round}[\mathrm{Random}(N_{chr}, N_{bit})] \tag{6.36}$$

Here, the "Random (N_{chr},N_{bit})" command generates a matrix of size $N_{chr}xN_{bit}$ with elements uniformly spread between 0 and 1. Such standard random number generation functions are available in any computer software. The "Round[ ]" command here rounds the resulting random numbers to the nearest integer (0 or 1). Each row of this matrix corresponds to a chromosome.

Then the target value is calculated by placing the variables in the target function. The large solution space allows for nice random sampling in GA. In Table 6.4, the

Table 6.4 Initial population chromosomes

Sequence number	Chromosomes	Target values	Rank	Vigor degrees
1	10111000101010	−11818	6	0.059508
2	00111000101010	−3626	21	0.018258
3	10111000101011	−11819	5	0.059513
4	10111001101010	−11882	4	0.059830
5	10101000101010	−10793	7	0.054347
6	11111000101010	−15924	2	0.080183
7	01110000100010	−7202	17	0.036265
8	01010000100010	−5154	19	0.025952
9	01110000101010	−7210	11	0.036305
10	01110110100010	−7586	12	0.038198
11	01110100100010	−7458	15	0.037554
12	01000110100010	−4514	20	0.022730
13	00110110100010	−3490	22	0.017573
14	01110010101010	−7338	16	0.036950
15	01100100100010	−6434	18	0.032398
16	01110110100010	−7586	14	0.038198
17	01110110101110	−7598	13	0.038259
18	01111000110010	−7730	9	0.038923
19	11111110101010	−16298	1	0.082067
20	01110101011101	−7517	10	0.037851
21	00101110001010	−2954	23	0.014874
22	00101011001010	−2762	24	0.013908
23	10000110101010	−8618	8	0.043395
24	11101110110100	−15284	3	0.078961

initial population of the chromosomes consisting of the latitude and longitude genes mentioned in the previous section is randomly shown with $N_{chr} = 24$ target values. When each gene has 7 digits, there are $2 \times 7 = 14$ digits in the chromosomes since there are two decision variables. In the 4th column of the same table, the order of the target values (considering the order from smallest to largest) and the vigor degrees are calculated according to Eq. (6.32) and shown in the last column.

GAs work with many decision points that can be a solution at the same time, all these decision points are coded according to the binary number system and form the population. The question that should be asked at first is how many chromosomes (solution alternatives) should there be in the initial population? The answer to this is only given by the observations as a result of the studies, but it is appropriate to take at least 10 – 15 chromosomes at the beginning. One can also increase this to 50 – 60. The reader who wants to learn GA with a real understanding should try to have an idea about how many chromosomes will need to be taken by trying to solve the same problem with simple optimization examples and different population numbers.

In each GA preparation, each of the decision variables of the problem is considered as a string of binary numbers. The chromosome containing the decision variables is obtained by arranging the whole array of decision variables one after the other.

The next step is how to choose the numbers (0 or 1) that will be put into the chromosomes that make up the population. For this, completely independent uniform random number generation methods are used. Since the number of chromosomes in the initial population is N_{chr}, it is understood that there is a need for as many digits as $N_{gen}xN_{bit}xN_{btop}$ in the whole population. Here, uniform and completely random number generators are used to fill the digit positions in this number with the numbers 0 or 1. If there are no clues about the solution of the problem, it is appropriate to initially bombard the decision space completely randomly. If there are clues about the approximate location of the desired optimization point, it is appropriate to randomly select the chromosome values around such a solution point, but in a narrower decision space.

There is no direction, even partial, that should be followed during GA work. Points move from their positions to their new positions by random evolutionary processes. With the establishment of the initial population, there are three consecutive GA operations based on probability principles in the later stages. The first of these is selection from the population, and the latter are the crossover and mutation (number-changing) operations to further strengthen the chromosomes in that population. Selection is the process of choosing the ones that can go even better in the next stage among the chromosomes that are solution options in the population. During this selection process, care should be taken to select chromosomes with high vigor. A person with a high degree of vigor is more likely to play a role in the next population. Since in this type of selection work, the one with the highest degree of vigor can get the largest share of the cake, elections are then made by comparing the roulette wheel described in Sect. 6.1 to a cake.

Although there is one best solution, which we can call the only one, in all the classical methods, there is a population that is ready to be the solution in GAs. Only one of them is the best; others are likely to be the best in their decision variable region. Since GAs are approached with a population to the best solution, it is not possible for them to be stuck in local best solution points as in classical methods.

6.9.8 Selection Process

The purpose of this is to select the chromosomes to be considered in order to be subjected to GA processing in later populations within the current population. The best solution may be missed due to fast approaches in obtaining vigor degrees with linear scaling given in Eq. (6.33). In this selection, a chromosome can be replaced with a decimal number many times, but it must be converted to an integer. Chromosomes that can enter the next population directly or after exposure to at least one of the GA treatments should be randomly selected from the population. In order to make this type selection, it can be used with the different method described previously in Sect. 6.4.1.

The solution points (chromosomes) that will constitute the next population that will evolve in the form of further improvement between the population decision points are randomly selected according to their vigor (approximately to the best solution). Such a selection process ensures that population evolves and develops for the better.

The population, which can be defined as the gene pool, is the pool where the suitability of the chromosomes is evaluated, and the selection process is made. In binary GAs, 0 or 1 represents a digit, their mixed sequences represent a gene, and the sequence of genes represents chromosomes, (Sect. 6.4.1). For example, if the chromosomes 00111, 11100, and 01010 have the target function values t(00111) = 0.1, t(11100) = 0.9, and t(01010) = 0.5, respectively, the vigor values are calculated according to the Eq. (6.32) as v(00111) = 0.0667, v(11100) = 0.6000, and v(01100) = 0.3333. During the GA process, chromosomes with high vigor values are selected and copied onto chromosomes with low vigor. In the next steps, the chromosomes are subjected to crossover and mutation (digit change) processes as will be explained in the next sections.

The aim here is to eliminate the chromosomes with small vigor and to generate a new population by making the larger ones feel their weight. The selection process of chromosomes from a population can be done by rewriting the large chromosomes over the small ones and/or using the roulette wheel. Characteristics of chromosomes are determined according to their vigor values. In this way, the chromosomes in a population with N_{pop} are divided into two groups as N_{good} and N_{bad}. In practice, it is generally taken as $N_{good} \approx N_{bad} = N_{pop}/2$.

It is also useful to change the N_{good} number of chromosomes in GA studies. If one calls the crossover ratio R_c, to determine the number of chromosomes to be considered for this pairing the following expression is used.

Table 6.5 GA population

Sequence number	Chromosome	Target values	Rank	Vigor degree
19	11111110101010	−16298	1	0.082067
6	11111000101010	−15924	2	0.080183
24	11101110110100	−15284	3	0.078961
4	10111001101010	−11882	4	0.059830
3	10111000101011	−11819	5	0.059513
1	10111000101010	−11818	6	0.059508
5	10101000101010	−10793	7	0.054347
23	10000110101010	−8618	8	0.043395
18	01111000110010	−7730	9	0.038923
20	01110101011101	−7517	10	0.037851
9	01110000101010	−7210	11	0.036305
10	01110110100010	−7586	12	0.038198

$$N_{good} = R_c N_{pop} \tag{6.37}$$

If the crossover rate is high, it is not possible for the population chromosomes to accumulate in a single chromosome. On the other hand, it will not be of much use in examining the target surface at low rates.

Other than the criteria given above, methods not found in the literature can be developed and used. For this reason, do not perceive what is written even in this book as dull and stereotyped. The most suitable 12 of the initial population in Table 6.5 were selected by considering their vigor values are given in the table.

The first 6 of them form two equal parts of N_{pop}, with N_{good} behind but N_{bad}. For these newly formed GA population chromosomes to become even more vigorous by mating among themselves, crossover and digit swapping (mutation) GA operations should be applied with random selections. Here are a few paths that can be followed.

1. The first of these is to leave the N_{good} chromosomes in Table 6.5 as they are, and to randomly select the partners among the remaining N_{bad} ones,
2. To apply GA crossover and digit swapping operations by matching between all chromosomes without making such a distinction.

6.9.9 Selection of Spouses

In the previous section, the chromosomes selected are matched among themselves, and the chromosomes are paired in pairs in order to obtain N_{pop} healthy chromosomes in the new population. Two new chromosomes are obtained by matching two randomly selected chromosomes from a GA population. The mating continues until the bad chromosomes excluded in the population are replaced by more vigorous new ones. There are different pairing methods.

1. *Top-Down Spouse Selection*: In a population, matching is done starting from the top of the chromosomes arranged from top to bottom (Table 6.5), according to the vigor values, until the chromosomes come down. In this approach, C_{2i-1} and C_{2i} continue until i = 1, 2,. . . . Although this does not model nature well, it is preferred because it is useful and practical. Recommended for GA employees,
2. *Random Spouse Selection*: Here, uniform random values are used to select chromosomes. First, the chromosomes are sorted from 1 to N_{good} according to their vigor values (Table 6.5). Two random numbers are generated to select the first two chromosomes. This is to show the good chromosome number of producing N_{good}, is determined by the following rule.

   Spouse = Round up [N_{good}xRandom]

   Here, the meaning of the "round up" command is that the decimal number is taken equal to the first integer after it. For example, let's say there are 6 random numbers 0.1535, 0.6781, 0.0872, 0.1936, 0.7021, and 0.3933. Multiplying these by 6 and rounding up gives the chromosome selection numbers 1, 5,1, 2, 5, and 3. So, C_1 matches C_5, C_1 matches C_2, and C_5 matches C_3,
3. *Weighted Random Spouse Selection*: Pair selection is done by assigning probability values to a good number of chromosomes in the pool. The odds are calculated based on the target values of the chromosomes. The chromosome with the lowest target value (for minimization) has the greatest probability. A random number indicates which chromosome will be selected. Matching that works with this type of weighted average is commonly referred to as the "roulette wheel" approach (Sect. 6.4.1.2). Here, too, there are two different techniques. These are rank and target weights.

   (a) *Order Weights*: Regardless of the optimization problem, it ensures that the probabilities are found according to the order of these chromosomes, not according to the chromosome value are generally determined from the formula.

$$P_n = \frac{N_{good} - n + 1}{\sum_{i=1}^{N_{good}} i} \tag{6.38}$$

For example, if there are 6 chromosomes in a good population, the calculation is made from the formula $P_n = (6 - n + 1)/(1 + 2 + 3 + 4 + 5 + 6) = (7-n)/21$. Thus, the probability of each chromosome and the sequential addition of them is the cumulative probability of each. These consecutive sum probabilities must eventually reach the value 1. After that, a random number between 0 and 1 is generated on the computer. Here, the chromosome corresponding to the probability of successive chromosomes slightly larger than this number is selected for mating. Likewise, if there are 6 chromosomes and the generated 6 random numbers are 0.1535, 0.6781, 0.0872, 0.1936, 0.7021, and 0.3933, it is understood that C_1 must match C_3,

C_1 must match C_1, and C_3 must match C_2. If a chromosome matches itself, there are different options. First, it means that if we let it go, the next generation will have three chromosomes. Second, we can randomly pick another chromosome,

(b) *Target Weights*: Here probabilities are calculated from target values, not chromosome levels. For each chromosome, the lowest target value of the excluded chromosomes, $t_{Ngood+1}$, (i.e., the lowest target value among the N_{bad} chromosomes) is subtracted from this chromosome target value, T_i, the standard target value is found.

$$T_i = t_i - t_{Ngood+1} \tag{6.39}$$

All standard targets are marked with a plus sign. The probability is calculated as,

$$P_i = \left| \frac{T_i}{\sum_{j=1}^{N_{good}} T_j} \right| \tag{6.40}$$

This method assigns greater probability to the top chromosomes if there is a large difference between the vigor of the top and bottom chromosomes. In Table 6.5, $tN_{good+1} = -10793$. On the other hand, if the chromosomes have similar vigor, the probability values are quite uniform. Using 6 random numbers as before, K1, K3; K1 K2; and in K3, K1 spouses are selected,

4. *Tourniquet Spouse Selection*: First, a small number of chromosomes are randomly selected among the N_{good} chromosomes, and the first among them with the lowest vigor is selected. Similarly, other spouses are chosen. Considering 6 tourniquets between two randomly selected chromosomes, the most suitable chromosomes are selected for mating. With the application of the turnstile selection method, for example, C_1 is paired with C_3, C_3 is paired with C_4, and C_1 is paired with C_4.

6.9.10 Crossing

The emergence of new products with the merging of spouses is called the crossover stage of GA. Here, the target surface is first investigated in GA. This is called the discovery phase because GA still uses chromosome values. Provided that the chromosome number remains the same, GA approaches the solution of optimization.

During the solution search, two new solution points (decision points) that are better than the previous ones, that is, closer to the optimal solution, are revealed by a cross between two different decision space points that are candidates for the solution.

These two new resolution points are inherited from the previous points. Thus, the decision points in the solution society evolve for the better, creating a new population. Here, the numerical values of the two solution points cross some digits between them, causing new solution points to emerge. This is accomplished in a similar way to the crossover in the DNA of biological organisms. In order to further increase the vigor of the chromosomes, there is also another option such as a mutation (number change).

It is a process that allows chromosomes to exchange genes with each other. First, the chromosome pairs to be crossed over are randomly selected. Then, the genes from which these chromosome pairs will be cut are again determined by random selection. Crossing genes are exchanged between chromosome pairs. The aim here is to obtain different populations (generations) from the existing population. In this process, the chromosome pairs to be crossed and the genes from which they will be crossed are determined randomly. Different operations in this regard are made by the following four operations. The decimal numerical values represented by each chromosome are shown on the right side of the arrows.

6.9.10.1 Single-Cut Crossover

Here, two matching chromosomes give birth to twins, keeping the chromosome number the same. First, a crossover point is randomly determined somewhere between the first and last digits of the two chromosomes. All digits to the left of the crossover point of one chromosome are replaced with the other chromosome as they are. First, all digits to the left of the cross point on one of the first two parent chromosomes pass on to the new child's chromosome. Similarly, those to the left of the crossover point of the other chromosome pass on to the other newborn chromosome. The digits to the right of the first primary chromosome crossover point then move to the right of the previous digits of the second newborn. Thus, the chromosome of the first newborn is complete. Similarly, digits to the right of the crossover point of the second primary chromosome are also placed to the right of the first newborn. Thus, the two newborn chromosomes partially contain the digits of the previous ones. The main chromosomes should cause a total of N bad new births so that the population size remains constant (N_{pop} = constant). These are called simple or single crossover point deliveries. In this process, chromosomes are cut from one place at random and then replaced with related genes.

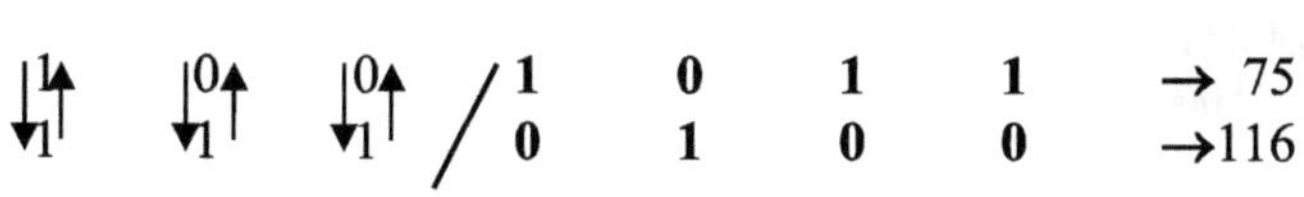

The chromosome pair above was cut right after the 3rd digit and the dark or light genes were replaced to obtain the following new chromosomes. As can be understood from the explanations above, a number corresponds to each chromosome, and these numbers change with the change of even one of the genes in the chromosomes.

1 0 0 0 1 0 0 →68
1 1 1 1 0 1 1 →123

With a single cut crossover, only two new chromosomes can be obtained that are different from each other.

6.9.10.2 Double-Cut Crossover

The same process as before is repeated, with the cut made in two places along the chromosomes.

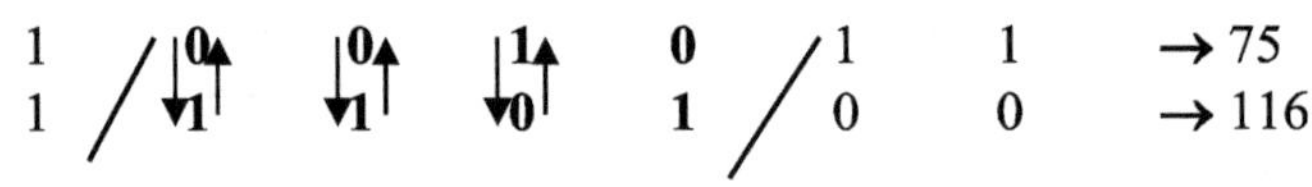

Here, the chromosome pairs are cut from two different places and crossover is applied. Dark genes have been crossed over.

1 1 1 0 1 1 1 →119
1 0 0 1 0 0 0 →72

As can be seen, this time, the numbers they represent have changed again as a result of the newly shaped chromosomes. In two-point crossover, each chromosome is divided into three parts. From these parts, 6 different chromosomes are obtained by crossing over both. Three parts of the chromosomes exposed to the two-point crossover are shown below: the straight, italic, and bold parts.

000*010111***000110**
0001*11100*0**01001**

Crossing over between the plain, italic, and bold parts, respectively, the different chromosomes between the straight parts.

001*0101110***00110**
000*1111000***01001**

Between the italic parts

000*1111000***00110**
001*0101110***01001**

and finally, between the bold parts there are 6 in shape.

000*0101110***01001**
001*1111000***00110**

It is not necessary to use all the new chromosomes. In this selection, either the plain, italic, or bold parts can be replaced and as many of the 6 new chromosomes as desired can be randomly selected to participate in the population.

6.9.10.3 Multi-section Crossover

It is achieved by replacing the genes by randomly cutting the chromosomes in more than two places diagonally.

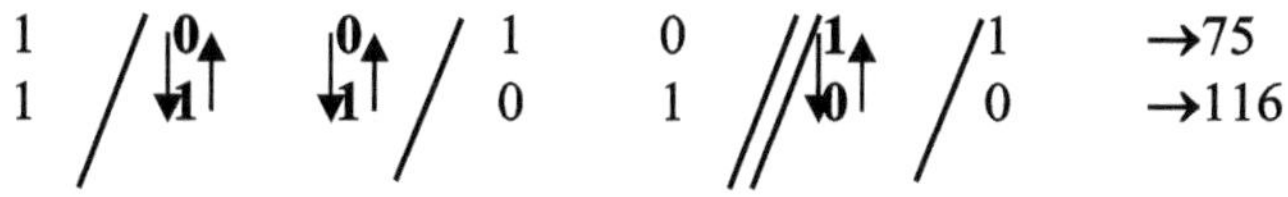

As a result of this crossover, the new chromosomes and numbers are as follows.

$$\begin{array}{ccccccccc} 1 & 1 & 1 & 1 & 0 & 0 & 1 & \rightarrow 121 \\ 1 & 0 & 0 & 0 & 1 & 1 & 0 & \rightarrow 70 \end{array}$$

Here, too, many new chromosomes are obtained by crossing over between opposing parts. If n represents the number of fragments, n^2 new chromosomes can be obtained.

6.9.10.4 Uniform Crossover

The essence of this process is the swapping of random digits between two chromosomes.

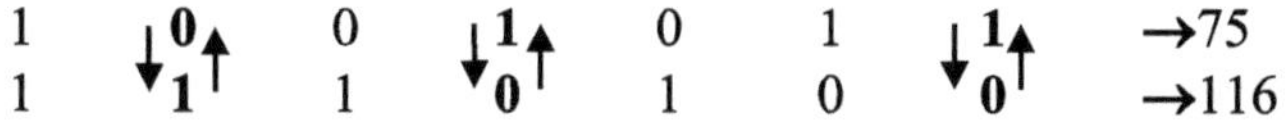

Here, a coin is tossed for each digit. For example, if heads come then crossing applied, in cases of tail crossing is not applied. For the dark-colored genes, crossover is applied for these genes and the new shapes of the chromosomes are as follows.

$$\begin{array}{ccccccccc} 1 & 1 & 0 & 0 & 0 & 1 & 0 & \rightarrow 100 \\ 1 & 0 & 1 & 1 & 1 & 0 & 1 & \rightarrow 93 \end{array}$$

In uniform crossover, the first two chromosome digits are looked at randomly. After the random digit is determined, the numbers are exchanged. For this, a random base is generated first. This base is a sequence that randomly contains 0s and 1s and has the length of the starting chromosomes. If the digit corresponding to a tick in the footer is 0, the digit in the corresponding chromosome digit is passed to the first chromosome. If the bottom digit is 1, the corresponding digit on the chromosome is passed to the second new chromosome. An example of this is given below.

00101011000110
01111100001100
Base ***00110110001110***
00111101001100
01101010000110

A uniform crossover would be a cut crossover for the substrate given below.

Base ***00000011111111***

Similarly, on the substrate, which can cause two-point crossover is in the form.

Base ***00000011110000***

From this, it is understood that the uniform crossover is a generalization of the single, double, and other crossovers. Numerous studies have shown that uniform crossover is more effective than single or double crossover. In fact, more specifically, it was concluded that the two-point crossover is also more efficient than the one-point crossover.

In the multiple crossover, only the randomly selected digits were exchanged. However, in the uniform crossover operation, each digit is left equally open to crossover.

With the uniform crossover, the two intermediate chromosome numbers of the new chromosomes are formed by taking the numbers corresponding to the first chromosome instead of the numbers 1, and the values corresponding to the numbers 0 in the second chromosome. In the formation of the second new chromosome, the numbers of the second chromosome are taken to the 1s of the intermediate chromosome and the numbers of the first chromosome are taken to the 0s of the second chromosome.

6.9.10.5 Inversion

Inversion, which is completely different from the previous processes, is the chromosome that occurs when it is read with the gene sequence from right to left instead of from left to right. Here, too, a completely different number is obtained from the number represented by the previous chromosome. For example,

$$1\ 1\ 0\ 0\ 0\ 1\ 0\ \rightarrow 98$$

If a chromosome like this is reversed, the following new chromosome is obtained.

$$0\ \ 1\ \ 0\ \ 0\ \ 0\ \ 1\ \ 1\ \ \rightarrow 35$$

6.9.10.6 Mixed Crossover

First, in a single-point crossover, instead of replacing the first parts of the chromosomes with numbers in the same order between them, the numbers are filled with 0 or 1 from the beginning with a completely independent and uniform random number generator, but if the second parts of the main chromosomes remain the same, it is called mixed crossover. Thus, the beginning parts of new chromosomes are completely independent of each other. For example,

1	0	0	/	**1**	**0**	**1**	**1**	→ 75
1	1	1	/	**0**	**1**	**0**	**0**	→116

as a result of mixed crossover of chromosomes are obtained.

0	1	0	/	**1**	**0**	**1**	**1**	→ 43
1	0	1	/	**0**	**1**	**0**	**0**	→ 84

6.9.10.7 Interconnected Crossover

If it is desired to produce other chromosomes close to a chromosome, then new and desired chromosomes can be produced by taking the appropriate parts from the two chromosomes and bringing them together according to the equation below. The two main chromosomes, C_1 and C_2, are mixed by an α, resulting in a new member form is obtained.

$$C_y = C_1 \alpha (C_2 - C_1) \tag{6.41}$$

According to the studies, it should be chosen as a random number between $-0.25 < \alpha < 1.25$. During assembly, Eq. (6.41) should be applied for the opposing genes of both chromosomes, while α value should be chosen randomly again. The positions of the new chromosomes produced in this way geometrically in the decision space cannot be too far from the parent chromosomes (see Fig. 6.32). These affinities vary according to the value of α.

6.9.10.8 Linear Associative Crossover

This is like the previous crossover, but all coupling states are obtained using a fixed value of α. As shown in Fig. 6.33, new chromosomes are obtained on the line connecting the first chromosomes in the decision space. This is expressed in the form of a straight-line right according to the following equation.

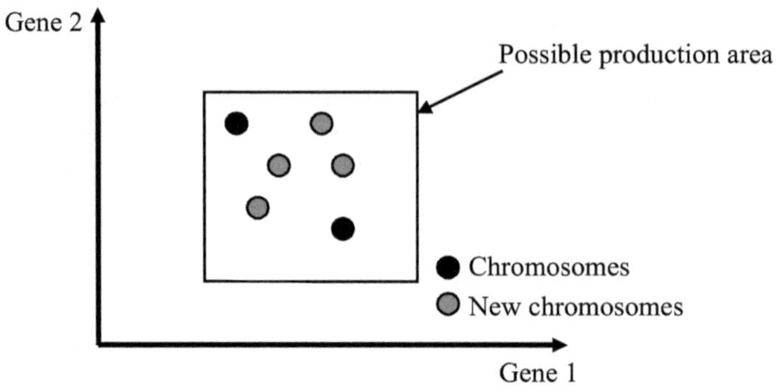

Fig. 6.32 Intermediate components cross-over decision space

Fig. 6.33 Linear union members

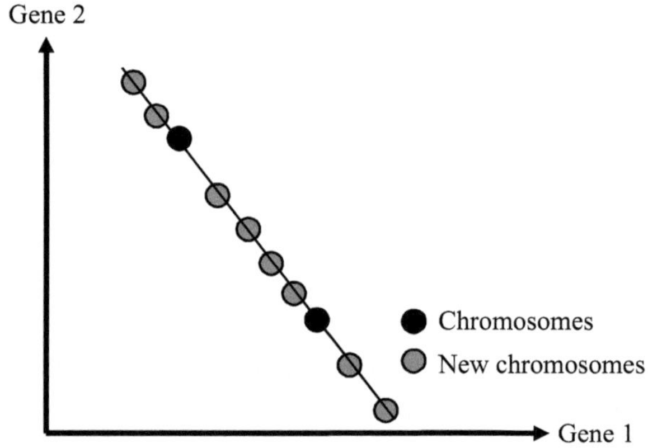

$$C_y = C_1 + \alpha(C_2 - C_1)$$

6.9.11 Mutation (Number Change)

Digit change is a GA operation performed on a single chromosome. Thanks to this process, during the search for the best solution, GAs also tries a much better or worse solution away from the next solution search area. Even if the algorithm goes bad, it can pick up and get closer to the best as quickly as possible with a series of number substitutions in the next step. Such an opportunity is absent in classical methods, because they all take the next step in the decision space within a predetermined system. Although such breakthroughs are not always successful, it may not be a waste of time as the algorithm renews itself randomly. Digit change (mutation) is the process of making 0 value 1 or 1 value 0 in a digit of the chromosome. This operation can be done in one digit of the chromosome or in more than one digit. Thus, a very different number emerges with the change of even one number.

This process occurs when the digit in one or more digits of the same chromosome is replaced with the opposite digit type. With digit shifts, GAs explores the target function surface of the second type. Thus, by obtaining new chromosome types that were not found at the beginning, it also prevents the GA from approaching the solution rapidly in the optimization process. If a one-digit change is applied to a chromosome then the number ‘1’ is converted to the number ‘0’ or vice versa. The digit location where the digit change will be made is randomly selected from the $N_{pop}xN_{bit}xN_{gen}$ total digit population. Increasing the number of digit change points enables the solution space that GA scans naturally to be examined in parts that cannot be touched with the classical GA approach. In general, 1% to 5% of the digits are exchanged in iterations. In the last iteration, the digits are no longer changed. In general, even in the best solutions, digit changes are not allowed. For example, 5%, or 0.05%, of numbers is changed among those other than the best chromosomes. It is mentioned earlier (Sect. 6.5) that the population is shown as a two-dimensional matrix. In this case, in the example given above, seven pairs of random numbers should be generated so that the rows and columns of the places where the digits will be changed can be determined. For example, if the first random pair is (4,11), the 4th row and the 11th row need to be digitized. The number “0” in this position is converted to the number “1.” This is shown in bold in the chromosome below.

0001011000**010** → 00010110001010

In addition, 6 more-digit changes are required. Their position can be (9,3), (2,2), (2,1), (5,14), (8,10) and (5,8) random number pairs. Most of the number changes made increase the target values of the chromosomes. It occurs from time to time in cases such as target value decreases. Digit change chromosomes search at points outside the solution path, which is usual in GA, and cause the solution surface to be investigated in more detail.

Digit swapping, unlike crossover, is the process of replacing randomly determined digits on a single chromosome with opposite numbers instead of two chromosomes. The goal here is to search for the best possible outside of the local best ones by making the digits 0 to 1 and the numbers 1 to 0 with a given value of change of digits. The digit change value, i.e., the number of digits to be exchanged, is kept small because randomness is not desirable. For example,

1 0 **0** 1 0 1 1 → 75

If the digit change in the chromosome is applied to the third digit, the new form of the previous chromosome is possible.

1 0 **1** 1 0 1 1 → 91

When examining digit substitution during a GA operation, two situations should be considered. These are the digit change types and rate. How much should the digit change be? Changing one of the digits in a number may oppose changing the corresponding variable value by 50%. The expected digit changes of a digit, whose digit is changed, can be calculated with the formula.

$$E(G) = \frac{1}{n}\sum_{i=1}^{n}\frac{1}{2^i} \tag{6.42}$$

Accordingly, using the 0.25 ratio in a 4-digit gene, it can be expected as (1/4)(0.5+0.25+0.0125+0.0675) = 0.23563 as a number change. On the other hand, the amount of change in 2-digit changes expected in an 8-digit gene is $(2/8)(1/2+1/2^2+1/2^3+1/2^4+1/2^5+1/2^6+1/2^7+1/2^8) = 0.24902$. In general, changing the digit based on the number of digits in a gene does not cause much change in the variable value. GA is very sensitive to the choice of the rate of change of digits.

After the position of one of the digits in a chromosome is randomly selected, a new chromosome is obtained if the digit in that digit is converted to the opposite digit. This jumps randomly to another part of the decision space. Since the digit change process is rarely done during GA operations, it is applied by considering low probabilities. Practical studies have concluded that this probability value varies between 0.01 and 0.001.

6.10 Probabilities of GA Transactions

The simplest of GA operations is crossover. This is the exchange of parts of the same length of two chromosomes among themselves. For example, let two sets of chromosomes be,

$$1\,0\,1\,1\,1\,0\,0\,1 \quad \rightarrow \quad 185$$

and

$$0\,0\,1\,0\,1\,0\,1\,0 \quad \rightarrow \quad 42$$

If one denotes their chromosome length by L (here L = 8), then one must first select one of the numbers from 1 to 8 with uniform randomness for the crossover operation.

Assuming 3 in the crossover change selection, one gets two new chromosomes with the crossover change of the first 3 genes from two chromosomes, as follows.

$$0\,0\,1\,1\,1\,0\,0\,1 \quad \rightarrow \quad 57$$

and

$$1\,0\,1\,0\,1\,0\,1\,0 \quad \rightarrow \quad 170$$

So, with the crossover, one can see where the previous numbers jump from. The arbitrarily chosen 3-digit crossover above should generally be done with a chromosome percentage (probability). If one shows the probability of finding the crossover site in a chromosome with P_C, its value must be determined beforehand for GA operations. For example, in the previous example, $P_C = (3/7) \sim 0.472$. This probability value refers to the initial randomness in GAs.

It is not necessary to apply the crossover operation to all members of population. For this, it is decided how many members in the population will be crossed in the percentage (proportion, probability) to be shown with the P_C. This means that if there are N members in the population, NxP_C should be selected for crossover. The selection of these members must be made completely at random from among all members. Thus, it is understood that the second possibility choice in performing GA operations is P_C.

The concept of third possibility in GAs is digit mutation, and it works in changing the value of any digit in the chromosome. A 0 becomes 1 or 1 becomes 0 in a digit change. In order to decide in which digit in the chromosome this number change will be made, the probability of a digit change, which is shown with P_D, should be chosen randomly as a number between 0 and 1. For example, if the above chromosome is going to change the digit corresponding to the probability of $P_D = 0.23$, first, since the number of genes in the chromosome is 8, the place of the digit change is $8 \times P_D = 8 \times 0.23 = 1.8 \sim 2$, starting from the left. Decimal results are rounded up to an upper integer if the decimal number is 5 or higher, otherwise a lower integer. Accordingly, if a number change is made in the 2nd digit of the chromosome then the following case is obtained.

$$1\,1\,1\,1\,1\,0\,0\,1 \quad \rightarrow \quad 249$$

The most important reason for the digit changes during the GA method is to enter the possible sub-solution spaces in the solution space and to search for solutions there. Thus, it is ensured that the algorithm reaches absolute (global) solutions instead of local solutions in optimization operations.

Above, we have understood how the members are coded and how they become values by crossover and mutation operations. The benefit of returning to the decimal system during these explanations is to take the place of the chromosome values in the target function and first determine the target values (in case of negative target value) and if necessary, vigor degree determination. As the processes continue, it is understood that successive chromosomes travel through the solution space quite randomly. In order to understand this, it is useful to look at how the crossover and mutation operations above change by leaps and bounds in the x_1 and x_2 decision space. For this, the stages described above are shown in Fig. 6.29 in the Cartesian coordinate system. If one uses other crossover and digit swapping operations, one sees that the point jumps to different parts of the decision space randomly and travels almost every part of the whole decision space. In response to this change in the decision variables, the value of the target function also changes. It is an advantage over other methods to catch the absolute optimization target point in a short time, since the decision variables will be circulated almost everywhere in the decision space with GA operations with random mechanisms. While traveling the decision space in this way, it reviews the different values of the target function. There must be

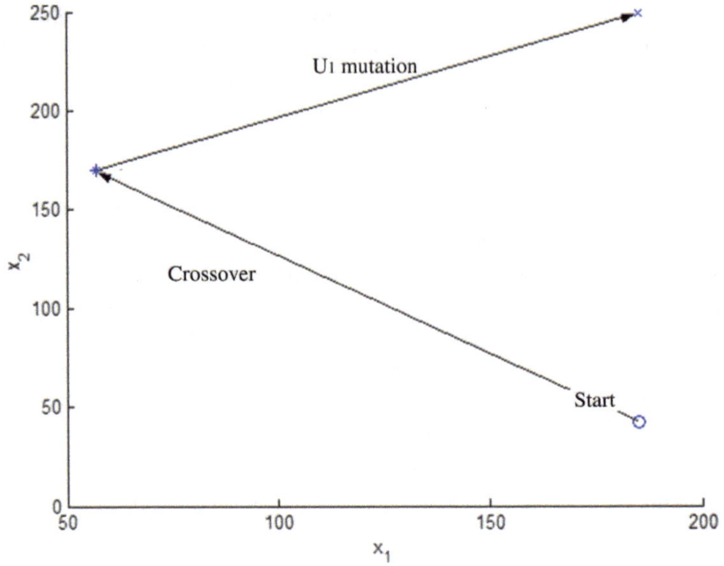

Fig. 6.34 Chromosome location change on decision space

a criterion to end this random circulation of the GA. It is desirable that the relative error α between the last two consecutive smallest target function values, T_1 and T_2, be less than 5 (5%).

$$\alpha = 100\frac{|T_i - T_{i-1}|}{T_i} \tag{6.43}$$

If $\alpha < 5$, GA operations are terminated. Otherwise, it is tried to reach different points in the decision space by crossing and mutation of numbers in new populations and within them (Fig. 6.34).

6.11 Gray Codes

Variable representation with classical binary numbers can slow down the speed of GA analysis. For example, let us consider (in bold numbers) the crossover gene, COG, which falls in the middle of the middle genes of two different chromosomes given below.

. . . **10000000** . . . . . . 128 . . .

. . . **01111111** . . . . . . 127 . . .

Here, it denotes the middle gene of **10000000** chromosomes. By encoding this gene according to the decimal number system, the number 128 on the right is found. The values corresponding to the newly formed chromosomes COG crossover between digits 3 and 4 are given as follows.

$$\ldots 10011111 \ldots \quad \ldots 159 \ldots$$
$$\ldots 01100000 \ldots \quad \ldots 96 \ldots$$

The partners are the better chromosomes, but the target values of the chromosomes resulting from the crossover process are further away from the previous ones. These binary number representations are the opposite of each other, as opposed to the very close target values of 128 and 127 of the previous ones. As a result, those born from these chromosomes are quite different from each other. In this case, the parameter values are 159 and 96. While it is expected that the variable values will approach each other, this is not the case at all. Although this is an extreme situation, it can be encountered in GA analysis. The problem can be exacerbated by increasing the digits in the variable representation (Fig. 6.35).

In order to avoid these situations, the variables must be gray coded, which defines binary numbers such that the Hamming distance between them is equal to 1. The Hamming distance is defined as the difference of the digits in the opposite digits of the two chromosomes. Let us consider the previous example as gray coded instead of binary number. If one defines two base-number encodings as 01000000 to represent 127 and 11000000 to represent 128, the COG becomes the following state.

$$\ldots \mathbf{11000000} \ldots \quad \ldots 128 \ldots$$
$$\ldots \mathbf{01000000} \ldots \quad \ldots 127 \ldots$$

Chromosome is cleaved by COG. Thus, it is possible to write the new chromosomes from COG between the numbers 3 to 4 as follows:

$$\ldots \mathbf{11000000} \ldots \quad \ldots 128 \ldots$$
$$\ldots \mathbf{01000000} \ldots \quad \ldots 127 \ldots$$

Chromosomes are the good solution. From here one can conclude that the new chromosomes will be fine, because gray coding is used. The process to gray encodes a binary number is simply shown in Fig. 6.36.

In Table 6.6, binary and gray coded opposites of integers between 0 and 7 are given.

Note that in gray coding, the Hamming distance of each neighboring integer is equal to 1. However, in encoding the integers 3 and 4 in the binary number system, the Hamming distance is equal to 3. The solution of a problem is approached more

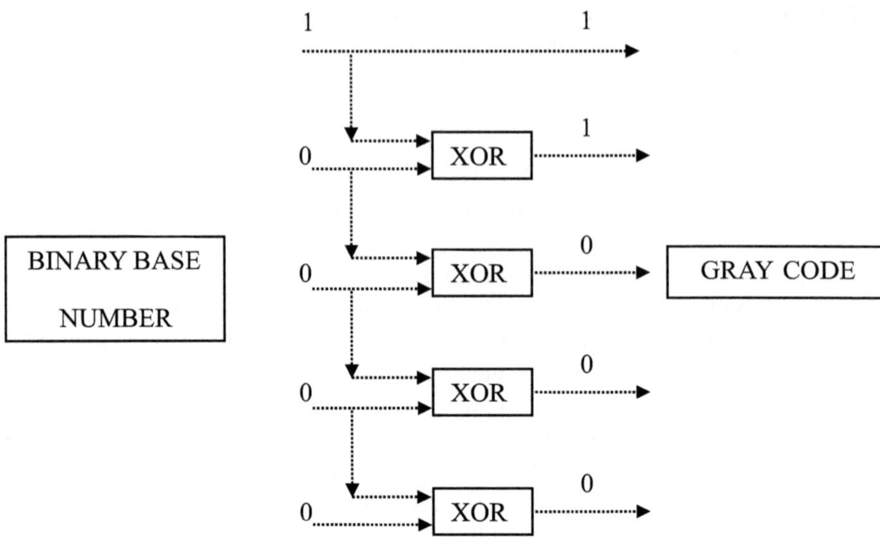

Fig. 6.35 Conversion diagram of binary number to gray code

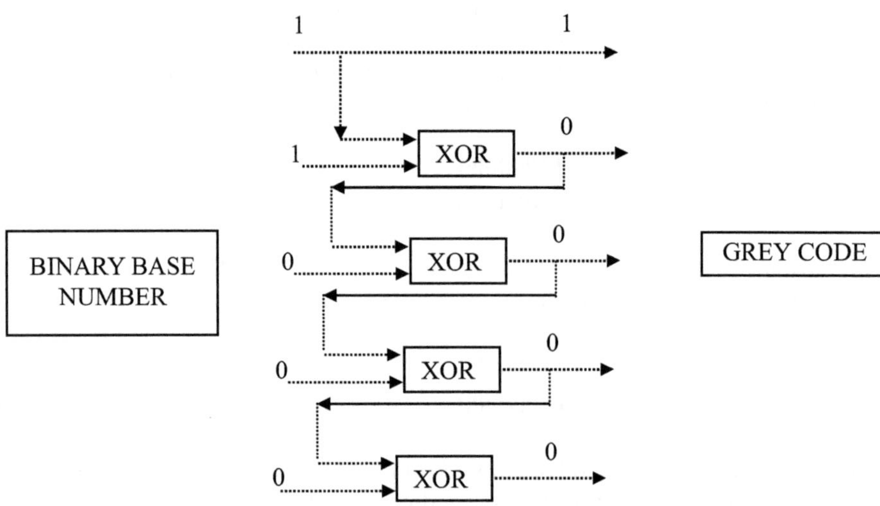

Fig. 6.36 Conversion diagram og grey code to binary number

quickly in GA operations with a gray coded system. In the studies conducted by Caruana and Scheffer (1988) and Hinterding et al. (1989), it was observed that the gray coded GA saves between 10% and 20% of time.

Table 6.6 Binary and gray coded numbers

Integer number	Binary number	Gray
0	000	000
1	001	001
2	010	011
3	011	010
4	100	110
5	101	111
6	111	100
7		

6.12 Important Issues Regarding the Behavior of the Method

The success of encoded GA information in genes comes in different forms from these combinations and selection of the most successful results, especially through crossover. Some important features of GAs can be summarized as follows.

1. Chromosome can be considered as a collection of genes that carry the information necessary for the solution of the problem. While the high number of chromosomes increases the working time, the low number of them eliminates the diversity,
2. Digit change is a process used to reach the most ideal solution. It allows the system to examine different areas by making jumps. Keeping the number change rate high will increase the randomness and make it difficult to reach the ideal point,
3. Normally the crossover is done at one point. It has been seen that multipoint crossover gives better results in some problems,
4. Encoding according to the binary number system is widely used,
5. Each population arises in connection with the previous population,
6. In general, although the first populations are random in GA, if the physics of the event are known, the first populations can also be selected from chromosomes close to the solution. In this case, the computation time is reduced,

The general operation of the GAs explained in detail above can be explained step by step as follows.

1. First, the chromosomes of the population are randomly selected. If the physics of the event is known, the chromosomes that make up the population can be selected as desired, considering the expert opinions. This choice makes it easier to reach the goal,
2. The vigor of the chromosomes is calculated by Eq. (6.32),
3. By looking at the vigor of the chromosomes, those who have adapted to their environment transfer their characteristics to the future population, while those who cannot adapt to their environment are eliminated, (selection operator),

4. Chromosomes constituting the new population are matched by the crossover process,
5. Random changes are made on some chromosomes (number exchange process),
6. The vigor of the new chromosomes obtained is calculated, if they are at the desired level, the processes are terminated. If the desired level is not reached, the good ones are selected and the bad ones are eliminated, and the crossover and mutation processes are continued. These processes are continued until the chromosomes reach the desired level or until a specified iteration is completed.

6.13 Convergence and Schema Concept

The optimization solution made with GA is obtained by staying within the desired error limits or by reaching the predetermined number of iterations. After a while, all chromosomes and their target values will be close to each other. Here, the change in the number allows this approach to decrease even more healthily. Population statistics are also kept during the operation of GAs. These include arithmetic means, standard deviations, and minimum target values. These then serve to gauge whether the solution has been approached in some way. To reach the solution, the target function calculation is determined by adding the following numbers.

$$\text{Initial population} + (\text{Crossover} + \text{Mutation}) \times \text{Iteration number}$$
$$= \text{Total number of transactions}$$

This means that the solution space has been explored in approximately 100 × Sum/(SumxSum) percentage.

Do GAs come close to a solution? There is no mathematical proof or guarantee that they will find the general smallest point. There is a handy proof by Holland (1975) called the schema theorem. A diagram contains randomly located digits and digits according to the binary number system. Apart from 0 and 1, there also exists a character, which means "empty." For example, in a diagram like 10ee00, the middle two blank digits can take either 0 or 1. By filling in the blanks, four different genes emerge. These are the 100000, 100100, 101000, or 101100 genes. The schema theorem states that "short schemas with better than average vigor is more frequent in the next population" or vice versa, "schemas with values below average vigor are less frequent in the next population." This means that "the most suitable schema survives in the next population." By following the best schema throughout the GA solution life, the best chromosome is approached. For example, suppose a given c_t chart is in a population at time t. The number of schemas that can be found in the next (t+1) chart can be given as,

$$c_{t+1} = c_t P_{cs}(1 + P_{ct})R_t \tag{6.44}$$

Here, P_{cs} denotes the probability of the chosen schema to survive in the next generation, the probability of the schema chosen to mate, and the probability of the schema that cannot be destroyed by crossover or change in R_t. If $P_{cs}(1 + I_{ct}) < 1$, then the schema disappears. The living schema propagates according to the following rule.

$$c_{t+1} < \left(P_{ct}P_{c(t+1)} \ldots P_{c1}\right)\left[(1 + P_c)\left(1 + P_{c(t-1)} \ldots (1 - P_{c1}\right]\left(P_{ct}P_{c(t-1)} \ldots P_{c1}\right)c_1$$

Here it is assumed that the probability of the schema in transition from one generation to the next changes. For the initial population, $P_{cs}(1 + P_{ct})$ can be >1. In the long run, this type of schema will hardly survive. For the schema to survive, $P_{cs}(1 + P_{ct}) > 1$ for all generations. When should this algorithm be stopped in practice? This is where the fuzziness in using GAs is. Some of the best possibilities are given below.

1. *Correct answer*: This might sound silly or simple. By looking at the best chromosome, it is necessary to make sure that this is the best solution. If this is the best solution, stop.
2. *The situation where improvement is not possible*: If the populations continue with the same best chromosome for X time, they should also stop. This means that the algorithm has either reached the best point or a local least vigorous point. Here one must be a little careful. Sometimes the solutions that seem to be the best results continue for several populations, and then the best solution can be obtained later, by crossing over or swapping numbers.
3. *Statistics*: If the mean or standard deviation of the population target function value reaches a certain level, the algorithm should be stopped. This means that the values will no longer change,
4. *Predetermining the number of iterations*: In case the algorithm cannot be stopped due to one of the above, the algorithm is run up to a predetermined maximum number of iterations. If this is not done, the algorithm can continue indefinitely.

If the algorithm does not come close to a good solution, the population number and digit change rate of the GA are changed. Maybe another crossover or switching from a continuous variable GA to a binary algorithm might give a better solution. Although rare, it can sometimes lead to weaker and slower results than other classical methods.

6.14 GA Parameters Selection

The selection of the size of the GA, such as population size and rate of change, is quite problematic as it presents a multi-choice situation. GAs involve initial randomness in the selection of population size and crossover and digitization rates. On the other hand, there are different types of crossover and mutation. Aging of chromosomes and gray coding are possible. It may be too time-consuming and impossible to determine the best one by testing all these options separately. After various tests, De Jong (1975) said that there are two criteria for the best parameter selection of GAs. One of them is to use the average of vigor of all populations from the beginning population to the last one. The GA is penalized for too many weak solutions, while the lowest vigor is rewarded. Second, the best vigor value found to the population in any iteration is considered. Binary numbers GAs are of different complexity.

1. Simple GA: Roulette wheel selectivity, simple crossover, and mutations are used,
2. An elitism is added to the previous algorithm,
3. The expected number of new chromosomes born from a chromosome,
4. The previous 2nd and 3rd steps are considered together,
5. Choosing a crowding factor,
6. Selecting a small fraction from a random population.

6.14.1 Gene Size

The time required for a GA to approach the solution and the health of the solutions depend on the size of the digits or genes that will be used to represent the variables. When solutions are made, computer users unnecessarily pursue extreme precision. For this, they consider many decimal places, even if they are meaningless after the comma. For each decimal digit in decimals, about 3.3 digits are needed in a gene. Therefore, if accuracy up to 8 digits is desired, the variable gene should consist of $8 \times 3.3 = 26.4$ or 27 digits. For example, if there are 10 variables, the chromosome length should be $27 \times 10 = 270$ digits. From here, it only brings more harm than good in GA analysis, due to the excess of desired sensitivity. For this, the necessity of the degree of accuracy to be expected from the solution of the problem should be considered beforehand.

6.14.2 Population Size

It is important in deciding the population size in the initial and later stages of GA. Here, too, there is a trade-off between population size and the number of iterations required for the GA's approximation. In the worst case, the initial ensemble iteration of calculating the required target values to find the solution must be at least as follows.

$$N_{pop} + M_{iter}\left[N_{bad} + \mu\left(N_{good} - 1\right)\right] \tag{6.45}$$

Here, M_{iter} indicates the number of iterations (population) required for GA's approximation, and μ is the mixing ratio of good and bad chromosomes. In this equation, it has been considered that the number change will only occur in the number of chromosomes and that there will be one COG in each of them. How many GA variables should be considered so that the target function calculations are minimal?

In small populations, the GA approaches a local minimum point in a very short time. Because COG such a society cannot exemplify the solution area as much as necessary. On the other hand, in large ensembles the time required to find the best solution becomes very long. According to a rule put forward by Goldberg (1989), very small ensembles are useful for solving problems connected in series, while large ensembles are useful for solving problems connected in parallel. Syswerda (1991) recommends determining the size of the population according to the following rule.

General rationality indicates that large ensembles will approach solution at much slower rates but will eventually arrive at better solutions than small ensembles. The experiences obtained at the end of the studies have pointed out that this rough rule is not always correct. The best population size should be adjusted to the behavior of the problem at hand. How the startup population will sample the target surface is also important. This depends on the FGG and gene cleavage rates used. For a better understanding of population size, let us first consider the population size of a bivariate problem, which is 16. The solution target surface can be sampled in different ways. These may be in the form of different examples, on the following explanations.

1. *Uniform*: Here, the solution surface is sampled at equal intervals along both parameters. Thus, all sides of the solution area are scanned with equal dots (Fig. 6.37),
2. *Random*: One side of the solution surface may have more sampling points than the other (Fig. 6.38),
3. *Partial Random*: Here the parameters in both directions are sampled with an equal number of random numbers (Fig. 6.39),
4. *Inverse Random*: Here, half of the chromosomes are randomly sampled, and the other half is taken as the complement of the previous one (see Fig. 6.40).

Completing here means changing 1's to 0 or 0's to 1. The application of this technique will be better understood by the following example.

$$\text{Random}\begin{cases} 0 & 1 & 1 & 1 & 1 & 1 & 0 & 0 & 0 & 0 & 1 & 1 & 0 \\ 0 & 1 & 0 & 1 & 0 & 0 & 1 & 0 & 1 & 0 & 0 & 0 & 0 \\ 0 & 0 & 1 & 0 & 1 & 0 & 0 & 1 & 0 & 0 & 0 & 1 & 0 \\ 0 & 0 & 1 & 0 & 1 & 0 & 1 & 1 & 0 & 0 & 0 & 1 & 1 \end{cases}$$

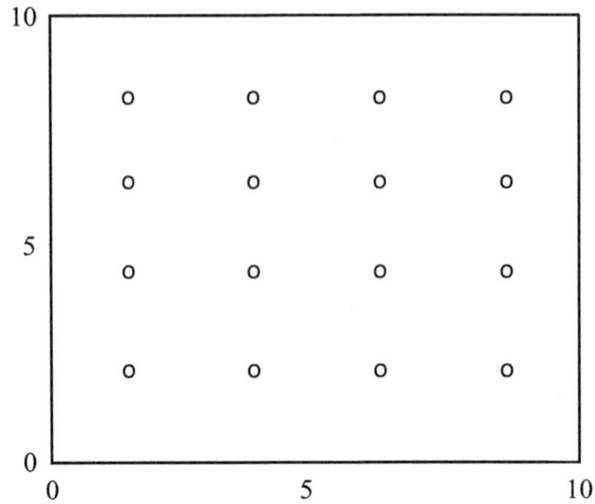

Fig. 6.37 Iso-range sampling

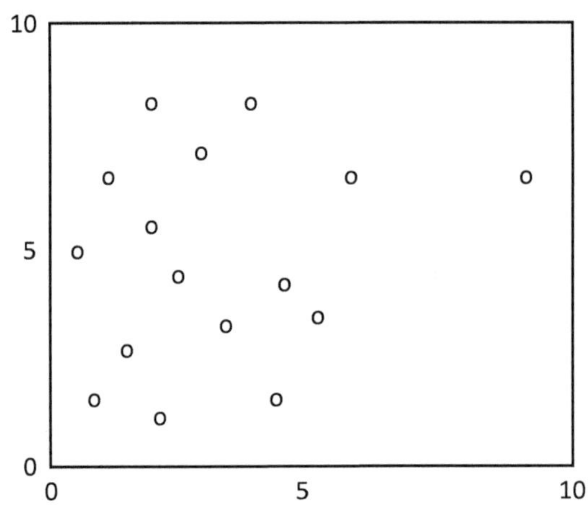

Fig. 6.38 Random sampling

$$\text{Complementary}\left\{\begin{matrix} 0 & 0 & 0 & 0 & 0 & 0 & 1 & 1 & 1 & 1 & 0 & 0 & 1 \\ 1 & 0 & 1 & 0 & 1 & 1 & 0 & 1 & 0 & 1 & 1 & 1 & 1 \\ 1 & 1 & 0 & 1 & 0 & 1 & 1 & 0 & 1 & 1 & 1 & 0 & 1 \\ 1 & 1 & 0 & 1 & 0 & 1 & 0 & 0 & 1 & 1 & 1 & 0 & 0 \end{matrix}\right.$$

Especially by providing enough sampling of the solution surface of the algorithm at the beginning, it is ensured that the approach time to the solution is reduced during future populations (iterations). The size of the initial population N_{ipop} is greater than the size of each subsequent population, N_{pop}. This means that after the solution surface is thoroughly sampled first, sub-samples of this first stage sampling are studied in other stages (populations). Changing the COG and the digits increases the sampling space of the algorithm. If the ensemble is large enough, each sampling method samples the solution surface almost equally frequently. For this, it is enough to compare 100 sampling points with the sampling methods in part of the same figures as in the Fig. 6.41.

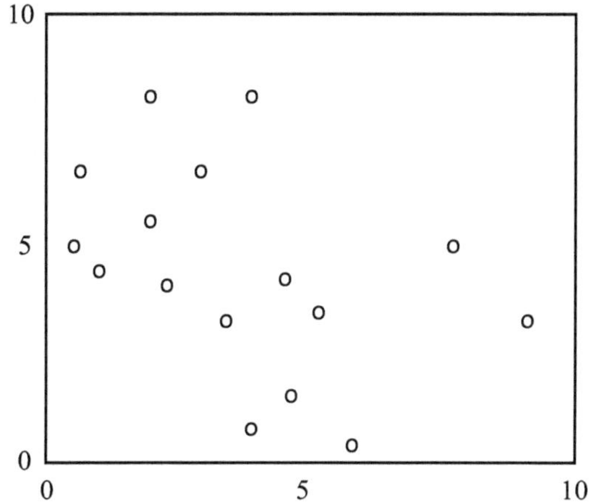

Fig. 6.39 Partial random sampling

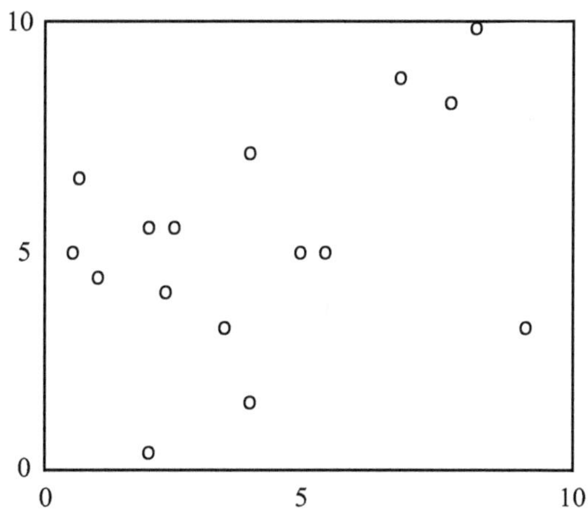

Fig. 6.40 Reverse random sampling

A simple control can be used for the best population size. The arithmetic average of the GA that optimizes different functions and its 100 independent usage results are taken. $N_{ipop} = N_{pop}$ and $M_{iter}xN_{pop} =$ are kept constant as shown in Table 6.7.

The size of the population also varies from population to population. Apart from this, it is possible to consider the aging of chromosomes (Michalewicz 1992). The lifetime of a chromosome is the number of populations (generations) in which it remains unchanged. The healthier chromosomes live longer than the weaker ones. Thus, with the exclusion of deceased chromosome numbers from the population, the population size varies from population to population. There are three approaches to extending the age of a chromosome: proportionality, linearity, and bilinearity. Their definitions are given in the table below. Definition of variables are as follows.

Fig. 6.41 100 point sampling

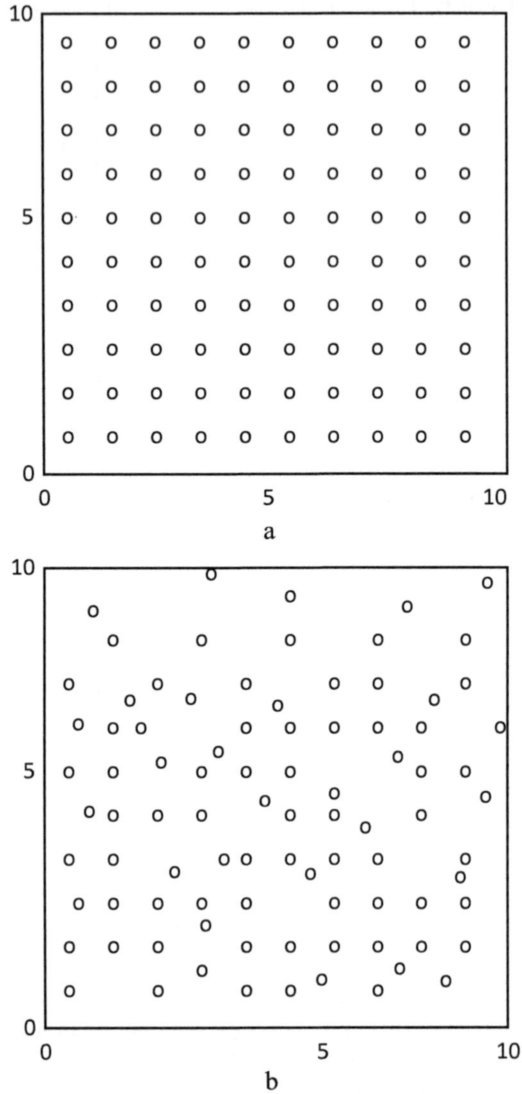

Table 6.7 Required numbers to obtain the same target function value (Haupt and Haupt 1998)

Usage (run)	1	2	3	4	5	6
M_{iter}	160	80	40	20	10	5
N_{pop}	4	8	16	32	64	128

Table 6.8 Increasing the lifespan of chromosomes (Haupt and Haupt 1998)

Allocation	Life time$_i$	Benefit
Proportionality	smallest{lifetime$_{max}$, lifetime$_{max}$ − ηtarget/B{target }}	lifetime α(1/st.)
Linearity	lifetime$_{max}$ − 2η(target$_i$-target$_{min}$)/(target$_{max}$ − target$_{min}$)	like the best target
Binary linearity	target > B{target} for lifetime$_{min}$ − 2η(target$_{eb}$-target$_i$)/(target$_{max}$ − target$_{min}$) target < B{target} for 0.5(lifetime$_{min}$ + lifetime$_{max}$) + 2η(B{target}-target$_i$)/(B{target} − target$_{min}$)	Good consiliator

Table 6.9 The ability to test with GA, which has a long-lived chromosome (Haupt and Haupt 1998)

Number	Function	Boundaries
1	$-x\sin(10\pi) + 1$	$-2 < x < 1.0$
2	integer(8x)/8	$0.0 < x < 1.0$
3	xsgn(x)	$-1.0 < x < 2.0$
4	$0.5 + \{[\sin^2(x^2 + y^2)]^{0.5} - 0.5\}/[1 + 0.001(x^2 + y^2)]^2$	$-100 < x < 100$

1. Lifetime$_{min}$ = smallest lifetime,
2. Lifetime$_{max}$ = greatest lifetime,
3. $\eta = 0.5(\text{lifetime}_{max} - \text{lifetim}_{min})$,
4. Target$_{min}$ = smallest targeted chromosome in the population,
5. Target$_{max}$ = largest targeted chromosome in the population,
6. B{Target} = expected value of the target.

It has been understood that for different vigor functions shown in Table 6.8, the varying population size gives better results than the fixed population size. The results here are given for 20 independent runs. The linear approach gives the best average (arithmetic) the lowest vigor. For this, the maximum number of functions is calculated. The binary-linear approach has the least number of function calculations, while the worst mean is the additional small target values. The principle of proportionality has a situation in between.

Is it always good to retain half of the chromosomes in one generation? This average is suitable for rough vigor functions, but for those with more roughness more chromosomes are needed and fewer chromosomes for those with less roughness. In practice, the principle of preserving half of the chromosomes when moving from one population to another yields very favorable results (Table 6.9).

One wonders if the solution is obtained more quickly by seeding the GA well at first. One can answer that this is not always the case. In almost every optimization method, the better the first insemination is done, the faster the solution is approached.

6.14.3 Example of a Simple GA

What is the peak (maximum) value of x in the range, 0 – 31 of a function $f(x) = x^2$? Since one will search for solutions according to GAs, some points in the decision solution space (here $0 < x < 31$) will be represented by a 5-digit binary number system. For GA, the following operations are performed.

1. *Initial*: At this stage, the decision variable solution candidates must be randomly assigned. Here, the solution points in the initial population should be decided. If the population size is taken as 4, 4 random decision variable values should be assigned. Let these 4 decision variables be 13, 24, 8, and 19, within the boundaries of the decision space ($0 < x < 31$). Their representations in the 5-digit binary number system are 01101, 11000, 01000, and 10011, respectively,
2. *Evaluation*: The target function values of these solution candidates are calculated. These are $13^2 = 169$, $24^2 = 574$, $8^2 = 64$, and $19^2 = 361$,
3. *Vigor*: At this stage, each of these values, which are candidates for solution, is compared among themselves and the degree of vigor is given according to the one closest to the solution. The percentage value of each solution point in the population is probability, so the vigor is calculated. According to this, since the sum of the four decision points is 1170, the vigor degree of each can be calculated from Eq. (6.21) 169/1170 = 0.144 (14.4%), 574/1170 = 0.492 (49.2%), 64/1170 = 0.55 (5.5%), and 361/1170 = 0.309 (30.9%), respectively. In that case, the most vigorous, that is close to the solution, among these solution candidates, is 19 according to the decimal number system and 10011 according to the binary number system, with a degree of vigor of 30.9%,
4. Now if a random number r between 1 and 1000 is chosen, according to these percentages of candidate solutions, the first decision point to enter the next population should be chosen as follows.

   If $r < 144$ then 01101 should be selected,
   If $145 < r < (144+492 = 636)$, 11000 should be selected,
   If $637 < r < (637 + 55 = 692)$, 01000 should be selected,
   If 693 $r < (636 + 309 = 100)$, 10011 should be selected.

   Using random numbers, 01101, 11000, 11000, and 10011 are selected. From this, it is understood that the 01000 series cannot enter the new population by dying. Thus, a new population emerges,
5. *Crossover*: Two decision points are randomly selected from this new population, and then one of the binary digits of this decision point is randomly selected. This second random selection indicates the location of the crossover in the sequence. The part preceding this crossover point is exchanged between both decision points, resulting in two new chromosomes. At the end of the random selection, the values of the 01101 and 11000 decision points are randomly selected from the population at 4 crossover points are 01100 and 11001 according to the new two-base number system. If the crossover point is 2, the strings 11011 and 10000 appear from the strings 11000 and 10011,

6. *Digit Change (Mutation)*: This is the counter-number change made in one digit of the chromosome of a decision point. It is 0 if the number in the change digit is 1, and 1 if it is 0. Changing the digits is done with a low probability of one in a thousand. In this example, the number has not been changed.
7. *Reevaluation*: After the crossover in the fourth step, since the decision points of the new society are 01100, 11001, 11011, and 10000 in binary, their decimal equivalents are 12, 25, 27, and 16, respectively.
8. *Iteration*: Finally, by going to the second step from here and repeating the next steps, new populations and better decision points are obtained in the transition to each society.

The example given above is very simple. The values calculated from the target function have a monotonic structure. However, this is not always the case. Then the percentage strength grades used above cannot be used. Digit swapping is a random search method, but it does not always occur. The effectiveness of the search method depends on the population and the size of the chromosome representing the decision points. The larger the population, the more the representative decision points in the original population and they are distributed over the decision space. The principle that the random-vigorous survives is just as valid. Also, in the case of the large population, multiple iterations increase the relevance of the solution and lead to better solutions.

6.14.4 Decimal Number-Based Genetic Algorithms

In the Sect. 6.3.7, it is said that when converting decimal numbers to a binary number system, approximation is made by taking the middle value of the intervals, and this affects the results. Although such approximations are not important in most practical work, in some sensitive cases such approximations may not be necessary. In this case, it may be desirable to make calculations directly with the decimal number system by disabling the conversion between decimal and binary number systems for better results. In this case, the decimal number system should be used directly, abandoning the representation of each variable according to the binary number system. If the binary number system is used, the chromosome lengths must be very large for reliable approaches.

When the variables are continuous, the decimal number system and fractional numbers are used. Since the precision that can be used with binary numbers is limited, the precision that computers allow can be achieved with the decimal number system. In addition, since the amount of memory required by decimals is less, computer optimization is also involved. On the other hand, since it is not possible to convert decimal numbers to binary numbers and vice versa, as in the previous sections, the computer time spent is even less.

Fig. 6.42 Decimal parameter GA stages

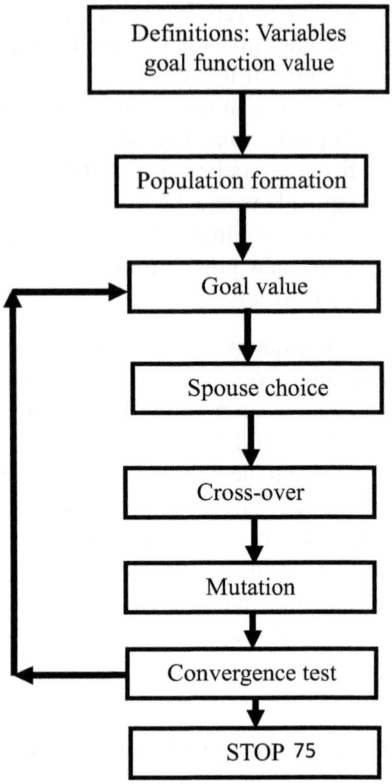

6.14.5 Continuous Variable GA Elements

Figure 6.42 shows the successive steps for GA operations with decimal number variables. Each of these stages will be explained in detail later.

6.14.6 Variables and Goal Function

Like the GA method, which is based on the binary number system, the smallest value of the target function, which varies according to the set of variables, is sought. If d_1, d_2, ... , N_{var}, then there are N_{var} number of variables, including dN_{var}, they are $1 \times N_{var}$ sized chromosomes as follows.,

$$C = [d_{1,} d_2, \ldots, d_{Nvar}] \tag{6.46}$$

Here, each of the variables is represented as a decimal number. From the goal function g( ) given the target value corresponding to a chromosome,

$$G = g(C) = g(d_{1,}d_2, \ldots, d_{Nvar}) \tag{6.47}$$

Defining the limits of the ranges of variation of these two equations and the variables represent the solution space of the optimization problem. In order to explain the decimal variable GA method, as an example, the target function is considered as follows,

$$G = g(x, y) = x\sin(4x) + 1.1y\sin(2y) \tag{6.48}$$

Let the limits of the variables be $0 < x < 10$ and $0 < y < 10$. In this case, the chromosome in general, with the x and y variants in the chromosome is,

$$C = [x, y] \tag{6.49}$$

This means $N_{var} = 2$. The equivalent line map of the target function is shown in Fig. 6.43.

6.14.7 Parameter Coding, Accuracy, and Limits

As with the base two-number GA, how many digits are required for the representation of the variables is not considered here. Variables are represented as decimals with numbers falling between the ranges of variation. Although decimal variables can accept any number, their digits are now represented by the digits of the computer processor. Here, they convert decimal variables with the finest precision by utilizing the binary numbers found in the computer's own processor. Now the precision of the variables is equal to the rounding error of the computers. They use the $2^4 = 16$-digit binary number system for single precision calculations. Their double precision is $2^5 = 32$ digits. On some supercomputers this can be up to $2^6 = 64$ digits of precision.

Since GAs are probing methods, it is necessary to determine the space boundaries of the decision variables of each problem well. If the boundaries are not known, the best solution is searched by keeping the decision space quite wide.

6.14.8 Initial Population

Chromosomes containing as many elements as N_{ipop} must be selected to initiate the GA optimization process. For this, a matrix with N_{ipop} rows is considered, each row

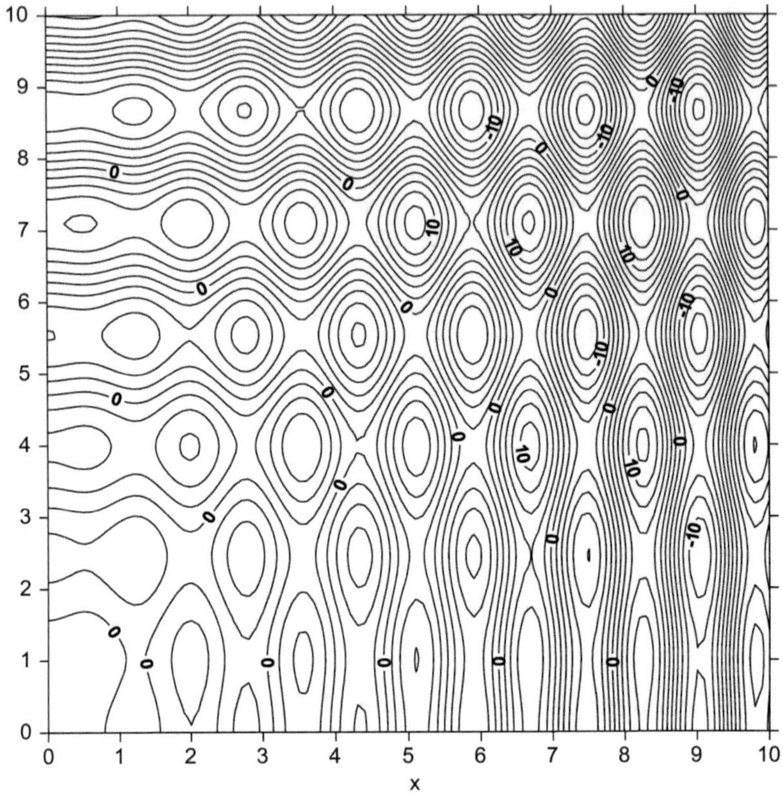

Fig. 6.43 Goal function map

of which has a chromosome of size $1 \times N_{var}$. Initially, random numbers are generated by giving N_{ipop} number of population chromosomes, and the $N_{ipop}xN_{var}$ sized matrix is constructed according to the following rule.

$$IPOP = (d_{max}-d_{min}) \times Random[N_{ipop}, N_{var}] + d_{min} \tag{6.50}$$

Here IPOP represents the initial population, the minimum and maximum values of the variables in the order of d_{min} and d_{max}, Random [ ], random values from a uniform distribution in the dimension $N_{ipop} \times N_{var}$ with values between 0 and 1. The 'Round' order is used to get only one of the values 0 and 1 for the random variables produced in the previous section. Here, there is no need for such rounding as GA operation will be performed with decimal numbers. For example, to produce a population of 8 chromosomes with 4 genes in each row, first a matrix of random numbers with 4 × 8 elements between 0 and 1 is created. Such a matrix can be obtained with any suitable off-the-shelf software. This form is obtained as,

$$\text{Random}[4,8] = \begin{bmatrix} 0.4451 & 0.8462 & 0.8381 & 0.8318 \\ 0.9318 & 0.5252 & 0.0196 & 0.5028 \\ 0.4660 & 0.2026 & 0.6813 & 0.7095 \\ 0.4186 & 0.6721 & 0.3795 & 0.4289 \\ 0.3046 & 0.3028 & 0.3784 & 0.4966 \\ 0.1897 & 0.5417 & 0.8600 & 0.8998 \\ 0.1934 & 0.1509 & 0.8537 & 0.8216 \\ 0.6822 & 0.6979 & 0.5936 & 0.6449 \end{bmatrix}$$

If the minimum and maximum values for the variables of the question under consideration are $d_{min} = 3$ and $d_{max} = 18$, this matrix is calculated by means of Eq. (6.50) and it becomes as,

$$\text{IPOP} = (18-3)\text{Random}[4,8] + 3 = \begin{bmatrix} 9.6764 & 15.6933 & 15.5718 & 15.4769 \\ 16.9772 & 10.8773 & 3.2946 & 10.5422 \\ 9.9899 & 6.0397 & 13.2192 & 13.6421 \\ 9.2797 & 13.0821 & 8.6922 & 9.4334 \\ 7.5693 & 7.5415 & 8.6756 & 10.4483 \\ 5.8448 & 11.1251 & 15.9002 & 16.4965 \\ 5.9015 & 5.2631 & 15.8048 & 15.3244 \\ 13.2333 & 13.4685 & 11.9034 & 12.6737 \end{bmatrix}$$

Here, each element in the last matrix shows the numerical values (genes) of the variables from the decision space between 3 and 18 and considering the 4 numbers in each row at the same time shows the chromosomes that are candidates for the optimum solution of the problem.

There is no democracy as one understands between chromosomes. The priority of each chromosome is according to the amount of goal value. According to the example given in the previous section (Sect. 6.3), $d_{min} = 0$ and $d_{max} = 10$. Since the goal function in Eq. (6.48) is very complex, the number of chromosomes should be increased. The softer the target surface in the solution space, the smaller can be the number of chromosomes. In the application here, by taking $N_{ipop} = 24$, the size of the population initial matrix is determined as 24 × 2, like the previous matrix formation.

$$IPOP = \begin{bmatrix} 8.1797 & 3.4197 \\ 3.4119 & 7.2711 \\ 6.6023 & 2.8973 \\ 5.3408 & 3.0929 \\ 8.3850 & 3.7041 \\ 5.4657 & 6.9457 \\ 5.6807 & 7.0274 \\ 4.4488 & 6.2131 \\ 7.9482 & 5.2259 \\ 1.7296 & 2.7145 \\ 9.5684 & 8.8014 \\ 9.7975 & 2.5233 \\ 8.7574 & 1.3652 \\ 8.9390 & 2.9872 \\ 7.3731 & 0.1176 \\ 1.9914 & 6.6144 \\ 2.8441 & 0.6478 \\ 5.8279 & 5.1551 \\ 4.6922 & 9.8833 \\ 4.2350 & 3.3395 \\ 4.3291 & 5.7981 \\ 5.2982 & 2.0907 \\ 2.2595 & 7.6037 \\ 6.4053 & 3.7982 \end{bmatrix}$$

Each row of this matrix is a solution chromosome for Eq. (6.48). The first column represents x and the second column represents the variable y. Chromosome (x, y) and target function values corresponding to this population are given in Table 6.10.

There is a relationship between the number of iterations required for a GA to reach a solution and the number of chromosomes in the initial population. In Fig. 6.44, the locations of some points in the initial society are shown with asterisks.

From here, it is seen that the chromosome locations are scattered all over the solution area.

6.14.8.1 Natural Selection

Now it must be decided which of the starting chromosomes will be sufficiently strong to pass on to the next population. For this, N_{ipop} chromosomes are sorted from

Table 6.10 Chromosome and target values

Number	x	y	Goal
1	8.1797	3.4197	13.7214
2	3.4119	7.2711	5.8807
3	6.6023	2.8973	1.1969
4	5.3408	3.0929	−1.8813
5	8.3850	3.7041	9.3703
6	5.4657	6.9457	3.5177
7	5.6807	7.0274	−3.5299
8	4.4488	6.2131	10.0463
9	7.9482	5.2259	13.0192
10	1.7296	2.7145	7.7105
11	9.5684	8.8014	8.9454
12	9.7975	2.5233	−1.7783
13	8.7574	1.3652	−8.6221
14	8.9390	2.9872	−0.8747
15	7.3731	0.1176	−9.7865
16	1.9914	6.6144	4.0073
17	2.8441	0.6478	−7.7977
18	5.8279	5.1551	6.9498
19	4.6922	9.8833	−8.3377
20	4.2350	3.3395	8.4416
21	4.3291	5.7981	−4.8037
22	5.2982	2.0907	−12.7438
23	2.2595	7.6037	10.6310
24	6.4053	3.7982	2.8903

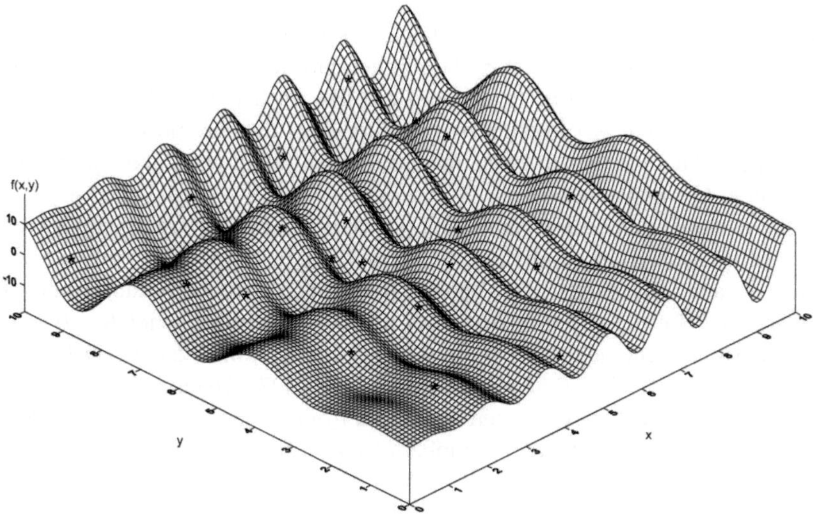

Fig. 6.44 Goal function surface and location of 20 chromosomes (Fig. 6.43)

the largest to the smallest according to their target values. One passes on the first N_{pop} of them to the next population. It is assumed that the rest of them died by being excluded from population. This process should be applied in all subsequent iterations (populations) of the GA so that better chromosomes can continue their lives, approaching the best solution after iterations based on target function. In all subsequent iterations, the chromosome number is taken as N_{pop}.

Some of the remaining N_{pop} numbers of chromosomes are then subjected to spouse selection and crossover operations. A good number of these are reserved for this work, while the rest, N_{bad}, are reserved for crossover and mutation operations to generate new species. This means,

$$N_{pop} = N_{good} + N_{bad} \tag{6.51}$$

In the example here, the arithmetic mean of the goal values of 48 chromosomes in the starting population is 1.8902, and the best target value is −13.92. By excluding half the number of chromosomes, N_{good}, the arithmetic mean of the population of chromosomes is – 4.27. Thus, N_{pop} = 24 chromosomes will be used for GA operations. N_{good} = 12 of these are used for spouse selection and crossover, but the others die. Chromosomes remaining after natural selection and their target values are given in Table 6.11.

6.14.8.2 Crossover

Among the chromosomes, N_{good} = 24 best ones were selected as described above. Of these, 12 chromosomes, which can be considered as 6 mothers and 6 as fathers, are randomly selected and crossed. Each pair generates two new pairs with remnants of themselves. In addition, the parents stay in the population to participate in the next population development. The more the mother and father resemble each other, the more the chromosomes born from them are like each other. There are different methods to produce new offspring by breaking apart the chromosomes. These will not be repeated here as they are described as three different ways in the previous two base number GA explanations.

In the example here, the weighted target value selection is used, whose weights are probabilities, and the results are shown in Table 6.12.

Mothers and fathers are rarely chosen from the grassroots, because their target valuesare low. Let us suppose that with a random number generator, 6 random number pairs are generated as (0.4679; 0.5344), (0.2872; 0.4985), (0.1783; 0.9554), (0.1537; 0.7483), (0.5717; 0.5546), and (0.8024; 0.8907). The maternal and paternal chromosomes selected to be crossed at random from the last column of the previous chart established by these random numbers are as follows.

Table 6.11 N_{pop} chromosomes sorted

Point	x	y	Goal value
1	9.0465	8.3097	−16.2555
2	9.1382	5.2693	−13.5290
3	7.6151	9.1032	−12.2231
4	2.7708	8.4617	−11.4863
5	8.9766	9.3469	−10.3505
6	5.9111	6.3163	−5.4305
7	4.1208	2.7271	−5.0958
8	2.7491	2.1896	−5.0251
9	3.1903	5.2970	−4.7452
10	9.0921	3.8350	−4.6841
11	0.6059	5.1941	−4.2932
12	4.1539	4.7773	−3.9545
13	8.4598	8.8471	−3.3370
14	7.2541	3.6534	−1.4709
15	3.8414	9.3044	−1.1517
16	8.6825	6.3264	−0.8886
17	1.2537	0.4746	−0.7724
18	7.7020	7.6220	−0.6458
19	5.3730	2.3777	−0.0419
20	5.0071	5.8898	0.0394
21	0.9073	0.6684	0.2900
22	8.8857	9.8255	0.3581
23	2.6932	7.6649	0.4857
24	2.6614	3.8342	1.6448

Table 6.12 Chromosomes in the waiting pool and their values

Sequence number, n	Probability, P_n	Cumulative sum probability, $\sum_{i=1}^{n} P_i$
1	0.2265	0.2265
2	0.1787	0.4052
3	0.1558	0.5611
4	0.1429	0.7040
5	0.1230	0.8269
6	0.0167	0.8637
7	0.0308	0.8945
8	0.0296	0.9241
9	0.0247	0.9488
10	0.0236	0.9724
11	0.0168	0.9892
12	0.0108	1.0000

$$\text{Mother} = [3 \quad 2 \quad 1 \quad 1 \quad 4 \quad 5]$$

and

$$\text{Father} = [3 \quad 3 \quad 10 \quad 5 \quad 3 \quad 7]$$

From here it is understood that C_3 will match C_3, etc.

From here it is understood that C_3 will match C_3, etc.

6.14.8.3 Waiting Pool

As in the GA based on binary numbers described in the previous section, the two new chromosomes that result from the fusion of two chromosomes will have partly the remnants of the previous chromosomes. Different approaches have been proposed for the detection of crossover points in the decimal parameter GA method. The simplest methods take one or more points on the chromosome as the crossover point. The parameters between these points are exchanged between the mother and father. For example, let us think two different chromosomes as follows.

$$\text{Spouse}_a = [d_{a1}, d_{a2}, d_{a3}, d_{a4}, d_{a5}, d_{a6}, \ldots, d_a N_{var}]$$

and

$$\text{Spouse}_b = [d_{b1}, d_{b2}, d_{b3}, d_{b4}, d_{b5}, d_{b6}, \ldots, d_b N_{var}]$$

Let a represent the mother and b represent the father from the subscripts here. If the in-between (here 2nd is the 4th position) is exchanged after the crossover point is randomly selected, then the following cases are obtained.

$$\text{Child}_1 = [d_{a1}, d_{a2}, \mathbf{d_{b3}}, \mathbf{d_{b4}}, d_{a5}, d_{a6}, \ldots, d_a N_{var}]$$

And

$$\text{Child}_2 = [d_{b1}, d_{b2}, \mathbf{d_{a3}}, \mathbf{d_{a4}}, d_{b5}, d_{b6}, \ldots, d_b N_{var}]$$

In the extreme case, as many as N_{var} points are selected, which chromosome to give its element to each case is also randomly chosen. By going to the base of the chromosome list and randomly choosing each variable, it is decided whether to exchange information between the two spouses. This is called uniform crossover as already explained earlier. For example, when the 1st and 6th parameters are swapped, the following two cases are arrived.

$$\text{Child}_1 = [d_{a1}, d_{a2}, \mathbf{d_{b3}}, \mathbf{d_{b4}}, d_{a5}, d_{a6}, \ldots, d_a N_{var}]$$

and

$$\text{Child}_1 = [d_{b1}, d_{b2}, \mathbf{d_{a3}}, \mathbf{d_{a4}}, d_{b5}, d_{b6}, \ldots, d_b N_{var}]$$

In this kind of crossover, no new information can be introduced, continuously variable values chosen randomly in the initial population enter the next stages of reproduction (population, iteration) with different combinations.

Instead, more efficient mixing methods have been developed. Here, new children are obtained by mixing the information found in the parents. The one-child variable value, the d_{new}, is obtained from the mixture of the chromosomes of the two spouses proportionally as follows.

$$d_{new} = \alpha d_{mother} + (1 - \alpha) d_{father} \quad (6.52)$$

Here α is the weight variable, which assumes a value between zero and one, and is the n-th variable in the maternal and paternal chromosomes in d_{mother} and d_{father}, respectively. The same variable of the second child is directly complementary to the first (i.e. the sum of α and 1−α equals 1). If $\alpha = 1$, d_{mother} dissipates in itself, but d_{father} dies. Conversely, if $\alpha = 0$, d_{father} propagates within itself and d_{mother} dies. However, when $\alpha = 0.5$, the child result is the arithmetic mean of the parents.

Determining which variable selection is appropriate for a good mix is another challenge. Sometimes this linear mixing is done to the right or left of the crossover point for all variables. Each variable can be mixed for the same α value or different α's can be selected for each. All these mixtures do not allow the emergence of values that may be beyond the extreme values found in population. For this, an outer extension (extrapolation) method must be applied. The simplest of these are linear crossovers. In this case, three different children from the mother and father occur. These are again the sum of their weights is equal to 1.

$$\left.\begin{aligned} d_{new1} &= 0,5 d_{mother} + 0,5 d_{father} \\ d_{new2} &= 1,5 d_{mother} - 0,5 d_{father} \\ d_{new3} &= -0,5 d_{mother} + 1,5 d_{father} \end{aligned}\right\} \quad (6.53)$$

Any variable out of bounds is excluded alongside the other two. The further spread of the two best children is then selected. In such a method, the 0.5 variable is not the only variable that can be used. It can be used in crossovers that conducive to different understanding. For this, α is randomly selected in the range [0,1] and the child parameters are defined as.

$$d_{new} = \alpha(d_{mother} - d_{father}) + d_{mother} \quad (6.54)$$

In another form of this method, it is possible to select a certain number of variables and to assign different α values for each variable. In this approach, the two parent couples cause the emergence of new children who are outside of them. Sometimes the parameters may overflow beyond the allowed solution area. In such a case, the child is neglected, and the operations are repeated with a value of α so that they remain within the solution area. After selecting α values for the mixture, crossovers, the distances outside the boundaries of the two main pairs are determined. This approach enables the GA method to consider the situations that are out of the way without deviating too much from the solution path. In practice, the joint use of extrapolation and crossover is often preferred here as well. This opposes crossover, starting with the random selection of a variant from the first main chromosome pair.

$$\alpha = \text{Upward round}\{\text{RandomxN}_{\text{var}}\} \tag{6.55}$$

One can obtain from here the following.

$$\text{Spouse}_1 = [d_{a1} d_{a2} \ldots d_{a\alpha} \ldots d_a N_{\text{var}}]$$
$$\text{Spouse}_2 = [d_{b1} d_{b2} \ldots d_{b\alpha} \ldots d_b N_{\text{var}}]$$

Consideration of these selected variables' combination gives rise to new children.

$$d_{\text{new1}} = d_{\text{mother}\alpha} - \alpha[d_{\text{mother}\alpha} - d_{\text{father}\alpha}]$$

and

$$d_{\text{new2}} = d_{\text{father}\alpha} + \alpha[d_{\text{mother}\alpha} - d_{\text{father}\alpha}]$$

The final stage should be completed in the form of crossovers in chromosomes, as explained earlier.

$$\text{Child}_1 = [d_{\text{mother1}} d_{\text{mother2}} \ldots d_{\text{new1}} \ldots d_{\text{father}} N_{\text{var}}]$$
$$\text{Child}_2 = [d_{\text{father1}} d_{\text{father2}} \ldots d_{\text{new2}} \ldots d_a N_{\text{var}}]$$

If the first variant of the chromosomes is selected, the variants to the right of this variant are exchanged. If the last variant of the chromosome is selected, the variants to the left of this variant are exchanged. With this method, the parameter does not appear outside of the basic covariates until $\alpha > 1$.

In the example given here, the first sets of spouse are identical and produce their own parts. The second pair of chromosomes are given as,

$$C_2 = [5.2693; 9.1382]$$
$$C_1 = [9.1032; 7.6151]$$

A random number selects d_1 as the crossover point. A second random number is $\alpha = 0.7147$. In this case, new children are available as,

$$\text{Child}_1 = [5.2693 - 0.7147 \times 5.2693 + 0.7147 \times 9.1032; 7.6151] = [8.0094; 7.6151]$$

$$\text{Child}_2 = [9.1032 + 0.7147 \times 5.2693 - 0.7147 \times 9.1032, 9.1382] = [6.3631; 9.1382]$$

If these processes are continued with the other four spouses, eight more children will emerge.

6.14.8.4 Mutation (Number Change)

If not paid attention to, it quickly approaches a point in the GA target values range. There is no problem if this place is in the smallest area overall. Like the goal values function used, in many cases there are too many minimization portions. If they do not find a solution to this quick approach, GA optimization may end up in an undesirable local area. To avoid this, mutations are needed for the comparative analysis of the GA by exploring other parts of the solution space. These number changes should be made by a random method. In binary number GAs this is achieved by changing the digit at random locations either from "1 to 0" or from "0 to 1." In GAs with decimal variables, the rate (probability, percentage) of number change between 1% and 20% is useful, as in the other ones. If one multiplies the rate of number change by the total number of variables, s/he finds the number of variables to be changed. After that, the row and column of the matrix are selected to determine the variables to be changed, and the variable that will change the number is determined with a new random variable. Here, the rate of gene change of 4% (0.04) was used. Since there are 12 chromosomes before crossover and then 24 chromosomes, 2 variables need to be digitized ($0.04 \times 24 \times 2) \sim 2$. The p_1 variable in C_7 and the p_2 variable in C_{22} have been digitized. Thus, while the variable p_1 in C_7 is deleted, it is replaced with a new random variable between 0 and 10 limit values,

$$C_7 = [2.7271; 4.1208]$$
$$\downarrow$$
$$C_7 = [3.1754; 4.1208]$$

The chart below shows the order of the target values of the second-generation variables from best to worst, with number changes after new children. The arithmetic mean of this ensemble is –8.51 (Table 6.13).

After seven populations (generations), the smallest goal value is found to be –18.5 with this GA method.

Table 6.13 Second generation of decimal parameters

x	y	Cost
9.0465	8.3128	−16.2929
9.0465	8.3097	−16.2555
9.1382	5.2693	−13.5290
,...	,...	−,...
,...	,...	− ,...
4.1208	3.1754	−2.6482

6.15 General Applications

The use of the principles and rules explained in the previous sections and the applications of the GA method are exhibited in this section. Each of the examples presented here constitutes a basis for optimization transactions on different subjects.

6.15.1 Function Maximizing

Function maximizing or minimizing is one of the most used areas of GAs. The optimization of the derivatives of continuous functions can be easily done by analytical methods. Here, an example of how such functions can be maximized without derivation will be presented. For this the following cubic function will be used.

$$f(x) = x^3 \tag{6.56}$$

For a simple GA application, the following steps are required.

1. Let us determine, for example, 6 completely randomly determined x values and convert these x values to binary (0 and 1). From this, it is understood that the GA initial population was selected with 6 members (chromosome). Since there is only one decision variable, does the software in base two corresponding to each x number represent a chromosome? In this simple example, each chromosome is equal to one gene. Let the lengths of these chromosomes be chosen as 6. In other words, there are 6 digits (genes) in each chromosome. The said x values and their chromosomes are shown in Table 6.14.
2. Let's calculate the sum of the y numbers found by substituting the X numbers by each chromosome in the function, and then the vigor values (percentages) of each in this total are calculated from Eq. (6.32). The results obtained are shown in Table 6.15.

   Each of these percentages shows the probability values of the numbers represented by the chromosomes in Table 6.14 within the chromosome ensemble. In other words, these percentages are a measure of the vigor (strength) of the

Table 6.14 GA first population

Decision variable	Chromosome
If x_1= 5 then coding is	0 0 0 1 0 1
If x_2=13 then coding is	0 0 1 1 0 1
If x_3= 1 then coding is	0 0 0 0 0 1
If x_4=7 then coding is	0 0 0 1 1 1
If x_5= 16 then coding is	0 1 0 0 0 0
If x_6=21 then coding is	0 1 0 1 0 1

Table 6.15 GA goal and vigor values

Decision variable	Goal value	Vigor degree
x_1= 5	125	0.00780
x_2=13	2197	0.13712
x_3= 1	1	0.00006
x_4=7	343	0.02141
x_5= 16	4096	0.25563
x_6=21	9261	0.57798
Total	16023	1.00000

Table 6.16 Chromosome selection

Decision variable	Chromosomes
x_1= 21	0 1 0 1 0 1
x_2= 21	0 1 0 1 0 1
x_3= 16	0 1 0 0 0 0
x_4= 13	0 0 1 1 0 1
x_5= 21	0 1 0 1 0 1
x_6= 21	0 1 0 1 0 1

chromosomes in their struggle to survive in the population. For example, the 6th chromosome is the strongest and the 3rd chromosome is the weakest.

3. Consider a roulette wheel with these percentage values as detailed in Chap. 1, Sect. 1.4.1.2. By turning the roulette wheel 6 times, the chromosomes given in Table 6.16 are selected. The point to be noted here is that in such a selection, the weak (small vigor) chromosomes could not pass the random roulette wheel selection. A regenerated GA population of 6 members was obtained, again in Table 6.16, with mostly strong chromosomes. In such a roulette game, the 6th chromosome, that is, the number 21 is expected to occur with the highest probability.
4. Chromosomes occurring in Table 6.16 are first crossed over with each other in order to develop over time. For example, the chromosome pairs to be crossed at the beginning and then the gene numbers to be crossed are randomly determined and presented in the Table 6.17.

   The chromosomes matched in the first column of the table and the number of digits to be crossed over is shown in the second column. For example, the cross between two chromosomes such as x_4 and x_5 can be seen in detail below.

Table 6.17 Crossover operations preliminaries

Chromosome pairs for crossover	Crossover digits
x_6 and x_2 chromosomes	3
x_5 and x_4 chromosomes	2
x_1 and $_3$ chromosomes	5

Table 6.18 Crossed GA population

Chromosomes	Decision variable
0 1 0 0 0 1	x_1=17
0 1 0 1 0 1	x_2=21
0 1 0 1 0 1	x_3=21
0 1 1 1 0 1	x_4=29
0 0 0 1 0 1	x_5= 5
0 1 0 1 0 1	x_6=21

Chromosome	Digit number 1	2	3	4	5	6
x_5:	0	1	0	1	0	1
x_4:	0	0	1	1	0	1

As a result of these processes, the newly formed chromosomes take the following shapes.

x_5 : 0 0 0 1 0 1

x_4 : 0 1 1 1 0 1

The results obtained after the completion of all crossover operations are shown in Table 6.18.

Comparison of this crossed population with the previous Table 6.16 shows that the population chromosomes in Table 6.18 become stronger.

5. Finally, let us apply the mutation (digit change) operation. Again, the digit of the chromosome that will undergo a number change is chosen randomly. For this, the random integer selection made between the numbers 1 and 6 is 1. It is seen that the chromosome in which the number will be changed is 3 by being randomly selected from the population in Table 6.19. Accordingly, the digit that will change the digit is the 1st digit of the 3rd chromosome. With the completion of this process, the above chromosome group takes its new form.

As can be seen from this chromosome ensemble, the value of chromosome 3 reached 53 and the function value reached 53^3 after the number change. Whereas, the best chromosome at first was 21, and thus the function was 21^3. Thus, all the steps required by GA have been completed. If the steps so far are repeated with similar operations for a while, it is seen that the best chromosome to be reached by GA [1 1 1 1 1 1] (63) and the best function value will be 63^3.

Table 6.19 Changed GA population

Chromosomes	Decision variable
0 1 0 0 0 1	x_1=17
0 1 0 1 0 1	x_2=21
1 1 0 1 0 1	x_3=53
0 1 1 1 0 1	x_4=29
0 0 0 1 0 1	x_5= 5
0 1 0 1 0 1	x_6=21

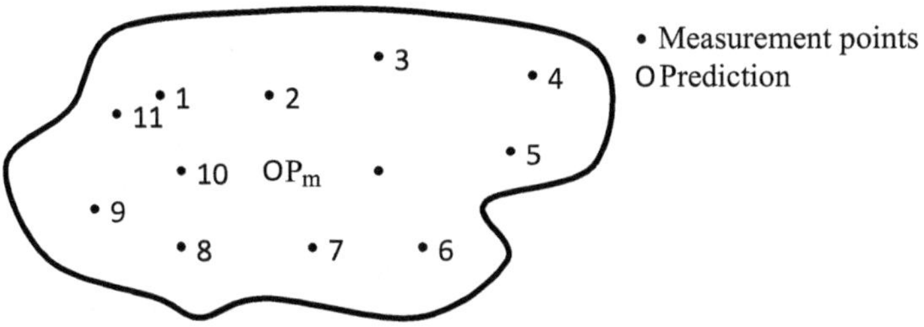

Fig. 6.45 Irregular sampling of regional variable

6.15.2 *Geometric Weight Functions*

Meteorological, earth sciences, hydrological, ore evaluation, oil exploration, etc., have regional variability. The information about the events is determined by the numerical information obtained from the measuring devices placed in a region in irregular locations or from the wells drilled. For economic reasons, a preliminary modeling is necessary to increase the value of the regional variable or the drilling of wells with the available information. Here, it is necessary to make the best estimation of the regional variable on a subject called the forecast point and for which there are no measurements. As shown in Fig. 6.45, the locations of the stations from which information was obtained are randomly distributed.

The problem here is to estimate the value at the prediction point, P_m, with a calculation that includes the weighted average of the measured variable values at a set of sites. Weight values with n number of measurement points are $w_{i,P}$, (i = 1, 2, 3, . . . , n). Hence, the prediction point value, P_m is calculated as,

$$P_m = \frac{\sum_{i=1}^{n} w_{i,P} P_i}{\sum_{i=1}^{n} w_{i,P}} \tag{6.57}$$

where P_i's are the measurements at each site i. Various researchers have proposed solutions to this problem of regional prediction, by considering weight functions deterministically that are functions of inter-station distances. Such functions depend only on the geometrical layout of the stations. However, regional values measured are not considered in these functions. They are shapes that have been obtained only with logic and geometry. Therefore, the weight coefficients found cannot be expected to reflect actual regional dependence.

Thiebaux and Pedder (1987) suggested the weighting coefficients in general according to the following expression without considering the actual variation of the variables.

$$w_{i,T} = \begin{Bmatrix} \left(\frac{R^2 - r_{i,T}^2}{R^2 + r_{i,T}^2}\right)^{\alpha} & \text{for} \quad r_{i,T} \leq R \\ 0 & \text{for} \quad r_{i,T} \geq R \end{Bmatrix} \tag{6.58}$$

Here R indicates the radius of influence and α the smoothing coefficient. This last equation is the general version of the classical Cressman (1955) method, where the smoothing coefficient is equal to 1. The radius of influence, R, and the softening coefficient, α, parameters are determined subjectively by personal experience and ability related to the problem under investigation. Another α number given in the literature is 4 (see Fig. 6.46). In practical studies, difficulties are encountered in determining α. Sasaki (1960) and Barnes (1964) substituted Eq. (6.58) by another form of the geometric weight function suggestion as follows.

$$w_{i,P} = \exp\left[-4\left(\frac{r_{i,P}}{R}\right)^{\alpha}\right] \tag{6.59}$$

where $r_{i,p}$ is the distance between the prediction point and i-th measurement point.

The weight functions should reflect the characteristics of the regional variable and, in addition, the geometry of the station's configuration. Şen (2004) suggested the cumulative semi-variogram technique as another alternative for regional dependency functions like Fig. 6.46.

GA was used to find R and α parameters in Eqs. (6.58) and (6.59) by using the experimental weight functions obtained (Şen 2004). These equations are made to represent regional variability with the help of GAs. For this application, 16 chromosomes and 16 genes are used. As the goal function, the sum of the squares of the differences (errors) is considered between the experimental weight values and the weight functions given by Eqs. (6.58) and (6.59). The flow of the application was made by executing all the steps in the frame given in Sect. 6.14.5. The results obtained at the end of the calculations are shown in Fig. 6.47.

The application can automatically give R, α and the estimation error, E. It has been observed that the Barnes equation represents the monthly precipitation data recorded in the Marmara Region of Turkey better than Cressman (1955) method. The whole results are presented in Table 6.20.

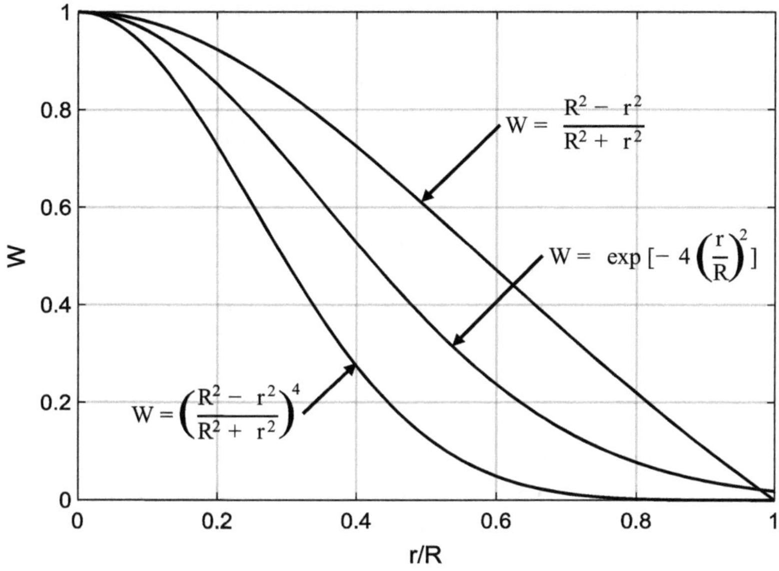

Fig. 6.46 Various weighting functions

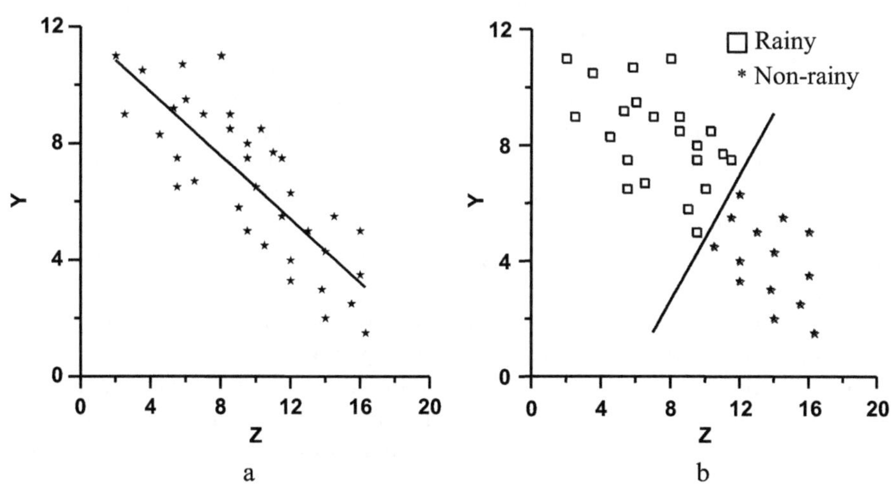

Fig. 6.47 GA-based weighting function (January)

6.15.3 Classification of Precipitation Condition

It is very important to divide the natural events into two classes as the presence or absence of natural events in the branches of science that examines natural events. For this, in general, logistic regression (Chap. 7) and separation methods are used in

Table 6.20 Calculated softening coefficients (α) and radius of influence (R).

	Cressman			Barnes		
Months	α	R	Error (10^{-2})	α	R	E (10^{-2})
January	2.47	0.98	6.65	2.28	0.79	4.35
February	16.23	2.32	8.68	2.28	0.75	6.97
March	30.81	2.75	8.56	1.89	0.73	8.00
April	28.74	2.68	7.74	1.85	0.75	6.91
May	25.35	3.14	6.17	1.82	0.95	4.72
June	22.70	3.23	4.05	2.00	0.96	4.04
July	28.38	2.95	7.19	2.12	0.75	6.63
August	1.50	0.80	7.70	2.42	0.79	3.84
September	2.07	0.93	7.97	2.31	0.81	4.90
October	1.25	0.75	24.30	3.15	0.69	5.15
November	1.52	0.76	5.28	2.26	0.78	3.71
December	1.69	0.77	21.70	3.03	0.63	5.30
Average	**13.50**	**1.83**	**9.66**	**2.28**	**0.78**	**5.37**

Note: The error expression represents the sum of the squares of the errors

statistics. However, these require a set of assumptions to be provided by the data. For example, these include the constancy of variance (homoscedasticity), stationarity, homogeneity, the distribution according to the normal (Gaussian) distribution, etc. On the other hand, there are no such assumptions in the application of GAs. These methods process the numerical values of the data according to a certain set of rules, resulting in a mathematical expression at the end, allowing the classification to be made accordingly. For example, dew point temperature and vertical air velocity play a role in the classification of precipitation or its absence. Statistical methods divide the scatter diagram into two parts by attaching the words rainy and non-rainy to the points of these two variables on the scatter diagram without giving details of the physical relations between these variables.

The representation of the classification to be made in general on a two-dimensional Cartesian axis set is given in Fig. 6.48.

Here, the classification of an event denoted by x depends on two other variables, y and z; accordingly, depending on the y and z values, the x value belongs to either 1 (rainy) or 0 (non-rainy) classes. The steps required for GA to make classification are given below.

(a) Scatter diagram of some of the y and z data is shown in Fig. 6.48a. This is called a training scatter diagram. If it is desired to find a relationship between only these two variables, the classical regression approach is used for this. However, this cannot make an accurate classification,
(b) Each point in the scatter diagram is written according to the verbal value of x (rainy or non-rainy). Thus, even before the application of the method, the scatter diagram is divided into two classes by means of observations. In Fig. 6.48b, the squares show the rainy stars.

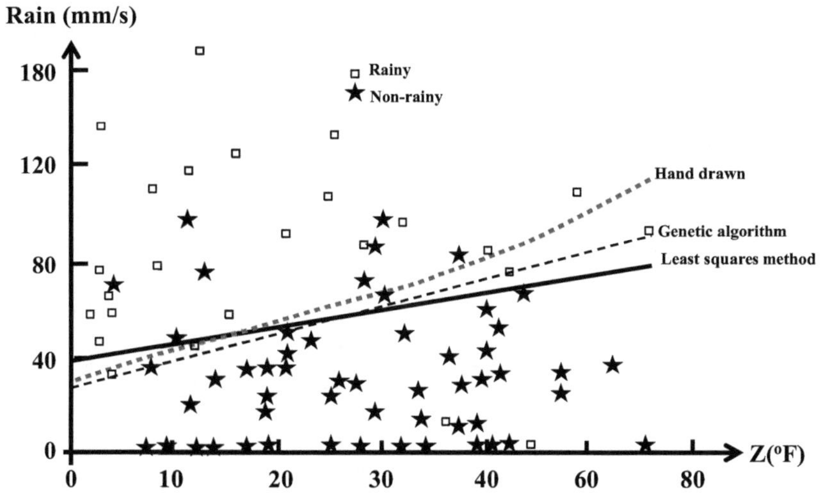

Fig. 6.48 Scatter diagram (**a**) regression line, (**b**) separator line

(c) A line or curve can be drawn which can separate the points thus divided into two groups. It is used as a straight separator according to the separation method. Finding the separator, which is quite difficult with classical methods, can be found more easily with GAs, whether it is curved or correct. The goal here is to have the smallest error in the number of classifications.

6.15.4 Two Independent Datasets

Here, the GA application is shown for two independent datasets. The first of these is the 24-h precipitation data from the literature (Panofsky and Brier 1968). Here the classification is based on the vertical velocity, Y (mm s^{-1}), and z, the dew point temperature ($^{\circ}$F). In the example used here, 28 days out of 91 were rainy (see Table 6.21). Their scattering diagram is given in Fig. 6.49.

The steps in the GA application required to find the line to separate the rainy and non-precipitated situations are as follows.

(a) Since there is a line expression sought, there are two decision variables, constant, a and the slope, b. First, two random number sequences are generated on the computer, which are thought to represent the cut-off point and slope of the line. Let the sequence a_1, a_2, ..., a_n represent the intersection point of the dividing line sought, and the sequence b_1, b_2, ..., b_n represent the slopes. In this case, the population size to GA is n^2,

(b) The reciprocal elements of these two sequences, a_i and b_i, are encoded as decision variables with a binary number system each unit length m (digits). If there are two variants, i.e., genes, there will be 2 m long chromosomes. During

Table 6.21 Rainy and non-rainy conditions data

Dew point temperature	Vertical velocity	Precipitation	Dew point temperature	Vertical velocity	Precipitation
7,08	155.88	1	20.80	40.55	2
3,00	136.64	1	20.86	49.95	2
24,99	134.46	1	21.86	52.28	2
15,93	125.17	1	22.93	46.64	2
11,10	117.84	1	24.86	24.13	2
8,05	110.53	1	25.94	28.59	2
6,12	97.47	1	26.95	28.60	2
2,96	76.42	1	29.01	17.51	2
8,12	79.21	1	31.03	58.65	2
24,92	107.49	1	32.00	49.67	2
20,67	92.18	1	33.12	26.45	2
15,06	58.57	1	33.18	14.42	2
3,96	65.72	1	37.06	26.79	2
4,02	58.35	1	38.13	28.73	2
1,99	57.41	1	36.13	40.83	2
3,02	46.31	1	39.14	30.76	2
31,93	97.06	1	41.23	32.80	2
28,09	87.82	1	40.07	44.11	2
40,18	84.02	1	41.22	52.71	2
42,20	76.97	1	40.13	60.07	2
48,29	108.96	1	43.29	67.58	2
55,34	93.08	1	36.56	11.23	2
12,10	45.30	1	39.12	14.69	2
16,99	48.79	1	47.19	24.58	2
20,98	53.18	1	47.13	34.49	2
4,14	32.68	1	52.22	37.07	2
36,00	11.52	1	7.13	0.79	2
44,33	1.52	1	9.25	0.61	2
11,17	99.04	2	12.12	1.25	2
4,12	71.68	2	13.54	1.06	2
13,16	76.74	2	14.00	0.46	2
10,07	47.30	2	16.85	1.00	2
8,08	36.06	2	17.37	1.01	2
12,14	46.31	2	18.99	1.13	2
30,10	98.55	2	24.93	1.50	2
29,10	85.61	2	27.91	1.43	2
28,04	71.05	2	31.75	1.47	2
30,06	65.11	2	32.37	1.78	2
37,12	83.68	2	32.98	1.59	2
12,12	20.35	2	33.56	1.70	2
14,04	32.69	2	34.24	1.80	2
16,88	35.66	2	39.05	1.76	2
18,89	36.39	2	40.22	2.18	2
18,85	23.76	2	42.05	1.89	2
18,92	16.99	2	5528	2.25	2
20,93	35.50	2			

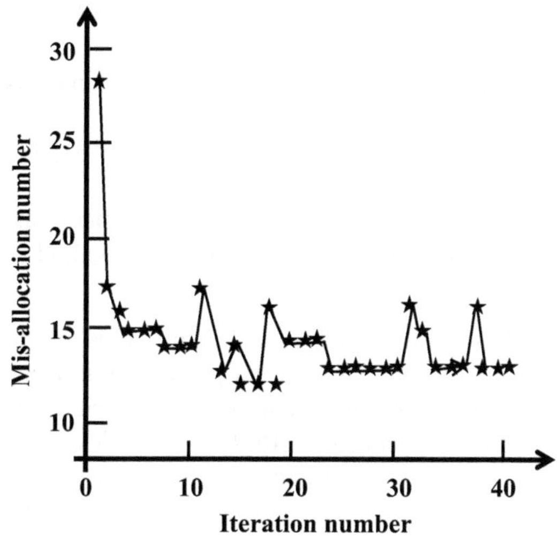

Fig. 6.49 Precipitation event and regression separation straight-line. (Panofsky and Brier 1968)

the whole GA process, it should be kept in mind that half of a chromosome represents the segment of the dividing line and the other half represents the slope,

(c) The importance of each chromosome in the population is determined by means of an error function. As will be explained below, the definition of percent error is used for each chromosome. With the completion of this and the previous two steps, the GA initialization process is also completed,

(d) Thus, the chromosomes in the population are ready for GA operations. After that, the chromosomes are subjected to selection, crossover, and mutation (number change) processes. Here the selection is made by the roulette wheel method. The smaller the error, the greater is the probability that that chromosome will enter the next society. The aim here is to exclude bad chromosomes or to combine them with other chromosomes to produce healthier chromosomes. For this, the crossover operation is used (Sect. 3.10),

(e) A margin of error of ±5% should be accepted during this series of operations. In case of error falling between these error limits, the GA operations are stopped, and the result is found,

(f) Finally, the match from a and b sequences arises in such a way that the GA error is the smallest. According to the explanations above, the data in Fig. 6.49 for the GA application is represented by a chromosome of $n^2 = 100$ ($n = 10$) and $m = 16$ in length.

(g) Naturally, according to step (b), each chromosome has 32 digits with a population size of 100. In this example, the probability values of crossover and digit change operations are taken as 0.80 and 0.01, respectively. After the completion of the above steps, the GA and separation line were determined by Öztopal (1998).

$$y = 27.83 + 1.12z \tag{6.60}$$

However, as a result of statistical linear analysis, Panofsky and Brier (1968)

$$y = 37.53 + 0.75z \tag{6.61}$$

In Fig. 6.49, the solid line is the separation lines found by Panofsky and Brier (1968), and the dotted line is the GA method. Comparing the two lines reveals the following differences.

(a) GA resolution has steeper slope. This means that as the dew point temperature z approaches zero, a lot of precipitation occurs,
(b) The regression separation line resulted in 15 misclassifications. Accordingly, the error percentage is 15/91 = 0.16. However, the GA discrimination line only gives rise to 12 misclassifications. Accordingly, the error percentage is 12/91 = 0.13. Accordingly, GA gives better results than the statistical discrimination line,
(c) The GA method, which reduces the number of misclassifications, performs sequential renewal operations. The sequential error reduction of the GA is shown in Fig. 6.50. It is not possible to follow such sequential error reduction in the regression discrimination line method.

The same data is now considered in two separate parts. Of these, the first 45 data were randomly selected from the data as training data for GA. Separation lines are shown in Fig. 6.51 with the data. There are 5 misclassifications of GA here. Statistical discrimination line can only make 7 wrong classifications.

Figure 6.52 shows the discriminant lines (GA and regression) of the remaining 46 data after the training.

Here again, precipitation and non-precipitation predictions can be made with less error than the GA method.

For example, in order to show the effectiveness of GA in predicting precipitation, GA modeling of precipitation at Van Lake station is presented. For this, according to the mathematical expression given in the equation below, the precipitation of a month is generally calculated in a linear model in terms of precipitation of the preceding 3 months is expressed as,

$$Y_t = aY_{t-1} + bY_{t-2} + cY_{t-3} + \varepsilon \tag{5.6}$$

Here, a, b, and c are the model parameters, and *a* is a random term with no internal or external dependencies. After optimization with GA using the data of the first 346 months (28 years), the model parameters are found as a = 0.49, b = 0.15 and c = 0.14. Then, using the expression in Eq. (5.6), precipitations of other months are estimated and shown in Fig. 6.53 together with the measurements. From this figure, we see how the estimation and measurement values follow each other closely.

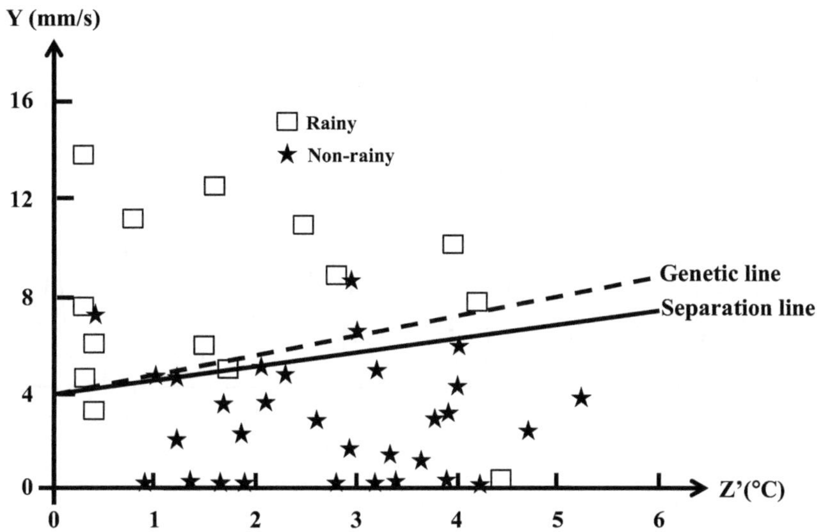

Fig. 6.50 GA error percentage reductions

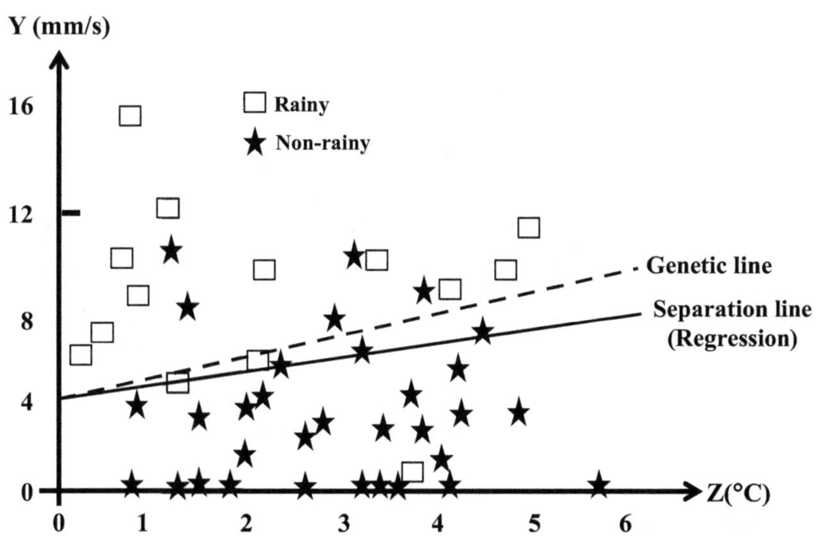

Fig. 6.51 Training data and separation straight-lines

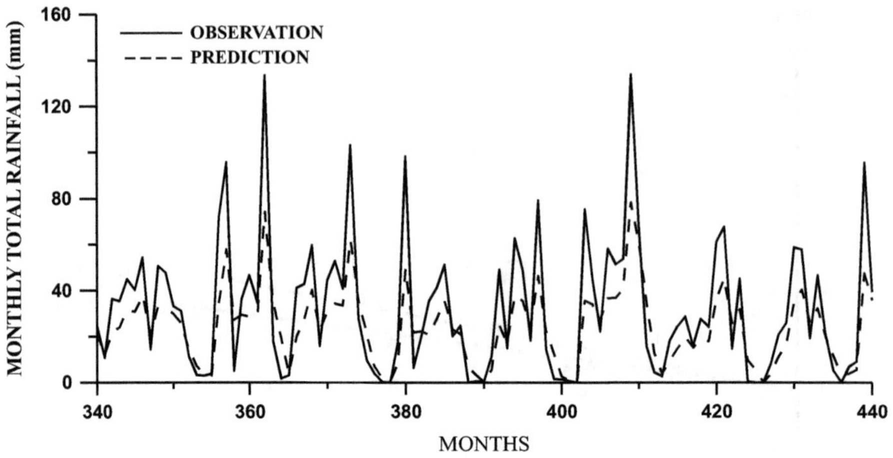

Fig. 6.52 GA error percentage decrease pattern

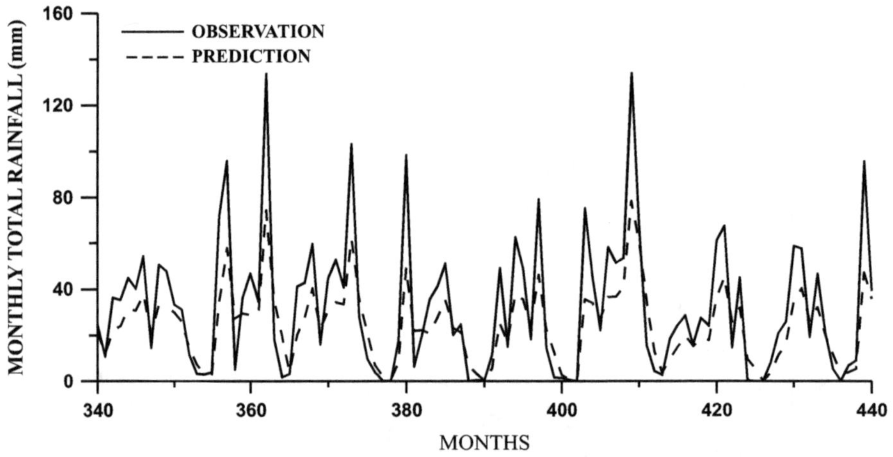

Fig. 6.53 Prediction and observation monthly precipitation data

6.16 Conclusions

One of the most important questions in research is whether optimization principles, efforts and studies can produce reliable results. This is because the available opportunities (materials, time, and space) are limited and significant economy is provided in the form of optimization by the service. In addition to systematic (deterministic) methods (Chap. 5) which provide the solution of various problems with exact equations (without any uncertainty), there are also systematic, but probabilistic and statistical methods (Chap. 4) that involve a certain degree of uncertainty. The Genetic Algorithms (GA) model provides the opportunity to reach a solution by eliminating the deficiencies in classical methods. Although there are random elements in the method itself, GA can reach the absolute optimization solution in the shortest time. The GA method is easy to understand by everyone as it can reach the result with only arithmetic calculations without requiring detailed and heavy mathematics. All the basic principles for GA modeling applications as its components, their combination, and finally practical examples are also given.

In this chapter, the principles, logic, and similarities of GA philosophy with other classical methods are explained and the reader's love for the subject and self-development principles is taken into consideration by giving necessary clues.

References

Adeli H, Hung SL (1995) Machine learning- neural networks, genetic algorithms and fuzzy systems, John Wiley & Sons, Inc.

Barnes SL (1964) A technique for maximizing details in numerical weather map analysis. J Appl Meteor 3:396–409

Buckles BP, Petry FE (1992) Genetic Algorithms, Washington: IEEE Computer Society Press, Technology Series

Caruana RA, Schaffer JD (1988) Representation and hidden bias: Gray vs. binary coding for genetic algorithms, paper presented at Fifth International Conference on Machine Learning, Univ. of Mich., Ann Arbor

Cressman GP (1955) An operational objective analysis system. Mon Wea Rev 87, 10:367–374

De Jong KA (1975) Analysis of the behavior of a class of genetic adaptive systems. Ph. D. Dissertation, The University of Michigan, Ann Anbor

Goldberg DE (1989) Genetic algorithms in search optimization and machine learning. Addison-Wesley, Reading

Haupt RL, Haupt SE (1998) Practical genetic algorithm. Willy, 177 pp

Hinterding, R Michalewicz Z, Peachey TC (1989) Self-adaptive genetic algorithm for numeric functions. Modifications and Extensions of Evolutionary Algorithms Adaptation, Niching, and Isolation in Evolutionary Algorithms. International Conference on Parallel Problem Solving from Nature PPSN 1996: Parallel Problem Solving from Nature — PPSN IV: 420–429

Holland J (1975) Genetic algorithms, Scientific American, July, pp 44–50

Ifrah G (1994) İslam Dünyasında Hint Rakamları. Rakamların Evrensel Tarihi (Indian Numerals in Islamic World. Evolutionary history of nembers). (in Turkish) TÜBİTAK, 159 pp

Kirkpatrick JB (1983) An iterative method for establishing priorities for the selection of nature reserves: An example from Tasmania. Biological Conservation 25(2): 127–134

Lorenz EN (1963) Deterministic non-periodic flow. J Atmos Sci 20:130–141

Michalewicz Z (1992) Genetic Algorithms + Data Structures = Evolution Programs. Springer, Third Edition, 383 pp

Öztopal A (1998) Genetik Algoritmaların Meteorolojik Uygulamaları.(Genetic Algorithm Applications in Meteorological Applications) M. Sc. Thesis, Istanbul Technical University, (in Turkish)

Panofsky HA, Brier GW (1968) Some application of statistics to meteorology, Pennsylvania State University Press, 224 pp

Sasaki Y (1960) An objective analysis for determining initial conditions for the primitive equations, Tech. Rep., (Ref. 60-16T). Texas A/M University, College Station

Syswerda G (1991) Schedule optimization using genetic algorithms. In: Davis L (ed) Handbook of genetic algorithms. Van Nostrand Reinhold, New York, pp 332–349
Thiebaux HJ Pedder MA (1987) Spatial objective analysis. Academic, 299 pp
Şen Z (2002) İhtimaller Hesabı Prensipleri. (Probability Calculation Principles) (in Turkish) Bilge, Kültür ve Sanat yayıncılık, İstanbul, 147 pp
Şen Z (2004) Genetik Algoritmalar ve Eniyileme Yöntemleri (Genetic Algorithms and Optimization Methods). (in Turkish) Su Vakfı, İstanbul, 142 pp

Chapter 7
Artificial Neural Networks

7.1 General

Mankind has lived with nature since its creation and learned many solutions from it through inspirations. One tried to fight against natural disasters as much as possible, along with using natural resources to meet needs. In addition, men tried to understand the events occurring in nature with physical or imaginative pure thoughts, mind activities, and feelings. They tried to scrutinize the cause-effect relations of the events as much as the knowledge and technology of the period. During these studies, they developed many methods.

The methods studied have made unprecedented advances in the last several decades as a result of the rapid development of computer numerical computation possibilities after 1950. Some of the developed methods were inspired by living organisms. Artificial neural networks (ANN) and genetic algorithms (GA) can be given as examples of the methods that emerged by trying to express the functioning of human organisms with mathematics.

It is a fact that technological developments are very rapid in the age we live in. Especially, the developments in the computer world have become dizzying. These developments are speed in calculations, examination of the smallest (nano) and largest (macro) realms, acceleration of information production, etc. These have helped humanity in scientific and technological ways. The basis of this lies in the fact that computers process the input data very quickly.

On the other hand, the human brain is very successful in seeing, speaking, error correction, and shape recognition in a noisy environment and with incomplete information. The human brain can produce much more efficient results in such processes than a computer. While nerves, known as electrochemical brain processing elements, respond to a process in milliseconds (10^{-3}), and today's electronic technology products respond in nanoseconds (10^{-9}). Even though electronic technology products work 10^6 times faster, the reason why the brain works more efficiently than computers in terms of processing with incomplete information and

Z. Şen, *Shallow and Deep Learning Principles*,
https://doi.org/10.1007/978-3-031-29555-3_7

recognizing shapes has been a matter of curiosity for a long time. This curiosity has led people to examine the working system of their brain.

As a result of the examination of the brain and its tissue, it was determined that the difference was caused by the information processing system. The striking result obtained showed that the nerves in the human brain process information in parallel. Based on this fact, scientists have accelerated their studies so that information is processed in parallel, as in the human brain, so that it is as close to human behavior as possible. Studies have focused on the design of methods, devices, and machines with the capabilities. Finally, methods such as ANN and GA have been developed for the establishment of intelligent systems that can model human behavior. Especially with the help of these methods, expert systems, which have attracted great interest, have become used in many fields.

ANN resulting from the modeling of the nervous system shows the characteristics of the biological nervous system in terms of parallel working and learning abilities. In addition to other features, the fact that it can process information quickly due to its parallel operation and that its hardware can be implemented easily makes ANN more attractive than other methods. That is why ANN is still used for classification, pattern recognition, control, prognostic (understanding the inner workings of an event) calculations, image processing, voice recognition, etc. It is used extensively in many fields. Although ANNs have just started to be used in many areas, their usage areas are increasing day by day.

7.2 Biological Structure

The nervous system in humans consists of cells called nerves. Nerves are the smallest units in which the vital activities of living things are carried out. The nerves, which are estimated to have about 10^{10}–10^{11} in a person, are spread not only in the human brain, but throughout the body on the central nervous system. The task of the nerves that make up the communication system of the brain is to receive signals and process and transmit electrochemical signals within the neural networks.

ANNs are nothing but the application of technology and scientific research methods by neglecting the details of some of the findings known today regarding the biological nervous system. ANN consists of a system that has layers that communicate with each other in parallel, and each layer contains enough nerve cells. Communications between these layers and their cells, which can be considered quite complex, are provided by weight coefficients determined according to the nature of the phenomenon under investigation. As stated by Lippman (1987), the principle of "back propagation of errors" is generally used as the modeling criterion in ANN. Figure 7.1 shows the structure of a biological nerve cell. As can be understood from this, a cell consists of the cell body, snaps (connections), dendrites (entrances), and axons (exits). Although there is a nucleus inside the cell, the size of the cell body varies between 5 and 130 microns. There are also larger cells. The fundamental neural system units are neurons, and they transmit information to different parts of the human body.

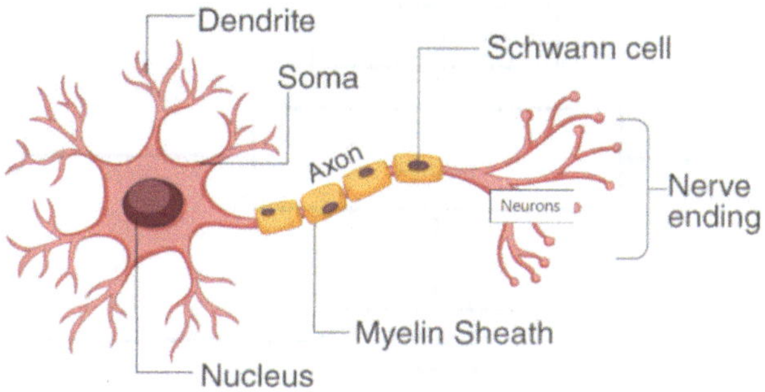

Fig. 7.1 A biological nerve cell

Snaps are connections that enable communication between a neuron output and a neighboring neuron input. Dendrites are responsible for carrying the information received from the snaps to the body. When the information collected in the cell body opens the excitation threshold, the cell is stimulated, and signals are sent to other cells with the help of axons, which are transport lines. Axons and dendrites can be compared in terms of their functions. The axon is longer than the dendrites. In addition, they have a regular structure because they have fewer branches. Dendrites, on the other hand, resemble a natural tree more, because they have a more irregular surface and many branches (Freeman 1995).

7.3 ANN Definition and Characteristics

The artificial neuron model, inspired by the structure of biological neurons in the 1940s, showed that logic conjunctives such as AND, OR, NOT could be modeled numerically (Chap. 3). Thus, the examination of the biological nervous system and the development of ANN models that work like them have become the common work of researchers from different branches.

ANN, which can be interpreted as information processing means, can also be described as a detailed black box model that produces outputs against the given inputs. Although ANN is defined as a similarity to the nervous system formed by a group of nerve cells that provide information flow with the axons, it has also been defined as networks that are formed by the intense parallel connection of simple elements that are generally adaptive (repeatable, iterative) (Kohonen 1988; Kung 1993). ANN differs from conventional information processing methods due to its features. In terms of some features, it gives more reliable results than many methods. Some of these features can be defined as parallelism, fault tolerance, learnability, and ease of implementation. Since the processing of information in ANN is carried out in parallel, the transmitted information is independent of each other. In addition, since there is no time dependence between the connections in the same layer, they can work simultaneously, thus increasing the speed of information flow (Bkum 1992). Due to the parallel working principle, the error occurring in any unit does not cause a significant error in the whole system. There is an effect only in proportion to the weight of the cell. Thus, the overall system is minimally affected by local faults.

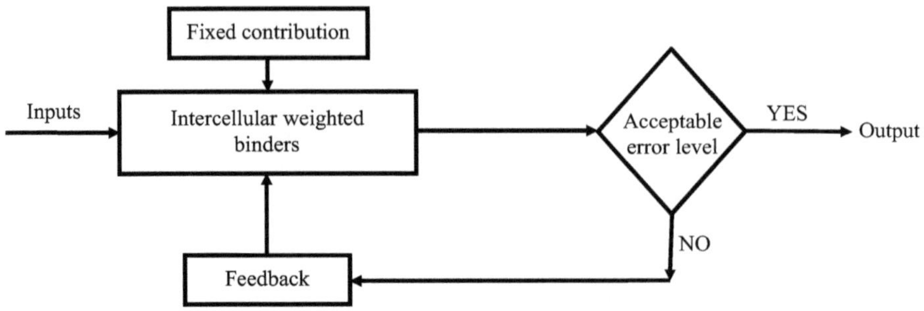

Fig. 7.2 ANN general structure

Learning in ANN takes place in the form of renewal of connection weights, where learning information is stored. In this way, it is possible to store the information obtained for a long time. In addition, learning ability allows solving problems that are not fully defined with the help of ANNs. A parallel running ANN model is preferred in solving many problems because it contains simple operations instead of complex functions and has an uncomplicated architecture.

ANNs are composed of simple nerve cells that operate in parallel. These elements are inspired by the biological nervous system. Like nature, the communication between the elements is provided by a network. By representing the connection activities between the elements in such a network with weights, they are trained according to the input and output data, and the contribution rates of the outputs are calculated separately.

After deciding on the type of data and the desired target for an ANN network, unknown connection values in this network are determined by training with sequential approaches in order to obtain the expected outputs from the inputs. The general structure of an ANN is shown in Fig. 7.2. Here, after the outputs obtained in the first calculations are compared with the expected outputs, the training of the ANN is terminated when they show an approximation to each other within acceptable error limits. Otherwise, the training continues. Thus, there is a network flow that can be driven by learning and training.

Grid connection weights and constant contribution are changed by training through feedback according to the error amounts between the output data (expected values) and the ANN outputs. The successive renewal and iterative execution of each training step to improve the previous one will be explained in detail in Sect. 7.5.1. This type of sequential improvement is sometimes called "instant" or "refreshing" training. Today, models that are more complex than classical computers and humans can solve can be done with ANN (Chap. 9).

In addition to using teacher training methods, unsupervised training techniques or direct design methods are also used in the training of some ANN networks. For example, unsupervised education approaches are used to group data. Some linear networks and especially ANN designs developed by Hopfield (1982) are trained without a teacher, that is, unsupervised training. From this, it can be said that there are different training and direct use opportunities in the development and use of ANN models.

ANN network can be examined in two parts, one is its architecture (structure), and the other is the mathematical functions that enable this architecture to work. In general, the architecture consists of nerve cells in the input, hidden (intermediate) and output layers, and the connections between them and a cell with a constant contribution. Inside the hidden cells, there is an inner working called the activation function (processor), that is, the trigger processor. One can think of ANN operation as two mathematical functions as internal and external operations. The inner workings of an ANN are provided by the activation functions in the hidden layer. Its external operation is the random assignment of connection values between successive layer cells and then the feedback of the output prediction error, etc. It is provided as a result of minimizing and renewing with transactions. These mathematical operations undertake the tasks of learning, training, remembering, and constantly detecting new information and renewing the network connections of the ANN.

ANN serves to obtain the desired outputs in parallel, by first separating all the inputs into simple components and then combining them, not sequentially as in time series and statistical data processing methods. We can compare the ANN architecture to a prism that can separate the incoming light into its components. In ANN network, it divides the data coming with its architecture and mathematics into parts of different simplicity and then recombines them in accordance with the desired outputs. In a way, ANN is like the plan, construction, operation, and benefit of a bridge that is planned to reach another point on the opposite shore from a construction point of a river. The information on the starting coast represents the input data, the information on the opposite coast represents the outputs, and the bridge between these two represents the ANN architecture. How are columns, beams, cables, slabs, etc. in the bridge? There are units, and if each of them relates to certain mathematical and physical rules, ANN tries to provide such a connection between layers and cells.

7.4 History

It is accepted that the first work on ANN started in 1943. McCullogh and Pitts (1943) first developed the cell model by describing the artificial nerve. In this study, nerve cells are modeled as logical elements with fixed threshold values. This model, which has fixed connection weights of $+1$ in its structure, is also called arithmetic logic calculation elements.

In 1949, when studies on learning intensified, scientists increased their efforts to model the human learning process. Hebb (1949) developed a rule that can be considered as the starting point for learning in ANN. This learning rule, which was put forward, gave an idea about how a neural network could perform the learning task at that time, but also formed the basis of many of the learning rules that are still valid today. Rosenblatt (1958) made the second breakthrough in ANN with the perceptron (single linear sensor) model and learning rule. This model also became the basis of today's machine learning algorithm (Chap. 8). Perceptron model is a learning machine that can be used in complex iterative behaviors. The biggest drawback of perceptron model is that it can only solve problems that can be separated by a linear multidimensional plane. It cannot produce solutions to nonlinear (curvilinear) problems.

Windrow and Hoff (1960) as a new approach developed ADAptive LINear Element (ADALINE) model in the early 1960s, when the studies accelerated. It can be called also as a single recursive linear neural network. Along with this, the Widrow-Hoff learning rule was put forward as a new and powerful learning rule. The most important feature of this rule is that it aims to minimize the total error throughout the training.

The increase in interest in ANN was only with the making of some new studies in the 1960s. The fact that the first methods were too weak to solve complex computational problems, therefore, hindered the progress of studies on this subject. In 1969, the anti-sensory publications by the mathematician Minsky and Paperts (1969) greatly diminished the interest in ANN and especially in perceptron model. Because, in their book called "Perceptron," they mathematically proved that its architecture cannot perform AND/OR (XOR) and many other logic functions (Chaps. 3 and 6). Demonstrating that perceptron model is insufficient in many aspects stopped the studies on ANN for a while. In fact, these developments have even caused the disappointment of many researchers. This stagnation continued from the early 1970s to the 1980s. Between these dates, although some scientists continued their studies with their own personal efforts, no significant progress could be made.

With the studies carried out in the early 1980s, ANN became widespread again. The first stirrings began with the development of nonlinear (curvilinear) networks by Hopfield (1982). He focused his work especially on the architecture of associative ANN networks. However, as a result of the works conducted by Kohonen (1982) and Anderson (1983), the studies gained momentum again with the development of unsupervised learning networks. Thus, these years were the years when the interest in ANN was revived and studies were intensified.

Rumelhart et al. (1986) developed a training algorithm called "back propagation" for multilayer perceptron-type networks. While this algorithm was powerful, it was based on very complex mathematical principles. In addition, the ability of this algorithm to enable effective learning has completely attracted attention. The introduction of this algorithm, which is still one of the most used teaching systems, broke new ground in the field of ANN. Today, studies on ANNs continue at a rapid pace in the world along with deep learning (Chap. 9). Frequently, new proposals are made about different learning algorithms and network architectures.

7.5 ANN Principles

ANNs with different architectural structures are used in practice to achieve one or more of the processes of learning, establishing relationships, classification, generalization, and optimization by utilizing the available data. There is a parallel flow of information from the input layer to the output in the architectural structure. Such a flow is provided by cells placed in parallel.

When the input data is given to an ANN, after the first training, the renewal (intensification or weakening) of the weights with the "steepest descend" method can be done by minimizing the mean error squares (MES) between the output data and the values expected to correspond to them (Chap. 6). If the MES is not less than a desired level, the weights in the ANN network are renewed by the feedback method. Feedback and network connectivity values are continued to be refreshed until the MES reaches a satisfactory level. There are two steps between ANN calculations: one is to transform the forward inputs into outputs, and the other is to regenerate the weights backward to reduce errors. As a result, it is desired that the model approaches the absolute minimum value in the error space, but this is difficult to guarantee. ANN can also reach a local minimization point, causing the illusion that the best case is achieved. The only way to avoid this is to run the same ANN several times with different initial weight assignments. If the same minimum error value is reached every time, it can be concluded that this corresponds to the absolute smallest error sought. If individual local minimum errors are obtained each time, the smallest local error position between them should be taken as the best solution.

For an ANN model to make reliable future predictions, it must be tested from different angles. For example, such a small error value can be reached during training that it may be nothing more than a random situation. For this reason, the data was entered in a random minimum error area before single training, the other during testing, and the third before testing. Hence, it should be divided into parts called harmony. Smith (1982) suggested that this separation should be 40%, 30%, and 30%, respectively.

7.5.1 ANN Terminology and Usage

For a subject to be perceived well, some basic concepts about it must be understood clearly. It is useful to use words with the same meaning in a cluster. In many works, the same concepts, methods, or interpretations are expressed with different words and special terminologies among them. If it is not known that some terminologies are used in the same sense, the reader may feel as if he is learning something different from the method he knows.

First, ANNs have architecture. There are two important elements in this architecture, the boxes and the connections between them. However, the presence of these two elements cannot be said to constitute the ANN model of that architecture. For example, in Fig. 7.3, different architectures are given very simply.

Figure 7.3a shows the operation of multiplying three variables—x, y, and z. The explanation for this can be made by adding three different variables together, multiplying them, and finding the result. Thus, the links between the three input boxes and the single operation (output) box point to bringing the multiplication operation to a common place (box) without the inputs being changed. If the connections move their inputs to the other side without making any changes, none of the architectures can be ANN models. Similarly, there is an architecture that

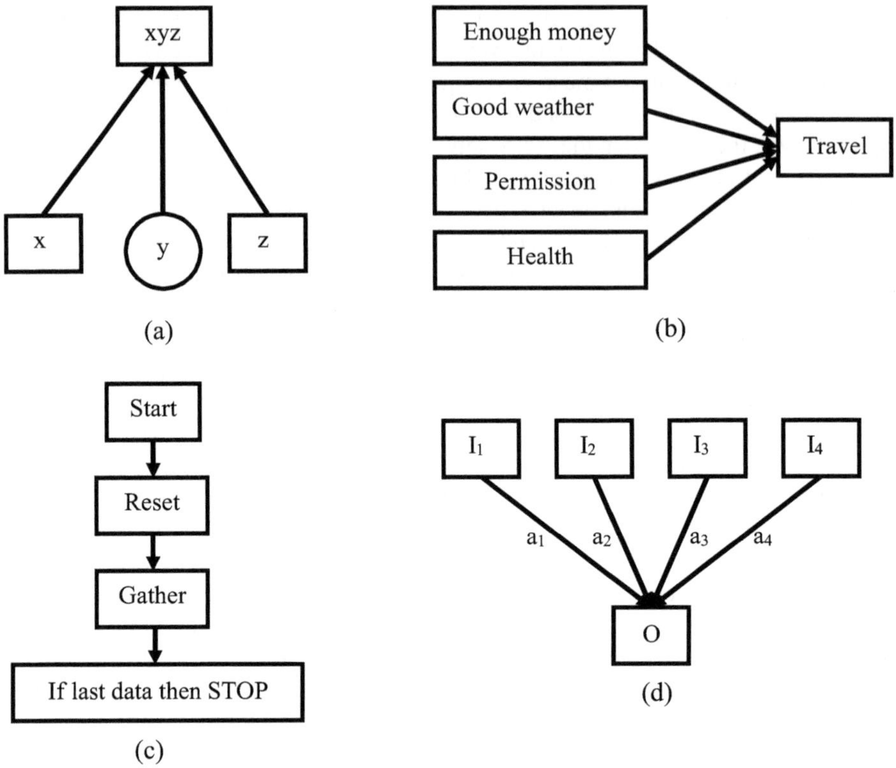

Fig. 7.3 Different architectures

explains the necessity of certain conditions to travel in Fig. 7.3b. Here, too, the links do not have numerical values and are only shown for information flow. Therefore, ANN model is not architecture. In Fig. 7.3c, which is a flow diagram, the connections cannot be perceived as ANN model architecture because they only show the flow directions. However, in the form of Fig. 7.3d, the connections between the input boxes and the output have weight values, and they play an active role in transferring the input data to the output. Only in this case can one look at the architecture as ANN architecture. As a result, architectures without weight values in their connections are not ANN models. In Fig. 7.3, the boxes are deliberately shown in the form of sometimes rectangular and sometimes circular. Although some publications say that there are slight differences between them, rectangular boxes will be used in this book in general. After knowing what the boxes mean, there is no harm in researching them to be made in triangles, trapezoids, or any other desired shape. However, rectangles and circles are widely used in the international literature. It is implied that direction is not important in the architectures in Fig. 7.3.

Each of the boxes in ANN architecture is given different names, such as node, unit, cell, and artificial nerve. However, the equivalent of all these terminologies in biology is nervous. In this book, the word cell is used in general.

7.5.2 Areas of ANN Use

Although the first ANN studies started about 50 years ago (Chap. 2), their extensive use and continuous development have been observed in the last 20 years. However, the principles, mathematical methods, terminologies and rules of many control systems, and optimization methods have been established in such a way that they have not changed for many years. ANN methods, together with fuzzy logic (FL) and GA methods (Chap. 6), constitute a very successful basis for higher education, research, industry, and technological developments. Each has applications in many different fields. Meanwhile, ANN has a simplicity of understanding and operation that can be adapted to applications in a short time.

For ANN applications to be carried out successfully, its principles and working processes must be understood very well. Unfortunately, most of the studies carried out in such cases cannot pass only a mechanical application. In mechanical studies, all parts of the ANN method, up to the mathematics, are written correctly from other basic sources. Wrong, incomplete, and sometimes illogical words are used in the interpretation and evaluation of the application results.

The ANNs can be used in almost every discipline and field of science; their uses are explained in relation to each methodology as follows.

1. *Classification*: In scientific studies, our daily classification habit continues, and one can make comments by dividing the examined events into classes in a way that does not overlap according to bivalent logic (Chap. 3). For example, a classification such that countries with more than 10,000 m^3 of water per capita per year are rich, those between 10,000 and 5000 m^3 are considered less rich, those between 5000 m^3 and 1000 m^3 are considered poor, and those with less than 1000 m^3 are considered very poor. Under the light of such a classification, it is easy to determine into which class a country falls if its annual per capita water consumption is given. Such a classification can also be achieved by ANN, and such a classification falls within the supervised learning procedure.
2. *Clustering*: Although the boundaries are determined and given beforehand in classification, there is no need for such information in clustering, because it means search for patterns (input data) that are like each other in the same group. ANN can be trained in such a way that it can first separate it into different subsets with a given set of patterns (data, vector). One can then decide in which cluster the patterns that will be given and that are not among the training data. Each cell in the output layer of ANNs can be developed to represent different clustering. Such ANNs achieve self-organizing clustering. From further explanation, it will be seen that the similarity for clustering will generally depend on the Euclidean distance between two data sets, or the angle between the vectors representing these two data, and hence, the concept of correlation in statistics is applicable (Chap. 4).
3. *Vector digitization*: Here, the process of separating many data into a small number of data becomes important with the same characteristics. In classical studies, there are acceptances or practical procedures related to this. For example,

in modeling and solving differential equations with finite elements, it is assumed that the properties of each finite element in the space in which it is valid are the same at all points in this space. Thus, since the property of millions of points is represented by a number, the number of data is significantly reduced. With the digitization process, for example, an area can be divided into nonoverlapping sub-domains. Looking at the map of European countries, it is seen that each country has its own borders. These are the process of separating into sub-domains with vectorial digitization. It is possible to have ANNs do such operations, like what is thought by the human mind. As will be explained in the next sections, special methods and algorithms have been developed for this (Sect. 7.6.2).

4. *Pattern compatibility*: As a result of overlaying a pattern given here as corrupt, worn, obsolete, or incomplete with a full and complete pattern previously stored in the memory of ANNs, ANN can output regular pattern to the corrupted input. This process is called pattern search and registration process. This can be likened to extracting a smooth curve from a noisy time series, for example, using Fourier analysis. It is this kind of human thought that shapes very close to a circle are idealized and accepted as a circle.
5. *Functional approach*: There are many shapes whose mathematical expression is unknown. In this case, the desired shape can be approached by overlapping other simple and regular shapes. It is possible to approach a given function in some basic ways through ANNs.
6. *Forecasting*: In general, the future can be predicted after understanding the internal and external workings of the event as a result of examining the behavior of past data. Forecasts of stock market oscillations, weather events, floods, etc. are always this type of function. Although it is not possible to make perfect estimations, it is possible to make the best estimation from the available data with ANN models. For example, according to Lee et al. (1989), it was possible to predict the classically known occurrence of sunspots every 11 years with ANN models. In fact, the estimation is like the functional approach mentioned in the previous step.
7. *Control problems*: There are internal relationships that are too complex to be compared with mathematical models for the operation of many tools and devices manufactured in the industry. Human modeling of these relationships is called automatic control or automation. Control can be defined as obtaining the desired output behavior from the given input values. For this purpose, ANN models have been used successfully in different studies. It is also possible to perform iterative (adaptive) checks, including changes over time.
8. *Optimization*: Maximizing (or minimizing) the target of the event examined in many commercial and scientific subjects under the given constraints is known as the optimization process (Chap. 6) (Şen 2004). Although many classical methods have been developed in previous studies for optimization, doing this with ANN modeling is useful, at least in the absence of limiting mathematical assumptions. In Chap. 6, various GA optimization procures are presented.

9. *Search studies*: It is also possible to achieve search methods, which are widely used in artificial intelligence (AI) applications with ANNs. Almost all the points mentioned before are search operations.

Currently, there is a tendency to be accepted as a scientific article, but even in the world, almost whatever is written on this subject, and most of the articles are doomed to remain as nonpermanent articles that will not be cited. Since the number of employees in these matters is low even in the world, there is a situation in the sense that whatever is written is almost printed. In order to benefit from this book and enter into prestigious scientific studies, please do not pass every line of scientific analysis (with filtering through science philosophy, logic and reason). Among the main benefits of ANN models are the following points

1. Absence of data-related assumptions required for statistics and other modeling techniques.
2. Nonlinear multi-input-multi-output (MIMO) systems can be easily modeled.
3. Automatic conversion of variables.

However, ANNs also have disadvantages that should be noted, which are the following major points

1. If it fits too well, it may be random.
2. It is not possible to know the binary relations of the input and output variables. Each ANN model is a detailed black box, as will be explained in the following sections. It is not possible to give meanings to the correlation values that could be related to the physics of the event.
3. Many data are needed to obtain reliable results.

7.5.3 Similarity of ANNs to Classic Methods

Although ANNs have a unique architecture, revealing their similarities with classical methods, concerning classification, control, forecasting, etc., will lead to a better understanding of the subject. Modeling has been done in many ways. These include matrix and vector calculations, regression and correlation concepts, Kalman iterative (adaptive) filter real-time adaptations, stochastic processes and time series analysis, multiple regression approaches, spectral analysis, and others. ANN includes some aspects like each of these, albeit partially. In this section, the relationship of ANNs with them will be mentioned, and it will be possible to better understand what similarities there are in their architectural structures and functioning. In order to determine the most suitable alternative, the validity of the methods and application conditions will be specified together with the necessary data.

ANNs are also called linked architecture, parallel distributed processing, and cell-structured systems. These are frequently interconnected, parallel structured information processing frameworks inspired by mammalian brain functions. In addition, thanks to the mathematical activation functions in their architecture, they transform the information they perceive into useful outputs, like the behavior of the biological nervous system. The most distinctive aspects of these systems are their information processing capabilities. The fact that these cells, which are related to each other by interconnectors, exchange weighted information with the connectors presents a situation like the information flow in the living body. ANNs learn the information they perceive through training by making mistakes, just as people obtain their knowledge by repeating their knowledge through training and learning to minimize mistakes over time and then using this knowledge in the face of new situations. After successfully passing the training, ANNs now decide to accept or reject the new information they perceive by testing. During training, the weights in the links are constantly changing and improving with good adjustments, complementing itself with training and post-testing like the same biological perception. How do babies learn by making mistakes after receiving some signals with their sense organs before they start to speak and by adjusting the whole nervous system according to their perceptions and errors, after a certain period by making no mistakes in some cases and little or practically insignificant mistakes in others. ANN functions ensure that their information is used for more useful purposes. It is also possible with ANNs to achieve reliable results in the future by accumulating information after mistakes and repetitions made in connections.

ANNs were mentioned for the first time in 1950, but due to the difficulties encountered, their use in practical and technological applications increased only after 1980. Today, their use continues to develop and increase in all disciplines in terms of number and complexity of operations. With these, patterns can be recognized, unbiased classifications can be made, and even if the information is somewhat lacking, generalizations can be made to reach the full result. In modeling an event with ANNs, all the information in the data is used in the most detailed way. For example, if classical methods are to be used even when searching for the simplest relationship between x and y data, it is encountered in many applications that the following assumptions are provided by the data, and even if not, models are made as if they satisfy these assumptions

1. The requirement that the terms remaining from the data or the relationship between them must conform to the normal (Gaussian) probability distribution function (PDF) (Chap. 4).
2. The necessity that the statistical parameters in the data do not change as the number of data increases, which is the mean, variance, etc. assumes that the parameters are constant (Chap. 5).
3. Post-model errors are linearly independent of each other.
4. There must be a certain mathematical pattern in the scatter diagrams of the data.
5. The measurements made are free of errors.

However, there are no prerequisites or acceptances in modeling the same problem with ANN, which can be used directly even in cases where there is no classical analysis algorithm or in solving very complex problems. To do ANN modeling, it is not necessary to understand the physics of the event beforehand. As a result, ANN models generally provide a multi-input multi-output relationship.

In recent years, there has been a tremendous increase in ANN applications. As a competitor to statistical methods, it can be used very effectively in modeling and then making predictions (Chap. 4). It should not be concluded that it will replace all statistical models. If necessary assumptions are made with statistical methods, one can make models by considering the physical structure of the phenomenon under investigation. For example, statistical mechanics, developed with Boltzmann's efforts at the end of the nineteenth century, was used in classical physics to reveal the internal dynamics of previously unknown events (temperature, pressure, energy, etc.). In addition, "particle physics" (quantum physics), which started to emerge after 1926 and causes many technological developments as current physics, continues to function with probabilistic and statistical rules.

7.6 Vector and Matrix Similarity

ANNs have at least three consecutive layers in information processing. These are the layer of sensing information from the external environment, which is called the input layer; the information processing layer, which is in the middle; and finally, the output layer that gives the information out from the ANN environment in a way that can be understood by humans. There are also cells working in parallel in each layer (see Fig. 7.4). Here, the weights of the inputs I, the outputs O, and the required number of ties connecting them are also indicated by $a_{i,j}$ and c_{ij}. Since each ANN can generally correspond to a function in mathematics, one can look at the input variables as independent variables to indicate that they are not affected by anything, the outputs as dependent variables since they are based on inputs, and finally, each of the ties as the weights (constants, parameters) of the mathematical expression connecting these two variables.

Each ANN corresponds to the expanded state of a function such as O = f(I). One can say that for each ANN, it provides an explanation of the relationship between dependent and independent variables. With such a relationship matrix notation takes its shape as follows:

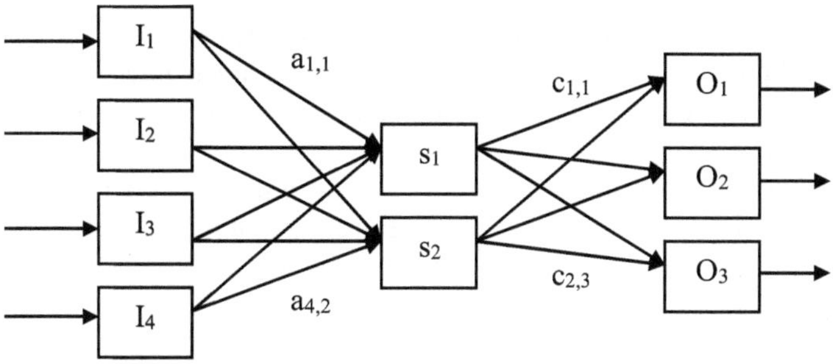

Fig. 7.4 Typical ANN

$$\begin{bmatrix} O_1 \\ O_2 \\ \vdots \\ O_m \end{bmatrix} = \begin{bmatrix} a_{1,1} & a_{1,2} & \dots & a_{1,n} \\ a_{2,1} & a_{2,2} & \dots & a_{2,n} \\ \vdots & \vdots & \vdots & \vdots \\ a_{m,1} & a_{m,2} & \dots & a_{m,n} \end{bmatrix} \times \begin{bmatrix} I_1 \\ I_2 \\ \vdots \\ I_n \end{bmatrix} \tag{7.1}$$

It is the input vector I with n elements, the output vector O with m elements, and the coefficients matrix with (nxm) elements in between showing the connection between these two vectors. This matrix incorporates the weights in ANNs. Equation 7.1 is equivalent to an ANN architecture that does not have an intermediate layer and consists of only input and output layers. In this representation, there is only a linear relationship between the input and output variables. In case of only one output, this equation can be written as:

$$[O] = [a_{1,1} \;\; a_{1,2} \;\; a_{1,3} \;\; \dots \;\; a_{1,n}] \begin{bmatrix} I_1 \\ I_2 \\ I_3 \\ \vdots \\ I_n \end{bmatrix} \tag{7.2}$$

This matrix structure corresponds to the single linear detector (perceptron), which will be explained in Sect. 7.8. In Eq. 7.2, the mathematical structure is purely linear, and its classical mathematical form can be written as follows:

$$O = a_{1,1}I_1 + a_{1,2}I_2 + a_{1,3}I_3 + \dots + a_{1,n}I_n \tag{7.3}$$

In this vector calculation, it is nothing but the scalar product of two vectors, namely, as Is and As, $a_{i,j}$'s [in Eq. 7.2 where I is the column and A is the row vectors]. So, the structure of linear perceptron is the scalar product of two vectors that can be displayed as:

$$O = \vec{A} . \vec{I} \tag{7.4}$$

If these vectors are normalized, Eq. 7.4 gives the linear correlation coefficient in statistics. If the result is zero, these vectors are fully independent; if +1 (−1), then they are completely dependent; between 0 and +1 (−1), they are partially positively (negatively) dependent.

In perceptron, the output is either 1 or 0, as will be explained in Sect. 7.8.3 Accordingly, perceptron makes two different classifications. After all these explanations, it is understood that the perceptron structure has a structure with a single input layer and an output cell as in Fig. 7.5.

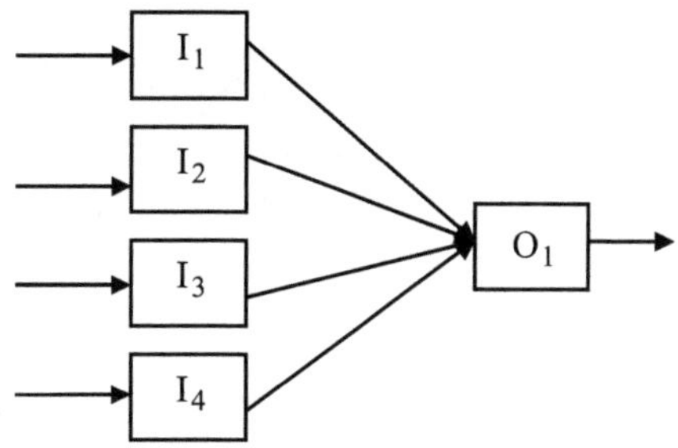

Fig. 7.5 Perceptron structure

The two most important features of ANN architectures, which will be explained in Sect. 7.11.1, which differ from perceptron, are: first, there is a third intermediate (hidden) layer between the input and output layers, and second, the mathematical operations are nonlinear. From this, it can be concluded that ANNs with a structure like the above matrix will only perform linear operations and have at most two layers. An ANN with an input and output layer and only one cell in the output layer is called perceptron. In matrix mathematics, the matrix that acts as a converter between variable vectors is called the matrix of coefficients. However, in ANNs, this is called the weight matrix. Here, the first simple neural network put forward in 1950 is nothing, but the form of Eq. 7.1 converted into bilayer cells (blocks) as follows with weights, a_{ij} ($i = 1, 2, \ldots, n; j = 1, 2, \ldots, m$) (Fig. 7.6).

With the consecutive coefficient's matrix, matrix equations that will form the basis of three or more layered ANNs can also be written. For example, where there are two coefficient matrices A and C, the architectural structure takes the following form.

$$\begin{bmatrix} O_1 \\ O_2 \\ \vdots \\ O_m \end{bmatrix} = \begin{bmatrix} a_{1,1} & a_{1,2} & \ldots & a_{1,k} \\ a_{2,1} & a_{2,2} & \ldots & a_{2,k} \\ \vdots & \vdots & \vdots & \vdots \\ a_{m,1} & a_{m,2} & \ldots & a_{m,k} \end{bmatrix} \times \begin{bmatrix} c_{1,1} & c_{1,2} & \ldots & c_{1,n} \\ c_{2,1} & c_{2,2} & \ldots & c_{2,n} \\ \vdots & \vdots & \vdots & \vdots \\ c_{k,1} & c_{k,2} & \ldots & c_{k,n} \end{bmatrix} \times \begin{bmatrix} I_1 \\ I_2 \\ \vdots \\ I_n \end{bmatrix} \tag{7.5}$$

There are three layers with input and output layers, one of which is the intermediate layer. However, this new three-layer ANN will not be able to avoid linear computation. At this point, researchers in the world thought a lot and could not make any developments due to the obsession that only linear operations can be done with ANN for 25–30 years.

The matrix structure given in Eq. 7.5 can be made more detailed with an intermediate operation by remembering that the operation flow is from right to left. First, by multiplying I input by the C weights, we can write a new intermediate (hidden) variable (vector) S or the intermediate outputs linearly as follows.

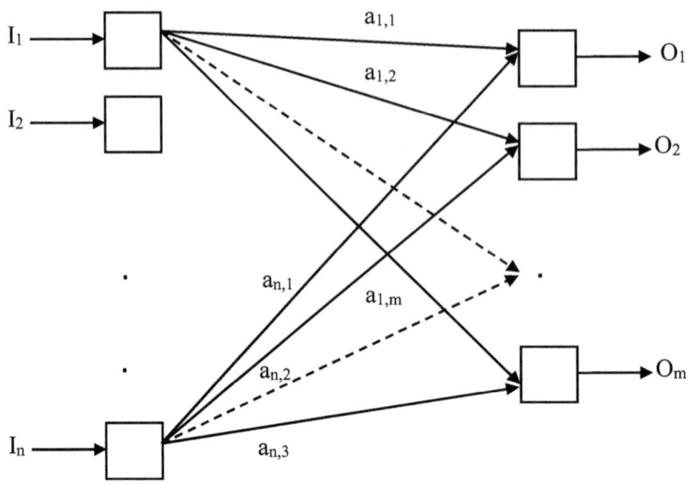

Fig. 7.6 ANN indication of matrix model

$$\begin{bmatrix} H_1 \\ H_2 \\ \vdots \\ H_k \end{bmatrix} = \begin{bmatrix} c_{1,1} & c_{1,2} & \cdots & c_{1,m} \\ c_{2,1} & c_{2,2} & \cdots & c_{2,m} \\ \vdots & \vdots & \vdots & \vdots \\ c_{k,1} & c_{k,2} & \cdots & c_{k,m} \end{bmatrix} \times \begin{bmatrix} I_1 \\ I_2 \\ \vdots \\ I_m \end{bmatrix} \tag{7.6}$$

Its ANN architectural structure is presented in Fig. 7.7.

By consideration of Eq. 7.6, Eq. 7.5 takes the following shape with H_i, ($i = 0, 1, 2, \ldots, k$) representing intermediate hidden layer elements.

$$\begin{bmatrix} I_1 \\ I_2 \\ \vdots \\ I_n \end{bmatrix} = \begin{bmatrix} a_{1,1} & a_{1,2} & \cdots & a_{1,k} \\ a_{2,1} & a_{2,2} & \cdots & a_{2,k} \\ \vdots & \vdots & \vdots & \vdots \\ a_{n,1} & a_{n,2} & \cdots & a_{n,k} \end{bmatrix} \times \begin{bmatrix} H_1 \\ H_2 \\ \vdots \\ H_k \end{bmatrix} \tag{7.7}$$

Its simple ANN architectural structure can be given as follows (Fig. 7.8).

The output of the previous ANNs in these last two figures is the same as the input of the latter. By overlapping them, the unified architectural structure given in Fig. 7.9 is obtained.

Thus, three-layer ANN architecture has emerged by transforming matrix algebra equations into ANN structure with logical approaches. However, there is a problem here. If the H cells in the hidden (intermediate) layer follow the same matrix calculations described above, generating outputs from inputs will remain a linear operation. To avoid this, either the coefficients must be changed nonlinearly, or the H cells or both must be skewed (put nonlinear mathematical operators). The easiest of these three options in practice is to structure the low number H cells with a

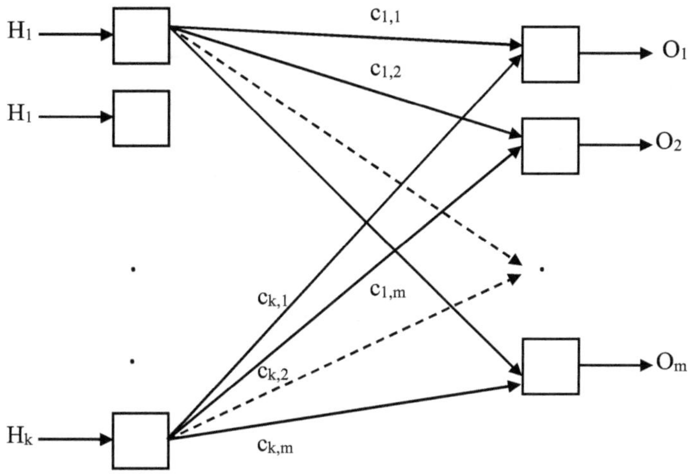

Fig. 7.7 Left-hand side ANN structure between the hidden and output layers

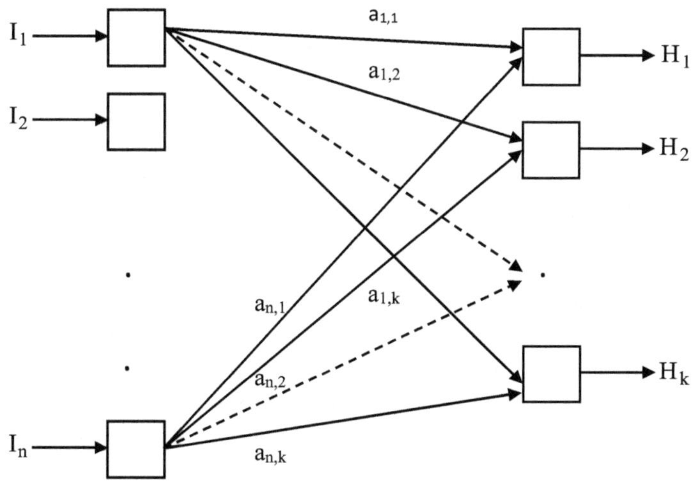

Fig. 7.8 Left side is ANN structure between the hidden and output layers

nonlinear mathematical processor inside each. For this option, the question is what will be the mathematical expression of the activation function to be placed in the cells in the hidden (intermediate) layer. One can immediately see that H cells have two functions. The first of these is to collect the weighted parts coming from each of the input layer cells, that is, an adder operator on the left of the cell, and then another activation function should be placed on the right side of the same cell as the nonlinearity operator (activation function) of this collected information.

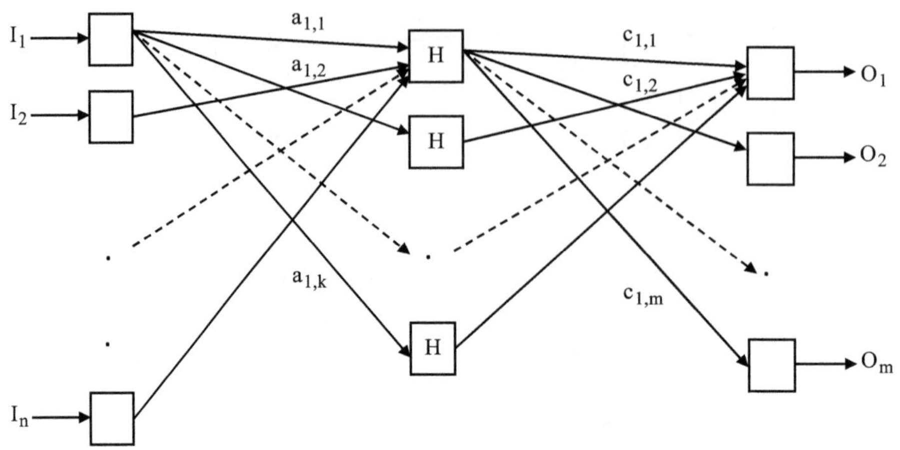

Fig. 7.9 Superimposition of two ANN

The task of this activation function is to curve the linear information it receives from the adder activation function and leave it to the output of the cell. From there, this information is shared between the output cells by means of the weights between the intermediate (hidden) layer and the output layer. Therefore, it is necessary to have a linear activation function in each output cell that will detect and collect only the signals coming to it. Accordingly, while there are no processors in the cells in the input layer of an ANN, there are two adder and curvilinear activation functions in the intermediate layer cells and linear adder processors in the output cell. After all this has been said, ANN architecture (topology) is defined (see Sect. 7.11.1).

For linear activation functions, there is no problem. These are units that perform simple arithmetic addition. However, the curvilinear operator (this is called the activation function in ANNs) needs to be determined. Here, it is possible to put different activation function functions in each hidden layer cell, and they are all placed in the same activation function to provide speed and simplicity in calculations. This subject will be explained in detail in Sect. 7.11.1 in due course.

7.6.1 Similarity to Kalman Filters

Kalman (1960) filters are a dynamic and repeatable linear version of regression analysis. As a general structure, it has two parts: one is the system, and the other is the measurement equations. Both parts are expressed separately by matrix equations. Among them, the system equation expresses the interdependence and inner workings of the state variables in the system between successive times. If the state variables of a system are represented by the D (d_1, d_2, ..., d_n) vector, the system equation is the relationship between two consecutive times takes the following form.

$$\begin{bmatrix} d_1 \\ d_2 \\ \vdots \\ d_n \end{bmatrix}_t = \begin{bmatrix} a_{1,1} & a_{1,2} & \cdots & a_{1,n} \\ a_{2,1} & a_{2,2} & \cdots & a_{2,n} \\ \vdots & \vdots & \vdots & \vdots \\ a_{n,1} & a_{n,2} & \cdots & a_{n,n} \end{bmatrix} \times \begin{bmatrix} d_1 \\ d_2 \\ \vdots \\ d_n \end{bmatrix}_{t-1} + \begin{bmatrix} \varepsilon_1 \\ \varepsilon_2 \\ \vdots \\ \varepsilon_n \end{bmatrix}_t \tag{7.8}$$

Here, ϵ_i's show the errors of the system at time t. The weight matrix is in the form of a square matrix. This equation in matrix form has ANN architecture like the expansion shown in Fig. 7.4. The measurement equation for Kalman filters is written in matrix form as follows.

$$\begin{bmatrix} m_1 \\ m_2 \\ \vdots \\ m_m \end{bmatrix} = \begin{bmatrix} 1 & 1 & 1 & 0 & 0 & 1 \\ 0 & 1 & 0 & 1 & 1 & 0 \\ 0 & 0 & 0 & 0 & 0 & 0 \\ \vdots & \vdots & \vdots & \vdots & \vdots & \vdots \\ 1 & 0 & 0 & 0 & 0 & 0 \end{bmatrix} \times \begin{bmatrix} d_1 \\ d_2 \\ d_3 \\ \vdots \\ d_n \end{bmatrix} + \begin{bmatrix} \varepsilon'_1 \\ \varepsilon'_2 \\ \vdots \\ \varepsilon'_n \end{bmatrix} \tag{7.9}$$

In this expansion, M indicates measurement (output) values and measurement errors in $\acute{\varepsilon}_i$'s. Here, the coefficients matrix of size (mxn) consists of 1 and 0 elements. Of these, 1s indicate which state variables directly contribute to the output values. Zeros indicate no such contribution. The full architectural structure of Kalman filter under the light of these points is shown in Fig. 7.10.

In this figure, each element of the coefficients matrix in the measurement equation represents the correlation values between the state variables and the measurements. There are no curvilinear operators in each of the cells in the hidden and output layer of the Kalman filter, and only random independent errors are added. Thus, the error terms ε_i's in the system equation and $\acute{\varepsilon}$ in the measurement equation are also included in Fig. 7.10 structure. However, with their additions, the linear operation of the Kalman filter does not change but becomes random. Another point to note in Fig. 7.10 is that the connection weights between the hidden and output layers in Kalman filters are completely determined. Comparing this Kaman filter architecture with Fig. 7.9 reveals how similar they are to each other.

7.6.2 Similarity to Multiple Regression

If there are many variables related to a problem, the most common method used to express a variable that is desired to be estimated among them in terms of the remaining estimator variables is multiple regression analysis (Chap. 4). In its most general form, one can write as follows:

$$O = a_1 I_1 + a_2 I_2 + a_3 I_3 + \cdots + a_n I_n + \varepsilon \tag{7.10}$$

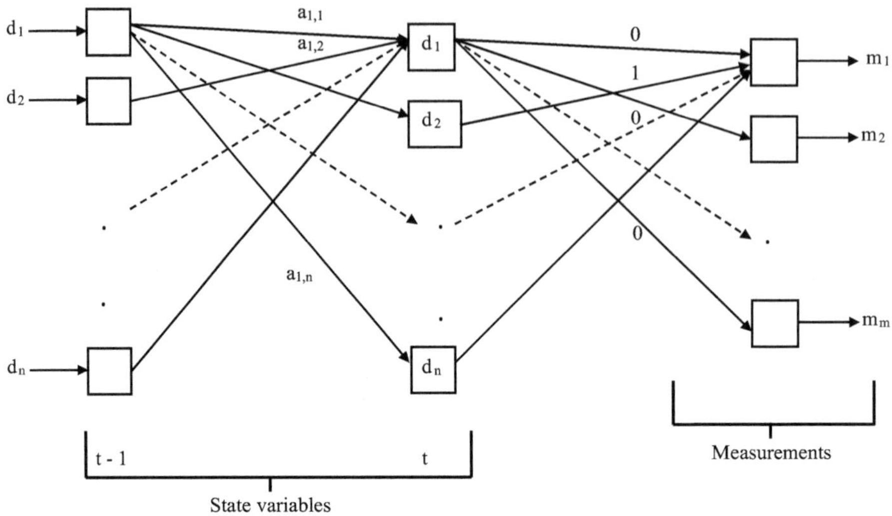

Fig. 7.10 Kalman filter ANN architecture

Here, I_i ($i = 1, 2, \ldots, n$) are called independent variables in mathematics and predictive variables in control theory. On the other hand, three are dependent or predicted output, O, variable. As can be understood from the mathematical expression, the I_i's must be completely independent of each other for this equation to be valid. In terms of ANN architecture, there must be only one connection between any input and output. In this case, the architecture of Eq. 7.10 should be as in Fig. 7.11.

It turns out that the regression architecture is equivalent to the perceptron architecture, which will be explained in Sect. 7.8. However, there are differences in connection values. In the multiple regressions, the connection values (correlation coefficients) are equal to the value of the connection between the inputs and the output. However, for these correlation coefficients to be valid for multiple regressions, the data must comply with some important assumptions.

1. It is assumed that the input variables are completely independent among themselves. This is an almost rare occurrence in practice. For this reason, the input variables are subjected to some transformations so that the correlation coefficient between them becomes zero. These include principle components analysis (PCA) and Fourier analysis methods. So, the structure given in Fig. 7.11 is also valid for the principle components and the transformed input variables in the Fourier analysis.
2. The error term ε in the model should fit the Gaussian (normal) PDF.
3. The data should be stationary, that is, they should not contain trends, periodic fluctuations, jumps, or spikes.
4. The variances of the data should be constant (homoscedasticity).

As a result, one can say that the multiple regression approach is like the modeling architecture with ANN, if all these assumptions are provided by the input and output data.

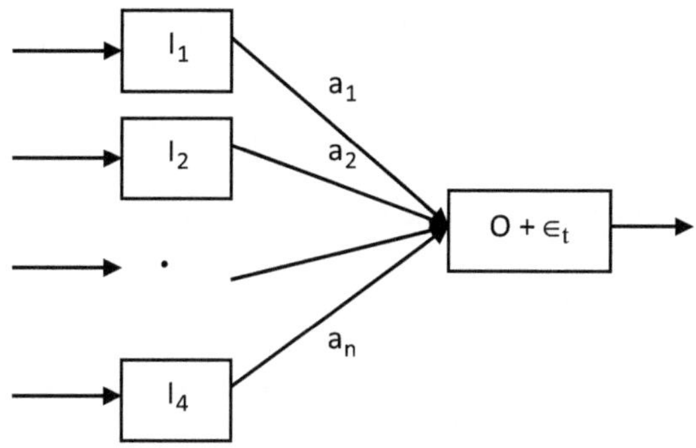

Fig. 7.11 Multiple regression ANN

7.6.3 Similarity to Stochastic Processes

We can say that stochastic processes, which are frequently used in the modeling of many natural and artificial events, and especially in signal and image processing areas, show similar situations to the ANN structure. In general, the stochastic process consists of two parts, one deterministic and the other random. The mathematical expression for a kth-order Markov process, the simplest of stochastic processes, is in the following form:

$$d_t = a_1 d_{t-1} + a_2 d_{t-2} + a_3 d_{t-3} + \cdots + a_k d_{t-k} + \varepsilon_t \tag{7.11}$$

This equation linearly connects the value of a state variable at time t to all its historical values (d_{t-1}, d_{t-2}, ... , d_{t-k}) up to time t−k. Here, ε_t denotes the uncertainty part. In stochastic processes, some assumptions must be valid from the theory. Among these are the conditions of being stationary and ergodicity. In addition, there are assumptions that the errors have Gaussian (normal) PDF and that they do not have periodicity in their structure. Equation 7.11 in terms of vectors can be written as follows:

$$[d_t] = \begin{bmatrix} a_1 & a_2 & a_3 & \ldots & a_k \end{bmatrix} \begin{bmatrix} d_{t-1} \\ d_{t-2} \\ d_{t-3} \\ \vdots \\ d_{t-k} \end{bmatrix} + \varepsilon_t \tag{7.12}$$

Its ANN-like architecture is also like the perceptron architecture given in Fig. 7.5.

7.6.4 Similarity to Black Box Models

Although ANNs were biologically inspired by the cells of the human nervous system in the brain, it is possible to explain this in terms of engineering modeling. Any natural or social phenomenon has a processor that converts inputs to outputs. In its simplest form, one can represent this as a black or closed box model, as shown in the figure below.

With this structure, it has been used in much different environment, society, engineering, and psychological modeling, and reliable results have been achieved. The essence of this is the principle in the philosophy of science that every event has consequences under certain reasons (Chap. 3). In Fig. 7.12, the fertile part is named the processor. The black box model has been particularly successful in examining events that have one input and one output. In terms of inputs and outputs, one can collect black box models in four main groups as single-input-single-output (SISO), single-input-multi-output (SIMO), multiple-input-single-output (MISO), and multiple-input-multi-output (MIMO). The perceptron model is the most common type in practical applications. It is more natural to have a wide variety of processors in other types. Since there is a processor box in accordance with the SISO model in Fig. 7.12, one can call it a three-box model. The first box here contains the properties of a single input variable, connecting it to a single processor box. The inputs that it processes uniformly in this box eventually yield a single output. With a rational inference from here, it is understood that if the input, processor and output are multiple, for example, 2-box, 3-box, 15-bok, etc., multi-box models emerge.

Before proceeding to the explanation of other multiple models, it should be noted that the model in Fig. 7.12 has one input, the other output, and a breakout or hidden box between them in terms of shape. Thus, even in the simplest univariate case, there are three different boxes, or three layers with one cell each, analogous to the ANN architecture. The input box is the measured input, the output box is the predicted output variable, and the processor box is a hidden (intermediate) box whose processor is not yet known. It is possible to measure outputs like input variables. It turns out that the operation of the unknown processor in the hidden box is not known for output fertility, but its behavior must be determined from the input and output data. One of the methods developed in this direction is the least square regression method. Here, by fitting the most appropriate straight line or curve to the scattering diagram that emerges in the scatter made by showing the input variables on the horizontal axis of the Cartesian coordinate set, and the output variables on the vertical axis, the productive processor can be determined (see Fig. 7.13). Other methods (Fourier, Walsh, wavelet, waves, spectral, stochastic processes, multiple regression, PCA, etc.) can reflect the behavior of the processor. In this direction, spectral and wavelet analysis are the most widely used approaches today.

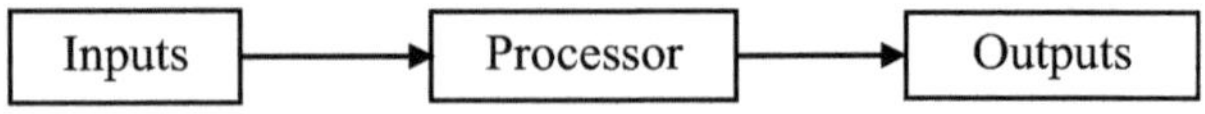

Fig. 7.12 Black box model

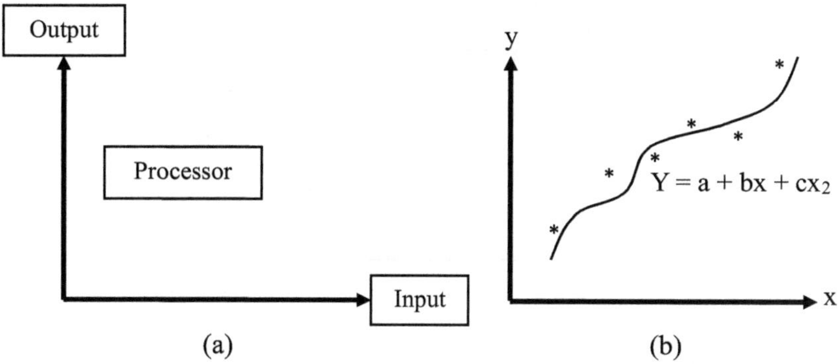

Fig. 7.13 Cartesian coordinate system

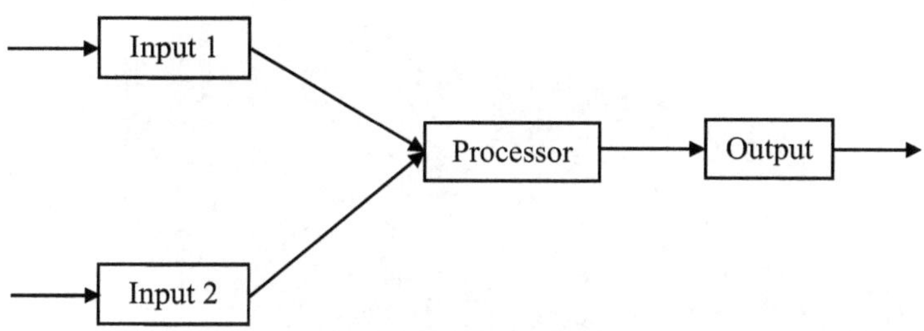

Fig. 7.14 2I1O box model

Now, if one considers the cases where there is more than one input, hidden or output boxes, it is understood that the three-layered but one-celled (boxed) parts given in Fig. 7.12 will not be enough. For example, if there are two input-single-output (2ISO), one box will be insufficient in the input part. If two boxes are placed here, as shown in Fig. 7.14, a slightly more detailed box model emerges.

In this model, arrows (links) that carry information from the input boxes to the hidden (processor) box are shown. Since no processing is done to the data in the input boxes, these arrows transmit the output of the boxes to the hidden layer as they are. It should be noted that each link has a value here. If there were more input data in the input layer instead of two boxes, each would transmit the data to the processor box at a certain rate. Thus, it is stated once again that no action is taken in the input boxes. In this model, the processor processes the messages from the two input boxes in some way and converts them into an output variable. This is just like triple regression. In this case, the dependent output variable is expressed in terms of the independent input variables. This means that part of the output variable comes from one input variable, while the other part comes from the second input data. Thus, the hidden box mixes the inputs in varying proportions, giving the output. If there are too many boxes in the input layer, the most general version of the model in Fig. 7.13, the 2ISO model, is obtained. A model of triple regression is shown in Fig. 7.15.

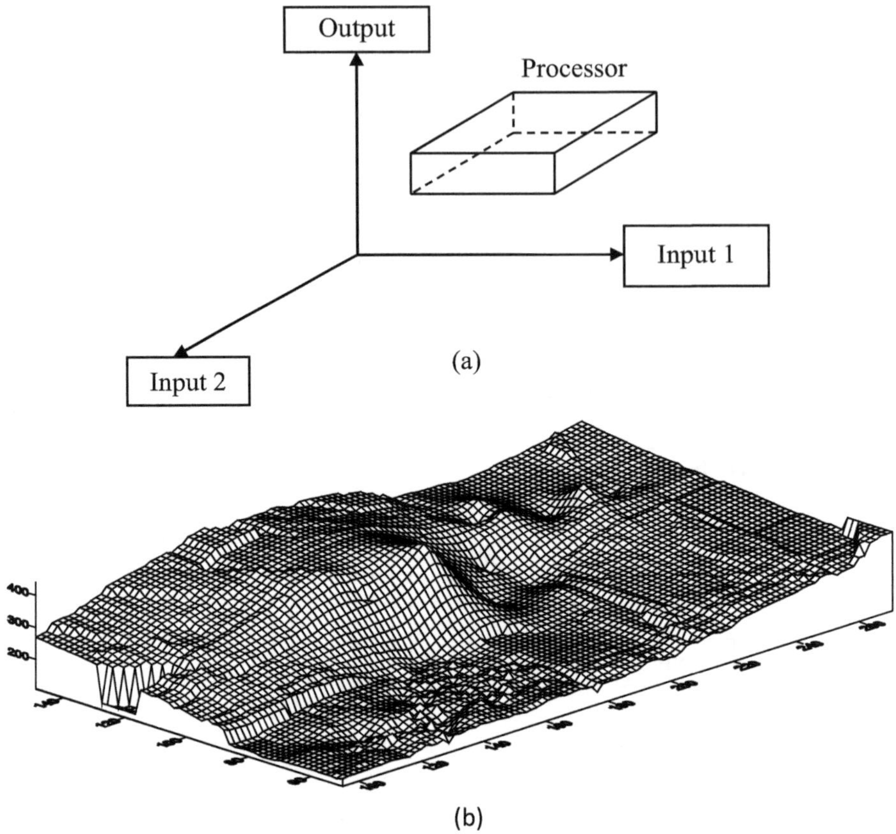

Fig. 7.15 Triple relationships

Although there are two input variables in the previous model, their mixing is done in a single box. If one divides the hidden box into two, as a result, the multiprocessor (MP) status of the 2ISO model emerges. The structure of this modeling is shown in the Fig. 7.16.

In this model, some of the inputs are transmitted to the lower processor of the hidden layer and the other part to the upper processor. The output can be perceived as the sum of the quantities coming from the individual processor boxes. So, in this model, the output value is found by adding the box on the output layer. The difference of this model from the previous one is that some operations are performed in the input and output layer boxes. It should not be concluded that the hidden boxes should be as much as the input boxes. For example, in Fig. 7.17, the hidden box is less than the input boxes. Generally, in practical studies, the number of boxes in the hidden layer is less than in the input layer.

In the most general modeling, the number of boxes in the input layer is equal to the number of input variables, and those in the output layer are as many as the number of variables to be predicted as output. The number of boxes of the hidden

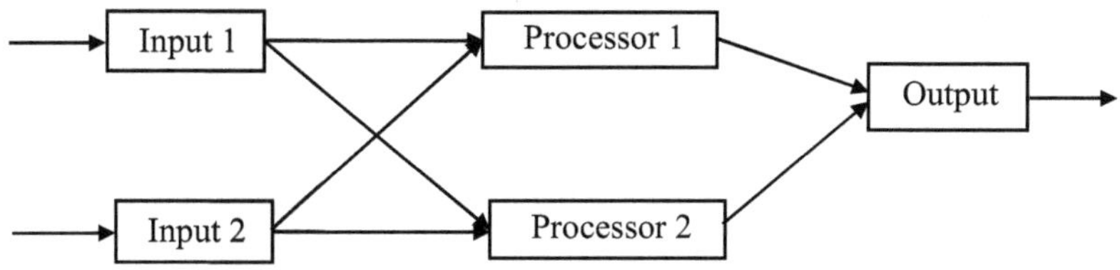

Fig. 7.16 2I1O binary model

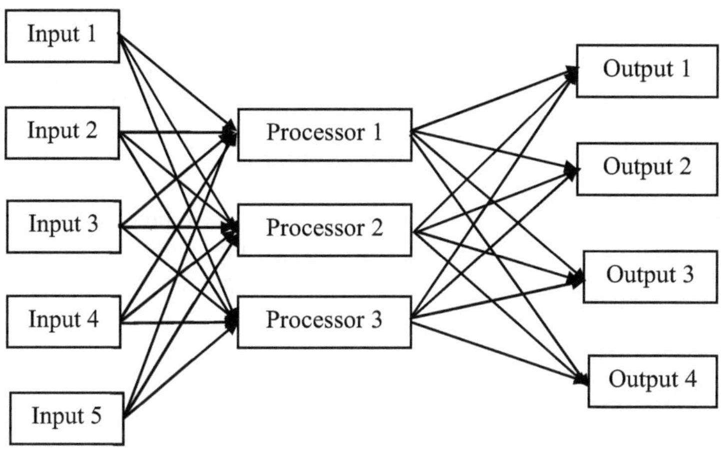

Fig. 7.17 5I4O ve triple model

layer is not known exactly, but it should not be taken too much. As a rule of thumb, the number of boxes in the hidden layer should be less than the number of boxes in the input layer. Again, in practical studies, the maximum number of hidden layer boxes should be 3–5. In many practical applications, the number of boxes in the hidden layer is initially taken as three. Now, let us show the model in its most general form, for example, with five input data, four predictive, and three intermediate boxes, as in the figure below.

As can be seen in this figure, the model becomes more complex as the number of boxes in different layers increases, but its basic principles are very simple as described above. From the explanations so far, the structure of the model has been presented. The models that emerge in this way are very similar to the architecture of ANN models. In the model shown in Fig. 7.17, it is a matter of curiosity what each of the processors in the boxes constituting the interlayer elements are. As explained earlier, this processor can be detected by a simple least squares (regression) analysis in models with SISO. As the model gets more complex, there is no complexity in the processor. The processors in the interlayer boxes generate the outputs by processing the inputs received by them with a kind of mathematical processing. If one thinks about this process a little closer, it will be understood that there must be two kinds of processors inside the processor boxes in the hidden layer, one for linear addition of the inputs and the other for sending them to the output layer boxes with a curvilinear

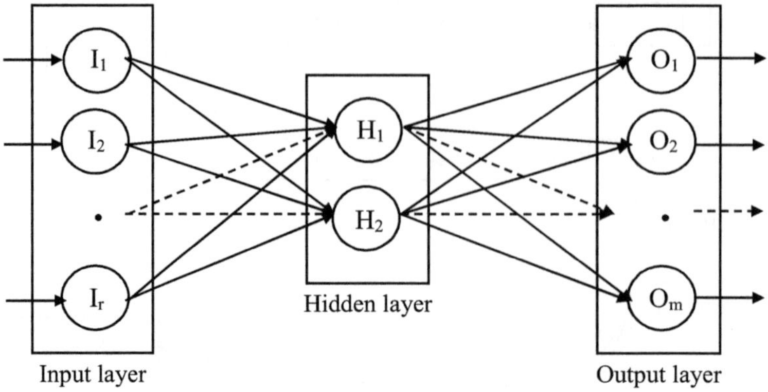

Fig. 7.18 ANN fundamental architecture

(nonlinear) processor. So, the first of these processors is to process the incoming by linear aggregation and distribute the outgoings to the output boxes. Already, the output boxes collect the ones that come to them and predict the outputs.

7.7 ANN Structure

After the similarities with the classical method and box models described above, let us focus on what needs to be done to switch to the ANN models in the literature. There is a one-to-one agreement between the above models and ANN approaches. In ANN terminology, circles, which are said to represent cells (see Fig. 7.18), can be used instead of boxes. In the figure below, a model with multiple inputs and two hidden layer cells and multiple outputs, which is the basis of classical ANN model architectures, is shown, inspired by the box models described so far. In this type of modeling, the following questions arise.

1. Incoming connections representing nerve cells do not operate in the ANN model.
2. Each cell has only one output value, which is linked to the others, namely, to the cells of the next layer, in different ways.
3. All input data also reaches the cells in the input layer and disperses from there without delay.
4. The information coming to the hidden layer cell passes through a curvilinear (nonlinear) processor (activation function) and forms the outputs of these cells.
5. Each cell of the output layer linearly collects the information from the hidden layer cells and gives the expected outputs.
6. Differences between expected and measured outputs are recognized as errors, and if this is not less than a desired limit, the errors are distributed backward to each link as described in Sect. 7.13.5.
7. With similar procedures, the training of the ANN is continued until the desired error limit is reached.

7.8 Perceptron (Single Linear Sensor)

It is explained that linear modeling can be done in most of the classical examples given in the previous sections. Considering these basic information, first, ANNs that can perform clustering by linear operations will be explained in detail. Understanding these will help to understand the more complex types of ANNs that are the subject of the next chapters for deep learning methodologies (Chap. 9). In this section, the basics of the types of single linear sensor, perceptron, will be presented, which is the first type of ANNs that are single-limited, have a simple architecture, and can make linear cluster separations. In English, such sensors are called "perceptron" for short, as mentioned earlier.

7.8.1 Perceptron Principles

Before going into the mathematical details and ANN architecture, a mind experiment will explain what the basis of perceptron is. The first step in the use of perceptron is to divide data into clusters. It will be useful to understand what kind of clustering will go from a picture given to human beings. For this, let us look at a pattern consisting of two different points (stars and circles) in the two-dimensional space given in Fig. 7.19.

Whenever possible, data should always be transformed into scatter diagrams that show the relationships between the input and output variables. In this way, people can try to make inferences away from mathematics by visualizing with their own understandings, thoughts, and interpretations. The following visual interpretations can be made by looking carefully at the scatter diagram

1. There are two different patterns: one is a star, and the other is a circle.
2. Stars are 5, and circles are 6.
3. Stars and circles have x and y values. Their coordinates are given in Table 7.1. Otherwise, by reading from this figure, the coordinates can be obtained even if they are approximate.
4. The star pattern is narrower on the left, and the circle pattern is wider on the right.
5. With the eye, we can separate the two pattern areas with an approximate line without any calculations as follows (see Fig. 7.20).
6. If desired, the approximate equation of this line can also be extracted. After a short approximate calculation, the separation line equation is obtained as:

$$x – 0.22y – 0.3 = 0 \tag{7.13}$$

7. Is this the only line of separation? The answer is immediately no. If thought wisely and thoroughly, it is seen that there are other straight lines (even broken

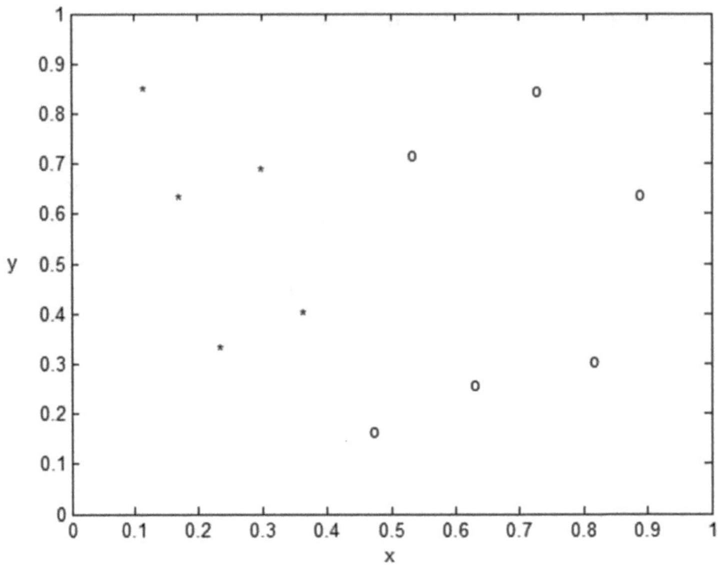

Fig. 7.19 Two set pattern

Table 7.1 Pattern type, number, and coordinates

* class		o class		* class		o class	
x	y	x	Y	x	y	x	y
0.12	0.85	0.47	0.16	0.12	0.85	0.47	0.16
0.17	0.63	0.53	0.72	0.17	0.53	0.53	0.72
0.23	0.33	0.63	0.26	0.23	0.63	0.63	0.26
0.30	0.69	0.72	0.84	0.30	0.72	0.72	0.84
0.37	0.40	0.82	0.31	0.37	0.82	0.82	0.31

lines) that separate these two different patterns. For example, the dashed straight lines indicated by the numbers I, II, and II in Fig. 7.21 can distinguish these two patterns. One can increase the number of such dashed lines as much as possible.

8. Thus, is there an area that must remain within the dividing lines that one can call an infinite number? The answer to the question is yes. It is concluded that all lines falling within the area indicated by the thick lines on the right and left sides in Fig. 7.22 can be used as separation lines.

   In that case, if the x and y coefficients in the line expression given by Eq. 7.13 change, if they remain in the area between the thick borders, each of the straight lines can be accepted as valid separation lines at least as much as the others. Then, the general mathematical expression of an infinite number of separation lines can be written as:

$$a_1x + a_2y + b = 0 \tag{7.14}$$

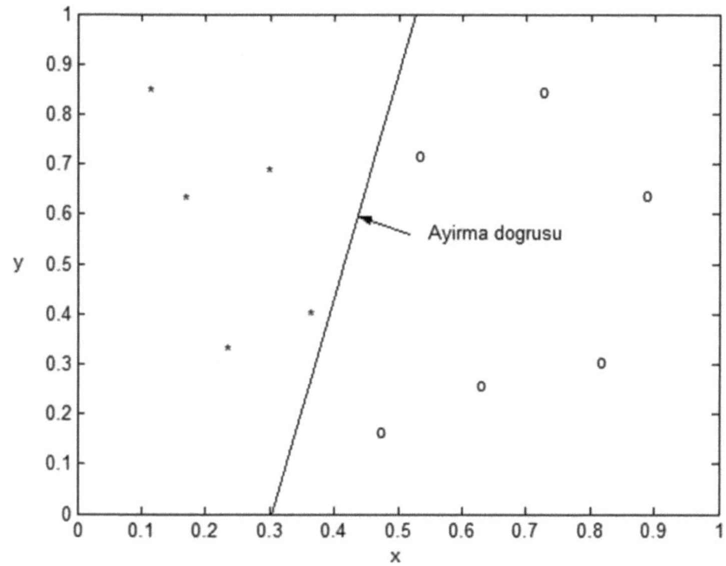

Fig. 7.20 Discrimination straight line

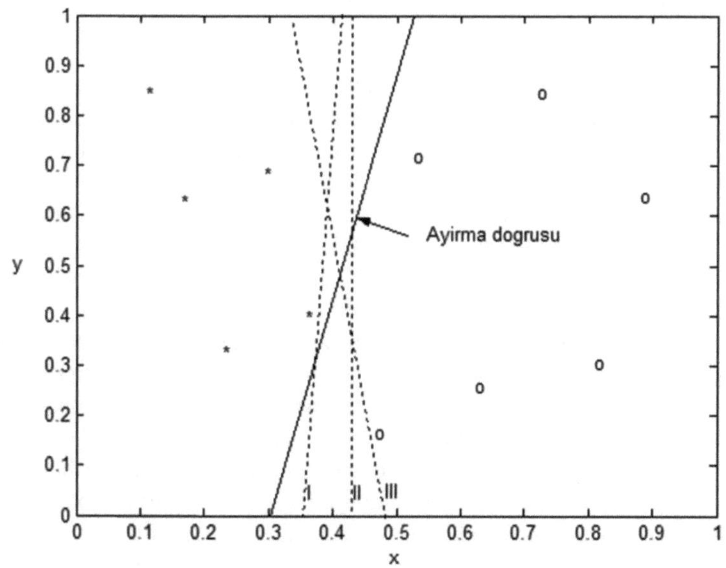

Fig. 7.21 Various discrimination straight lines

If one interprets the coefficients in this equation, he can say that the contribution of the variable x to a_1 is the weight of the variable y, like a_2. Finally, b can be called a constant or a bias constant, which ensures that the straight line does not pass through the beginning of the coordinate (if $b \neq 0$). Equation 7.14 can be

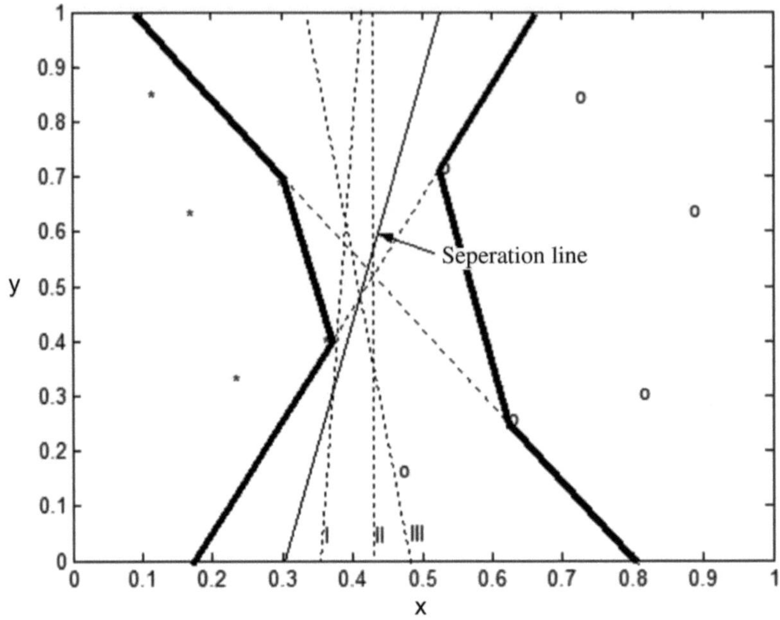

Fig. 7.22 Necessary boundaries within the discrimination straight lines

established by discarding the coefficients arbitrarily. However, one can see whether this line equation can separate the given patterns by drawing on Fig. 7.22. If the line falls within the area indicated by the thick lines, it is accepted as the separation line; otherwise, these coefficients should be discarded, and other lines should be tried. The person who is an expert on the subject makes head throws with a certain consciousness. Let us not forget that the saying "getting out of your head" for humans here means using "random numbers" when working with computers.

9. The struggle to find the most suitable one among the infinite number of separation lines is like fishing. The patient one reaches his goal by catching a large or small fish. Why not one should have such porter operations done by our computer, which is calculation porter? Since computers cannot have an unsystematic thinking like humans, they absolutely need a system and an algorithm (method) that can achieve calculations step by step by performing similar operations. At this point, perceptron methods, which are the simplest of ANNs, come to the aid. In the light of the above discussions, one can develop a system and method called perceptron. The most important point to be emphasized in the explanations is that the separator will be a straight line with constant coefficients, but the coefficients are changed randomly. Let us keep this in the back of our minds for now,

10. What kind of method can one suggest using the abovementioned procedure? If one is asked, he can answer the consideration of Eq. 7.14 and Fig. 7.22 without knowing any method.

    (a) First, the coefficients a_1, a_2 and b are randomly thrown out of the head, and this selected line is drawn on Fig. 7.22 using Eq. 7.14.
    (b) If this straight line does not stay within the valid area defined in Fig. 7.22, its coefficients are increased or decreased one by one or by two or all three at a random rate or some are increased and some are decreased.
    (c) If the new straight line does not stay in the desired area, it is continued to change the coefficients in a similar way.
    (d) Continuing in this way, a line that will remain in the valid area after a certain effort is captured, and thus a separation line is found, and all these steps are called perceptron method.

11. After all these operations are done, the constants of Eq. 7.14 are obtained according to the given calculation procedure. This equation is then used to decide which pattern the new points will belong to. Here, too, two paths can be followed:

    (a) When the x and y coordinates of the new pattern point are given, if the point is marked using these, it is decided that if it is on the right of the separation line, it will belong to the star pattern if it is on the left of the circles,
    (b) Since one has the equation, he does not need to mark the point. Substitute the x and y values of this point in the separator equation, and the result is either plus (>0), minus (<0) or equal to zero. If it is plus, the point will belong to the circle, if it is minus, it will belong to the star pattern. In practice, even if it is very rare, in case of zero, the researcher can decide which cluster it belongs to by flipping a coin. However, if the star pattern was added in the first zero case, it should be adhered to in all other zero cases.

12. Circle from the last part of the previous step too? Or is it a star? While deciding, the researcher can call one of them white and the other black, as well as coding the final decision with the number 1 or 0 so that computers can understand it like humans and separate patterns according to crisp (binary) logic (Chap. 3).

7.8.2 Perceptron Architecture

Considering what was said in the previous section, since Eq. 7.14 resembles a two-variable regression model, its architecture is given in Fig. 7.23, considering x and y as variables as x_1 and x_2, respectively.

If Eq. 7.14 is subjected to a mathematical synthesis after the perceptron architecture, which is structured generally in this way, one can understand that the following points are verbally hidden in this equation.

1. There are two mathematical operations in the equation, one for addition and one for multiplication (between constants and variables). The priority is in multiplication operation.
2. The variable values x_1 and x_2 should be given as inputs to the same equation, so that the numerical value of the equation can be calculated as the output.
3. In the same equation, in addition to the terms with variables x_1 and x_2, there are also constants (a_1, a_2 and b).
4. This equation consists of two parts, one variable and the other constant.
5. There are obvious variables (x_1 and x_2) as input, and the others are random constants a_1 and a_2 weights and bias, b.
6. Since there are two inputs, as indicated in Fig. 7.23, if two boxes (cells) are taken as input, x_1 and x_2 values are entered in the boxes. Then, the first structure of Fig. 7.24 appears like Fig. 7.23, in order to indicate the variable part of Eq. 7.14 as multiplications first and then as the sum of these two multiplications. This is called the architectural structure of the variable part.
7. After the variable part is thus arranged by adding a constant, b, a function f is obtained (see Fig. 7.24).
8. The X inside the input boxes indicates that the inputs will be multiplied by the weights in the perceptron box output. In the next sections, a multiplication sign will not be used in the input cells. The + sign in the first circle indicates that the inputs multiplied by the weights will be added. The part up to this point represents the variable part of Eq. 7.14.
9. With the inclusion of the fixed part, the left side is now represented by a simple architecture $a_1x + a_2y + b$. The equation value calculated in the previous step is not expected to be equal to zero most of the time. Therefore, the result,

$$f = a_1x + a_2y + b$$

Now the cluster assignment will be made according to the value of f. As stated before, since the star (circle) will be represented with the binary values of 1 (0), the given data series (patterns) of f must also be divided into two parts. For this, an output unit called a processor is placed in the output unit. This is actually nothing more than a function that simply converts incoming inputs to either 1 or

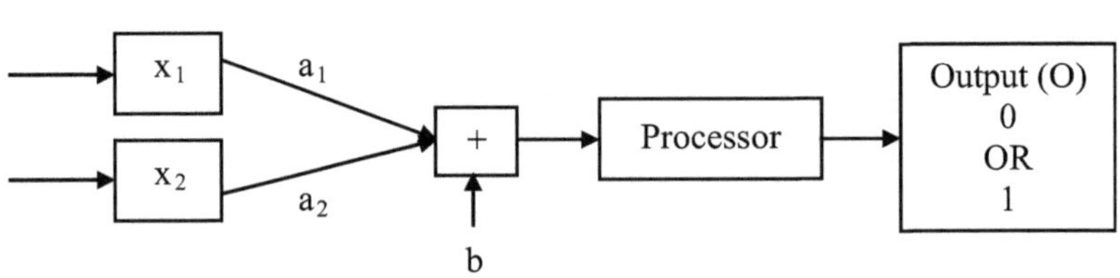

Fig. 7.23 Perceptron architecture

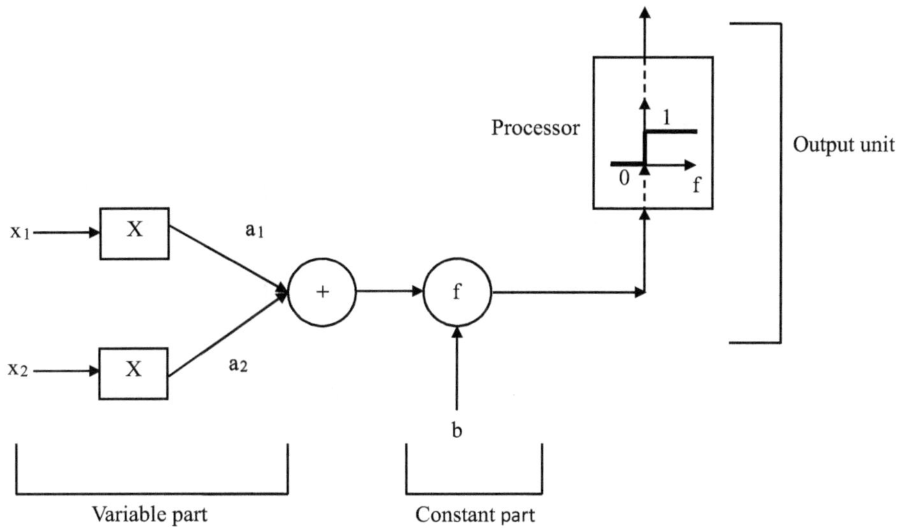

Fig. 7.24 Fundamental neural cell

0. To achieve this, the plus threshold processor shown in Fig. 7.25a is placed in the output unit in Fig. 7.25. Thus, an output of 1 for f > 0 or 0 for f < 0 is valid. Similarly, considering the opposite polarity threshold operator in Fig. 7.25b, the value +1 or −1 is obtained.

Thus, the problem is solved when a perceptron inserts the input pattern introduced to it into one of the clusters in its output. In binary clusters, 1 represents a suitable clustering, while 0 represents another clustering.

10. Since the input patterns are known as stars or circles and are encoded as 1 and 0 by the researcher, the perceptron must provide a known output for every known input. If it is provided, the procedures are continued in a similar way by referring to another entry.
11. If not provided, it means that the code 1 (0) on the circle (star) pattern appears as output. In this case, since the inputs x_1 and x_2 cannot be changed, new straight lines are searched for by changing the coefficients a_1 and a_2 and/or the constant b. Thus, by changing these numbers and re-clustering, the process continues until the output code is obtained as suitable for the input pattern. This is iterative training and eventually perceptron learns to cluster. Information on how to change the coefficients will be given below.
12. These processes are continued until all known data (pattern) are exhausted. Thus, perceptron has now adjusted itself (its weights, constants) according to the data. All the steps up to this point are called the perceptron training or shallow learning phase. Perceptron can now automatically find patterns whose set is unknown.

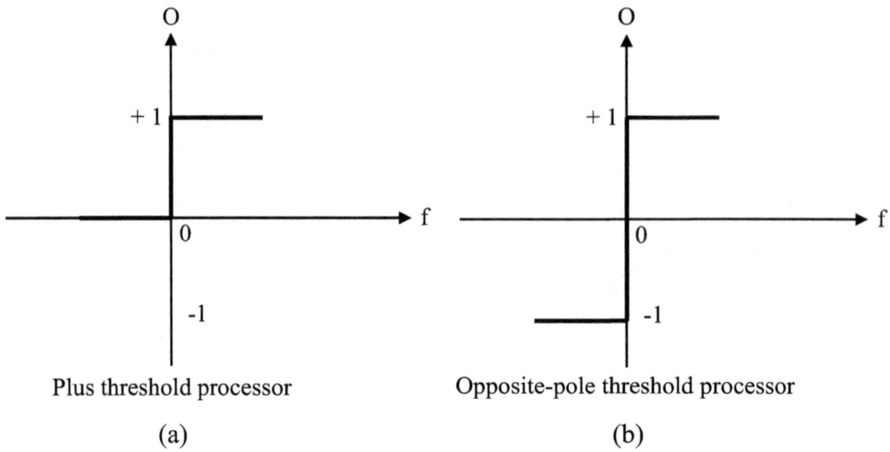

Fig. 7.25 Splitter processors into two classes

The explanations above are given for linear separator equations with two input variables. If the number of variables is generally n, the separator equation takes the following form

$$a_1x_1 + a_2x_2 + a_3x_3 + \cdots + a_nx_n + b = 0 \tag{7.15}$$

Under the philosophy of what has been said, the architecture of this equation can be drawn similar to Fig. 7.24. The processor in Fig. 7.23 and the processor that divides the incoming values in the output unit in Fig. 7.24 into two classes is called a hard or threshold processor.

Opposite polarity threshold processor divides an entered pattern into two pattern sets according to 1 and -1 encodings for plus ($f > 0$) and minus ($f < 0$) values, respectively. The plus-threshold processor encodes the plus input values as 1 and the minus values as 0, as indicated in Fig. 7.24. These processors are also called strict threshold functions in some publications.

7.8.3 Perceptron Algorithm

Here the mathematical operation of perceptron, whose basic and architecture is be presented according to the explanations in the previous section. For this, the following steps must be executed in order:

1. *Assignment of initial weight values*: Small random numbers are assigned to the fixed value of the whole weight and the dividing line.

2. *Calculation of the variable part*: When a pattern is detected by the perceptron, the variable part value as a result of adding the inputs by multiplying with the assigned weight values are calculated as:

$$\sum_{j=1}^{n} a_{ij} x_{ij} \tag{7.16}$$

Here, the first index (i) indicates the sequence of data (pattern), and the second index (j) represents the j component in the i-th data. For example, the fifth pattern (1, 3.8, 0.3) is $a_{51} = 1$, $a_{52} = 3.8$ and $a_{53} = 0.3$.

3. *Fixed value addition*: Perceptron value that will enter the processor when such a constant value is added to Eq. 7.16 as:

$$f = \sum_{j=1}^{n} a_{ij} x_{ij} + b$$

A fixed value is initially assigned as a small random number.

4. *Processor output*: Clustering by entering the value in the previous step to one of the threshold processors (opposite or positive) results in:

$$F_h(f) = \begin{cases} 1 & f \rangle 0 \\ 0 \ (\text{or} - 1) & \text{otherwise} \end{cases} \tag{7.17}$$

5. *Checking*: Since it is known which set the entered pattern belongs to beforehand, it is checked whether this result is the same with it. If it is the same, the pattern perceived by the perceptron is placed in the cluster that is suitable for it. In this case, similar operations are performed by going to the next pattern without making any changes in the weight and constant values. If perceptron puts the detected pattern in the reverse cluster, the weight values selected for this pattern should be changed,
6. *Weight training*: The randomly selected weights will now need to be adjusted until they put the detected pattern in the appropriate cluster. Random assignment is no longer possible here. There is a formal adjustment of the weights according to the perceived data (pattern structure). Formality requires the new weights to be in a certain connection with the previous ones. The most plausible here is to slightly change (increase or decrease) the previous weight values. In general, if the increment is shown as Δaij, the following expression can be writable

$$a_{ij}(k+1) = a_{ij}(k) + \Delta a_{ij} \tag{7.18}$$

Here, k + 1 and k are counters showing two consecutive training stages (next and previous). The question arises of how to calculate Δa_{ij}.

7. *Error correction*: Since the contribution of each element of the pattern detected by the perceptron to the output will be different, the weight increment values must be different. If the output does not hold the expected set, the error will be $h = 1 - 0 = 1$ or $h = 0 - 1 = -1$, so the weight increment of the x_{ij} element appears in the following form.

$$\Delta a_{ij} = hx_{ij} \tag{7.19}$$

By substituting this in Eq. 7.18, the error renewal pattern is obtained as:

$$a_{ij}(k+1) = a_{ij}(k) + hx_{ij} \tag{7.20}$$

This means that in a simple perceptron ANN, all the weights are added as either plus or minus sign as their input.

8. *Cluster iteration*: The simplest training of perceptron as described above is continued until all patterns are clustered correctly. A very important point to be noted here is that after each clustering error occurrence, all designs are reintroduced into the perceptron training system with their renewed weight values. Changing the weights for a pattern does not mean that the patterns that were clustered correctly before that will stay in the same cluster without being replaced by these new weights. In general, the clusters of some of the previous patterns have changed. Then the training process should be continued until all the patterns are converged into suitable clusters with the same weights and constant values,
9. After a perceptron has assigned all the patterns to their known suitable clusters as described in the previous step, it is now ready to be used for clustering of patterns whose set is unknown.

7.8.4 Perceptron Implementation

In order to perform clustering with perceptron, it is necessary to explain the difference between the concepts of data and pattern. Data means not only the numerical values of a variable but also groups of more than one variable. For example, precipitation, humidity, and evaporation variables in a place represent a data set. It is considered as data, say as (620, 0.80, 700) with the knowledge that the average precipitation in Istanbul is 620 mm, the humidity is 80%, and the evaporation is 700 mm. Sequences of such a data group in different years also represent the data series. However, this can happen not only in time but also in space. The consideration for the ensemble of cities also indicates another set of data.

The concept of pattern is rather understood as the numerical values of very small parts of an area called pixels. These take the values 0 or 1 according to the two number system. For example, if the 26 pixel positions in an 8×8 area are 1 (black) and 0 (white) in the rest, a pattern emerges.

7.8.4.1 Data Clustering

In order to fully understand the steps described in the previous section, it is useful to perform the application step by step with a data series. Let the components in the Cartesian coordinate system of three data, two of which belong to the set 1 and three to the set 0, be (1.3, 0.5), (2.5, 0.6), and (1.8, 3.4). Let us consider the first element of each data as the input i_1 and the second as the input i_2. Suppose we have information that the first data belongs to the clusters 1 and the others belong to the clusters 0. All calculations according to the perceptron principles described above are obtainable below step by step.

1. First, let's look at how they are scattered in the Cartesian coordinate set. Since these data are in two-dimensional space, one has the opportunity to display them in the Cartesian coordinate set. Otherwise, it is not possible to make such a representation if there are more than three data components. According to the data, the scatter diagram of the three data is shown in Fig. 7.26, the first of which is "*" and the other two are "o" signs.

   By looking at all the points at once, one can immediately see where the dividing line will pass through according to the steps described in Sect. 7.6.1. However, since computers do not have such a capability, it should be ensured that they see each point one by one. This takes a lot of time compared to humans, but humans can never perceive data that is too large in size.
2. Since the architecture of the perceptron to be used here is given in Fig. 7.23, two weight values should be assigned as random small numbers to start the calculations. These are taken as $a_{11} = 0.2$ and $a_{12} = 0.7$. In addition, $b = 1$ is taken randomly as a fixed value. Here, the first index represents the first data, and the second index represents the resultant of the data. It is also given that this data belongs to the one coded cluster. Then, one should expect the value of 1 to be obtained as a result of the perceptron processing with the given data and weights. The value given by Eq. 7.20 in the previous section is $f_1 = 0.2 \times 1.3 + 0.7 \times 0.5 + 1 = 1.61$. Since $f_1 > 0$, $O_1 = 1$, and the weights for this pattern do not need to be changed because perceptron has assigned this data to the correct cluster.
3. Now let's get the second data to be detected by perceptron. Its set is known to be 0. So $f_2 = 0.2 \times 2.5 + 0.7 \times 0.6 + 1 = 1.92$ with the same weight coefficients. Since the result is greater than 0, it is concluded that its set is 1. However, the expected set is 0. Accordingly, the error is $e_2 = 0 - 1 = -1$. In that case, the weights need to be reduced by the input values according to Eq. 7.20. Thus, the new weights are $a_{21} = 0.2 - 2.5 = -2.3$ and $a_{22} = 0.7 - 0.6 = 0.1$. At the end of

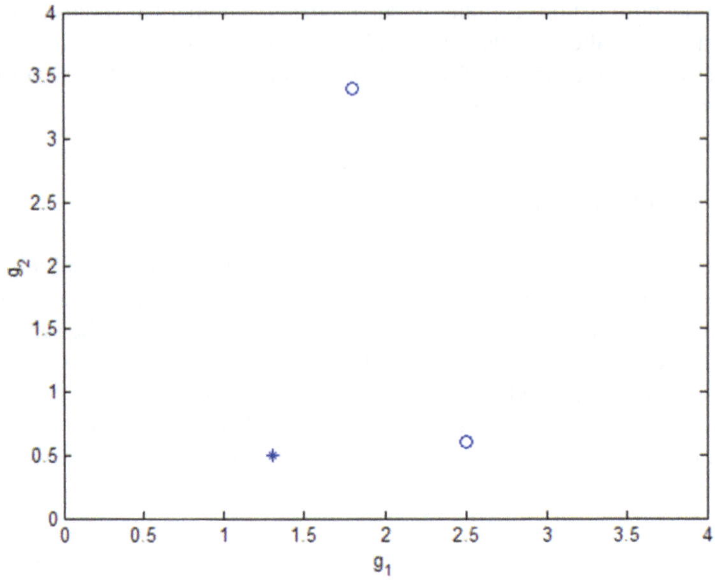

Fig. 7.26 Data scatter diagram

the recalculations, since $f_2 = -2.3 \times 2.5 + 0.1 \times 0.6 + 1 = -4.69 < 0$, the clustering is 0, and the second data is put in the correct cluster with these weights. However, before going to the third data, it should be checked whether the previous data remains in the same cluster with these new weighting coefficients. For this, since the previous data gives the value $f_1 = -2.3 \times 1.3 + 0.1 \times 0.5 + 1 = -1.94 < 0$ with the new weight coefficients, the set with these weights is distorted. So, to bring it back to the set, if the new weight coefficients are increased this time by the inputs according to Eq. 7.20, taking into account $e_1 = 1 - 0 = 1$, $a_{11} = -2.3 + 1.3 = -1.0$, and $a_{12} = 0.1 + 0.5 = 0.6$. Accordingly, the same pattern comes back to the cluster since $f_1 = -1.0 \times 1.3 + 0.6 \times 0.5 + 0.1 = 0.001 > 0$ with these weight coefficients. One wonders if the new weights left the second pattern in the set? For this, since the value of the second set with the same weights is $f_2 = -1.0 \times 2.5 + 0.5 \times 0.6 + 1 = -1.2 < 0$, it is understood that the set has not changed. Thus, the first and second data found the correct clusters with the same weight values.

4. Now if we consider the third data, its expected set is 0. Clustering could not be done correctly since the calculation of this with the latest weights is $f_3 = -1.0 \times 1.8 + 0.6 \times 3.4 + 1.0 = 1.24 > 0$, $o_3 = 1$ and $e_3 = 0 - 1 = -1$. Then, if the weights are renewed according to Eq. 7.20 by decreasing the components of this last data, $a_{31} = -1.0 - 1.8 = -2.8$ and $a_{32} = 0.6 - 3.4 = -2.8$ are obtained.

With these new weights, the latest data is placed in the correct cluster since $f_3 = -2.8 \times 1.8 - 2.8 \times 3.4 + 1.0 < 0$. One wonders, if these new weights bothered previous sets? For this, the first cluster calculations are $f_1 = -2.8 \times 1.3 - 2.8 \times 0.5 + 1.0 < 0$, so $f_1 = 1 - 0 = 1$. Unfortunately, the pattern has changed clusters again. The weights need to be replenished similarly to the previous ones by increasing and $a_{11} = -2.8 + 1.3 = -1.5$ and $a_{12} = -.8 + 0.5 = -2.3$. With these weight values, f_1 will still have a minus sign, so it cannot come back to the dataset. Once again, the weights need to be increased by the amount of data components. So $a_{11} = -1.5 + 1.3 = -0.2$ and $a_{12} = -2.3 + 0.5 = -1.8$. With these, it is in the first dataset since $f_1 = -0.2 \times 1.3 - 1.8 \times 0.5 + 1.0 = 0.36 > 0$. Again one wonders if it remained in the second and third datasets that were previously placed in their correct set. To check this, the second data calculation with weights $f_2 = -0.2 \times 2.5 - 1.8 \times 0.6 + 1 > 0$ changed the cluster and new weights as $e = -1$ $a_{21} = -0.2 - 2.5 = -2.7$ and $a_{22} = -1.8 - 0.6 = -2.4$. With these, the second set returns to the set since $f_2 = -2.7 \times 2.5 - 2.4 \times 0.6 + 0.1 < 0$. One wonders what the first data did in this situation? The calculation has changed again since $f_1 = -2.7 \times 1.3 - 2.4 \times 0.5 + 1.0 < 0$ with new weights. So the weights must be renewed by increasing $a_{11} = -2.7 + 1.3 = -1.4$ and $a_{12} = -2.4 + 0.5 = -1.9$. Since the first data with these coefficients still cannot return to the set, the new weights are $a_{11} = -1.4 + 1.3 = -0.1$ and $a_{12} = -1.9 + 0.5 = -1.4$. It returns to the first dataset with these weights, but what about the second data? Since $f_2 = -0.1 \times 2.5 - 1.6 \times 0.6 + 1.0 < -0.21$, it returns to the set. These weights keep the first and second data in their set, but what about the third data? So, if one looks at the third data with the weights, since $f_3 = -0.1 \times 1.8 - 1.6 \times 3.4 + 1.0 < 0$, the weights of the trained perceptron will be the result, since the same weights keep the third data in its set.

7.8.4.2 Pattern Separation

Perceptron is also used for pattern detection and clustering. For this reason, after perceptron detects the patterns given as input, it should be able to distinguish which cluster the outputs will yield. As a very classical example, the patterns (+1 +1), (−1 +1), (+1 −1), and (−1 −1) are divided into two groups. Let there be in the first group (+1 +1) and (−1 +1) in the second group (+1 −1) and (−1 −1). Let a perceptron be developed so that when the second group catches [(+1 −1) and (−1 −1)] it yields +1 and −1 for the other. Although the superiority of perceptron in such a simple separation process is not well known, their superiority becomes evident when the number of patterns is large. Perceptrons can distinguish many patterns in a much shorter time compared to other classical separation methods. The number of patterns a perceptron can distinguish is called its discrimination ability.

To understand the linear discrimination capabilities of perceptron, one can use the (+1 +1), (−1 +1), (+1 −1), and (−1 −1) patterns, for example, +1 for (−1 +1) and the remaining 0 in the patterns (+1 +1), (+1 −1), and (−1 −1). It is possible to represent these four patterns on a two-dimensional plane with the variables x_1 and x_2. Thus, the four patterns show the four points symmetrical from the start (see Fig. 7.26). In general, the two-dimensional plane space is

$$x_2 = mx_1 + b$$

It can be divided into two parts by a straight line. This distinction can be made so that each part contains only one of the patterns. If a perceptron processor can do this, we can call it a pattern separator.

Accordingly, it is understood that the coefficients m and b must be represented as a function dependent on perceptron. If a perceptron with parameters m = 1.5 and b = −1.1 can be developed, it will be a function that can separate four given patterns, that is, indicates the truth.

$$x_2 = 1.5x_1 - 1.1$$

Since this shows a very simple application, there are also many solutions as there will be many m and b coefficients that can separate the four given patterns. In general, the solution is geometrically presented in Fig. 7.27.

In order to make the above distinction with perceptron, the operations are as follows. Let us consider a design with two input neurons as in Fig. 7.23, where the result of perceptron is 1 when a critical threshold value is exceeded and 0 because it is not exceeded. One can represent the threshold value,Θ_i, with such an output from perceptron with the following expression:

$$O_i = x_1 a_{i,1} + x_2 a_{i,2} - \Theta_i = 0 \tag{7.21}$$

Here, $a_{i,1}$, $a_{i,2}$, and i-th pattern show the two weights and constant values, respectively, in perceptron architecture. From Eq. 7.21, by performing the necessary simple operations, one can find

$$x_2 = -\left(\frac{a_{i,1}}{a_{i,2}}\right)x_1 + \frac{\Theta_i}{a_{i,2}} = ax_1 + b \tag{7.22}$$

Here, $a = -a_{i,1}/a_{i,2}$ is the slope of the dividing line, and $b = \Theta_i/a_{i,2}$ the point where this line intersects the x_2 axis. From these last two expressions, the parameters Θ_i, $a_{i,1}$, and $a_{i,2}$ can be found. If a = 1.1 and b = 0.8, then $\Theta_i = 0.7$ threshold value then $a_{i,1} = 0.875$ and $a_{i,2} = -0.9625$. It is explained above what the slope and intercept (constant) values of the separation line mean and how they depend on the pattern values, and calculation of pattern is shown by giving slope and constant values. However, in order to find the slope and constant from the pattern values, the steps must be repeated similar to the data series described in the previous section.

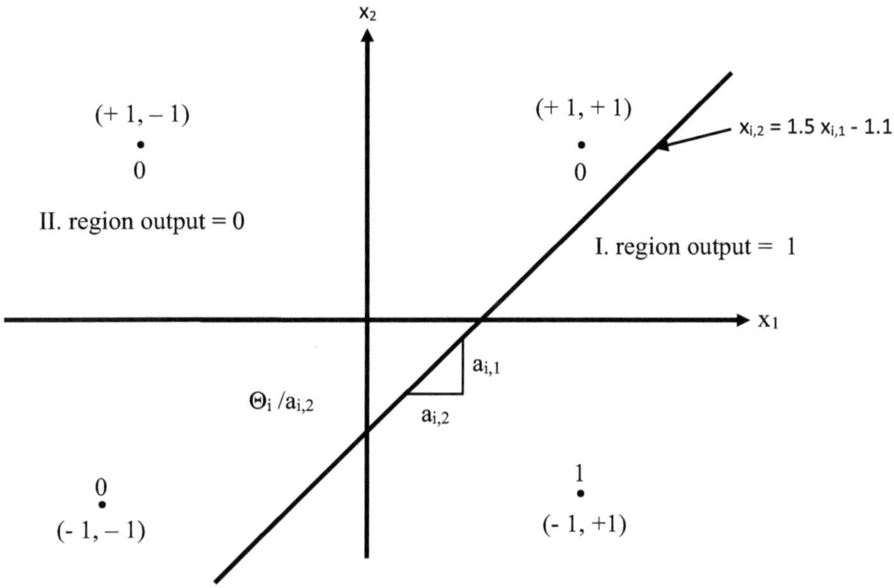

Fig. 7.27 Graphical indicator of a straight line separator

7.9 Single Recurrent Linear Neural Network

This structure is called ADAptive LINear Element (ADALINE) for short in English. It is similar to the simple linear neural network developed by Widrow and Hoff (1960) and is used as a neural network that can repeat itself and is slightly more advanced than perceptron. It has one processor, and its learning rule is based on the principle of minimization of the mean squares of errors (MSEs), and its general structure is given in Fig. 7.27. It should not be overlooked that its architecture is similar to perceptron. The difference between the two is only in the learning rule. Another important difference of this algorithm from perceptron is that even if the pattern stays in the correct set, the minimization of MSE affects the change in weights.

The output of Fig. 7.28, with a given i-th pattern ($I_{i,1}$, $I_{i,2}$, ... , $I_{i,n}$) and the corresponding output if O_i is the weights a_{ij}'s as:

$$(NET)_i = \sum_{j=1}^{n} a_{ij} I_{i,j} + \Theta_i \tag{7.23}$$

The expected output value is O_i, and the square of the errors, E, is calculated as:

$$E_i = \left[O_i - (NET)_i\right]^2 \tag{7.24}$$

The derivative of this with respect to the weight coefficients leads to:

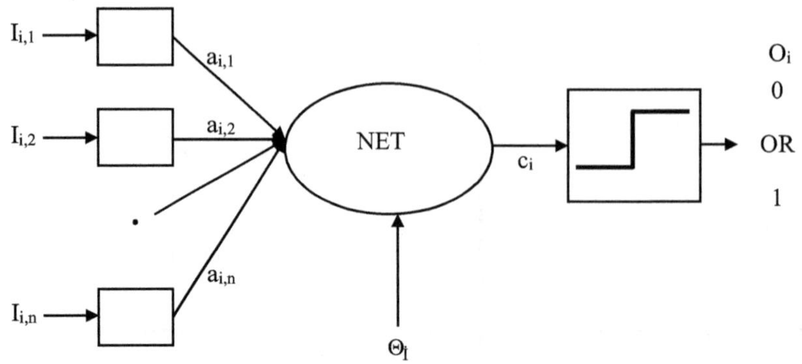

Fig. 7.28 ADALINE neural network

$$\frac{\partial E_i}{\partial a_{ij}} = -\left[O_i - (NET)_i\right] I_{ij} \tag{7.25}$$

From here, the increment value of the weights is a certain percentage of this error slope, α, which is defined as:

$$\Delta a_{ij} = \alpha\left[O_i - (NET)_i\right] I_{ij} \tag{7.26}$$

The NET value found as a result of increasing the weights in this way is finally passed through a threshold processor and clustering results as:

$$C_i = \begin{cases} 1 & \text{if} \quad NET \geq 0 \\ 0 & \text{if} \quad NET \leq 0 \end{cases} \tag{7.27}$$

When the expected output of the network is represented by the symbol O (0 or 1), the forward calculation error, e_1, result is:

$$e_1 = O_1 - C_1 \tag{7.28}$$

The weights should be renewed in such a way that this error is minimized. For weight regeneration to reduce the error, the following relationship is valid.

$$a_{ij}^{new} = a_{ij}^{old} + \alpha(O_i - C_i) I_{ij} \tag{7.29}$$

Here, α is defined as the learning coefficient. If this coefficient is 1, ADALINE turns into perceptron. Similarly, with a constant coefficient, when the initial value is taken as 1:

$$\Theta^{new} = \Theta^{old} + (O - C_O) \tag{7.30}$$

7.9.1 ADALINE Application

If an automatic separator is considered so that the slaughtered chickens and roosters that come to a factory do not mix with each other, we come across two types of patterns. Let us assume that one has components that represent these patterns and that there are measurements such as weight, length, shape, and sizes among these components. By measuring each of these, the numerical values of the pattern components can be obtained. Let us represent the clusters of chicken and rooster species with $o_c = 1$ and $o_r = -1$, respectively. From what has been said, we understand that in an ADALINE structure, there will be two input cells and an output cell representing the cluster, among which there will be an intermediate processor as in Fig. 7.28.

Let the given patterns be predetermined as (1, 0) for the chicken and (0, 1) for the rooster. To solve the problem, first the initial weights, $a_1 = 0.4$, $a_2 = 0.2$; suppose the constant $b = 0.3$ and the learning rate $\alpha = 0.5$. Of these, the weight and constant coefficient are assigned as small random variables. If there is knowledge from previous experiences, the learning coefficient should be chosen accordingly, or if there is no knowledge, it can be taken as 0.5, as if a situation with half-teaching and half-non-teaching (partly with a teacher) can be considered. Practical studies are taken at $\alpha < 0.5$ and preferably around 0.2 with increasing expertise. One can start the iterations by taking advantage of all this data.

1. *Rooster pattern*: If one of the patterns in *ADALINE* architecture and, for example, (1, 0) is detected (rooster), the net value (Eq. 7.21) is $o_c = 1$ since $NET = 0.4 \times 1 + 0.2 \times 0 + 0.3 = 0.7 > 0$. However, since the value of the rooster expected cluster is -1, the error in between is $e = -1 - 1 = -2$. The weights must be renewed according to Eq. 7.21. By doing this,

$$\begin{bmatrix} a_1 \\ a_2 \end{bmatrix}_{new} = \begin{bmatrix} 0.4 \\ 0.2 \end{bmatrix}_{old} + 0.5 \times (-2) \begin{bmatrix} 1 \\ 0 \end{bmatrix} = \begin{bmatrix} -0.6 \\ 0.0 \end{bmatrix}$$

The constant coefficient is calculated as:

$$\Theta^{new} = 0.3 + 0.5(-2) = -0.7$$

Could the rooster be clustered correctly with these new weight and constant coefficient values? If asked, from Eq. 7.23, since $NET = -0.6 \times 1 - 0 \times 0 - 0.7 < 0$, ADALINE placed the rooster in the correct cluster this time. Then one can go to the next pattern,

2. *Chicken pattern*: Since the chicken pattern is (0, 1), the NET value is the last weight values and $NET = -0.6 \times 0 - 0.8 \times 1 - 0.7 < 0$, since $o_c = -1$ and the number of chickens is +1, there is an incompatibility between them. The error is $e = 1 - (-1) = 2$. Therefore, the weights must be renewed. For this, from Eq. 7.30 one can obtain the following:

$$\begin{bmatrix} a_1 \\ a_2 \end{bmatrix}_{new} = \begin{bmatrix} -0.6 \\ 0.0 \end{bmatrix}_{old} + 0.5 \times (2) \begin{bmatrix} 0 \\ 1 \end{bmatrix} = \begin{bmatrix} -0.6 \\ 1.0 \end{bmatrix}$$

The constant coefficient is calculated as follows

$$\Theta^{new} = -0.7 + 0.5(2) = 0.3$$

With these, we find NET = $-0.6 \times 0 + 1.0 \times 1 + 0.3 > 0$ from Eq. 7.21 to see if the chicken pattern has changed back to the coop. This means that $o_c = 1$. These final weights and constants are valid as they coincide with the given number of clusters, 1. One wonders if the roosters stay in the cluster with these? With the latest values for this

$$\text{NET} = -0.6 \times 1 + 1.0 \times 0 + 0.3 < 0$$

is found, and therefore, $o_r = -1$. Since this coincides with the number of rooster coops -1, these final weights and constant values help to distinguish between chickens and roosters.

As a result, ADALINE's parameter values for dividing chicken and rooster patterns into appropriate clusters are $a_1 = -0.6$, $a_2 = 1.0$, $\Phi = 0.3$, and the learning rate is $\alpha = 0.5$.

7.9.2 *Multi-linear Sensors (MLS)*

In this case, perceptron again separates three or more patterns whose positions are defined by two variables (x and y). As shown in Fig. 7.29, there must be more than one separator straight line equation with different weights and constants.

Here, A_1, A_2, and A_3 separator lines are passed visually in order to separate the patterns. Similar to the steps of the binary pattern (perceptron) described in the previous section, it is entirely up to the reader to make the verbal explanations first and then converting it into mathematically correct equations. Readers who really want to grasp the essence of the matter should think about this issue so that they can easily make applications in the future, explain these issues to others easily, and even open new horizons on these issues.

7.9.3 *Multiple Adaptive Linear Element (MADALINE) Neural Network*

ANN with two or more ADALINE infrastructures in their architecture are given this name (Fig. 7.30). Here, the outputs obtained from each ADALINE system are passed through an inactive cell with AND (ANDing) or OR (ORing) (Chap. 3); the result output is obtained as +1 or −1.

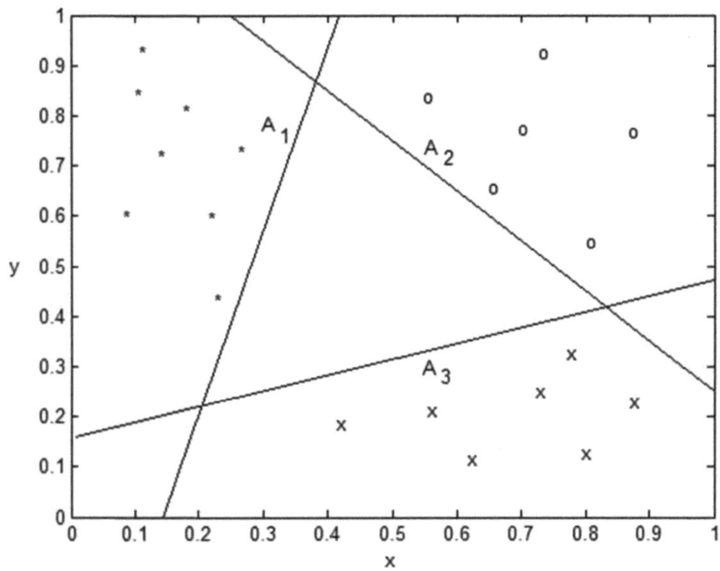

Fig. 7.29 Three-pattern straight line separators

The teaching rule for this is the same as for the ADALINE unit. Finally, the two ADALINE outputs are combined with ANDing or ORing to get the output of MADALINE. According to the bivalent logic rule, the truth tables of AND OR combiners are given below.

7.9.4 ORing Problem

All the simple ANNs presented in this section have a linear structure both in terms of the operations they perform while moving data or patterns from the input layer to the output and in terms of discrimination in the pattern space. Therefore, they cannot solve some problems. The most famous of these is the ORing operation. One of the most important rules of logic is ORing (Chap. 3) if two separate pieces of information are given. An example of this is given in Table 7.2. We can look at the ORing in this chart as if given four patterns. Let these patterns be (1, 1), (0, 0), (0, 1), and (1, 0) looking at the situation in the Cartesian coordinate set before distinguishing between the clustering of the first two, with the output 0 and the others 1. If those in cluster 1 are "o," the situation is presented in Fig. 7.31. Let's first interpret this verbally with the mind. Star patterns occupy the corners of a square, staying on two axes in circles in a 45° line. What is desired is that these two sets are separated by a straight line.

No matter how hard one tries as human beings, it is not possible to pass a straight line to separate these two sets. For this reason, ANN studies have come to a dead end for a long time (about 30–35 years) since they could not achieve this in the

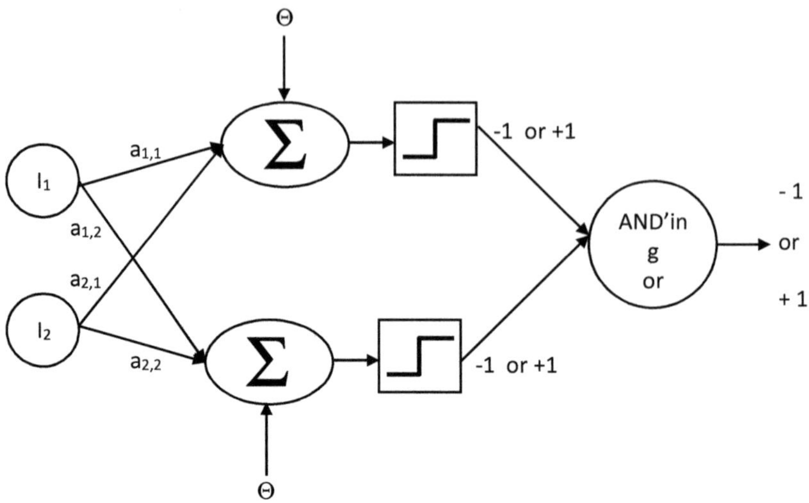

Fig. 7.30 MADALINE neural network

Table 7.2 MADALINE output table

	ANDing	Output		ORing	Output
ADALINE	ADALINE	MADALINE	ADALINE	ADALINE	MADALINE
1	1	1	1	1	1
1	0	0	1	0	1
0	1	0	0	1	1
0	0	0	0	0	0

architectures presented in this section and constitute the foundations of ANNs. However, the two sets can be easily separated by a curve rather than a straight line. In this respect, it was necessary to develop architectures that can make curvilinear separations for the development of ANNs. Clustering can be done easily by ORing the ANN architectures that will be presented in the next sections.

7.10 Multilayer Artificial Neural Networks and Management Principles

After explaining the basics, structure, and usage areas of the simplest neural network in the previous sections, in this section, detailed information will be given about the structures of multilayer ANNs that can model wider curvature (nonlinear behavior). These have a multilayered and multicellular architectural structure that emerges as a result of the combination of many simple nerve cells described in the previous

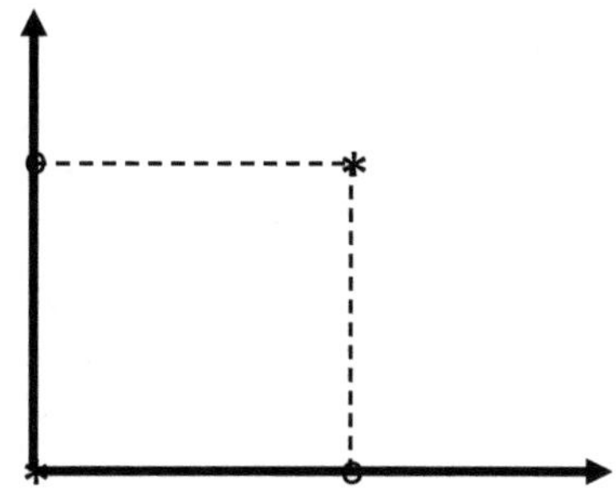

Fig. 7.31 ORing procedure

sections. One of the most important differences from the previous ones is that there is at least one more layer called hidden or intermediate between the input and the output. For this reason, they are called multilayer ANNs or multilayer sensors. In such an embodiment, the same processor is generally used in all cells of the hidden layer. Unlike the previous threshold processors, these processors not only vary between −1 (and 0) and +1 but also constantly change. Those with output values ranging from −1 to +1 are tangent hyperbolic, and those with output values ranging from 0 to +1 are sigmoid. The commonly used processor is sigmoid. A linear processor can also be used in a multilayer ANN. So far, it has been observed that multilayer ANNs, which are called standard and only have a single hidden layer, can represent many engineering problems.

It is not possible to precisely determine the number of cells in the hidden layer. What was said about the number of interlayer boxes in the box model described in Sect. 7.5.3 also applies here? In general, in the figure, there are input, hidden, and output layers consisting of n, k, and m cells, respectively. Of these, the input layer cells are connected to the hidden layer cells, and the hidden layer cells are connected to the output layer cells with a set of weight coefficients. The first connection set is shown with the letters a_{ij} and the second with the letters c_{ij}. From these explanations, we understand that the flow of information naturally goes from left to right, that is, forward.

7.11 ANN Properties

From the geometry of the network, the general architectural structure (topology) and the determination of the connection type are understood. Here, determining the number of cells in the hidden and other layers in general is important problems. The layers in Fig. 7.32 have the following properties:

1. *Input Layer*: The boxes in this layer are called input units or cells. These represent the network's perception of information given as input to the network. Each input unit has values of the input data at a given moment. Input units do not process the data and simply distribute the information contained in them to the hidden layer units.

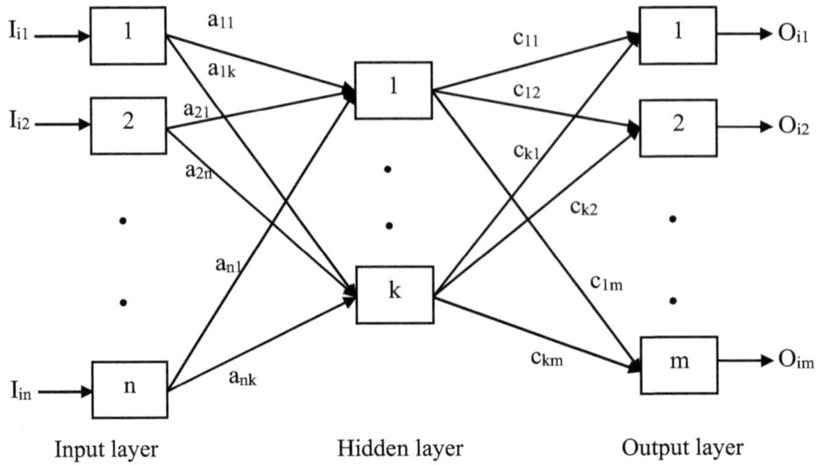

Fig. 7.32 General ANN architecture structure

2. *Hidden Layer*: Boxes here contain hidden units or cells. These cannot be observed directly. These cells have processors that provide curvilinear (nonlinear) behavior to the network.
3. *Output Layer*: The boxes here are also output units or cells. They represent expected values as output for any data (pattern). For example, each output unit can have a class of input data.

The weight connecting the cells of successive layers can be forward or backward fed symmetrical or asymmetrical. Their definitions are as follows:

1. *Feedforward networks*: All weights provide the correct flow of information from input to output.
2. *Feedback networks*: These represent feeds (information flow), either reversible or in loops.
3. *Symmetrical connections*: Information flows from one cell to the next and again to the first cell. If their weights in both directions are equal, they are called symmetrical connections.
4. If the connection is not symmetrical as said in the previous step, they are called asymmetrical connections.

Multilayer ANNs serve to establish mathematical parallel processing bridge (relationships) between input and output cells. Small changes in the connection (weight) values, which can be considered as a parameter during their operation, can sometimes lead to important and even uncontrollable results. This means that the operation of the ANN may exhibit nonstationary or chaotic behaviors for certain parameter values. We can list the important points in ANN modeling as follows:

1. The architecture of the network
2. The number of layers in the network

3. The number of nerve cells in each layer
4. Selection of teaching algorithm
5. The number of iterations to be made during the training
6. The number of calculations in each iteration
7. Group or pattern recognition speed
8. The performance of the network
9. ANN validity after learning
10. Whether here is a need for a constant term
11. Selection of the processor function
12. Selection of weight values as well as possible
13. The results are stationary

7.11.1 ANN Architectures

The structure that occurs as a result of connecting a certain number of nerve cells with each other in a way that makes information flow possible is called ANN topology, structure, or architecture. ANN architecture includes nerve cells, connection weights, and processors. This resulting topology can have smooth or irregular geometries.

Just as the nerves in the human brain die and their new connections are established, it is possible for ANN to renew its architectural structure. For ease of calculations, it is assumed that the cells are connected to each other with a smooth geometry, and there is no structural change.

Although a topological structure selected at the beginning of the study remains constant throughout the study, it provides great convenience in terms of calculation, but it brings some limitations. Because in this case, only the internal weights of the artificial network can be changed during the training, but there is no structural change. Thus, the solution is tried to be reached with the network that was originally designed. This situation, which is called the structure problem by Lee (1991), sometimes leads to problems such as not being able to reach a solution or reaching a solution as a result of too much training. The reason for this is that the solution is tried to be reached without changing this structure with the architectural structure determined by the researcher who designed the network. All these problems can be eliminated only if the researcher designing the network chooses the most suitable architecture.

There are many options in the selection of networks that contain many artificial neurons in their structure, as shown in Fig. 7.33a–d. In these figures, cells are shown with a circle. In this book, square or rectangular is preferred.

From this figure, it is understood that there can be single or multi-layer architectures. In general, there is no problem in determining the number of artificial neurons in the first (last) layer representing the input (output) variables. The main problem is how many cells are in the middle layer, which adds curvature (nonlinearity) to the operation of the multilayer ANN architecture. In addition, it is necessary to consider how many hidden layers will be. In general, one hidden layer is taken in the first studies. In ANN, modeling should pay attention to the selection of architecture and elements, since researcher wants to make evaluations by understanding nature. The following points should be considered in the mathematical analysis of an ANN.

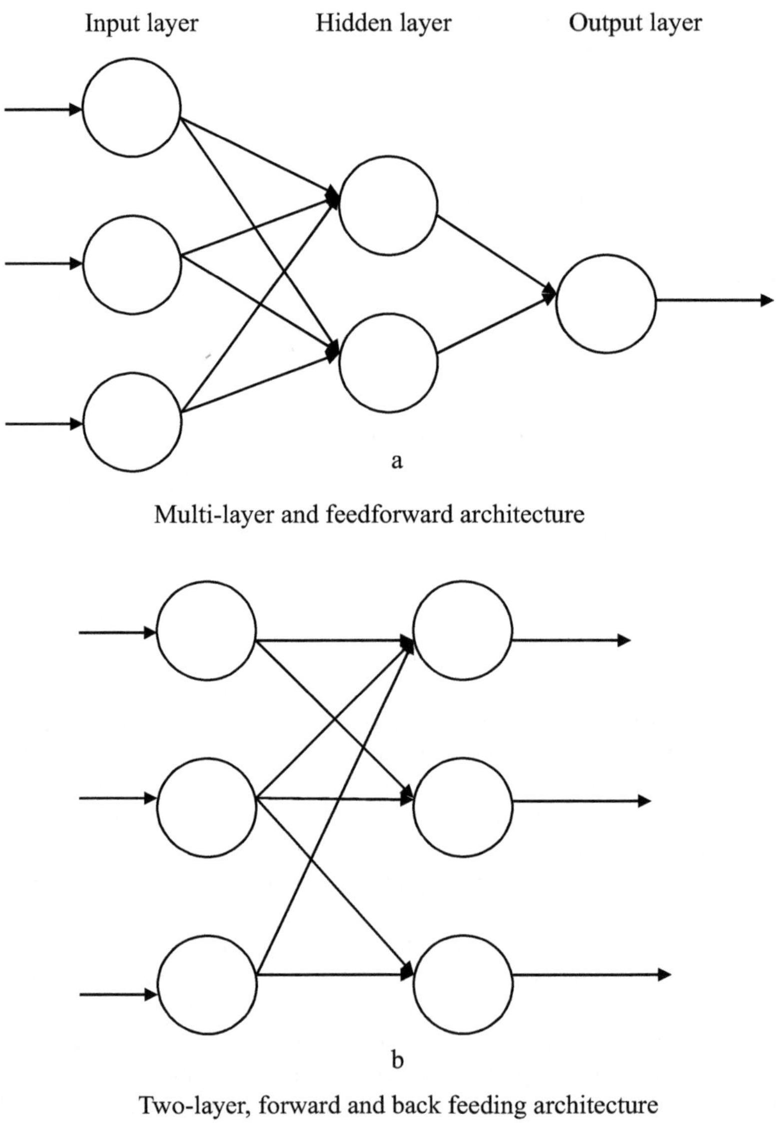

Fig. 7.33 Self explanatoryANN architectures

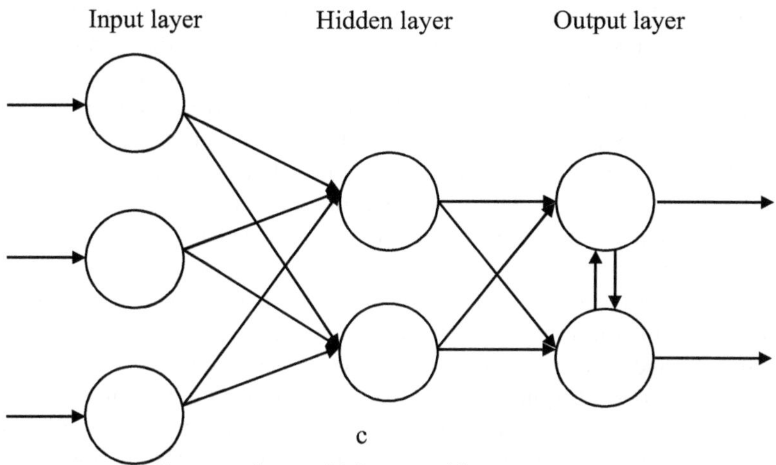

c
Compatative multi-layer architecture structure

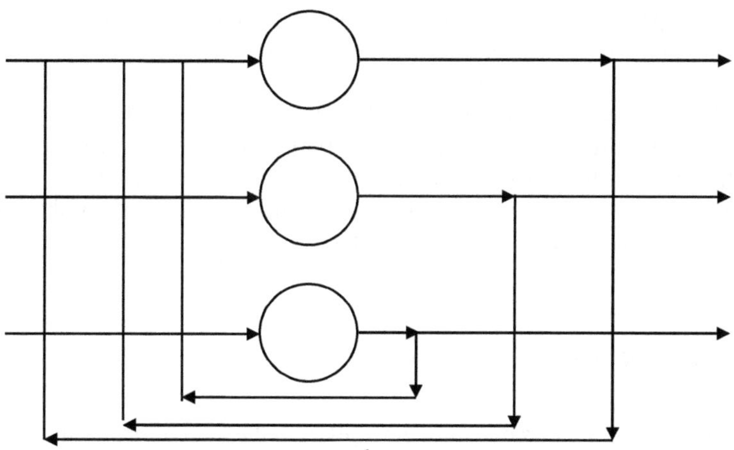

d
Single-layer mixed feed architectural structure

Fig. 7.33 (continued)

1. How much detail the ANN should have according to the problem to be solved?
2. How much information will be stored in an ANN and used later? Considering the importance of the variables that affect the problem from the information to be stored here, it should be considered in the modeling.
3. How should the most suitable ANN architecture be selected in applications?
4. How to provide the best ANN behavior?
5. How fast should ANN learn?
6. How much time delay after submission of inputs should information about outputs be available?
7. Under what conditions can an ANN learn wrong?

If the input data is discrete, each of them can be represented as a single input unit. The number 1 corresponds to the processor value "yes" or "true," and the number 0 corresponds to "no" or "false." If the processor value falls between 0 and 1, it represents the probability of that value. Continuous input data can be represented in different ways. In practice, continuous variables are generally set to be between 0 and 1. Discretization of the continuous variable is done by dividing it into adjacent nonoverlapping subintervals. For example, if one considers the continuous variability between a and b, the range of variation is divided by k + 1 points into k subintervals. These subranges are m_0, m_1, m_2, ... , m_{k-1}, m_k; then $m_0 = a$ and $m_k = b$. In the case of input data such as x as a path in data representation, if $m_i < x < m_{i+1}$, then in unit encoding, this range has a processor value of 1, while others are expected to have a value of 0. In another approach, it is possible for a unit to have a processor level equal to 1 if and only if the lower limit of this range is equal to or less than x as $\overline{x} < x$. The second approach will also trigger one or more units to encode a single value. Since the second approach is more dispersed, it is understood that it will give less noise (noise, error) than the first.

In a third approach, a unit is processed in proportion to the closeness of the respective subranges to the value. This is again a distributed processor and is called a knowledge-based ANN. On the other hand, it is possible to represent continuous value with binary number system instead of decimal (Şen 2004). For example, 8 is represented by four digits like 0111 and 7 in 1000. In this system, it is understood that since the numbers that are close to each other do not appear close to each other, a physical meaning cannot be given. In this way, it constitutes the basis of the concept of representation according to the binary system.

7.11.2 Layered ANN

As explained before, ANN architecture emerges with the help of one or more layers. The layer consists of a certain number of nerve cells connected to each other with a smooth or irregular geometry. The choice of a layered structure for problem solving is decided by considering the characteristics of layered ANNs described above. In this section, brief information about two- and multi-layer ANNs will be given.

7.11.2.1 Two-Layer ANN

A two-layer ANN emerges from a single input and output layer as seen in Fig. 7.34. Due to the absence of a middle layer to provide curvature, this architectural structure is mostly used for linear depictions (see Sect. 7.8).

As can be seen here, the weight values connecting each cell in the input layer to the cells in the output layer are called weights (a_{11}, a_{12}, a_{1m}, ... , a_{n1}, a_{n2}, a_{nm}). The i-th pattern components coming to the input layer are multiplied by the connection

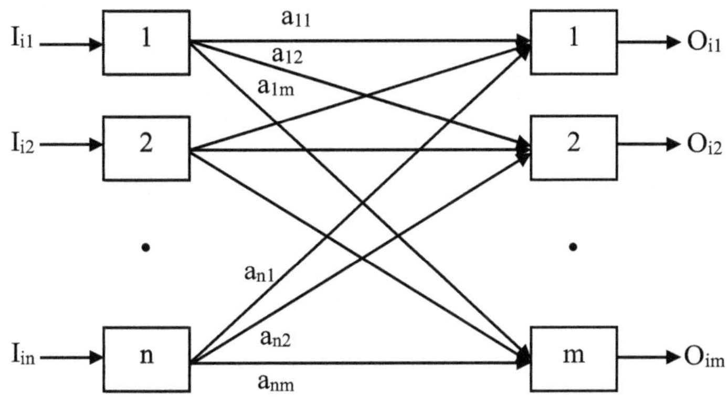

Fig. 7.34 Two-layer ANN model

weights I_{ij} (i = 1, 2, ... ,n) and delivered to each cell in the output layer. The information collected in each cell is passed through an adder function and outputs O_{ij} (j = 1, 2, ... ,p) are obtained.

A remarkable point here is that while there is no flow of information from the nerve cells in the same layer to each other, information is transmitted from each nerve cell in the input layer to the nerve cell in the output layer. The value of the carried information is found by multiplying and summing the weight values between the layers, like the previous one.

7.11.2.2 Multi-layer ANN

It is a feedforward network with at least one intermediate layer between the input and output layers. The intermediate layers consist of nerve cells called hidden cells, which are directly connected to the nerve cells in the input and output layers. These layers, which are formed by hidden cells, are also called intermediate or hidden layers. The ability of multilayer ANN is due to its layered structure and the use of nonlinear processors in the hidden layer nerve cell outputs.

An example of multilayer ANN is shown in Fig. 7.35. Hidden layers only carry information between the input layer and the output layer. In the figure, although the nerve cells in the hidden layer only transmit information to the nerve cells in the next hidden layer, information transfer between the layers is also possible.

It would be wrong to give a definite answer whether placing a hidden layer between the input and output layers or increasing the number of hidden layers will increase the efficiency. However, the general opinion is that bilayer ANN gives good results in linear and multilayer ANNs in curvilinear (nonlinear) depictions. It is possible to implement unknown or unexpressed nonlinear functions in dynamic models with the help of multilayer ANN architecture.

The question of how many layers an ANN should be used to solve any problem may come to mind. Although it is wrong to generalize about the number of layers,

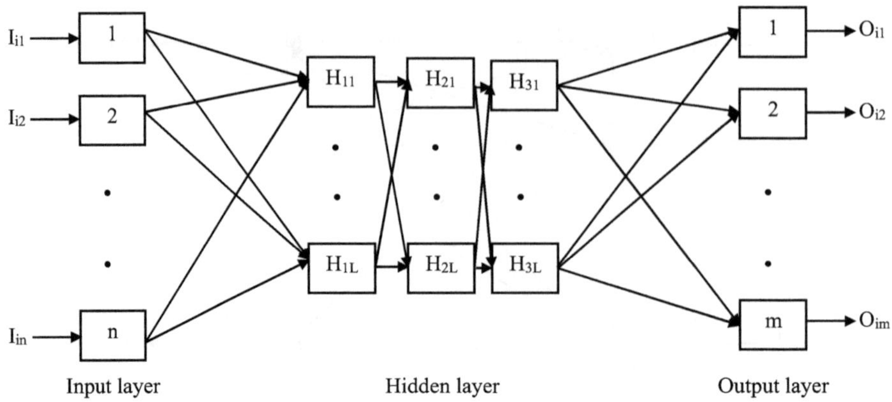

Fig. 7.35 Multi-layer ANN model

Simpson (1992) stated that a three-layer ANN using a single hidden layer successfully performs many nonlinear representations. It is useful to consider the following steps in the discovery of an ANN:

1. *Determination of network properties*: These are the geometric shapes of the network, connection types, connection sequences, and determination of the change intervals of weights.
2. *Cell characteristics*: Processor limits and type.
3. *Determination of system dynamics*: It is the determination of initial weight values, processor calculations, and error distribution training.

7.11.3 System Dynamics

After ANN architecture is determined, the connection between the layers and the fixed input values and the determination of the processor shapes to be placed in the cells in the hidden layer are considered as determining the system dynamics of that ANN. With the detection of these, the ANN can now detect the inputs to be given to it and perform forward operations.

Selecting the initial weight values according to a decided ANN architecture requires some special considerations. Generally, these values are taken as random small numbers. The most important step in an ANN is the learning rule, which will be explained in Sect. 7.5.1. This rule provides how link weights should be changed to optimize the network. It shows how to make link weight corrections at each step. Here, all the operations performed with the learning rule until the output values are close to the previously known values are called ANN training. While ANN is used to solve a problem, the solutions are the values reached after the processor and training in the output units. For example, suppose an ANN is used to separate fruits into oranges, apples, and lemons. Here, the network should have three units to represent each fruit type. A fruit whose type is unknown is placed in this network, and it is asked to determine what its type is. After the information is detected by the input unit, it is forwarded. If the unit corresponding to the apple in the output unit reaches the largest processor, it is concluded that the fruit type as a class is apple. From this example, it is understood how the processor levels should be adjusted within the network. Generally, when training an ANN, the difference between the value in an output unit and the measurement is calculated as an error. Correction of connection weights is done by taking these errors into account.

All the links between the different layers are called interlayer links or weights. Connections between cells in the same layer are also called interlayer connections. Links linked from a cell to itself be called self-links. Connections between layers with at least one layer between them, not sequential, are also called layer-hopping connections. In case of full connection, each cell in one layer is connected to all cells of the next layer. Otherwise, partial connection is mentioned. A higher-order link is a link that combines information from more than one cell. The number of input data determines the number of connections. The rank of a network is equal to the rank of the highest link in it. ANNs are always first order unless otherwise stated. Connection weights can be integers or decimal numbers. These change during the training of the network. Some of these can be kept constant depending on the situation.

With the above definition, the layers in the ANN, the cells in each layer, and their connections that transmit information from one layer to the next, seem to form an information network. As shown in Fig. 7.35, such a network contains parallel layers with cells within them and communication paths that provide cascading connections between them. For example, three-layer ANN architecture is shown in Fig. 7.32. Here, three parallel layers contain a certain number of cells. If each of these layers is represented by the indices I, H, and O, layer I is called "input," layer H is called "hidden or intermediate," and layer O is called "output" layer. Here, by analogy with "black box" models (see Sect. 7.6.4), the initial information that causes the input layer outputs to occur; the interlayer, the inner parts of the process, which adjust their connection with the output; the output can be visualized as the layer that gives the desired information to the researcher, for example, in the form of predictions or as artificial values.

The weighted connections between successive layers in the mesh in Fig. 7.32 are shown by (a_{i1}, a_{i2}, a_{i3}, . . . , a_{in}) and (c_{i1}, c_{i2}, c_{i3}, . . . , c_{in}). In such a network, although the input and output values are known, the weight coefficients in the ANN are trained, and the internal structure suitable for these inputs and outputs is developed with sequential approaches. During such a process, after the necessary steps, which will be presented step by step, are targeted by the ANN architecture in such a way that the errors are minimal, the future values are found with the architecture that is now trained and ready for predictions. In general, the method used was described by Rumelhart et al. (1986) as the generalized delta rule (see Sect. 7.5.1).

During the operation of ANNs, first the number of variables that can be input and accordingly the number of cells in the input layer are determined. For example, if the problem is to predict the future time series value, first, it is necessary to decide how many past time data to predict the future. Is this step a time interval (lag) just like stochastic processes? two? or how many time intervals? Oppose his decision. If it is decided to use the past three records in estimating the future value, then the number of cells in the input layer will be three. The input for each cell will be these three historical precipitation values. The training of the ANN is carried out by using the measured output value. In general, there is only one cell in the output layer. How many cells are required in the intermediate layer is decided in the light of the knowledge, skills, and experience of the ANN designer. The more knowledge, skills, consultation, and experience there are, the more appropriate number of cells will be used in the establishment of the ANN architecture. This provides economy in the number of communication lines and weight coefficient connecting cells in successive layers. Thus, with the available information, preferably an ANN with a small number of cells is used. This means that it is possible to calculate the desired communication weight coefficients correctly and to minimize the time required for training. In such an embodiment, the sum of the values that will come from i-th input data array (pattern) to the m-th cell of the next layer is calculated through the following expression.

$$(\mathrm{NET})_m = \sum_{j=1}^{L} a_{ij} I_{ij} + \Theta_i \tag{7.31}$$

Here, I_{ij} represents the j-th component of i-th data array, the total number of cells in that layer is L, and a constant value as the intrinsic contribution. The cells in each intermediate and output layer process the input information in the form of Eq. 7.31 and produce output value. The output values take their final form after the information collected in the cells with the f(.) activation function.

$$O = f(\mathrm{NET}) \tag{7.32}$$

The f(NET) activation function used here can be represented by different mathematical functions, according to the study.

7.12 Activation Functions

The description (mapping) between the input and output data in the ANN is provided by the mathematical function of the activation function, which are also referred to as the activation functions. Therefore, the selection of the activation function is of great importance in order to make the most appropriate description. First, it should be decided whether the depiction will be linear or not. In order to reach the best solution in the studies, it is necessary to choose the most suitable type of description. In multilayer ANNs, there are several activation functions to be used in hidden layer cells. Different activation functions can be used in different cells of a hidden layer, but in practice, the same activations are always used unless otherwise necessary to eliminate some problems with their selection at the outset.

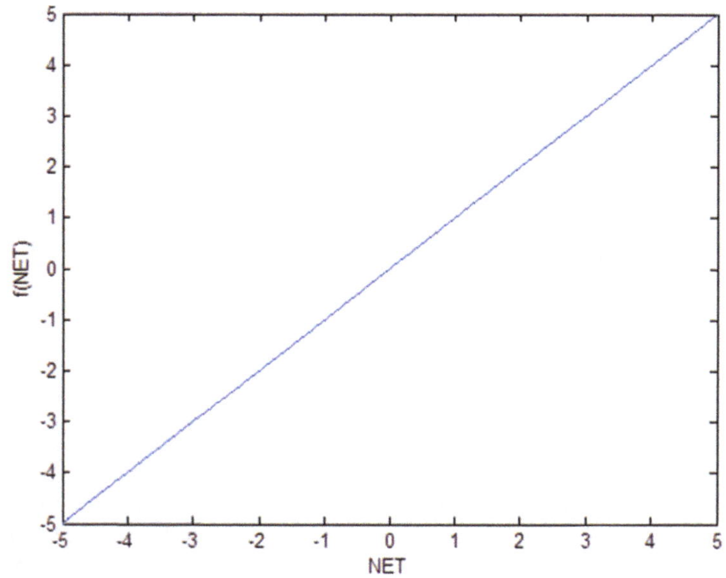

Fig. 7.36 Straight line activation function

7.12.1 Linear

The linear function shown in Fig. 7.36 corresponds to any input value I, where α is a scalar number as:

$$f(NET) = \alpha NET \tag{7.33}$$

If $\alpha = 1$, the information coming to the hidden layer cell remains unchanged at the activation output. Thus, since no curvature operation is performed, the ANN returns to the MADALINE architecture described in the previous section. On the other hand, if $\alpha \neq 1$, the inputs to the cell are output by only changing the scale (increasing or decreasing) but remaining linear.

The reader should compare the concepts of linear activation function with linearity as earlier mentioned in this chapter. For example, linearity in perceptron is related to the linearity separating the patterns. The operator of such a perceptron is one of the activation (threshold) functions. The operator of such a perceptron is one of the activation functions, which will be explained below. There is no limitation on linear activation function output here. Output values change constantly, not 0 and 1 (or −1). It can output values less than zero or greater than one.

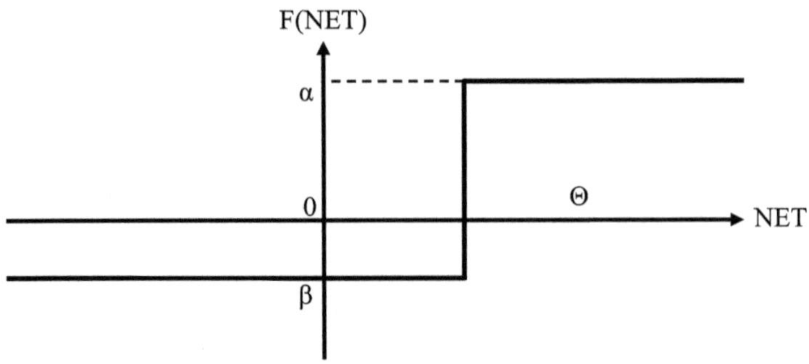

Fig. 7.37 Threshold activation function

7.12.2 Threshold

The most important feature of this is that it produces only two kinds of output against all input value (Fig. 7.37). If the input value, NET, exceeds the threshold value, Θ, it takes the constant value α as the output, otherwise β. The mathematical function of the activation takes the following form:

$$f(\mathrm{NET}) = \begin{cases} \alpha & \text{if} \quad \mathrm{NET} \geq \Theta \quad \text{then} \\ \beta & \text{if} \quad \mathrm{NET} \leq \Theta \quad \text{then} \end{cases} \tag{7.34}$$

In an ANN, α, β, and Θ values can be determined according to the type of study. Until now, in many studies, $\beta = 1$ and $\delta = 0$ or $\beta = 1$ and $\delta = -1$ with $\Theta = 0$ (Simpson 1992).

7.12.3 Ramp

It occurs when the two previous activation functions are together (Fig. 7.38). Here, it shows the linear operator in the range and the threshold operator in the range. The mathematical function of the ramp activation function has the following form:

$$f(\mathrm{NET}) = \begin{cases} \alpha & \text{if} \quad \mathrm{NET} \geq \alpha \\ \dfrac{\gamma - \delta}{\alpha - \beta}\mathrm{NET} + \dfrac{\alpha\delta - \beta\gamma}{\alpha - \beta} & \text{if} \quad \beta \leq \mathrm{NET} \leq \alpha \\ \delta & \text{if} \quad \mathrm{NET} \leq \beta \end{cases} \tag{7.35}$$

Such activation functions are mostly used in binary sets (Chap. 3). Also, γ and δ values are called the upper and lower saturation values. For each and values, the activation function takes only γ and δ values, respectively. In practical studies, the values of γ and δ are generally $\gamma = +1$ and $\delta = 0$ (or -1).

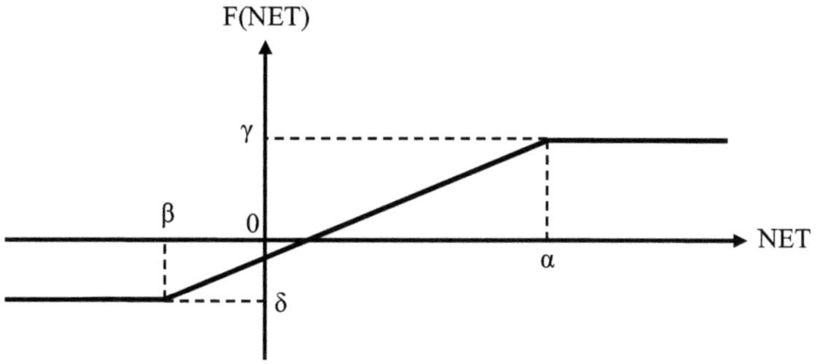

Fig. 7.38 Ramp activation function

7.12.4 Sigmoid

The sigmoid function is the continuous state of the ramp function. Since the function is in S-shape, it is often called an S function. Its mathematical form is given as:

$$f(\mathrm{NET}) = \frac{1}{1 + e^{-\alpha \mathrm{NET}}} \tag{7.36}$$

Depending on the value of α, this activation function appears in different curves as shown in Fig. 7.39. It gives outputs in the range of sigmoid function expressed as Eq. 7.36. If desired, the outputs of the function can be adjusted as in the figure or to fall within the range. Since the sigmoid operator represents a continuous function, it is most commonly used in nonlinear representations. The reason for this is that the derivative of the NET variable should be easily taken according to the parameters (Eq. 7.23). This will be seen in Sect. 7.5.1, in the section on back propagation of errors.

7.12.5 Hyperbolic

The tangent hyperbolic operator function (Fig. 7.40) has no parameters. The mathematical expression for this is given below:

$$f(\mathrm{NET}) = \frac{e^{\mathrm{NET}} - e^{-\mathrm{NET}}}{e^{\mathrm{NET}} + e^{-\mathrm{NET}}} \tag{7.37}$$

This activation function can be differentiated easily. However, since it has no parameters, it is not as elastic as the sigmoid activation function.

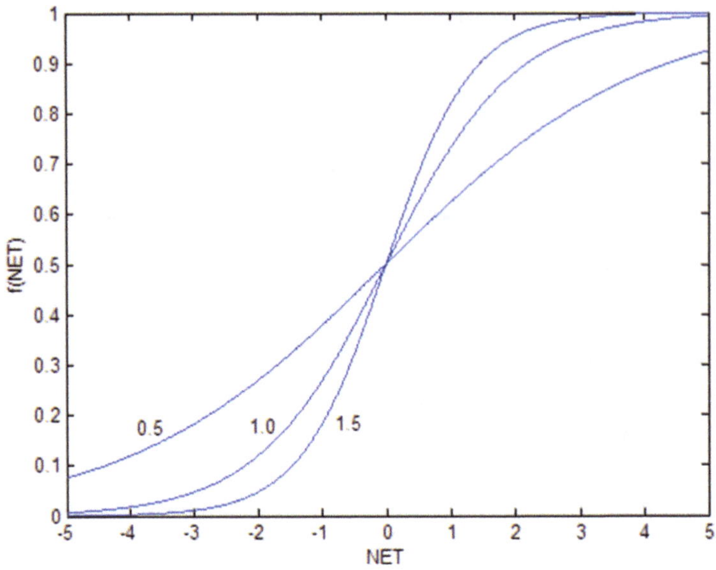

Fig. 7.39 Sigmoid activation function

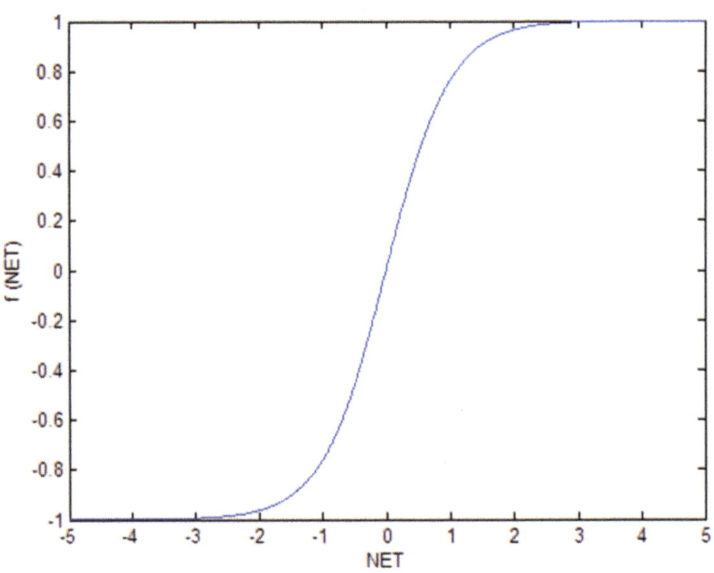

Fig. 7.40 Sigmoid activation function

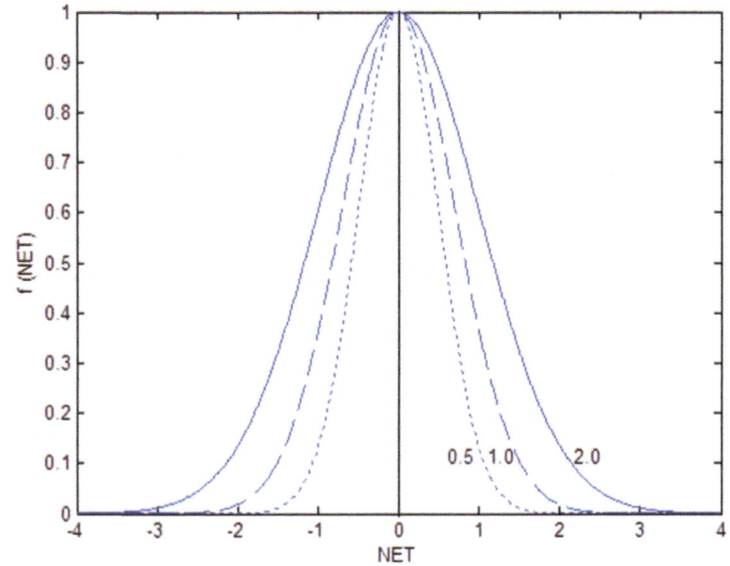

Fig. 7.41 Gaussian activation function

7.12.6 Gaussian

As can be seen in Fig. 7.41, the Gaussian activation function has a variance value greater than zero and symmetric with respect to the axis parallel to the f(NET) axis passing through its top. Mathematical expression with variance σ, is in the following form:

$$f(\mathrm{NET}) = \exp\left(\frac{-\mathrm{NET}^2}{\sigma^2}\right) \tag{7.38}$$

Since the last three activation functions explained above are continuous, they are used in analytical studies that need to be differentiated. One of them can then be used to develop the necessary application equations according to the least square's principle in the back propagation of errors. In general, it has become customary to use the sigmoid activation function.

7.13 Key Points Before ANN Modeling

Considering the simple ANNs explained in the previous sections, preparations should be made by thinking in detail about the event and data properties to be examined at the following points before the extensive modeling studies to be carried out

1. *Data collection*: It is necessary to collect data about the problem to be examined and decide which ones will be inputs and outputs. This data quality is not the numerical values and the number of data, but primarily the investigation of how many variables will be input and output. Thus, it will be decided how many cells will be in the input and output layers in the ANN architecture to be designed.
2. *Splitting the data into sub-data*: Some of the data will be used in training the ANN and some of it will be used in training for investigating its suitability and the rest in testing. Accordingly, the data is divided into three subgroups of 40%, 30%, and 30%. Randomization of this separation yields good results if the serial (internal order dependency) properties of the data set are not important. Let's not forget that the assignments of the ANNs weights are made randomly, just as in the generation of genetic algorithms (GA) (Chap. 6) in the formation of the first population, in the use of techniques such as Monte Carlo.
3. *ANN architecture*: Generally, there is an intermediate (hidden) layer between the input and output layers specified in step 1 to ensure the curvature. The number of cells in this layer should also be decided. As a recommendation here, it would be appropriate to include at least three cells at the beginning. Thus, the ANN architecture is completed.
4. *Mathematical software*: The mathematical functions should be decided for the activation function to be used in the hidden layer. The fact that the activation function is of the differentiable type is useful for making calculations. Therefore, sigmoid, tangent hyperbolic, or Gaussian operators are attractive (see Figs. 7.39, 7.40, and 7.41). It is also useful to predetermine the learning rate and memory coefficients to be used during ANN training (Sect. 7.5.1). There are no strict rules for their selection. However, by gaining experience over time or benefiting from previous studies, the researcher can have an idea about how much these coefficients should be. Even if they are given random values at the beginning, they can be changed according to the situation during the training if it is deemed appropriate. It even needs to be changed in some cases, which will be explained later.
5. *Initial weight values*: It is appropriate to assign the connections between the cells of the input-hidden and hidden-output layers as small random numbers. Also, for fixed threshold strict activation functions, the limits are taken as 0 and 1 or −1 and 1.
6. *Forward calculations (feedforward)*: With the completion of the above steps, it can now be used with ANN architecture and software. First, after the data is inserted into the input layer, aggregations with the connection weights (Eq. 7.31) and the necessary curvilinear (nonlinear) transformations are made

with the activation function in each hidden layer cell (Eq. 7.32) and constant values, Θ_i, are added, if any. Thus, the final values in each cell of the hidden layer are transmitted to the output layer cells with their assigned weights. Output values are calculated after summing in each output cell.

7. *Back-calculation (feedback)*: In case the output values are different from the values known as pre-measured data, the connection weights of the ANN need to be renewed. For this, an error value is obtained by taking the arithmetic average of the squares of the errors that occur in each output cell. This error must be distributed proportionally to all weights with their input data sizes. This is also called the feedback process, the details of which will be explained later.
8. *Iteration*: The renewed connection values obtained after feedback are repeated with the same input values by performing feedforward operations again. If the mean of the squares of the errors between the newly expected output and the corresponding measurement values is not smaller than a previously accepted limit, this time the feedback step is returned and the processes are repeated.
9. *Pause*: Until the calculated error is below a certain level, the forward-feedback phases, that is, the training, are continued. When the desired level is achieved, the latest ANN structure training is complete.
10. *Control*: With the completion of the above steps, the importance of the training data, which is defined as 40%, is over. This final version of the ANN (weight coefficients) is retrained with the fitness data (30%). Here, if there are no significant differences in the errors of the outputs, it is concluded that the ANN is ready for use with the absolute smallest error. If there are differences, this time the same ANN calculations are renewed with the fitness data as above,
11. *Usage*: After all the above steps, ANN can now predict unknown outputs from inputs.

7.13.1 ANN Audit

In the evaluation of any ANN architecture, mathematics, and outputs, the answers to the following questions should be sought:

1. How does ANN learn the given data series and what results does it give afterward? Here, the speed, quality, and quantity of learning and its function are important.
2. What are the responses of the ANN to the data that were not used during its training? Can satisfactory outputs be achieved?
3. How much computer memory and time will be used by ANN? What are the difficulties during the operation of the algorithm application? How can these be made more understandable and easier?
4. Is it possible to do the work of ANN with simpler existing classical methods and get more valid results?
5. Considering the investigated event, can comments and inferences be made in ANN architecture and mathematics?

In order to increase the reliability of the results obtained at the end of the use of an ANN, it is necessary to carry out some tests. The following tests can be used for better confirmation depending on the ANN status:

1. *Euclidean distance test*: The expected output, b_i, and the ANN output, the error, E, value obtained by calculating the Euclidean distance between the O_i ($i = 1, 2, \ldots, n$) data (pattern, vector) can be taken as a basis.

$$E = \sum_{i=1}^{n} (b_i - O_i)^2 \tag{7.39}$$

The more this error can be reduced, the better are the results. This error criterion has always been taken as a basis in the previous explanations.

2. *Absolute distance test:* This is defined as the sum of the absolute values of the differences of the reciprocal elements of the input and output patterns.

$$E_M = \sum_{i=1}^{n} |b_i - O_i| \tag{7.40}$$

3. *Relative error testing*: The previous error definitions are made without making any distinction between the output cells of an ANN. However, each cell has its own fault. The fact that the relative error of the largest of these is less than a certain percentage (±5 % or ± 10 %) indicates that modeling is reliable and satisfactory. If the biggest error is in the j-th output cell, its relative value is calculated as:

$$E_j = 100 \times \frac{|b_j - O_j|}{b_j} \tag{7.41}$$

If $E_j < 5$, the desired result has been achieved.

4. *Classification error test*: Unlike the above, the success in the classification made by ANNs the classification error is calculated as:

$$Ec = \frac{W_c}{T_c} \tag{7.42}$$

Here, W_c is wrong, and T_c is the total number of classifications. This counters the possibility of miscalculation. The lower the probability, the better the ANN works. In practical applications, it is desirable to keep it less than 0.05.

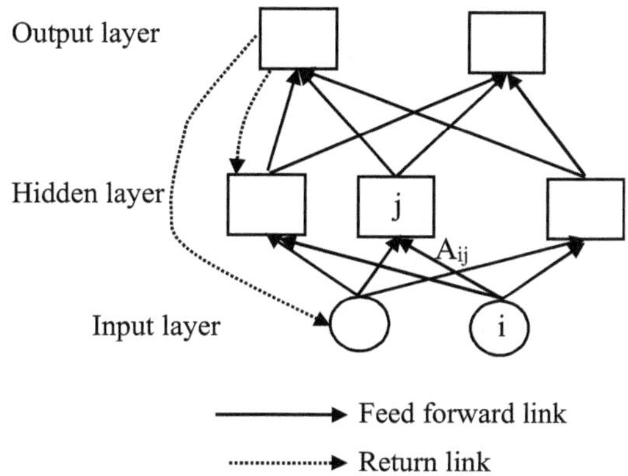

Fig. 7.42 ANN computation model

If ANN gives good results with especially given data, but undesirable results with new and independent data, it cannot be mentioned as generality. An ANN developed for an event should show similar behavior in all data related to that event.

7.13.2 ANN Mathematical Calculations

ANNs are a network that emerges when many cells are connected to each other in parallel distribution. There are weight values that show the effect of this connection from one cell to another. The following figure is given to explain the simplest structure and behavior of ANNs. Here, there are exactly three layers, one for the input, and the other for the output, and hidden between these two (Fig. 7.42).

So far, different ANNs and their training, learning, renewal, iteration, and forward and backward computation processes have been explained. In order to reach better and more reliable results in a short time, some points have been passed closely. Here, ANN architecture and some subtleties that can be found in mathematical hardware will be presented, with emphasis on them. For this, the following questions and answers are given:

1. How are weights assigned as initial values? After the input data is detected by the ANNs, the weight coefficients of the connections are needed at the first stage of calculations. As has always been said earlier, these are randomly assigned as small numbers. However, it is not always necessary to do so. If the reader gives arbitrary initial values to the weights, the ANN model will still work, but it may take a long time. The most reliable approach in assigning initial values is the opinions of experts in the subject under study. With expert opinions, it is possible to significantly shorten the processing time of an ANN. In the absence of such a view, the initial values are generally assigned randomly, staying between -1 and $+1$. At this point, the reader should pay attention not only to the ANN outputs with ready-made software but also to how the weights change step by step, at

which values they reach stationarity, after how many training cycles (iterations) this is achieved, how many layers there are in the ANN, and how many cells are in each one and make the necessary comments. Otherwise, it is not possible to specialize with the use of mechanistic software, and the dilemma of many ANN modelers and even article writers arises from not paying enough attention to these points. Considering these, inferences and comments about the mechanism of the event will increase in the light of the expert opinion on the one hand and the outputs on the other hand. While assigning the initial values, the researcher does not need to be an expert, but making numerical assignments by taking the opinions of the experts saves time and improves experience. It is beneficial for activation functions that reach saturation at very large and small values, such as sigmoid, where the initial values are slightly far from the -1 and $+1$ limits. In such activation functions, extreme values can bring the ANN to saturation immediately and cause subsequent transactions not to develop properly. For this reason, some researchers have proposed to choose initial weight values between -0.5 and $+0.5$.

If the input data are not too different from each other, it is appropriate to take initial values with a uniform random distribution. There is a huge difference between the data, and for example, if one or two data values are too large compared to the others, the uniform random selection may go in a biased direction, giving more importance to large input values.

2. Is the weight refresh after each new data string is presented to the input layer? Or should it do it after certain regular or irregular intervals? A general answer to this is that if the elements in the data array are not too different from each other, it may be appropriate to refresh the weights at certain intervals, but in the case of complex data, the weights should be renewed after each data series (pattern) is detected by the input layer. A minimum error state can be reached after a certain training cycle (iteration) with the refreshes made after each pattern entry or ANN training stops after the execution of the largest initially allowed number of training cycles without this attainment. If the order of the data is not important, and similar patterns are known at the beginning and approximately how many clusters they fall into, it may be necessary to renew the weights so that successful results can be obtained in a short time. The most time-consuming part in an ANN calculation is the renewal of the weights.
3. What should be the value of the learning rate? The learning rate, η, is a measure of how quickly the ANN learns from the inputs during training. This ratio shows that the net change in the weights at each step is included in the calculation of weight replacement, reduced by a certain scale adjustment. If a near-continuity course is desired in renewal, this should be kept very small. Keeping this ratio high (small) causes the ANN to learn fast (slowly). It is necessary to find a middle way by avoiding both situations. Theoretically, it is unthinkable for this ratio to be greater than 1 and less than 0. Since it is necessary to avoid both limit values, it is constant with experience in all studies conducted in the world, and this ratio has been taken as $0.1 < \eta < 0.8$. In general, every ANN is fast in the beginning, but as it gets to know the data, it moves toward the ability to learn more slowly. For this

reason, instead of a constant learning rate that takes a lot of time in ANN training, initially large, but decreasing rates gradually approaching 0.1 should be considered. Although there is no general rule in the selection of this ratio, the following points stand out in the studies carried out so far.

(a) *Reduced testing approach*: The value of η, which is kept large at the beginning of the training, can be decreased continuously as time goes on. The general rule to consider here is that the increases in weights should not be excessive. Otherwise, weight regeneration may be oscillating. These oscillations may increase, especially when approaching a local minimum error point.
(b) *Incremental and detrimental testing approaches*: If the ANN behavior improves after each design, η is increased relative to its previous value, but decreased if the behavior is not good,
(c) *Double Increment*: The learning rate is doubled until the ANN errors get worse,

4. What should the value of the memory coefficient be? This coefficient is a measure of how much of the previous weight values will be reflected in the new weight values when the weights are renewed. If there are successive local minimum points that differ from each other in the variation of the error amount with the training circuit (number of iterations) during the ANN training, the slope of the error term should also be included in order to avoid a local minimum point during the weight renewal. Thus, the weight increments at time (t + 1) come with the arrival of a term with the memory coefficient, α, in addition to the back-propagation errors to those at time t, the following expression is used.

$$a_{ij}(t+1) = \eta E_j I_j + \alpha \Delta a_{ij}(t) \tag{7.43}$$

It is very difficult to predict the value of the parameter in ANN calculations. However, an idea can be obtained by running ANN with trial and error processes. Similar ways of learning rate can be applied. The smaller (larger) the memory parameter, the less (more) the effect of previous weight values on weight regeneration.

5. *Generalizability*: If an ANN model is overtrained, it is not possible to benefit from it as generalized. The simplest criterion of overtraining is that ANN cannot recognize additional data or recognize it very poorly, although it gives good answers to training data series and hides them in its body. For this reason, Hecht-Nielsen (1990) suggested that ANN should be developed with training patterns and mathematics, but it should be checked continuously with test data errors. In some cases, even if the continuous decrease of the ANN error is hidden by the training data, the training is terminated even though the smallest error point has not been reached. Starting from the larger error level with the test data to be used later, if the ANN errors are gradually reduced, it can reach a minimizing point. In this case, the ANN is optimized according to the test data.

Especially when there are many architectural elements in layers and cells in the layers, the training data may be insufficient for training ANN mathematics. In this respect, as a rule, ANN architectures with fewer elements can be preferred. In some cases, it is possible to artificially add errors to the same training data and renew the weights with them to check the generality of the ANN. Depending on a study, how many layers and cells and how many training data arrays are needed to train, this architecture can only be answered with a trial and error approach. By paying attention to the details of such studies, it may be possible to obtain ANN architectures that can work in a short time with less effort by gaining expertise for similar studies,

6. *How many hidden layers and how many cells should be considered?* In order to answer this, special and additional trial testing should be done. For this, the most appropriate layer and cell numbers are obtained in the case of the smallest error by reducing the elements one by one, respectively, from the architectures with many elements, and by comparing the resulting ANN error with the previous architectural structure, continuing the element reduction until the smallest error is reached. It is also possible to do the opposite. In other words, starting from an architectural structure with very few elements and increasing the number of elements one by one, following the error again, the most suitable architecture and its elements can be obtained. Unfortunately, since these are very tedious, time-consuming and uneconomical to do, many researchers are content with training the ANN with the first element and their configuration and the data they have, and even without looking at the faults, it gives satisfactory results. The reader of this book is advised to arrive at an appropriate ANN design in a comparative manner by working with different element assemblies, even if it is time-consuming, by being very patient. According to the data at hand and the desired output type, there is no problem in the number of cells in the input and output layers. The problem is the number of hidden layers and the number of cells in each hidden layer. In practical studies, only one hidden layer is taken unless there is a logical necessity. In this case, one only must find the optimal number of cells in this hidden layer. Here, again, as a rule of thumb, it is useful to keep in mind that ANN architectures, which usually have three or five hidden layer cells, have been used in most of the previous studies. Here, as shown by Lippmann (1989) in Fig. 7.43, each cell in the first hidden layer stores information from the input layer and works as a separation line. Thus, the number of hidden layer cells can be determined according to the expected number of classes in classification problems.
7. How many data sets (patterns) are needed for a good training? It is quite difficult to answer this. Generally, expect to have five to ten times as many data strings as the number of weight links. By Baum and Haussler (1989), N_t is the number of training data, N_w is the number of weights to be renewed, and the expected confidence number in α is stated as:

$$N_t \geq \frac{N_W}{1 - \alpha} \tag{7.44}$$

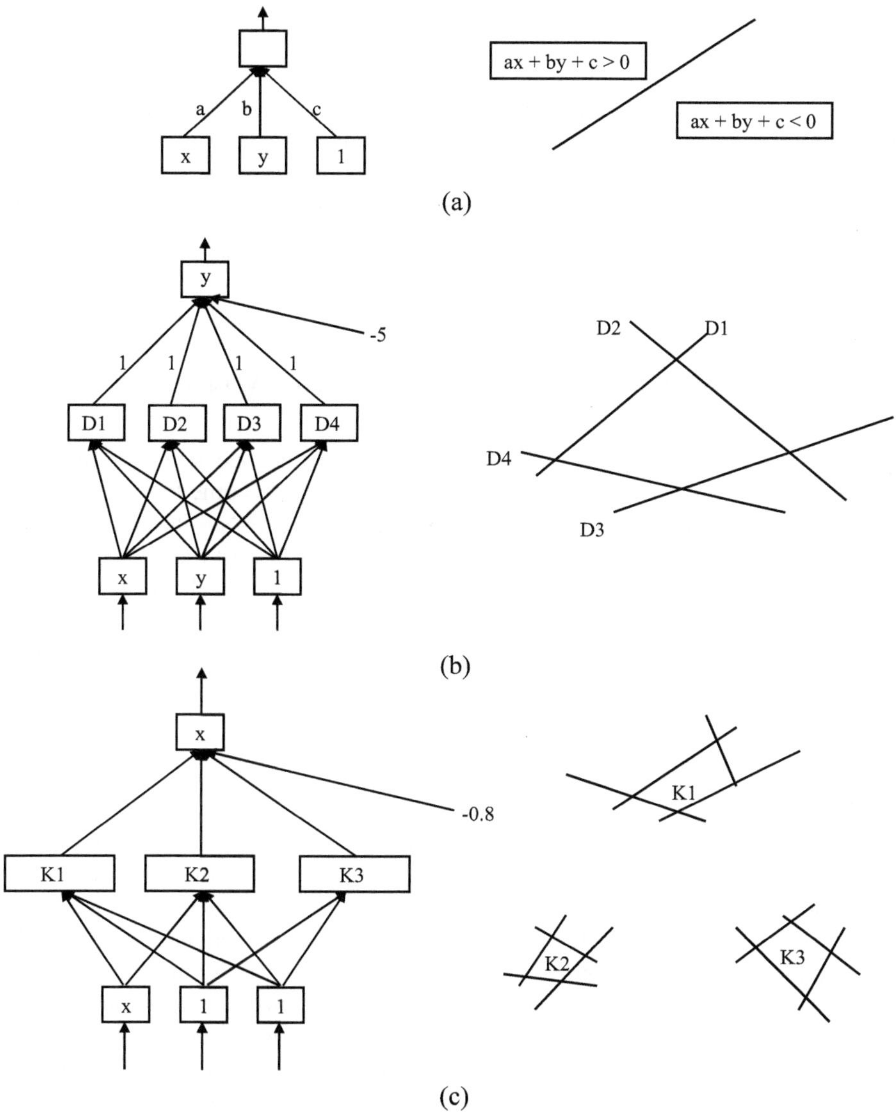

Fig. 7.43 Hidden layer cell and seperation straight line relationship, (**a**) single cell threshold process, (**b**) single hidden layer and convex region, (**c**) Two hidden layer and three convex region union (Mehrotra et al. 1997)

For example, if there are 36 connections in ANN architecture and it is desired to obtain results with a reliability of 0.95, the number of data required to be used in training should be at least 720. With the inclusion of the number of cells, n, into the work, it is recommended to use the following formula:

$$N_t \geq \frac{N_w}{1-\alpha} \log \frac{n}{1-\alpha} \tag{7.45}$$

If there are eight cells in the example given above, the number of training data must be at least 1587. All these theoretical calculations cause the theory to be inconsistent with practice, just as the account at home does not match the market. Because collecting and processing this much data in practice is a separate problem. For this reason, the number of training data is selected by the method at best or the available ones are satisfied.

7.13.3 Training and Modeling with Artificial Networks

In the previous sections, ANN basic, architectural structure, and activation functions are explained. After all the initial preparations are over, it is time to train the ANN and then test it. One of the most distinguishing features of ANN is its ability to learn. Learning is defined as the calculation of link weights between available samples (data, patterns, vectors) that can make the structure behave well. In the previous section, learning in ANN was also defined as changing the connection weights between cells to realize the most appropriate description between input and output information. These changes are mainly as follows:

1. Making new connections
2. Replacing existing weights
3. Destroying some link weights

ANN stores the information obtained during learning as connection weights between nerve cells. These weight values contain the information necessary for the ANN to process the data successfully (Freeman 1994). Since the information is stored in the entire network, the connection value of a single network does not make sense on its own. For a meaning, a group of connection weights must be considered together (Özmetel 2003). However, all intercellular connections must have appropriate values for the ANN to display an intelligent behavior. Artificial nerves are like the human brain in terms of needing information during the learning period and storing information through the connection weights between the nerves.

One of the important points in ANN learning is the selection of the training set that will enable learning. The misconceptions in this selection are that the larger the training set is chosen, the better the trainer will be. However, the training set should be chosen to provide the best learning with the least amount of information (Sönmez and Şen 1997). While generating the training set, choosing different and independent information rather than close information provides a more efficient learning. Already, the training set is expected to give reasonable outputs against the previously untrained input values. This feature sought in learning is called generalization.

For example, the data from I_1 to I_5 show the actual data value with the input and the corresponding outputs. The curves in Fig. 7.44a, b are the same, but the sampling

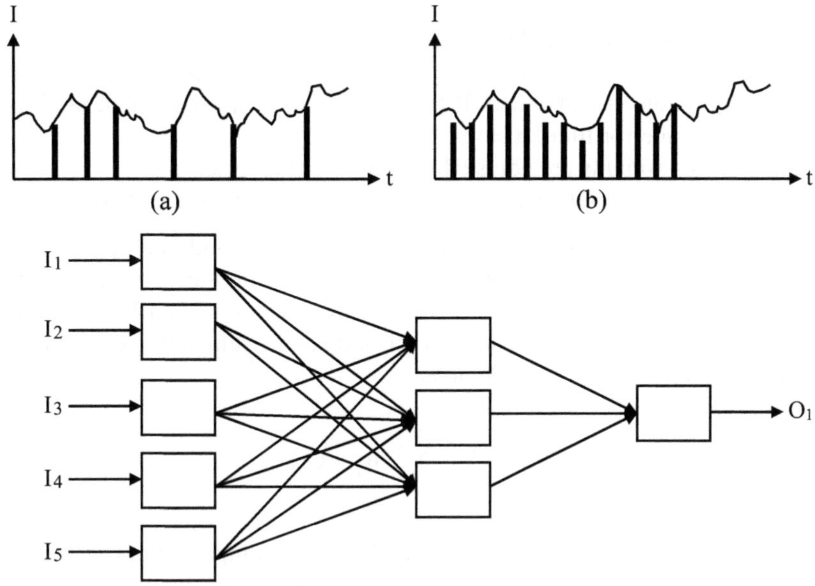

Fig. 7.44 Training set with generalisation variation

for ANN training is different. Whereas the first sampling is random, the sampling in the second is formal. Random sampling should be preferred for the best training of ANNs. It includes different features in terms of taking the changes in the structure of the input data randomly. This type of sampling is not biased.

7.13.4 ANN Learning Algorithm

In order to develop an artificial intelligence (AI) program with ANN, the following steps must be followed in order:

1. Selecting a suitable ANN architecture according to the situation of the problem
2. Selection of ANN characteristics (activation functions, connection weights) in accordance with the application
3. Training the system with data according to the selected model and given characteristics
4. Using the trained system to make inferences, that is, testing. If proper results cannot be obtained, one or more of the previous steps must be rearranged.

If there are n sequential training sequences (patterns) in a problem, the desired results can be obtained by training the ANN after completing the following processes:

1. In the first step ($i = 1$), the initial weights are selected.
2. The input layer detects the data valid for $i = 1$.
3. By using the inference algorithm (forward feeding), the activation function levels of the output units are obtained. If the ANN predictions give the desired results within a certain error limit, it stops.
4. Otherwise, the weights are adjusted with the learning rule.
5. If $i = n$, $i = 1$ again. Otherwise increment i by 1 and go to step two.

In order to understand whether the selected model and the application are satisfactory, it is necessary to optimize the ANN by minimizing the sum of the errors between the actual data and the output layer data obtained from the model. One of the criteria used for this is the minimization of the sum of the squares of error. The sum of squares of error is written as:

$$E_T = \sum_{i=1}^{n} \sum_{j=1}^{m} \left(b_{ij} - O_j\right)^2 \tag{7.46}$$

Here, b_{ij} and O_j represent the expected and network output values at the time i (or when entering the i-th data string) in output unit j, respectively. Moreover, it is tried to optimize the ANN with cross-entropy, which is expressed by the formula (Hinton 1989):

$$-\sum_{i=1}^{n} \sum_{j=1}^{m} O_j \log_2\left(O_j\right) + \left(1 - b_{ij}\right) \log_2\left(1 - O_j\right) \tag{7.47}$$

The following steps must be completed in order to make inferences from the given training sequence:

1. Entering the input data into the ANN input layer units.
2. Calculating the activation function levels in each of the units in the hidden and output layers of the ANN.
3. In the feedforward ANN, the process is terminated after the activation function levels of all output layer units are calculated. In a feedback situation, processing is terminated if the activation function levels in each of the output layer units reach near constant values. Otherwise, it goes back to step 2. If the ANN cannot reach a stable state, then the ANN architecture is not successful.

7.13.5 ANN Education

Since there is no certainty in all ANN trainings, some philosophical approaches are in question. During the calculation of ANN weights, all the sequential forward-backward feeding processes are called training. It is desirable that this training ends with the minimization of the error. One can check the quality of the training with a

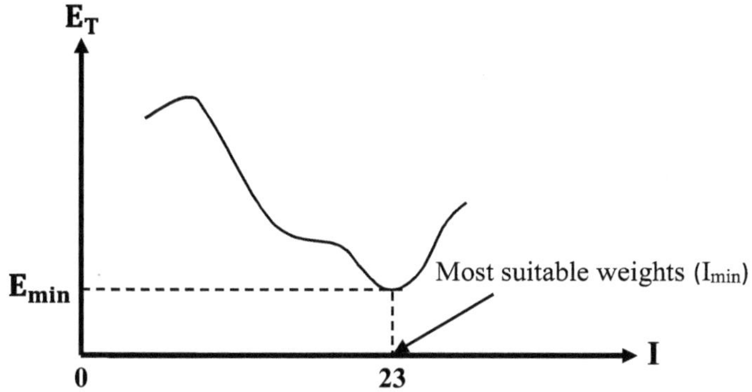

Fig. 7.45 Error-training period number

graph showing the variation of errors with the number of forward feeding iteration numbers. Such a graph is presented in Fig. 7.45. In general, the arithmetic mean of the sum of the squares of the errors is large, since the weights are chosen randomly at the beginning. During training, this is expected to decrease with the iteration number, I, of advanced feeding (training).

From this figure, it is understood that the situation with the smallest (minimum) errors is reached only after the 23rd training. If the relative error percentage, α, is less than ±5%, the training cycle is terminated. The mathematical expression for this is calculatable as follows:

$$\alpha = 100\frac{|E_{i-1} - E_i|}{E_{i-1}} \tag{7.48}$$

In this case, the weights have the most optimized values. This was achieved after I_{min} took several training cycles. The following questions can be answered by considering the relative error calculations.

1. *Question-1*: Is the smallest error achieved a local situation? Or absolute (global) error has been reached? The first of the quantities that affect the calculated error and that one can change is the selection of the initial values of the weight's coefficients. For this reason, the ANN is retrained by choosing other weight values that are completely different from the coefficients that started the previous training circuit and leaving the other conditions the same. Thus, a minimum error value is reached again according to Eq. 7.48. If this new minimum error value is practically no different from the previous one, it is concluded that the weight coefficients are obtained according to the absolute minimum error at the end of the training. In this case, there are no significant differences between the newly trained weight values and the previous ones. However, the new smallest error may have been reached at the end of a different number of training cycles. That is to be expected as well. If the new smallest error and the previous one is very different from each other, it is understood that neither of them reaches the

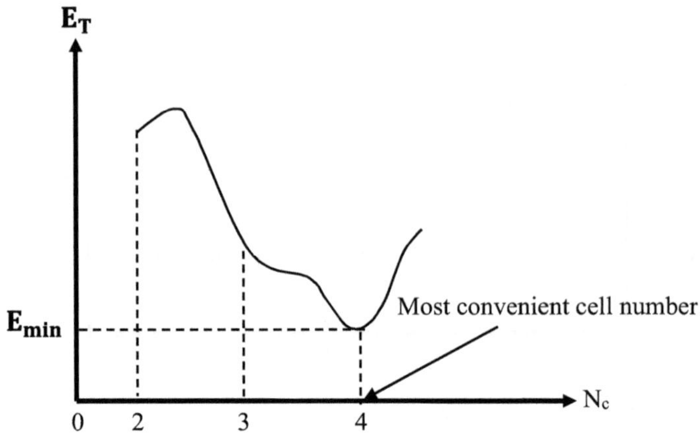

Fig. 7.46 Error cell number variation

absolute minimum. Retraining begins by selecting a third set of weight initial values. In this way, at the end of several different trainings, it is decided whether the absolute minimum has been reached or not.

2. *Question-2*: Another factor affecting the errors is the number of cells in the hidden layer. Can one reduce the error if we reduce or increase the number of them? The answer to this is yes, one can reduce it. So, in order to find the best number of hidden cells, this time, it is necessary to look at the variation of the errors with the cell number, N_c. One can show this graphically as in Fig. 7.46. However, if one wants to reach this result with calculations on the computer, the sequential relative error is calculated using Eq. 7.48. These calculations, which are based on the number of cells in the interlayer, are started with two cells, and the smallest error is obtained, then similar calculations are made by increasing the number of cells one by one. For each hidden layer cell number, a predetermined minimum error is found. When the relative error between consecutive cell numbers is less than $\pm 5\%$, the most appropriate hidden layer cell number is found for the available data. It is seen from Fig. 7.46 that the optimal number of interlayer cells is 4.

   The number of cells in the input and output layers of ANNs is determined according to the character of the problem under investigation. However, it is more difficult to determine the number of cells in the hidden layers, and there is no specific method in this regard. By sequentially changing the number of cells in the hidden layer, the best results can be found by investigating the cases where the squares of the errors are the least, using the trial and error method. However, too much computer time is needed.

3. *Question-3*: Could another factor affecting the errors be the activation functions? It could be some part of the answer. In that case, the minimum error is reached by making separate training circuits for different activation functions. Among these functions, the activation function that gives the minimum error is selected for ANN.

4. *Question-4*: Could it influence the learning coefficient if one considered before the ANN training? The answer to the question may be yes. So, one can similarly determine this by minimizing the errors based on Eq. 7.48. Theoretically, it is, but since this is a number between 0 and 1, it is practical to choose it as an expert opinion without going through such tedious error calculations. However, for the ANN, provided that the other conditions remain the same, only the smallest error change for different learning coefficients can be found, and the learning coefficient corresponding to the minimum error among them can be obtained. The same is true for the memory coefficient. In practice, it is preferred to include it in the ANN as an expert opinion at the very beginning.

Before using an ANN, whose architecture is given in Sect. 7.11.1, it must be immune to the event examined with the available data. This means that the weight coefficients, which play a very important role in the information communication between the cells of the successive layers, are estimated by using the data for the investigated event; in other words, the appropriate values are found. Such a process is called the training phase of the ANN. After giving the desired input data to the cells of the input layer in a formal way in the training, a person who has no skills and experience about the event can give random values to the weight coefficients if desired for the first time. Thus, as explained in the previous section, the predictions in the output cell or cells are performed by operating the ANN.

The sum of the squares of the errors is found numerically resulting from each of the estimations made for the n data sets available at hand. Here, the set of weights corresponding to the case that the sum of the squares of these errors is the smallest constitutes the ANN architecture weights of the event under investigation. In general, E_i is represented as the total error value among the expected (o_{i1}, o_{i2}, o_{i3}, . . . , o_{im}) ANN model outputs for the i-th data value, and the measurement (O_{i1}, O_{i2}, O_{i3}, . . . , O_{im}) values are found with the help of the following expression as:

$$E_i = \sum_{j=1}^{m} \left(o_{ij} - O_{ij}\right)^2 \tag{7.49}$$

Then, the sum of these errors calculated for p data inputs and outputs form is obtained as follows:

$$E_T = \sum_{k=1}^{p} E_k \tag{7.50}$$

All weights and thresholds are chosen randomly, for example, between −1 and +1. This has no drawbacks during ANN operations. As a natural consequence of this, it is easily understood that the estimation result obtained by ANN will be different from the data value. Thus, the errors corresponding to the weight values obtained by the "back error propagation" of the sequential improvement represent a point in the space formed by these weights. Here, there are peaks and troughs on the surface of the errors generated by the weight variables in an ANN. The depressions in this surface represent the places where the errors are smallest. The "steepest gradient descent" method is used in numerical and systematic research to reach the trough points (Şen 2004).

This opposes the method in mathematics, where the derivative of a function is zero. In the process of searching for the consecutive smallest sloped point, a location close to the vertex is initially selected. After that, it is progressed toward the lowest point by making research in the form of decreasing errors in succession of the weight set connected to each other. This process is displayed mathematically as:

$$\Delta a_{ij} = -\eta \frac{\partial E}{\partial a_{ij}} \tag{7.51}$$

Here, Δa_{ij} is the error variation between different layers; η represents a proportionality coefficient called the learning rate, and $\partial E / \partial a_{ij}$ the slope of the error surface at that point. The learning rate is kept small so that changes in weight coefficients do not oscillate during training. The training time can be shortened by taking high values of this. After these processes are completed, the weights are updated according to the following expression.

$$a_{ij}^{new} = a_{ij}^{old} + \Delta a_{ij}^{old} \tag{7.52}$$

In practical applications, in Eq. 7.51 weight increments are calculated according to:

$$\Delta a_{ij}^{old} = \delta_i E_T O_j \tag{7.53}$$

Here, δ_i indicates the error term of i-th cell, and O_i is for the value coming from the previous layer to this cell. Rumelhart et al. (1986) proposed δ_i to take values according to whether the cell under consideration is in the output or hidden layer.

7.14 Description of Training Rules

There are several rules that an ANN can benefit from during its training. The first of these rules was presented in the simplest way by Hebb (1949). The later ones have been developed a little more accordingly. Most of these rules are named after the first bidder. There are several different training rules that are generally in use. These depend on the mathematics that will be used to refresh the weight values and change them in a way that prepares the ANN for decision-making. Human knowledge of the training rules of natural neural networks is limited, not exhaustive. The learning rules presented here for ANNs are necessarily much simpler than the natural ones. This subject is open to research. The main rules that have been developed and used so far are as follows:

1. *Hebb's* (1949) *rule*: This is the first and best training rule. If a neuron receives information from another cell, and if both are quite active (mathematically the same sign), the link value between these two cells should be strengthened, that is, the weight value should be increased. Otherwise, it is necessary to weaken the weight value. In summary, this rule is based on changing the sign (+ or −) of the weight values with a predetermined fixed magnitude (see Sect. 7.8.3).
2. *Hopfield* (1982) *rule*: This rule is like the previous one. The difference between them is that in this rule the weight values are changed in magnitudes other than sign. Accordingly, if the signs of the desired input and output cells are the same, that is, active or both are not active, the weight values are increased by the teaching rate (coefficient), η; otherwise, it is decreased by the same amount. The rule here, unlike the previous one, also includes a training coefficient, which is a magnitude chosen by the ANN designer, generally between 0 and 1 and often between 0.2 and 0.4. In Hebb's rule, this ratio is equal to 1.
3. *Delta rule*: This is the most widely used rule, which is also based on the Hebb rule. Here, it is a learning method by constantly changing the previous weight values a little (we can call it delta ,δ, difference, inspired by mathematics) until the ANN outputs of the connection values are close to the output data values. It is based on the process of minimizing the mean square errors of the ANN outputs and restoring the weights. The errors are distributed between the two successive layers, starting from the last layer, toward the first layer in the connection weights. This is called the feedback process, which means the backscattering of errors. This process is continued until the first layer (input layer) is reached. Architectures trained with this type of learning can be called error distributed feedback ANNs. This rule is also called Windrow-Hoff (1960) or least mean squares learning rule in the literature.
4. *Kohonen* (1988) *rule*: This rule was inspired by the learning rule in biological systems. Here, nerve cells take care of learning and regenerating weights within their own body. A cell with the largest output is declared as the winning cell. While this cell affects its neighbors, it does not affect the others at all. Only the winning cell outputs, but in the renewal of connections, neighboring cells are involved by renewing their weights. The Kohonen rule does not require an output data array. For this reason, he completes his training with a teacher-free training (unsupervised). For this, there is the central cell effect and proximity to the neighbors. Thus, a self-regulating ANN emerges. Algorithms for these rules will be explained in Sect. 7.13.5.

7.14.1 Supervised Training

For this type of training, the output data of the cells in the ANN output layer should be known numerically. This cognition allows the entire ANN to learn the network and even memorize the outputs in such a way that the inputs are exactly against the outputs. During this training, ANN outputs a numeric value. Since the data value of this is known, it is as if a teacher compares the outputs with what should be and decides whether the result is acceptable or not. In this decision, the closer the value obtained by the ANN is to the data value, the higher the acceptability. Just as there are differences in the grades of each teacher, the teacher here defines an error limit according to himself, and if the difference between the ANN output and the data values falls between these limits (say, $\pm$ 5 %), the teacher accepts such errors. Thus, the training of the ANN is terminated.

Comparing the results obtained with the output data values (measurements), the differences are observed as errors. By calculating the neural connection (weight) values in the ANN structure to minimize the sum of the squares of these errors, the outputs are approached with the smallest error. Here, there will be a forward flow from the inputs to the outputs, and if the error term is not within the desired limits, there will be a backflow (feedback) from the outputs to the inputs. However, in these forwards and backwards, the input variables will never change their values, but the output variables will always change their values closer to the measurements. ANN operation will be terminated if the relative error term of these forwards and backwards is less than the desired value, for example, $\pm 5\%$ or $\pm 10\%$ to be selected. Thus, ANN can be used to make predictions by calculating the outputs from the next input data. In this process, learning is done by minimizing the sum of the squares of the errors.

For example, when learning a foreign language, one repeats the word after hearing it from the teacher. If the sound heard from the teacher remains in the mind, one continues to use it by repetition (by going back and forth) even if the pronunciation is not the same as the pronunciation one heard, but if it is very close, one is convinced that he has pronounced that word correctly. Thus, in supervised training, there should be a student and a teacher.

In training with the help of the supervisor, there is a teacher who interferes with learning from the outside. This process is completely under the control of the trainer. The teacher decides to determine the training set, to what stage the training will be continued, and how it will be carried out. The most important feature of supervised training is the use of real values during training (Kosko 1990). Supervised training can be examined in two classes: training where the teacher gives the right result and training where the teacher gives only reward-punishment.

7.14.1.1 Training Where the Teacher Tells the Correct Result

Here, first, the trainer gives only the login information to the ANN. Allows ANN to generate output information. Next, the teacher also gives the desired exit information (actual data). Thus, the ANN renews its connection weights in order to reach the desired output values by comparing the output data values. This process is continued under the supervision of the teacher and the training of the ANN to reach the minimum error.

7.14.1.2 Education in Which the Teacher Gives Only Reward-Punishment

During the training, the teacher only states that the result found by the ANN is true or false, instead of giving the exact answer. ANN is trained by changing the connection weights in line with these warnings.

7.14.2 Unsupervised Training

In ANN modeling, which has no exit information, there is no longer a teacher to control the quality of outputs. ANN only takes the information from the input layer and divides them into clusters after processing it according to itself. Here, the user is free to make own clustering. It is also called self-learning because there is no need for a teacher. In this training, only login information is given. ANN is provided to generate output. Since the desired outputs are never given to the network, the error is not considered (Kosko 1991).

The entered information is processed by the ANN and analyzer. The goal in decoupling is to identify as many different clusters as possible. Therefore, the link weights only change depending on the input data. The separation criteria may not have been previously known. In such cases, the network must develop its own classification rules. However, it should be noted that this type of learning can be applied to a limited number of ANN models.

In this type of training, the student must do something by himself without a teacher. Similarly, there is no information about the outputs of the ANN, that is, the situation on the opposite side, which is likened to a bridge for crossing from a known point of a river. In such a case, ANN architecture takes some input values and separates them into some groups or patterns after parallel operations. If there is an output status later, the group or pattern is given its new status with the help of this. In the absence of such information, a group or pattern may have been obtained, which may even be completely arbitrary. If one gives soil to someone, that person can divide it subjectively into three classes as gravel, sand, and silt. If he is given another sample later and if it falls into one of these three groups, he assigns it to that group, but if not, he establishes a new group. Each new information leads to a classification that is made to one of the previously existing groups or new groups (clusters) are revealed.

Although unsupervised training does not require output data, it is useful to know some tips about output. In the absence of such clues, unsupervised training may fail as well as succeed. Just as a babies can classify, compare, and aggregate their knowledge by time, without the need for a teacher, with the progression of age in months and the visual and auditory information they receive from the environment, similarly, they can make clusters, inferences, and predictions by arranging themselves by using the data given to them in ANNs. ANNs that have working principles

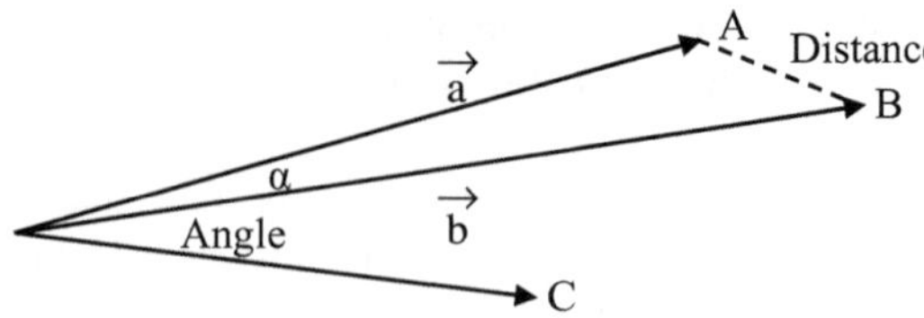

Fig. 7.47 A and B data vectors

in this way are called unsupervised ANNs. The most important feature of these predecessors is that they do not even require special activation functions in the hidden layer. With these ANNs, who can be trained unsupervised? Many clustering operations that can be done with human thought and some previously known classical methods can be done in a different and more effective way. Especially in the clustering process, the idea of keeping the similar data together as much as possible and keeping the dissimilar ones as far away as possible is dominant (Chap. 9). By making use of a given data (vector, pattern) sequence set, different ideas can be reached in finding its subsets. One can list the main ones as follows:

1. *Dependency principle*: As much as two data sets are dependent on each other, they should be put in the same cluster. Two data arrays with the same number of elements, say n, represent two points in n-dimensional space. As shown in Fig. 7.47, $(x_{11}, x_{12}, \ldots, x_{1n})$ and $(x_{21}, x_{22}, \ldots, x_{2n})$ data can also be represented as vectors with the same beginning. The closer they are to each other, the smaller the angle between them, and the more similar they are. Thus, they can be put in the same cluster.

   From the vectorial calculation of the cosine of the angle, α, between the two vectors, their scalar product can be written as:

$$\cos \boldsymbol{\alpha} = \frac{\vec{a} \cdot \vec{b}}{\left|\vec{a}\right| . \left|\vec{b}\right|} \tag{7.54}$$

   The denominator of this expression represents the strength of the vectors. In the case of the standardized vector, Eq. 7.54 takes the following form, since the magnitude of each vector is equal to one.

$$\cos \boldsymbol{\alpha} = \vec{a} \cdot \vec{b} \tag{7.55}$$

   Since Eq. 7.55 verbally means the sum of the products of the reciprocal components of the two vectors, clearly can be written as:

$$\cos \boldsymbol{\alpha} = \sum_{i=1}^{n} x_{1i} x_{2i} \tag{7.56}$$

This is called the correlation coefficient in classical statistics (Chap. 4). Its value ranges from −1 to +1, and if the value is close to +1, the two can be included in a set by saying that the vectors are in nearly the same direction and close to each other. In fact, Eq. 7.56 can be viewed as carrying the input data (pattern) with weights from that input layer to a cell in the next layer (hidden or not). To see this, for example, as a result of the displacement of the vector x_{21} by the weight vector elements, a_i becomes:

$$\cos \boldsymbol{\alpha} = \sum_{i=1}^{n} x_{1i} a_i \tag{7.57}$$

This is like Eq. 7.31, and it is possible to write as:

$$\text{NET} = \sum_{i=1}^{n} x_{1i} a_i \tag{7.58}$$

A very important conclusion that can be drawn from here is that if the input and weight vectors are close to each other, the weight vector can be taken instead of the input vector. Considering this principle in Kohonen (1988) ANNs, a set of input patterns can be assigned to different clusters by self-learning (Sect. 7.6.1).

2. *Distance principle*: Again, it can be considered that the similarity of the two data series mentioned in the previous point is measured by the distance between them. The distance between points A and B is shown in Fig. 7.47. It is understood that the distance criterion will be healthier than the angle criterion. Because even if the angle is zero (parallel vectors), the points determined by the two vectors may not overlap. However, if the distance is zero, the points fall on top of each other. It is, therefore, necessary for the calculation method to provide a criterion for the distance. Although there are different distance criteria, the Euclidean distance definition is used in order to be differentiable. Accordingly, the distance between the two data series (pattern, vector) is calculated as:

$$U = \sqrt{\sum_{i=1}^{n} (x_{1i} - x_{2i})^2} \tag{7.59}$$

Here, again, if one of the vectors is replaced by a weight vector the following is obtained:

$$U = \sqrt{\sum_{i=1}^{n} (x_{1i} - a_i)^2} \tag{7.60}$$

This is used for clustering in many ANN calculations (see Sect. 7.6.1).

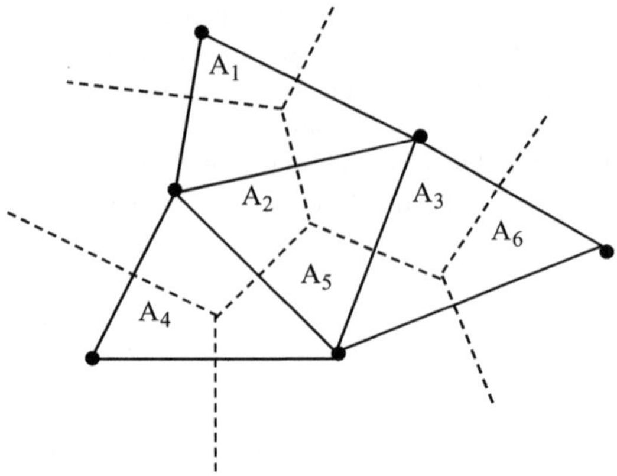

Fig. 7.48 Subarea calculation method

3. *Common digitization principle*: In another process that one can define as clustering, it is the determination of domains. Such studies use Voronoi regions in different disciplines, areas covered by finite elements in numerical analysis with finite polygons (Şen 1998). As shown in Fig. 7.48, all events in the sub-areas determined by the middle perpendiculars of the lines connecting the measuring stations with similar areas are included in the calculations as if they have the same behavior. Without such a transaction, for example, a clustering operation would be performed, if the average tax payment of that city is the same at every point of the area within the borders of each region.

   The detection of the effective fields finds its counterpart in the numerical decomposition method of vectors in ANNs (Sect. 7.6.1). With this method, excessively large data can be represented by very few clusters. For example, geological maps made by classifying rocks as igneous, precipitate, and metamorphosed rocks in earth sciences not only divide the properties of rocks at an infinite point into three groups but also facilitate human understanding and interpretation.
4. *Efficiency principle*: Here, a kind of clustering is done by determining which of the different data series collected about the investigated event, not the spatial activity described in the previous step, is more effective, that is, priority. From the collected data, one can also deduce the elimination of those that are not very effective with the event examined. If there is a language factor among the variables used in the comparison of different nations, this variable will lose its effectiveness only in the examination of Turks, and it will not be considered in the examinations because every Turk speaks Turkish. While examining Turks from Turkey, Turkey is not taken into consideration as a field, and thus, some variables are eliminated, and variables that are effective in putting Turks from Turkey into different clusters (social, economic, cultural, origin, religion, etc.) are considered. Thus, the number of input variables is reduced, and easier

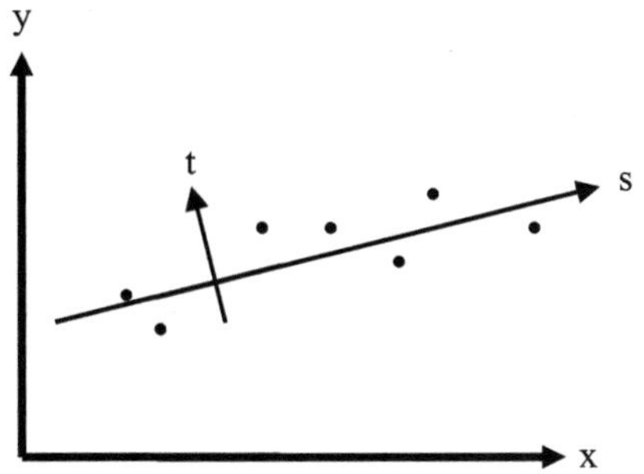

Fig. 7.49 Principle components

modeling can be made.

5. *Principal component*: Unlike the previous one, there is also the use of some transformations to reduce the number of variables. For example, in the analysis of prime components, data obtained by reducing them to a smaller number of variables with a transformation is used instead of real data. The scatter diagram of two variables, x and y, is given in Fig. 7.49.

   Instead of these two independent variables, it can be expressed with a variable in the direction indicated by s in the figure. The state can be thought of as an ANN that connects the inputs x and y to the output s. In fact, there are projections of points in the t-direction perpendicular to s, but since their ratio is much smaller than in the s-direction, only the s-direction is considered. By adjusting the weight values, the two variables become common. In general, there is also the separation of the n-variable state into $m < n$ prime variables.

7.14.2.1 Hamming Networks

Here, some basic ANN principles will be included that are necessary to achieve clustering and similar operations with ANNs. In ANNs, each cell represents important features of the data used in learning. Thus, each cell in the network is individually important. They both preserve the efficiency principle and significantly reduce the size of large numbers of data. Here, untrained education ANN units will be explained, which are put forward by using very simple methods. These are Hamming, Biggest, and Simple competitive learning networks.

In most of the ANNs that can be trained unsupervised, distance calculations and their comparison with each other are in question. The Hamming distance is a measure between data strings whose elements are 0 or 1. In the Hamming ANN shown in Fig. 7.50, there are the input and output layers and the weight connections between them.

These weight links are for storing the different data sets given to the input layer. For example, the i-th of k 0–1 input arrays of size n is $I_{i1}, I_{i2}, \ldots, I_{in}$. Accordingly,

New data entry, i3

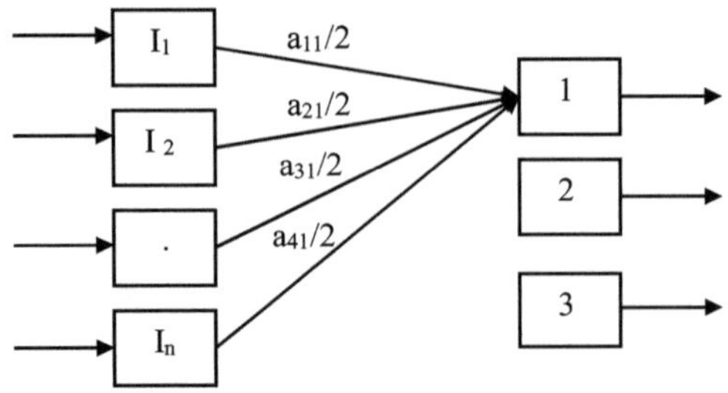

Fig. 7.50 Hamming ANN

there should be n and k cells in the input and output layers, respectively. The weight correlation value between the j-th cell of the input layer and the i-th cell of the output is $a_{i,j} = I_{ij}/2$. Elements of the weight and threshold vectors, assuming a threshold value – n/2 in each of the output layer can be written as:

$$A = \frac{1}{2}(I_1, I_2, \ldots, I_k) \tag{7.61}$$

and

$$E = -\frac{n}{2}(1, 1, \ldots, 1) \tag{7.62}$$

When a data series such as I is detected in an ANN that has been digitized in this way, it is possible to write the output value as:

$$O = AI + E = \frac{1}{2}(I_1.I - n, I_2.I - n, I_3.I - n, \ldots, I_k.I - n,) \tag{7.63}$$

Since $I_i.I$ represents the scalar product of two vectors as the sum of the products of their respective components are calculated as follows:

$$\sum_{j=1}^{k} I_{ij} I_j \tag{7.64}$$

7.14.2.2 Application

How Hamming vectors, $I_1 = (1, -1, -1, 1, 1)$, $I_2 = (-1, 1, -1, 1, -1)$, and $I_3 = (1, -1, 1, -1, 1)$ are stored in the ANN? Since there are three patterns and five elements in each, the architecture of the Hamming ANN appears as two layers with five input and three output cells. From here, it is seen that the weight and threshold values appear with the following matrices:

$$A = \frac{1}{2}\begin{pmatrix} 1 & -1 & -1 & 1 & 1 \\ -1 & 1 & -1 & 1 & -1 \\ 1 & -1 & 1 & -1 & 1 \end{pmatrix}$$

and

$$E = -\frac{5}{1}\begin{pmatrix} 1 \\ 1 \\ 1 \end{pmatrix}$$

When this ANN first detects data, the outputs in three output cells become available as:

$$\frac{1}{2}\times 1 - \frac{1}{2}\times 1 - \frac{1}{2}\times 1 - \frac{1}{2}\times 1 - \frac{1}{2}\times 1 - \frac{5}{2} = -4$$

$$-\frac{1}{2}\times 1 + \frac{1}{2}\times 1 - \frac{1}{2}\times 1 - \frac{1}{2}\times 1 + \frac{1}{2}\times 1 - \frac{5}{2} = -3$$

$$\frac{1}{2}\times 1 - \frac{1}{2}\times 1 + \frac{1}{2}\times 1 - \frac{1}{2}\times 1 + \frac{1}{2}\times 1 - \frac{5}{2} = -2$$

These are the minus sign values of the Hamming distance between the input and the hidden vectors. From here, one can find which of the hidden patterns is closer to the input pattern from the largest value of the outputs. For this, it is necessary to add the largest ANN after the output layer.

7.14.3 Compulsory Supervision

In this training style, there should be a few nerve cells at the output. Thus, it is decided whether the output is the same as the existing nerve cells. An output is obtained with ANN for each given input. The supervisor says this output is successful or unsuccessful, without saying that it falls into the following category. Thus, the binary logic rule is used. Upon the supervisor's statement that it is unsuccessful, the ANN continues the training until it takes the decision that it is successful by processing the input again and again.

7.15 Competitive Education

Although this is like unsupervised training, the characteristic operations and architecture in ANNs are slightly different. There are some artificial nerves at the output here. When there is an input, the ANN mechanism tries to have the output at least close to one of the targets. This output nerve now becomes a major group, while others do not matter. In case of another input, another output neuron becomes active so that work can be continued. One can say that competitive training is like clustering. As will be seen in Sect. 7.6.1, Kohonen ANNs receive competitive training. The basic principle here is that the winner takes all, that is, leaves nothing to the others. In the normalized state of the data, the winning cell in competitive training is the one with the largest activation functions output. In the absence of normalization, the winning cell is determined according to the distance principle (Euclidean distance) explained in Sect. 7.14.3.

7.15.1 Semi-teacher Training

If the output values are not known numerically in an ANN modeling, the previous training is abandoned, but one of the two options should be chosen. One of them is to have verbal information about the outputs or not to have any information at all. Verbal knowledge can be acquired in semi-directed training. In such knowledge, there are generally good-bad, beautiful-ugly, yes-no, right-wrong, etc. with binary logic. Since the teacher does not have full knowledge here, he can only partially guide the ANN. After this training, the teacher helps the ANN to obtain the outputs divided into two classes.

7.15.2 Learning Rule Algorithms

As a result of various studies, many learning rules have been developed as explained in Sect. 7.14.1. When the desired behavior could not be achieved in the studies, in some cases, the rules of supervised and unsupervised training were used together (Hsieh and Chen 1993). Since learning in ANN is possible by renewing the connection weights, the training takes place step by step. Here, only the most known and used perceptron, Widrow-Hoff, Hebb, and delta learning rules will be mentioned.

Some notations in the training algorithms that will be mentioned are as follows. The vector $A_i = [a_{i1}\ a_{i2} \ldots a_{in}]^T$ is the link weight vector of the n inputs to the i-th cell in the network. Likewise, vector $I = [I_1, I_2 \ldots I_n]$ shows the input data (pattern, vector). Also, O_i and B_i are the expected output value with the output obtained from the i-th cell, respectively. Finally, η is the learning rate, indicating a fixed number.

7.15.2.1 Perceptron Learning Rule

As mentioned earlier, perceptron proposed by Rosenblatt (1958) is a linear ANN. The perceptron learning rule is a supervised algorithm that tries to find the vector of link weighting coefficients between input and output. Output value can be expressed as:

$$O_i = \mathrm{sgn}\left(A_i{}^T I\right) \tag{7.65}$$

On the other hand, considering O_i and B_i, the weight values in the (k + 1) iteration are renewed by using the k-th as follows:

$$A_i(k+1) = A_i(k) + \eta\ \left\{B_i(k) - \mathrm{sgn}\left[A_i{}^T(k)\ I(k)\right]\right\} I(k) \tag{7.66}$$

Note that the change of weights is only valid in the case of $B_i \neq O_i$. That is, when the wrong value is obtained as the output value, the weights are renewed. Otherwise, since the parentheses will be zero, $A_i(k + 1) = A_i(k)$.

7.15.2.2 Widrow-Hoff Learning Rule

Widrow-Hoff (1960) training rule is also a supervised rule. The most important feature of this rule is that it works independently of the activation functions. The rule is only based on the principle of minimizing the error between the actual output value and the output given by the ANN. The weight coefficients are renewed according to the following expression:

$$A_i(k+1) = A_i(k) + \eta\left[B_i(k) - A_i{}^T(k) I(k)\right] I(k) \tag{7.67}$$

This is considered as a special case of the delta learning algorithm, which will be explained in detail in Sect. 7.14.3.

7.15.2.3 Hebb Learning Algorithm

The Hebb algorithm is an unsupervised training method. The basic idea here is to increase the effect by enlarging the weight coefficients of the connections that cause the cell to generate abruptly. The weights are restored as follows:

$$A_i(k+1) = A_i(k) + \eta f\left[A_i{}^T(k)\ I(k)\right] I(k) \tag{7.68a}$$

In other words, it can be displayed as:

$$A_i(k+1) = A_i(k) + \eta O_i(k) I(k) \tag{7.68b}$$

In Eq. 7.68b, if the product of the output and the input values is positive, an increase in the vector A_i will occur; otherwise, a decrease will occur. This learning rule has been applied to the perceptron architecture in Sect. 7.8.

7.15.2.4 Delta Learning Rule

This is a supervised learning rule and is valid only if the derivative of the activation functions is taken. In this learning rule, the derivatives are included in the calculation when the weights are renewed. Although it is most used, another name is the slope fall rule. The connection weights are renewed as:

$$a_{ij}(k+1) = a_{ij}(k) - \boldsymbol{\eta} \frac{\partial H_i}{\partial a_i(k)} \tag{7.69}$$

Here, the total error of the k-th iteration is calculated as follows:

$$H(k) = \frac{1}{2}\,[B_i(k) - O_i(k)]^2 = \frac{1}{2}\,\left[B_i(k) - f\left(a_i^T(k)I(k)\right)\right] \tag{7.70}$$

The variation of the error according to the connection weights is expressed as:

$$\frac{\partial H(k)}{\partial a_i(k)} = -\,[B_i(k) - O_i(k)]\,\frac{\partial O_i(k)}{\partial a_i(k)} \tag{7.71}$$

Here, the a_i change takes place in the opposite direction of the slope of the error surface. Thus, it is aimed to go to the place where the error is the lowest on the surface formed by the error function.

7.15.3 *Back Propagation Algorithm*

Minsky and Paperts (1969) showed that two-layer feedforward networks remove many limitations compared to a single-layer perceptron, but they could not offer a solution on how to change the link weights between the layers. Since no solution could be found to this problem, which emerged with the training of ANN, no progress could be made in the studies conducted for many years. Therefore, ANN, which entered a period of recession, maintained its silence until the 1980s. Rumelhart et al. (1986) proposed the back propagation algorithm leading to a revival of interest in ANNs. Although the proposed algorithm is a formal method, it is based on a complex mathematical basis. In addition, this algorithm has provided a better solution to many problems such as input-output descriptions, clustering, and generalization, which were not successful before. Thus, the back propagation algorithm, which is considered a turning point in ANN, has made it possible to solve many problems and enabled many successful applications.

Back propagation algorithm is a powerful learning algorithm used in hidden layer ANNs. The basis of this is the changes in an ANN consisting of subsystems that can be calculated completely and effectively, which makes it possible to use ANN to learn the relationship between complex, nonlinear, and process parameters (Verbos 1994).

There are two basic flows in the back propagation algorithm. The first is forward information flow over networks, and the second is backward error propagation. In the forward flow, inputs versus outputs are obtained using weighting coefficients (see Sect. 7.12). In the backward flow, the error calculated from the difference between the model output and real data is propagated backward, and the weights are changed.

As in all learning methods, the purpose of the back propagation algorithm is to obtain the connection weights that will provide the most appropriate description between the input and output data. The back propagation algorithm on the network, which has a topology as in Fig. 7.32, takes place step by step as follows:

1. First, in order to determine the topological structure of the ANN, the layer and the number of cells in each layer are determined.
2. The values of the constant parameters in the expressions above are assigned.
3. The values of A and C, which are the weight relationships between the layers, are randomly assigned.
4. Using the assigned connection weights, where n is the number of measurements, c_{kj} ($i = 1, \ldots, n$; $j = 1, \ldots, m$), output is obtained for each input vector.
5. c_{ij} values, starting from the connection weights between the output layer and the hidden layer, to the back propagation of the errors is renewed as:

$$a_{ij}^{new} = a_{ij}^{old} - \eta \frac{\partial E}{a_{ij}} \tag{7.72}$$

where η is a nonzero numerical value. In order to renew the weight coefficients a_{ij}, first, the derivatives of the error value and the error value should be calculated according to the a_{ij} connection weights.

6. The output values O_{kj} and a_{kj} given by the ANN are calculated using the model output values o_{ij}, and the total error value is calculated as follows:

$$E_T = \sum_{i=1}^{n} \sum_{j=1}^{m} \left(O_{ij} - o_{ij}\right)^2 \tag{7.73}$$

7. In order to renew each weight coefficient, by taking the derivative of E_T according to the connection weights one by one:

$$\frac{\partial E_T}{\partial a_{ij}} = \frac{\partial}{\partial a_{ij}} \left[\sum_{k=1}^{n} \sum_{j=1}^{m} \left(O_{kj} - o_{ij}\right)^2 \right] \Rightarrow \frac{\partial E_T}{\partial a_{ij}} = 2 \sum_{k=1}^{n} \left(O_{kj} - o_{ij}\right) \times (-1) f'\left(o_{ij}\right) y_i \tag{7.74}$$

Since the sigmoid function is used as the activation function, its derivative can be expressed as:

$$f'(o_{ij}) = o_{ij}(1 - o_{ij}) \tag{7.75}$$

Thus, after all operations are completed, a_{ij} connection weights are renewed.

8. The error must spread over c_{ij} in the same way as it spreads over the connection weights a_{ij}. For this reason, the renewal of the c_{ij} connection weights is like Eq. 7.72 and carried out as follows:

$$c_{ij}^{new} = c_{ij}^{old} - \eta \frac{\partial E}{\partial c_{ij}} \tag{7.76}$$

9. With the help of the chain rule, the variation of the E_T term with respect to the cij connection weights is shown below:

$$\frac{\partial E_T}{\partial c_{ij}} = \frac{\partial H_T}{\partial o_{ij}} \frac{\partial o_{ij}}{\partial a_{ij}} \frac{\partial a_{ij}}{\partial c_{ij}} \Rightarrow \frac{\partial E_T}{\partial c_{ij}} = 2\sum_{i=1}^{n}\sum_{j=1}^{q}(O_{ij} - o_{ij})$$
$$\times(-1)f'(o_{ij})a_{ji}f'(a_{ji})a_{ij} \tag{7.77}$$

Thus, since the connection values of A and C are completely renewed, they will have different values from the randomly assigned values at the beginning, so the training of the ANN is completed once. ANN can be trained as desired by repeating the same operations in a similar way. Note that the error value will decrease as the total error is always propagated back as a result of training. The decrease in the error value after each training shows the appropriate weighting coefficients are approached, which will perform the description between the input and output data.

For the training of the network to be carried out in a reliable way, the randomly assigned connection weights at the beginning are very important. These weights determine from where one will start the training. Whether this starting point is very close or far from the real solution depends entirely on the values one will determine at the beginning. Alexander and Norton (1992) stated that if the training process starts with assigning equal values to the weights, the training process will not take place. The reason for this is that when the same error value is propagated toward the link weights with the same value, the link weights will not take different values from each other.

One of the important points in ANN is deciding how long the education will continue (see Sect. 7.5.3). There are two options for completing the training process. The first of these is to risk a certain amount of error tolerance and to continue training until it reaches a lower error value. So, in this case, the amount of error is more important than the number of training. Here, the error tolerance must be within reasonable limits. Otherwise, it will be difficult to reach error values lower than desired, no matter how much the number of training is increased. The other option is to choose a fixed training number. Here, the trainer accepts the error that will be obtained as a result of the specified training number. The trainer decides from the beginning how the trainer will be carried out according to the work to be done.

After all, when Eqs. 7.72 and 7.76 are examined, the derivative of the error according to the connection weights is seen on the right side of the equations. Since the error value will decrease at the end of training, these derivatives will gradually decrease, and it is expected to reach zero value for an ideal situation. In this case, no matter how long the training process is continued, there will be no change in the a_{ij} and c_{ij} values, that is, the connection values will become stationary. For such ideal cases, the training stop phase is the point at which the error term reaches zero. However, it is practically impossible for the error term to reach zero, except in special cases. It is not possible to describe with zero error between input and output variables, especially in natural events where uncertainty is also effective. The error surface is usually a surface that has both local points and an overall minimum point. The purpose of training the ANN is to reach the general minimum point on this surface.

It should not be expected that the error value will always decrease during the training. Sometimes, it can be observed that the error increases. This is often a sign that one is moving away from the solution. If training continues while a local minimum point is reached on the error surface, the error value increases by a certain amount until the local minimum is eliminated. Then, when it moves toward another smallest, the error value starts to decrease again. However, the fact that the error value increases without decreasing is an indication that this point is a general minimum point. Despite all this, is the point reached local minimum since one does not have any information about the error surface during the solution? or is it a general minimum point? It is hard to decide what it is. Therefore, in many studies, the training process is terminated as soon as the error value starts to increase.

Another important parameter in ANN training is the selection of the learning rate, η. This ratio controls the amount of change in link weights. It is important to choose the learning rate appropriately in order to choose an efficient learning. While an appropriate η value can lead to a solution with a small amount of training, an unsuitable η value will not cause a solution to be reached only after too much training.

The learning rate can be expressed as the size of the step taken while approaching the minimum on the error function. If this ratio value is set small, the change in connection weights and error value will be small, so one will need to keep the number of training high in order to reach the real solution. In such cases, since the amount of change is very low, it is often possible to stop the trainer with the thought that the smallest point has been reached. Haykin (1994) proposed that if the value of η is chosen large, it will show sudden and rapid changes in the connection weights, and this will lead to instability.

In order to determine the learning rate in the best way, attention should be paid to the slope at the point on the error surface. If the slope is small, increasing the steps, that is, choosing a large learning rate, will be beneficial in terms of reaching the smallest point. Since the change in the error surface is very low, there is no harm in taking a big step. If the slope is high, this time the steps should be taken as small as possible. While going toward the smallest, choosing a high step where the slope is high may cause instability as it will cause jumps.

As mentioned above, it is difficult to assign the value of η according to the slopes, since there is not much information about the error surface. For this reason, a value between 0 and 1, preferably between 0 and 0.2, is assigned in the studies. In addition, the value of η can be determined by control according to the indecision situations that will arise during the training. It should be noted that choosing an appropriate value of η can perform the description with less training.

Although the back propagation algorithm is a powerful learning algorithm and is widely used, it has some drawbacks. Among them, for example, there is no guarantee that the network can be trained. It can be thought that it would be enough to enlarge the network in order to carry out the training. Keeping the network large does not guarantee how difficult it will be to learn. Since the enlargement of the network will bring more processing load, the probability of being trained in a finite time frame decreases (Morgan and Scofield 1991).

7.15.3.1 Unsupervised ANN

Apart from the common ANNs described in the previous sections, ANNs with vector separation, recursion oscillation theory and radial basis function ANNs will be mentioned in this section. These ANNs do not generally have a hidden layer or, if present, have a special activation function unlike the previous ones. In this section, linear vector segment, regenerative oscillation theory, and radial basis activation function ANNs will be explained, which are widely used in practice.

7.15.3.2 Linear Vector Piece

While discussing the comparison of ANN and other models in Sect. 7.2, we said that the simple ANN system functions are in the form of the following matrix:

$$\begin{bmatrix} O_{i1} \\ O_{i2} \\ \vdots \\ O_{im} \end{bmatrix} = \begin{bmatrix} a_{1,1} & a_{1,2} & \cdots & a_{1,n} \\ a_{2,1} & a_{2,2} & \cdots & a_{2,n} \\ \vdots & \vdots & \vdots & \vdots \\ a_{m,1} & a_{m,2} & \cdots & a_{m,n} \end{bmatrix} \times \begin{bmatrix} I_{i1} \\ I_{i2} \\ \vdots \\ I_{in} \end{bmatrix} \tag{7.78}$$

This matrix can also be explained in detail in the form of parts in vector. In terms of ANN, one can talk about three different vectors:

1. The i-th input vector in the rightmost column, I_{ij} (j = 1, 2, . . . , n).
2. The i-th output vector in the leftmost column, O_{ij} (j = 1, 2, . . . , n).
3. In each row of the coefficient's matrix, a_{ij} (j = 1, 2, . . . , m) m weight vectors (row vector), connecting the input vector elements to each of the output cells.

Thus, there are a total of (m + 2) vectors, two of which are columns, and m are rows. It should be noted that the input and weight vectors are of the same size. Also, the coefficient matrix can be thought of as column vectors, but these are not important for the ANN operations described in the previous sections. From these explanations, one can conclude that the ANN architecture is a collection of vectors. Each of these vectors performs linear operations, and there are no curvilinear operations. According to the values of the weight vector elements, the inputs are spread over the output cells separately. A question arises as to which of the output layer cells would take the largest share in such a spread.

The weight vector with the largest weight array elements moves the input vector to the corresponding output cell with the greatest weight. Accordingly, the output cells are classified as the most important; the second most important ones can be sorted accordingly. In that case, whichever of the weight vectors carrying the information in the input data to the output cells can make this transport amount the largest, the importance of most of the cells in the output decreases accordingly. The weight values in the weight vectors are not already known values. Since the values are determined with random assignments at the beginning and the input information is moved to the output cells, is it not possible to adjust the weights of these weight vectors in such a way that one of them carries the most? The ready answer is "yes," but what method should be developed so that this can be done formally? Assuming this is done, one can represent the input vector, for example, by an output layer with two outputs (cells). Thus, the input vector is divided into a small number of weight vectors to obtain the state in the output cells. The method with this philosophy is called linear vector quantization (LVQ).

7.15.3.3 Linear Vector Piece ANN

Since it was first developed by Kohonen (1988), this ANN method is also called Kohonen ANN. The basic concept is the splitting of the input vector into sub-vectors as described above. In this ANN, the weights must be trained in such a way that the vector indicated by the weights is closest to the input vector. The learning rule to be considered in training such an ANN is the development of the weight vector, which will allow closer convergence to the input vector at each weight renewal. At the end of this approach, if a weight vector is very close to the input vector, since one can count both of those from the same category, the output cell of it, for example, can be given the code 1; thus, a cluster is opened. If the same weights or slightly different ones approximate other input data, it is decided that they can also be put in the same category. On the contrary, if the weight vector is closest to the input vector, but the distance between them is important, then a new cluster is opened by assigning the

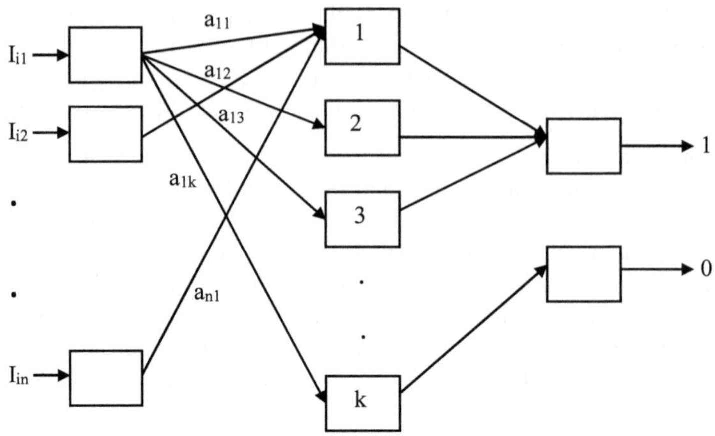

Fig. 7.51 LVP YSA

value 0 to the output layer cell, which is the output of this weight vector. Thus, a two-cell output layer is obtained, whose output is known as 0 or 1. After all this has been said, it is concluded that there are two layers in an LVQ as ANN architecture. The output layer has no other function than to show the aggregated state of the results. Input layer perceives all functions, and information comes to the hidden layer, and there is a competition between the cells of this layer in terms of proximity to the input vector indicating that this layer has an active function. LVQ ANN architecture is given in Fig. 7.51. The hidden layer in this architecture is called the Kohonen layer.

In these ANNs, the correlation is only between the input and the hidden layer. The relations between the Kohonen and the output layers are all equal to 1. Each cell in the Kohonen layer is described as 1 and 0, respectively. There will be as many weight vectors as the number of cells in the Kohonen layer. The operation of LVQ ANN is according to the following steps:

1. Data collection and quality control.
2. In this network, by determining the number of cells only in the Kohonen (intermediate, hidden) layer, the ANN architecture emerges like Fig. 7.51.
3. For the link weights to be renewed during training, a learning coefficient must be determined.
4. The initial values of the weights should be assigned as small random numbers.
5. Inserting a set of data into the input layer for detection.
6. Calculation of known weights and input data, separately for each cell in the Kohonen layer.

$$\mathrm{NET_i} = \sum_{j=1}^{n} a_{ij} I_{ij} \tag{7.79}$$

7. Determining the cell with the largest NET value in the Kohonen layer.
8. Renewing the binder weights connecting this cell to the input layer cells, ignoring all cells except this cell.
9. After all data sets are entered the LVQ ANN, applying the above steps for each (starting from step 5).

As a result of training all data sets, important reference vectors (connections of cells of the input-Kohonen layers) should be determined.

7.15.3.4 Learning Rule of LVQ ANN

As a training rule from the previous explanations, after finding the cell in the Kohonen layer that represents the input vector the most, the weights of the connections to this cell should be adjusted to be close to the input vector. Here, proximity is understood as minimizing the distance between two vectors (see Chap. 5, Sect. 7.14.1). If the i-th cell in the Kohonen layer represents more of the input vector than the others, the Euclidean distance between the weight vector a_{ij} ($j = 1, 2, \ldots, n$) and I_{ij} ($j = 1, 2, \ldots, n$) is calculated as:

$$u_i = \left\| A_{i,j} - I_i \right\| = \sqrt{\sum_{j=1}^{n} \left(a_{ij} - I_{ij}\right)^2} \tag{7.80}$$

When the similar process is completed for all the cells in the Kohonen layer, one has n distance values (u_i; $i = 1, 2, \ldots, n$). Here, the weight values of the cell corresponding to the smallest distance between them represent the first reference vector. From now on, only the weights of that Kohonen cell will be considered to make this smallest distance even smaller. There are two sets of states for this winning cell as "true" (1) or "false" (0). The difference between these two situations requires training. Considering that the winning cell is in the correct cluster first, the weights should be moved a little closer to the input vector. In the renewal of the weights during the training, α being the learning coefficient the following expression is used.

$$a_{ij}^{new} = a_{ij}^{old} + \alpha\left(I_{ij} - a_{ij}^{old}\right) \tag{7.81}$$

For the zoomed vector not to diverge again after getting very close to each other, it is necessary to decrease the learning coefficient toward zero over time. On the other hand, if the winning cell is in the wrong set, the weight vector must be removed from the input vector. When the same data set is detected by the ANN again, the weight vector is removed from the input vector so that the same cell does not win. For this, the minus sign of the previous equation should be used during training.

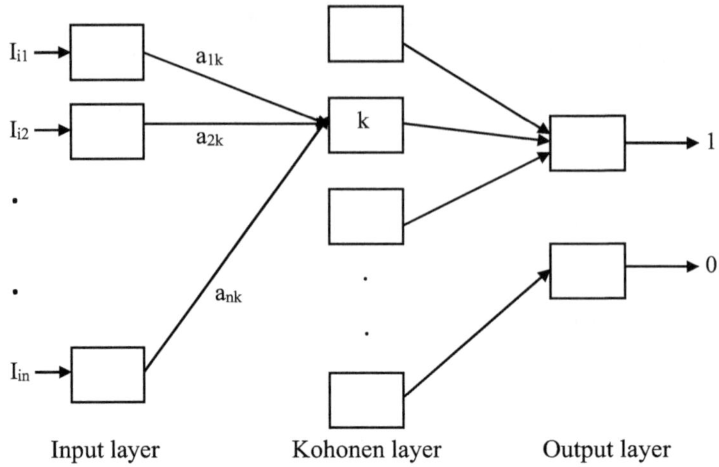

Fig. 7.52 A reference vector

$$a_{ij}^{new} = a_{ij}^{old} + \left(I_{ij} - a_{ij}^{old}\right) \quad (7.82)$$

Here, too, it is necessary to bring the learning coefficient closer to zero over time. The main drawbacks of such ANNs are as follows:

1. Difficulties in adjusting the learning coefficient so that it decreases to zero over time.
2. Since the same reference vector always wins in the solution of some problems, the flexibility of the ANN is lost.
3. Difficulty in deciding objectively in which class to include vectors that are the same distance from two boundaries, if there are two classes as a result. Various LVQ ANNs have been put forward during the studies carried out to eliminate these drawbacks.

7.15.3.5 Training of LVQ ANN

The purpose of training these networks is to determine the weight reference vector that is closest to the input vector. The vector formed by the weights coming into each cell in the Kohonen layer is called the reference vector. For example, in Fig. 7.52, only the weights coming from the top to the second cell is shown in the Kohonen layer.

Kohonen (1988) tried to make some changes to overcome the drawbacks of the standard LVQ ANN. In order to decide on which class to include vectors that fall very close to the cluster boundaries, the same LVQ ANN architecture has made some changes only in training. Let the two weight vectors close to I vector be A_1 and A_2 as shown in Fig. 7.53. Likewise, it is shown as a standard distance to be considered in classification.

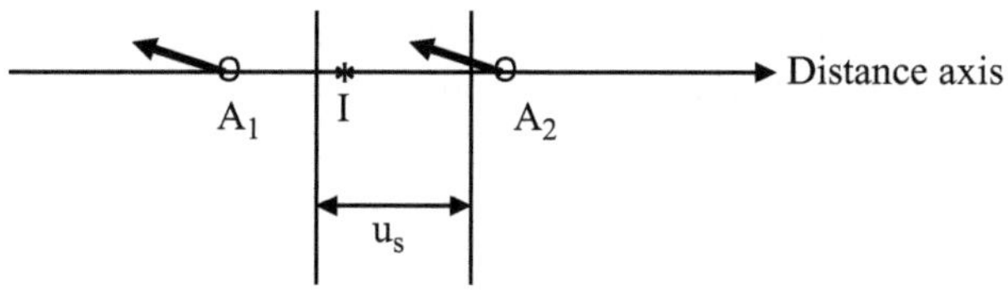

Fig. 7.53 New LVP ANN traning

In this way, A_1 is the closest to the input vector, and A_2 is the next closest weight vector. A_1 is in the wrong class, and A_2 is in the correct class. The input vector also lies within a centrally determined range between vectors A_1 and A_2. In both cases, the weights of both vectors A_1 and A_2 must be changed during the same training period. New weights like the previous teaching rule are calculated from the expressions:

$$a_{i1}^{new} = a_{i1}^{oldi} + \alpha\left(I_{i1} - a_{i1}^{old}\right) \tag{7.83}$$

and

$$a_{i2}^{new} = a_{i2}^{old} + \alpha\left(I_{i2} - a_{i2}^{old}\right) \tag{7.84}$$

It must also be decided whether the input vector is between two weight vectors. For this, if u_1 and u_2 are the distances of the vectors A_1 and A_2, respectively, from the input vector then:

$$s = \frac{1 - u_s}{1 + u_s} \tag{7.85}$$

After the variable is defined, the input vector is between the two weight vectors.

$$\min\left(\frac{u_1}{u_2}, \frac{u_2}{u}\right) \geq s \tag{7.86}$$

While solving this problem, the standard LVQ ANN training should be used first, and then this new training should be passed. Accordingly, standard LVQ ANNs determine the clusters, and the training here should be used to separate the vectors in these boundaries into correct clusters.

Another problem in implementing standard LVQ ANNs is that some weight vectors overlap too often, making the flexibility of the network dull. With the effect of this, other weight vectors that can be reference vectors cannot gain this qualification. While different weight vectors should represent different aspects of the input vector, this cannot be the case. In order to get rid of this, it is possible to penalize the winning weights during the training. Thus, successive gains of the same weights are prevented. A penalty is achieved by adding the weight vector a distance d from the input vector. The d value here should be determined based on how many times the weight vector wins the approach. For example, d_i amount to be added to the distance of the weight vector of i-th cell from the input vector is calculated as:

$$d_i = C\left(p_i + \frac{1}{n}\right) \tag{7.87}$$

Here, C represents a user-specified constant the probability of i-th cell winning is p_i. The term 1/n in parentheses indicates the initial probability of each of the n cells in the Kohonen layer. Regeneration of p_i in later training circuits is calculated according to the following expression:

$$p_i^{new} = p_i^{old} + \beta\left(y_i - p_i^{old}\right) \tag{7.88}$$

where β is a user-assigned coefficient to avoid any fluctuations in the data. This does not have a definite value, and the assignment is made based on experience as a result of trial and error. In Eq. 7.88, y_i represents output values (0 or 1):

$$y_i = \begin{cases} 1 & \text{Cell gains the race} \\ 0 & \text{Cell cannot gain the race} \end{cases} \tag{7.89}$$

During the punishment-driven LVQ ANN training, the d value is added to the distance between the input and weight vectors is calculated as:

$$u_i^{new} = u_i^{old} + d_i \tag{7.90}$$

The competition between Kohonen cells should be done with these new distance values in mind.

7.16 Renovative Oscillation Theory (ROT) ANN

All ANNs described in the previous sections needed either a full or partial teacher during their training. Regenerative oscillation theory (ROT), which is the subject of this section, divides ANNs into appropriate classes by processing the input data (patterns) as its own teacher, without the need for a teacher. These were first developed by Goldberg (1989) under the name of Adaptive Resonance Theory (ART) networks. In multilayer ANNs, two sets 1 and 0 play a role as a partial teacher in the output layer of LVQ ANNs. No teachers are needed here. Only when the input pattern enters the ROT ANNs network, it is placed in the appropriate cluster by internal processes.

In order to keep in mind why these two-layer networks are called "repetitive oscillation," during a study that starts with the input data during operation, the data is placed in the appropriate cluster in the output layer with the oscillations between the two layers. During each oscillation (round-trip), the shape of the data is renewed, and the result is approached. Another feature of these ANNs is that they keep some information in their memory for short-term, forgetting others by making short-term recalls that were not present in the previous ones. If one interprets this as the forgetting feature of ROT ANN, he can understand that it has a functioning very close to human nature. Man is a forgetful creature, but he does not forget all that he has learned. This means that after keeping some information in the human memory for a short time, it is forgotten because it is not emphasized when its importance is over. Another humanlike aspect of these ANNs is that they learn by repetition. In fact, the word "oscillation" means that it kneads the information repeatedly between the input and output layers. Ultimately, perceiving it in its final form implies clustering and increasing the number and quality of clusters as a preliminary preparation for subsequent inputs. These oscillations (round-trip) show detailed learning by itself through repetition. The situation of these ANNs is like the learning story of a baby called "Hayy bin Yekzan," who stayed in a deserted place by himself in an article written by the great Islamic thinker İbn-i Sina (Avicenne) (950–1030). After his birth, this person falls on a deserted island, where a deer takes care of him. By taking information from him and his natural environment and forgetting some in a short time and keeping some in his memory for a long time, oscillations in understanding how to cook without a teacher, burying the dead, clustering those who are similar to what they see in nature in different ways, educating oneself with new information, according to their place (repeats) by doing, he learns without a teacher by always getting better. He then uses them at different stages of his life. In fact, this philosophical (theory) article of Ibn Sina is nothing but the explanation of ROT ANNs.

From all these explanations, ROT ANNs have short- and long-term memories, train themselves without a teacher (unsupervised), after making the most appropriate clustering of the given data with the round trips between the input and output layers, putting that data in the appropriate cluster in order to compare it with other input data in the future. One understands that it has a continuous process to work with the input data. It compares the new incoming data with its short-term memory but determines the last appropriate clusters with its output long-term memory. Like human behavior, it is now immune to that pattern during oscillations, and if data like it comes after that, it immediately places it in its cluster with oscillation and in some cases without oscillations. ROT ANNs can consider even the smallest details. Thus, it also can separate very finely. For these reasons, one witness that these ANNs are widely used in the clustering of even the most complex events and expect that it will be increasingly used in the future (Chap. 9).

7.16.1 Differences of ROT ANN and Others

These ANNs are very suitable for data clustering, classification, and recognition (Carpenter and Grossberg, 1988). In addition to the features described above, ROT ANNs have many features that distinguish them from other ANNs, and therefore, they have attractiveness. These can be listed as the following points:

1. As they have fast and detailed clustering capabilities, ROT are a fast solution for ANNs clustering real-time data.
2. Although other ANNs must get support from the outside environment to make decisions, ROT ANNs are connected to the outside only through input data. Apart from that, it continues its function without the need for any external information. Because it sets it outputs on its own instead of getting it from a teacher.
3. Finally, it makes a self-stable clustering and quickly decides whether new data can enter one of these clusters. If the new data cannot enter a cluster determined by the ANN beforehand, the same ANN opens a new cluster for it, this time causing the ANN to renew itself over time.
4. Detecting even small details in clusters that are like each other as soon as possible and quickly places them in one of the classes that are in the long-term memory and determined beforehand.
5. For the oscillation process, if the two layers of this ANN are considered one under the other, the classification operations are sent from right to left, and if there is no suitable cluster to be placed on the right side, the classification operations are continued until it finds a suitable cluster in the long-term memory by making additional operations on it. There are two sets of connection networks due to these two types of commuting. The ones from left to the right determine the weights by comparing the clusters in short-term memory with the input data. Right-to-left linkage weights represent each long-run set. Accordingly, the information from left to right is compared with the information above. If two knowledge populations are the same or similar, then the pattern is placed in its appropriate set. Otherwise, if it cannot be included in a suitable long-term memory class after different round-trips (oscillations), a new cluster is opened for this data.
6. In the structure of these ANNs, there should also be a unit called "gain" in order to prevent entries that are not in the same class from entering that cluster in comparison with short-term memory information. Its job is to prevent clustering weights from being affected by a re-incoming data entry. Thus, the previously made clustering weights are preserved. This provides ROT ANNs with the ability to continuously learn by them.
7. After these ANNs generate enough clusters, if there are any similar clusters in the long-term memory from the incoming input data, it places that data in its class in a very short time and without any oscillations (returns). However, if it does not fit any of the clusters in the long-term memory in quite different data, it continues to oscillate for a while to determine a new cluster for it. After all, it puts it in a new cluster with self-learning.

7.16.2 ROT ANN Architecture

As it was said before, there are two layers in these ANNs: input and classification (Fig. 7.54). In each of these layers, there is short-term memory (STM) and long-term memory (LTM) that connect the two layers.

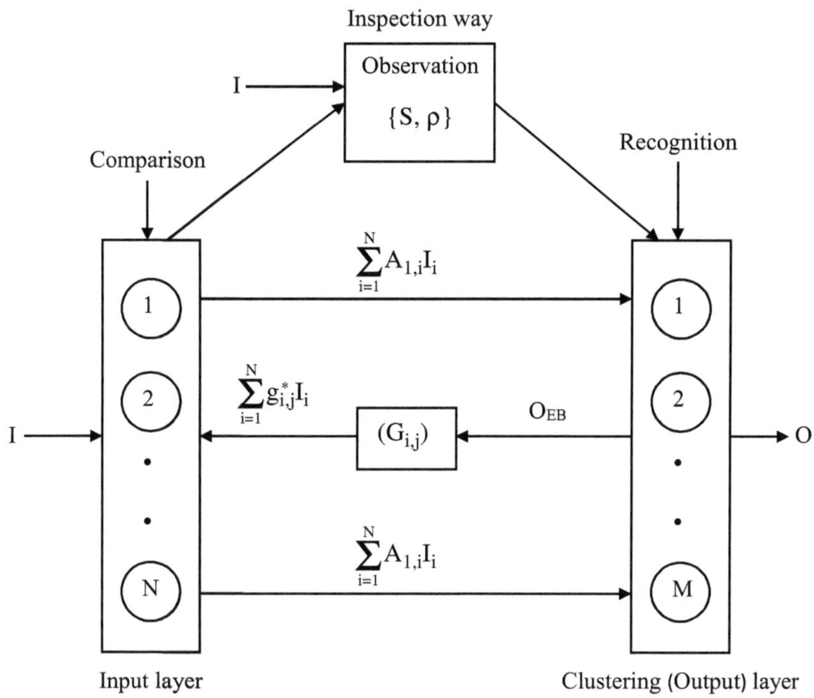

Fig. 7.54 ROT ANN structure

During the operation of these ANNs, there are consecutive oscillations, first from left to right and then from right to left. I input (pattern) are first given to the input layer. With the STM of the input layer, an active data is generated with the weights of this cluster first.

ROT ANNs benefit from clustering of data arrays given in varying numbers of clusters. The difference of clustering with these from the others is that after starting clustering with the first data, if the later data do not have similarity with it within a certain measure, it gives rise to new clusters. The setting of whether a given data is like the clusters previously made with the ANN. ANN is within a measure called the "surveillance parameter." There are two layers in the structure of this ANN. The first of these performs both input and comparison operations. This is called the input or benchmark layer. The second layer contains different clusters and is called the recognition or output layer. ROT ANN has neither middleware nor the activation function described in Sect. 7.11.1. The number of elements required to detect the data given in the input layer has N cells. The number of cells in the output layer, on the other hand, is M, the number of different clusters that will develop and appear until all the training data is finished. In order to decide which of the pre-existing clusters an input data will be like, there are oscillations between the input and output layers. If the data entering during these oscillations is like one of the clusters in the recognition layer by providing the surveillance factor, this pattern is included in that cluster and a new cell is not opened due to this data in the recognition layer. Otherwise, if the input data is not like any of the predetermined clusters in the recognition layer, a new cell is determined in the recognition layer, and this data is added to the recognition set to be used in further studies.

After the input layer detects data, a similar pattern is presented from the recognition layer to this layer from each of the existing clusters, respectively. Thus, it is decided whether the two data are similar or not, considering the observation parameter of the incoming and outgoing data in the comparison layer. If the data from the recognition layer is like the input data, there is a fit. If no conformity is observed, oscillations (round-trip) are continued until a correspondence is obtained between the input data and the data coming from the other cells of the recognition layer between the recognition and comparison layers. If no match is obtained, a cell is opened in the recognition layer to represent this data and its future counterparts.

7.16.3 ROT ANN Education

The departure and arrival weights between the comparison and recognition layers are adjusted by processing the learning of ROT ANNs with training. During these round-trips, the ANN weight coefficients are renewed according to the rules given below, and they are continued until they become stationary. In a ROT ANN that becomes stationary, the connection coefficients do not change anymore, and the most suitable cluster data is selected from the recognition layer for each data that is made to be detected by the comparison layer from the outside world. In further studies, the number of cells in the recognition layer is increased if the ROT ANN architecture cannot assign a new input to any of the clusters in the recognition layer with its pre-trained connections. Thus, this ANN shows improvement over time.

In practical studies, two similar education systems, ROT1 and ROT2, were introduced. The first of them is binary, and the other works according to the decimal number-based system. Before these, the training of ROT1 will be explained.

From the round-trips described in the previous section, it is deduced that both outbound and return weights will be found. Of these, the input (from the restrictive layer to the recognition layer) weight coefficients will be different from the return weight coefficients. The going coefficients show the weights required for the input data to reach the recognition layer. The assignments of their departure starting values, I_{ij}, are taken as the reciprocal of the number of elements in the input data, N, one more.

$$I_{ij} = \frac{1}{1 + N} \tag{7.91}$$

The initial values of the rotation weight coefficients d_{ij} are always taken equal to 1.

$$d_{ij} = 1 \tag{7.92}$$

The point to be noted here is that the outgoing and return weights are not equal ($I_{ij} \neq d_{ij}$). Provided that the number of clusters in the recognition layer is M then $i = 1, 2, N$ and $j = 1, 2, \ldots,$ is M. As said earlier, M can change over time. Of these, a ROT ANN with weights I_{jl} is used to organize the learning from the input layer to the output, and d_{jl}'s from the output layer to the input layer. When an X data is given to the input layer, it is given to this recognition (output) layer as y_j with the weights based on the following account:

$$y_j = \sum_{i=1}^{N} I_{ij} x_i \tag{7.93}$$

This is called the similarity coefficient because it is a measure of the similarity between vector I_{ij} ($i = 1, 2, \ldots, N$) and vector x_j. At the end of these calculations made for each of the cells in the output layer, the cell with the largest amount of similarity is determined as the cluster that best represents the input data in the recognition layer. The final decision of the data belonging to this cluster depends on the value of the surveillance parameter. For this, a return data is obtained by using the return weights from the cluster that is thought to represent this data best in the recognition layer elements and they are obtained as:

$$z_l = d_{l,j} x_l \tag{7.94}$$

Here is the observation parameter, p, which is used to compare this return data with the input data; it is calculated as:

$$p = \frac{\sum_{l=1}^{n} z_l}{\sum_{l=1}^{n} x_l} \tag{7.95}$$

If this parameter value is greater than a predetermined observation value, it is judged that the two data are similar. In this case, the I_{ij} and d_{ij} weights are changed so that the return and input data are as similar as possible.

7.16.3.1 Application

For a better understanding of ANNs, the application given below should be followed step by step. Suppose that binary number-based clustering is given as follows for ROT ANNs.

$$(1, 1, 0, 0, 0, 0, 1)$$
$$(0, 0, 1, 1, 1, 1, 0)$$
$$(1, 0, 1, 1, 1, 1, 0)$$
$$(0, 0, 0, 1, 1, 1, 0)$$
and
$$(1, 1, 0, 1, 1, 1, 0)$$

As additional information, if the observation parameter is also 0.7, let us find out how many different clusters these patterns will enter without any other information. ROT ANN operation can be done according to the following steps:

1. The number of pattern elements is N = 7 cells in the input layer of the bilayer ROT ANN architecture, but there are no cells in the output (recognition) layer because no clustering has been done yet. But to start, it is assumed that there is one cell in the recognition layer, (M = 1).
2. Initial values of the outgoing and return connection weights between the two layers according to Eqs. 7.91 and 7.92, at time zero $I_{1,1}(0) = 1/(7 + 1) = 1/8$ and $d_{1,1}(0) = 1$ is taken.
3. After the first pattern (1, 1, 0, 0, 0, 0, 1) is detected by the input layer, it is moved to a single cell in the recognition layer by means of Eq. 7.93 leading to:

$$y_1 = \frac{1}{8} \times 1 + \frac{1}{8} \times 1 + \frac{1}{8} \times 0 + \frac{1}{8} \times 0 + \frac{1}{8} \times 0 + \frac{1}{8} \times 0 + \frac{1}{8} \times 1 = \frac{3}{8}$$

The number 3 in the nominator is the sum of the 1s in the pattern. Since there is no other cell in the recognition layer, the current cell with this value is respected as the winner.

4. Using return weights, in this case, the return pattern is the same as the input pattern, because the return weights are all equal to 1.
5. Observation parameter is calculated according to Eq. 7.95, p = 3/3 = 1 is obtained. Since this value is greater than the 0.7 observation value initially given, it is concluded that the input pattern will be clustered with the first cell in the recognition layer.
6. In this case, the outgoing and return connections are renewed according to their statements, respectively:

$$I_{j,1}(k) = \frac{d_{l,j}(k-1)x_l}{0.5 + \sum_{l=1}^{n} d_{l,j}(k-1)x_l}$$

and

$$d_{l,j}(k) = d_{l,j}(k-1)x_l$$

Here, k represents the number of steps. According to the available data, the renewed outward and return weight coefficients are briefly shown as G(1) and D(1) are obtained as:

$$G(1) = \begin{bmatrix} \frac{1}{3.5} & \frac{1}{3.5} & 0 & 0 & 0 & 0 & \frac{1}{3.5} \end{bmatrix}$$

and

$$D(1) = \begin{bmatrix} 1 & 1 & 0 & 0 & 0 & 0 & 1 \end{bmatrix}$$

Thus, the first pattern is placed in the cluster, and the return weights are renewed to reach this cluster from the input layer and to return from there.

7. It is time to detect the second pattern and start the clustering process. First, it is necessary to determine whether it belongs to the above-specified cluster or not. For this, (0, 0, 1, 1, 1, 1, 0) elements must be transferred to the recognition layer with their input weighting coefficients. After adding the input coefficients of these pattern elements and their reciprocal products, $y_1 = 0$ is obtained. Multiplying the reciprocal components of the last input pattern by the return weights gives the comparison to the input layer as the components of the return pattern (0, 0, 0, 0, 0, 0, 0). At first glance, it appears subjectively that the return pattern is completely different from the entry pattern. However, using Eq. 7.94 objectively, $p = 0$ is obtained, and it cannot be included in the cluster in the pattern recognition layer, because it is less than the predetermined observation value of 0.7. Since a cluster will be determined in it, the second cell is opened in the recognition layer. If the outbound and return weights of this cell are made according to the formulas given in step 6, then one can find:

$$G'(2) = \begin{bmatrix} 0 & 0 & \frac{1}{4.5} & \frac{1}{4.5} & \frac{1}{4.5} & \frac{1}{4.5} & 0 \end{bmatrix}$$

and

$$D'(2) = \begin{bmatrix} 0 & 0 & 1 & 1 & 1 & 1 & 0 \end{bmatrix}$$

8. Since there are two cluster cells in the recognition layer, the weights of going and returning to these cells are collectively in the form of a single matrix as:

$$G(2) = \begin{bmatrix} \frac{1}{3.5} & \frac{1}{3.5} & 0 & 0 & 0 & 0 & \frac{1}{3.5} \\ 0 & 0 & \frac{1}{4.5} & \frac{1}{4.5} & \frac{1}{4.5} & \frac{1}{4.5} & 0 \end{bmatrix}$$

and

$$D(2) = \begin{bmatrix} 1 & 1 & 0 & 0 & 0 & 0 & 1 \\ 0 & 0 & 1 & 1 & 1 & 1 & 0 \end{bmatrix}$$

9. Since there are two cluster cells in the recognition layer, the weights of going and returning to these cells are collectively in a single matrix can be displayed as follows:

$$G(2) = \begin{bmatrix} \frac{1}{3.5} & \frac{1}{3.5} & 0 & 0 & 0 & 0 & \frac{1}{3.5} \\ 0 & 0 & \frac{1}{4.5} & \frac{1}{4.5} & \frac{1}{4.5} & \frac{1}{4.5} & 0 \end{bmatrix}$$

and

$$D(2) = \begin{bmatrix} 1 & 1 & 0 & 0 & 0 & 0 & 1 \\ 0 & 0 & 1 & 1 & 1 & 1 & 0 \end{bmatrix}$$

10. When the third pattern (1, 0, 1, 1, 1, 1, 0) is detected in the input layer, if the components in each row of the G(2) matrix are multiplied and added together to transfer it to the output layer, the new pattern can belong to two cluster cells in the recognition layer. Accordingly, since $y_2 > y_1$, the second cluster cell is the winner for this last pattern. Will this cluster accept the new pattern? In order to answer the question, if the components in the second row of the rotation matrix given in the previous step are multiplied by the corresponding components of the input pattern, the rotation pattern (0, 0, 1, 1, 1, 1, 0) is obtained. At first glance, when this pattern is compared with the input pattern, it is concluded that all the others are the same except for the first two elements, and therefore, it may subjectively belong to the second set. To make this objective, if the observation coefficient is calculated from Eq. 7.95, $p = 4/5 = 0.7$ is obtained. Since this value is equal to the pancake value given at the beginning, it is decided that the entered pattern will belong to the second set. In this case, since the input and return weights given in the previous step do not change, they are automatically $G(3) = G(2)$ and $D(3) = D(2)$.
11. The same ROT is achieved by detecting the fourth pattern (0, 0, 0, 1, 1, 1, 0) in the ANN and performing similar operations to the previous step. Again, the second cluster cell is the winner, and the return pattern are found as (0, 0, 0, 1, 1, 1, 0). Accordingly, since the observation parameter $p = 1 > 0.7$, it can be included in the second cluster in this design. As a result of the renewal of the outgoing and return weights of the second set according to the weight renewal operations given in step 6, the new matrices are calculated as:

$$G(4) = \begin{bmatrix} \frac{1}{3.5} & \frac{1}{3.5} & 0 & 0 & 0 & 0 & \frac{1}{3.5} \\ 0 & 0 & 0 & \frac{1}{3.5} & \frac{1}{3.5} & \frac{1}{3.5} & 0 \end{bmatrix}$$

and

$$D(4) = \begin{bmatrix} 1 & 1 & 0 & 0 & 0 & 0 & 1 \\ 0 & 0 & 0 & 1 & 1 & 1 & 0 \end{bmatrix}$$

12. At the end of similar operations with the introduction of the last pattern (1, 1, 0, 1, 1, 1, 0) to the input layer, the winning cluster cell is again the second one, because $y_2 > y_1$. Will this cluster cell accept the new pattern? To answer to the question, first find the return set (0, 0, 0, 1, 1, 1, 0) to be obtained from here. Since the pancake parameter is $p = 3/5 = 0.6 < 0.7$, the second and winning cluster cell cannot accept this new pattern. It is necessary to ask, can the first cluster cell in the recognition layer accept the new pattern? By using the rotation elements of the first cluster cell, the rotation pattern (1, 1, 0, 0, 0, 0, 0) is obtained. Since the observation parameter calculated from here is $p = 2/5 = 0.4 < 0.7$, the input pattern cannot be a member of this set. Then it will be necessary to open a new recognition cell.
13. When the travel and return weights of this new and third cluster cell are calculated according to the equations in step 6, one can obtain:

$$G'(5) = \begin{bmatrix} \frac{1}{5.5} & \frac{1}{5.5} & 0 & \frac{1}{5.5} & \frac{1}{5.5} & \frac{1}{5.5} & 0 \end{bmatrix}$$

and

$$D'(5) = \begin{bmatrix} 1 & 1 & 0 & 1 & 1 & 1 & 0 \end{bmatrix}$$

Then, after all the patterns are entered, the ROT ANN travel and return matrices take the final shapes as follows:

$$G(5) = \begin{bmatrix} \frac{1}{3.5} & \frac{1}{3.5} & 0 & 0 & 0 & 0 & \frac{1}{3.5} \\ 0 & 0 & 0 & \frac{1}{3.5} & \frac{1}{3.5} & \frac{1}{3.5} & 0 \\ \frac{1}{5.5} & \frac{1}{5.5} & 0 & \frac{1}{5.5} & \frac{1}{5.5} & \frac{1}{5.5} & 0 \end{bmatrix}$$

and

$$D(4) = \begin{bmatrix} 1 & 1 & 0 & 0 & 0 & 0 & 1 \\ 0 & 0 & 0 & 1 & 1 & 1 & 0 \\ 1 & 1 & 0 & 1 & 1 & 1 & 0 \end{bmatrix}$$

14. Finally, all patterns are re-entered into ROT ANN, and calculations are made. The first pattern (1, 1, 0, 0, 0, 0, 1) is the winner, as the recognition layer reaches the cluster cells with the values $y_1 = 3/3.5$, $y_2 = 0$ and $y_3 = 2/5.5$. It is understood that there is no change in the outgoing and return weights due to this pattern after the necessary actions are taken. Entering the second pattern (0, 0, 1, 1, 1, 1, 0) yields the values $y_1 = 0$, $y_2 = 3/3.5$ and $y_3 = 3/5.5$, where the second cluster cell is the winner. Again, it is seen that there will be no change in weights. As a result of entering the third pattern (1, 0, 1, 1, 1, 1, 0), the second cell of the recognition layer is the winner. Because the similarity numbers of the input pattern are $y_1 = 1/3$, $y_2 = 3/3.5$ and $y_3 = 4/5.5$. In this case, based on the second cluster cell, since the observation parameter is $p = 3/5 < 0.7$, the input pattern cannot be included in the second cluster, and in this case, its status relative to the other cells in the recognition layer (1 and 3) should be examined. The third cell is the winner, because $y_3 > y_1$. Since the corresponding monitoring parameter is $p = 4/5 = 0.8 > 0.7$, the input cluster is included in the third recognition layer cluster cell. Accordingly, as a result of the renewal of the outgoing and return weights, the following matrices are obtained, respectively:

$$G(8) = \begin{bmatrix} \frac{1}{3.5} & \frac{1}{3.5} & 0 & 0 & 0 & 0 & \frac{1}{3.5} \\ 0 & 0 & 0 & \frac{1}{3.5} & \frac{1}{3.5} & \frac{1}{3.5} & 0 \\ \frac{1}{4.5} & 0 & 0 & \frac{1}{4.5} & \frac{1}{4.5} & \frac{1}{4.5} & 0 \end{bmatrix}$$

and

$$D(4) = \begin{bmatrix} 1 & 1 & 0 & 0 & 0 & 0 & 1 \\ 0 & 0 & 0 & 1 & 1 & 1 & 0 \\ 1 & 0 & 0 & 1 & 1 & 1 & 0 \end{bmatrix}$$

After that, it is left to the reader to see that nothing will change with the introduction of the fourth and fifth patterns into the ROT ANN architecture with this new weight matrix. Thus, it is decided that the ANN architecture and mathematics developed above have become stationary, and the necessary clustering of similar patterns that will come after that continues according to this architecture.

7.17 ANN with Radial Basis Activation Function

In all previous ANNs, the activation functions in the cells of the hidden layer are either bipolar (threshold) or of the ever-increasing type according to the entered values. These are mathematical functions that assign large values to large and small operators to small ones. Such activation functions are used when there is no clustering or classification in the input data. However, if there is some clustering in the data or patterns, as a reflection of this, hidden layer activation functions are expected to have such a structure. Such clustering operations are frequently encountered in practical applications. For this reason, radial basis activation function (RBP) ANNs is used for ANN hidden layer activation functions to work with pre-clustered data. In general, a vector with n elements (x_1, x_2, ... , x_n) represents a point in n-dimensional space. If this is three-dimensional, for example ($x_1 = 3.1$, $x_2 = 1.2$, $x_3 = 2.5$), its representation is given in Fig. 7.55.

Vectors correspond to many points, and points can be grouped according to their proximity and distance. For example, in Fig. 7.56, we can see that the 12 points obtained from 12 vectors visually fall into three clusters named A, B, and C.

When attention is paid, the distances between the points falling in each cluster are very close to each other, but the distances between the clusters are much larger. If we need to derive a logical rule from here, the points in each set are very close to the point indicated by the arithmetic mean of their vectors. So, in order to cluster the points, first, the arithmetic mean points from which their distances can be calculated must be known. In this, first, it should be known in advance how many clusters the point collection will be divided into. In visual situations like Fig. 7.56, one can immediately understand that there will be three clusters. However, one cannot see into how many clusters the hundreds of points entered into a computer will be divided. There is a need for a method to understand this. After the method is determined, it can be automatically performed on computers by transferring it to the computer through convenient software.

7.17.1 K-Means

The simplest method for dividing data into clusters is the so-called k-means. Separation into two clusters occurs by extracting the smallest of the distances of each of the data from at least two predetermined fixed points. To explain this in the simplest way, let there be seven points along the x horizontal axis—1.3, 7.2, 4.5, 6.8, 15.2, 11.7, 0.3. Since at least two clusters are required to start the clustering process, for example, when the numbers 3 and 10 are cluster centers, the first thing to do is to calculate the distances of each of them to the numbers 3 and 10 when it is desired to divide these seven numbers into two clusters. Since the distance in one axis is the absolute value of the differences, they are obtained as:

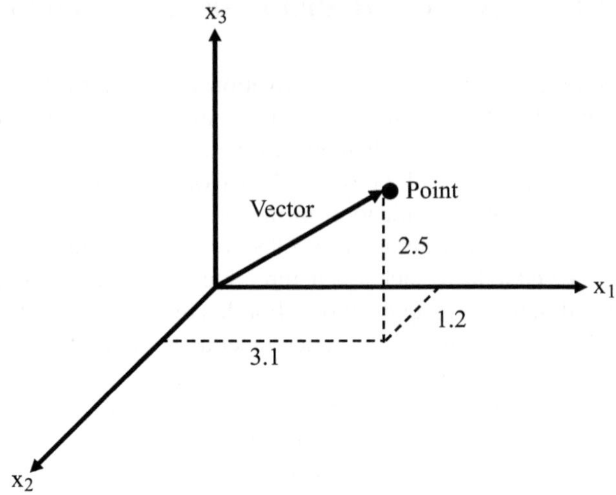

Fig. 7.55 Vector point transformations

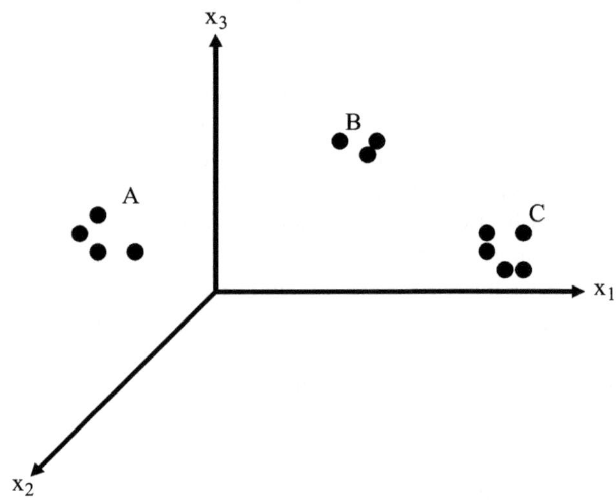

Fig. 7.56 point clusters

Distances of 7 points from 3: 4.3, 4.2, 1.5, 3.8, 12.1, 8.7, 2.7
Distances of 7 points from 10: 11.3, 3.8, 5.5, 3.2, 5.2, 1.7, 9.7

According to the minimization of the reciprocal elements of these two sequences, we can find which elements of the seven number sequences will participate in the three-center clustering. Because of this comparison, it is understood that the first, third, and seventh elements will be included in the three-center set because they are closer to the number 3. Others (7.2, 6.8, 15.2, 11.7) are included in the cluster with 10 centers.

For points that can be represented on two or more axes, the distance is generally measured by the Euclidean distance. If there are two vectors as $(x_{11}, x_{12}, \ldots x_{1n})$ and

$(x_{21}, x_{22}, \ldots x_{2n})$, then the distance, D, between these two points can be calculated from Eq. 7.59. The literal meaning of this is to find the square root of the sum of the squares of the difference of their reciprocal elements, the way to measure the similarity (closeness) of two vectors or patterns. However, when many patterns are given, it has not been determined how they will be clustered. For clustering, it is necessary to determine the cluster centers of the same size as vector (pattern) beforehand, even if it is approximate. In the k-means method, it starts with the principle of knowing the cluster numbers in advance. In order to develop the K-means method, it would be appropriate to define some notation first. If the number of n-dimensional vectors (patterns, data) to be clustered is N, let the i-th pattern be denoted by $x^{(i)}$ $(i = 1, 2, \ldots, N)$. After determining the number of clusters as K, let the j-th of n-dimensional cluster centers be represented by $z^{(j)}$ $(j = 1, 2, \ldots, K)$. Accordingly, we can list the steps of the k-means method as follows:

1. After the number of clusters K is selected, K cluster centers with n dimensions are chosen randomly, if they remain within the range of the input values given at the beginning. Since these initial values will be renewed in the next steps, the values of the cluster center vectors in the l-th step is $\{z_1(l), z_2(l), \ldots, z_K(l)\}$.
2. According to this determined cluster center, the first clusters are established by dividing the patterns among K clusters according to their smallest distances. Here, for $x^{(i)}$ to join the j-th cluster, it is required that the distance from that cluster center is smaller than the distances from the other cluster centers of the same pattern. Its mathematical software shows the distance between the D_{ik} i-th pattern and the center of the k-th cluster according to the following expression:

$$D_{ik} < D_{ij} (j = 1, 2, \ldots, k)$$

Then pattern $x^{(i)}$ belongs to the k-th cluster. In distance comparison, i is changed from 1 to K, provided that $i \neq k$ condition is valid. Here, the $P_k(l)$ notation is an ensemble containing the patterns belonging to the k-th set in the l-th step. Thus, there will be K communities.

3. To calculate the new cluster centers, the arithmetic means of the patterns found in each community is calculated as:

$$z_k(l + 1) = \frac{1}{N_k} \sum_{i=1}^{N_k} x^{(i)} \tag{7.96}$$

Here, N_k denotes the number of patterns found in the k-th cluster obtained in the previous step.

4. As a final step, it is checked whether the sequential calculations of the cluster centers are stationary. Theoretically complete stationarity condition sought is possible according to the following expression:

$$z_k(l + 1) = z_k(l) \quad k = 1, 2, \ldots, K \tag{7.97}$$

However, in practical studies, it is desirable to keep this approach smaller than a certain percentage of relative error α:

$$E = 100\frac{z_k(l + 1) - z_k(l)}{z_k(l + 1)} \tag{7.98}$$

If E < α, the calculations are terminated, and thus, the N patterns given at the beginning are divided into K clusters. In this regard, the error percentages used in practice should be ±5% or ±10% at the maximum.

The two most important features of the K-means clustering method are that sequential training is unsupervised and that it can organize itself in its structure. More explanation of the K-means is presented in Chap. 9.

7.17.2 Radial Basis Activation Function

In order to determine the value at an unmeasured point in an area, for example, in a region where meteorological stations are located, the meteorological variable at that point is tried to be estimated by considering the measurements at the nearest three to five stations. In order to work on such clustered or limited time or space problems, it is useful to set the activation functions in the hidden layer of the ANN to include the desired clustering at the very beginning. For this reason, instead of an ever-increasing activation function in the hidden layer, it is preferred to use activation functions with a certain domain, as shown in Fig. 7.57.

Such activation functions can be triangular, trapezoidal, semicircular, etc. It can be in similar shapes, or it is more suitable for subsequent theoretical and practical calculations to have a kind of curve that can be derived mathematically. For this reason, a Gaussian (bell curve) activation function is generally preferred. The expression of this is given as:

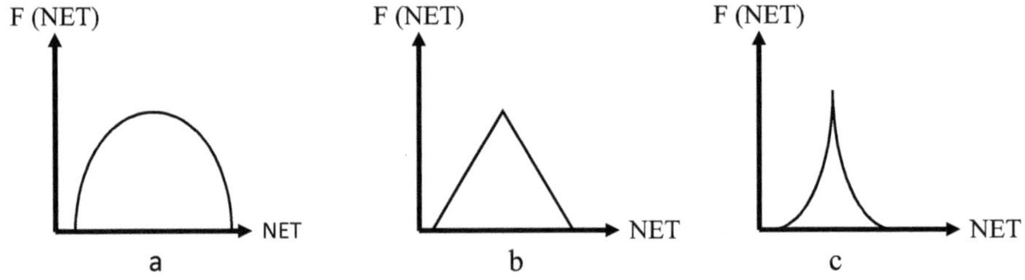

Fig. 7.57 Clustering processor

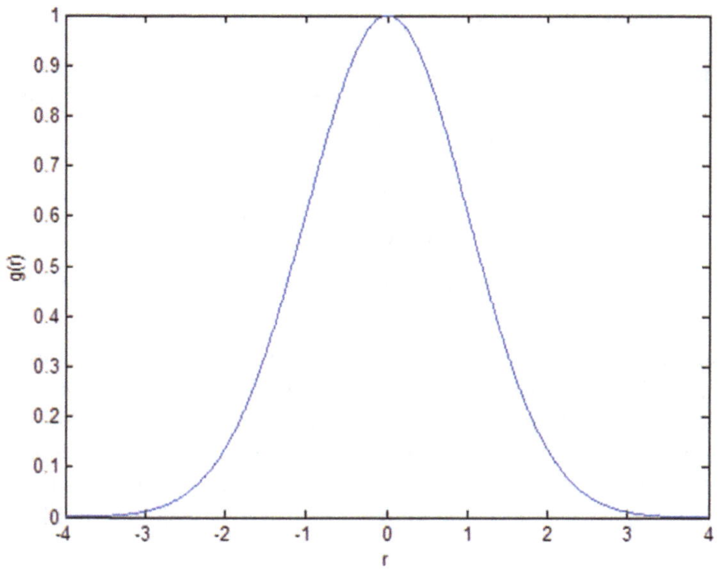

Fig. 7.58 RBP basic activation functions

$$I(r) = e^{-\frac{(r-\mu)^2}{\sigma^2}} \tag{7.99}$$

Here, r gives the data values of a variable in applications, and σ gives the radius of the mean domain of that variable around the peak, μ. Among the parameters here, μ is called the center, and σ is called the distribution parameters. The radial activation function gets its maximum value when r = μ and I(μ) = 1. As one moves away from this value in all directions, the value of the radial activation function decreases. Here, μ denotes the center of the cluster, which has been predetermined and given to the activation function. The radial activation function then compares the known cluster center (μ) with the unknown r value (data). Logically, the closer r is to μ, the more the pattern falls into this cluster. The standard ($\mu = 0$, $\sigma = 1$) Gaussian curve is shown in Fig. 7.58. If the parameters are not standard, similar bell curves are obtained, narrowing or expanding on the horizontal scale. These are called radial basis functions (RBFs) in ANN terminology because the Gaussian value of each equal distance in the radial direction is the same. From here, one can understand that it has a circular effect from the phrase radial basis.

To see even more clearly why a clustering activation function is needed, let's look at the two-cluster scatter diagram in Fig. 7.59. Here, the cluster of small empty circles is surrounded by the filled circles.

With the previous ANN architecture and mathematics, at least four correct separators are needed to distinguish the two clusters. Considering that each separator corresponds to an operator in the hidden layer, it is concluded that at least four cells would be required in a classical ANN hidden layer. However, if a radial basis

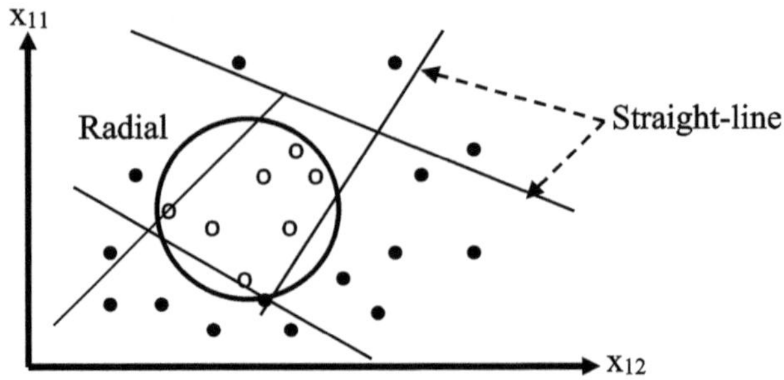

Fig. 7.59 Linear and radial basis separations of the two-set pattern

activation function is considered such that the apex is in the middle of the cluster of small empty circles, it alone can separate the two clusters. From this, it is understood that if there is one cell in the hidden layer and its activation function is RBF, the clustering problem can be solved simply. This example shows how economical the radial basis activation functions are. Especially in space with more than two variables, this becomes even more important. Here, ANNs with radial activation functions in the hidden layer are called RBF ANNs.

Studies like RBFs are frequently found in the classical literature. The main problem here is to estimate the values at the points where no measurement is made by using the values taken from the measurement points in the sampling of the investigated event. The question is, how far away from the point estimation point will the measurement points be considered? For example, in the field of meteorology, this problem was suggested by Cressman (1959) that it is necessary to consider all stations located in a circular area of 500 km or 750 km around the forecast point. Thus, he determined the radius of the circle of influence mentioned above. This is a very subjective opinion and does not match the facts. Instead, different views have been put forward by different researchers.

In the estimation of geology and mineral deposits of earth sciences, the radius of influence can be determined by means of semivariograms by Matheron (1963) and total semivariograms by Şen (1989). All the efforts mentioned above are for estimation with the locally best (optimum) interpolation method. The basic equation of the best interpolation is given in Fig. 7.60, considering the configuration of the measurement and prediction points is made as:

$$P_N = \frac{a_1 m_1 + a_2 m_2 + \ldots + a_n m_n}{a_1 + a_2 + \ldots + a_n} \tag{7.100}$$

Here, P_N is the prediction, and m_i ($i = 1, 2, \ldots, n$) represent the measurement values made at the nearest n points. Each a_i indicates the contribution (weights) of m_i measurement to overall prediction value. This equation can also be written as:

$$P_N = \alpha_1 m_1 + \alpha_2 m_2 + \ldots + \alpha_n m_n \tag{7.101}$$

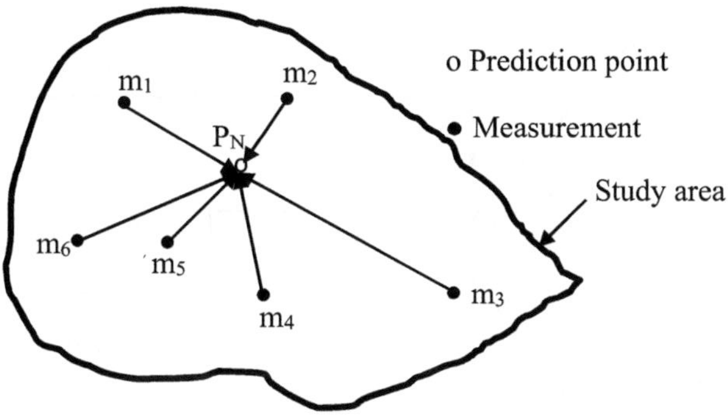

Fig. 7.60 Estimate and measurement point configurations

Here, the α's are interpreted as the percentage contributions of each measurement to the predictive value, and their sum must necessarily equal 1.

$$\alpha_1 + \alpha_2 + \ldots + \alpha_n = 1 \tag{7.102}$$

From Eq. 7.100, it is seen that the prediction (output) value consists of a linear combination of measurement (input) values.

The n points that enter this calculation and are close to the point where the prediction is desired form a cluster. Naturally, the closer (farther) the measuring point is from the prediction point, the larger (smaller) the weight value will be. If the closeness in this sentence is obtained from this curve according to the Gaussian curve and the distances in the weights, estimates can be made by using activation function in Eq. 7.99. With this logic, radial basic activation functions ANNs are established.

7.17.3 RBF ANN Architecture

Although the ANN with the radial basis activation function given in Fig. 7.61 is seen as having three layers, it can be viewed as a two-layer ANN since the training operations pass between the hidden layer and the output layer.

It can be said that this ANN hidden layer outputs as a combination of radial basis activation function in its cells. Every data sequence or pattern given to the input layer is transferred directly to the hidden layer, and an output is generated from here. Thus, each hidden cell has its own output. These outputs are numbers between 0 and 1. RBF plays the role of activation function in hidden layer cells of classical ANNs. In the mathematical formula of RBF given in Eq. 7.99, σ^2 is related to the spread of the Gaussian curve. The practical meaning of this is to take its value as small (large) for clusters spread over a narrow (wide) area. If one shows the output of the i-th radial basis activation function in the hidden layer with O_{ri} and the j-th output in the output layer with O_j, if the weights between the hidden and output layer are A_{ji}, then among them, the following relationship is valid:

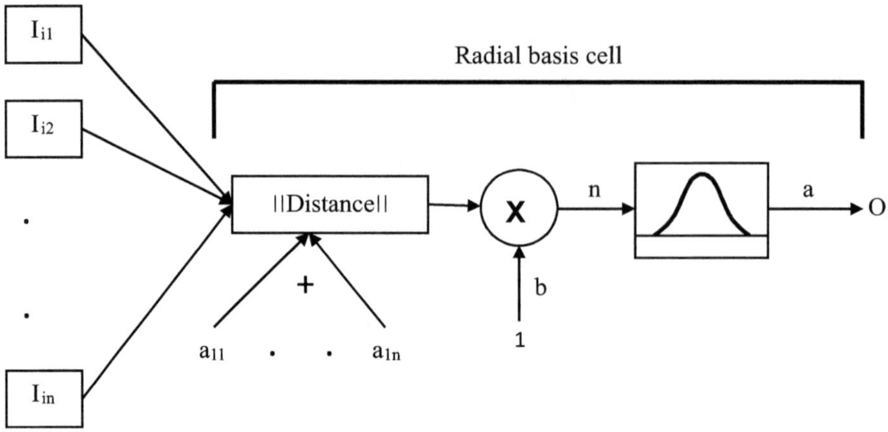

Fig. 7.61 ANN with radial basis processor

$$O_j = \sum_{i=1}^{n} a_{ji} O_{ri} \tag{7.103}$$

This means that it is a linear combination of the outputs nonlinear RBF, like Eq. 7.101. Thus, it is concluded that the RBF ANN is not linear, and this is provided by the RBF.

7.17.4 RBF ANN Training

One can divide the training process of RBF ANNs into two groups, one of which is the training of the parameters μ and σ of the radial basis operators in the hidden layer according to Eq. 7.99, and the second is the training of the weights between the hidden layer and the output layer. The first of these is unsupervised training, and the second is supervised training, as explained in Sect. 6.7.1. The basic questions to be answered for this training are as follows:

1. How many cells should be in the hidden layer? This means how many clusters one will need to divide the data into.
2. How to find the values of the central parameters, μ, of the radial activation function?
3. How to find the distribution parameter, σ, of the radial operator?

There are different approaches for the determination of RBF parameters. The most primitive of these is the determination of μ and σ in the light of the information obtained by utilizing those who have expert opinion on the pattern or data.

Objectively, the k-means clustering algorithm described above can be used to determine the center parameters. With this algorithm, the distribution parameter is the arithmetic mean of the squares of the deviations around the center, after the parameter μ is found, the σ value is computable as:

$$\sigma^2 = \frac{1}{n}\sum_{i=1}^{n}(r_i - \mu)^2 \tag{7.104}$$

This expression is equal to variance in statistical data processing method. After the weights are assigned as random small numbers, the entire numerical structure of the RBF ANN is determined. Thus, after assigning initial values for both the radial activation function parameters and the weights, these values can be further improved with one of the supervised training methods (e.g., back propagation algorithm, Sect. 7.14.3) to further improve the network. From the above explanations, one can deduce that the following steps are necessary in the training of an RBP ANN:

1. Hidden layer parameters are initially determined by clustering operations, and weights are found by randomly assigning small numbers.
2. By entering the data (pattern) in the input layer, the output of each cell of the hidden layer is calculated.
3. Outputs in the output layer (Eq. 7.103) are calculated from the outputs of the radial basis activation function.
4. The weights are the result of the calculation of the weight increments over time, Δa_{ij}, and they are calculated according to the following expression:

$$a_{ij}(t+1) = a_{ij}(t) + \Delta a_{ij}$$

5. Weight changes are based on the back propagation principle under the light of the expression as follows:

$$\Delta W_{ij} = \eta \delta_j C_i$$

Here, η is the learning rate, and δ_j is the error in the j-th output cell, that is, the difference between the output and the observation. This error is defined as:

$$\delta_j = I_j - O_j$$

6. The above steps are continued until convergence is achieved.

7.18 Recycled Artificial Neural Networks

All the ANNs described in the previous sections had feedforward training, and many had a subsequent feedback training circuits. However, none of them had a separate unit for recycling or exchange between cells of the same layer. One of the questions that come to mind is, "Cannot the capability of ANN be increased by forward feeding in addition to the input layer cells by recycling the values from the hidden layer cells?" On the other hand, one wonders if the cells of the same layer cannot communicate with each other. ANNs that challenge this are known as competitive training networks. ANNs that allow such an additional input are called reversible ANNs. In Chap. 9, the most advanced ANN alternatives are presented related to this point.

7.18.1 Elman ANN

The type of ANN, which has the whole multi-layer ANN structure and additionally contains the outputs of the hidden layer as a parallel input layer, is called Elman (1990) ANN. Its architecture is shown in Fig. 7.62. In the first study of this ANN, it starts with the input layer that only receives the input data. Elman starts the ANN by returning the first values from the hidden layer and inserting them into the additional input layer. Since the return is delayed, the said additional layer will be called the "lagged input layer" from now on.

Recycling can be one time-lapse, or they can be two or more intermittent. Accordingly, this ANN is called one-time-lagged, two-time-lagged, etc. It is like interval (lag) modeling in the stochastic processes with 1, 2, 3 etc. lags. The tasks of each layer in the Elman ANN in Fig. 7.62 are as follows:

1. In the input layer, there are cells that make as much detection as the number of data elements.

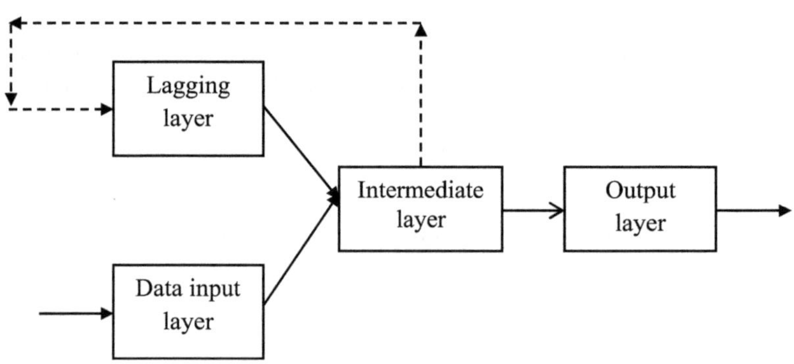

Fig. 7.62 Elman ANN architecture

2. In the hidden layer, there is an activation function that collects the information from the input cells, and then a curvature (converting linear information into nonlinear form).
3. In the output layer, there are as many cells as the number of elements in the output data, and inside each cell, there is an adder activation function that linearly collects the information it receives.
4. There are as many cells as the hidden layer cells in the lagging layer. They only have detection feature. The connection of each cell of the hidden layer to the corresponding cell in the lagging layer is by one-weight connectors. This means that the lagging layer cells perceive the information in the hidden layer cells as they are. However, it is also possible to weaken the returns by giving values less than 1 to the weight coefficients.

7.18.2 Elman ANN Training

Philosophically, the training of this ANN is the same as the multi-layer ANNs described in Sect. 7.11.1. In addition, there will be some additions in education due to the presence of the lagging layer. At the beginning of the training, in addition to the random selection of the connection weights between the input and hidden layer cells in small numbers, the connection weights, $a_{ij}^{\text{input-hidden}}$, between the lagged, and hidden layer cells are assigned randomly. If the one-time-delay Elman ANN is shown as the first return values, $I_{ij}^{\text{input}}(t-1)$ then the output amount in one of the intermediate layer cells at time t is calculated as:

$$\text{NET}_i(t) = \sum_{j=1}^{n} a_{ij}^{\text{input-hidden}} I_{ij}(t) + \sum_{j=1}^{k} a_{ij}^{\text{input-hidden}} I_{ij}(t-1) \tag{7.105}$$

Here, n and k represent the number of cells in the input and lagging layers, respectively. The output of this ANN is obtained by linear summing of the information coming to the output cells. If one denotes the weights between the intermediate and output layers as, $a_{ij}^{\text{hidden-output}}$, then the output of i-th output cell is calculated as:

$$O_i(t) = \sum_{j=1}^{m} a_{ij}^{\text{hidden-output}} \text{NET}_i(t) \tag{7.106}$$

In case these outputs do not match the expected outputs, O_i, all forward weights are renewed in the network, just as in multilayer ANNs (see Sect. 7.5.1), by way of feedback by taking advantage of the error amounts. In general, the error at time t is given as:

$$e_i(t) = o_i(t) - O_i(t) \tag{7.107}$$

If the output function is sigmoid, the amount of error to be distributed to the weights at time t, using Eq. 7.75, with the notations at hand is possible as:

$$\delta_i(t) = O_i(t) - [1 - O_i(t)]e_i(t) \quad (7.108)$$

Calculations to be followed in feedback should be done exactly as described for multilayer ANNs (Sect. 7.5.1).

Recyclable ANNs are very useful in modeling dynamic events, because they consider time lags. With these features, they can replace the stochastic process types such as Markov, ARIMA (Auto regressive integrated moving average), etc. models. Although there are some assumptions arising from data and modeling in stochastic modeling, there are no presuppositions in reversible ANNs.

7.19 Hopfield ANN

ANNs, which are frequently used in same-fit classification, were first developed by Hopfield (1982). These ANNs are mostly used for optimization operations. In this ANN, inputs and outputs are made by cells in a single layer. Cells work according to binary (crisp, two-value) logic as open (+1) and closed (−1). It can be considered in two parts according to whether the activation function in the cells is continuous (sigmoid and tangent hyperbolic functions) or discontinuous (threshold functions). The output of each cell is used as input by returning to other cells. Therefore, the Hopfield ANN is also a reversible ANN, (see Fig. 7.63). The previous outputs are distributed as inputs to the cells with weights.

There is a weight coefficient for the correlations from each entry point in front of the cells. In general, $a_{ij}(t)$ shows the weight of the relation from i-th point to cell j. If this type is represented as the input of the ANN at time t as I, then:

$$I_i(t) = \sum_{j \neq i}^{n} a_{ij} O_{ij}(t-1) - \Theta_i \quad (7.109)$$

Here, Θ_i denotes the constant of i-th cell. The output of the same cell is either discrete or continuous depending on the activation function, respectively, as:

$$O_{ij}(t) = \mathrm{sgn}\left[I_{ij}(t)\right] \quad (7.110)$$

or

$$O_{ij}(t) = \mathrm{sigmoid}\left[I_{ij}(t)\right] \quad (7.111)$$

Since sgn is a sigmoid, which is defined as a threshold function, the open software of Eq. 7.110 is like Eq. 7.35 and can be written as:

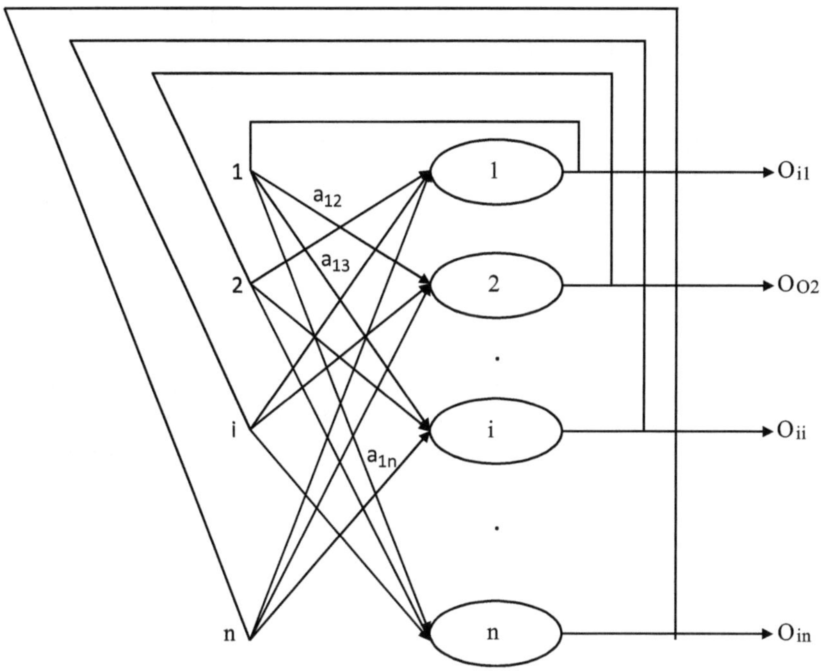

Fig. 7.63 Hopfield transformable ANN

$$O_{ij}(t) = \begin{cases} +1 & \text{if} \quad I_{ij}(t) \rangle \Theta_i \\ -1 & \text{if} \quad I_{ij}(t) \langle \Theta_i \\ O_{ij}(t-1) & \text{if} \quad \text{otherwise} \end{cases} \tag{7.112}$$

Here, Θi indicates the threshold values to be assigned by the designer. In the training of Hopfield transform ANN, there are two phases, such as regenerating weights (feedback) and obtaining outputs (forward feeding), like multilayer ANNs:

$$a_{ij}(t) = \begin{cases} \frac{1}{n} \sum_{k=1}^{n} m_{ik} m_{jk} & \text{if} \quad i \neq j \\ 0 & \text{if} \quad \text{otherwise} \end{cases} \tag{7.113}$$

Here, n represents the number of sample sets to be detected in the ANN, and m_i represents the measurement value of i-th element in a sample data. In order to ensure stability in a balance way, an example is given to the ANN that it did not perceive before. The deficiencies of this sample, which may be missing data, should be filled by Hopfield ANN. After this sample is inserted into the network, it is expected that the network becomes stationary at the end of successive iterations. Iteration of the network is like Eq. 7.109:

$$O_i(t+1) = \mathrm{sgn}\left[\sum_{j \neq i}^{n} a_{ij}(t) O_{ij}(t) - \Theta_j\right] \tag{7.114}$$

For this, the inputs of the network must be assigned as initial values. The energy formula proposed for the network to become stationary is as follows:

$$E(t) = -\frac{1}{2}\sum_{i=1}^{n}\sum_{j \neq i}^{n} a_{ij}(t) O_{ij}(t) O_{ij}(t) + \sum_{i=1}^{n} O_i(t) \tag{7.115}$$

During network operation, this formula must either be stationary or decrease. Replenishment of weights is done according to Hebb's rule. Another advantage of other ANNs is that there is no need for much iteration in renewing their weights. There are two types of them, discrete and continuous, as explained below.

7.19.1 Discrete Hopfield ANN

Here, the input pattern elements have values of -1 or $+1$. The number of cells in the layer is equal to the number of elements in the pattern. It ensures that the cell outputs are -1 or $+1$ by subjecting the inputs coming from the outside and other cells with weights to a threshold operator in each cell. K input patterns, each be shown as with n dimensions, $I_1, I_2, \ldots, I_K$. On the other hand, let's denote the output of the j-th cell after t time with $O_{kj}(t)$. Let the correlation coefficient (weight value) from cell j to cell i be a_{ij}. If the k-th pattern is entered in the i-th cell, D_{ki}, it shows the external inputs, the cell output value at the time of t + 1 can be written as:

$$O_{kj}(t+1) = \mathrm{sgn}\left(\sum_{i=1}^{n} a_{ji} O_{ki}(t) + D_{kj}\right) \tag{7.116}$$

If $x \geq 0$ then $\mathrm{sgn}(x) = +1$, for $x < 0$ then $\mathrm{sgn}(x) = -1$ and $\mathrm{sgn}(0) = 0$. Another feature of classical Hopfield ANNs is that the outputs of the cells are renewed asynchronously. This means that one cell output is renewed at each time interval; which cell is considered for renewal is done by random selection. In order to obtain reliable results, each cell should be given an equal chance in the selections. Sorting the cells to be renewed can be done at the beginning and put in order. Simultaneity can also be done alternately according to the distinction between odd and even cell numbers. Simultaneous regeneration of all cells is also possible.

7.19.2 Application

Consider a four-cell Hopfield ANN with the patterns (1, 1, 1, 1) and (−1, −1, −1, −1) in memory. If a distorted pattern such as (1, 1, 1, −1) is presented to it, the outer inputs of the four cells are $D_1 = 1$, $D_2 = 1$, $D_3 = 1$, and $D_4 = -1$. If the second cell is chosen randomly, its net input is according to Eq. 7.116.

$$a_{21}I_1 + a_{21}I_1 + a_{23}I_3 + a_{24}I_4 + D_2 = 1 + 1 - 1 + 1 = 2$$

Accordingly, its output is sgn(2) = 1, and thus, its state does not change. Secondly, if the fourth cell is thought to be randomly selected for the refresh, the state of this cell has changed from −1 to 1 since its input value will be 1 + 1 + 1 − 1 = 2 and sgn (2) = 1. Thus, since the pattern (1, 1, 1, 1) was brought from the memory instead of the corrupted pattern, the disorder was eliminated.

7.19.3 Continuous Hopfield ANN

If the cell outputs can take any value between −1 and +1, continuously then Hopfield ANNs are in question. There are the following differences for these ANNs from the discrete Hopfield ANN:

1. The cell operator should be a continuous function rather than a discrete threshold function.
2. Since it is continuous in time, the inputs and outputs of the cells are constantly renewed.
3. In very small-time increments, very small changes are not abruptly seen in the cell outputs. In this respect, variation over time (derivative), $\partial O(t)/\partial t$, is important in continuous Hopfield ANNs. Instead of Eq. 7.116, the following expression is used:

$$\frac{\partial O_k(t)}{\partial t} = \eta f\left(\sum_{i=1}^{n} a_{jk}O_i(t) + D_k\right) \tag{7.117}$$

Here, η represents the learning percentage, and f(.) represents one of the continuous activation function (sigmoid, tangent hyperbolic, etc.).

4. Cell output values are limited. It reaches saturation for very large and small values. In most of the studies, it is seen that the cell outputs are confined between −1 and +1. Accordingly, the renewal process is done using the following equation:

$$\frac{\partial O_k(t)}{\partial t} = \begin{cases} 0 & \text{if} \quad O_k = 0 \text{ ve } f\left(\sum_{i=1}^{n} a_{jk}O_i(t) + D_k\right) \geq 0 \\ 0 & \text{if} \quad f\, O_k = 0 \text{ ve } f\left(\sum_{i=1}^{n} a_{jk}O_i(t) + D_k\right) \leq 0 \\ \eta f\left(\sum_{i=1}^{n} a_{jk}O_i(t) + D_k\right) & \text{otherwise} \end{cases}$$

7.20 Simple Competitive Learning Network

In these ANNs, there are n-cell input and m-cell output layers and weights connecting them. Unlike previous ANNs, connections can also be found between cells in the output layer (see Fig. 7.64).

Here, in order to determine the weights iteratively, it is necessary first to decide what is being looked at from such an ANN. For example, if it is desired to cluster the pattern sequences presented to the input layer of each cell in the output layer, each of the output cells is desired to represent some of the collection of patterns. The number of cells in the input layer is equal to the number of elements in the pattern. The same number of connections comes from the input layer to each cell of the output layer, and each of them has a weight value. The distance between these two vectors can be looked at as the output cell weight vector resembles the input data. There are weight connections from the same input pattern to each output cell. Therefore, whichever of these weight connections is closest to the input pattern, then output cell is first declared as the winner and its output is assumed equal to 1. Thus, the weight values

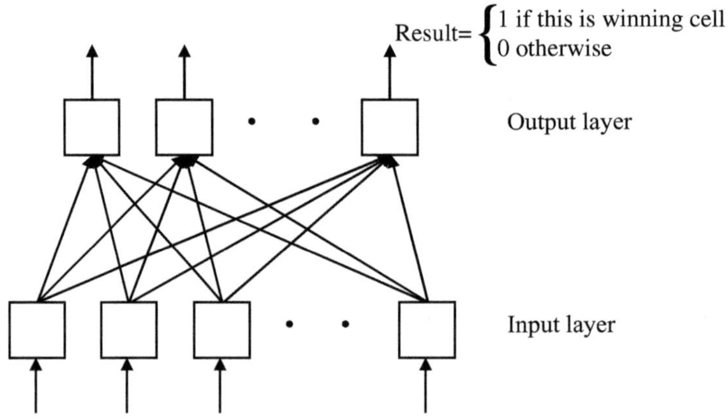

Fig. 7.64 Simple competitive ANN

of these winning output cells are renewed according to a certain rule, and the connection vector is tried to be as close as possible to the input vector. This means trying to minimize the distance between two vectors (weight and input pattern). For the renewal of the weights, η(t) the learning rate is made according to the following formula:

$$a_{ij}(t+1) = a_{ij}(t) + \eta(t)\left[I_{ij}(t) - a_{ij}(t)\right] \tag{7.118}$$

The weight renewal process is continued until the calculations become stationary. Since this can take a long time, in practice, the processes are terminated either when successive weight differences are smaller than a certain percentage of error or after a predetermined number of iterations.

7.20.1 Application

Five training patterns are given as $I_1 = (1.1, 1.7, 1.8)$, $I_2 = (0, 0, 0)$, $I_3 = (1, 0.5, 1.5)$, $I_4 = (1, 0, 0)$ and $I_5 = (0.5, 0.5, 0.5)$. In this case, let's start with the input and output layer, which has three cells each, namely, the cells of the output layer as A, B, and C. Suppose one gets the initial values of their weights to the input layer as random numbers between 0 and 1 as follows:

$$A = \begin{pmatrix} 0.2 & 0.7 & 0.3 \\ 0.1 & 0.1 & 0.9 \\ 1.0 & 1.0 & 1.0 \end{pmatrix}$$

1. Detection of the first data set to the input layer and the weight connections to cell A and the Euclidean distance between this input from Eq. 7.59 is found as:

$$D_{A1} = \sqrt{(1.1-0.2)^2 + (1.7-0.7)^2 + (1.8-0.3)^2} = \sqrt{4.1}$$

   and similarly

$$D_{B1} = \sqrt{4.4}$$

   and

$$D_{C1} = \sqrt{1.1}$$

Since the smallest of these is the value of cell C, it is the winner, and its output are taken equal to 1. As a result of iteration of the weights with a learning rate of $\eta = 0.5$, the new weight matrix becomes:

$$A = \begin{pmatrix} 0.2 & 0.7 & 0.3 \\ 0.1 & 0.1 & 0.9 \\ 1.05 & 1.35 & 1.4 \end{pmatrix}$$

2. As the second data (pattern, vector) (0, 0, 0) is detected by the input layer, similar distance calculations result as $D_{A2} = \sqrt{0.6}$, $D_{B2} = \sqrt{0.8}$ and $D_{C2} = \sqrt{4.9}$. From here, since the smallest distance belongs to A, the new weight matrix with the renewal of the weights become:

$$A = \begin{pmatrix} 0.1 & 0.35 & 0.15 \\ 0.1 & 0.1 & 0.9 \\ 1.05 & 1.35 & 1.4 \end{pmatrix}$$

From here, it is seen that only one of the rows of the weight matrix changes with each new pattern entry.

3. As the smallest distance, $D_{B3} = \sqrt{0.5}$ value is found with the detection of the third pattern (1, 0.5, 1.5), the values in the second row of the weight matrix change and takes the following shape:

$$A = \begin{pmatrix} 0.1 & 0.35 & 0.15 \\ 0.05 & 0.3 & 1.2 \\ 1.05 & 1.35 & 1.4 \end{pmatrix}$$

4. Since it is found as the smallest distance $D_{A4} = \sqrt{1.0}$ by entering the fourth data, the weight matrix by renewing the first-row weight values becomes:

$$A = \begin{pmatrix} 0.55 & 0.2 & 0.1 \\ 0.05 & 0.3 & 1.2 \\ 1.05 & 1.35 & 1.4 \end{pmatrix}$$

5. By inserting the last data string into the input layer, cell A wins and the weight matrix takes the following form:

$$A = \begin{pmatrix} 0.5 & 0.35 & 0.3 \\ 0.05 & 0.3 & 1.2 \\ 1.05 & 1.35 & 1.4 \end{pmatrix}$$

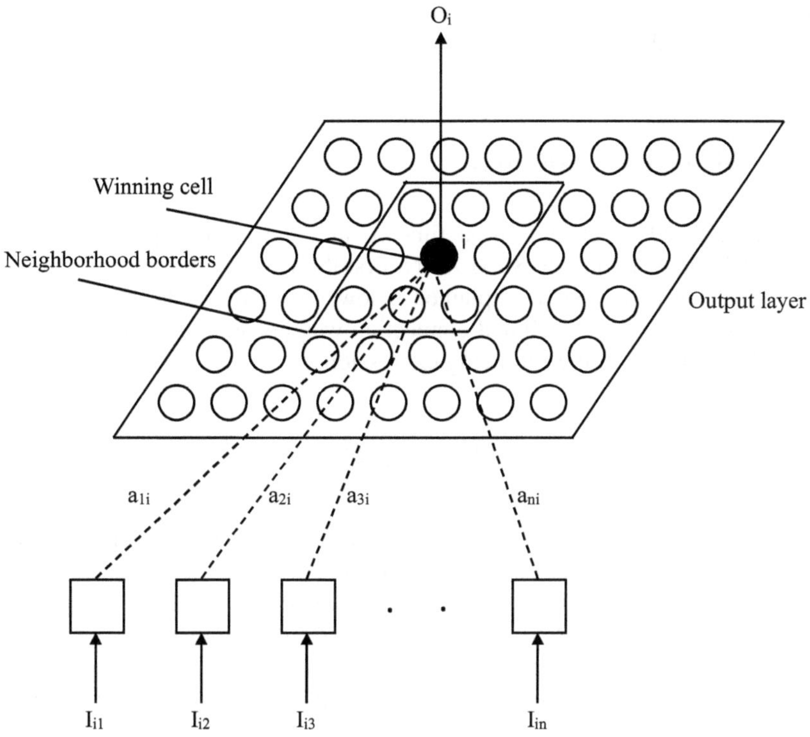

Fig. 7.65 SOM YSA architecture

7.21 Self-Organizing Mapping ANN

This approach proposed by Kohonen (1988) is called self-organizing mapping (SOM). One needs to clearly understand what each word here means. In SOM ANNs, the input layer is one-dimensional as in the previous ones, but the output layer is two-dimensional. The cells of the input layer are arranged in a row, while the cells of the output layer are arranged in a plane (see Fig. 7.65). This ANN can be called a bilayer network. The connections from the input cells to each of the output cells in the plane are exactly there.

Only the connections from the input layer cells to an output cell (j) are shown in the figure. First, the presence of the word map in this ANN is due to the map drawn on two-dimensional planes. The most important feature of this ANN is that when the data is detected on the input layer, there is no need for a teacher to train it, nor does it need to give real data to the output layer. For this reason, it is used to solve problems whose outputs are unknown. This is equivalent to saying that this ANN works according to unsupervised training. The cells on the output plane want to win by competing. Finally, when one of them wins, it takes the cells closest to it in every direction. Thus, when a cell wins, for example, the set of cells consisting of nine cells, shown in Fig. 7.65, and eight of them are closest to the winning cell. The winning cell will output 1 with all other cells 0. However, during the training, the weights are trained in the neighboring cells.

7.21.1 SOM ANN Training

At the normalized input and for a time, t, and the randomly assigned weight coefficients' entrance into the SOM ANN, first the winning output cell is found by using one of the two ways as explained below. Primarily in the two-dimensional output layer, input of each cell and connection weights are multiplied and summed up linearly according to the following expression.

$$O_i = \sum_{j=1}^{n} I_{ij} a_{ij} \quad (7.119)$$

The cell with the greatest output value is determined as the winning cell. The output of such a cell is taken as 1 with all others as 0.

Another way to determine the winning cell is, first as already mentioned in Sect. 7.6.1 for the Kohonen ANN, the closest weighted cell to the input pattern is determined by Euclidian distance calculation. For this, the distances between each weighted pattern and the input layer is calculated according to the following expression:

$$u_j = \left\| I_i - A_{i,j} \right\| \quad (7.120)$$

The output cell with small distances will win the competition. After the determination of the winning cell, it is necessary to update the weights between it and the closest neighbors. For this, the following equation is used:

$$A^{old}(t) = A^{new}(t) + \alpha k(i, l) \left[I_i(t) - A^{old}(t) \right] \quad (7.121)$$

Here, α denotes a learning coefficient that needs to be reduced over time. On the other hand, according to the k(i,l) neighborhood function is expressed as Gaussian distribution function:

$$k(D) = \exp\left(-\frac{D^2}{2\sigma^2} \right) \quad (7.122)$$

Here, D denotes the distance between the winning cell, i, and one of the neighboring cells, j. In this equation, σ denotes the width of the neighborhood area, which must be reduced over time after initial assignment by the researcher. If the vectors showing the positions of these two neighbors are Pi and Pj then the difference between them is calculated as:

$$D = \|P_i - P_j\| \tag{7.123}$$

There are two methods for determining a winning cell and its neighbors whose weights will change together. The first of these is to enclose the neighbors around the cell in a square or rectangle. Such a situation is shown in Fig. 7.65 in the simplest way. A second method is to determine the neighbors in a regular polygon with the winning cell in the center (see Fig. 7.66).

7.22 Memory Congruent ANN

The ANNs to be explained here help to find out which data (pattern) is suitable for which of the data (patterns) in the memory. Although the input pattern is partially missing or corrupt, these ANNs determine the closest regular pattern in memory. This is like how one perceives what a broken Turkish speaker means as proper Turkish. It is in our memory in the form of proper Turkish patterns. Any separation from these patterns, that is, partial deficiency or defect, can be perceived as smoothed by the person immediately. Here, the identification or matching of two slightly different patterns in perceiving an incomplete and/or corrupted Turkish sentence properly is called memory congruent (MC) ANN. From this short explanation, it is understood that there will be two stages in MC ANNs, one of which is the

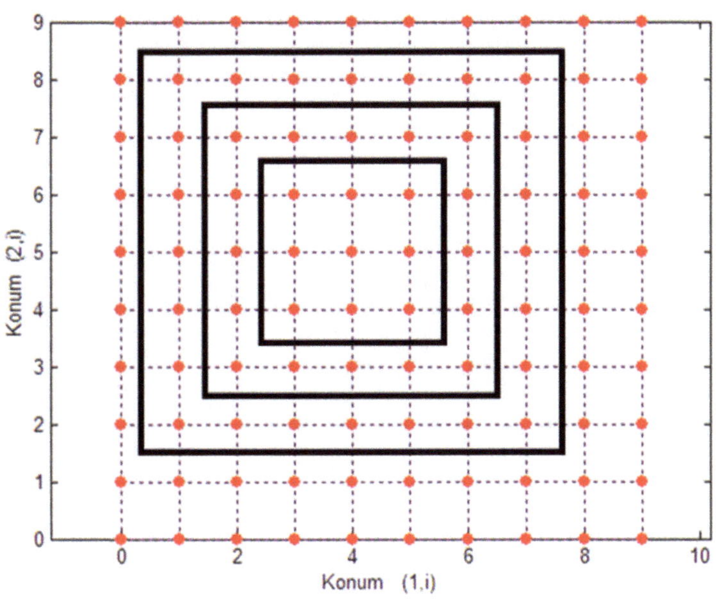

a

Fig. 7.66 Polygon neighborhood areas (**a**) rectangle, (**b**) hexagon, (**c**) random

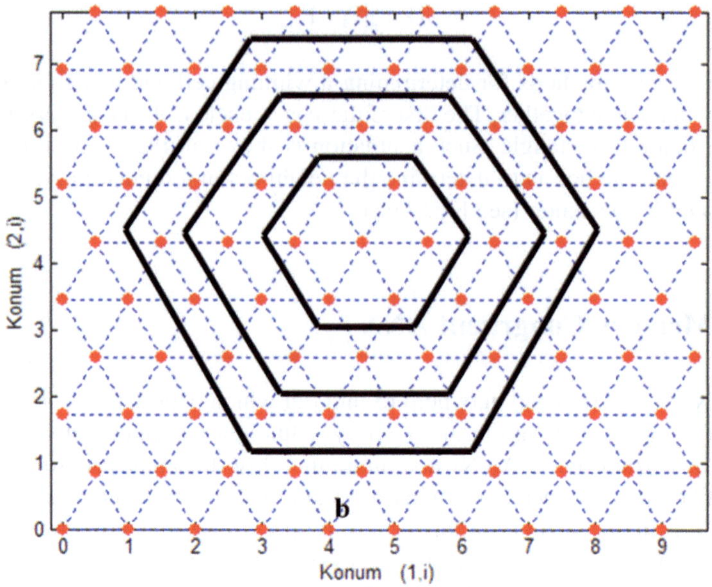

b

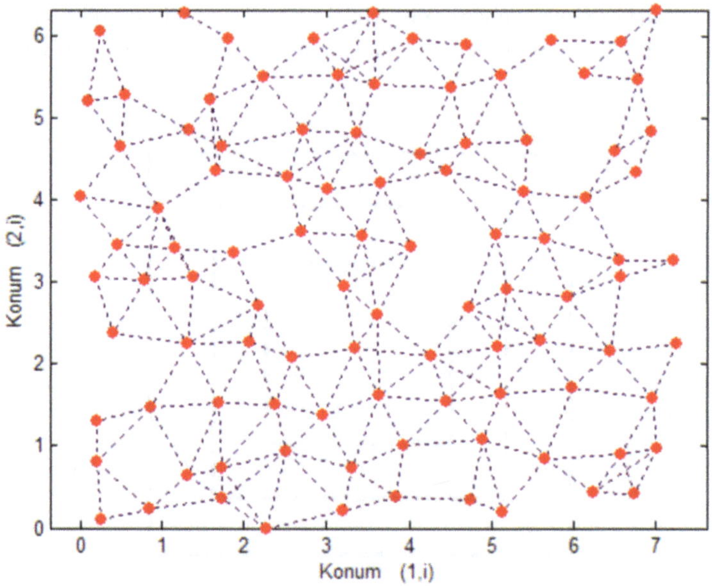

c

Fig. 7.66 (continued)

assignment of weights for placing regular patterns in the memory and thus the presence of many weight sets in the memory, and the other one is searching for which a pattern made to be detected by this ANN matches which one in the memory. The first of these is called pattern storage in the memory and the second is called pattern extraction from memory.

There are two types of fit in MC ANNs as different-fit and same-fit. The types of patterns in the input and memory in different harmony are different from each other. For example, when translating from Turkish to English or vice versa, the harmony between the patterns falls under the category of different-harmony. However, the matching of Turkish defects or deficiencies with Turkish, that is, the same type of patterns, falls into the same-congruity class. MC ANNs are used in cases, where there are irregularities in the pattern and there are several regular patterns that may correspond to a broken pattern. They can also be used to determine if the given pattern does not fit any of the ones in the memory.

7.22.1 Matrix Fit Memory Method

The weight matrix obtained by multiplying the input and output patterns matrices is important here. Weight matrix elements are calculated by Hebb's rule as:

$$a_{jk} = \sum_{i=1}^{n} I_{ji} O_{ki} \tag{7.124}$$

In short, the software for this in matrix form is to multiply I matrix containing the input patterns line by line and the O matrix containing the output patterns line by line by the transpose:

$$A = IO^T$$

7.22.1.1 Application

Finding a matrix fit memory for two input and two output patterns, each with two elements, is straightforward. For example, the weight matrix for (1, 1) and (1, −1) input and (−1, 1) and (−1, −1) output patterns is obtained as:

$$A = \begin{pmatrix} -1 & -1 \\ 1 & -1 \end{pmatrix} \begin{pmatrix} 1 & 1 \\ 1 & -1 \end{pmatrix} = \begin{pmatrix} -2 & 0 \\ 0 & 2 \end{pmatrix}$$

When the first input pattern is given to an ANN equipped with these weights, then the output pattern is reached as follows:

$$O = \begin{pmatrix} -2 & 0 \\ 0 & 2 \end{pmatrix} \begin{pmatrix} 1 \\ 1 \end{pmatrix} = \begin{pmatrix} -2 \\ 2 \end{pmatrix}$$

By passing it through a threshold operator, the (−1, 1) pattern is found. This output (1, 1) is the memory matched and correct pattern found for the input pattern. Similar operation can be done in the second entry pattern. If a new pattern (−1, −1) is given to this ANN, then:

$$O = \begin{pmatrix} -2 & 0 \\ 0 & 2 \end{pmatrix} \begin{pmatrix} -1 \\ -1 \end{pmatrix} = \begin{pmatrix} 2 \\ -2 \end{pmatrix}$$

where the pattern (1, −1) is obtained as a result of passing it through the threshold operator. This is a pattern that does not exist in memory.

7.22.1.2 The Least Squares Method

Here, the weight matrix is found by minimizing the squares of the differences between the input and memory patterns. If A is the weight matrix, then the k-th output pattern is derived from the input pattern I_k by matrix multiplication:

$$O_k = AI_k \tag{7.125}$$

Accordingly, the error is:

$$e = \sum_{k=1}^{n} (d_k - O_k)^2 \tag{7.126}$$

Here, the weight matrix is calculated to minimize this error definition. For this, it is enough to set the derivative of e with respect to A then equate to zero. Details of the inference are presented in Sects. 7.5 and 7.14.3. Finally, the weight matrix becomes available as:

$$A = \frac{\left(\sum_{i=1}^{n} d_i I_i^T\right)}{\left(\sum_{i=1}^{n} I_i I_i^T\right)} \tag{7.127}$$

Here, the letter T denotes the transpose of the pattern. That is, it converts the row shaped data vector to column shape. In Eq. 7.127, numerator means the sum of the products of a pattern in memory and the corresponding elements of the input pattern. The denominator is the sum of the squares of the elements of the input pattern.

7.22.1.3 Application

Find the least squares weight matrix from the two input and output patterns given in the previous section. Since, $I_1 = (1, 1)$, $I_2 = (1, -1)$, d1 = (−1, 1) and d2 = (−1, −1):

$$a_{11} = d_1 I_1^T = -1 \times 1 + 1 \times 1 = 0$$

$$a_{12} = d_1 I_2^T = -1 \times 1 + (-1) \times 1 = -2$$

$$a_{21} = d_2 I_1^T = (-1) \times 1 + (-1) \times 1 = -2$$

and

$$a_{22} = d_2 I_2^T = (-1) \times 1 + (-1) \times (-1) = 0$$

So, the final form of the weight matrix becomes as:

$$A = \begin{pmatrix} 0 & -2 \\ -2 & 0 \end{pmatrix}$$

7.23 General Applications

In the previous sections of this book, some applications of ANNs are given in the appropriate sections. The applications in this section require computer software. In the following, the data types and the ANNs architectural structure are explained in detail in table and figure forms.

7.23.1 Missing Data Complement Application

In this example, the analysis of data contributes to solving the problems of earth sciences like in many branches of science. For this, it is required that the data be complete and continuous before the application stage. However, there may be deficiencies in the data between certain periods due to the failures arising from situations such as missing measurements. These deficiencies may be for a single variable, or there may be deficiencies in several of them in the same period.

In case of missing data, they must be completed first in order to carry out the study. In this study, it was assumed that there were deficiencies in the data of the temperature parameter after a certain date at a measurement station, and these missing data were tried to be completed with the ANN approach.

The selected meteorological parameters include temperature, atmospheric pressure, water vapor pressure, relative humidity, and wind velocity. The daily average values of these between March 1 and April 20, 1996, are shown in Table 7.3.

It is assumed that the temperature data have not been recorded since March 28, since the measurement could not be made or recorded. Thus, 27-day data values of the temperature parameter are known within a 50-day period, but the next 23-day data is incomplete, but there is no deficiency in the data of other variables.

In order to complete the missing data, first, an architectural design is made between the temperature and other variables with the help of ANN methodology. For this, the data of the first 27 days is used, in which it is fully recorded. A three-layer ANN model consisting of input, output, and a single hidden layer is chosen for ANN architectural model construction as in Fig. 7.67. In the model, Z and W are for input-hidden layer and hidden layer-output connection weights.

In the input layer, four cells are used for the air pressure, relative humidity, wind speed and water vapor pressure data, and in the output layer, a single cell was used for the temperature records. Three cells are placed in the hidden layer. The sigmoid activation function is used in the exits of the hidden layer cells.

In the training of the ANN model, the learning method is applied in which the trainer tells the correct result, which is a branch of learning with the help of the trainer (Sect. 7.15.3) and back propagation algorithm (Sect. 7.14.3). The learning rate value is chosen as 0.04 in the model, and a generalization approach is used between the temperature and other datasets with 9000 iterations.

In Fig. 7.68, the temperature measurements and the temperature model values are shown after the proposed ANN model application and training completion. As can be seen from the figure, temperature patterns follow each other closely.

The absolute error value between the temperature measurements and ANN model values can be calculated as follows:

$$E = \sum_{i=1}^{50} \left| T^{i}_{Observation} - T^{i}_{ANN} \right| \tag{7.128}$$

The resulting value is 15.82 °C, which is expressed as the total absolute error; when divided by the number of data, the average absolute error value is 0.62 °C. To be able to analyze the error values closely, the scatter diagram of the measured and ANN model predicted values are shown in Fig. 7.69.

The correlation between the actual and predicted temperature values because of the regression analysis yields:

$$T_{ANN} = 1.01\ \text{Observation} + 0.07 \tag{7.129}$$

The results show that reliable predictions can be made for the temperature from other variables. The question is how accurately the model to complete the missing temperature values?

In order to find the answer to this question, the data belonging to the first 27 days are selected as the training set. Then, the ANN model in Fig. 7.67 is trained, and after

Table 7.3 Göztepe/Istanbul meteorology station daily data

Temperature (°C)	Atmospheric pressure (mb)	Water vapor pressure (mb)	Relative humidity (%)	Wind velocity (m/s)	Temperature (°C)	Atmospheric pressure (mb)	Water vapor pressure (mb)	Relative humidity (%)	Wind velocity (m/s)
3.6	1008.0	5.8	82.3	4.6	5.7	1016.6	7.3	77.7	1.2
2.9	1004.1	5.3	68.3	1.5	8.1	1014.2	7.8	80.0	1.7
2.4	1005.3	5.6	77.0	2.4	12.6	1012.5	10.8	70.7	2.0
3.3	1012.5	4.9	65.0	1.6	11.7	1002.3	11.8	72.3	1.8
4.2	1010.2	5.3	71.7	3.9	8.2	1004.0	9.7	91.0	1.5
3.5	1012.0	5.0	64.7	5.3	12.9	1006.9	9.6	88.7	2.7
3.3	1016.9	6.1	74.7	4.3	8.9	1012.0	8.7	81.7	3.6
2.3	1021.6	6.4	82.5	3.6	9.2	1016.3	8.5	73.6	3.0
2.3	1020.1	6.7	84.3	3.2	11.9	1014.6	9.4	63.7	2.3
2.4	1016.0	6.7	95.3	4.1	7.9	1010.9	9.1	82.3	5.1
5.1	1017.1	6.6	81.3	3.7	7.1	1004.8	9.8	92.7	2.2
6.5	1014.5	7.8	74.0	4.9	6.6	1003.6	8.8	94.7	2.1
6.3	1012.4	7.6	80.0	3.5	6.2	1009.1	9.0	92.7	1.8
6.2	1013.7	7.9	79.3	4.3	5.7	1014.3	7.7	87.3	2.4
6.3	1014.9	7.8	83.3	3.5	6.4	1016.0	7.9	85.7	1.6
5.1	1012.4	8.1	82.3	3.4	9.1	1012.4	7.6	65.7	1.6
3.1	1007.2	6.8	92.0	3.1	8.8	1007.5	8.9	79.3	3.5
3.2	1011.9	5.9	89.0	2.8	8.6	1006.4	7.5	69.0	3.2
4.5	1013.5	6.8	77.3	2.7	7.3	1010.9	7.4	69.7	1.5
5.5	1012.7	7.1	80.7	2.9	11.2	1010.1	7.2	52.7	1.2
5.1	1013.4	7.0	77.7	3.2	11.2	1003.2	9.3	70.7	1.6
4.7	1014.9	7.3	78.3	4.5	8.7	1005.7	7.7	69.0	1.8
4.7	1015.5	7.1	83.0	3.8	8.2	1002.8	7.9	77.7	2.0
5.3	1018.8	7.1	72.3	3.6	5.8	1008.2	8.6	92.7	2.2
7.8	1013.3	10.1	92.7	2.1	8.4	1013.6	9.6	79.3	1.6

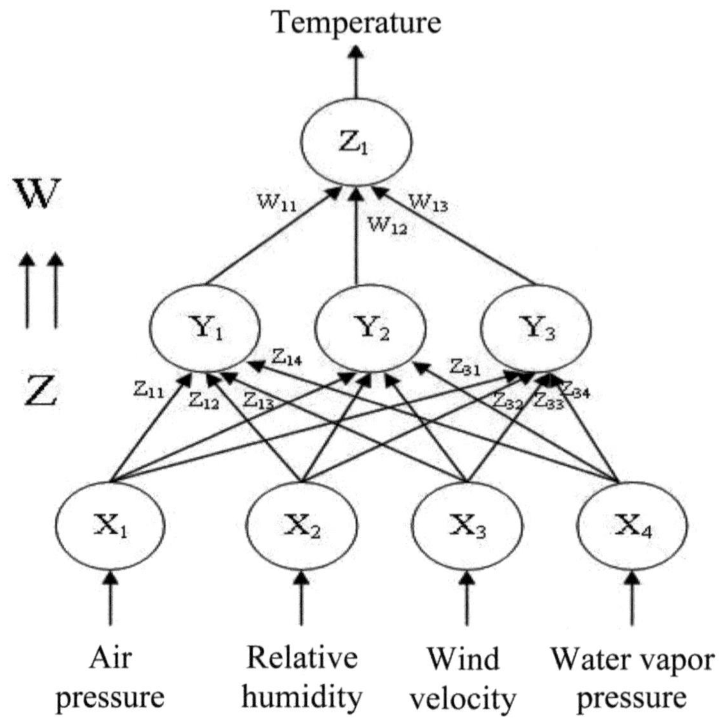

Fig. 7.67 Example ANN architecture

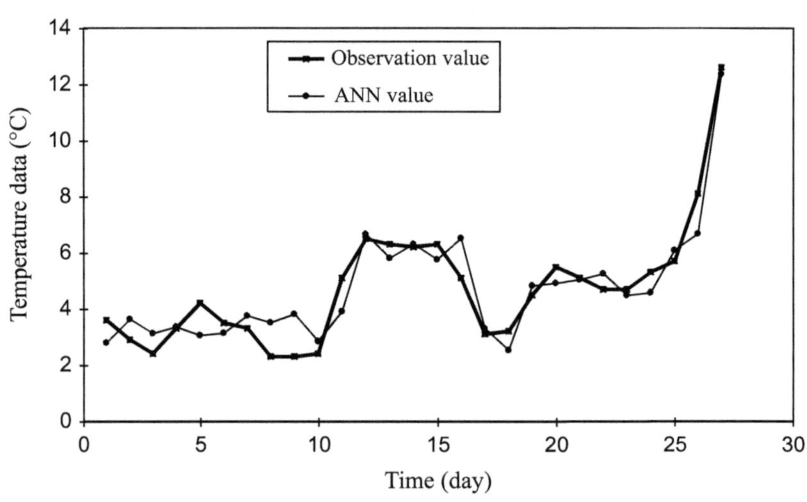

Fig. 7.68 Observation and ANN data variations

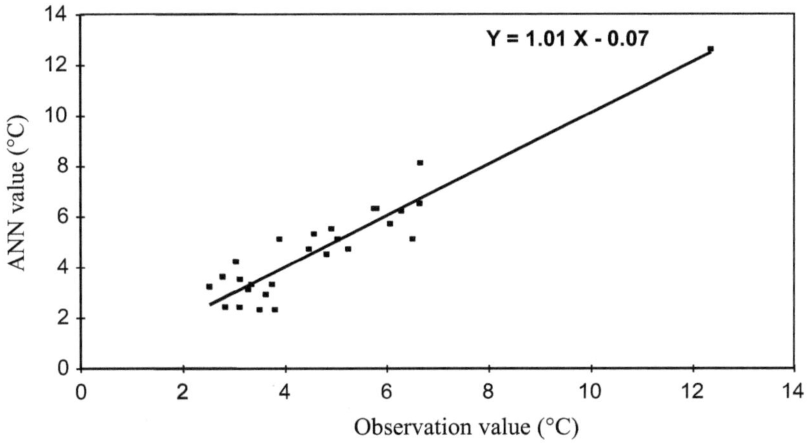

Fig. 7.69 Scatter diagram between the observation and ANN data

the training completion, the variables other than the temperature after March 28 are given to the ANN as input data. The output value for each input data set is the daily temperature value of that day. Thus, ANN values are obtained for the missing temperature data in the desired time interval. Figure 7.70 shows the values produced by the ANN model with the training set to complete the missing data.

In this figure, a result close to the graph in Fig. 7.68 is obtained. Although the general trend is maintained, the error values increase only at the maximum and minimum points. In addition, sharper transitions are observed at some turning points compared to the previous approach. This time, the average absolute error only for the test region is 0.64 °C. The scatter diagram of the actual versus ANN model temperature values is given in Fig. 7.71.

Likewise, with the help of ANN, it is possible to complete the missing information also spatially. For example, the generalization made above depending on the time series can be completed with the help of ANN by using the measurements at a set of irregularly located stations in an area.

7.23.2 *Classification Application*

Humankind's work on classification dates to ancient times, which was made for the first time to classify animals, plants, sounds, etc. Today, studies on the classification problem continue in different fields and in dimensions with numerically different methods. Intensive studies on classification have led to the introduction of many mathematical models. With the technological developments, classification studies provided help to machines and especially to machine learning technologies (Chap. 8). Machine classifications provide possibility to solve many problems that were difficult to do so before.

Many statistical methods are used to decompose different data groups. For example, two different event groups belonging to variables such as temperature,

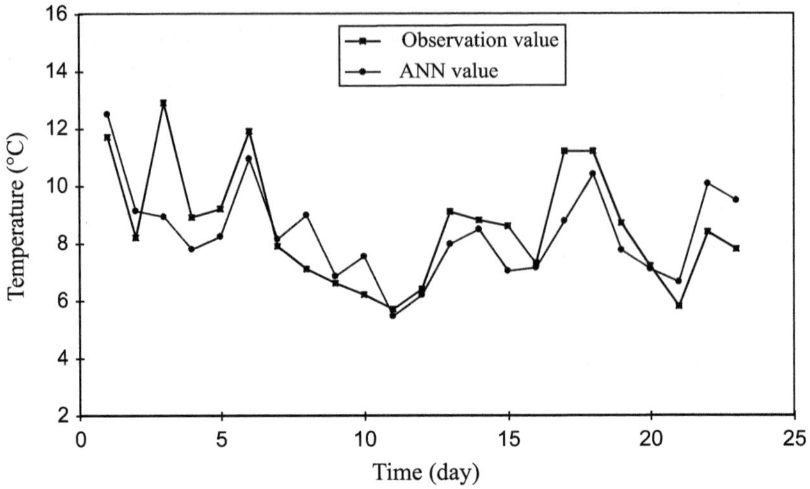

Fig. 7.70 Observation and ANN value variations

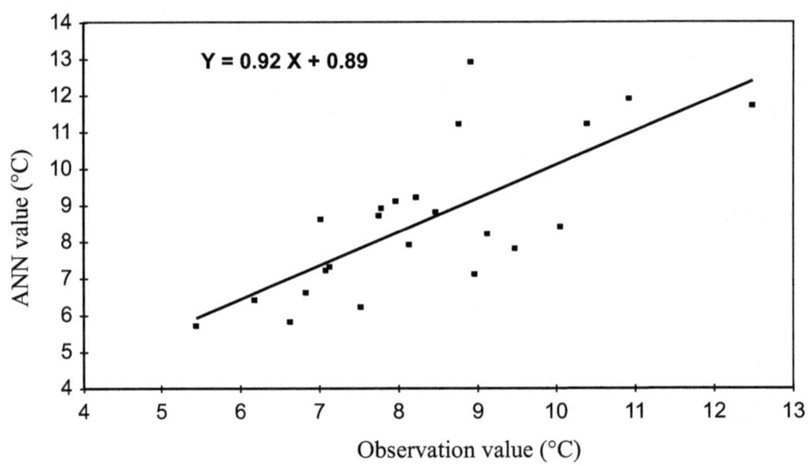

Fig. 7.71 Scatter diagram between the observation and ANN values

pressure, and humidity can be successfully separated with the help of classical or logistic regression analysis. Classification can be performed by means of a linear or nonlinear "separation function." With X_1 and X_2 being the variables in the study, a linear parser, P, can be written generally as:

$$P = b_o + b_1X_1 + b_2X_2 \tag{7.130}$$

Panofsky and Brier (1968) conducted a classification study using data on vertical velocity, dew point temperature depression, and precipitation. Of the 91 measurements made in Albany, New York, 28 had precipitation, and 63 had no precipitation

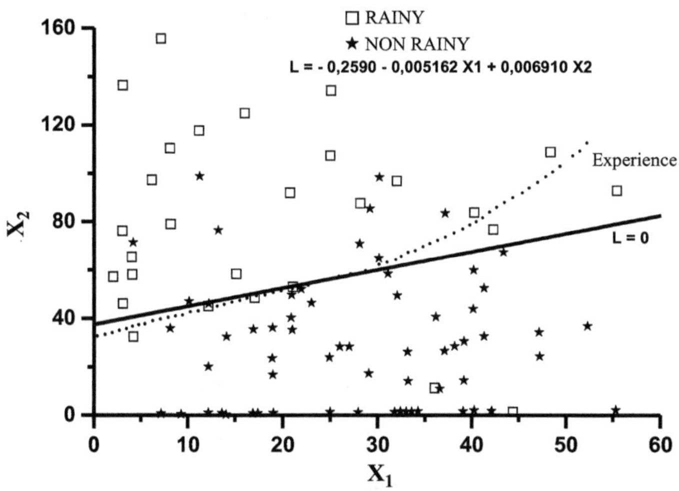

Fig. 7.72 Linear separations by L straight line

data. The scattering diagram of the data of vertical velocity, dew point temperature depression, and precipitation are shown in Fig. 7.72. By trying to obtain the lowest error rate, two different data groups, with and without precipitation are separated correctly.

$$P = -0.2590 - 0.005162X_1 + 0.006910X_2 \tag{7.131}$$

In response to any vertical velocity and dew point temperature depression values, it represents the rainy (and non-rainy) region for $P > 0$ ($P < 0$). As a result of the research, a total of 15 points are included in the wrong group; nine of them are in the rainy zone, and six of them are in the non-rainy zone.

Lippmann (1989) states that in classification problems, ANN is very different from statistical approaches and gives better results due to its features such as processing speed, trainability, and easy tool. Although the perceptron can only linearly separate two different classes, Tattersall et al. (1989) state that multilayer ANN models can successfully decompose more than two different classes. First, while the classification process is performed, only unclassified data is processed instead of processing all data. In addition, being able to follow the classification step by step is another privileged feature (Antognetti and Milutinavic 1991).

In the study, by using the same data, the classification problem is solved with the help of three-layer ANN architectural model as in Fig. 7.73. In the input layer, two nerve cells are used, namely, vertical velocity and dew point temperature depression. There are eight neurons in the hidden layer and a single neuron in the output layer. The sigmoid is chosen as the activity function for the cell outputs in the hidden layer and the output layer. The output value greater than 0.5 for each input pair is included in the rainy group, and the small output values are in the non-rainy group.

As a result, training is provided with 2000 iterations using the back propagation algorithm. Thus, values ranging from 0 to 1 are obtained for each vertical velocity

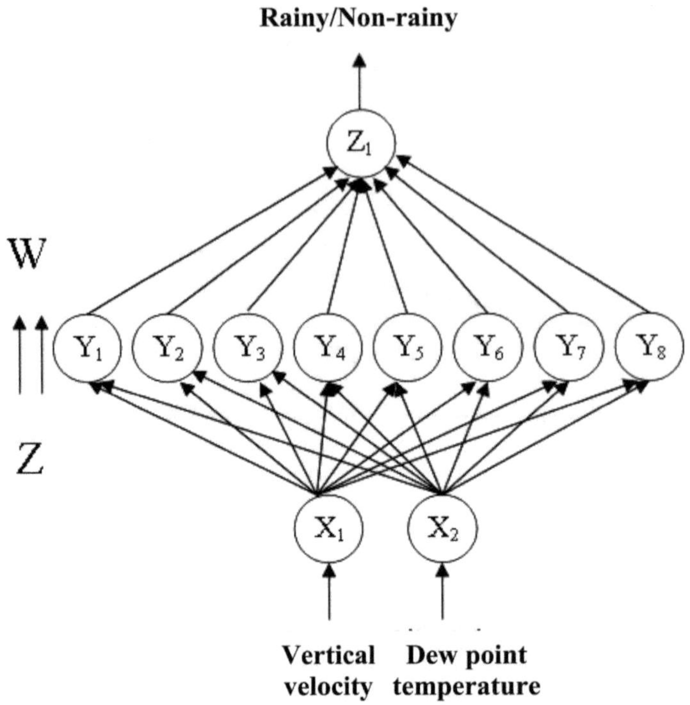

Fig. 7.73 ANN model for classification

and dew point temperature depression point. Considering the values at all points, contour maps are drawn for the study area, but only 0.5 contour is shown in Fig. 7.74. In the upper and lower parts, values greater than 0.5 are obtained.

In this way, two different data groups are separated with a 0.5 contour, with rainy part at the top and no-rainy at the bottom. As a result of this separation, the number of points in the wrong groups is reduced to 12, 10 of them are in the rainy region, and two of them in the non-rainy region. When the decomposition made with the straight line and the decomposition with the ANN is compared, the decomposition made with the ANN contains fewer erroneous points. The use of a nonlinear parser instead of a linear parser in the separation of data groups has been effective in this result. The use of a curvilinear parser can give better results, especially in data groups that cannot be separated linearly. Since nonlinear functions can be used at cell outputs in ANN, it becomes possible to separate data groups with a nonlinear parser.

7.23.3 Temperature Prediction Application

Atmospheric phenomena have a direct impact on some of our daily needs, such as clothing and transportation. Because of these needs, mankind has been wondering

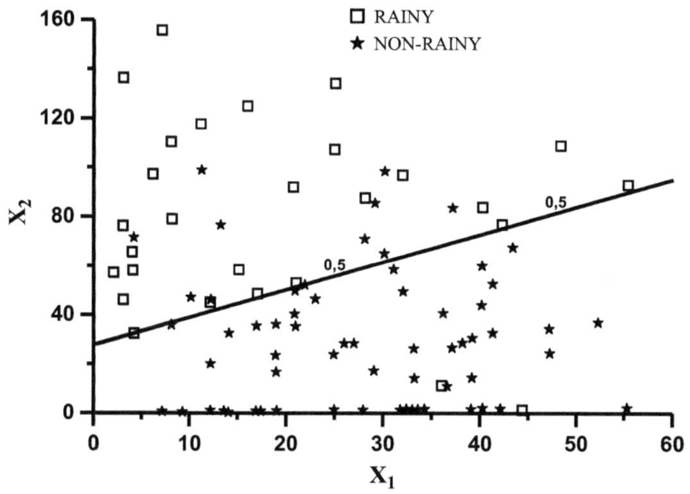

Fig. 7.74 ANN and curve separation

about the next steps of atmospheric events since ancient times. Therefore, the problem of predicting meteorological parameters has survived to the present day without losing its importance starting with the first humans. In the early days when it was not possible to make measurements, human beings were content only with observing the atmosphere. Therefore, the first predictions were made based on the experience gained as a result of these observations. With the acceleration of technological developments, many models and prediction methods have been developed based on numerical calculations.

In this study, it has been tried to predict the temperature variable, which is one of the meteorological parameters. In other words, it has been investigated how accurately the temperature value of the next day can be predicted with the help of the temperature values observed in the past days.

In the study, the temperature measurements are considered at 14:00 in the afternoon of May and June of 1966 belonging to the province of Ankara. The data for the 61 days are shown in Table 7.4.

Out of the 61-day data, 40 are reserved for training and remaining for testing. Using the training data, it is determined how many days ago data were needed in order to predict the data at the time of t + 1, and finally, it is tested on 21 test sets. In the application, a three-layer ANN architecture model is considered, as shown in Fig. 7.75.

Back propagation algorithm is used as the training algorithm. First, the interval number (N) is taken as 1 and the data at time t = 0 is used to predict the data at time t + 1. The model was trained by finding t = 0 as input data to the ANN model, and data at time t + 1 as output data. After the model is trained with 10,000 iterations, the retraining set as the test set is given as an input to the model. Thus, it is aimed to obtain how well the ANN generalizes between the data at t = 0 and (t + 1). For this, the mean absolute error value between the measurements and the ANN model output values is calculated with the help of the following expression where L is the lag number.

Table 7.4 Ankara temperature data

Sequence	Temperature (°C)	Sequence	Temperature (°C)	Sequence	Temperature (°C)	Sequence	Temperature (°C)
1.	18.4	17.	19.7	33.	26.6	49.	25.0
2.	19.7	18.	21.8	34.	31.0	50.	24.0
3.	16.2	19.	23.8	35.	24.8	51.	23.6
4.	14.2	20.	17.0	36.	21.1	52.	26.4
5.	17.4	21.	19.0	37.	24.2	53.	28.0
6.	21.0	22.	16.8	38.	26.8	54.	25.8
7.	22.4	23.	19.7	39.	27.6	55.	24.3
8.	23.4	24.	11.3	40.	27.0	56.	24.8
9.	23.1	25.	18.0	41.	23.8	57.	27.6
10.	18.8	26.	18.0	42.	24.6	58.	26.1
11.	18.4	27.	21.5	43.	19.2	59.	18.6
12.	23.6	28	24.0	44.	17.0	60.	24.6
13.	18.8	29.	22.4	45.	20.2	61.	30.0
14.	18.8	30.	21.8	46.	22.6		
15.	17.0	31.	21.4	47.	25.5		
16.	18.0	32.	23.4	48.	27.4		

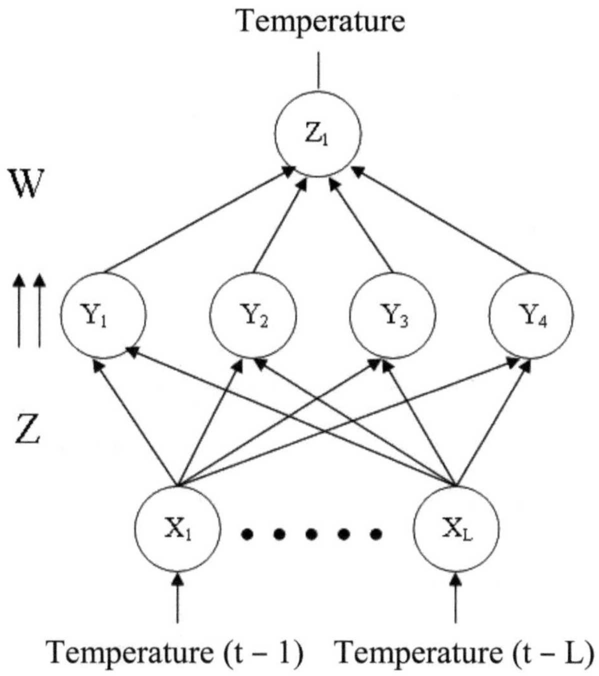

Fig. 7.75 ANN model for prediction

$$E_N = \frac{1}{40 - L} \sum_{i=L+1}^{40} \left| T^i_{Observation} - T^i_{ANN} \right| \tag{7.132}$$

The same operations are repeated for N = 2, 3, . . . , 7, and the mean absolute error value according to the number of intervals is obtained with the help of the Eq. 7.132. Figure 7.76 shows the oscillation between the number of intervals and the mean absolute error value.

It should also be noted that the number of cells in the input layer changes for each new value of N during training. For example, for N = 3, since a prediction will be made between the temperature value at (t + 1) and the values at t = 0, (t − 1) and (t − 2), three cells are used in the input layer. The average absolute error value decreases as the number of intervals increases, and at the fifth interval the average error value reaches the lowest value as 8.28, and the average error value starts to increase again with the increase in the number of intervals. Hence, ach of the values belonging to five steps ago should be used for prediction in order to predict the temperature value at the time t + 1.

After the interval number is determined as 5, the model is tested on 21 temperature values, which are previously reserved as the test set and not subjected to any processing. Here, for example, 36th, 37th, 38th, 39th, and 40th data are entered the ANN model in Fig. 7.77 in order to predict the 41st data.

Likewise, data of 37-days, 38-days, 39-days, 40-days, and 41-days, which are observed by measuring the true value, are used for the temperature prediction of the

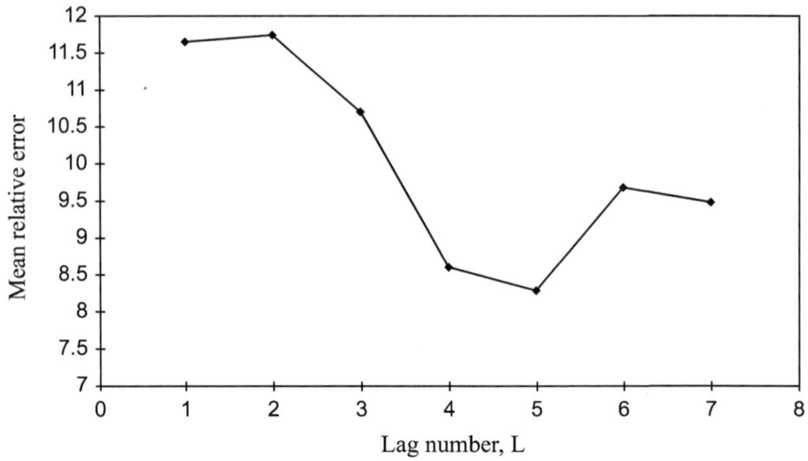

Fig. 7.76 Variation of mean relative error by lag number

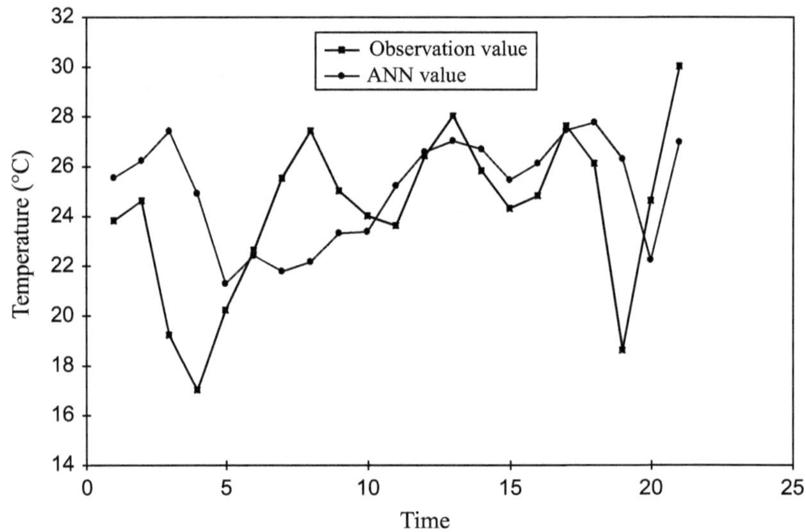

Fig. 7.77 Observations and ANN prediction variations

42-day. By repeating the processes, the forecasts for 21-days are calculated one by one. The results are close to the measurement values. The model and measurement patterns are close to each other.

Because of the calculations, the average absolute error amount between the real and ANN prediction values is 2.52. Considering that the trainer in the study is made with a small number of data such as 40 and the prediction is achieved using only the temperature values measured in the past, the results are reasonable.

7.24 Conclusions

In early engineering and technology applications, a model of an event, a behavior pattern like human acquired data is sought. Parallel process modeling methods are inspired by the brain working system and original and very shallow learning field artificial neural networks (ANN) methods have taken their place in the literature. Its specific properties are explained by regression and stochastic methods, which have been widely used until recent years and are still used in many parts of the world. There are no assumptions about the event or data at the outset. In this chapter, first, similarities between the two modeling types are explained by explaining the similar aspects of classical methods even if the reader not knows or does not know beforehand. In fact, it may be possible for a reader modeling according to classical systems to get used to ANNs and make applications in a short time. After a brief philosophical discussion of the presentation style of ANN model, explanations about the reasons and necessity of ANN model, considering its similarity to previous methods and the philosophical, logical and rationality rules. Efforts have been made to better understand it with the given applications. For the implementation of ANN, at least three layers must be established together with the cells, one being the input, the other the output, and the third hidden (intermediate) layer. For this, it is necessary to develop an architecture that will not model mathematical rules, but only the action and response variables that control the event and the reactions that may occur in it.

References

Alexander I, Norton H (1992) An introduction to neural computing. Chapman and Boundary Row, pp 137–142

Anderson JA (1983) Cognitive and psychological computation with neural models. IEEE Trans Syst Man Cybernet SMC-13(5):799–814

Antognetti P, Milutinavic V (1991) Neural networks, concepts, applications and implementation. Prentice-Hall, Englewood Cliffs

Baum EB, Haussler D (1989) What size net gives valid generalization? Neural Computation. For completeness, we reprint this full version here, with the kind permission of MIT Press

Bkum A (1992) Neural networks in C++. Wiley, New York

Carpenter GA, Grossberg S (1988) The ART of adaptive pattern recognition by a self-neural network. Computer (March):77–88

Cressman GP (1959) An operational objective analysis system. Month Weath Rev 87(10):367–374

Elman JL (1990) Finding structure in time. Cogn Sci 14:179–211

Freeman JA (1994) Simulating neural networks. Addison-Wesley Publishing Company, pp 8–13

Freeman WJ (1995) Mass action in the nervous system. Academic, New York

Goldberg DE (1989) Genetic algorithms. Addison-Wesley, Reading

Haykin S (1994) Neural networks: a comprehensive foundation. Prentice Hall PTR, Upper Saddle River

Hebb D (1949) The organization of behavior. Willey, New York

Hecht-Nielsen R (1990) Neurocomputing. Addison-Wesley, Reading

Hinton GE (1989) Connectionist learning procedures. Artif Intell 40:185–234

Hopfield JJ (1982) Neural networks and physical systems with emergent collective computational abilities. Proc Natl Acad Sci 79:2554–2558
Hsieh KR, Chen WT (1993) A neural network model which combines unsupervised and supervised learning. IEEE Trans Neural Netw 4:357–360
Kalman RE (1960) A new approach to linear filtering and prediction problems. Trans ASME Ser D J Basic Eng 82:35–45
Kohonen T (1982) The self-organized formation of topography correct feature maps. Biol Cybern 43:59–69
Kohonen T (1988) An introduction to neural computing. Neural Netw 1:3–6
Kosko B (1990) Unsupervised learning in noise. IEEE Trans Neural Netw 1(March):44–57
Kosko B (1991) Stochastic competitive learning. IEEE Trans Neural Netw 2(September):522–529
Kung SY (1993) Digital neural computing. PTR Prentice-Hall, Inc./Simond and Schoster Company, pp 1–5
Lee LS, Stoll HM, Tackitt MC (1989) Continuous-time optical neural associative memory. Opt Lett 14:162
Lee TC (1991) Structure level adaptation for artificial neural networks. Kluwer Academic Publishers, pp 3–10
Lippman R (1987) An introduction to computing with neural nets. IEEE ASSP Man 4:4–22
Lippmann RP (1989) Pattern classification using neural networks. IEEE Commun Magaz 27 (November):47–64
Matheron G (1963) Principles of geostatistics. Econ Geogr 58:1246–1266
McCullogh WS, Pitts WA (1943) A logical calculus of the ideas immanent in nervous activity. Bull Math Biophys 5:115–133
Mehrotra K, Mohan CK, Ranka S (1997) Elements of artificial neural networks. MIT Press, Cambridge, MA
Minsky M, Paperts S (1969) Perceptrons. MIT Press, Cambridge, MA
Morgan DP, Scofield CL (1991) Neural networks and speech processing. Kluwer Academic Publishers, pp 141–145
Özmetel E (2003) Yapay Sinir Ağları (Artificial neural networks). Papatya Yayıncılık Eğitim (in Turkish), 232 sayfa
Panofsky HA, Brier GW (1968) Some applications of statistics to meteorology. Earth and Mineral Sciences Continuing Education, College of Earth and Mineral Sciences, University Park
Rosenblatt F (1958) The perceptron: a probabilistic model for information storage and organization in the brain. Phsychol Rev 65:386–388
Rumelhart DE, Hinton GE, Williams RJ (1986) Learning representations by back propagation error. Nature 32:533–536
Şen Z (1989) Cumulative semivariogram model of regionalized variables. Int J Math Geol 21:891
Şen Z (1998) Average areal precipitation by percentage weighted polygon method. J Hydrol Eng ASCE 3:69–74
Şen Z (2004) Genetik Algoritmalar ve En İyileme Yöntemleri (Genetic algorithms and optimization methods). Su Vakfı Yayınları (in Turkish), 142 pages
Simpson P (1992) Foundations of neural networks. In: Sanchez-Sinencio E (ed) Artificial neural networks. IEEE Press, pp 1–25
Smith BC (1982) Linguistic and Computational Semantics. Xerox, Palo Alto, Ca.
Sönmez İ, Şen Z (1997) Yapay sinir ağları yardımıyla debi öngörüsü (Discharge prediction by artificial neural networks). Meteorolojik Karakterli Doğal Afetler Sempozyumu, Ankara, pp 445–457 (in Turkish)
Tattersall GD, Foster S, Linford P (1989) Single layer look-up perceptrons. In: First IEE international conference on artificial neural networks, Conference Publication no 313
Verbos PJ (1994) The roots of backpropagation. Wiley-Interscience Publications, New York, pp 169–173
Windrow G, Hoff ME (1960) Adaptive switching circuit, IRE Western electronic show and convection. Convection Record, pp 96–104

Chapter 8
Machine Learning

8.1 General

Computer program writings of what one thinks of about a phenomenon provide the basis of machine learning (ML) principles that try to deduce the most distinctive internal characteristics of large datasets in the forms of features, labels, clusters, or classifications purposes. Deductive inference system of ML is like AI deductions. Distinctions of ML approaches are in their representations and adaptations to reach the final label output products. The property of storage of learned information in knowledge representation form is an inductive hypothesis, which is in the form of a model. The two main duties of ML are the task of classification and regression for prediction of the target from input datasets in the forms of features and labels. If the data type is in the forms of nominal or ordinal format, then classification procedure is the convenient ML procedure; otherwise, regression analysis is the role of ML procedure. Both classification and regression procedures are supervised learning aspects based on a previously labeled set of training alternatives. The suggested model structure learns the best underlying pattern within the large dataset. Furthermore, clustering procedure is an unsupervised learning task, which puts into groups of the similar objects. Hence, there are similar and dissimilar clusters. This is quite equivalent to cluster analysis in statistics.

Under the umbrella of AI are ML, GA, ANN, and deep learning (DL). It is possible to state that shallow learning is a sub-branch from machine learning and machine learning is a sub-branch of deep learning. It is not possible to make crisp distinction between the ML and DL; ML has more algorithms than the other. ML is an attractive scientific research methodology, but its use is not straightforward without having active shallow information given in Chaps. 1, 2, 3, 4 and 5.

Computational algorithms provide the bases of ML, which helps to emulate human natural intelligence (NI) through vivid education system or one's own ambiguous desire to work with large datasets. ML techniques are successfully usable

Z. Şen, *Shallow and Deep Learning Principles*,
https://doi.org/10.1007/978-3-031-29555-3_8

in diverse disciplines such as computer vision, pattern recognition, space research, engineering, entertainment, medicine, finance, sociology, etc.

Carbonell et al. (1983) stated that "learning is many faceted phenomena, which includes acquisition of new declarative knowledge, the development of motor and cognitive skills through instruction or practice, the organization of new knowledge, effective representation, the discovery of new facts and theories through observations and experiments. Since the inception of the new computer era, researches have been striving to implement such capabilities in computers. Solving this problem remains as a most challenging and fascinating long-range goal in AI works. The study and computer modeling of learning processes in their multiple manifestations constitute the subject of machine learning." Among AI procedures, ML is dependent on computer programming science based on the large dataset usage and imitation of human algorithms by machines even approximately, which can be improved gradually toward better accuracy.

At its most basic level, ML refers to any type of computer program that can "learn" by itself without explicitly programmable by a human. The phrase (and its underlying idea) has its origins decades ago—all the way to Alan Turing's 1950 paper "Computing Machinery and Intelligence," which featured a section on his famous "Learning Machine" that could fool a human into believing that it's real. Today, ML is a widely used term that encompasses many types of programs that one can run across big datasets concerning analytics and data mining. At the end of the day, the "brains" powering most predictive programs—including spam filters, product recommenders, and fraud detectors—are ML algorithms.

Data scientists are expected to be familiar with the differences between supervised and unsupervised ML procedures as well as ensemble modeling using a combination of techniques and semi-supervised learning means, which combine supervised and unsupervised approaches. At its most basic level, ML refers to any type of computer program that can "learn" by itself without having to be explicitly programmed by a human. ML technology is not new, because it is accumulation of several years' works, but it became attractive as a result of AI and digital innovation studies. By time, it became quite different from AI and is bound to play significant role in the future with DL procedures.

In general, any data point is characterized by various properties in the groups of features and labels. Although identification of features by ML automatization is comparatively easy than labels, whose identifications cannot be achieved without human expert inferences. ML seeks prediction by learning from the features of especially two-dimensional large datasets. ML procedures integrate available large datasets with model and loss functions through successive trial and error procedure until the most suitable data representation is reached through computer software applications. It is necessary to have a continuous adaptation of a hypothesis (model) about the phenomenon, where the data come from. The final product is referred to as the label of a data point. A hypothesis map is fitted to low level properties of the data, which is referred to as data feature. The researcher should be aware of the possible hypothesis maps for the convenience to the datasets. The measure of the correspondence between a hypothesis map and the data is the loss function, which should have the least value for the trial and error procedure ending. As will be explained later in this chapter, there are different types of loss function for various matching properties between the hypothetical map and the available dataset. The best result can be achieved with the most harmonious combination improvement of dataset source with model and loss function. Two very frequently used methodologies in the ML procedure are the DL procedures as will be explained in the next chapter and linear or nonlinear regression models. The following figure shows the harmonic consistent achievement of dataset with model and loss function.

It is obvious from Fig. 8.1 that there are two trial and error cycles, one for the single model, which might be thought by the expert researcher that it is the most suitable one in the light of similar previous experiences, so the first cycle solves the problem at hand. If there are several model possibilities, then the ML procedure revolves with multitude of models and finally comes as the best result among alternatives choices with overall minimum loss function value.

Scientific principle of "trial and error" plays important role in ML methods' implements. The methods validate and refine continuously a model based on the loss function minimization that generates data predictions in the best match with observations (dataset, measurements).

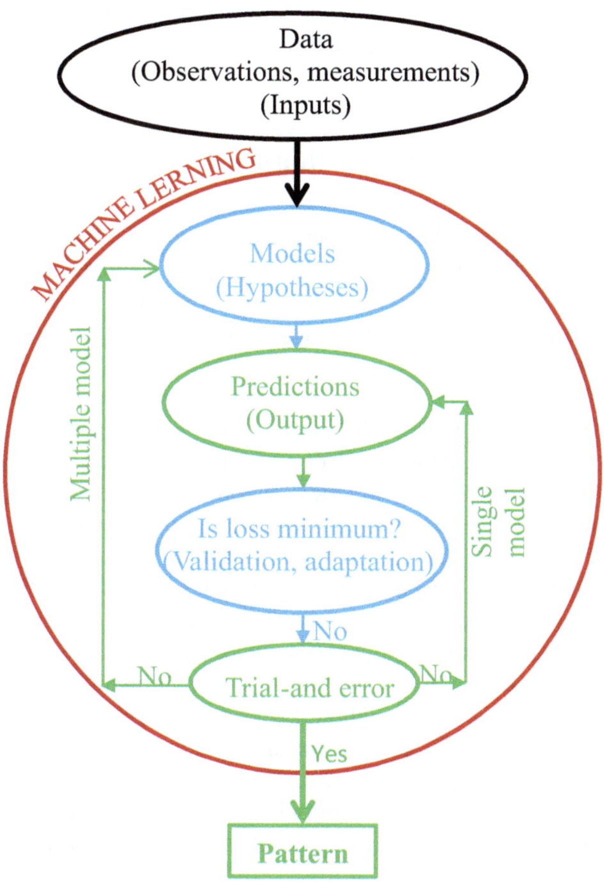

Fig. 8.1 ML combines three main components: data, model and loss function

Under the ML umbrella, there is basically a set of shallow learning techniques as explained in Chaps. 1, 2, 3, 4 and 5, which concentrate on tools and techniques that help for computers to learn and after various training procedures they adapt on their own.

Shallow learning methodologies in terms of logical, uncertainty and mathematical bases, shallow ANN, and GA procedures are explained in the previous chapters. DL techniques became more popular in recent years as a result of high-performance facilities provided by computers. DL procedures can deal with unstructured large number of features to identify from a given datasets with higher power and edibility. The ability of AI and DL methodologies is to pass the dataset through several hidden layers each of which is capable for extraction of features progressively and their passage to the next layer. Initial layers extract shallow level features, which are then combined to form a complete representation for better feature identifications.

The main purpose of this chapter is to expose how it works, what are its connections with shallow and deep learning processes. ML is dependent on a set of algorithms mentioned in the previous chapters including numerical data and developed along many years due to experience enhancements in addition to explicit computer programming language advancements. To perform certain tasks, ML procedures requite different but convenient algorithm to the problem at hand for dataset structure pattern detection purposes.

8.2 Machine Learning-Related Topics

There are varieties of numerical and verbal methodologies within the ML application possibilities that are based on the shallow learning procedures, majority of which are explained in the previous chapters. It is assumed that the reader is familiar with many of these basic approaches through the undergraduate and possibly postgraduate training periods.

Learning is an effective training procedure for any individual ambitious to develop cognitive skills based on rationally digested instructions and their practical applications, new knowledge acquisition in a declarative manner, organization, and presentation of new practical, theoretical, or patent gainable discoveries of new and enhanceable shallow and especially deep aspects that are automatically useful for society, in general, and individuals in particular. With the advent of computer hardware and especially software possibilities, such activities gained continuously increasing acceleration since 1950. ML provides a domain for challenge fascinating problem solutions automatically based on measurements in the form of datasets, observations that provide verbal information and personal or teamwork experiences with the use of AI methodologies as already explained in the previous chapters.

Whatever are the shallow or DL knowledge contents, the final ML product boils down to computer software, which is the most essential topical knowledge used for ML and AI problem solutions. In this manner, old day voluminous hand programming problem-solving procedures became more complex, and thus, human intelligence gained fascination challenging characteristics to transit from shallow learning to DL procedures.

With the advent of ML, originally engineering-oriented procedures entered numerous topics in divergent disciplines, such as engineering, medicine, biology, psychology, human learning, and social sciences. To produce acceptable productions in these areas, it is necessary to care for knowledge acquisition that leads to skill refinement in any concerned topic. Thus, physical world behavior problems can be predicted with better sensitivity and put into functionality for the service of men. The transition from shallow learning domain into DL is possible by polishing rationally present level of information toward more accurate, optimum, and beneficial end products.

In today's educational system, whatever is the topic, there are variety of learning procedures as mentioned at different chapters of this book including rote, instructive, analogical, exemplary, observative, experimental, experience, discovery, and critical review.

As for the analytical and empirical procedures, among ML techniques are parameters in algebraic and scientific mathematical expressions, which may require linear algebra, computer science, optimization, simulation, information content, probability, statistics, decision tree, formal linguistical grammar, graphs, figures, flow charts, science philosophical and rational logical principles, frames, schemes, taxonomy, labels, scatter diagrams, datasets, conceptual analytical and numerical models, and basic calculus principles. Even though they are interconnected, ML has different aspects than AI, in that ML is comparatively easier than deep AI methodologies, because it is the predecessor. The far most difference is that ML provides mathematical layouts as mentioned in Chaps. 3, 4 and 5, whereas AI is for mimicking and imitating human behaviors with the help of ML. Additionally, ML principles support AI procedures to communicate with human, understand language, and conduct conversation. In this manner, AI continuously improves with experience accumulation based on the background of ML. Since with the usage of computers in numerical calculation starting from 1950 onward, human began to learn from available dataset the behavior of phenomenon concerned by means of various mathematical, probabilistic, statistical, and AI methodologies. At this stage, the important role of fuzzy logic principles and rules must not be forgotten to communicate with computers, and hence, machines. Human learning from dataset was at the shallow learning levels, but AI raised this to higher levels tremendously by means of DL procedures with ML solution algorithms.

ML technology might be very intelligent, but its final productions may not be acceptable perfectly. Human intervention helps to improve learning methodology for better pattern recognitions. For example, the learning methodology may not overcome some parts of the dataset and may get stacked, but human intervention supports the algorithm for better recognition, and hence, at the end, more reliable output pattern can be obtained. This point implies that ML technology cannot understand as the human understanding level. Interference of human activation assistance to ML techniques improves the precision of final label identification, and in this manner, ML algorithms become more self-aware of the training procedure. Hence, intimidating instance can be avoided by human interferences. One must not expect perfect performance from ML-based robots, because the best combination is achievable when robotic performances are coupled with human aids. Human assistance improves the accuracy of final products by data labeling for ML model feed or by correction of inaccuracies in predictions. Thus, cooperation of ML and human intelligence increases the efficiency of final products in an adaptive manner.

8.3 Historical Backgrounds of ML and AI Couple

ML has started with algorithm descriptions that motivated the desire to solve rather difficult and complex problems by quite simple automation principles with were initially concerned with pattern, number, and letter recognitions. Along the same direction, time series temporal, spatial or spatiotemporal simulations, and then onward future prediction methodologies all supported with innovative ML technique evolutions. In the past, many methodological algorithms and preliminary computer programmers and scientists were not aware of ML terminology.

In 1959, the Turing test by Alan Turing brought a wonderful idea as to whether computers were capable to simulate human intelligence. Samuel (1953) wrote the first computer program; hence, ML implementation started to enter scientific literature. By playing the same game on the computers, the relevant algorithm improved the communication between human and machine, and hence, winning moves are identified. Rosenblatt (1958) proposed the first neural network computer software design by considering human mind simulation possibility. The nearest neighbor algorithm facilitated rather primitive but illuminating basic pattern recognition possibility in computers. Later, Dejong (1981) proposed explanation-based learning procedural principles, which allowed computers to train and test data, and, hence, suggestion of dataset analyses and of insignificant data. About 4 years later this time, rather than individual researchers, a team of ML research group, namely, AT&T (1985), made a series of ML meetings, which dispersed early ML representation and further awareness among the researchers with the support of experts' systems enhancing the machine and human communications through computer programs. In this way, real-world problems rather than theoretical approaches took place for characterization and identification of hidden patterns from the noisy datasets. The first time in speech recognition, Cox et al. (2000) proposed speech recognition automatization, where Markov and especially hidden Markov models played initial role. Vapnik (1998) proposed statistical learning theory including support vector machine technique, which provided large-scale classification predictions. Haffner (1998) and researches from AT&A team proposed the first form of convolution neural network (CNN) architecture for speedy pattern recognition works with large data availability (Chap. 9). Subsequently of all the abovementioned developments, until 2005 started to be mentioned under the name of "deep learning." In the meantime, decision trees methodology treated unstructured data applications. In 2006, DL concept promoted significantly with the power and accuracy increments in neural networks. In 2011, deep neural networks in addition to a set of new algorithms provided possibility of training a model after trial and error study of numerous examples, which were superior over all well-known previous techniques.

The ML model procedures are for imitation, generalization, and prediction of large datasets. There are a variety of models, algorithms, and techniques that are used in ML procedures in order to reach at a desirable goal. These procedures learn from datasets for pattern recognition. ML is not for fitting a given pattern to the dataset, but identification of the best pattern even though there are some uncertainties in the dataset structure. There is not a certain concrete rule or methodology for ML procedures, but the researcher must choose the most convenient model and algorithm depending on the purpose of problem solution. In such a decision, the type, number, and extent of the dataset play primary role in addition to the supervised or unsupervised structure of the dataset. In case of likely pattern or event identification problems, supervised learning is convenient for solution in human-labeled data. This is the case when input and output data are known. For this purpose, the algorithm learns certain features of input dataset from output data, and after several forward feeds and back propagation iterations, the algorithm identifies the best solution within practically acceptable error limits, which are in practical studies either $\pm 5\%$ or $\pm 10\%$. One tries to reduce the error percentage after several trials. For example, classification into known groups is achievable by supervised learning after trying to fit the output to input approximately. For learning, the algorithm must be informed about possible image, text, or speech. Increase in the number of classes causes thousands of trial to get acceptable accuracy. On the other, unsupervised learning works with unlabeled data, where the machine identifies a set of patterns without any priory information. Unsupervised training works well in case of clustering to find similar groups from the dataset.

Powerful learning systems can be established through deep neural networks (DNN), which are successors of ANN (Chap. 6). They add several hidden layers for immediate representation extraction. These networks entered the DL scene especially after 2010 due to the parallel hardware structure appearance in addition to easy open source software availability (Sze et al. 2017). Furthermore, recurrent neural networks (RNN) provide sending feedback signals to each hidden layer in the architectural construction. Convolution neural networks (CNNs) are suitable for visual or image recognition purposes as feed forward ANN alternatives (Chap. 9).

8.4 Machine Learning Future

Although there are many futuristic development possibilities in ML and AI, they depend on overall methodological and algorithmic procedures that are available at present time. There is suspicion about the smartness of ML systems. The level of smartness can be improved by time. Improved dataset structural perceptions by ML techniques are expected toward more accurate and precise predictions and pattern identifications. In the meantime, more expertise increments are also expected, which will add to the better organization of ML techniques with human aid. Various data types and larger data possibilities are bound to lead for better algorithmic alternatives and accuracy improvements.

In order to gain expertise in ML, a strong foundation, and to achieve ML in the best possible manner, the following five steps must be completed intellectually:

1. *Science philosophy and logic:* These are the foundation key stones for appreciation of any problem in terms of linguistical prepositions and basic subjects for rational approximate reasoning principles (Chaps. 1, 2 and 3).
2. *Shallow learning skills:* Mathematical, probabilistic, and statistical certainty and uncertainty principles must be either established in the light of previous step by encountering with their symbolic presentations, and then their philosophical and logical backgrounds must be translated into rationally intelligent statements (Chaps. 4 and 5),
3. *Computer coding skills:* Knowledge and information obtained from the previous steps are for intra-intelligence of human beings, but their speedy digestions by computers require software programming abilities. Such abilities do not provide competence in software writing, but also translation of any software of interest to human communication and vice versa. This stage is basically for communication with machines, which are computers at large.
4. *Machine learning theory:* ML is an intensive mathematical discipline, so if one plans to modify ML models or generate new ones from scratch, familiarity with basic mathematical concepts is essential for the process (Chap. 5). Knowing the basics of ML theory provides a foundation to build on and helps one troubleshoot when something goes wrong.
5. *Project design and road map:* Gaining hands-on experience with ML is the best way to test one's knowledge, so not to be afraid to dive in early with a simple collaboration or tutorial to get some practice.

ML methodologies are active tools in many social, art, medicine, finance, science, and engineering topics. As mentioned earlier, there are three components in harmony for the success of ML methods, which are data, method, and uncertainty as loss. Their successful combinations are possible by trial-and-error successive activations in improvement with computationally efficient calculations. It is essential to know about the data generation mechanism for continuous adaptation to arrive at the desired goal. There is relevant hypothesis in the ML procedures to obtain reliable and rational predictions for the phenomenon concerned. The method part of ML is already explained in several previous chapters as shallow learning procedures, and the loss function will be explained in Chap. 9. Hence, here, only dataset part is explained briefly.

8.4.1 Dataset

The most important input element of the ML is dataset; in general, they are in the form of two dimensional and full of individual data points in the form of pixels as atomic units, each one of which contain information different from others. Data points may appear as time series recorded by sensors, text, photographs, pictures, random variables, maps, forests, and sceneries and alike. In different disciplines, different datasets are available. For example, for image processing works, each data point, that is, pixels, is considered as images. As a preliminary approach, many ML methods provide estimates for a quantity of interest simply as a prediction or forecast by averaging a certain area data points by overlapping or nonoverlapping moving window of certain dimensions such as 2 × 2, 3 × 3, 5 × 5, or in general, nxn. These are referred to as sample size or subsample size within the whole dataset. It is well known from statistics that the more the data, the better are the averages. In most applications, one may not have access to every simple microscopic data points. Different groups of data points have different data properties. These are either features or labels. Features are low-level properties of a data point, which can be either measured or computed easily by some automotive means. Features are referred to by various terminological keywords such as input variable, independent variable, explanatory variable, and covariate. On the contrary, labels are high-level data point properties with some quality of request. For their determination, expert human interference is necessary. Some synonyms for label data points are response variable, target, and output variable. There is no clear-cut distinction between features and labels; they are fuzzy with some overlapping parts. For this reason, the same data point may be used for features, whereas as labels for other modeling applications. The fuzziness between the features and labels becomes obvious in cases of missing data. In general, each data point may be characterized by several properties. Each property is candidate to be used as a feature. However, some of the properties may be missing at some data points, and hence, there is not completeness of each property for each data point. The data points with missing property may be defined as label points. These properties may be defined from the feature dataset points that are available for the whole data points. Determination of missing property values is referred to as imputation (Rogier et al. 2006). In image processing as a result of hardware failures, some of the pixels may have corrupted or missing data. It is possible to estimate the color intensity of such a pixel from neighboring pixels. In some cases, a patch of data field may have corrupted or missing property; it is then possible to define each data pixel in the patch and, finally, adapt central pixel as the label of such a patch.

8.5 Uncertainty Sources and Calculation Methods

Estimating and reducing as much as possible the uncertainties in any dataset is an important issue. These should be made for the input dataset and the general trends in them, and calculations should be made with reliable methodological approaches. The following points gain importance in uncertainty studies.

1. The uncertainties of each variable data must be determined separately.
2. Calculation of the uncertainty that may arise from all of them if more than one variable is used.
3. General trends in the data (identification of uncertainties in trend components).

4. Determining the uncertainties in order of importance and making improvements in data collection studies. It is possible to have uncertainties other than those that can be excluded statistically. For example, these include omissions, double spellings or enumerations, omissions, or ambiguity in terms. These should also be considered in modeling.

After excluding unreliable data outside of the 95% confidence interval, the theoretical probability density function (PDF) best matches the dataset. Before using data in a model or calculations, it is useful to learn about its types as for stationary, homogeneous, independent, homoscedastic, trend free, or not. Among the main sources of uncertainty, the following points are worth to consider:

1. *Uncertainty of lack of integrity:* This is because a measurement method related to the subject of interest has not been developed or a process for its functioning has not yet been defined. Therefore, there is an incomplete understanding of the dataset. As a result, a biased view may arise, and the researcher may remain in a completely random ambiguity.
2. *Model uncertainty:* Models can often be complex for better representation of the dataset at hand. Biased or random uncertainties regarding PDF calculations and deductions can arise as a result of a variety of reasons:

   (a) Models often approximate a simplified operating mechanism of real structures and are therefore incomplete. Among these, time and/or space features cannot be represented exactly, the desired resolution cannot be achieved, and the stability situations cannot be obtained in the numerical solution of complex equations.
   (b) Since each model works with estimations (interpolation) in the data ranges of the input variables, approximate values are obtained according to the accepted method in calculating the intermediate values, and therefore, a source of uncertainty arises.
   (c) Uncertainties arise because models often work on external estimates (extrapolation) or extensions (projection), and their validity cannot be checked outside the dataset range.
   (d) There is another source of uncertainty, since different results can be obtained using different formulations in a mathematical model.
   (e) The use of some parameters related to the model inputs by making estimations with a limited number of data reveals other uncertainties in addition to the model uncertainties. For example, the number of data plays a significant role in the statistical parameter calculations.

3. *Data availability:* In some problematic approaches, dataset may not be available for a problem. In this case, it is possible to use data found elsewhere in similar situations. For this, internal (interpolation) or external (extrapolation) calculation methods can be used with another sort of uncertainty.
4. *Lack of representative data:* There is another uncertainty source, when there is dataset that does not fully comply with local requirements. The best example of this is that if a facility has data in full operation, this data does not match the data

in that facility's start-up (warm-up time) or deceleration states. In these cases, there is uncertainty in the form of bias.

5. *Measurement errors:* These errors can be regular (systematic) or random. These can occur in measurement, recording, and communication situations. For example, when measurements are made according to a device, replacing that device with more sensitive alternatives gives rise to another uncertainty source between the previous measurements and the new ones. The use of methods based on approximate assumptions about measurements or their processing also includes uncertainty.
6. *Statistical random sampling error:* A finite dataset obtained is a part of its whole (population), and the arithmetic mean becomes constant as the number of data increases, but it requires more data to fix the variance. Therefore, an uncertainty type arises due to the data number limit. Here, it is also necessary to understand the difference between volatility and uncertainty. The longer the number of data, the more constant values are achieved in the calculation of the parameters, thus eliminating the uncertainty in the parameter estimations, but variability is always present in the datasets.
7. *Incorrect registration or misclassification:* The uncertainty here includes data based on incomplete, inconsistent, or incorrect definitions. Such uncertainties cause to biased results.
8. *Missing data:* Measurements may not be possible for sometimes or locations in data series for different reasons. In this case, estimations are tried to be made by applying some methods to complete the missing data. Among these methods, one can mention the following types.

   (a) If only one data is missing, the arithmetic mean of two adjacent data is taken.
   (b) In case of consecutively more than two missing data:

      (i) After obtaining the theoretical PDF of the dataset at hand, random numbers are generated according to that PDF and the missing data filling process is completed in this manner by generation of random numbers,
      (ii) In case of interdependent datasets, missing data are filled after the sequential behavior patterns of the data by considering the Markov or auto-regressive-integrated-moving-average (ARIMA) stochastic processes.

8.5.1 Reduction of Uncertainties

Before applying a model, the above-described uncertainties in the datasets should be reduced as much as possible for better representation. Priority should be given to variables that are effective in the representation of the studied event, and it would be appropriate to neglect some of the less influential effects, such as those that have an impact of less than 5% in applied studies. It is recommended to pay attention to the following points in reducing uncertainties.

1. *Further development of concepts:* Efforts should be made to eliminate the ambiguity of the effective variables in the structure of the model with better concepts. For example, working with refined temporal or low-resolution spatial data hides detailed concepts, features, labels, etc.
2. *Improvement of the model:* It is useful to try and improve the model structure and the parameters in order to eliminate regular (systematic) and random uncertainties as much as possible. Thus, understanding of the modeling philosophical and logical backgrounds leads to more detailed insight possibilities, and thus, uncertainties are greatly reduced,
3. *Representational improvement:* There are benefits to doing detailed and piecemeal studies for this purpose. For example, it would be beneficial to make such improvements in image processing, classification, regression and clustering prediction, simulation, and identification studies. For example, uncertainty can be reduced further by making continuous measurements to reduce sampling errors and gain additional insight about the event and its modeling structure improvement.
4. *Use of more precise measurement methods:* By avoiding simple assumptions and using more precise measurement methods and tools, a significant reduction in measurement errors can be achieved,
5. *Collecting more measurement data:* It is beneficial to increase the number of data in order to reduce the uncertainties in random sampling. Thus, bias and random uncertainties are reduced,
6. *To eliminate the risk of deviation uncertainty:* The measuring devices used for this should be placed in well-represented positions, and their settings should be checked regularly.
7. *Improving the state of knowledge:* In measuring and examining an event and its dataset, trying to improve the missing states can be reduced by a better understanding of the event and its parts.

8.5.2 Probability Density Functions (PDF)

The following points gain importance in order to make the necessary inferences by examining the behavior patterns of the datasets, which forms the basis of the modeling studies.

1. Determination of arithmetic mean, variance, standard deviation, and skewness coefficients as statistical parameters for dataset general behavior appreciation.
2. Determination of internal dependence (correlation) coefficients since the dataset is in the form 1D temporal series or 2D spatial image, map, feature, and pattern forms. Thus, more reliable statistical parameter values can be determined from the data by increasing a limited number of simulations according to Markov or ARIMA stochastic processes (Chap. 4), if necessary.
3. In order to perform the risk and reliability analysis, the theoretical PDF of the dataset must be identified and examined concerning its location, scale, and shape parameters.

Each of these points can be applied to any dataset structure that may contain uncertainty. It is necessary to determine the theoretical PDFs, especially in terms of cumulative distribution functions (CDFs) since they provide visual and numerical floor for identification of reliable and unreliable data values. In practice, normal (Gaussian), log-normal, two-parameter Gamma, general extreme value, and Weibull CDFs are in use for the reliability analyses. In general, given datasets do not fit normal PDF. In Appendix, the Matlab program is program is given for identification of the most suitable PDF to given dataset. Mathematical expressions and explanations of these PDFs are available in different textbooks (Feller 1967; Kendall and Stuart 1974; Benjamin and Cornell 1970). It is recommended to follow the steps below for theoretical PDF calculations.

1. First, simply the dataset is ranked in ascending order, and the risk, R, is calculated according to the following expression:

$$R = \frac{r}{n+1} (0 < R < 1) \tag{8.1}$$

   Here, r is the rank of each data, and n is the number of data. It must be remembered at this stage that risk has synonymous meaning with exceedance probability.
2. Plot ranked dataset versus risk values on the vertical axis and, hence, a decreasing (or increasing) scatter of points are obtained.
3. The best fitting theoretical PDF to this scatter diagram is identified by minimization of errors, for example, by least square methodology.
4. Based on this theoretical CDF, the largest possible dataset values are obtained for a set of risk levels.

Execution of these steps yields to a representative theoretical CDF and the scatter points as in Fig. 8.2 with a set of risk level values in the dataset.

8.5.3 Confidence Interval

It is recommendable that datasets within the 95% confidence interval are reliable. This means that 2.5% areas at the right and left sides of Fig. 8.2 present outliers. For example, the consideration of reliable dataset with exclusion of outliers may be one of the feature dataset input into the model considered. When the confidence interval was 95%, the unreliable regions were taken as 0.025 for low data and 0.975 for high data values. In this figure, the points that have importance are the scattering of the data (* score), the theoretical CDF curve, and the vertical lines drawn to the left and right sides according to the 0.975 and 0.025 significance levels. Between the vertical lines are reliable datasets and outside the outlier data values appear.

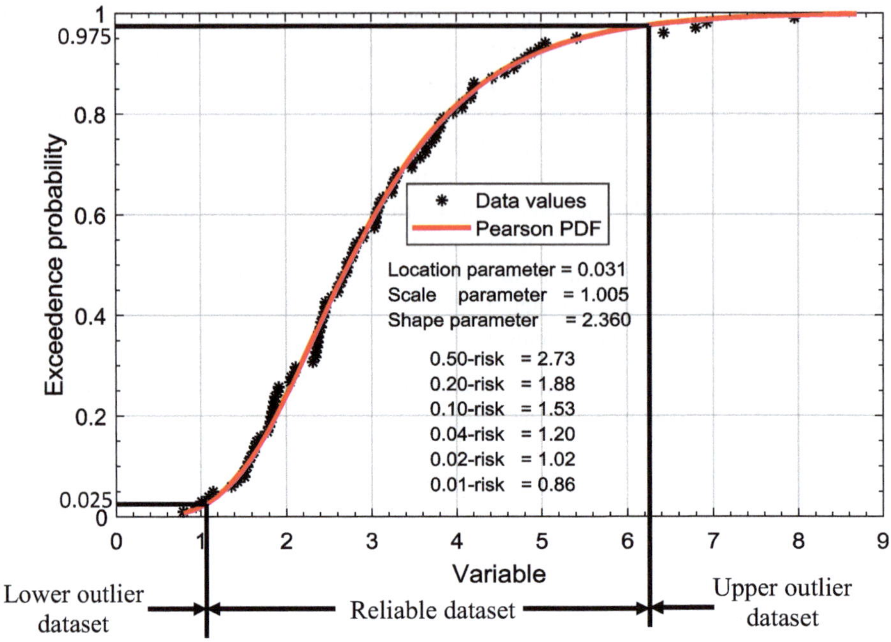

Fig. 8.2 CDF example

8.5.4 Uncertainty Roadmap

After all what have been explained in this section, in Fig. 8.3, a roadmap is proposed to ensure the reliability of the data before it is considered as input dataset.

Each part of this flowchart must be implemented by considerations of what have been explained in this section.

8.6 Features

They are like data points in terms of design properties for ML applications and can be counted among shallow learning elements that can be computed easily, but their decision choice is quite challenging prior to ML applications. For example, in Fig. 8.4a, time series, X_i ($i = 1, 2, \ldots, n$), is given in time domain. Its equivalence in the relative frequency domain appears on the right-hand side (Fig. 8.4b).

Apart from frequency indication, it is possible to convert a given dataset into various other numerical features such as point minimum, maximum, mean, standard deviation, some percentages, and alike. The features can be stuck as data points in one- or two-dimensional vector and matrix forms. In most of the application cases, they are treated by means of linear algebraic procedures, which may occur in the

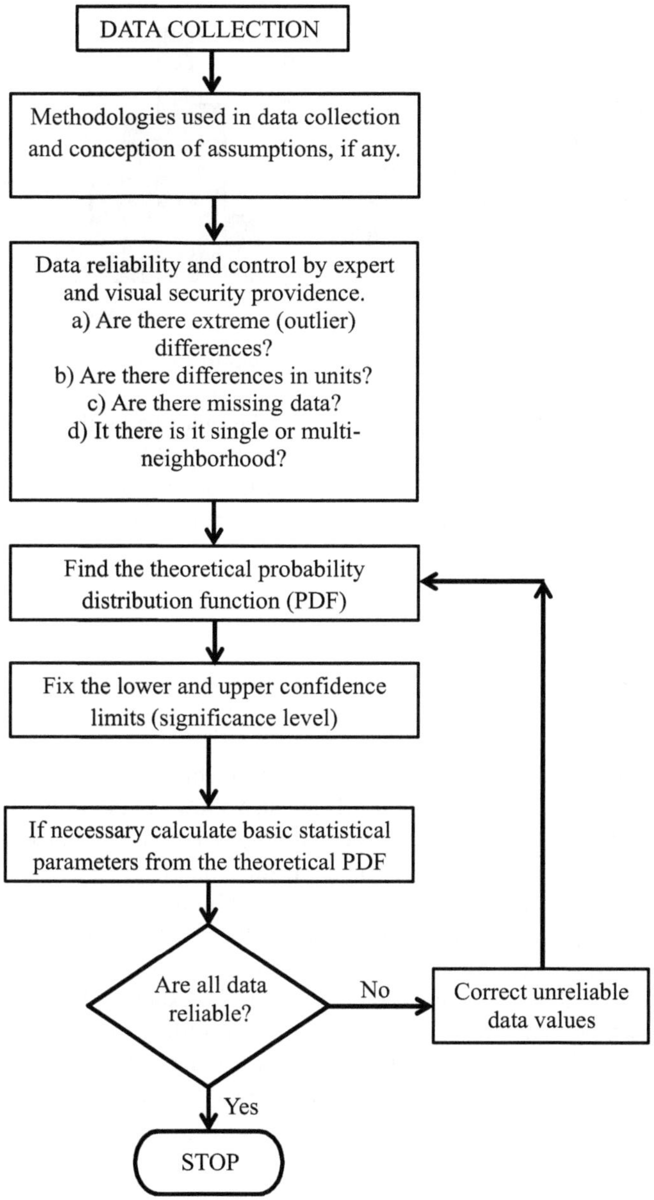

Fig. 8.3 Data evaluation flow chart

form of geometric features. Algebraic and especially geometric features in the forms of patterns provide efficient search possibility for element searches with desirable characteristics. One of the difficult problems is to manage model errors efficiently for

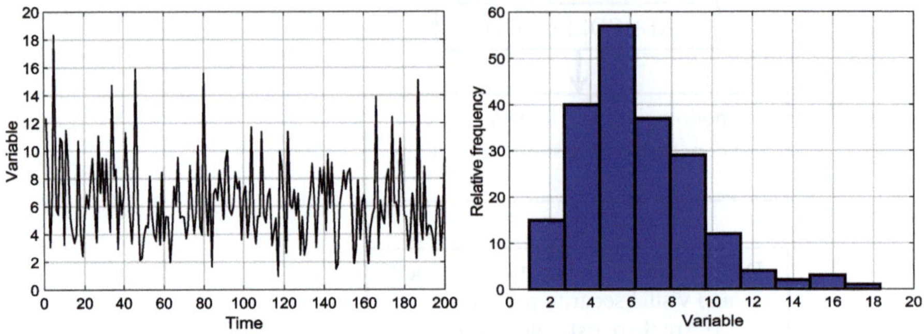

Fig. 8.4 Data point visualization, (**a**) time series line plot, (**b**) spectrograph

ML convenience in different applications. On the other hand, the question of how many features can be coupled with ML is another problem. Only an expert can provide a convenient number of features for ML application. As a rule, label data number points, n_L, is larger than the numerical feature n umbers, n_F, ($n_L > n_F$), in any ML model training.

8.7 Labels

Apart from data feature patterns, there may be different kind of properties within each data point, and these properties may represent a higher-level facts or quantities, which are referred to as labels. The labels are also shown in the form of vectors like feature vector forms. Frequently, ML methodologies try to find efficient approximate prediction for each data points based on the features in the same dataset body. Label properties can be determined with the aid of expert human interactions, and they are not separable crisply from the features, because there is fuzzy interference between the two properties (Chap. 3). A prominent example for the link between regression and classification is logistic regression (Chap. 7). For example, human representing data points can be defined by its label as 0 or 1 according to bivalent (crisp) logic principles. Logistic regression methodology is a binary classification application where the separation between categories is achieved by linear regression.

8.7.1 Numeric Labels: Regression

To apply the ML methodology, the data points of the label field must contain all possible label values. The possibility of numerical label estimation with ML methods has been introduced in the previous chapters, especially in Chap. 4.

8.7.2 Categorical Labels: Classification

In some ML applications, data point labels may imply either categorization of classification possibilities, in which case classification methodologies are used for application. The simplest classification procedure is binary, that is, bivalent (crisp) logic classification, where each data point belongs in one of the two different classes either 1 or 0. There are also multi-classification procedures, where each data point may belong to more than two classes based on fuzzy logic principles (Chap. 3). Data points can also belong simultaneously to several categories. Multi-label classification considers predictions such that a given input may belong to more than one label. Here, the classifications are mutually inclusive contrary to numerical labels where the classes are mutually exclusive.

8.7.3 Ordinal Labels

These are values in between numeric and categorical labels from an ordered finite set. It is proposed by Nam et al. (2014) that like numeric labels, ordinal labels take on values from an ordered set. For example, such an ordered label space represents rectangular areas of size 1 km by 1 km. The features x of such a data point can be obtained by stacking the RGB pixel values of a satellite image depicting that area. Beside the feature vector, each rectangular area is characterized by a label y where,

- y = 1: The area contains no trees.
- y = 2: The area is partially covered by trees.
- y = 3: The area is entirely covered by trees.

Verbally, it is possible to say that label value y = 2 is "larger" than label value y = 1 and label value y = 3 is "larger" than label value y = 2. It is to be noted that descriptions such as "larger" are inclusive in fuzzy logic domain as explained in Chap. 3. In many ML applications, although feature data points can be determined easily, but their labels are known for few points only, because such data is quite scarce. Even though there may not be any point data label, ML teaching procedure can extract relevant information from feature data. These are unsupervised ML methods (Chap. 7). Scatter of dataset and its probabilistic model have been explained in Chap. 4.

8.8 Learning Through Applications

Theoretical imaginative ideas are necessary for mind triggering and critical thoughts in search for innovative knowledge generation but as one of the Muslim thinker Abo'l-Iz Al-Jazari's (1136–1206) pioneer of innovative robotic and other mechanical device designs, stated that

Information without any application remains between falseness and truthiness.

After messy information scraps and previous thinker parlances classification, better and simpler scientific inferences are achievable. The scientific affairs may confront with difficulties, but they are handbill by means of systematic categorizations.

In this way, he advised that any work must not remain in the theoretical form but must be converted to application for useful technological services. The knowledge must be turned to applications and technology so that apart from the truthiness or falseness, they become serviceable and helpful to humanity. All his advices are valid even today, because he has said many things that went over centuries and did not lose their validity (Şen 2022).

ML is based on computer algorithms and software application that are improvable with experience and observations after several trial and error procedures, if necessary.

8.9 Forecast Verification

Prediction validation relates to temporal or spatial quality measures. There are differing opinions among many scientists as to what constitutes a good guess (Murphy 1993). All prediction validation procedures attempt to compare somehow predicted variables with observations on error or loss function minimization principle. In general, according to some criteria, the closer the prediction values to the observations, the better the prediction. Among these criteria, there are many subjective and objective estimation measures. For example, personal judgments and probabilistic, statistical, parametric, or non-parametric bases may be chosen as key criteria, depending on the nature of the estimation problem at hand. Regardless of the criteria, an expert's opinion and comments may seem invaluable, but they are complementary and supportive information. In short, any prediction validation procedure necessarily involves comparisons between matching pairs of predictions and the observations to which they relate. Spatial and temporal estimation should be compared with improvement in estimation procedure for reduced uncertainty and increased precision in the form of incremental trends. It is also important to compare the estimation result from different sources, especially in 2D studies. Therefore, it is possible to evaluate the relative merits of competing estimators or estimation systems via any computer software. Experts can provide additional feedback on the performance of forecasts in the hope of better future forecasts.

The goal of forecast verification is to identify weak points in any forecasting procedure and improve them with day-to-day knowledge and technological developments. The inclusion of up-to-date information and diagnostic checks on the actual appearance throughout the forecast and time reference can lead to better forecasts. In the prediction validation procedure, the most important variable that will guide the estimator for improved predictions is the series of errors quantified by probabilistic and statistical methodologies and their uncertainty characteristics.

Whatever the improvement, it should be kept in mind that it is impossible to consistently approach acceptable minimum prediction error, but another prediction error may occasionally occur. Real improvement has to do with overall behavior, not instantaneous or partial reductions in the error amount. Therefore, general behaviors can be represented most often either by statistical parameters or practically by relative frequency distributions of error sequence. This means that any error in forecasts is a component of the forecast's overall performance.

8.9.1 *Parametric Validation*

As described earlier in the regression straight-line technique (Chap. 4), the error sequences must have a zero arithmetic mean. A logical consequence of this statement is that there should be insignificant random positive and negative deviations over a long period of time.

8.9.1.1 Frequency Distribution

It is stated earlier that the frequency distribution of any uncertain variable provides not only probabilistic but also statistical parametric behavior of the variable (see Fig. 8.2). In a successful estimation procedure, the frequency distribution functions of the prediction values and the observations should be close to each other. The frequency distribution in prediction validation can be implemented in two different ways.

1. The joint distribution of forecasts and observations is of fundamental importance for the verification of the forecast.
2. The frequency distribution of the prediction error sequence, which is the combined expression of prediction and observations, the difference between the two.

In practical forecasting procedures, both forecasts and observations are separate variables. The prediction by Y_i data value for the implementation can take any of the valuesfrom the sequence Y_1, $Y_2, \ldots$, Y_m and the corresponding observations as X_i, which can take the values of X_1 $X_2, \ldots$, X_n. With these notations, the common PDF, $P(X_i, Y_i)$, of predictions can be expressed as follows (Chap. 4):

$$P(X_i, Y_i) = P(X_i/Y_i)P(Y_i)(i = 1, 2, \ldots, m); (j = 1, 2, \ldots, n) \quad (8.2)$$

By associating a probability with each of the mxn possible combinations of predictions and observations, the bivariate PDF can be derived from the given datasets. Eq. 8.2 was named as calibration-improvement factorization by Murphy and Winkler (1987). Part of this factorization consists of a set of m conditional distributions, $P(X_i/Y_i)$, each of which is made up of probabilities for all n outcomes of X_i given one of the predictions Y_i. This conditional PDF indicates how often each possible event occurs when a single Y_i forecast is published, or how well each Y_i forecast is calibrated. However, the unconditional probability part, $P(Y_i)$ in Eq. 8.2, specifies the relative frequencies of each predictor value, Y_i, or how often each of the m possible predictive values is used. It is also possible to write Eq. 8.2 equivalently in the following form:

$$P(X_i, Y_i) = P(Y_i/X_i)P(X_i) \tag{8.3}$$

This is known as likelihood-based ratio factorization (Murphy and Winkler 1987). The term conditional probability on the right represents the probability that each of the allowed forecast values is given before the observed event X_i.

The relative frequency distribution function of forecast errors is also a fundamental quantity for evaluating forecast quality. The smaller the variance of this distribution function, the better is the estimate. In addition, this PDF provides a basis for the reliability of the estimates. Using the relative frequency distribution function of the prediction errors, it is possible to derive the confidence of the prediction value (Fig. 8.2 and Chap. 4).

8.9.2 Prediction Skill

When defining the estimation skill, it is necessary to have two different bases for comparison. Prediction skill refers to the relative accuracy of a set of predictions. Two baselines shorten estimates and a set of standard control or reference estimates. In practical estimation procedures, three reference estimation values are usually chosen as follows.

1. In permanent forecasts, the values of the forecast and the previous time period.
2. Random forecasts according to the dataset relative frequencies of forecast events.
3. A quantitative measure of estimation skill is the skill score (SS), defined as the percent improvement over the reference estimates:

$$SS = 100 \frac{A - A_r}{A_p - A_r} \tag{8.4}$$

This formulation gives consistent results whether the measure of accuracy has a positive (larger values of A is better) or negative (smaller values of A are better) orientation. The SS reaches its maximum value of 100% when $A_p = A$ and the lowest SS value of 0% when $A_r = A$.

8.9.3 The Contingency Table

Comparison of two or more categorical data is performed in the form of tables in which the absolute frequencies or counts of the dataset, for example, mxn possible combinations of estimates and observations, are displayed in one mxn contingency table. The contingency tables contain the relative frequencies at each categorical position in the table, as value divided by the sample size of each tabulated entry. Such tables have similarities with the conditional and combined probability quantities mentioned earlier in this section on the validation of relative frequency estimation. Table 8.1 shows a one-to-one agreement between a probability table and the joint distribution of estimates and observations for the case m = n = 2. A two-way contingency table provides information about the counts of two categorical variables. In such a table, the rows (columns) are for one (another) variable.

The total number of prediction/event pairs in the dataset is equal to n = a + b + c + d. Common probabilities consist of the relative frequencies at which each of the possible prediction/event pairs occur and are obtained from the contingency table of raw counts divided by sample size. The table is labeled "yes" and "no," so the relative frequency of a successful "yes" guess would be P(X,Y) = a/n.

It is also common practice to include marginal rates in contingency tables. These are simply row and column totals, in which case the number of occurrences of each "yes" or "no" prediction or observation, respectively.

On an assumed-perfect prediction, non-diagonal terms in a 2 × 2 contingency table will have zero values. This corresponds to all "yes" predictions for the event after the event and not after all "no" predictions for the event. However, in the real case, the cross-diagonal values are expected to be as small as possible for accurate predictions. Hit rate or rate effect calculations are the simplest measures of objectively validation of forecasts from a given contingency chart. This is a simple part of the prediction case, in which the categorical prediction accurately predicts the next event or none (Wilks 1995). With the illustrations in Table 8.1, the hit rate H is defined as:

$$H = \frac{a + d}{n} \tag{8.5}$$

This definition equally credits correct "yes" and "no" predictions. Also, the hit rate punishes both types of errors equally. Worst hit rate equals zero, and best hit rate equals one. It is possible to determine the percentage of prediction error by multiplying Eq. 8.5 by 100.

Table 8.1 A contingency table

	Yes	No	Total
Strong	a	b	a + b
Mild	c	d	c + d
Total	a + c	b + d	N = a + b + c + d

Another predictive validation measure based on a contingency table is the critical success index (CSI), which is useful when the event to be predicted occurs significantly less frequently than a nonoccurrence event. It is defined by considering all entries in the 2 × 2 contingency table, except for the "no" situation in the observed and predicted values. In fact, the CSI is the percentage of correct "yes" predictions among all inputs except observed and predicted "no" situations. Therefore, it is defined by the following notation:

$$\mathrm{CSI} = \frac{\mathrm{a}}{\mathrm{a} + \mathrm{b} + \mathrm{c}} \tag{8.6}$$

This can be considered a hit rate for the predicted quantity after excluding the correct "no" guess from the evaluation. It also varies between zero and one, including the critical success index.

Yet another useful measure of prediction validation from the 2 × 2 contingency table is the probability of detection (POD), defined as the fraction of cases where the forecast event occurred and was also predicted. This is another expression of the probability expression given in Eq. 8.3. Considering the notation entries in Table 8.1, the POD can be given as:

$$\mathrm{POD} = \frac{\mathrm{a}}{\mathrm{a} + \mathrm{c}} \tag{8.7}$$

In case of perfect prediction, POD is equal to one. It is also possible to make another useful definition for prediction validation measurement based on the conditional probability expression considering P(X/Y). This measure is called the false alarm rate (FAR) and is defined as follows:

$$\mathrm{FAR} = \frac{\mathrm{b}}{\mathrm{a} + \mathrm{b}} \tag{8.8}$$

Contrary to previous definitions of prediction validation, the smaller the FAR, the better is the prediction. Based on the contingency tables, it is possible to make another measure to compare the mean forecast with the mean observation, and this is called the contingency table bias (B), defined as:

$$\mathrm{B} = \frac{\mathrm{a} + \mathrm{b}}{\mathrm{a} + \mathrm{c}} \tag{8.9}$$

It is simply the ratio of "yes" predictions to the number of "yes" observations. In the case of unbiased prediction, which is the required quality, $\mathrm{B} = 1$ indicates that the event is predicted the same number of times it was observed. It is possible for B to take values greater than one, indicating that the event is predicted more often than observed, and this is called overestimation. Likewise, a bias value of less than one is observed less frequently than predicted, which corresponds to an underestimation.

Although 2 × 2 probability tables help to understand the underlying philosophical system in its interpretation and use, categorical predictions are not limited to 2 × 2 contingency tables.

8.10 Learning Methods

Scientific researches and technological innovations gained speed with the ML methodological principles. In ML method, generally, there are three types of learning procedures, unsupervised, supervised, and reinforced supervised alternatives. These learning systems are based on computer technology through software with interdisciplinary theoretical aspects such as probability, statistics, and algorithm developments in order to strengthen and AI attributes. Various shallow learning training possibilities are explained already in Chap. 7 including unsupervised and supervised trainings and their different versions.

8.10.1 Supervised Learning

Supervised learning is attached with a label association depending on the problem concerned, which is supposed to answer to the question at hand. In discrete label cases, the task is classification problem, and in the case of revaluated labels, regression methodological solution is the focal point.

Among the methodologies of ML, supervised learning is relatively shallow (basic) learning technique (Chap. 7). In its structural establishment, relevantly convenient rational verbal and logical statements must be set down by researcher or engineer before learning. The machine depends on the information technological learning principles in a supervised manner to execute the relevant statements either through an architectural structure between input and output components or to use mathematical formulations. General learning abilities of the machine can be gained basically with supervised learning by complete stimulation of the problem at hand. In this way, the supervised learning method reaches at systematic organization and management levels, and hence, it can be employed for different methodological solutions if there is dataset at hand leading to instant classification, simulation, prediction, estimation, and regression type of methodological solutions ready under the hand.

Supervised machine learning is counted among the shallow learning principles, because it represents a relatively basic learning methodology. Prior to its use in computers, the human must identify at least verbal learning principles. It is necessary to collect basic information about data numerically and linguistically with goal consideration. Its comparison with other learning methods indicates that it can simulate common ML learning. The ML researcher can solve some classification and regression problems as already explained in Chaps. 3, 4 and 5. Among the classical learning approaches are k-nearest neighbor (KNN), support vector machine

(SVM), and Bayesian network (BN) methodologies. In this learning procedure, the learning content is more systematic and there is certain regularity.

8.10.2 Unsupervised Learning

This is another shallow learning alternative to supervised learning procedure. In this methodology, the machine does not know the content during the whole learning procedure. The machine itself adjusts the learning principle extraction from available data. In this way, the machine has complete authority over the data information analysis. During the learning process, machine does not depend on certain directions for completion of the work. Operational methodologies help machines to learn the basic content of the problem, its concepts, and, hence, a series of content learning in freedom as fundamental principles. ML improves continuously through successive learning stages and as a branch of AI aims to mimic human intelligence abilities and skills by machine. ML deals with understanding context inductively for presenting hidden pattern within a large dataset after the application of statistical principles through AI methodologies. The main purpose of unsupervised learning is to deduct hidden pattern regularities, for example, in the form of clusters, to detect local and global anomalies in the dataset. More information about unsupervised learning is presented already in Chap. 7.

8.10.3 Reinforcement Learning

As for the reinforcement learning (RL), it is based on psychological inspiration to identify characteristic behavior of data. An ideal outcome can be identified by a series of trial and error work. During the trials, certain actions are chosen for output desirable pattern. Loss function is frequently used in ML procedures for evaluation and comparison of different hypothesis. If the future data points generation predictions are influenced by a hypothesis, then RL procedure is used in applications. This learning-type application provides data points, which represent the programmable system states at different time instances. Like unsupervised ML, RL methods often must learn a hypothesis without having access to any labeled data point. Opposite to other learning methodologies, RL cannot evaluate loss function for different hypotheses. From mathematical point of view, loss function evaluation is possible on point-wise for the current hypothesis for the most advanced prediction.

Sample datasets are basic gradients of supervised learning procedures, whereas reinforcement learning (RL) has interaction with environment (Mnih et al. 2015). This is also referred to as semi-supervised learning. There is not a concrete loss function in the RL system. There is not straightforward access to the function for optimization, and therefore, there are queries about the interaction. On the other hand, if the state is integrated with the concerned environment, the input requires the preceding actions (Al-Dulaimi et al. 2019). For RL, execution scope of the problem

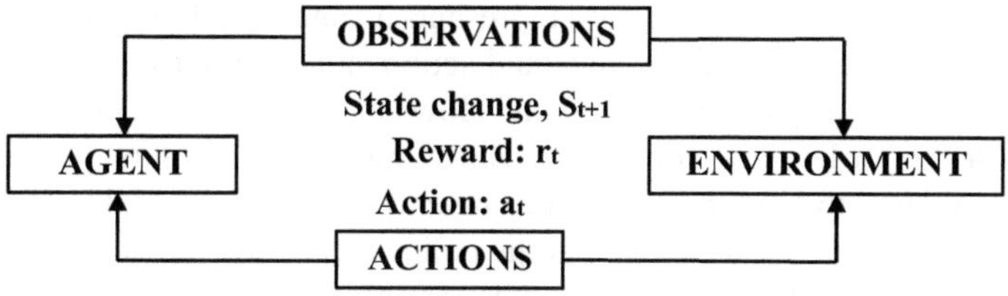

Fig. 8.5 RL components

is important and supportively also the space. This approach is useful when in the case of many parameter optimization. The parameters influence the speed of computation in learning. There are some specific characteristics for the use of RL procedure, which provides identification of the best action over a longer period; it helps to identify the action requirement; for far better solutions, it enables to find the best approach. Its use is not recommended in case of insufficient dataset availability, because it may give rise to heavy computation and time consumption. Figure 8.5 indicates the key concepts in a RL system.

Action is equivalent to a move in the environment within action space, A, which is a possible set of movements in the environment. The movements are discrete. Observations are possible to understand according to the actions that take place in the environment. Observations are sent back and perceived in the form of states. The state changes by time and space, and its stages are shown, for example, state at time t, s(t), and at time t + 1, s(t + 1). Agent's action can be measured as success or failure as feedback, which is referred to as reward. Sometimes, the reward may be delayed without instantaneous response. The total reward, R_t, (return) that the agent can record can be expressed as the summation of all the possible rewards according to the following simple mathematical expression:

$$R_t = \sum_{i=t}^{\infty} r_i$$

There is also discounted total reward as a return, R_d, according to the following mathematical expression:

$$R_d = \sum_{i=t}^{\infty} \gamma^i r_i$$

where γ is a discount number. Its importance is to make future rewards less that present ones, and therefore, it is a number between 0 and 1. It is quite similar to discount rate in economics. In the reinforcement learning, there is also another concept of Q-function, which reflects expected R_t an agent in states after execution of a certain action, a. The valid expression in its implicit form is:

$$Q(s_t, a_t) = E(R_t/s_t, a_t)$$

This is equal to saying that Q-function represents a conditional reward, R_t; at a given state, s_t; and action, a_t, at the same state and action characteristics. How can one take action provided that a Q-function is given? How can one use this Q-function to have better regards? In order to decide on this aspect, one needs a policy, p(s), depending on the current state to have best action to take as its state, s. For this purpose, it is necessary to have a firm strategy so that the policy can choose an action to maximize future reward. To get the best reward, there should be a set of states for the same action that one can enter the Q-function, and thus, for each stage, there is a return according to the Q-function, and the desired result is the one that maximizes the Q-function as follows:

$$P'(s) = \max_{a} Q(s, a)$$

where $P'(s)$ is the optimum solution policy. Deep reinforcement learning algorithms include two blocks of assessment as value learning and policy learnings as shown below.

Value learning	Policy learning
Find Q(s,a)	Find p(s)
$a = \max_{a} Q(s, a)$	Sample $a \sim p(s)$

Q-function for value learning

8.11 Objective and Loss Functions

Loss functions teach to machines how to learn. It helps to evaluate the suitability of a specific algorithm model to a given dataset. Some of the loss functions are given in the following items in cases of regression problem solutions.

1. *Mean absolute error (MAE):* It is one of the robust loss functions used for regression problems, which may be non-Gaussian form in cases of extreme dataset values. This loss function does not consider unrealistically high or low values. Its definition is arithmetic average of absolute differences between model output, M_i, and measurement, m_i datasets ($i = 1, 2, \ldots, n$) where n is the number of data:

$$AME = \frac{1}{n} \sum_{i=1}^{n} |M_i - m_i| \tag{8.10}$$

2. *Mean squared error (MSE):* This is the most frequently used loss function in practice. It is the mean of squared differences between the model and measurement values, and the mathematical definition is given below:

$$\mathrm{MSE} = \frac{1}{n}\sum_{i=1}^{n}(M_i - m_i)^2 \tag{8.11}$$

In perfect match, which is not possible in real modeling approaches, MSE is equal to zero.

3. *Mean bias error (MBE):* This is for the average bias calculation between the hypothesis model and the available measurement data. It shows overestimation (positive value) or underestimation (negative value) cases. The model can be updated depending on the final MBE value. Its formulation is given below:

$$\mathrm{MBE} = \frac{1}{n}\sum_{i=1}^{n}(M_i - m_i) \tag{8.12}$$

This is not frequently used loss function because positive and negative errors can cancel each other.

4. *Mean squared logarithmic error (MSLE):* In order to reduce huge differences according to the previous loss functions, one can consider MSLE conception. Its calculation is like MSE, where instead of genuine values, their logarithms are considered as follows:

$$\mathrm{MSLE} = \frac{1}{n}\sum_{i=1}^{n}[\log(M_i) - \log(m_i)]^2 \tag{8.13}$$

8.11.1 Loss Function for Classification

There are problems of a given dataset classification into few or several classes for identification and comparison purposes. The data is divided into different and unique classes based on different parameters. The following loss function can be used for classification problems.

8.11.1.1 Binary Cross Entropy Loss (BCEL)

Entropy is the measure of randomness that occurs in any information content including datasets, which is the difference of randomness between two random variables. This methodology is usable for binary classification tasks, which are capable to answer a question with only two alternative choices, 0 and 1. In cases of two classes, this is the most frequently used loss function. As the predicted probability diverges away from the actual label, the BCEL increases leading to high loss value. If the log loss is equal to zero, the model predicts perfectly the dataset pattern. The mathematical expression of BCEL is given by the following expression:

$$J = -\sum_{i=1}^{n} M_i \log[h_\theta(m_i)] + (1 - M_i)\log[1 - h_\theta(m_i)] \tag{8.14}$$

where M_i is the true label and $h_\theta(m_i)$ is the prediction post hypothesis. Since binary means that the classes take on two values as 0 and 1, if $M_i = 0$, the term does not exist, and in case of $M_i = 1$, then the $(1 - M_i)$ term is zero.

8.11.1.2 Categorical Cross Entropy Loss

This is the extension of binary cross entropy loss as explained in the previous subsection to multivariate variable case. In this loss function, only one element will be nonzero as other elements in the vector are multiplied by zero. This property is extended to an activation function called soft-maximum.

On the other hand, hinge loss (HL) is conveniently used in cases of support vector machines (SVM) to calculate the maximum margin from the hyperplane to the classes. It penalizes wrong predictions, and hence, the target label score is greater than the sum of all the incorrect labels by a margin at the least one. HL is used for maximum-margin classification, most notably for SVMs. The loss function is not differentiable in the convex form, and it is used in ML procedures. Its mathematical form is as follows:

$$\text{SVMLoss} = \sum_{j \neq M_i} \max\left(0, s_j - s_{M_i} + 1\right) \tag{8.15}$$

See: https://www.section.io/engineering-education/understanding-loss-functions-in-machine-learning/#introduction

8.12 Optimization

Optimization is an integral part of ML based on mathematical procedures to reveal unseen data features by numerical computation of model parameters for a decision-making system. The learning problem leads to a set of parameters that are the optimal with respect to a given dataset. ML design principle needs optimization of mathematical formulation procedures. For example, the weather prediction problem below can be formulated as an optimization (minimization) problem. Mathematical and statistical computational frameworks of optimization techniques are among straightforward application means of optimization. Figure 8.6 shows the main differences between the optimization and loss function.

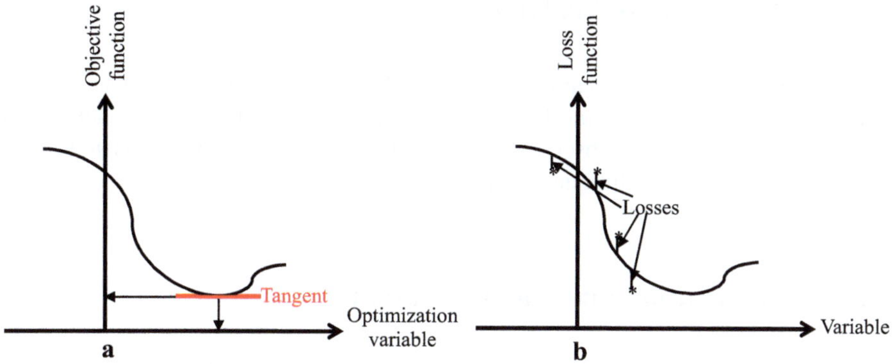

Fig. 8.6 (**a**) Objective function for optimization, (**b**) loss function

The curve in Fig. 8.6a shows a simple optimization problem with the minimum optimization value and objective function value. Figure 8.6b is represents ML method learning to find a hypothesis by loss function minimization. It should be considered that the average loss is a noisy version of the ultimate objective, which is equivalent to expectation in the case on large number of data, and its PDF is identifiable.

8.13 ML and Simple Linear Regression

The linear or multiple statistical regression methodologies are employed in ML supervised training for categorization of given dataset. Its main structure aims at prediction-dependent variable from a set of independent input variables through the statistical least squares' methodology. In the classical regression analysis, there is a set of assumptions as follows:

(a) Input variables are regarded as independent from each other.
(b) Input and output variables have uncertain, vague, random, and incomplete components, and their PDFs are assumed as normal (Gaussian) PDF.
(c) A convenient mathematical function is fitted to the scatter of plots on the condition that the sum of the squares is subject to minimization.
(d) Regression methodologies require numerical data and help to predict or explain a numerical value based on a set of prior data.

Linear regression has the simplest mathematical expression, which relates the output variable, O, to inputs, I_i ($i = 1, 2, \ldots, m$), where m is the number of input variables and a_i's are unknown parameters:

$$O = a_0 + a_1 I_1 + a_2 I_2 + \ldots + a_m I_m \quad (8.16)$$

8.13.1 Least Square Technique

The essence of many statistical models is based on the principle that the sum of the squares of forecast error should be minimal. Here, the prediction error is defined as the difference between the observed and predicted value. A very brief revision of the statistical least square technique is explanation in Chap. 4.

8.14 Classification and Categorization

Classification is also among the ML methodologies that deal with prediction outputs from a set of input data. In general, the output is in the forms of bivalent logic "yes, 1" or "no, 0" format although it is not limited with two exclusive alternatives only. For example, in a geological map, there are three rock types: "volcanic," "sedimentary," and "metamorphic," and thus, a convenient classification method application can identify these three types explicitly as "0," "1," and "2," or with three different colors, respectively. Logistic regression is the simplest classification method, which separates different classes with linear lines (Chaps. 6 and 7). The adjective regression here does not mean statistical regression, but rather probabilistic one based on inputs probability to estimate output.

For instance, if there is a set of water samples sent to laboratory for quality evaluation, each one will have different values, and a set of numbers appear, and the logistic regression can take these values as inputs to estimate the probability that the water is of "good" or "bad" quality according to standard base values for each chemical component. Hence, the logistic regression outputs are either 0 or 1. If the base probability is fixed as 0.3, then the water samples with greater score than 0.3 can be regarded as "good," and the remaining are in "bad" classifications.

Categorization is a classification procedure of a given dataset into several groups each with the similar objects so that later new data can be recognized in one of these groups automatically. It has been already explained in Chap. 3 that in case of bivalent logic, there are groups that are mutually exclusive and exhaustive. As for the fuzzy logic, the groups are mutually inclusive non-exhaustive but partially nested. As already mentioned in Chap. 3 in any language, categorization begins with words like "low," "medium," and "high" as opposed to the classical two-way logic of "to be or not to be" within certain limits.

8.15 Clustering

This keyword is considered especially in unsupervised learning procedure in search for a structural composition in a set of unlabeled data. For example, as explained earlier, k-means procedure, and c-means will be explained later in this chapter; both

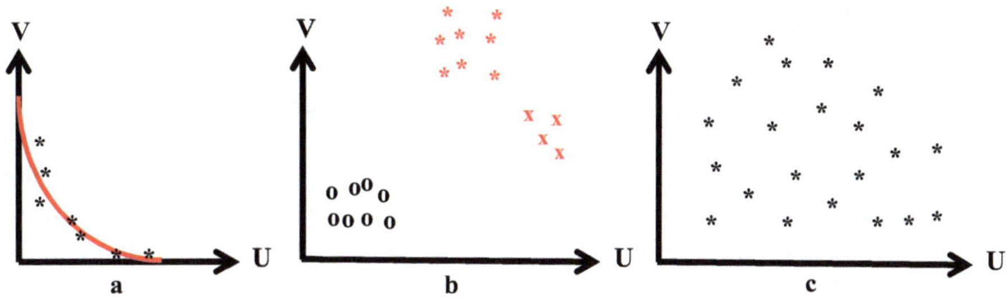

Fig. 8.7 (**a**) Mathematical single cluster, (**b**) visual cluster, (**c**) procedural cluster

of them work with unsupervised learning procedure for clustering. Clustering procedure divides dataset into several groups, and in each one, there are elements that are similar according to a criterion. In a way, it is a procedure that collects in groups similar feature properties and in others dissimilar objects. Clustering can be defined simply as "the process of organizing objects into groups whose members are somewhat similar." Figure 8.7 provides a visual impression for such clustering procedures that can be grouped primarily by visual inspection prior to any computer software application for procedural application.

Figure 8.7a has scatter points that fall around a mathematically expressible pattern, and its expression can be obtained by regression analysis (Chap. 4) based on the least squares methodology. The scatter points may not have such a definitive feature, but obviously and visually identifiable clusters as in Fig. 8.7b. This pattern may not even need the application of a technique for clustering, because there are three clearly visible groups shown by o, x, and * signs.

Finally, Fig. 8.7c does not provide obvious and visual pattern, and therefore, its division into a set of clusters is dependent on researcher and his computer program for clusters and its number identification. For this purpose, in this book, the researcher is advised to start clustering with two groups and then apply, k-means or c-means procedure as convenience requires (Chap. 7). If for any logical and rational reason more than two clusters are necessary, then the same procedure can be applied with group number increment one by one. The researcher then can decide which cluster number provides the best characteristic of the dataset. The simplest form of similarity criterion is called distance-based as mentioned earlier in Chap. 6.

Another clustering procedure is named conceptual alternative, where objects belong to the same cluster if they have a common concept for all that objects. This means that the objects are not classified according to objective similarity measures, but according to their descriptive concepts. For example, people can be put into "young," "middle," and "old" age categories according to the epistemological implication of each word for classification. In any clustering procedure, two basic principles are applicable as mentioned earlier in this chapter.

1. Cluster centroids must be as far as possible from each other.
2. Each cluster center must gather all the closest elements.

8.15.1 Clustering Goal

The main goal is to identify clusters of an unlabeled dataset into intrinsic groups. In such a procedure, one cannot expect to obtain best criterion in clustering that could be independent of the final purpose. Initially, one wishes to cluster the dataset into several groups with the satisfaction of a certain criterion even though it may be purpose dependent. In scientific literature, there are procedures such as homogeneity, outlier detection, data usefulness, stationarity, data reduction, trend identification, and alike. It is by now well known that clustering is applicable in different disciplines like marketing for finding similar client groups, biology in plant and animal grouping, insurance for identification of policy holder graduation, city planning for house groups, earthquake in epicenters identification for dangerous zones, geology for fissure, fracture and fault identification, etc.

In the identification of clusters, there are requirements that should be considered before the application of any procedure. Among such requirements are attribute differences, dimensionality, noise, outlier, independence, decision on input parameters, cluster shape determination, scalability, etc. One must not take clustering as granted, because there are several problems that should be taken into consideration before the application start. It is necessary to consider whether the cluster procedure is capable to address adequately all requirements by the researcher; one must be careful in case of large datasets availability as for its proper treatment by means of clustering procedures, determination of distance criterion for similarity measures, and interpretation of the clusters must be evaluated from different viewpoints.

8.15.2 Clustering Algorithm

In general, there are four clustering groups as explained in the following subsections. Among these are exclusive, overlapping, hierarchical, and probabilistic alternatives. The first alternative is concerned with clustering in an exclusive manner that each object belongs to a single group. A simple example for this case is given in Fig. 8.8. A straight line separation can be applied as described for logistic regression methodology in Chaps. 6 and 7. In the exclusive clustering, there are no differences, and all the points in each group have MD equal to 1.

As for the overlapping clustering method, the clusters are mutually inclusive, and therefore, clusters overlap with each point belonging at the same time to more than one cluster. In other words, each object in the dataset can belong to two or more clusters with different MD. Figure 8.9 shows such a simple fuzzy clustering.

Two clusters, A and B, overlap, and the red objects belong to both groups with different MD. Of course, each one of the points has a MD according to fuzzy logic principles (Chap. 3).

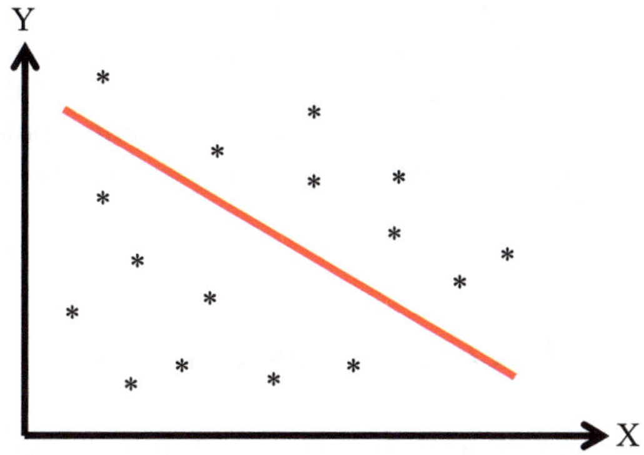

Fig. 8.8 Exclusive clustering

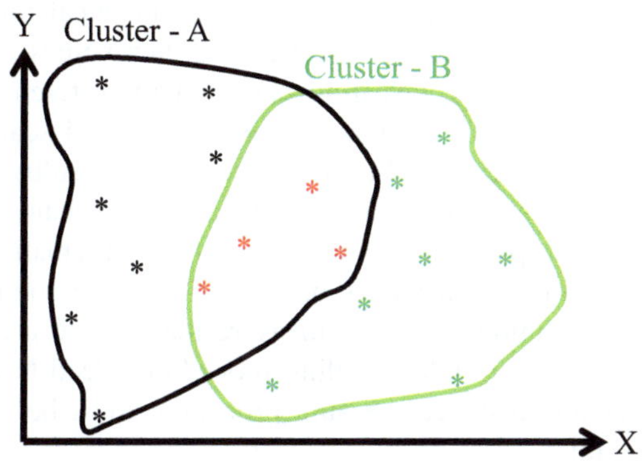

Fig. 8.9 Overlapping clustering

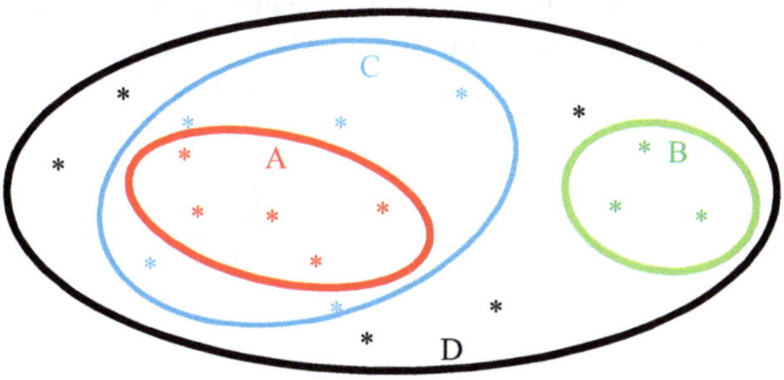

Fig. 8.10 Hierarchical clustering

Hierarchical clusters have the union between the nearest two clusters. Initial condition is that each datum is set as a cluster. Thus, the final clustering can be reached after some iteration. These are generated as having a predetermined ordering from top to bottom. Figure 8.10 is for such a clustering example shown by Venn diagram concept.

In this figure, the initial clusters are A and B, where cluster A is inside cluster C, and all of them are under the umbrella of dataset without clustering.

The forth clustering is completely probabilistic, and the most widely used alternatives are k-means and c-means, which are explained in detail in the subsequent sections. This clustering belongs to one of the previous three clustering types. Briefly, k-means is a special clustering procedure, but fuzzy c-means is an overlapping clustering algorithm, because each object in the cluster has more than one MD.

8.15.3 Cluster Distance Measure

Distance measure between any two data points or between a data point and the proposed cluster center is the main component in each clustering procedure. In cases of the same datasets, the distances can be calculated directly; otherwise, the datasets can be converted to standardized data values as already explained in Chap. 4. Of course, different scaling can lead to different clustering. Figure 8.11 illustrates comparatively clusters in the original and scaled down datasets. Thus, a scaling operation may lead to biased clustering results.

It is obvious graphically that there is change in the clustering after scaling, but another problem comes from the mathematical formulation procedure used to combine the distances between the single components of the data feature vectors into a unique distance measure that can be used for clustering purposes, because different formulas (scaling procedures) lead to different clustering results. Thus, a suitable distance measure procedure must be adapted for convenient and reliable results.

In many applications, Euclidean distance measure is used, which expresses the distance as the length of hypotenuse in a right-angle triangle. This simple expression has been generalized by Minkowski as distance measure, $D(X_i, X_j)$, between two sets of data X_i and X_j (i,j = 1, 2, . . ., n) as follows:

$$D(X_i, X_j) = \left(\sum_{k=1}^{n} |X_{i,k} - X_{j,k}|^p \right)^{\frac{1}{p}} \tag{8.17}$$

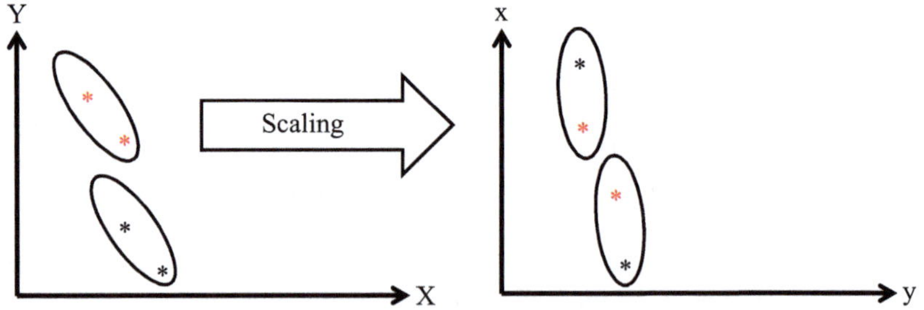

Fig. 8.11 Scaling and clustering

This distance measure reduces to Manhattan metric for $p = 1$ and to Euclidean distance measure for $p = 2$. There may be any number between $1 < p < 2$, but there is no guidance as for which value to adapt in distance calculations in a practical problem-solving.

Clustering methods do not need for output data, but output is identified automatically from the input dataset. The performance of these methods as for the quality of output can be followed by visualization. There are two clustering methodologies that are frequently used in clustering studies. One depends on crisp logic and referred to k-means and the other on fuzzy logic and known as c-means.

8.16 k-Means Clustering

This is the simplest unsupervised learning procedure to separate a given dataset into several clusters (MacQueen 1967). Its preliminary explanation is given in Chap. 7. The cluster number must be prefixed according to the experience of the researcher or expert. Hence, the number of clusters, k, is predetermined. The main goal is to find the cluster centers that are representative and fixed according to the data scatter points at hand. The first guess of cluster centers is suggested either under the light of expert view or in case of no preliminary suggestion; they are allocated randomly. The next step is to consider each point's closeness to the nearest cluster centroid. After the clusters are determined in this way, the arithmetic mean of the points falling into the same cluster is calculated and revised. If the new cluster centers do not fall on the previous centroids, then the same principle is recalculated, and hence, another new cluster elements are identified. The repetition of the similar calculations can be done within a do loop until the new cluster centers stabilize or the relative error difference between the final and previous step cluster centroid locations remains within ±5% limits. At the end, k-means algorithm aims to minimize an objective function (distance closeness) in the form of least square procedure. The objective function, J, expression can be written as follows:

$$J = \sum_{j=1}^{k} \sum_{i=1}^{n} \left\| X_i^j - C_i \right\|^2 \tag{8.18}$$

where $\left\| X_i^j - C_j \right\|^2$ is a distance measure between a data point X_i^j and the cluster center C_j. The following execution steps are necessary for the completion of k-means methodology.

1. Decide about the number of clusters depending on the previous experience or take advice from an expert in the area.

2. Choose the first group of cluster centers either based on plausible reasons about the phenomenon otherwise allocate randomly.
3. Calculate the distance of each dataset point to each cluster center and allocate the point to the nearest cluster. Complete the same procedure for all the data points available.
4. Recalculate the cluster center new positions by calculating the arithmetic mean values for each cluster.
5. Repeat steps 2–4 until the new center points do not move or use the percentage error criterion as explained earlier.

After the execution of all the steps although the procedure will end up with no moving or acceptable error limit criterion cluster centers, but this does not mean that optimum configuration is found. One must not forget the sensitivity of the result on the initial random cluster center allocations. To avoid such a situation, it is recommended to run the k-means procedure few times with different cluster center randomization. In the following is a simple procedural application of k-means. Let us suppose that there is a dataset as X_i (i = 1, 2, . . ., n) with k, (k < < n) number of cluster search possibility, in general. Initial cluster centers are C_k (k = 1, 2, . . ., k). Any point, X_i, is in cluster C_j belongs to cluster j provided that $\|X_i - C_j\|$ is the minimum distance of all k distances. Furthermore, it is necessary to find the mean of the points that fall within k cluster. After the procedural step execution as explained above through steps 1–4, one should care for the following items:

1. Calculate arithmetic mean cluster centroids after each round; hence, there are m_1, m_2, . . ., m_k cluster centers.
2. Estimated means are used for classification of the samples into new clusters.
3. Continue until the final point according to the criteria mentioned earlier.

Figure 8.12 presents hypothetically the successive progress of the cluster centers and the division line between them. Here, the linear line is just the piece of holistic cluster boundary.

As a result, one may notice that the new centroids change their location step by step until no more significant changes appear, which implies that centroids do not move significantly any more. Here, the significance must be defined objectively through some criterion. For instance, if the last centroid positions are less than 5% relative error distance from the previous centroid locations, then the looping process of finding new centroids can be stopped.

Fig. 8.12 Cluster center successive movements

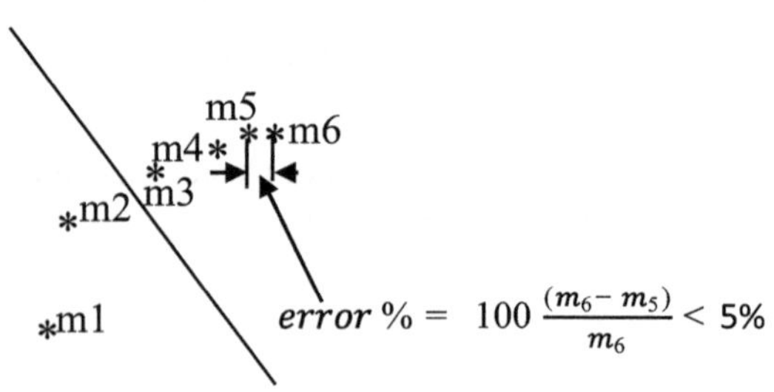

The weakest point of the k-means procedure is that it is not clearly obvious to have a general solution right from the beginning to decide about the number of clusters. As mentioned earlier, the guarantee is to run the same procedure with different initial random cluster centroid allocations and to try the same procedure with different cluster numbers starting from two onward until the optimum cluster number is obtained.

8.17 Fuzzy c-Means Clustering

This method of clustering allows a dataset element to belong to more than two or more cluster groups with different MDs, the summation of which is equal to 1 for each data point. The basis of the fuzzy clustering technique is proposed by Dunn (1973) and Bezdek (1981), and it is used quite frequently in pattern recognition applications. Mutually inclusiveness brings into the k-means procedure another dimension in the optimization, which requires that the summation of each point's MDs for each cluster be equal to 1. For this reason, it is referred to as the fuzzy c-means by Bezdek (1981). It is an iterative clustering approach like the k-means. However, c-means uses fuzzy membership functions instead of hard values of 0 and 1. The objective function is also referred to as the loss function and is given like k-means procedure as follows:

$$J_m = \sum_{i=1}^{n} \sum_{j=1}^{c} u_{ij}^{m} \left\| X_i - C_j \right\|^2 (1 \leq m < \infty) \tag{8.19}$$

where m is any real number greater than 1, u_{ij} is the MD of x_i in the cluster j, X_i is the i-th of multidimensional measured data, c_j is the multidimension center of the cluster, and ||*|| is any norm expressing the similarity between any measured data and the center. Fuzzy clustering necessitates iterative optimization of the objective (loss) function along the following expressions. During iteration successive adaptation of MDs, u_{ij} and cluster centers, C_j are obtained:

$$u_{ij} = \frac{1}{\sum_{k=1}^{C} \left(\frac{\left\| X_i - C_j \right\|}{\left\| X_i - C_j \right\|} \right)^{\frac{2}{m-1}}} \tag{8.20}$$

and

$$C_j = \frac{\sum_{i=1}^{n} u_{ij}^{m} X_i}{\sum_{i=1}^{n} u_{ij}^{m}} \tag{8.21}$$

The completion of iteration steps is possible when $\max_{ij}\left\{\left|u_{ij}^{(k+1)} - u_{ij}^{(k)}\right|\right\} < \varepsilon$ criterion is satisfied with condition that $0 < \varepsilon < 1$ after k iterations. The valid applicable steps for the fuzzy c-means can be summarized as follows.

1. Initialize MD values randomly or systematically as $U^{(0)}$ matrix containing values such that $0 < u_{ij} < 1$ (i = 1, 2, . . ., n) and (j = 1, 2, . . ., k). The exponent m is taken conveniently equal to 2.
2. After k iteration the cluster centers are calculated according to Eq. 8.21.
3. Update $U^{(k)}$ until above stated ε condition is satisfied.
4. Calculate the maximum difference among $u_{ij}^{(k)}$ and $u_{ij}^{(k+1)}$ as ε and if $0 < \varepsilon < 1$ is not satisfied then return to step 2.

8.18 Frequency-k-Means-c-Means Relationship

Histogram is for classification whereas k-means and c-means for clustering. To distinguish the concepts between them, let us consider the simple dataset given in Fig. 8.13.

In all graphs in Fig. 8.13, two clusters are identified near the two cluster centers. A first glance on Fig. 8.13a indicates that there are two visual clusters, A and B without application of any ML procedure. Computers are not capable to sense like human mind visually, but the visual inspection can be thought to computers through AI procedures. Let us consider first the concept of frequency diagram as explained in Chap. 4, which is a classification methodology. For this purpose, the range, R, variation of dataset projections on the horizontal axis is divided into five classes as in Fig. 8.13b. For k-means and c-means applications, the projections of the dataset on the horizontal axis is taken into consideration. In k-means algorithm, bivalent logic is used with all data points belonging degree equal to 1 (see Fig. 8.13c). As for the c-means approach in Fig. 8.13d, the same dataset does not belong to crisp clusters as in Fig. 8.13c, but MDs take place along smoother line to indicate that some datasets may belong to two (or several) clusters with different MDs.

8.19 Dimensionality Reduction

This methodology corresponds to the elimination of some relatively insignificant (redundant) dimensional variables from the data. The simplest example is the linear regression line. Figure 8.14 indicates the scatter of points and the best straight-line fitted by least squares principle (I-I direction). For many researchers, this is the end of regression analysis, but one point is missed from the consideration. The following two points are important for further interpretation of this straight-line from dimensionality point of view.

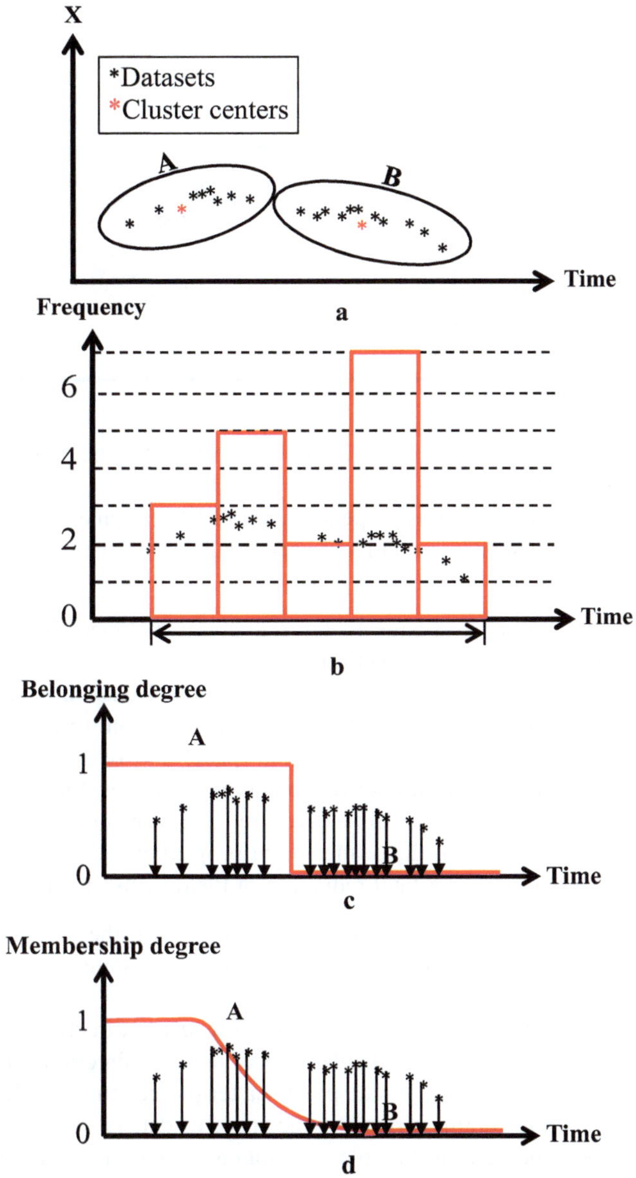

Fig. 8.13 (**a**) visual clustering, (**b**) frequency classification, (**c**) k-means clustering, (**d**) c-means clustering

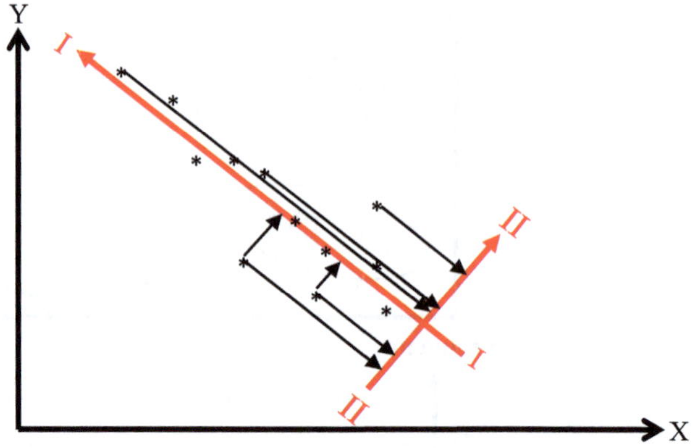

Fig. 8.14 Dimensional reduction

6. The dimension along the straight-line in I-I direction has the most scatter of data points projection.
7. The second direction, II-II, is perpendicular to I-I direction along which the data point projections' scatter is relatively very small and, hence, may be considered as negligible.

The regression line corresponds to the reduction of dimensionality from two (x and y) directions to along I-I direction axis only.

In practical applications, there are many dimensional data in several columns, which are collectively called as "features." If there are n columns, each row of the data corresponds to a point in an n-dimensional space. Like two-dimensional scatter diagram as in Fig. 8.14, there are scatter points in n-dimensional space. Like the classical regression analysis, the dimensionality of n can be reduced to several directions with significance of the scatter point projections on different directions. The following steps lead to the final reduction of the dimensionality.

1. In n-dimensional space, find the direction along which the scatter of data projections is the maximum. Since the maximum scatter is sought, the best measure is the variance, $V_{I\text{-}I}$.
2. After having identified the maximum projection scatter direction as the primary direction, there are many perpendicular directions to I-I direction and identify the next maximum variance, $V_{II\text{-}II}$, data point scatter projection along II-II direction.
3. These two directions define a plane in the space of data scatter; hence, the third direction is perpendicular to this place and then finding the next maximum data point projection scatter along this direction by variance, $V_{III\text{-}III}$, value defines the third dimension, III-III.
4. It is possible to continue like the previous steps for the fourth ($V_{IV\text{-}IV}$), fifth ($V_{V\text{-}V}$), and other additional directions, and hence, after each step application, a new direction is identified, and the number of dimensions increases,
5. It is necessary when to stop the dimensionality reduction search procedure, which depends on the percentages of total variances of dimensions to the total variance,

V_D, of the data; for this purpose, the successive variance ratio percentages are calculated, and when this ratio reaches to more than 90%, then the number of dimensionality at this step is the reduced dimension.

6. The calculations continue according to the following sequence of variance ratio, V_R, percentage calculation:

$$V_{R_1} = 100\frac{V_{I-I}}{V_D}; V_{R_2} = 100\frac{V_{I-I} + V_{II-II}}{V_D}; V_{R_3} = 100\frac{V_{I-I} + V_{II-II} + V_{III-III}}{V_D} \ldots \text{and so, on}$$

All these percentages theoretically remain between zero and one, but as the new dimension is added to the previous ones, the total variance increases steadily.

7. The first time when the final variance ratio calculation exceeds 90%, then this number, say $m < n$, dimensions are important for the explanation of the data feature, and the remaining dimensions become redundant.

It is possible than an image can include thousands of pixels, which requires the reduction of the total number. Such a huge dataset may become manageable with dimensionality reduction. The steps that are explained above are referred to as the principle component analysis (PCA) in the literature for dimension reduction. The PCA methodology can reduce the number of dimensions of original data dramatically without losing too much information. Another effective methodology that is available in the literature is the t-Stochastic Neighbor Embedding (t-SNE), which helps to dimension reduction with nonlinear dimensionality.

8.20 Ensemble Methods

The meaning of ensemble is a collection of a group of individual items that are viewed. It is possible by the ensemble methods to combine several alternative methods for the solution of the same problem through supervised machine learning methodology for getting higher quality prediction than each of the individual alternatives. For example, random forest algorithm is such an ensemble methods that combine many decision trees trainmen with different samples of the dataset. Another example is the synthetic generation with convenient stochastic modes for simulation of a natural event. The stochastic methods provide a set of alternatives, and the combination of these alternatives by a procedure leads to the final solution.

In a single machine learning, reduction of the variance and bias can be achieved by ensemble methodology applications. In this manner, the quality of precision can be balanced out. Availability of several alternatives in an ensemble provides to adapt the one with a certain risk or reliability level.

8.21 Neural Nets and Deep Learning

Neural networks are composed of at least three successive layers as input, I_1, I_2, I_3, (data) and output, O, (prediction, classification) layers connected by a hidden layer with a set of nodes, which provides nonlinearity functionalities as in Fig. 8.15. According to the required solution accuracy, the hidden layer number can be increased to few more each with connection weights between successive hidden layers, which is the primary foundation of deep learning architectural neural networks of different types (convolution, recurrent).

A lot of data even though very complex and a lot of computer power are required for the best deep learning performance. Deep learning neural network architectural structures provide many self-tuning parameters for the completion of the task. In the assessment of audiovisual, text and image classification executions for detailed and finer identification of hidden patterns in the complicated data deep learning techniques are extremely successful.

8.21.1 Learning

There are numerous real-world applications where data mining and machine learning methods are used frequently in the forms of shallow learning (Chaps. 3, 4, 5, 6 and 7). In such quite traditional methodologies, a set of rational assumptions are taken into consideration for data training, testing, and probabilistic and statistical feature detections. It is noticed later that some of the well-known assumptions are not valid in ML studies, for instance, assumptions of homogeneity, stationarity, homoscedasticity, normality, and independence structure of datasets. Especially when training data is rare, expensive, and not readily obtainable, the ML procedures cannot be applied smoothly. It is always desired to generate high-performance trainers with more easily obtained data and their relationship with other meaningful data for support. Weiss et al. (2016) named all these methodological approaches as transfer learning. There are difficulties and expensiveness in obtaining matching data training in order to match the feature pattern of data characteristic predictions. It is therefore necessary to generate a high-performance learner for a target domain training from a related datasets, and thus, transfer learning procedure can be satisfied.

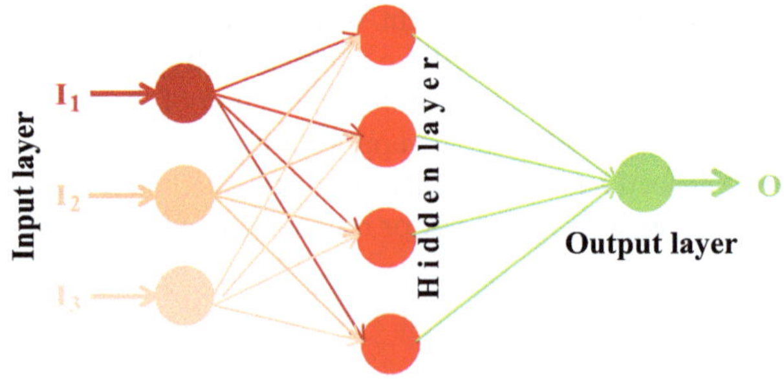

Fig. 8.15 Basic neural network architecture

8.22 Conclusions

The main purpose of this chapter is to provide basic information about machine learning (ML) procedures and artificial intelligence (AI) models that can be managed by computer software, capturing what the expert thinks about a phenomenon by trying to extract the most obvious internal properties of large amounts of data for patterns, classifications, or label description end products. The relationship is presented between AI, ML, deep learning (DL), and artificial neural network (ANN). Three ML stages are described regarding data, model, and loss function that minimizes for best model calibration and validation. A good explanation of the historical link between the ML and AI is given. It is possible to say that ML is the backbone that connects shallow learning procedures with deep learning methodological approaches. The most important learning components of ML are given as philosophy and logic of science, data collection roadmap, shallow learning (Chaps. 3 and 4), and computer coding skills. A detailed explanation is presented for uncertainty type calculation methods and their applications in practice. Unsupervised, supervised, and reinforced supervised learning procedures are explained comparatively. Different types of loss (error) functions are discussed, and the best one is recommended as a measure of root mean square error. Bivalent (crisp) and fuzzy logic classification systems are presented as k-mean and c-mean procedures for classification, respectively.

It is recommended to read on for a solid understanding and practice of ML procedures with priority explanations in Chaps. 3, 4, 5, 6 and 7.

Appendix: Required Software for Data Reliability Analysis

```
function [SS, I] = ErrorCheckByCDF(D,Xtitle,StName,Unit)
% This program is written by Zekâi Şen
% This program provides identification of data sets that do not fall within 90% confidence limits and also
% prediction of data risk values at a set of risk levels.
% Istanbul Medipol University: e-mail:zsen@medipol.edu.tr
% This program considers Gamma, Log-Normal, Extreme Value (EV Gumbel) and Generalized
% Extreme Value (GEV, Pearson) probability distribution functions (PDFs)
% This program produces Intensity-frequency curve for any given time
% duration
% D          : Dataset
% Xtitle     : Name of dataset
% R          : Risk levels
% StName     : Title
% SS         : It is the least sum of squares of probability deviations
%              from the theoretical probability distribution
% I          : Probability distribution function (PDF number)
%              If I = 1 Gamma PDF
%              If I = 2 Log-Normal PDF
%              If I = 3 Extreme value (Gumbel)PDF
%               If I = 4 Generalized extreme value (Pearson III)PDF
%              If I = 5 Weibull PDF
Risk=0.001:0.001:0.999;
R=[1-Risk(500) 1-Risk(200) 1-Risk(100) 1-Risk(40) 1-Risk(20) 1-Risk(10) 1-Risk(4) 1-Risk(2)];
n=length(D);
DM=1.1*max(D);
Dm=min(D);
x=Dm:0.1:DM;
```

```
pp=(1:1:n)/(n+1); % Data probability in ascending order
p=1-pp'; % Data probability in descending order
SD=sort(D); % Sorted time series in ascending order
pgam=gamfit(D); % Gamma PDF parameters
ygam=1-gamcdf(x,pgam(1),pgam(2));
ptgam=1-gamcdf(SD,pgam(1),pgam(2));
ppt2gam=(p-ptgam).^2;
GTest=mean(ppt2gam);
plon=lognfit(D); % Log-Normal PDF parameters
ylon=1-logncdf(x,plon(1),plon(2));
ptlon=1-logncdf(SD,plon(1),plon(2));
ppt2lon=(p-ptlon).^2;
LNTest=mean(ppt2lon);
pevd=evfit(D); % Extreme value PDF parameters
yevd=1-evcdf(x,pevd(1),pevd(2));
ptevd=1-evcdf(SD,pevd(1),pevd(2));
ppt2evd=(p-ptevd).^2;
EVTest=mean(ppt2evd);
pgev=gevfit(D); % Generalized extreme value PDF parameters
ygev=1-gevcdf(x,pgev(1),pgev(2),pgev(3));
ptgev=1-gevcdf(SD,pgev(1),pgev(2),pgev(3));
ppt2gev=(p-ptgev).^2;
GEVTest=mean(ppt2gev);
pwbl=wblfit(D); % Weibull PDF parameters
ywbl=1-wblcdf(x,pwbl(1),pwbl(2));
ptwbl=1-wblcdf(SD,pwbl(1),pwbl(2));
ppt2wbl=(p-ptwbl).^2;
WBLTest=mean(ppt2wbl);
rgam=gaminv(R,pgam(1),pgam(2));
rlon=logninv(R,plon(1),plon(2));
revd=evinv(R,pevd(1),pevd(2));
rgev=gevinv(R,pgev(1),pgev(2),pgev(3));
rwbl=wblinv(R,pwbl(1),pwbl(2));
[SS I]=min([GTest LNTest EVTest GEVTest WBLTest]);
if I == 1
  yf=ygam;
  rf=rgam;
  pr=pgam;
  PR='Gamma probability distribution;
  LC=gaminv(0.025,pr(1),pr(2));
  UC=gaminv(0.975,pr(1),pr(2));
  elseif I ==2
    yf=ylon;
    rf=rlon;
    pr=plon;
    PR='Log-normal probability distribution;
    LC=norminv(0.025,pr(1),pr(2));
    UC=norminv(0.975,pr(1),pr(2));
    LC=exp(LC);
    UC=exp(UC);
    elseif I == 3
      yf=yevd;
      rf=revd;
      pr=pevd;
      PR='Gumbel probability distribution;
      LC=evinv(0.025,pr(1),pr(2));
      UC=evinv(0.975,pr(1),pr(2));
elseif I == 4
  yf=ygev;
  rf=rgev;
  pr=pgev;
  PR='Pearson probability distribution;
  LC=gevinv(0.025,pr(1),pr(2),pr(3));
  UC=gevinv(0.975,pr(1),pr(2),pr(3));
```

```
else
  yf=ywbl;
  rf=rwbl;
  pr=pwbl;
  PR='Weibull probability distribution;
  LC=wblinv(0.025,pr(1),pr(2));
  UC=wblinv(0.975,pr(1),pr(2));
end
UG=100*(rf(1)-UC)/rf(1); % Upper confidence limit percentage according to theoretical mean
AG=100*(rf(1)-LC)/rf(1); % Lower confidence limit percentage according to theoretical mean
figure
scatter(SD,p,'k*')
title(StName)
xlabel(Xtitle)
ylabel('Exceedence probability')
hold on
grid on
box on
plot(x,yf,'LineWidth',2,'Color','r') % Theoretical PDF plot
text(DM/2,0.80,['Location parameter = ' num2str(pr(1))])
text(DM/2,0.75,['Scale parameter    = ' num2str(pr(2))])
text(0.60*DM,0.60,[' 0.50  risk = ' num2str(rf(1)),'  ',Unit])
text(0.60*DM,0.55,[' 0.20  risk = ' num2str(rf(2)),'  ',Unit])
text(0.60*DM,0.50,[' 0.10  risk = ' num2str(rf(3)),'  ',Unit])
text(0.60*DM,0.45,[' 0.04  risk = ' num2str(rf(4)),'  ',Unit])
text(0.60*DM,0.40,[' 0.02  risk = ' num2str(rf(5)),'  ',Unit])
text(0.60*DM,0.35,[' 0.01  risk = ' num2str(rf(6)),'  ',Unit])
text(0.60*DM,0.30,[' 0.004 risk = ' num2str(rf(7)),'  ',Unit])
text(0.60*DM,0.25,[' 0.002 risk = ' num2str(rf(8)),'  ',Unit])
if I == 4
  text(DM/2,0.70,['Shape parameter   = ' num2str(pr(3))])
else
end
m=min(D);
M=max(D);
hold on
line([LC LC],[0 0.975],'LineWidth',2)
line([UC UC],[0 0.025],'LineWidth',2)
legend('Data values', PR, 'Lower confidence limit (0.025)',...
  'Upper confidence limit (0.975)','--', 'Location','Northeast')
text(m,0.05,['Percentage between upper confidence limit and mean value = ',num2str(AG)])
text(m,0.10,[' Percentage between lover confidence limit and mean value = ',num2str(UG)])
end
```

References

Al-Dulaimi A, Zabihi S, Asif A, Mohammadi A (2019) A multimodal and hybrid deep neural network model for remaining useful life estimation. Comput Ind 108:186–196

Benjamin JR, Cornell CA (1970) Probability, statistics and decision for civil engineers. McGraw-Hill, New York

Bezdek C (1981) Pattern recognition with fuzzy objective function algorithms. Plenum Press, New York

Carbonell JG, Michalski RS, Mitchell TM (1983) An overview of machine learning. Tiogo Publishing Company, Palo Alto, 23

Cox RV, Kamm CA, Rabiner LR, Schroeter J, Wilpon JG (2000) Speech and language processing for next-millennium communications services. Proc IEEE 88(8):1314–1337. https://doi.org/10.1109/5.880086

Dejong G (1981) Generalizations based on explanations. In: Proceedings of the seventh international joint conference on artificial intelligence. Morgan Kauffman, Vancouver, pp 67–69
Dunn JC (1973) A fuzzy relative of the ISODATA process and its use in detecting compact well-separated clusters. J Cybern 3:32–57
Feller W (1967) An introduction to probability theory and its applications, vol 1. Wiley, 499 pp
Haffner P (1998) Gradient-based learning applied to document recognition. Proc IEEE. http://yann.lecun.com/exdb/publis/pdf/lecun01a.pdf
Kendall MG, Stuart A (1974) The advanced theory of statisrics, vol 1. Hafner
MacQueen J (1967) Some methods for classification and analysis of multivariate observations. In: Proceedings of the fifth Berkeley symposium on mathematical statistics and probabilities, vol 1. University of California Press, pp 281–296
Mnih et al (2015) Human-level control through deep reinforcement learning. Nature 518(7540): 529–533. https://doi.org/10.1038/nature14236
Murphy AH (1993) What is a good forecast? An essay on the nature of goodness in weather forecasting. Am Meteorol Soc 8:281–293
Murphy AH, Winkler RL (1987) A general framework for forecast verification. Mon Weather Rev 115:1330–1338
Nam J, Kim J, Mencia EL, Gurevych I, Fürnkranz J (2014) Large-scale multi-label text classification—revisiting neural networks. In: Calders T et al (eds) ECML PKDD 2014, Part II, LNCS 8725. Springer, Berlin/Heidelberg, pp 437–452
Rogier A, Dondersa T, Geert, van der Heijdenc JMG, Stijnend T, Moons KGM (2006) A gentle introduction to imputation of missing values. J Clin Epidemiol 59:1087–1091
Rosenblatt F (1958) The perceptron: a probabilistic model for information storage and organization in the brain. Psychol Rev 65(6):386
Samuel AL (1953) Computing bit by bit or digital computers made easy. Proc IRE 41(10):1223. https://doi.org/10.1109/JRPROC.1953.274271. S2CID 51652282
Şen Z (2022) Scientific philosophy and principles in medicine. Bentham Science Publishers Pte. Ltd. (in print)
Sze V, Chen Y-H, Yang T-J (2017) Efficient processing of deep neural networks: a tutorial and survey. Department of Electrical Engineering and Computer Science, Massachusetts Institute of Technology, Cambridge, MA
Vapnik VN (1998) Statistical learning theory. Wiley Interscience
Weiss K, Khoshgoftaar TM, Wang D (2016) A survey of transfer learning. J Big Data 3:9. https://doi.org/10.1186/s40537-016-0043-6
Wilks DS (1995) Statistical methods in the atmospheric sciences, International geophysics series, vol 59. Academic, 464 pp

Chapter 9
Deep Learning

9.1 General

Deep learning (DL) is a mixture of artificial intelligence (AI) and machine learning (ML) concepts, which helps to pave the way to automatic scientific and technological developments based on shallow learning (SL) principle improvements. In general, SL principles are quite linear and based on descriptive conceptions leading to applicable formulations and algorithms, whereas DL procedures are comparatively abstractive, complex and based on data mining statistical procedures for predictive models. Especially, large amounts of data collection, analyses, and interpretation are dealt with DL procedures in easier and faster manners. Traditional ML algorithms are linear compared with DL methodologies, which have hierarchically increasing developments, improvements, and abstractive complexities. Original work about deep learning goes back to Ivakhnenko and Lapa (1965) who were working with algorithmic learning for supervised deep feed forward multilayer perceptron (Chap. 7). Later, Ivakhnenko (1971) described a deep network with eight layers, which is still in use in different disciplines.

DL procedures are heavily dependent on computer software than simple traditional soft learning methods. Its procedures follow an architectural structure composed of successive nonlinear transformations that lead to statistically final model output. As in the SL applications, the internal dynamism of successive layers continues through iteration rules until the final prediction remains within a certain domain of acceptable error limits such as ±5% or ±10%, as required. The word "deep" implies the number of processing layers through which dataset must pass until the final prediction is reached satisfactorily.

In classical ML, most often, the learning dynamism is guided by supervisor for the directions to reach at desired result. This procedure is for future extraction, which takes tedious time durations after laborious works provided that the supervision principles are applied in a sensible and accurate manner. On the contrary, DL advantageously builds the feature pattern itself without supervision. Unsupervision

Z. Şen, *Shallow and Deep Learning Principles*,
https://doi.org/10.1007/978-3-031-29555-3_9

provides faster processing with better accuracy. Most often, ready software are provided for training data in the forms of datasets, images, maps, features, patterns, or graphs. The software executes the internal program with training dataset information and generates a similar predictive feature within error limits. The software looks simply for pattern of pixels. Successive model iterations improve previous features predictively through complex but in more accurate manner. DL algorithms sort through numerous images for the one that is compatible with the input pattern feature within short time duration of few minutes. Big data, data mining, ML, and cloud computing procedures provide achievement of acceptable accuracy level in recognition of input pattern among a set of others by means of complex statistical models, which provide opportunity to reach at the output pattern after a number of iterations. Images, texts, and sounds are preliminary datasets for learning classification tasks with appropriate models via ML supported by deep learning principles. DL differs from surface learning in that SL has only a few layers as explained in Chap. 7, but in the case of deep learning, there are many hidden layers with different tasks.

The main purpose of this chapter is to flourish basic conceptual, rational, procedural, mathematical, logical, and structural composition of a given data set for identification of meaningful hidden patterns and features leading to output labels with acceptable reliability.

9.2 Deep Learning Methods

There are a set of methods that are usable to generate acceptable DL methodologies, which are based initially on SL procedures. In such approaches, there is learning rate decay, transfer learning among the layers, training from scratch and output stages.

Learning rate decay is a procedure that controls the system based on a set of assumptions and conditions for system operation prior to learning process. In this manner, through the model weight alterations, DL principles help to control how much change the model experiences in response to the estimation error after a number of iterations. Too high learning rates may cause instability in the model structure leading to suboptimal predictions. On the other hand, very small learning rates can cause sticking after a long training procedure. The easiest and most common adaptations of learning rate during training include techniques for reducing the rate of learning over time.

Information transfer among successive layers is referred to as the transfer learning, which seeks previously trained model improvement toward perfection in the next steps. After the introduction of dataset to the system and the necessary adjustments are made, then new task performances can be achieved with more specific categorization abilities. Much less data are required, and thus, computation time reduction is possible. The researcher should try to develop network architecture model for a large data set digestion. Such a training from starch can take days or weeks first to train the large set of data and then to make predictions. The over fitting problem, if any, can be solved by dropping units and systematic connections during the neural network training procedure. This is referred to as the dropout model that helps to improve neural network performance on supervised learning works.

9.2.1 Deep Learning Neural Networks

Artificial neural network (ANN) is an advances ML algorithm that furnishes the foundations of DL. In addition to the conventional ANN (Chap. 7), there exist different versions of ANN among which is the convolutional neural network (CNN), recurrent neural network (RNN), and feedforward neural networks (FNN). Their functionalities are almost similar in the way that after the entrance of input dataset and execution through a set of hidden layers, the output pattern is obtained within a certain range of acceptable error limits. In general, neural networks work by means of large amount of dataset and reach the final result after a trial-and-error iterative process. Consideration of the first few iterations by close expert guesses in a structural and systematic manners helps to obtain the final result in comparatively short time durations in a more sensitive and accurate way. DL methodology provides an effective manner of analysis to arrive at acceptable accuracy levels after training.

SL-based DL procedures have similarities to human mind, and therefore, these procedures can be applied to almost all problems provided that there are enough datasets. Among the most frequent application domains are graph, image and map recognition, natural language processing (NLP) including fuzzy logic principles, language translations, medical diagnoses, and stock market affairs. DL procedures provide extensive interaction between human and machine through intelligent software.

9.2.2 Limitations and Challenges

DL algorithms, procedures, and models are not effective in cases of missing observations and measurement datasets, because they infer what are available in the input dataset and, accordingly, provide estimations and predictions in the output results.

The biggest limitation of DL models is training stages through observations. One of the big problems in DL is the bias issue. Any dataset with some bias in it will be trained by a DL model alternative, but training procedure will assimilate the bias, and finally, the results will also include the bias effect to a certain percentage. Another problem in DL is the issue of learning rate. At high rates, the model converges rapidly but may produce suboptimum prediction estimations. Otherwise, the procedure may get stuck without even reaching to an acceptable result. Another possible limitation is concerned with the availability of hardware possibilities such as random access memory capacity and hard disk drive. In a DL procedure, execution efficiency improvement is preferable during short time durations. However, such requirements may cause to expensive investments in addition to large amount of energy consumption. As has been aforementioned, DL requires large amount of dataset and, hence, more effective parameter inclusion. It is also possible that after training procedure, DL model may get many stuck with inflexible and nonreturnable conclusions. Even a large dataset is available, SL principles must be laid down in a rational and systematic manner for effective solutions with acceptable accuracy limitations. This is tantamount to say that the user must not expect from DL software to yield acceptable and sensitively accurate results, if the internal structure of the model composition is not well constructed architecturally. Otherwise, any ready DL software program is at the service of user blindly without logical and rational fundamentals, which leads to similar application results in a manner of memorization, because the SL principles may not be taken into consideration effectively.

9.3 Deep Learning and Machine Learning

As for the problem-solving is concerned, DL remains as a subset of ML, which requires expert views for identification of most features concerning the problematic issue. One of the main differences is that DL tries to identify the problematic features incrementally without any need for expertise, and thus, DL procedure requires much longer training execution time within few minutes or hours. However, ML takes shorter time durations during testing. DL algorithms test time requirement increases with the size of data.

In general, scientists prefer traditional ML algorithms than DL for problem-solving in majority of datasets, because of comparatively simpler and superior interpretability of the algorithm internal structure and outputs, which provide more sensitive solution availabilities. ML procedural applications do not require large data amounts.

DL as a ML technique trains by validly developed software to identify after a set of layer executions to estimate, predict, or classify the input data into different pieces. Based on SL principles, DL procedures act similar to human brain, memory, and reasoning mind activities. DL is also a subset of AI domain, which implies learning similar to human brain functions, whereas ML provides computers to think and act with less human intervention.

9.4 Different Neural Network Architectures

The most detailed and general information about the neural networks is presented in Chap. 7 concerning artificial alternatives, which pave the way to CNNs. In the following sections, the reader must base the SL principles from that chapter in order to grasp and understand firmly the deep learning neural network procedures.

9.5 Convolutional Neural Network (CNN)

Preliminary basic information about CNN algorithm can be obtained from Hubel and Wisel (1959). After this year, several handwritten digital recognition procedures took place in the literature. These early trials paved the way to communicate with computers for fast computations. Original pattern recognition procedures were functional only by means of forward feeding operations, which required improvement inside the proposed methodology to reach the acceptable solution. Later, the internal model structural improvements started activation after forward propagation subsequent back-propagation works until practically acceptable solutions are obtained. Of course, back-propagation procedure is supportive in the case of supervised training as explained in Chap. 8. CNN procedure application was originally restrictive due to the insufficient computer facilities, finite data availability, training, and insufficient computer software. The application of CNN procedure became possible after computer hardware and software improvements due to large amount of data in addition to natural linguistical concepts toward pattern recognition.

CNN algorithm searches for useful classification, clustering, segmentation, and image processing procedures by inclusion of few convolution layers in their hidden layers part. Final optimum solution is achieved by convolution layers, which help to shift input dataset and provide systematic scanning possibilities for filtration. Thus, CNN architectural structure provides successive activation of shifting, scanning, and filtration operations. These DL algorithms guide to inspiration through visual inspections for short duration execution of the internal operations to reach better solution improvement. Another supportive feature in the CNNs is the language processing possibilities similar to fuzzy logic rule base as explained in Chap. 3. At initial stage, visual inspection and inspirations are important guidance tools based on the dataset features. It is advised that whoever plans to use CNN algorithm for the problem, first, the researcher must assess the data features through any convenient SL procedures as already explained in Chaps. 3, 4, and 5. Architectural structure of CNN technique is similar to SL ANN procedures explained in Chap. 7. As a digression from the classical ANN, the convolution layer as backbone of CNN provides refined input dataset pattern features through supervised learning procedure. Convolution layers have learnable filters in addition to several kernel functions. Such a structure provides opportunity to identify in detail different feature maps after heavy computation. Nonlinear activation functions similar to the ones explained in Chaps. 7 and 8 are used in the CNN hidden layers; thus, the input features are identified with the aid of kernels. These layers also help to reduce the complexities by parameter sharing procedures. The output from the convolution layer is taken as a subsample to generate an output. The most frequently used subsampling procedures in practice are maximum pooling, average pooling, and minimum pooling techniques that help to reduce the computational parameter number and network translation invariance. A small portion of convolution layer output is taken as input by pooling and subsampling procedures to down-sample and, thus, to produce a single output. The last part of the hidden layers in the CNNs architectural structure is referred to fully connected (FC) layers that are active during the feedforward procedure application; they operate on an input part, and each input relates to all neurons. The number of FC layer may be one or more depending on the output refinement request. The main role of FC layers is to transfer convolution layer outputs into CNN model final output. During the CNN algorithm application, as for the image classification, future extraction and classification parts are combined by CNN architecture. Convolution and classification parts and common activation are achieved by future extraction mechanism. Under the light of what have been explained above, one is able to expose the interconnected activities in a DL procedure as activation, normalization, objective (loss) function (Chap. 8), regularization speed, segmentation, detection, classification, and optimization.

A CNN architectural structure includes a single or multiple convolution-pooling layer blocks leading to again one or two FC layers prior to the output layer. The main concern is with the convolution layer, which tries to learn input feature representation and consists of several other learnable convolution layers each of which is used to compute different feature maps as filters between the input and output layers. Each convolution layer with feature map transfers knowledge to the next pooling layer. The convoluted results provide reductions by applications of nonlinear activation functions. The model complexity is reduceable by the parameter sharing property. The main role of pooling layer is to down sample the input feature from its previous convolution layer by dimensionality reduction. The last part of CNN has one or few FC layer, which typically exist in feedforward neural network. The final pooling layer output enters FC layer as input, and thus, CNN output products take shape in the form of feature maps. Although layer design is a significant initial procedure in CNN architecture, but for successful results, there are other different procedures such as activation functions, normalization, optimization (loss function), and satisfaction of computation speed. Among the three major fields that are treatable by CNN methodology are classification of images, detection of objects, and segmentation. In addition to these major fields, CNN applications became currently applicable in different disciplines such as text detection, language processing, question answering visually, etc.

9.5.1 CNN Foundation

As already mentioned in the previous section, the pioneering initial work is proposed by Hubel and Wisel (1959), who worked on animal visual cortex cells and took inspiration from.

Fukushima (1980) proposed the concept of neocognitron, which constituted the first theoretical CNN model foundations. Later, LeCun et al. (1990) suggested the CNN architectural framework for handwritten digit recognition. Further improvement in CNN architectural structure improvement is given by Krizhevsky et al. (2012). In general, a CNN model is composed of single or multiple layers of convolution and pooling in addition to subsequent one or multiple FC layers and, finally, an output layer (Fukushima 1980).

The CNN structure can learn objects that are highly abstracted features and converts them to more efficiently identifiable features. The simple block diagram conceptual model of CNN is presented in Fig. 9.1 under the light of the previous explanations.

For multiple layer CNN structure, the two layers of convolution and pooling can be repeated conveniently for the betterment of the output quality. As for the computer vision, CNN is preferable over the classical ANN due to the following reasons:

1. There is weight shearing feature that reduces trainable number of parameters in the system. This feature helps to avoid over fitting by generalization improvement.
2. The classification and future extraction layers learn jointly, and thus, the CNN model output appears more organized depending on the extracted features.
3. Facilitates the implementation of a large network, which is rather difficult to achieve by other neural networks.

9.5.2 CNN Network Layers

The sequence of layers as in Fig. 9.1 can be explained rather simply under the light of the following subsections.

9.5.2.1 Input Layer

As in many neural network procedures, the data is entered by this layer in the form of digits that represent the visual pattern. CNN model input is in the form of multichannel image, and in Fig. 9.2, a 3-channel type is given for grayscale image.

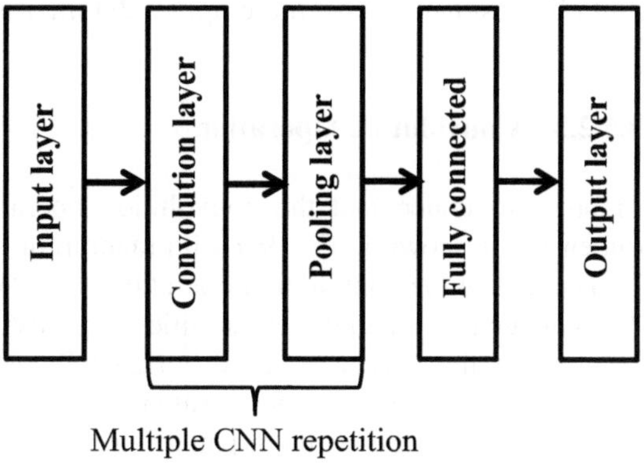

Fig. 9.1 Conceptual CNN layers

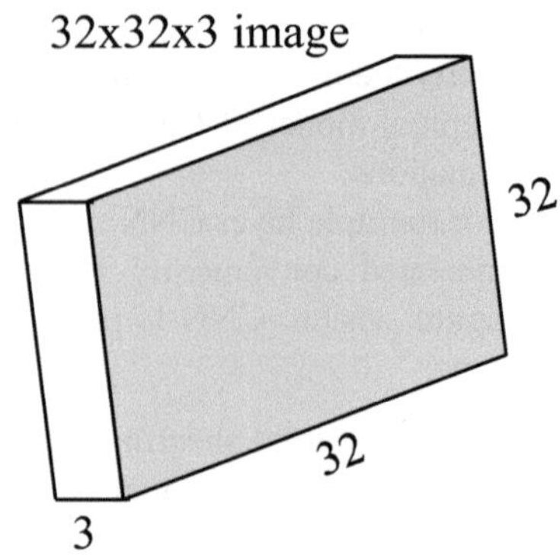

Fig. 9.2 RBG (red-blue-green) image

Fig. 9.3 2D kernels

-3	1
2	0

Fig. 9.4 Grayscale images 4 × 4

-3	0	2	1
1	4	3	2
-1	0	-1	1
2	0	1	0

9.5.2.2 Convolution Layer

It is composed of convolution kernels (filters), which convert the input data to an output feature generation map. A grid of discrete values constitutes a set of kernels with known weights as in Fig. 9.3. Initially, each kernel is value allocated by random numbers, which are then tuned to extract meaningful features. Although there are multidimensional filters, in the figure, 2D filter is shown:

9.5.2.3 Convolution Operation

In order to understand the convolution operation, a grayscale image of 4 × 4 dimension is shown in Fig. 9.4 with random initialization 2 × 2 kernel in Fig. 9.3.

The first of the convolution operation is to slide 2 × 2 kernels over the complete 4 × 4 image to scan all the possibilities. It is advised to follow a systematic way first by moving the kernel window horizontally and then vertically. Corresponding values are multiplier and then written on the final window pattern, which is 3 × 3 in this example. The resultant pattern grid is shown in Fig. 9.5 after 3 shifting and summation of corresponding cell multiplication values.

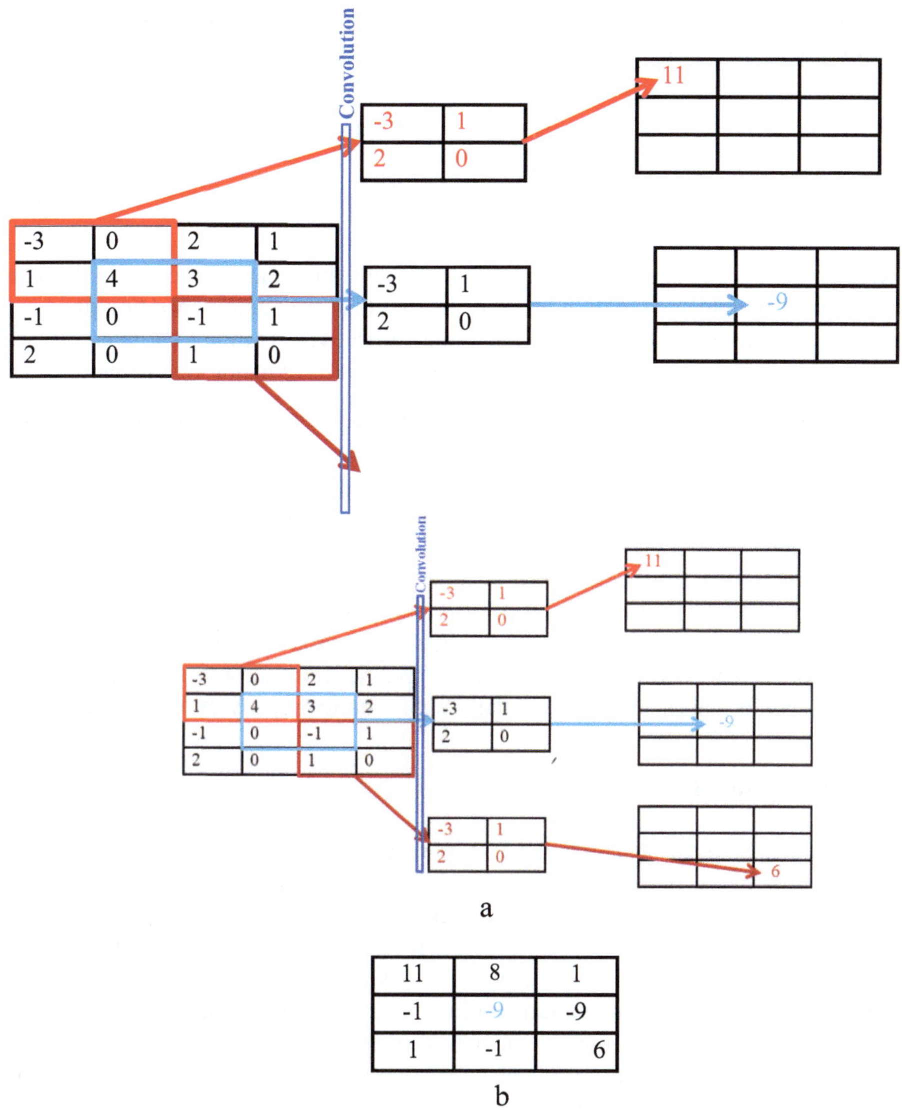

Fig. 9.5 Convolution calculations, (**a**) three examples, (**b**) complete result

In these calculations, one stride horizontal and vertical scans are used, but there are other stride values. It is obvious that increase in stride value will yield smaller grid results, which may not be in refinable scale. Image can be improved with more importance provided that the padding yields border size information; otherwise, the border side features fade away. Input image size can be increased by padding operation so that feature map at the output can also increase. For instance, the convolution operation can be achieved with zero padding and three-stride value similar to the convolution procedure shown in Fig. 9.5 with nonoverlapping shifting procedure horizontally and vertically by consideration of all possible alternatives.

After the convolution operation, it is possible to find the output feature map size according to the following formulation:

$$h' = \frac{h - f + p}{s} + 1 \tag{9.1}$$

$$w' = \frac{w - f + p}{s} + 1 \tag{9.2}$$

where h' and w' are height and width of the output feature map, respectively. Here, h and f are the height and width of input image; p is the convolution operation padding, and s is the stride of convolution operation, respectively. Among the main advantages of convolution layer, the following two points have major significance:

1. In traditional ANN, each neuron in one layer connects with each neuron in the next layer. In the CNN model, small number of weights play significant role between two layers through sparse connectivity property. This provides memory sufficiency, because small number of weights and connections are kept in the computer storage. Usage of dot multiplication rather than matrix multiplication provides swift computation in short time duration.
2. All neurons work with every pixel individually of the input matrix, and therefore, there is no need for dedicated weights between neurons of subsequent layers. It is possible to learn one set of weights for all input rather than learning new weights for each neuron. This is referred to as the weight sharing property of the CNN model.

9.5.2.4 Pooling Layer

These layers help to subsample the feature maps of large size and shrinks to lower scale feature maps. In each pooling step, the most dominant features are preserved. Pooling operation is applied after specification of pooled region size and convolution operation stride application like the previously explained content. There are different types of pooling such as minimum, average, maximum gated, tree polling, and alike. Among these operations, maximum pooling procedures are the most used one in the applications. Sometimes, the overall CNN performance is decreased due to the pooling layer. This is due to the fact that pooling layer spends time to find the specific feature in the given input image.

In the following example, a 4×4 matrix (Fig. 9.6a) is considered as initial input data. If a 2×2 filter is used by taking into consideration maximum pooling operator over the input matrix as maximum pooling procedure, then the output includes the results as in Fig. 9.6b. Maximum pooling means a sample-based discretion process as is obvious from Fig. 9.6c, d the main purpose is to down sample a given input dataset for its dimensional reduction and thus to allow some convenient assumptions about features in subregions. In a CNN pooling layer, the architecture is for subsampling, dimensionality, and computational reduction in computers for parameter estimates during execution. Pooling layers are useful in CNN procedure for feature consolidations after the convolutional layer feature learning pattern. There are three pooling operations.

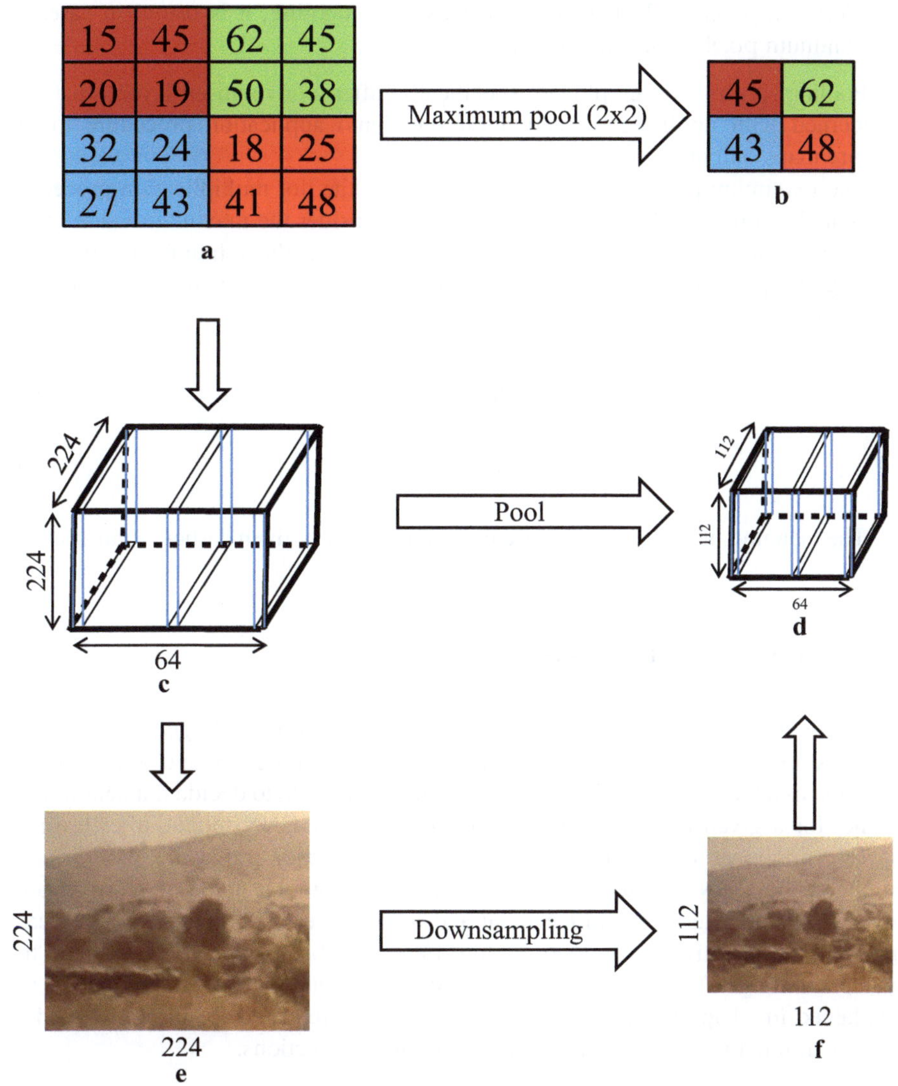

Fig. 9.6 Maximum pooling procedure examples

1. *Maximum pooling:* This is useful for extracting the most activated feature presence in the dataset. It helps in extracting low-level features like edges, points, etc. This is the most frequently used operator in CNN methodology than other pooling alternatives especially for image classification task.
2. *Average pooling:* This summarizes the average feature presence in the original dataset.

3. *Minimum pooling:* Just opposite of maximum pooling, it is useful in cases of minimum pixel value extractions.

Pooling layers are employed specifically after convolution layers that are concerned with nonlinearity (activation functions) application procedure to the feature maps output.

The maximum pooling procedure is similar to the dot multiplication procedure presented in Fig. 9.6. The only difference is to consider the maximum value of the grid shifting instead of dot multiplication. In such a case, the output feature map size finding is possible after the pooling operation according to the following formulations:

$$h' = \frac{h - f}{s} \tag{9.3}$$

$$w' = \frac{w - f}{s} \tag{9.4}$$

The meaning of each term in these two expressions is given earlier in this section.

9.6 Activation Functions

The main purpose of these functions is to map input to output. In general, the input value is obtained after weight sum calculation of neuron's input and sometimes by addition of a bias factor, if necessary. These functions help to decide if a neuron fires or not for a given input for output production. In any CNN model, after learning layer, a nonlinear activation layer is used for classification procedure. Nonlinearity supports the CNN architectural structure to learn complex patterns and provides nonlinearity managements in the input and output layers. Another very significant property of an activation function is to differentiate back propagation error in the model training. The most commonly used activation functions in any ANN are explained in Chap. 7, but herein, some of them are repeated for the CNN models with additional new information in the following subsections.

9.6.1 Sigmoid

Any range of input from large negative to large positive numbers is confined between 0 and 1. It has an S shape according to the following mathematical formulation and resultant shape in Fig. 9.7:

$$f(x) = \frac{1}{1 + e^{-x}} \tag{9.5}$$

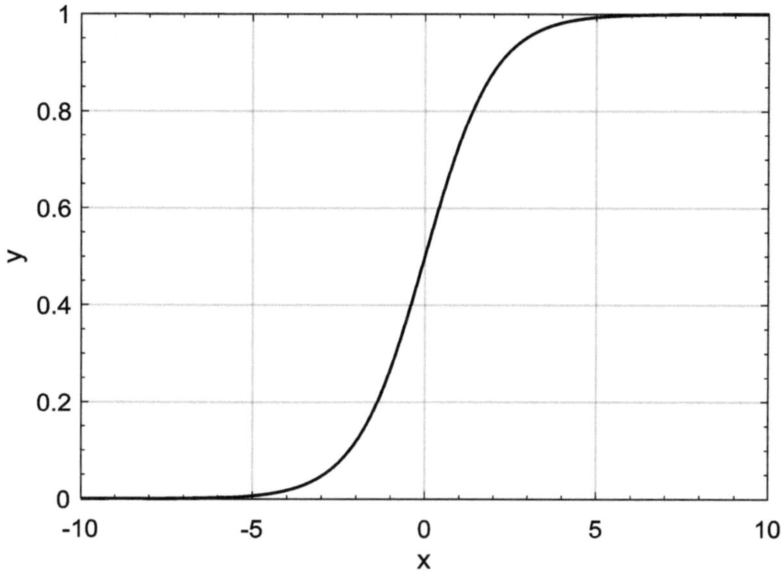

Fig. 9.7 Sigmoid activation functions

9.6.2 *Tanh*

This activation function binds real numbers within the confinement from −1 to +1 according to the following mathematical expression and its graph in Fig. 9.8:

$$f(x) = \frac{e^x - e^{-x}}{e^x + e^{-x}} \tag{9.6}$$

9.6.3 *Rectifier Linear Unit (ReLU)*

In CNN model, this is the most commonly used activation function, because it converts input values to positive numbers. One of its advantages is minimal computational time requirement than others. Its mathematical function and resulting graph are shown in the following mathematical expression and Fig. 9.9, respectively:

$$f(x) = \max(0, x) \tag{9.7}$$

ReLU activation function may give some major problems in cases of large gradient, and some of the neurons may not get fired again. This problem can be solved by the ReLU alternative activation function explained below.

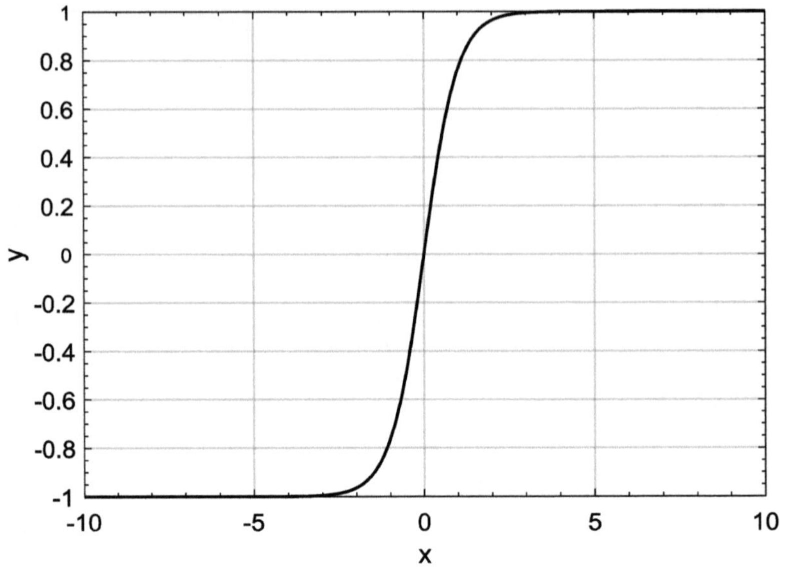

Fig. 9.8 Tanh activation function

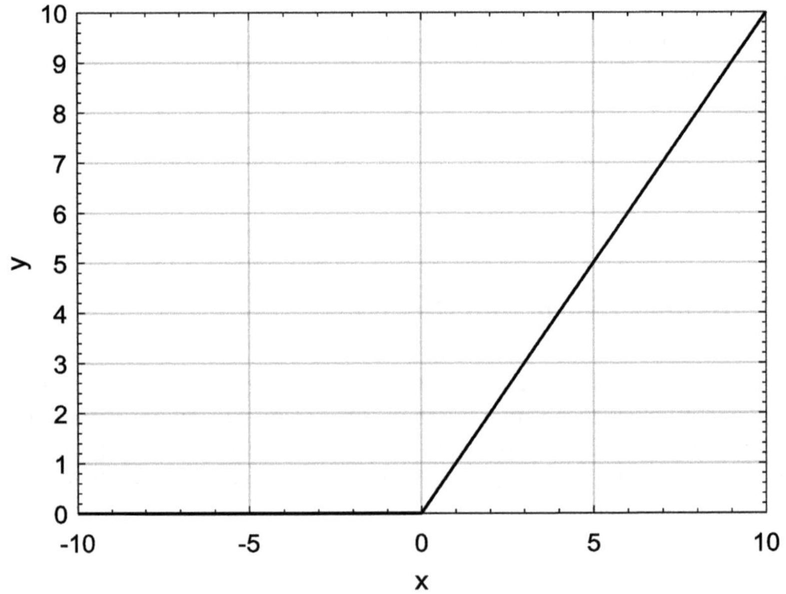

Fig. 9.9 ReLU activation function

9.6.4 Leaky ReLU

Negative input values are not ignored by this activation function completely, because it downscales these input values. Its mathematical expression and graphical presentation are given mathematically and graphically in Fig. 9.10:

$$f(x) = \begin{cases} x & \text{if } x > 0 \\ mx & \text{if } x \leq 0 \end{cases} \tag{9.8}$$

In the figure, m is adapted as 2, but different values can be taken into account, and the similar two-part linear lines can be obtained.

9.6.5 Noisy ReLU

This is the noisy form of ReLU activation function, which is loaded by noise according to normal (Gaussian) probability distribution function. Its mathematical expression and graphical representation are given below (see Fig. 9.11):

$$f(x) = \max(0, x) + \varepsilon \tag{9.9}$$

where ε is random varşable according to the standard (Gaussian) PDF with zero mean and unit variance.

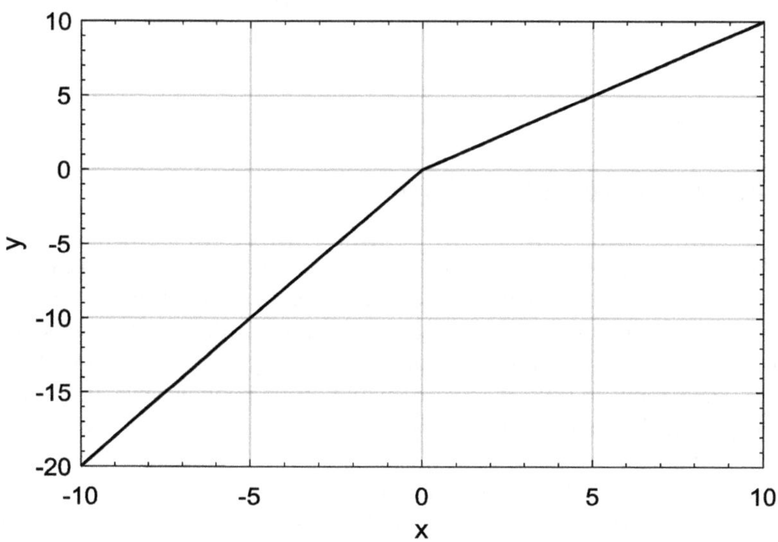

Fig. 9.10 ReLU activation function

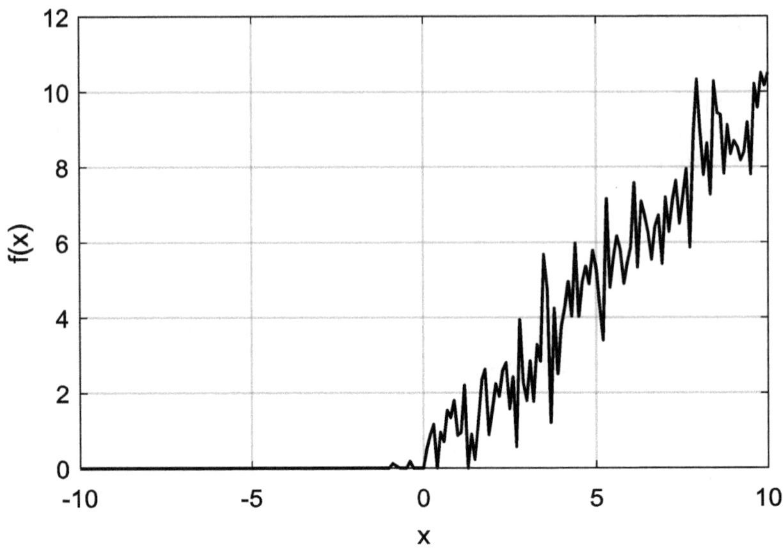

Fig. 9.11 Noisy activation functions

9.6.6 Parametric Linear Units

Even though it is very similar to Leaky ReLU activation function, the leak factor tuning is achieved during the training process. Its mathematical formulation is as follows, where αlearning weight is:

$$f(x) = \begin{cases} x & \text{if } x > 0 \\ \alpha x & \text{if } x \leq 0 \end{cases} \tag{9.10}$$

9.7 Fully Connected (FC) Layer

These layers constitute the last part of any CNN model architecture, and each neuron in this layer is connected with its previous layer neurons. The last FC layer is classification output layer. The FC layers operate feedforward wise similar to traditional ANN system following traditional multilayer perceptron (MLP) neural network. Their input is from the last convolution or pooling layer in the form of future maps, and the matrices are generated as a vector, which feeds the FC layer for final CNN output as in Fig. 9.12.

Two-dimensional pooled feature patterns are converted into single linear features as a linear vector, which is an input to the FC layer for image classification. It is a

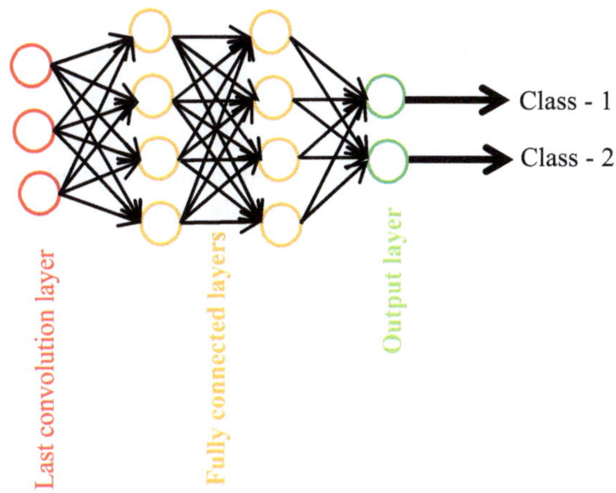

Fig. 9.12 FC layers architecture

component of feedforward neural network architectural structures that are at the last few layers in the CNN architectural structure and takes information from convolution layer followed by pooling layer and then FC layer, which enters the output layer for final pattern recognition.

9.8 Optimization (Loss) Functions

One of the most important stages in CNN model application is to check the validity of the final output production by measuring the efficiency of training operations to appreciate the quality of fitting by examining the prediction error, which should be optimized after a sequence of feedforward and backward training. For his purpose, there are different functions depending on the different loss functions that are available depending on the type of problem concerned. Although some of classical loss function alternatives are proposed in Chap. 8, in the following subsections, some others are explained briefly.

9.8.1 Soft-Max Loss Function (Cross-Entropy)

This measurement criterion is based on the probability, p, of the output, and it is frequently used as an alternative to square error function in the multiple classification problems. Herein, p is the probability for each output category, and y is the desired output. Hence, the probability of each class is defined according to the following expression:

$$p_i = \frac{e^{\alpha_i}}{\sum_{k=1}^{N} e^{\alpha_k}} \tag{9.11}$$

Here, N is the output layer neuron number, and e^{α_i} is for unnormalized output from the previous layer in the network structure. Accordingly, cross-entropy, H(p,y), is defined as:

$$H(p, y) = -\sum_{i=1}^{N} y_i \log(p_i) \tag{9.12}$$

9.8.2 Euclidean Loss Function

This is equivalent to mean square error (MSE) that is frequently used in statistical fitting controls. It is the difference between the prediction output, $p \in \mathbb{R}^N$, and actual output $y \in \mathbb{R}^N$ in each neuron of CNN output layer, which is defined as:

$$H(p, y) = \frac{1}{2N} \sum_{i=1}^{N} (p_i - y_i)^2 \tag{9.13}$$

9.8.3 Hinge Loss Function

It is used in cases of binary classification problems most frequently in vector support machines (VSMs). The difference between two target classes is maximized by optimization procedure. Its mathematical expression is as follows:

$$H(p, y) = \sum_{i=1}^{N} \max[0, m - (2y_i - 1)p_i] \tag{9.14}$$

Here, m is the margin, and it is set to 1 normally.

9.9 CNN Training Process

The CNN training process has the following four subsequently executable steps for completion:

1. Data pre-processing and augmentation
2. Initialization of parameters
3. CNN architectural structure regulation
4. Selection of optimizer

The pre-procession of data is necessary for raw data artificial transformation cleanness, more feature representative, better learnable, and in a convenient uniform format. This is achieved prior to data entrance to the CNN architectural structure. CNN performance is directly proportional with the amount of the training data, because efficiently good data improves the CNN model accuracy. The following stages are necessary for data pre-preparation:

1. The subtraction of the arithmetic average, $\overline{x}$, from the dataset provides new data sets, which varies around the zero value as in Fig. 9.13. The following expressions are for the zero centering procedure:

$$X' = X - \overline{X} \tag{9.15}$$

where

$$\overline{X} = \frac{1}{N}\sum_{i=1}^{N} X_1 \tag{9.16}$$

2. The next step is standardization of the zero mean data such that the standard deviation will be equal to unity. The mathematical expressions are as follows:

$$X'' = \frac{X'}{S} \tag{9.17}$$

Here, the standard deviation is defined as in the statistical calculations as follows:

$$S = \sqrt{\frac{1}{N} X'^2} \tag{9.18}$$

The standardization procedure is shown in Fig. 9.14.

Zero mean and unit standard deviation of the transformed data is also dimensionless; hence, CNN procedure is applied to such a transformed data set.

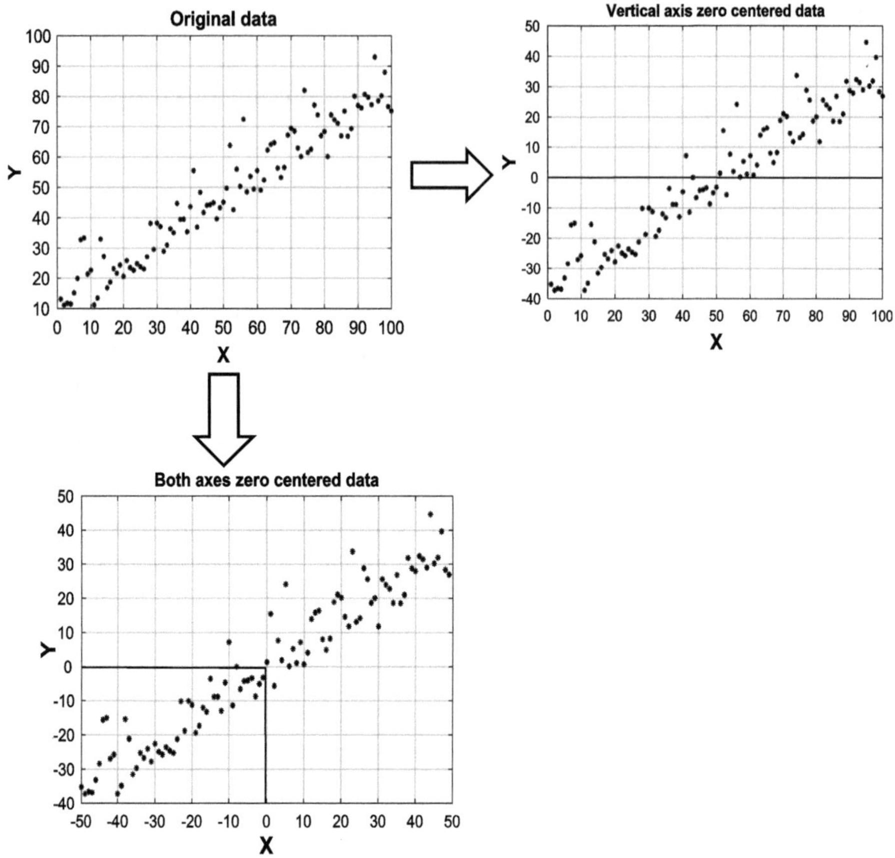

Fig. 9.13 Zero data centering

3. Artificially to expand the training, dataset size is referred to as the data augmentation technique, which can be achieved by the application of different operators to the sample dataset transformation as one or many new data samples for training process usage. There are different data augmentation procedures including scaling, contrast adjustment, translation, cropping, rotation, flipping, and alike (Chap. 5). Few of them can be applied individually or collectively. Data augmentation helps to avoid CNN model over-fitting problems.

9.10 Parameter Initialization

The most important initialization procedure in any ANN is first to establish and decide about its architectural structure and then its training for parameter learning. In general, the full procedure includes the steps of parameter initialization, optimization

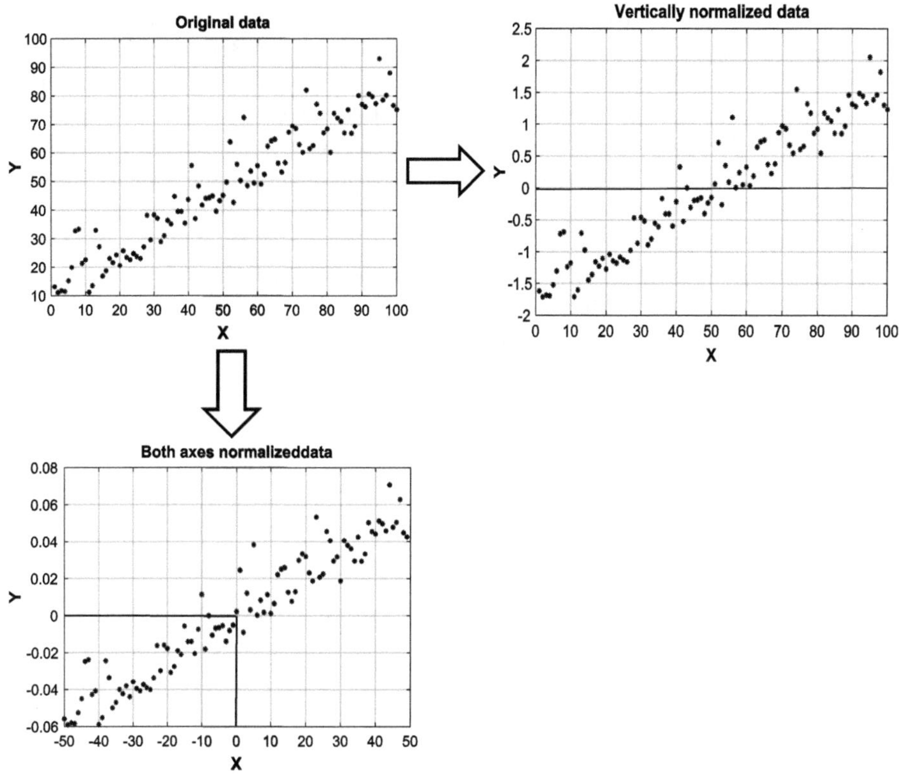

Fig. 9.14 Data standardization

algorithm in accordance with the problem solution purpose, and repetition of successive operations of forward propagation, cost (optimization) function computation, output label identification, and gradient computation based on cost function during back propagation for parameter renewal.

The initialization step can be critical to the model's ultimate performance, and it requires the right method. To illustrate this, consider the three-layer neural network below. You can try initializing this network with different methods and observe the impact on the learning.

CNN model fast convergence and final output accuracy are dependent on training process weight initializations. There are different techniques to initialize the weights:

1. *Random Initialization:* Convolution and FL layer weights are initialized randomly, where in general, each digit is different from others. For this purpose, normal (Gaussian) PDF (Chap. 4) or any other PDFs can be used but with very low standard deviation values as 0.1 or 0.01 and zero arithmetic averages. The main problem in random initialization is the possibility of vanishing or exploding gradients. Among some of the random initialization methodologies are Gaussian, uniform, and orthogonal procedures.

2. *Xavier Initialization:* This is valid in cases of sigmoid or tanh activation functions usage for neural network weight initialization (Glorot and Bengio 2010). Its main concept is random parameterization according to a uniform PDF that includes values depending on the number, n, of inputs to the node between −1/sqrt(n) and +1/sqrt(n). Another version is the standardized Xavier initialization to the traditional one depending also on the number, m, of outputs from the node as –sqrt(6)/sqrt(n + m) and + sqrt(6)/sqrt(n + m).
3. *He initialization:* He et al. (2015) provided an extended version of Xavier methodology by consideration of ReLU activation function, and the mathematical form of this procedure is achievable through the following expressions:

$$y^i = W_j^i \mathrm{ReLU}\left(x^j\right) + b_i \tag{9.19}$$

$$z^j = \mathrm{ReLU}\left(x^j\right) \tag{9.20}$$

$$\mathrm{Var}\left(y^i\right) = \sum_{i=1}^{n} E\left[\left(z^i\right)^2\right] \mathrm{Var}\left(W_j^i\right) = \sum_{j=1}^{n} \frac{1}{2} E\left[\left(x^j\right)^2\right] \mathrm{Var}\left(W_j^i\right) \tag{9.21}$$

The same authors advised that these formulas can be adopted in both forward and backward cases.

9.11 Regularization to CNN

A CNN model ends up with over-fitting in case of exceptionally well on training data, which is referred to as generalization. Otherwise, an under-fitting result generation comes into view, which happens if the model did not learn enough from the training data. The justified fitting results take place if model shows well performance by proper learning from the training and test data. By consideration of the following intuitive ideas, regularization process helps to avoid over-fitting results.

9.11.1 Dropout

Each training epoch neurons are dropped out randomly to distribute equally to all neurons during feature selection power: hence, the model is forced to learn several independent features. Any dropped neuron does not take active role during forward or backward propagation operations. However, in the testing process, the prediction is performed by means of the full-scale network.

9.11.2 Drop-Weights

Similar to the previous case, but here, instead of neuron dropouts, weights are dropped randomly during each training epoch.

9.11.3 The i^2 Regularization

The main idea behind this regularization is to fade out of the weights toward zero and addition of a penalty term equal to the squared magnitude of the loss function coefficient. Larger weight vectors are heavily panelized by addition of $0.5\lambda\ \|w^2\|$ to the objective function, where λ is a parameter, and it is concerned with penalization strength. In case of output layer N neurons, y_n and p_n prediction, and actual data values, respectively, then the following expressions are applicable with Euclidean objective cost function, C:

$$C = \text{loss} + 0.5\lambda\|w^2\| \tag{9.22}$$

and

$$C = \sum_{m=1}^{M}\sum_{n=1}^{N}(p_n - y_n)^2 + 0.5\lambda\|w^2\| \tag{9.23}$$

Herein, M is the training number. Finally, the weight increment rule based on the i^2 regularization can be expressed as follows:

$$\text{Weight increment} = \underbrace{\text{argmin}}_{w}\sum_{m=1}^{M}\sum_{n=1}^{N}(p_n - y_n)^2 + \lambda\|w\| \tag{9.24}$$

9.11.4 The i Regularization

This is almost similar to the previous one with the only difference that the absolute magnitude coefficient is considered instead of squared magnitude as penalty to the loss function:

$$C = \text{loss} + \lambda\|w\| \tag{9.25}$$

9.11.5 *Early Stopping*

In order to control stopping stage earlier, cross-validation process is employed with 20–30% of train dataset part for performance behavior evaluation. As the new trial worsens, one can stop.

9.12 Recurrent Neural Networks

These networks are useful as deep learning tools for sequential data modeling. In the sequential execution, a deep forward strategical procedure is employed with specific parameters for each sequence. For each element in the sequence, the same weight is attached, and hence, parameter number is reduced significantly. The RNN methodology is also applicable for two or more dimensional datasets in the forms of graphs and spatial patterns. The concepts of ML procedure are also valid in RNN such as feature types, classification, and similar training procedures. Feedforward subsequent back forward training trends are available through multiple hidden multilayer perceptron. Multiple-layer perceptrons are dealt with in Chap. 7 in addition to CNN in the previous sections, which are necessary with their very productive prediction capabilities concerning various tasks. In almost all tasks, word sequences are involved in the forms of natural language, speech recognition, and video captions.

It is necessary to follow carefully sequential data for successful strategy determination afterward RNN method application. In the strategy, there must be element ordering capacity task in addition to variable length sequence parameterization independent of sequence length. In an element-level model, all sequence elements share the same parameters. The contexture information in the sequence is captured; thus, an output is reached, which behaves as context for the subsequent element in the same sequence. After all these tasks, a final element-level output is reached, and hence, encoded information becomes useful for a category prediction. It is possible through another RNN procedure to decode for machine translation, speech recognition, and similar works.

9.12.1 *RNN Architecture Structure*

The main digression of RNN from CNN is that there is at least one feedback possibility in a hidden layer in the form of round loop for better model activation. For example, MLP is a standard part in the structure of RNN in addition to the loops, and thus, a nonlinearity is provided for better temporal predictions along the sequence, which are kept in a memory. There are structures where each neuron is connected to others with stochastically activation functions. In case of stochastic activation functions, simulated annealing procedure (Chap. 7) seems more appropriate. Figure 9.15 provides a simple RNN architectural structure.

A convenient time step must be chosen for the problem at hand, and it must correspond to real network operation procedure. A delay unit is necessary for activation function holding instruction at the next time step. If the input-hidden layer, hidden-hidden layer, and hidden-output layers' weights are shown as W_{ih}, W_{hh}, and W_{ho}, respectively, then one can write implicitly nonlinear hidden and output activation functions h(t) and o(t) as follows:

$$h(t) = f_h[W_{ih}x(t) + W_{hh}h(t-1)] \tag{9.26}$$

and

$$o(t) = f_{out}[W_{ho}h(t)] \tag{9.27}$$

Here, the state definition is by the hidden unit activations, h(t), set, and apart from input and output space in-between is the state space, in addition to the input and output spaces, there is also a state space. The dynamical system dimensionality is composed of the state space and the number of hidden units. After all what has been explained, there remains important questions concerning the RNN architecture model as in any other models about the stability, controllability, observability, and manageability? As for the stability, the important parts are neural network output boundless over time and possible small changes concerning inputs or weights as response of this dynamic system. Controllability of a RNN is concerned with initial state steerability to any state during a finite number of time steps. Observability, after stability and controllability checks deal with possibility to observe the results. Finally, manageability is possible after all the previous stability, controllability, and observability steps' checks.

Another feature of RNN systems is possible approximation to any desired accuracy without any restriction on the state space compactness with enough number of sigmoid hidden layers, which is also indicator of computational power of the

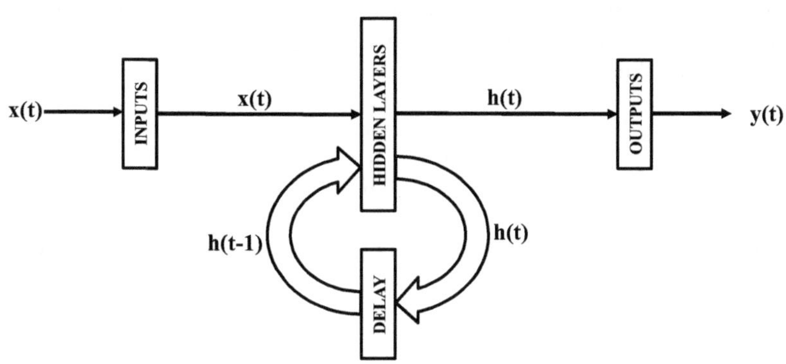

Fig. 9.15 A simple RNN architecture

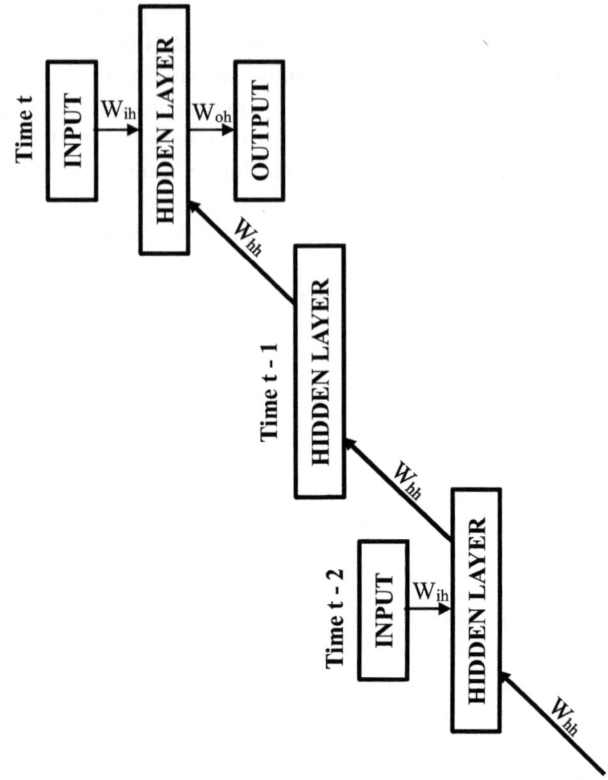

Fig. 9.16 Explicit structural time steps in RNN

system. It is possible to convert by unfolding over time, the recurrent networks into a feedforward network. In Fig. 9.16, more explicit structural components of a RNN is presented by considering each time step.

Common back propagation procedure version learning algorithm is valid in RNN system as back propagation through time coupled with gradient descent search for optimum solution to reach unfolded network completion. Provided that RNN training starts from initial time and ends after t time, then the total loss function (cost function, objective function) is equal to the summation of standard error function, $E_{sse/ce}(t)$, at each time step as follows:

$$E_{total}(0, t_1) = \sum_{0}^{t_1} E_{see/ce}(t) \tag{9.28}$$

It is well known that the gradient descent weight updates at each time step is calculated as follows (Chap. 7):

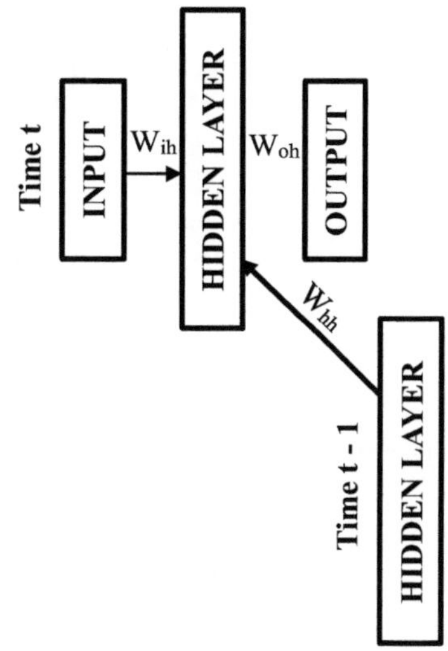

Fig. 9.17 Simple recurrent network (Elman system)

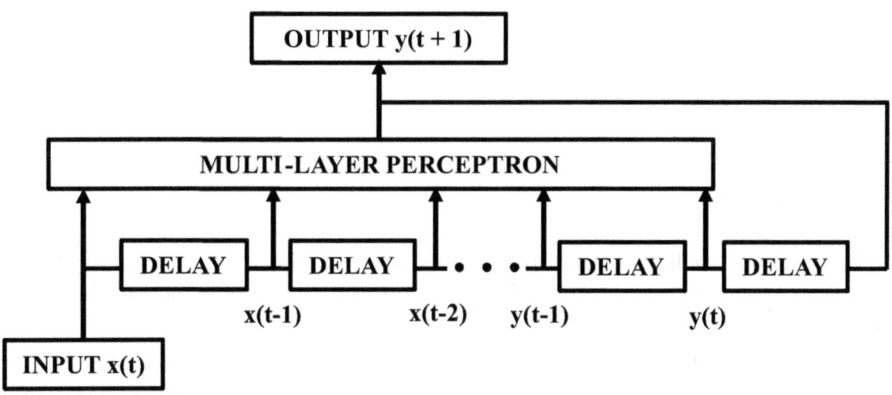

Fig. 9.18 NARX schematic representations

$$\Delta w_{ij} = -\eta \frac{\partial E_{total}(t_0), t_1}{\partial w_{ij}} = -\eta \sum_{t_0}^{t_1} \frac{\partial E_{sse/ce}(t)}{\partial w_{ij}} \tag{9.29}$$

In this last expression, $\partial E_{sse/ce}(t)/\partial w_{ij}$ is dependent on the input and hidden layer activation functions at previous time steps. In case of practically unacceptable error, it must not be back propagated through the network by time. Despite the fact of complexity in the neural system models like RNN, the back propagation through time procedure provides an effective learning algorithm.

In Fig. 9.17, truncation from RNN in Fig. 9.16 is shown as a simple RNN architecture for just one time step, which is quite equivalent to Elman network already discussed in Chap. 7.

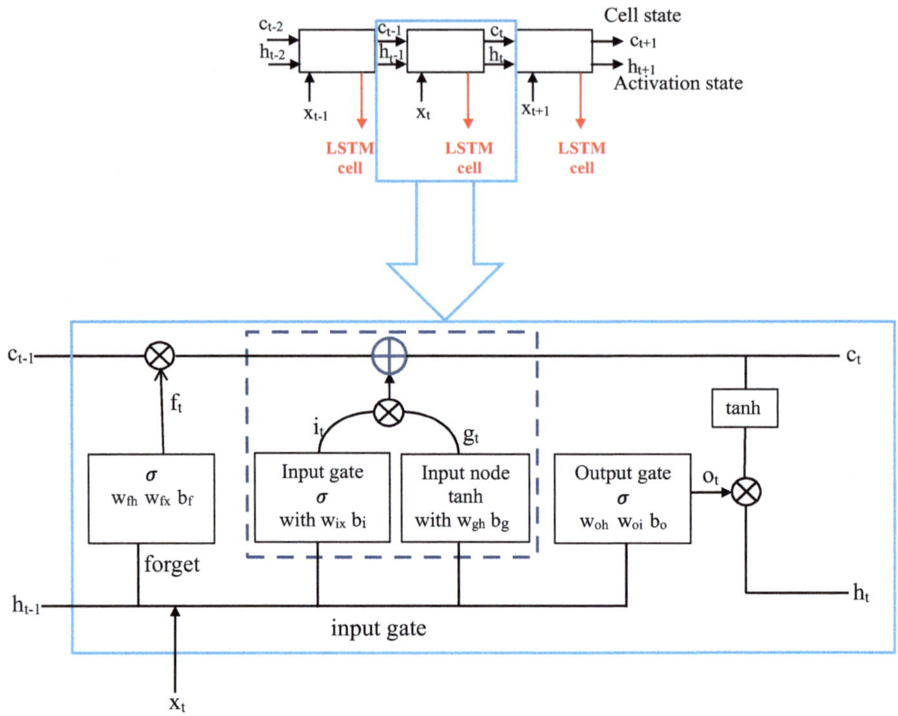

Fig. 9.19 System basic units

In this figure, each weight set appears only once, which indicates the possibility of partial back propagation through time procedure application by means of the gradient descent method.

There is also another alternative version of RNN known as the non-linear autoregressive with eXogeneous inputs (NARX) model with a single input and output that takes into account a delay line on the inputs with the outputs fed back to the input by another delay line, which is presented schematically in Fig. 9.18, where MLP unit plays significant role.

This structure is very useful for time series prediction for y(t + 1) from x(t + 1). The reader may compare this structure with the Kalman filter and various stochastic process structures mentioned in Chap. 6.

Glorot and Bengio (2010) proposed that RNN sensitivity to the exploding gradient and vanishing problems represent one of the main issues with this approach. The gradients may explode or decay exponentially during the training process due to the large or small derivative reduplications. Due to the long short-term method (LSTM) rectification procedure, the accuracy can be obtained over long time due to several storage capacity memory cells. Additionally, there are gated units for information flow control. The basic unit in such a system is shown in Fig. 9.19 with its most explicit form.

In the upper side of this figure is a condensed part from the RNN system, and its core appears in a larger scale at the bottom as shown by vertical arrow, where x_t is the input layer, h_t is the hidden layer, and c_t is the convolution layer representative variables. The necessary formulations about the system are presented in the following set of expressions concerning forget, input gate, and input node:

$$f_t = \sigma[(w_{fh}.h_{t-1}) + (w_{fx}.x_t) + b_f] \tag{9.30}$$

$$i_t = \sigma[(w_{ih}.h_{t-1}) + (w_{ix}.x_t) + b_i] \tag{9.31}$$

$$g_t = \tanh\left[\left(w_{gh}.h_{t-1}\right) + \left(w_{gx}.x_t\right) + b_g\right] \tag{9.32}$$

$$c_t^i = i_t.g_t \tag{9.33}$$

$$c_t = c_t^i + c_t \tag{9.34}$$

$$o_t = \sigma[(w_{oh}.h_{t-1}) + (w_{ox}.x_t) + b_o] \tag{9.35}$$

and

$$h_t = \tanh(c_t) + o_t \tag{9.36}$$

Four kinds of weights are used to complete LSTM method, which are given in the column matrix forms as follows:

$$w = \begin{bmatrix} w_f \\ w_i \\ w_g \\ w_O \end{bmatrix}; b = \begin{bmatrix} b_f \\ b_i \\ b \\ b_O \end{bmatrix}; h = \begin{bmatrix} w_{fh} \\ w_{ih} \\ w_{gh} \\ w_{oh} \end{bmatrix}; g = \begin{bmatrix} f_t \\ i_t \\ g_t \\ o_t \end{bmatrix}$$

9.12.2 RNN Computational Time Sequence

The same weight matrices are used at every time steps. In Fig. 9.20, loss function generation mechanism is shown by forward feeding and back propagation procedures subsequent iteration operations in an RRN system.

The initial procedure is to go forward across time by updating the cell state from the input and the previous state and thus generating an output. The following step is to calculate loss values at the individual time steps in sequence and then to sum each individual losses to get the total loss value. Although it is possible at a single time to back propagate errors through a single feedforward network, it is better to back propagate the total loss value through each individual time step toward the beginning, and hence, all the errors go back in time and this is the reason why it is referred to as back propagation.

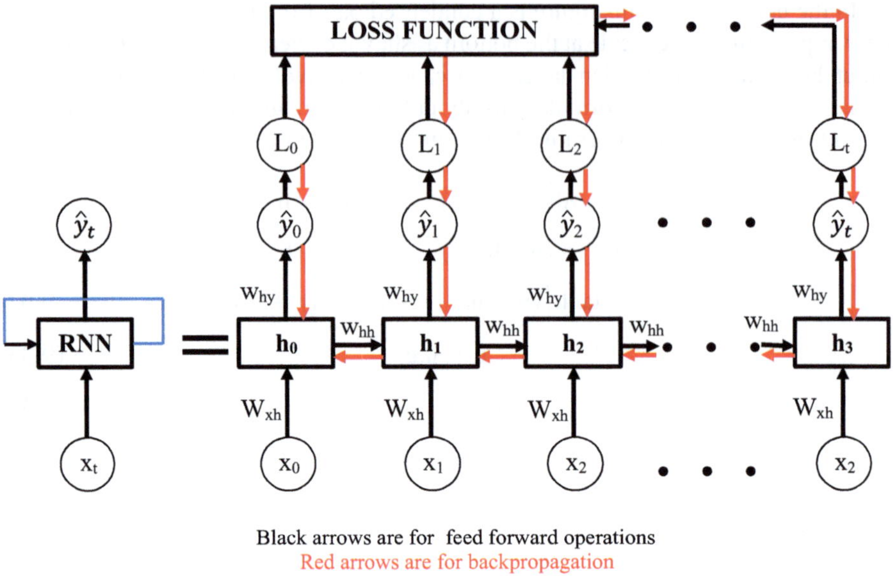

Fig. 9.20 Loss function operation mechanism

A closer look at how gradients are executed across the h_0, h_1, h_2, …., h_t box sequence (see Fig. 9.20), it is important to notice how gradient descend repetition slow down with each iteration. Between each step matrix multiplication must be executed involving w_{hh}. The initial cell value of h = 0 involves many factors also with respect to the weight matrix. For gradient calculation at time t, h_0 has many factors of W_{hh} and repetitive gradient computations.

In gradient calculations in some steps, the gradient may be greater (smaller) than one, which is referred to gradient explosion (gradient vanish); therefore, it is necessary to clip gradient. Especially, in cases of some gradients, continuously less than one cause to gradient vanishing problem, and it is necessary to solve it before forward step executions. This is possible by changing the activation function and network architecture adjustment and also by weight initialization. In RNN sequence modeling, one can consider the following three alternatives:

1. Multiple-input-single output, MISO (sentiment classification)
2. Single-input-multiple-output, SIMO (text generation, image captioning)
3. Multiple-input-multiple-output, MIMO (translation, forecasting, music generation)

Among the design criteria primarily, the following points are recommended to take into account for improved results:

1. Variable length sequence handling
2. Long-term dependencies

3. Information maintenance in order
4. Parameter share across the sequence

9.13 The Problem of Long-Term Dependencies

The gradient vanishing is due to successive multiplication of small numbers between 0 and 1, and thus, each back time steps will have successively smaller gradients. One should remember bias parameters that are useful in short-term dependence capture. First, it is necessary to determine short-term rather than long-term dependence. To capture long-term dependence, one can try to change the activation function by choosing the most convenient one between a ReLU, tanh and sigmoid alternatives. In practice, it has been observed that the ReLU derivative prevents gradient shrinkage when $x > 0$. Another way to avoid shrinkage problem is to initialize parameters as an identity matrix with initial zero bias values.

Finally, gated cells are used as a more complex recurrent unit gates to control the passing information. This operation captures quite rapidly the long-term dependencies in the data. The most frequently used method in RNNs is gated cell concept and its application. In the process of track information throughout many time steps, long short-term memory (LSTM) methodology depends on gated cell composition.

9.13.1 LSTM Network

LSLM networks have received close attention by researchers in different disciplines because of their effectiveness. The essential fundamentals of RNN and LSTM are explained by Sherstinsky (2020) based on signal processing concepts. This methodology relies on gated cell for information traction through many time steps, and it is very convenient to maintain long-term dependencies in the dataset for more

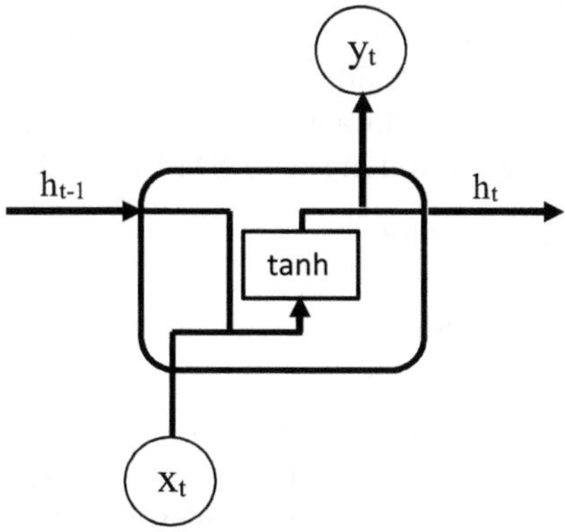

Fig. 9.21 Standard RNN unit cell

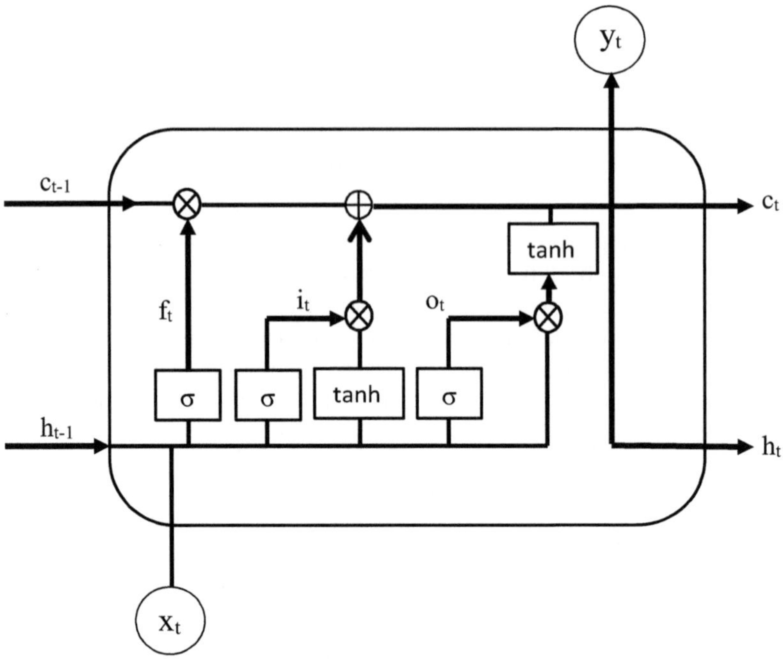

Fig. 9.22 Standard RNN chain structure

effective modeling of sequential data. These networks are very essential activity in the DL for most sequential modeling tasks. Let us consider a simple computation node of a standard RNN cell as in Fig. 9.21. This is a repeating module containing a simple computation node in a standard RNN. In this figure, the black lines capture weight matrix multiplication, and the activation function, tanh, shows a nonlinear activation function.

There is also the chain-like structure as in Fig. 9.22 with internal repeating module where recurrent unit is quite complex. It is the main duty of LSMT cells to track information through many time steps. In more refiner detail of LSTM, cell is given in the following figure, where information can be added or removed through cell gates, for example, via sigmoid neural network.

There are different interacting layers that are defined by standard neural network operations through nonlinear activation functions. They control the flow of information through such LSTM cells. The LSTMs work through four stages:

1. Forgetting
2. Storing
3. Updating
4. Outputting

The first step is to forget of irrelevant information from the previous state, which is achieved by consideration of previous state and to pass it through one of the sigmoid gates. After forgetting state, it is important to review available information and to decide which parts to keep and store, where i_t input at time comes into view in the previous figure. LSTMs store relevant information into the cell state. The final step is the output gate control as for the sent information to the next time step. This step is important for output y_t and h_t for the next step. Information addition or removal through this structure is called gates, which let information through a convenient activation function. After all what have been explained above about the LSTMs procedure, the following summery reflects their function briefly:

1. LSTMs from the outputted information maintain a separate cell state
2. Information flow is controlled by four gates as follows:
   (a) Forget gate for getting rid of irrelevant information
   (b) Relevant information storage from current output
   (c) Cell state updating selection
   (d) A filtered cell state version outputting
3. Back propagation is possible without interruptions in gradient flow computations

The completion of the previous explanations concerning LSTMs, the modeling by RNN, will not cause any problem even concerning the gradient descend. LSTMs constitute the backbone of successful RNN modeling work.

9.14 Autoencoders

Rumelhart et al. (1985) introduced autoencoders the first time as DL architectures based on unsupervised learning. This technique is useful for data compression so as to learn abstractions. They are helpful mainly to discover hidden featural structure of the data to produce better new data by minimizing noise in the input dataset. Coupled with CNNs, they are applicable to generic or image data. Hinton and Salakhutdinov (2006) suggested autoencoder procedure for data dimensionality reduction.

These are a type of feedforward neural networks with the same input and output features. The first step is to encode the input dataset into a lower-dimensional code representation and thus get the output reconstruction from this representation. The code is in the form of compressed input dataset, which is also referred to the latent-space representation. In its structure, there are three operation units as encoder, code, and decoder. In general, it has the structure as shown in Fig. 9.23.

In this figure, the first layer is for original dataset entrance input and the last layer as output is the layer that includes the reconstructed input. Autoencoder maps the

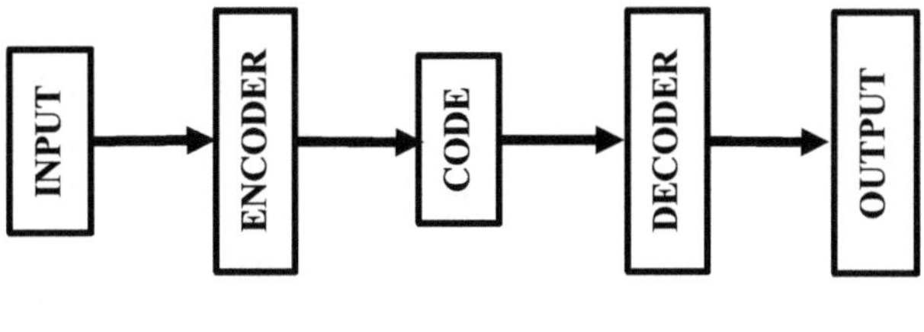

Fig. 9.23 Autodecoder

input dataset to a latent space by dimension reduction, and then latent representation is coded back to the input like output dataset. Reconstruction error reduction results after autoencoder compressive learning process. This is a procedure that is treated as unsupervised deep learning, which is also referred to in the literature as representation learning. Its concept is quite similar to principle component analysis (PCA) as discussed in Chap. 4. PCA is a statistical procedure that treats multitude of data set and puts them down into independent form of new variable dataset formation. Depending on the variance of each new variable, they are ranked ascendingly and successive summation of variances is expressed as percentage of the original total variance. Acceptance of a certain percentage, say, 90% the number of data, n, is reduced to m number of new independent data sets (m < n). In practical studies, most often m is less than even one third of n. Thus, essentially, PCA is a linear transformation opposite to autoencoders, which are nonlinear complex functional modeling procedures. PCA features are perfectly uncorrelated components, because they are projections on orthogonal basis.

Another type of unsupervised deep learning neural network type is autoencoder model with its hidden layers to decompose and then regenerate their input. Among their applications is dimensionality reduction, classification preprocessing, and filtering out any noise from the input dataset so as to identify essential input elements (Chap. 8). Autoencoders have similarities to PCA, and the simplest modeling background is presented in Fig. 9.24.

Fig. 9.24 PCA background

Fig. 9.25 Autoencoder architectural structures

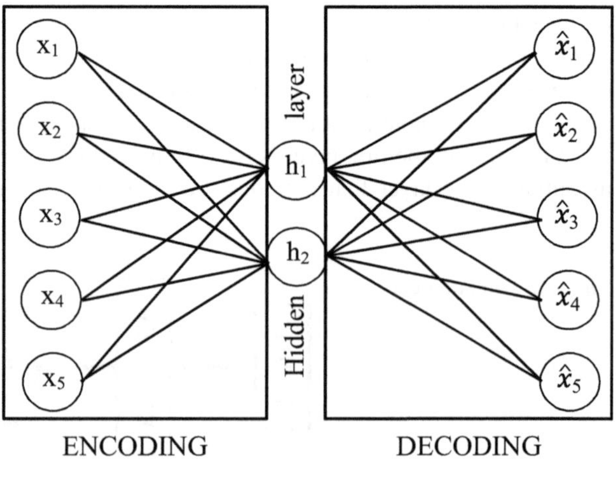

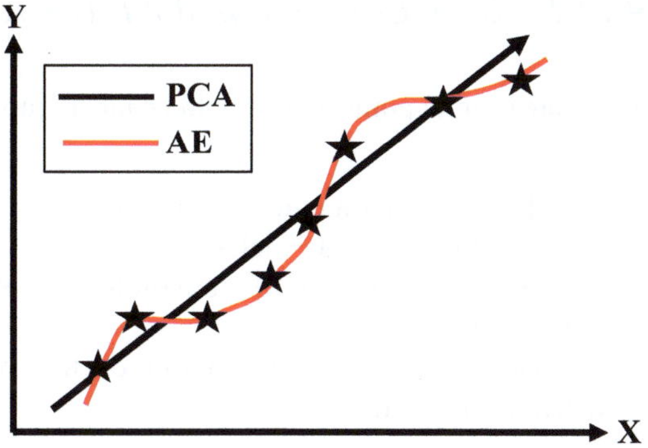

Fig. 9.26 PCA and AE comparison

PCA yields orthogonal directions of principle components with variance corresponding arrow lengths from the original datasets. Among the PCA deficiencies are the following items:

1. PCA features are derived from original data in the form of linear relationship, whereas some of them may have nonlinear trends.
2. There may be complex nonlinearity, and therefore, PCA may not yield the best lower dimensional representation, and finally, the best representation can be defined differently from the PCA variance-based definition in many other ways.

Figure 9.25 indicates briefly three parts of an autoencoder architectural structure similar to shallow learning ANN structure, as explained in Chap. 7.

There may be more than a single hidden layer. Encoding corresponds to original dataset, whereas decoding is the reconstruction of the original data perhaps with some irrelevant or insignificant points. Sı decoder is the mirror image of encoder or vice versa.

In contrast to PCA, autoencoders can perform nonlinear transformations through nonlinear activation function and multiple hidden layers. Autoencoder basic idea is to generate an architecture that has lower dimensionality representation of the original dataset. Autoencoders can encode and decode. Encoder is the mean of data compression into lower dimensionality domain, which is referred to as latent space. Representation learning needs the condition that all dimensional data should be independent from each other as it was in the case of PCA elements; otherwise, it is not possible to learn lower-dimensional representation. Figure 9.26 presents distinction between autoencoder (AE) and PCA methods.

It is possible to say that encoder is equivalent to PCA provided that it uses linear activation function. AEs can be trained by back propagation procedure. There is a loss (optimization) function that represents error, $E(x, \widehat{x})$, as the difference between the original data and reconstructed data. An ideal AE should be sensitive enough to input data during its reconstruction and insensitive to over-fitting possibilities. In order to avoid over-fit, the error data must be supported with regularization component as "$E(x, \widehat{x})$ + regularization."

9.14.1 Deep Convolutional AE (DCAE)

There are several points that should be taken into account for a successful application of AE:

1. DCAE has similar architecture to AE
2. Convolution layers are included
3. Encoder is composed of convolution + ReLU activation function + batch normalization
4. Decoder is composed of convolution transpose + ReLU activation function + batch normalization.

Latent space helps to keep the most important input data attributes and the leverage of the latent space leads to several interesting task performances. AEs can be useful in data generation, anomaly detection, noise avoidance, etc.

9.15 Natural Language Models

This model helps to identify the human language words by prediction through statistical tools. They are neural language processing models that are very useful for a variety of tasks such as speech recognition, sentiment analysis audio to text conversion, spell check, summarization, etc. An alphabet of tokens is represented by a probability distribution function (PDF) over a sequence as a language model. A language model can learn from examples among which are the suggested auto-completer in the forms of few characters or words, handwriting recognition for guessing a word from its etymologic or epistemological contents, speech recognition, spelling checking, and language translations. Although there were statistical language models, recently, a new trend came into existence as neural language models, which yield state of the art accuracy with the aid combination of two very recent neural network methodologies as deep and recurrent learning approaches.

A sequence of tokens, $t_1, t_2, \ldots, t_n$ is represented by joint PDF, $P(t_1, t_2, \ldots, t_n)$. The problem is to learn such probabilities from a training set of dataset. Let us consider the conditional probability of t_n given all previous data as $P(t_n/t_2, \ldots, t_{n-1})$. It is possible then to decompose the overall probabilities as in the following chain form of successive conditional probability multiplications (Chap. 4).

$$P(t_1, t_2, \ldots, t_n) = P(t_1)\,P(t_2/t_1)\,P(t_3/t_1, t_2)\ldots.P(t_n/t_1, t_2, \ldots, t_{n-1}) \quad (9.37)$$

Linguistic modeling has learning approach as the next token prediction problem, which corresponds to supervised learning. It is notable that the output is a fixed-length probability sequence, that is, probability vector. On the other hand, the input has not a fixed-length vector, because its length varies. Since there are thousands of probabilities, the problem provides a very difficult solution possibility. In order to search for solution, it is necessary to make assumptions, because without them the solution is intractable. For instance, if there are n = 10 data and alphabet has only two alternatives as 0 and 1, then the whole possible solution alternatives is equal to $2^{50} = 1.1259e+15$, which is almost impossible number for computational successive result. The solution is one of these numbers, but its search is complex and computationally almost impossible. Statistically, the most frequent assumption is Markovian sequence structure, which reduces the sequential serial memory length down to k instances, where k is a small number. Hence, $P(t_n/t_2, \ldots, t_{n-1})$ is replaced by $P(t_n/t_{n\text{-}k}, \ldots, t_{n-1})$. This implies that the next event probability depends on the previous k instances or tokens in this context. In practical studies k is fixed to a small number as 2 or 3 and at the maximum 5. There are convincing examples in the literature concerning this reductionist approach (Movshovitz-Attias and Cohen 2013; Bowman et al. 2015; Liu et al. 2019).

9.16 Conclusions

Deep learning (DL) procedures rely on a combination of machine learning (ML) and artificial intelligence (AI) methodologies for which necessary information is provided in Chaps. 2 and 8 for key components. DL procedures are similar to shallow learning artificial neural network (ANN) modeling principles, but with additional features for larger dataset processing for classification and regression works to reach the final request of the problem at hand in shorter time frames than conventional modeling styles. The most prominent AI DL methodologies are convolution neural network (CNN) and recurrent neural network (RNN). The first proceeds with the CNN method-specific hidden layers in terms of convolution and pooling pair series with the fully connected layer before the output layer. However, RNN has back propagation iterations within hidden layers that enable faster and nonlinear behavior capture in data.

Although ready-made software is used for problem solving, it is recommended that the reader try to understand the philosophical and logical internal structure of CNN and RNN. Without this type of internal audit, it is not possible to move toward changing of existing methods or replacing them with better improvements or innovative aspects.

References

Bowman S, Angeli G, Potts C, Manning CD (2015) A large annotated corpus for learning natural language inference. In: Proceedings of the 2015 conference on empirical methods in natural language processing

Fukushima K (1980) Neocognitron: a self-organizing neural network model for a mechanism of pattern recognition unaffected by shift in position. Biol Cybern 36:193–202

Glorot X, Bengio Y (2010) Understanding the difficulty of training deep feedforward neural networks. In: Proceedings of the thirteenth international conference on artificial intelligence and statistics, vol 9. PMLR, pp 249–256
He K, Zhang X, Ren S, Sun J (2015) Delving deep into rectifiers: surpassing human-level performance on ImageNet classification. In: Proceedings of the IEEE international conference on computer vision, pp 1026–1034
Hinton GE, Salakhutdinov RR (2006) Reducing the dimensionality of data with neural networks. Science 313(5786):504–507
Hubel DH, Wiesel TN (1959) Receptive fields of single neurons in the cat's striate cortex. J Physiol 148:574–591
Ivakhnenko AG (1971) Polynomial theory of complex systems. IEEE Trans Syst Man Cybern 4: 364–378
Ivakhnenko AG, Lapa VG (1965) Cybernetic predicting devices. CCM Information Corporation
Krizhevsky A, Sutskever I, Hinton GE (2012) ImageNet classification with deep convolutional neural networks. Adv Neural Inf Proces Syst 25:1097–1105
LeCun Y, Boser BE, Denker JS, Henderson D, Howard RE, Hubbard WE, Jankel LD (1990) Handwritten digit recognition with a back-propagation network. Adv Neural Inf Proces Syst 2: 396–404
Liu X, He P, Chen W, Gao J (2019) Improving multi-task deep neural networks via knowledge distillation for natural language understanding. arXiv preprint arXiv:1904.09482
Movshovitz-Attias D, Cohen WW (2013) Natural language models for predicting programming comments. In: Proceedings of the 51st annual meeting of the association for computational linguistics, Sofia, Bulgaria, August 4–9 2013. c 2013 Association for Computational Linguistics, pp 35–40
Rumelhart DE, Hinton GE, Williams RJ (1985) Learning internal representations by error propagation. Technical report, California Univ San Diego La Jolla Inst for Cognitive Science
Sherstinsky A (2020) Special issue on machine learning and dynamical systems fundamentals of recurrent neural network (RNN) and long short-term memory (LSTM) network. Phys D: Nonlinear Phenom 404

Index

Z. Şen, *Shallow and Deep Learning Principles*,
https://doi.org/10.1007/978-3-031-29555-3

S

T

U

MIX
Papier aus verantwortungsvollen Quellen
Paper from responsible sources
FSC® C105338
FSC
www.fsc.org

If you have any concerns about our products,
you can contact us on
ProductSafety@springernature.com

In case Publisher is established outside the EU,
the EU authorized representative is:
Springer Nature Customer Service Center GmbH
Europaplatz 3, 69115 Heidelberg, Germany

Printed by Libri Plureos GmbH
in Hamburg, Germany